International Association of Geodesy Symposia

Klaus-Peter Schwarz, Series Editor

Springer
Berlin
Heidelberg
New York
Barcelona
Budapest
Hong Kong
London
Milan
Paris
Tokyo

International Association of Geodesy Symposia

Klaus-Peter Schwarz, Series Editor

Symposium 101: Global and Regional Geodynamics
Symposium 102: Global Positioning System: An Overview
Symposium 103: Gravity, Gradiometry, and Gravimetry
Symposium 104: Sea Surface Topography and the Geoid
Symposium 105: Earth Rotation and Coordinate Reference Frames
Symposium 106: Determination of the Geoid: Present and Future
Symposium 107: Kinematic Systems in Geodesy, Surveying, and Remote Sensing
Symposium 108: Applications of Geodesy to Engineering
Symposium 109: Permanent Satellite Tracking Networks for Geodesy and Geodynamics
Symposium 110: From Mars to Greenland: Charting Gravity with Space and Airborne Instruments
Symposium 111: Recent Geodetic and Gravimetric Research in Latin America
Symposium 112: Geodesy and Physics of the Earth: Geodetic Contributions to Geodynamics
Symposium 113: Gravity and Geoid
Symposium 114: Geodetic Theory Today
Symposium 115: GPS Trends in Precise Terrestrial, Airborne, and Spaceborne Applications
Symposium 116: Global Gravity Field and Its Temporal Variations
Symposium 117: Gravity, Geoid and Marine Geodesy
Symposium 118: Advances in Positioning and Reference Frames
Symposium 119: Geodesy on the Move

Geodesy on the Move

Gravity, Geoid, Geodynamics and Antarctica

IAG Scientific Assembly

Rio de Janeiro, Brazil, September 3-9, 1997

Convened and Edited by
Rene Forsberg, Martine Feissel
and Reinhard Dietrich

Springer

Volume Editors
Prof. Dr. Rene Forsberg
Department of Geodynamics
Kort & Matrikelstyrelsen
Rentemestervej 8
DK-2400 Copenhagen NV
Denmark

Prof. Dr. Martine Feissel
Observatoire de Paris
61 Avenue de l'Observatoire de Paris
F-75014 Paris
France

Prof. Dr. Reinhard Dietrich
Institut für Planetare Geodäsie
Technische Universität Dresden
Mommsenstraße 13
D-01062 Dresden
Germany

Series Editor
Prof. Dr. Klaus-Peter Schwarz
University of Calgary
Department of Geomatics Engineering
2500 University Drive N. W.
Calgary, Alberta T2N 1N4
Canada

ISBN 3-540-64605-1 Springer-Verlag Berlin Heidelberg New York

Library of Congress Cataloging-in-Publication Data
Geodesy on the move : IAG Scientific Assembly, Rio de Janeiro, Brazil, September 3-9, 1997 / convened and edited by Rene Forsberg, Martine Feissel and Reinhard Dietrich.
p. cm. -- (International Association of Geodesy symposia ; symposium 119)
Includes bibliographical references and index.
ISBN 3-540-64605-1 (hardcover : alk. paper)
1. Gravity-Measurement-Congresses. 2. Earth-Figure-Congresses. 3. Geodynamics-Congresses. 4. Geodesy-Antarctics-Congresses. I. Forsberg, Rene. II. Feissel, Martine. III. Dietrich, Reinhard, 1949- . IV. Scientific Assembly of the International Association of Geodesy (1997 : Rio de Janeiro, Brazil). V. Series.
QB330.G45 1998 526'.1-DC21 98-28910 CIP

Printed in Germany

Typesetting: Camera ready by editor/author
Cover layout: design & production GmbH, Heidelberg

SPIN: 10654039 32/3020 - 5 4 3 2 1 0 - Printed of acid-free paper

Dedication

This volume is dedicated to the late Professor Herman van Gysen, an outstanding geodesist from South Africa, who as IAG Special Study Group president and co-convener took part in the preparation of the Rio Scientific Assembly. He had to cancel his participation in the meeting at short notice due to the unexpected discovery of cancer, which claimed his life on February 19, 1998, at a far too young age.

FOREWORD

Klaus-Peter Schwarz, IAG President
The University of Calgary

The Scientific Assembly of the International Association of Geodesy (IAG) was held from September 3 to 9, 1997 in Rio de Janeiro, Brazil, in conjunction with the 18th Brazilean Congress of Carthography. This was the first time that one ofthe major IAG meetings took place in Brazil. It provided an opportunity to showcase the progress of geodetic work in South America through campaigns such as SIRGAS. It also provided an opportunity for a large group of international experts to present the state of the art in geodesy and geodynamis and to interact with their hosts on possibilities of future cooperation. For the IAG, it continued a trend, started in Beijing four years ago, to hold major geodetic meetings outside of Europe and North America. The International Geoid School which was held in Sao Paulo following this meeting showed another facet of this growing internationalization of IAG activities and services.

The scientific program of the meeting consisted of three symposia and two special sessions, namely

Symposium1: Advances in Positioning and Reference Frames
Symposium 2: Gravity and Geoid
Symposium 3: Geodynamics
Special Session 1: IAG Services
Special Session 2: Geodesy in Antartica.

Papers presented at the first symposium are published in volume 1 of these proceedings, while papers of symposia 2 and 3 as well as special session 2 are contained in volume 2. Papers presented at special session 1 will be published separately. More details on the individual volumes are given in the prefaces written by the convenors.

The meeting was jointly organized by the IAG and the Brazilean Society of Cartography, Geodesy, Photogrammetry and Remote Sensing. The two local organizing committees worked closely together to economize on some of the organizational aspects and to guarantee a smooth running of two conferences in the same place. Thanks go to our Brazilean colleagues for their hard work and their warm hospitality. On the geodetic side, special thanks go to Professor D. Blitzkov, the national representative of the IAG who chaired the IAG Local Organizing Committee (LOC), and the dedicated group of individuals working working with him. Similarly, Professor W. Torge, past president of the IAG, who provided the liason between the LOC and the IAG Executive deserves a special vote of thanks. Finally, the symposium and session convenors who not only organized the scientific program but also took care of organizing the review rocess and editing these proceedings, are thanked for their outstanding efforts.

PREFACE

The International Association of Geodesy (IAG) held its Scientific Assembly in Rio de Janeiro, Brazil, 3-9 September 1997, in conjunction with the XVIII Brazilian Congress of Carthography. It was organized by the Brazilian Society of Carthography, Geodesy, Photogrammetry and Remote Sensing, with Prof. D. Blitzkow as the IAG national organizer. This was the first time the regular (4-year interval) IAG Scientific Assembly was held in South America. The meeting was attended by approx. 250 participants from around the world.

The meeting was difided into three main symposia: *Advances in Positioning and Reference Frames, Gravity and Geoid,* and *Geodynamics*. In addition, special sessions were scheduled on *IAG services* and *Geodesy in Antarctica.*

The present proceedings (Vol. 119) - cover the symposia *Gravity and Geoid, Geodynamics*, and the special session *Geodesy in Antarctica.* The title, *Geodesy on the Move - Gravity, Geoid, Geodynamics and Antarctica,* symbolizes the dynamics of geodesy, both in terms of movement in the rapid scientific development of the field, the constant movement of the Earth's crust, and the move in geodesy to cover the whole Earth geodetically - with Antarctica as the last frontier.

The proceedings were peer-reviewed in accordance with IAG rules. For *the Gravity and Geoid* symposium, the review process was organized by the session convenors: G. Boedecker, M.G. Sideris, N.K. Pavlis, and P. Holota, in cooperation with the symposium convenor (R. Forsberg). This is gratefully acknowledged.

The *Gravity and Geoid* symposium was divided into four sessions:
- Developments in static and kinematic gravimetry;
- Global gravity field from gravimetry and satellite altimetry;
- The geoid: theory and methods;
- Regional geoids and the gravity field in South and Central America;

These areas represent active areas of current research.

In the area of gravimetry, the presentations show constant development towards increased accuracies at the 10^{-9} level, with new technological developments such as superconducting and airborne gravimetry, which continue to improve. The global gravity field session covers the ongoing developments of global spherical harmonic reference models, such as the newest global gravity field model EGM96, including developments in recovering the marine gravity field from satellite altimetry, and the prospects of future gravity field missions to yield improved models. Within the theoretical and methodological developments in geoid determination and gravity field modelling, new advances include presentations on wavelets, cap sizes and modified kernels, and on terrain corrections and harmonic continuation.

The *Global Gravity Field* session was a special occasion to thank Prof. (Emeritus) R.H. Rapp, Ohio State University, who over the decades has been the pioneer and driving force in developing high-resolution global spherical harmonic models, with the EGM96 model being the crowning chievement, representing the combined efforts of space, military, and university groups. R.H. Rapp was presented a special memorial gift during the session. He gave an expanded overview of the topic, followed by presentations of many of his former students.

The *Geodynamics* proceedings included:
- Earth rotation and variations of the geopotential;
- Reference frames and global deformations;
- Combination of space observations with other observations of deformations.

The symposium presented an excellent opportunity to survey the main results of more than a decade of work in setting up global reference frames and organizing programs to monitor crustal deformations on a world wide or regional scale. During the 10 years of its existence, the International Earth Rotation Service (IERS) has monitored and provided access to the International Celestial and Terrestrial Reference Systems and to the Earth's orientation with an overall consistency equivalent to 1 cm on the surface of the Earth. The International Terrestrial Reference Frame (ITRF) provides positions and velocities for about 300 sites worldwide. This continuing effort is based on the close cooperation of laboratories and agencies making use of VLBI and satellite-geodetic techniques (laser ranging, GPS, DORIS).

Several global site motions solutions based on a single technique or on a combination of techniques were presented, showing that the motion of the major plates can be estimated with a precision of a few millimeters per year, thus allowing the study of some of the boundary zones. Plans or results were presented for a large number of crustal deformation programs based on space geodesy that concern many parts of the world: Asia-Pacific, Central Europe, the Middle-East, Egypt, the western Mediterranean Sea, the Andes, Central America. A number of contributions deal with deformation studies based on the integration of space-geodetic observations (GPS) with other types of observations, such as SAR interferometry, tide gauges, strainmeters/tiltmeters, and variations in the vertical or geological data. Exciting results showing seasonal and longer term geopotential and geocenter variations in the atmosphere-hydrosphere-solid Earth system, based both on space geodetic observations (SLR, DORIS) and atmospheric and oceanic data and models, were also presented.

The *Geodesy in Antarctica* session was a special "interdisciplinary" geodesy session, added to the program by the IAG executive committee, in order to stress the importance of this continent in modern geodesy, and taking advantage of the location of the IAG Scientific Assembly in the southern hemisphere. The importance of Antarctica for geodesy is increasing, as Antarctica holds the key to many challenging scientific questions regarding global change and geodynamics. Antarctica is also the last major void of terrestrial gravity field information on Earth, making an effort to collect additional data urgent in coming years, in order to support future gravity field space missions (planned missions will leave polar coverage gaps) and to improve global geopotential model development.

The Antarctica special session featured a broad spectrum of topics, ranging from overviews of global change and geodynamics, presentations relating to gravity field (geoid and airborne gravity), to presentations of more localized efforts in geodynamics and fundamental geodetic reference stations. Presentations on ice dynamics include measurements by GPS methods and SAR interferometry.

May 1998

Rene FORSBERG
Martine FEISSEL
Reinhard DIETRICH

CONTENTS

Regional geoids and the gravity field of South America
Convenors: R. Forsberg and D. Blitzkow

GEODYNAMICS
Earth rotation and variations of the geopotential
Convenors: Anny Cazenare and Christoph Reigber

Combination of space observations with other observations of deformations
Convenors: Paul Perdomo and Christoph Reigber

GEODESY IN ANTARCTICA
Convenor: R. Dietrich and J. Manning

The changing role of gravity reference networks

Wolfgang Torge
Institut für Erdmessung
Universität Hannover
Schneiderberg 50
D-30167 Hannover
Germany

1. Introduction

Gravity reference (or fundamental) networks realize the *gravity standard* through the gravity values of a *set of selected stations.* The networks may have global or regional extension, and consequently their establishment then follows either international conventions or national regulations. Obviously, the national networks should be tied to the global reference. High accuracy and reliability are demanded for the gravity values, fully exploiting the scientific knowledge and the technical skill available. The networks shall serve the needs of geodesy, geophysics and metrology over a longer time span, which requires monumentation and careful documentation of the stations, as well as a long-term local control.

The fundamental quantities of the gravity acceleration are *length* and *time*, which implies that gravity determinations, on principle, have to be based on measurements of these two quantities. During the last two centuries, gravity measurement and evaluation techniques developped significantly, increasing the accuracy by about four orders of magnitude (e.g. TORGE 1989). The need for establishing a *gravity reference system* was recognized about 100 years ago. The *classical approach* as developped within the frame of the International Association of Geodesy (IAG) consisted of providing a global gravity reference network, based on a few absolute gravity determinations, with subsequent interpolation by relative gravimetry, down to the national reference nets. This strategy had to be reexamined in the 1980's, when transportable free-fall gravimeters became operational. The resulting *modern approach* is based on the fact that these instruments realize the gravity standard at each individual gravity survey, and allow to independently establish gravity reference networks on local, regional or global scale. Obviously, now these *gravimeter systems* have to be carefully *controlled*, in order to guarantee a homogeneous gravity standard worldwide, a task again strongly supported by IAG.

In the sequel, we describe the main steps of this development, the state of the art at the modern approach at realizing the gravity standard, and the problems still to be solved in order to reach a 10^{-9} g accuracy and reliability.

2. The precursors

More accurate gravity measurements started around 1800, following the *wire pendulum* experiment in Paris by Borda and Cassini (1792), see LENZEN and MULTHAUF (1965). Bessel (1828) significantly improved this method, while Kater (1818) made the *reversible pendulum* operational. The lengthy procedure of these absolute measurements, as well as their large systematic errors led to the introduction of *relative techniques*, employing "*invariable*" *pendulums* (Kater 1819). Only gravity differences to some "reference" station were determined now, by observing the pendulum's oscillation time at the reference and the field stations.

Several *gravimetric campaigns* were carried out during the first decades of the 19th century, delivering around 100 gravity values. Apart from observations in Europe (Kater, Biot), British (Sabine, Hall, Foster), French (Freycinet, Duperrey) and Russian (Lütke) marine expeditions were extremely important. Connected to the reference stations London/Greenwich, Paris and St. Petersburg, they delivered a first insight into the global variations of the gravity field. The errors of the (few) absolute determinations performed at that period reached some 10^{-5} g, while the relative connections could be as good as 10^{-5} g or even better. The scarce data and their poor accuracy, at that time did *not yet* call for a common gravity reference.

The situation changed completely when after 1862 the IAG and its precursors included gravity measurements as well as gravity field modelling and interpretation into their scientific program (e.g. TORGE 1996). The construction of *transportable reversible pendulum* apparatus (Repsold/Bessel 1862, Brunner/Defforges 1887) nevertheless neither did significantly increase the number of gravity determinations nor their accuracy. The development of a *relative pendulum* device by Sterneck (1887), on the other hand, led to systematic surveys in many parts of the world, with about 1400 gravity stations available around 1900, carefully collected by IAG. The accuracy of those relative connections was 10^{-5} to 10^{-6} g, which by far exceeded the accuracy of the absolute "reference" stations. Now the need for a *common reference system* became obvious.

3. The classical approach

The classical approach for establishing a gravity reference system was developped mainly by Helmert, who became Director of the IAG Central Bureau in Berlin (later Potsdam), in 1886 (e.g. TORGE 1993a). The basic principle of this approach consists in defining the absolute gravity standard by the *gravity value of just one* well determined *gravity station* (later a few stations), and to refer all relative measurements of a *global network* to this fundamental value. We have three realizations of this approach.

The *Vienna Gravity System* was adopted by the IAG at its 13th General Conference in Paris 1900 (HELMERT 1901). It was based on the absolute determination in the Military Geographic Institute, carried out by v. Oppolzer (1884), and the transfer of an earlier absolute result obtained by v. Orff (1877) in Munich. Around 1400 gravity stations (most of them observed with relative pendulums) could be referred to the Vienna absolute value, with relative accuracies of ± 50 to 250 μms^{-2}. On the other hand, Helmert already supposed a systematic offset of + 150 to 200 μms^{-2} of the Vienna reference value, with respect to the absolute gravity at Potsdam.

As a consequence, Helmert initiated a new absolute determination in the Geodetic Institute Potsdam. It was carried out between 1898 and 1904 by KÜHNEN and FURTWÄNGLER (1906) using different reversible pendulums. The adjustment of the observation series resulted in a gravity value for the Potsdam pendulum pillar, with a standard deviation of ± 30 μms^{-2}, and a correction of - 160 μms^{-2} to the Vienna System. The IAG adopted the *Potsdam Gravity System* at its 16th General Conference in London 1909 (HELMERT 1910), following the proposal by BORRASS (1911) who presented a catalogue of 2700 gravity stations referred to the Potsdam gravity value, including 20 fundamental stations with adjusted results (± 10 to 20 μms^{-2}). The Potsdam Gravity System was well accepted by the user community, and *national gravity networks* were developped and tied to the Potsdam System (or even directly to the Potsdam station) until the late 1960's.

The Potsdam System had the advantage (like the Vienna System) not to suffer from calibration problems, but several *deficiencies* showed up rather soon. There ist only *one* "fundamental" gravity value, and no clear definition of a global set of "reference" stations. The catalogue's gravity values were rather different in quality, and most station descriptions were not satisfactory either. The Potsdam absolute value itself soon was suspected to be 100 to 200 μms^{-2} too high, and transfer errors of the same order of magnitude were found at national reference stations. At the late 1940's it became clear that the gravity standard as provided by the Potsdam System was no longer acceptable, and could be significantly improved (MORELLI 1946). New *absolute determinations* with reversible pendulums and free-fall devices confirmed the large bias of the Potsdam gravity value, while *relative pendulum* measurements experienced a slow but steady accuracy improvement to ± 10 μms^{-2} and better. More important became the use of elastic *spring type gravity meters* for long-range gravity connections (and not only for local interpolation) as indicated and demonstrated by WOOLLARD (1950) and WOOLLARD and ROSE (1963).

IAG took up this challenge by a dedicated "*World Gravity Net*" project, through its *International Gravity Commission* (set up 1951) and several Special Study Groups. The original concept (First Order World Gravity Net with few stations, connected by relative pendulum and gravimeter measurements, and tied to new absolute determinations, with the gravimeters calibrated on North-South-calibration lines) hat to be modified. This was due to the progress in absolute and relative gravimetry, and the reality of the survey projects (carried out in the 1950's and 1960's by different international groups) where for logistic reasons many more stations had to be occupied than scheduled, with many overlapping connections and loop closures.

The result of this enterprise was the *International Gravity Standardization Net 1971 (IGSN 71)*, adopted at the 15th IUGG General Assembly in Moscow 1971 (MORELLI et al. 1974).

It consists of the gravity values of 1854 stations (about 500 main stations, the rest being excenters or densification points along the calibration lines) derived from a least squares adjustment of 10 absolute values (Cook, Sakuma, and Faller with the first *transportable* free-fall instrument, HAMMOND and FALLER 1971), about 1200 relative pendulum ties, and 25000 gravity meter observations (mainly with LaCoste-Romberg gravimeters). The overall accuracy of IGSN 71 is estimated to be better than ± 1 μms^{-2}, with higher relative accuracies especially along the calibration lines. The mean correction to the Potsdam System is - 140 μms^{-2}, with large regional deviations.

With an accuracy increase of about two orders of magnitude compared to the Potsdam System, IGSN71 still serves the needs of most users in metrology, geosciences, engineering and navigation. It was immediately adopted as a standard for *national reference networks*, densified by *regional adjustments* (e.g. for South America, MCCONNELL et al. 1979), and finally also *extended* to regions as Eastern Europe, Russia, and China. The *quality* of IGSN 71 has been verified many times, especially through absolute gravimetry. We mention a comparison performed on 14 stations (gravity range 9.78 ms^{-2}: Caracas, to 9.83 ms^{-2}: Thule/Greenland) with the JILAG-3 absolute gravimeter operated by IfE Hannover since 1986 (TORGE et al. 1987). After restoring the constant tidal term to IGSN 71, no bias could be found, and the r.m.s. discrepancy is only ± 0.5 μms^{-2}. Slight scale non-linearities (several 10^{-5} over some 100 to 1000 km) have been indicated, which may be explained by calibration deficiencies of IGSN 71 at the polar regions, and systematic errors introduced through the relative pendulum data.

Starting in the 1970's gravimetry again experienced a drastic change. More *transportable free-fall gravimeters* had been developped and became operational with an accuracy increase (adjusted station value) to ± 0.05 ... 0.1 μms^{-2}, and even a few 0.01 μms^{-2} under good station conditions (MARSON and FALLER 1986, NIEBAUER et al. 1995). This triggered the modern approach for establishing gravity reference networks.

4. The modern approach

Absolute gravity meters are characterized by the fact, that the *standards* of *length* and *time* are fundamental parts of the system. By observing differences of length and time at a free-fall experiment, the gravity standard is realized independently from any external reference system, on local, regional or global scale. Obviously, this approach completely depends on the quality of the absolute gravimeter results, as expressed in terms of repeatability, accuracy, and reliability. While the accuracy strongly depends on the system's hard- and software, the *reliability*, i.e. the capability of detecting systematic errors, is difficult to judge. The *repeatability* depends on the system's control with time, and a sufficient modelling of changing environmental effects on gravity. Around 25 *transportable free-fall* or *rise and fall absolute gravimeters* are available today, among them are the (small) series of JILA (Joint Institute for Laboratory Astrophysics, Boulder, Col.) and FG5 (Micro-g Solutions Inc., Arvada, Cal.) instruments operating worldwide (MARSON and FALLER 1986, TORGE 1991). The JILA gravimeters (since 1983) have been the result of a long development by J.E. Faller and coworkers (FALLER et al. 1983), while the FG5 (since 1993) instrument significantly improved that system, especially by largely removing the influence of floor vibration and tilt on the gravity result (NIEBAUER et al. 1995).

The limit in accuracy of absolute gravimeters is given through the *standards* of *length* and *time*. Present-day technology employs either a polarization-stabilized ($\pm 2 \times 10^{-9}$ long-term stability) or an iodine-stabilized ($\pm 2 \times 10^{-11}$) laser as length standard, and a rubidium frequency standard ($\pm 1 \times 10^{-10}$) as time base. The test mass is free falling in high vacuum, and its motion is interferometrically measured with respect to a (quasi) inertial reference. The instrumental design thus incorporates a number of high-technology mechanical, optical and electronic components, including a data acquisition system and software for the on-line evaluation of the multiple (up to 200) position-time data pairs, collected at one drop. As a consequence, the instrumental error budget is rather complex, for the JILA and FG5 gravimeters estimates of ± 0.03 resp. ± 0.01 μms^{-2} are available. In addition to the uncertainties related to the gravity measurement itself, not sufficiently modelled *environmental effects* at different time scales have to be added to the error budget, with respect to accuracy and repeatability (TORGE 1991, TIMMEN 1994, NIEBAUER et al. 1995).

Acceleration effects due to *microseism* are strongly reduced through the instrumental design, and behave randomly over the observation time of 1 to 2 days, as demonstrated by the small standard deviations of station adjusted gravity values. On the other hand, *floor-gravimeter-system recoil effects* of periodic type may heavily disturb the result, if not removed through the gravimeter's construction as at the FG5 gravimeter. These site-dependent effect should be counted to the instrumental and site uncertainty. The *solid Earth* and *ocean tides* today can be modelled with an uncertainty less than 0.01 μms^{-2}, if the absolute data set is distributed over 1 to 2 days (TIMMEN and WENZEL 1994). Close to the coast and at border seas, these synthetic tides may be in error of several
0.01 μms^{-2}, which asks for dedicated gravimetric tidal observations. Very close to the sea, *swell* and *surge* effects may reach the same order of magnitude, and are difficult to model. *Atmospheric attraction* and *loading effects* generally are reduced by linear regression with local air pressure. This simple model and the use of a conventional regression coefficient may introduce errors of 0.02 μms^{-2} or more. Research is underway to reduce these errors by local regression effects and refined reduction models. *Ground water* and *soil moisture* variations play an important role at stations located on sediments. Short-term (rainfall) and seasonal effects may reach 0.1 μms^{-2} or more, modelling has been successful only at very few test stations. The effects of *long-periodic* tides and *polar motion* are small and can be modelled to 0.001 μms^{-2}, absolute gravimetry could even contribute to improve the models. *Abrupt* or *abnormal gravity variations* may be related to *earthquake* and *volcanic* activities, while *long-term* or *secular changes* could be due to *tectonic process, isostatic uplift, sedimentation*, and *natural gas, oil* or *water extraction*. These signals are part of actual research in geodesy and geophysics, understanding properly reduced gravity values as a time-dependent quantity with valuable information on recent geodynamic processes (TORGE 1993b). Taking the reduction uncertainty of environmental effects into account, *accuracy* and *repeatability* estimates vary between a few 0.01 and 0.1 μms^{-2}. This corresponds to the accuracy of relative gravimetry, over small resp. large gravity differences, if several well calibrated instruments are employed

(e.g. TORGE 1993b).

As a consequence of this technological progress, two strategies may be distinguished at the modern approach of realizing the gravity standard.

The first one is the *combination* of *absolute* and *relative* measurements in well-designed networks, using proper weighting eventually improved by an a posteriori error variance estimation. Improved linear and eventual quadratic calibration terms for the relative gravimeters may be derived from the network's adjustment, while periodic terms have to be determined in the laboratory and/or in dedicated calibration systems (KANNGIESER et al. 1983). Among the many examples for this strategy we have the (West) *German Gravity Base Net 1976* (21 stations, 4 absolute determinations by the IMGC absolute gravimeter, 656 gravity differences observed with 4 LaCoste-Romberg gravimeters, SIGL et al. 1981), and the *gravity base network* of *Venezuela* (29 stations, relative ties adjusted to 6 absolute values determined with JILAG-3 in 1988, DREWES et al. 1993).

Even more advanced (and with more instruments available in future certainly more and more employed) is the strategy to establish a gravity reference network through *absolute measurements only*. This will be applied especially at large-scale networks, where the accuracy of the relative ties even with great efforts may not reach the quality of the absolute gravity values. A global example is the *International Absolute Gravity Basestation Network* (IAGBN), proposed by IAG mainly for geodynamic investigations and already widely established (BOEDECKER and FRITZER 1986). Of regional extension is the *German Gravity Base Net 1994* (FRANKE et al. 1998). It comprises 30 stations arranged in clusters of 3 to 4 points, all of them observed with FG5-101 (BKG Frankfurt). Relative measurements were used only for a rough control of eventual systematic errors larger than 0.1 μms^{-2}, and for observing local differences (± 0.01 μms^{-2}). A comparison with JILAG-3 results (time span between the FG5 and the JILAG determinations less than 4 years) on 7 stations gave a bias of only 0.02 μms^{-2}, and a r.m.s. discrepancy of ± 0.04 μms^{-2}, which confirms the accuracy of ± 0.01 to 0.03 μms^{-2} claimed for the network.

As mentioned above, the modern approach of independently establishing gravity reference networks based on absolute gravimetry, requires a *continuous* and *thorough control* of the gravimeter system, and the reduction quality of environmental gravity effects.

5. Quality control of absolute gravimeter systems

The quality of the gravity standard realized through absolute gravimeter systems can be controlled in different ways. This control plays an essential part at the modern strategy of establishing gravity reference networks. Following mainly the experiences collected at IfE Hannover with the JILAG-3 gravimeter (TORGE 1993b, RÖDER 1994, TIMMEN 1994), and the results obtained with FG5 instruments (NIEBAUER et al. 1995, FRANKE et al. 1998) the following strategies should be applied:

- regular control of the *length* and *time standards* used in the gravimeter. At drift controlled polarization-stabilized lasers the residual uncertainties can reach a few 0.01 μms^{-2}, while iodine-stabilized lasers contribute only a few 0.001 μms^{-2} to the error budget. Rubidium frequency standards calibrated against Caesium standards are accurate to a few 0.001 μms^{2}
- evaluation of the multiple observations on an *individual station*. The drop-to-drop scatter gives an information on the random behaviour and microseism. The results strongly depend on the site conditions (JILAG-3: ± 0.1 μms^{-2} on very stable sites, e.g. underground stations in continental China, and ± a few μms^{-2} on sediments, close to the coast, or in a city environment. FG5: ± 0.05 and ± 1 μms^{-2}, respectively). The run-to-run (JILAG-3 one run consists of 300 consecutive drops) or set-to-set (at FG5-101 one set comprises 150 drops) scatter (short-term repeatability) also includes non-random parts of microseism and vibrations, as well as short-term residual environmental effects (JILAG-3: ± 0.01 to ± 0.1 μms^{-2}, FG5-101: < 0.01 ... < 0.1 μms^{-2}),
- repeated observations on a *reference station* are a measure for long-term *repeatibility*, indicating the time stability of the gravimeter system, and the modeling quality of environmental effects. For JILAG-3 wie have available time series over 10 years, for the reference station Clausthal (hard bedrock, low artificial microseism, r.m.s. scatter ± 0.04 μms^{-2}) and Hannover (sediments, city environment, ± 0.08 μms^{-2}). For FG5 gravimeters results of a *half year* control are available for bedrock stations (Table Mountain, Col.: ± 0.02 μms^{-2}, Wettzell/Germany: ± 0.03 μms^{-2}),
- comparison with *relative gravimetry* is an independent control, but of course not effective on the absolute level (provided that calibration is based on external standards). Tidal registration and relative ties between absolute stations can be used under this aspect. As an example, we have relative connections carried out with several LCR gravimeters between JILAG-3 absolute stations in China. For adjacent (twin) station, the "absolute" and "relative" differences agreed to ± 0.06 μms^{-2}, increasing to ± 0.09 μms^{-2} for large distances. This mainly indicates the high instrumental reliability of the absolute gravimeter, but also the quality of modelling environmental effects,
- simultaneous measurements of different *absolute gravimeter systems* at the *same place*, possibly even internationally agreed as reference station. Following the initiative and continuous activity of the late Professor Boulanger, IAG has organized comparison campaigns at the Bureau International des Poids et Mesures (BIPM) in Sèvres/France, in 1981, 1985, 1989, 1994 and scheduled another one for 1997 (MARSON et a. 1995). Through the (quasi) simultaneous measurements and the homogeneous reduction procedures, the results of these comparisons especially reveal systematic discrepancies between the gravimeter systems, and the scatter about a long-term mean value (a change for the BIPM has not been found). The comparisons show a continuous progress with respect to the instrumental scatter about the mean value of one epoch (from ± 0.1 to ± 0.03 μms^{-2}), but also systematic effects of individual instruments of 0.1 to 0.2 μms^{-2} and once even up to 0.7 μms^{-2}. This clearly shows the need of these comparisons, in fact the successful participation of an absolute gravimeter at those campaigns meanwhile is a certain quality certificate. For JILAG-3 the mean value from 17 determinations between 1986 and 1994 deviates by + 0.04 μms^{-2} from the mean value derived from 43 determinations with other instruments. At the 1994 campaign, this deviation of the

JILAG-3 value (1 determination) from the mean of the other instruments (11 determinations) is + 0.03 μms^{-2}.

The worldwide distributed and well designed IAGBN stations can also be used for comparisons of this type. As the time interval between the observations then generally is larger the resulting discrepancies will contain also residual uncertainties of the environmental reductions. For JILAG-3, we have comparisons of that type in Brasilia (JILAG-3 minus JILAG-4: - 0.08 μms^{-2}, time interval 3 years), Beijing (JILAG-3 - JILAG-5: - 0.04 μms^{-2}, same year, and Wettzell/Germany (JILAG-3 - FG5-101: - 0.02 μms^{-2}/3 years).

It should be mentioned, that a comparison with *other reference networks*, established by absolute (and relative) gravimetry can be done also by accurately connecting those networks through high-precision relative gravimetry. Those connections then will show eventual systematic differences (bias, scale), and thus again give information on the quality of the regional or local realization of the gravity standard. An example for this strategy is the United European Gravity Network 1994 (BOEDECKER et al. 1995).

Summarizing, we clearly recognize the need of a continuous thorough control of the absolute gravimeter systems and the evaluation and reduction software, especially through comparisons with independent gravity determinations. If we exclude gross and systematic errors larger than 0.05 to 0.1 μms^{-2} (which still occur through different causes), a station mean value can be determined now with an accuracy of a few ± 0.01 up to 0.01 μms^{-2}, mainly depending on site conditions and on the reduction of vibration induced effects through the systems hard- or software. Long-term accuracy may come close to ± 0.05 μms^{-2} or even better with most advanced instruments, and at stable and well-controlled sites.

6. Conclusions

We have demonstrated that the role of gravity reference networks realizing the gravity standard since about 100 years, is changing drastically. Previous strategy started with a global gravity network, based on only a few absolute determinations, and the bulk of the network stations determined by relative methods. National reference networks were tied to the global system, again by relative gravimetry. Due to the progress in gravimetric techniques, this classical approach experienced a remarkable accuracy and reliability increase by around two to three orders of magnitude, between the 1880's and the 1970's.

The development of transportable free-fall absolute gravimeters since the late 1960's, and their still increasing operational use since the 1970's, brought another quality improvement by one to two orders of magnitude, and triggered the modern approach for the gravity reference. Gravity reference stations nowadays can be established for any dedicated purpose, within the frame of local, regional or global networks, without the need for a global reference system. This new strategy, on the other hand, requires a continuous careful quality control of the absolute gravimeter systems and the procedures employed for the reduction of environmental gravity effects, to be performed at least on the accuracy level of 0.01 to 0.05

μms^{-2} which can be reached today.

The International Association of Geodesy engaged itself strongly at the classical and the modern reference network approach, by either establishing global reference systems, or nowadays organizing intercomparison campaigns for absolute gravimeter systems. Future progress can be expected by improved and more handy hardware, with a larger number of instruments available, and by software improvements especially at the reduction of instrument-floor recoil effects. While the 0.02 μms^{-2} accuracy may be reached at this instrumental part of the gravity standard's error budget, the uncertainties at the reduction of environmental effects are more difficult to reduce to that level. Corresponding efforts have to be concentrated especially on modelling earth and ocean tides close to the coast, the direct and indirect effects of air pressure changes as well as gravity variation produced by groundwater and soil moisture variations on sediments.

References

Boedecker, G., Th. Fritzer (1986): International Absolute Gravity Basestation Network. Veröff. Bayer. Komm. für die Internat. Erdmessung, Astronom. Geod. Arb., Nr. 47, München.

Boedecker, G., I. Marson, H.-G. Wenzel (1995): The adjustment of the United European Gravity Network 1994 (UEGN 94). In H. Sünkel and I. Marson (eds.): Gravity and Geoid. IAG Symp. Proceed. 113, 82-91, Springer, Berlin-Heidelberg-New York.

Borras, E. (1911): Bericht über die relativen Messungen der Schwerkraft mit Pendelapparaten in der Zeit von 1808 bis 1909 und über ihre Darstellung im Potsdamer Schweresystem. Verh. 16. Allg. Conf. der Internat. Erdmessung, London 1909, III. Teil, Berlin.

Drewes, H., W. Torge, R. Röder, C. Badell, D. Bravo, O. Chourio (1993): Geodynamic research by absolute and relative gravimetry in Venezuela. In W. Torge, A. González-Fletcher, J.G. Tanner (eds.): Recent Geodetic and Gravimetric Research in Latin America. IAG Symp. Proceed. 111, 202-214, Springer, Berlin-Heidelberg-New York.

Faller, J.F., Y.G. Guo, J. Geschwind, T.M. Niebauer, R.L. Rinker, J. Xue (1983): The JILA portable absolute gravity apparatus. Bur. Grav. Int., Bull. d'Inf. 53, 87-97.

Franke, A., E. Reinhard, B. Richter, H. Wilmes, W. Torge (1998): The German Gravity Base Net 1994 (DSGN 1994). In: IAG Symp. Proceed., Springer, Berlin-Heidelberg-New York (in press).

Hammond, J.A., J.E. Faller (1971): Results of absolute gravity determinations at a number of different sites. J. Geophys. Res. 76, 7850-7854.

Helmert, F.R. (1901): Bericht über die relativen Messungen der Schwerkraft mit Pendelapparaten. Verh. der 13. Allg. Conferenz der Internationalen Erdmessung, 1900, II. Teil: Spezialberichte und wissenschaftl. Mitteilungen, 139 - 385, Berlin.

Helmert, F.R. (1910): Bericht über die Schwerebestimmungen. Verh. der 16. Allg. Conferenz der Internationalen Erdmessung 1909. 1. Teil: Sitzungsberichte und Landesberichte über die Arbeiten in den einzelnen Staaten, Berlin.

Kanngieser, E., K. Kummer, W. Torge, H.-G. Wenzel (1983): Das Gravimeter-Eichsystem

Hannover. Wiss. Arb. Fachr. Verm.wesen, Univ. Hannover, Nr. 120.
Kühnen, F., Ph. Furtwängler (1906): Bestimmung der absoluten Größe der Schwerkraft zu Potsdam mit Reversionspendeln. Veröff. Königl. Preuß. Geod. Inst., Neue Folge Nr. 27, Berlin.
Lenzen, V.F., R.P. Multhauf (1965): Development of gravity pendulums in the 19th Century. Contr. from the Museum of History and Technology, Bull. 240, paper 44, U.S. Governm. Printing Office, Washington, D.C.
Marson, I., J.E. Faller (1986): g - the acceleration of gravity: its measurement and its importance. J. Phys. Earth Sci. Instr., 19, 22-32.
Marson, I., J.E. Faller, G. Cerutti et al. (1995): Fourth International Comparison of Absolute Gravimeters. Metrologia 32, 137-144.
Morelli, C. (1946): Compensazione della rete internazionale della stazione de riferimento per le misure di gravita relativa. Inst. Geofis. (Boll. Soc. Adr. Sc. Nat., Vol. XLI), Publ. No. 211, Trieste.
Morelli, C., C. Gantar, T. Honkasalo et al. (1974): The International Gravity Standardization Net 1971 (IGSN 71). I.U.G.G.-I.A.G. Spec. Publ. No. 4, Paris.
McConnell, R.K., P.J. Winter, R.F. Geller (1979): Latin American Gravity Standardization Network 1977 (LAGSN 77). Earth Physics Branch, Ottawa, Canada.
Niebauer, T.M., G.S. Sasagawa, J.E. Faller, R. Hilt, F. Klopping (1995): A new generation of absolute gravimeters. Metrologia 32, 159-180.
Röder, R.H. (1994): Zum Einsatz des Absolutgravimeters JILAG-3 in Präzisionsschwerenetzen. Wiss. Arb. Fachr. Verm.wesen, Univ. Hannover, Nr. 205.
Sigl, R., W. Torge, H. Beetz, K. Stuber (1981): Das Schweregrundnetz 1976 der Bundesrepublik Deutschland (DSGN 76), Teil I. Deutsche Geod. Komm., Reihe B, Nr. 254, München.
Timmen, L. (1994): Untersuchungen zur Modellbildung bei der Auswertung absoluter Schweremessungen, Wiss. Arb. Fachr. Verm.wesen, Univ. Hannover, Nr. 204.
Timmen, L., H.-G. Wenzel (1994): Worldwide synthetic gravity tide parameters available on INTERNET. Bur. Grav. Int., Bull. d'Inf. No. 75, 32-40, Toulouse.
Torge, W. (1989): Gravimetry. W. de Gruyter, Berlin-New York.
Torge, W. (1991): The present state of absolute gravimetry. Cahiers du Centre Europeen de Geodynamique et de Seismologie, Vol. 3, 9-22, Luxembourg.
Torge, W. (1993a): Von der mitteleuropäischen Gradmessung zur Internationalen Assoziation für Geodäsie - Problemstellungen und Lösungen. Z.f.Verm.w. 118, 595-605.
Torge, W. (1993b): Gravimetry and tectonics. In J. Kakkuri (ed.): Geodesy and Geophysics. Publ. Finn. Geod. Inst. No. 115, 132-172, Helsinki.
Torge, W. (1996): The International Association of Geodesy (IAG) - More than 130 years of International Cooperation. J. of Geodesy 70 (The Geodesist's Handbook), 840-845.
Torge, W., R.H. Röder, M. Schnüll, H.-G. Wenzel, J.E. Faller (1987): First results with the transportable absolute gravity meter JILAG-3. Bull. Geod. No. 61, 161-176.
Torge, W., L. Timmen, R.H. Röder, M. Schnüll (1994): The IFE Absolute Gravity Program "South America" 1988-1991. Deutsche Geod. Komm., Reihe B, Nr. 299, München.
Woollard, G. (1950): The gravity meter as a geodetic instrument. Geophysics 15, 1-29.
Woollard, G.P., J.C. Rose (1963): International Gravity Measurements. Geophysical and Polar Research Center, Univ. of Wisconsin, Madison, Wisconsin.

IMPROVING ACCURACY AND RELIABILITY OF AIRBORNE GRAVIMETRY BY MULTIPLE SENSOR CONFIGURATIONS

K.P. Schwarz and C. Glennie
Department of Geomatics Engineering
The University of Calgary
2500 University Drive, NW,
Calgary, Alberta, Canada T2N 1N4
e-mail: schwarz@ensu.ucalgary.ca

Abstract

In this paper, the question is investigated to what extent multiple sensor configurations could contribute to the accuracy and reliability of airborne gravimetry and whether such designs are economical. Because of the latter requirement, special attention is given to the use of off-the-shelf strapdown inertial technology. As a first step, the error models of the two approaches used in strapdown inertial gravimetry - the SISG approach (Strapdown Inertial Scalar Gravimetry) and the RISG approach (Rotation Invariant Scalar Gravimetry) - are analyzed with respect to their usefulness for optimizing multiple sensor configurations. A preliminary conclusion is that, for the same accuracy, the RISG approach is much more economical than the SISG approach. In a second step, a numerical analysis was done using a set of airborne gravity data obtained with two independent strapdown systems flown in parallel. Results of the numerical analysis show that the error spectra of all configurations are highly correlated when compared to the reference gravity field and that their RMS values are typically between 3.0-3.5 mGal (1 mGal = 10^{-5} m/s^2). When comparing results of different systems, however, the RMS values are considerably reduced, typically to a level between 1.2-2.0 mGal. Thus, the agreement between different airborne systems is considerably better than the agreement between the airborne systems and reference gravity. This bodes well for the design of highly accurate and economical multi-sensor systems based on the RISG method.

1. Problem Statement

Airborne gravimetry by strapdown inertial sensors and differential GPS (DGPS) implements a differencing concept, i.e. gravity is obtained by differencing the outputs of two data streams, one of which contains the effect of gravity while the other one does not, see for instance Schwarz and Li (1996). The difference, when expressed as acceleration, represents the sum of gravity and measurement noise. The signal-to-noise ratio is especially low in the high and low frequency parts of the spectrum. A reliable estimation of gravity is therefore only possible in a limited spectral band which is often called the spectral window of airborne gravimetry. To widen this window, major efforts have gone into the improvement of filtering methods in order to achieve the best separation of signal and noise possible. In parallel with these efforts, considerable work has been done reducing the amplitude and range of the noise spectrum either by system design or by operational procedures. An approach which has not been investigated so far, is the multiple observation of the required signal. Since the noise varies for each system while the signal remains the same, a better separation of signal and noise can be expected.

There is little question that a multiple sensor configuration will add to system reliability. By configuring the sensors as independent airborne gravimeters, the likelihood for a simultaneous breakdown is considerably reduced. Even if two independent systems are not feasible, augmentation of current systems by additional accelerometers could considerably improve reliability. In terms of accuracy, the situation is not as clear. If the remaining noise in airborne gravimetry is white noise, noise reduction by a second independent system would be about 30%. Combined with the increased reliability, this would be attractive in many cases. However, current results seem to indicate that the residual error spectrum does not have white noise characteristics and that the accuracy gain will most likely be much smaller than 30%. To analyze and compare the noise spectrum of multiple sensor configurations in theory and in practice is therefore a major objective of this study. The effect of sensor errors, filtering and modelling methods, aircraft dynamics, and accuracy of the reference model will be considered in the following. Sensor error propagation through the SISG and the RISG model will be studied in the next section. In section 3, some of the other effects will be discussed.

2. *Error Models and System Design*

Error models of the SISG approach and the RISG approach have been derived in Czompo (1994), Wei and Schwarz (1994, 1996). In both approaches, differential GPS is used to determine acceleration of the aircraft. Thus, GPS measurement errors and errors of modelling acceleration from carrier phase data have to be taken into account. In both approaches, specific force is measured by sensors of the inertial system, in one case by orienting one accelerometer along the vertical (SISG), in the other case by determining the magnitude of the specific force vector from measurements of three orthogonal accelerometers (RISG). Thus in the first case, accelerometer and orientation errors will affect the error budget, while in the second case only accelerometer errors will play a role. Measurements from both systems are differenced to obtain the gravity signal. Thus, synchronization errors will also enter into the picture. The major error sources to be considered are therefore GPS measurement and modelling errors, INS accelerometer and attitude errors, and INS/GPS synchronization errors.

The significant errors affecting the SISG approach are

$$d\delta g = -da_u + \dot{f}_u dT + \mathbf{A} d\mathbf{f}^b + f_n \varepsilon_E - f_e \varepsilon_N \tag{1a}$$

where $\{a_e, a_n, a_u\}$ are the three components of kinematic acceleration **a** as determined by DGPS in a local-level frame; $\{f_e, f_n, f_u\}$ are the three components of specific force **f** in the local-level frame while $\{f_x, f_y, f_z\}$ are the components of the same vector in the body-frame; ε_E and ε_N describe the non-verticality of the upward accelerometer due to alignment errors and gyro drift; and $\dot{f}_u dT$ is the effect of the synchronization errors. The superscript 'b' denotes the body frame. A lower case d in front of any of these quantities denotes an error. The row matrix **A** is of the form

$$\mathbf{A} = (-\cos\theta\sin\varphi, \sin\theta, \cos\varphi) \tag{1b}$$

where φ and θ are the roll and the pitch of the strapdown system as determined by gyro measurements.

The significant errors affecting the RISG approach are

$$d\delta g = -da_u + |\dot{\mathbf{f}}|dT + \mathbf{B}d\mathbf{f}^b - \frac{a_e}{f_u}da_e - \frac{a_n}{f_u}da_n \tag{2a}$$

where

$$f_u = \left\{|\mathbf{f}^b|^2 - a_e^2 - a_n^2\right\}^{1/2}$$

is the upward component of the specific force $\mathbf{f}$ in the local level system, and $|\dot{\mathbf{f}}|dT$ again denotes the synchronization error. The row matrix $\mathbf{B}$ is of the form

$$\mathbf{B} = \left(\frac{f_x}{f_u}, \frac{f_y}{f_u}, \frac{f_z}{f_u}\right) \tag{2b}$$

A comparison of formulas (1a) and (2a) shows a very similar structure, however, only the term da_u is identical in both equations. This means that the effect of errors in GPS-determined vertical acceleration is the same. The effect of a synchronization error dT and accelerometer errors $\mathbf{df}$ will only be the same for the case of a flight at constant velocity and constant height. This will be shown for the accelerometer term. Mathematically, the effect of $\mathbf{df}$ will be the same in equations (1a) and (1b) if $\mathbf{A} = \mathbf{B}$. This means that the following equations must be satisfied:

$$\frac{f_x}{f_u} = \cos\theta\sin\varphi \qquad \frac{f_y}{f_u} = \sin\theta \qquad \frac{f_z}{f_u} = \cos\theta\cos\varphi \tag{3a}$$

If pitch and roll are zero, i.e. $\mathbf{f}^b = \mathbf{f}^\ell$, where 'l' denotes the local-level frame, and if the effect of the deflections of the vertical is neglected, we obtain

$$\frac{a_e}{f_u} = 0 \qquad \frac{a_n}{f_u} = 0 \qquad \frac{f_u}{f_u} = 1 \tag{3b}$$

For these equations to be valid, no horizontal accelerations can be present. Thus, for a constant velocity flight at constant height, accelerometer errors will affect the RISG method in the same way as the SISG method, if the b-frame is mounted in such a way that it corresponds to the l-frame. Since horizontal flights with constant velocity are the desired flight scenario for airborne gravimetry, this situation is often approximated in practice. If pitch and roll are small, the small-angle approximations $\sin\alpha = \alpha$ and $\cos\alpha = 1$ can be used. Applying it to equation (3a) shows that rotational motion will only be a second-order effect in this case. However, horizontal acceleration can be considerable and would tend to make the effect of these terms different in the RISG system. Considering, however, that horizontal accelerations are rarely larger than 0.5 m/s^2 while g is 10m/s^2, the resulting effect is only 5% of the horizontal accelerometer errors, as long as the b-frame is close to the l-frame.

Using the same approximation as before, the remaining two terms in equation (1a) can be written as

$$f_n\varepsilon_e - f_e\varepsilon_N \approx a_n\varepsilon_e - a_e\varepsilon_N$$

These terms are not matched by the remaining terms in equation (2a). While the ε-terms in equation (1) denote an orientation error in the attitude matrix, the da-terms in equation (2as) denote errors in the DGPS acceleration determination. Considering the size of d**a**, the effect of this error in equation (2a) will always be small. The effect of the ε-terms in equation (1a) will only be small if the horizontal accelerations are small and short term.

This brief analysis leads to some important conclusions for multiple sensor configurations in airborne gravimetry.

1. The SISG and RISG methods will show a similar error behaviour for flight conditions which approximate a constant-velocity/constant-height scenario if the b-frame is roughly parallel to the l-frame. The RMS error will be of the same magnitude in this case. This means that the RISG approach, which is much more economical than the SISG approach, should be used for the design of multiple sensor configurations.
2. Using the SISG and RISG approach with data from the same system will be a reliable check on the data processing, but will result in highly correlated errors. Thus, an averaging of the two solutions will not increase accuracy.
3. Mounting the b-frame for the RISG design approximately parallel to the l-frame has the advantage that the noise contribution of the horizontal accelerometers becomes negligible. This means that total system performance can be improved by simply improving the accuracy of the vertical accelerometer.
4. Mounting a number of vertical accelerometers on the same platform seems to be the most economical way of increasing the accuracy of the RISG configuration, as long as accelerometer errors with white noise characteristics dominate the overall error budget.
5. If accelerometer performance is affected in a major way by aircraft dynamics, the errors of parallel accelerometers will most likely be strongly correlated. In this case, a more complex design may be needed.

3. Test Design, Instrumentation, and Expected Accuracy

The data analyzed in the following are part of a larger test which was flown in September 1996 over very rugged terrain in a part of the Rocky Mountains close to Banff, Alberta. More details on this test are given in Glennie and Schwarz (1997). The data were acquired on the first day of testing (Sept. 9) when an area of 100 by 100 km was covered with east-west flight profiles of about 10 km spacing at an average altitude of 4350 m. The height of the terrain varied from about 800 m to 3600 m and gravity anomalies at flight level from about -70 mGal to about +100 mGal. The data were acquired at night (24h to 6h) using a Conquest aircraft flying at an average velocity of 360 km/h.

To operate DGPS reliably and with sufficient accuracy, three GPS ground stations were used of which two were directly in the flight area; thus, the distance between ground stations and aircraft never exceeded 100 km. Various types of receivers were used for ground stations and aircraft (Trimble SSE and SSI, Ashtech Z12, NovAtel GPSCard). Since previous tests had shown that, for acceleration determination, there was no significant difference between different receiver types (Wei and Schwarz, 1996), a variety of combinations were used in the processing.

The inertial component of the system configuration consisted of two strapdown inertial systems. The first is a Honeywell Laserref III system which is equipped with a triad of dithered ring laser gyros (GG1342) and a triad of quartz flexure accelerometers (QA 2000), developed by AlliedSignal. The second is the Litton LTN-101 system which has a triad of undithered zero-lock ring laser gyros and a triad of accelerometers of the pendulous type (A4) developed by Litton. Both systems represent state-of-the-art technology for navigation grade inertial strapdown systems with performance characteristics of about 0.5 nm/h. The standard system output is 50 Hz. Table 1 gives the noise level (standard deviation) of the two vertical accelerometers under ideal conditions, i.e. as achieved in lab experiments, for filtering periods of 30, 60, 90, and 120 seconds. These are typical filter times used in airborne gravimetry. The noise spectra are not white but have some spikes in the low frequencies. The standard deviations represent the lower limit for the accuracy achievable with these systems and filtering periods under ideal conditions. The results are considerably better for the QA 2000 accelerometer used in the Honeywell system. This may be somewhat influenced by the difference in solving the quantization problem and may not reflect on the inherent accuracy of each sensor. By replacing the current accelerometers by sensors of higher accuracy, better results can be achieved. This is relatively simple for the Honeywell system where the QA 3000 can be used as a retrofit.

INS	120 s	90 s	60 s	30 s
LRF III	1.3	1.3	1.5	1.8
LTN-101	3.1	3.5	3.8	5.0

Table 1: Standard Deviation of Z-Accelerometer Noise in Static Tests (m Gal)

A second limitation on the accuracy of the results is the quality of the reference gravity field. It is computed as an upward continuation from ground gravity data and a digital elevation model (DEM) available in the area. The methods used to do this are described in Argeseanu (1995). In the area under consideration, point gravity anomalies are available at an average spacing of 10 km and point elevations are given on a 1 by 1 km grid. For a small part in the centre of the area flown, point gravity anomalies have an average spacing of about 3 km and the DEM is spaced at 100 m intervals. To take the differences in data coverage into account, a Poisson-integral approach was used for the upward continuation. Considering the data and taking into account the uncertainties in the global geopotential model used for the computations, the accuracy of the upward continued gravity is probably not better than 2 mGal for the whole area and 1.5 mGal for the central part. Thus, the best resolution at flight altitude to be expected from this data set is about 2.5 mGal for the Honeywell system and about 3.5 mGal for the Litton system, assuming that aircraft dynamics and other sensor errors do not play a significant role.

4. Analysis of Results

Table 2 gives an overview of the results achieved with the SISG and the RISG methods for different filtering periods and different inertial systems. They are shown for three representative profiles in the centre of the flight area where the reference gravity field is expected to be best. All results are root mean square differences (RMS) between filtered airborne gravity and reference gravity in units of mGal. Analyzing the table while keeping in mind the findings of section 3, leads to the following conclusions:

• For both systems, the RMS values are in the expected range for filtering periods of 90 and 120 seconds. They are much larger than expected for the 30-second filter. This shows that the phugoid motion, whose period is 55 seconds for this airplane, heavily influences error behaviour, see also Salychev and Schwarz (1995). The proximity to this critical frequency also explains the slightly poorer results for the 60-second filter.

• When comparing the two systems, results are very similar for the 90 and 120 second filter. Thus, for these filtering periods the higher noise level of the Litton accelerometers seems not to affect the overall results, while it does for the shorter filtering periods. This result is somewhat puzzling, although some of it can be explained by the large amplitudes of the Litton high frequency error spectrum.

• For both systems the SISG method generally outperforms the RISG method by 10% to 20%. This has been observed in an earlier analysis, see Wei and Schwarz (1996) and, at this point, cannot be fully explained.

Line	LRF III SISG				LRF III RISG			
Number	120 s	90 s	60 s	30 s	120 s	90 s	60 s	30 s
4	3.0	3.2	4.0	11.6	4.1	4.3	4.3	13.6
5	3.4	3.0	3.4	10.3	3.8	3.5	3.4	9.1
6	3.2	2.7	2.9	7.4	3.7	3.4	3.8	9.6

Line	LTN-101 SISG				LTN-101 RISG			
Number	120 s	90 s	60 s	30 s	120 s	90 s	60 s	30 s
4	3.2	3.3	4.5	29.1	3.6	3.9	6.1	34.4
5	3.4	3.4	4.3	26.6	3.9	3.9	5.7	31.6
6	3.4	3.3	5.4	33.8	3.4	3.6	6.8	39.4

Table 2: RMS Errors Between Filtered Airborne Gravity and Reference Gravity (mGal)

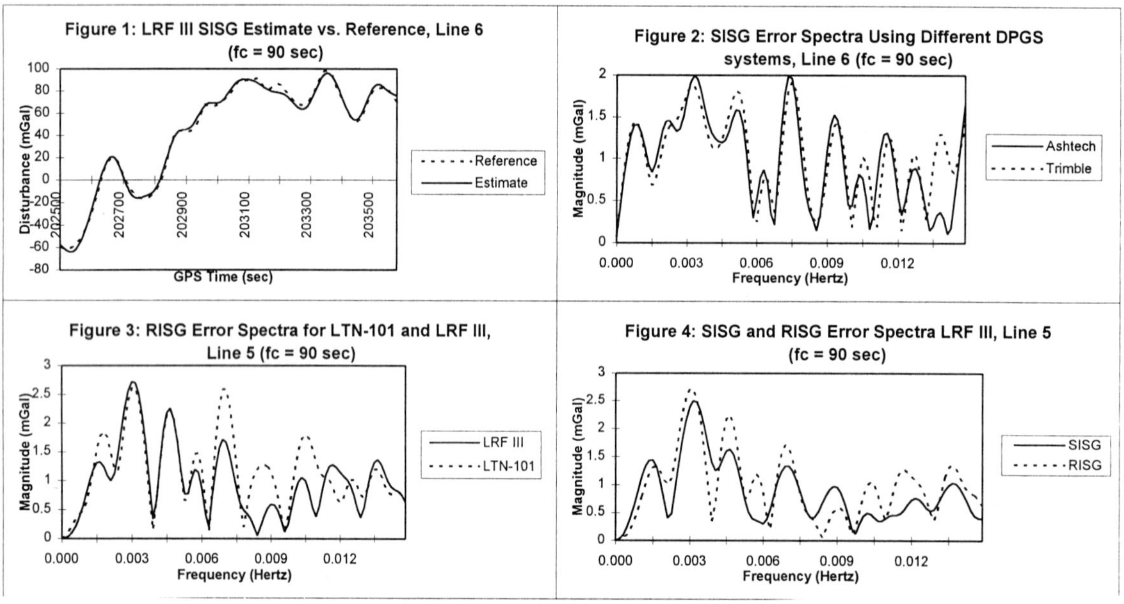

Figures 1-4: Error Estimates and Spectra

To further illustrate some of the above results, Figures 1 to 4 will be briefly discussed. Figure 1 shows a time domain plot of a typical flight profile. It not only shows the good agreement between airborne results and gravity reference, but also indicates that a 90-second filter has sufficient resolution to estimate the essential gravity field features at this altitude. Figure 2 was generated to illustrate that the choice of a specific GPS receiver does not have a major effect on the error spectrum. It shows the error spectra of two different DGPS pairs with the same inertial system. For the low frequencies, the deviations of the two spectra are less than 10% of the total error. Figure 3 explains why the RMS values of the two systems are so close for filtering periods of 90 and 120 seconds. The error spectra in this part of the spectrum are very similar and have little resemblance to the error spectra obtained in static mode. This will be further discussed in the next paragraph. Figure 4 shows the similarities in the error spectra of the SISG and the RISG method. They should be seen in light of the simplified error analysis in section 2 for constant velocity/constant height flights.

The results presented in Table 2 and Figures 1 to 4 indicate that the dominant frequencies in the error spectrum are the same for all system configurations when using filter lengths of 90 and 120 seconds. Possible reasons for this are systematic errors generated by the filtering approach, errors in the upward continued reference gravity field, or errors due to flight dynamics. To investigate this further, SISG and RISG solutions for the LRF III and the LTN-101 were directly compared. The RMS of their differences are given in Table 3. Results show that the RMS values are dramatically reduced for the 120s-filter, on average by 40-50%. In addition, the RMS values are between 1.2 mGal and 2.7 mGal which corresponds to the noise level of the accelerometers. For the 90s-filter the average improvement is 20-25%. These results indicate that the agreement between different airborne systems is considerably better than the agreement between the airborne systems and reference gravity. This either points to errors in the reference gravity field or to systematic errors in the filtered flight data which affect all systems in the same way. Based on the data currently available, it is not possible to decide on the cause of the errors, although errors in the reference gravity seem to be more probable. If this proves to be the case, a multi-sensor design with parallel vertical accelerometers will considerably improve the accuracy of airborne gravimetry.

	SISG		RISG	
Line Number	120 s	90 s	120 s	90 s
4	1.8	1.9	1.9	2.6
5	1.2	1.9	1.2	1.7
6	2.0	3.1	2.7	3.2

Table 3: RMS Errors between LRF III Results and LTN-101 Results (mGal)

References

Argeseanu, V.S. (1995). Upward Continuation of Surface Gravity Anomaly Data. IAG Symp. G4 on **Airborne Gravimetry**, IUGG XXI General Assembly, Boulder, Col., July 2-14, 1995, 95-99. Published by The University of Calgary.

Czompo, J. (1994). Airborne Scalar Gravimetry System Errors in the Spectral Domain. Ph.D. Thesis, UCGE Report No.20067, Dept. of Geomatics Engineering, The University of Calgary, Calgary, Alberta.

Glennie, C. and K.P. Schwarz (1997). Airborne Gravity by Strapdown INS/DGPS in a 100km by 100km Area of the Rocky Mountains. Proc. Int. Symp. on **Kinematic Systems in Geodesy, Geomatics and Navigation, KIS97**, Banff, Canada, June 3-6, 1997, 619-624.

Salychev, O.S. and K.P. Schwarz (1995). Airborne Gravimetric Results Obtained with the ITC2 Inertial Platform System. IAG Symp. G4 on **Airborne Gravimetry**, IUGG XXI General Assembly, Boulder, Col., July 2-14, 1995, 125-141. Published by The University of Calgary.

Schwarz, K.P. and Z. Li (1996). An Introduction to Airborne Gravimetry and Its Boundary Value Problems. IAG Int. Summer School on **Boundary Value Problems in Geodesy,** Como, May 1996; Springer Verlag, 1997, 312-358.

Wei, M. and K.P. Schwarz (1994). An Error Analysis of Airborne Vector Gravimetry, Proc. International Symposium on **Kinematic Systems in Geodesy, Geomatics and Navigation**, KIS94, Banff, Canada, August 30 - September 2, 1994, 509-520.

Wei, M. and K.P. Schwarz (1996). Comparison of Different Approaches to Airborne Gravimetry by INS/DGPS. Proc. of the IAG Symp. no.117 **Gravity, Geoid, and Marine Geodesy (GraGeoMar96),** Tokyo, Sept.30-Oct.5, 1996. Published by Springer Verlag, 1997, 155-162.

COMBINING AIRBORNE AND GROUND GRAVITY USING COLLOCATION

C.C.Tscherning
Dept. of Geophysics, Juliane Maries Vej 30, DK-2100 Copenhagen Ø, Denmark,
F. Rubek, R. Forsberg
KMS, Rentemestervej 8, DK-2400 Copenhagen NV, Denmark.

Abstract. In the Greenland Aerogeophysical Project (GAP) airborne gravity data have been collected covering all of Greenland. In addition the coastal areas are covered by ground gravity data. Improved gravity field modeling results may be obtained by combining the two data sources. Least Squares Collocation (LSC) may be used with advantage for this purpose. The method furthermore allows for (a) the correct modeling of the (filtered) observation functional, (b) the use of an error correlation function, (c) the determination of biases in the airborne data and (d) the estimation of the errors of the quantities calculated for the obtained gravity field models. We describe here the results obtained using LSC in the two areas with good ground data coverage. The basis for using LSC is the estimation and subsequent analytic modeling of the empirical covariance function. However, this modeling was only successful when a model error correlation function was subtracted from the covariance function of the airborne data. The variances calculated from the ground data were used to control that the variances derived from the airborne covariance had correct values. This made it possible to correctly downward continue the GAP data using LSC. When comparing observed and ground data predicted from the GAP data, biases were found. This is probably due to systematic errors in the airborne data or systematic errors from terrain effects. The bias was removed by assigning a bias-parameter to each track and adding (a part of) the ground data as new observations. The remaining ground data were used to verify that the bias has been removed. The results, however, were in both areas only marginally better than those obtained when using one ground gravity sub-data set to predict the other. On the other hand they were much better than results obtained using LSC or frequency domain collocation for the prediction of ground gravity from the GAP data only.

INTRODUCTION

An airborne gravity survey covering all of Greenland was carried out in 1991 and 1992 by the US Naval Research Laboratory in cooperation with Defence Mapping Agency, Naval Oceanographic Office and KMS. The Greenland Aerogeophysics Project (GAP) is described in e.g. Brozena (1992). The data have been analyzed in several publications, e.g. Forsberg & Brozena (1993), Forsberg & Kenyon (1994, 1995), Rubek (1997). Here we study the problem of optimally combining the GAP data with the ground gravity data, with the purpose of determining the best gravity model for Greenland.The method of Least Squares Collocation (Moritz, 1980) is ideal for this purpose. It not only permits the combination of the two types of data, but also the determination of systematic errors (e.g. in the form of biases), the estimation of prediction errors, the correct representation of the along-track filtered GAP data, and the use of individual and possibly correlated errors.

We have selected two areas in Northern Greenland with a good ground gravity coverage and digital elevation model, cf. Table 1 and Fig. 1 & 2. Despite all its nice properties, it is not possible to use LSC to create a gravity model while using all the GAP and the ground data in one run. A system of equations with as many unknowns as the number of data would have had to be solved, making solutions in smaller blocks necessary.

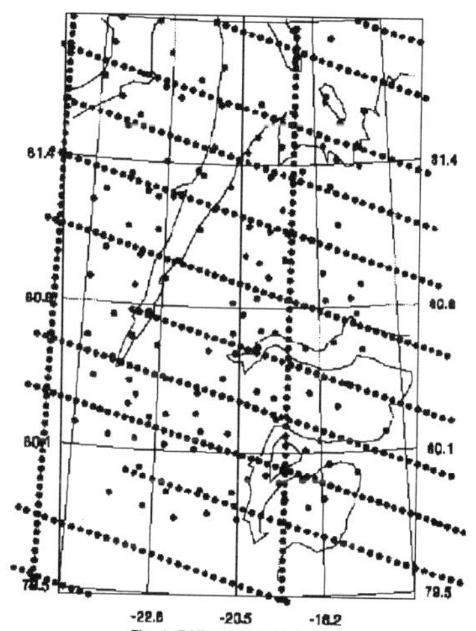

Fig. 1. GAP and ground gravity.

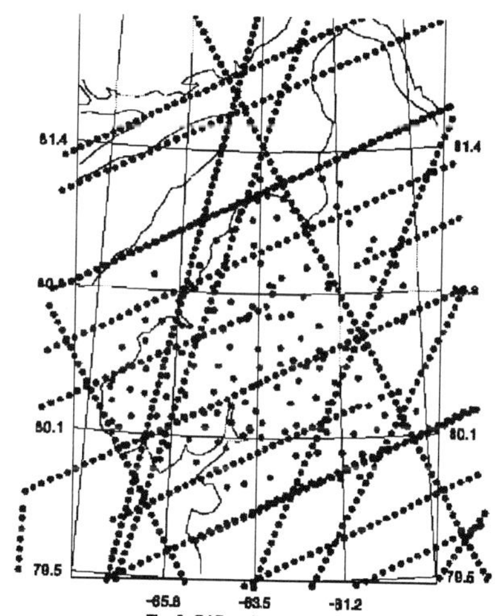

Fig. 2. GAP and ground gravity.

Table 1. Location of test areas, and number of data points.

Min Latitude, deg.	Max Latitude, deg.	Min Longitude, deg.	Max Longitude, deg.	Number of airborne data	Number of ground data
79.25	82.1	-26.0	-14.5	691	139
79.0	82.3	-70.0	-47.0	2126	151

DATA SELECTION AND PREPROCESSING

The airborne and surface gravity data have prior to the use of LSC been reduced by subtracting the contribution from EGM96 (Lemoine et al., 1996) and the terrain effects due to residual topography using standard procedures. A dense topographic elevation model of 500 m resolution obtained from KMS digital mapping projects was used. The model does not include information about the depth of the fjords, which are in the test areas. This might be a source of a bias, since the terrestrial gravity data are only measured on land, whereas the airborne data samples the land, fjords and the ice areas randomly. A typical Greenland fjord will have a large negative gravity anomaly due to the bathymetry.

When evaluating the topographic as well as the spherical harmonic contributions, it was taken into account that the airborne data are along-track filtered data. The filter used in the GAP processing is symmetric and has a resolution around 12 - 15 km (Brozena, pers.comm.). It has been implemented in the LSC procedure using 11 weighted point values to represent the filter in a 60 km filter window.

The analyzed GAP data were ordered by flight tracks. Values at or very close to track crossing points does not have to be equal, since they are weighted means along track. However, the values should in general be quite close, since a cross-over adjustment has been performed as part of the original GAP data processing. A calculation of root mean square values of differences between values within 1 km of track cross points gave values of from 6 to 15 mgal for equal-area blocks with sidelength 6°. For the two test areas the values were 11 mgal for the eastern block and 7 mgal for the western block. This indicated that there might be systematic errors present, larger than the standard error of the data of 4.6 mgal for the single track noise obtained from original cross-over adjustments (Brozena, pers.comm.), but it might also reflect that in the present study some additional aerogravity data from tracks with only Bell gravimeter data have been used.

COVARIANCE ESTIMATION

The use of the LSC method requires the estimation and analytic modeling of a covariance function. In an ideal situation ground data alone would have been used for covariance estimation, due to its low, uncorrelated noise. But for Greenland the data spacing is so large that only the rough shape of the covariance function can be estimated. Consequently the GAP data had to be used, but in such a way that only products of values located on the same track were used to calculate the covariances.

The resulting empirical covariances were unusually smooth for small distances, reflecting the inherent along-track filtering of the airborne data. Despite the correlated noise present we tried to determine an analytic representation of the two covariance functions, using a version of the program COVFIT (Knudsen, 1987). This was not successful, since the derived ground gravity variance became many times too large. The estimated radius of the Bjerhammar sphere in the used T/R-covariance model, (Tscherning & Rapp, 1974) was too close to the Earth mean radius.

When a 1-D covariance function (Sanso and Schuh, 1987) with 20 mgal^2 zero-lag value, was subtracted, values consistent with the ground gravity variance were obtained. However, the ideal procedure would have been the use of a covariance function derived from the ground data. The difference between the upward-continued (and filtered) covariance function and the GAP covariance function would in principle then have given an estimate of the error-covariance function of the GAP data.

NUMERICAL EXPERIMENTS

A number of numerical experiments have been carried out by (Rubek, 1997), but without taking the occurrence of correlated errors, or possible track-related biases into account. Both LSC and frequency domain collocation (FFT) were used, giving similar results. Here we only report on results obtained using LSC and taking the correlated errors into account. The results are summarized in Table 3.

First ground data were predicted from the GAP data. The mean (and also the standard deviation) of the differences between observed and predicted values are quite large, but it should of course be pointed out that estimating point values from along-track averaged airborne data will always be a noisy process. Furthermore a comparison of the prediction error with the absolute value of the differences observed - predicted showed that the percentage of differences larger than the error estimate was too large, see (Tscherning & Knudsen, 1986). This could indicate possible (track related) biases in the GAP data.

The LSC method includes the possibility for the estimation of parameters such as track biases, and this is implemented in the program GEOCOL (Tscherning, 1997). Consequently a bias parameter was associated with each track, and its value as well as the error was estimated. The 5 largest biases and their values as estimated from the GAP data only and from GAP data combined with ground data are found in Table 2. Note the large differences between the results with and without ground data, and the decrease in the error estimate when ground data are added. (At this point it would have been most correct to re-estimate the empirical covariance function from the LSC-filtered GAP-data, however we have aimed here to show the potential of LSC and not to produce the optimal result).

Table 2. Five largest biases and their standard deviations (units mgal) from the two test areas using GAP data only and GAP data combined with ground data.

Eastern area				Western Area			
Bias from GAP	Error	Bias from GAP + ground	Error	Bias from GAP	Error	Bias from GAP + ground	Error
-14.84	6.33	-27.01	4.66	16.12	7.44	17.17	7.43
-12.40	5.02	-16.17	4.88	-14.49	4.54	-19.70	4.46
-10.86	3.91	-12.24	3.81	-13.70	3.69	-19.37	3.61
-10.00	5.64	-22.72	4.04	-11.69	3.29	-16.22	3.19
-8.31	3.28	-10.87	3.05	-10.22	3.28	-16.03	3.05

The gravity prediction results, cf. Table 3, did also improve when ground gravity were added. Only every second of the ground data points were used in the form of observation data. The remaining values were used as control data. The biases have disappeared, but note that the result is not better than when the ground "observation" data set was used to predict the control data set. The error estimates, however, are 1 - 2 mgal higher than for those where the GAP and the ground data were combined.

Table 3. Mean and standard deviation (mgal) of observed minus computed ground gravity values computed from ground data only, from GAP data only, from GAP data only including bias-estimates and from GAP data and ground gravity including bias estimates.

Eastern area				Western area			
Number of data +parameters		Obs.-predicted		Number of data +parameters		Obs.-predicted	
GAP data	Ground data	Mean	Standard deviation	GAP data	Ground data	Mean	Standard deviation
0	70	0.76	6.07	0	76	2.89	7.07
691	0	5.45	9.09	2126	0	8.54	14.96
691+25	0	3.46	7.95	2126+62	0	7.52	13.82
691+25	70	0.31	5.96	2126+62	70	0.31	5.96

CONCLUSIONS

We have demonstrated how LSC may be used to combine airborne and ground gravity data, especially taking into account the along-track filtering inherent in the airborne gravity data. The LSC method, which includes the possibility for bias estimation, should be used if there is an indication that data contain biases. A traditional cross-over analysis will not always remove biases efficiently, and a bias-and-tilt procedure might actually degrade data in smaller sample areas. Estimation of biases in airborne data from surface data should, however, only be done with caution, due to the possibility of biases arising from terrain corrections, e.g. from insufficient digital elevation models or unmodeled effects of fjord depths and glaciers.

The use of a model for correlated errors is important, in order that downward continued values have a a correct variance. The combined use of the two data types much improves the result as compared to only using airborne data.

The challenge now is to find a procedure allowing large areas to be covered in one LSC run. In this way we can take advantage of ground data in the coastal regions to remove biases on tracks also passing areas without ground data. The solution here may simply be to only use every third or fourth value. This may give data sets of size 5000 - 8000 observations which are not difficult to handle. Naturally the computational effort will be large, considering that the numerical evaluation of a covariance involves the calculation of 11 x 11 point covariances.

REFERENCES

Brozena, J.: *The Greenland Aerogeophysics Project. Airborne Gravity, topographic and magnetic mapping of an entire continent.* In: O.Colombo (Ed.): From Mars to Greenland. Proc. IAG Symp. G3, Vienna, Austria, Aug. 1997, Springer Verlag, 1992.

Forsberg, R. and J.Brozena: *The Greenland Airborne Gravity Project - Comparison of Airborne and Terrestrial Gravity Data.* In: W.Torge (Ed.), Geodesy and Physics of the Earth, Proceedings IAG Symp. 112, Potsdam, Oct. 1992, pp. 171-175, Springer Verlag, 1993.

Forsberg, R. & S.Kenyon: *Evaluation and downward continuation of airborne gravity data - The Greenland example.* Proc. KISS94, Banff, Canada, pp. 531-538, 1994.

Forsberg, R. & S.Kenyon: *Downward continuation of airborne data.* Proc. Symp. On Airborne Gravimetry, IUGG XXI General Assembly, Boulder, 1995.

Knudsen, P.: *Estimation and Modelling of the Local Empirical Covariance Function using gravity and satellite altimeter data.* Bulletin Geodesique, Vol. 61, pp. 145-160, 1987.

Lemoine, F.G., D.Smith, R.Smith, L.Kunz, E.Pavlis, N.Pavlis, S.Klosko, D.Chinn, M.Torrence, R.Williamson, C.Cox, K.Rachlin, Y.Wang, S.Kenyon, R.Salman, R.Trimmer, R.Rapp and S.Nerem: *The development of the NASA GSFC and DMA joint geopotential model.* Proc. Symp. on Gravity, Geoid and Marine Geodesy, Tokyo, 1996.

Moritz, H.: *Advanced Physical Geodesy.* H.Wichmann Verlag, Karlsruhe, 1980.

Rubek, F.: *Comparison of three different methods for harmonic downward continuation of airborne gravity measurements.* Master Thesis, Dep. of Geophysics, University of Copenhagen, 86 p., 1997.

Sanso, F. and W.-D. Schuh: *Finite Covariance Functions.* Bulletin Geodesique, Vol. 61, pp. 331-347, 1987.

Tscherning, C.C.: *Current Problems in Gravity Field Approximation.* Proceedings 1. Hotine-Marussi Symposium, Rome, June 3-6, 1985, pp. 363-384, 1986.

Tscherning, C.C.: *GEOCOL - A FORTRAN-program for Gravity Field Approximation by Collocation.* Technical Note, Geophysical Institute, University of Copenhagen, 12. ed., 1997.

Tscherning, C.C. and P.Knudsen: *Determination of bias parameters for satellite altimetry by least-squares collocation.* Proceedings 1. Hotine - Marussi Symposium, Rome, June 3-6, 1985, pp. 833-852. Politecnico di Milano, 1986.

Tscherning, C.C. and R.H.Rapp: *Closed Covariance Expressions for Gravity Anomalies, Geoid Undulations, and Deflections of the Vertical Implied by Anomaly Degree-Variance Models.* Reports of the Department of Geodetic Science No. 208, The Ohio State University, Columbus, Ohio, 1974.

COMPARISONS BETWEEN ABSOLUTE (AG) AND SUPERCONDUCTING (SG) GRAVIMETERS

Martine Amalvict & Jacques Hinderer
Institut de Physique du Globe, 5 rue René Descartes, 67 084 Strasbourg, France
corresponding e-mail: martine@selene.u-strasbg.fr
Olivier Francis (ORB, Brussels, Belgium)
J. Mäkinen (Geodetic Institute, Helsinki, Finland)

Abstract

The French scientific community has recently (December 1996) acquired an absolute gravimeter (AG) , Micro-g FG5-206. This instrument is based at the Strasbourg gravity station, J9, where superconducting gravimeters (SG), model TT70 replaced in summer 1996 by a compact model C026, are operating since more than 10 years. First we check the instrumental specifications of the FG5 by comparing observations to the specifications given by the manufacturer. Secondly we perform several series of calibrations of the SG by the AG spanned over a time interval of 6 months. We provide results on the stability of the calibration factor in time from different experiments, as well as on the evolution of the factor and its uncertainty according to the duration of each experiment. We conclude that an accuracy of several parts per mill can be achieved in standard conditions.

I - Instrumental specifications

The Micro-g FG5 absolute gravimeter is a free-fall type of gravimeter. According to the manufacturer, its main specifications are as follows (see Niebauer et al., 1995):

- **accuracy** : ± 2 μGal*
- **precision**: ± 1 μGal
- **measurement time**: less than 2 hours for 1 μGal precision in a quiet site.

Set standard deviation

A possible way to approach the accuracy of the AG measurements is to use the Set Standard Deviation (SSD) which expresses the dispersion of the set mean values with respect to the session mean (mean of the set means). In fact, strictly speaking, one cannot constrain the accuracy from a single instrument alone as stated by Okubo et al. (1997); for instance, the SSD could be small but the mean gravity wrong due to a laser frequency shift for example.

At the moment, the FG5 has been operating in three sites: Strasbourg-J9, a site in the Vosges Mountains, named Welschbruch, and Membach (Belgium), the site of the C021

* 1 μGal = 1 x 10^{-8} m s^{-2}

superconducting gravimeter. The results are given in Table 1. The station J9 is known to be a medium quality site, with respect to microseismic noise (Freybourger et al., 1997) mainly because it is located on the sediments of the Rhine Graben. This medium quality is what we observe. Nevertheless, the value of the Set Standard Deviation is close to the one given by the manufacturer and much better than the best SSD obtained by the Jilag5 operating at this site. The Welschbruch is a station which is used for magnetic measurements, but no gravity measurements have been performed before this year. It is not yet a very well equipped station for gravimetry (we are testing it): there is no pillar or thermal stability. Nevertheless we can see that the results are very good; without any peculiar care neither in the measurements nor in their treatment, the SSD is smaller than the manufacturer's specification. Membach is known to be a very good site, located in a mine, very stable in temperature and very quiet. The set standard deviation is much smaller than the one given by the manufacturer, and very close to the value (1.35) of the FG5 from the ROB which was operating at the same time in the same site. More complete results about this experiment of two FG5s measuring in parallel can be found in Hinderer et al. (1997). There is a recent study in Japan (Okubo et al., 1997) which could also verify that the precision of the FG5 measurements is between 1 and 2 μGal.

	Manufacturer	J9 - Strasbourg (F)	Welschbruch (Vosges, F)	Membach (B)
Set Standard Deviation FG5 - 206	2.0	2.29	1.69	1.4
other AGs		3.6 Jilag 5		1.35 FG5 - 202

Table 1 - Set Standard Deviation of the AGs in different sites (in μGals).

Single set standard deviation

The precision of the AG measurements can be inferred from the Single Set Standard Deviation (SSSD) which expresses the dispersion of the drop by drop values in a set which is usually of short duration (~15 min).

The specified value seems easy to reach (cf. Table 2). Usually the SSSD is obtained from the Mean Standard Deviation (MSD), which expresses the single drop scatter, divided by the square root of the numbers of drops per set. The single set standard deviation we give for the FG5-206 was obtained in the Vosges site, Welschbruch, which seems to indicate a very low seismic noise as compared to the value obtained in Membach.

	Manufacturer	FG5 - 206 (Vosges)	FG5 - 202 (Membach)
Single Set Standard Deviation (μGal)	1.0	0.9	1.2

Table 2 - Single Set Standard Deviation of FG5s (in μGal).

Measurement time

This is the duration of measurements which is necessary to get a value of the mean gravity g, at a given precision. It will be, of course, depending on the quality of the site and on the weather conditions.

If we assume 200 drops per hour, and a 2-hour experiment and consider the Mean Standard Deviation obtained in the different sites, we can estimate the precision as: $\frac{MSD}{\sqrt{400}} = \frac{MSD}{20}$, assuming an ideal white noise distribution of the measurement errors. The precision obtained in the three sites, under the assumptions of this fictitious experiment, is given in Table 3 where the MSD values are typical for the considered stations.

Theoretically, we can conclude that two hour duration of measurements seems to be enough for getting a value of the mean gravity with a precision of 1μGal, and even better. Once more, the Welschbruch station appears to be a very promising site. The value of g at 1μGal precision could be reached in fifteen minutes.

	Manufacturer	J9 - Strasbourg	Welschbruch	Membach
MSD		17	7	9
precision	1.0	0.85	0.35	0.45

Table 3 - Mean Standard Deviation of the FG5 (in μGal).

Actually, in real experiments, one has to measure for much longer time, to stabilize the mean gravity measurement according to the dispersion of individual set mean values which is generally observed.

II - Calibration results

The first superconducting gravimeter (SG TT70) at the Strasbourg site J9 has been calibrated almost every year since 1989 with the Jilag 5 absolute gravimeter of the Finnish Geodetic Institute (Hinderer et al., 1991, 1995). In July 1996, it was replaced by a compact SG (SG C026) which has been calibrated for the first time in February 1997, in parallel by the Jilag 5 Finnish instrument, in order to insure the continuity of experiment, and by the new FG5 AG.

The calibration is performed by least squares fitting at every time t_i the values of the AG y_i (in μGals) to the feedback voltage of the SG, x_i (in volts):

$$y_i = bx_i + a$$

This provides the calibration factor b in μGal/volt, and the offset a in μGal (cf. Figure 4).

	Jilag 5/SG CO-26	FG5-206/SG CO-26
b : calibration factor (in (μGal/volt)	- 78.92 ± 0.31	- 79.03 ± 0.26

Table 4 - Calibration factor at Strasbourg - J9

The values of the calibration factor obtained with two AGs operating at the same site are in good agreement. The uncertainties are 1 σ error bars from the least squares adjustment. We used the mean gravity values for each set (25 drops every 15 seconds; 1 set every 15 minutes) in the inter-comparison.

More details can be found in Hinderer et al. (1997). We discuss below other aspects of the calibration problem.

Monthly calibration

We have repeated the calibration experiment, at J9, every month, and obtained a series of 6 values from February to July 1997. The values of the calibration factors are shown in Figure 1, the y-axis of which has a very dilated scale. The value of May is quite different from the other estimates but the quality of the experiment was bad because of noisy measurements (single drop scatter of 25 μGals). The uncertainty on the value is clearly under-estimated as can be seen from its error bar. The March experiment was long (6.5 days) while the duration of the other ones is between 2 and 5 days. The error bar for March is hence much smaller.

We conclude that 4 values are perfectly coherent, when taking into account the error bars. A fifth one (April) is easily included when taking 2 sigma uncertainty instead of 1 sigma.

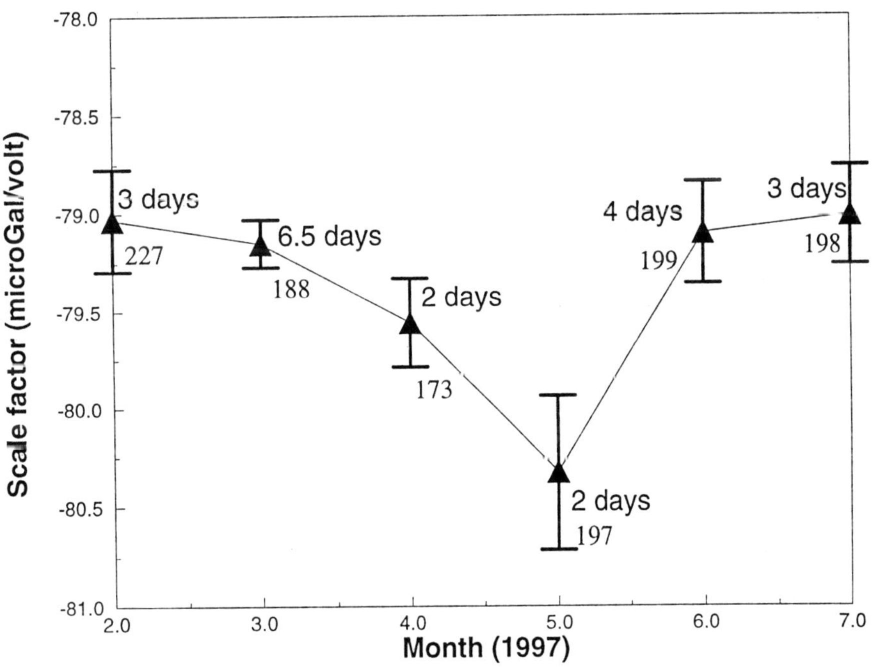

Figure 1 - Monthly calibration FG5-206/SG CO-26 in 1997.
The duration of the experiment and the amplitude of the tide are indicated.

Daily calibration

As the March experiment was long, we use it to study how the calibration evolves with increasing time (cf. Figure 2). First of all, we notice that the calibration factor is increasing with time, whereas the offset is decreasing; this is a verification that these two parameters are not independent. If there is an aperiodic instrumental process (e.g. drift, tilt, superspring), the model we are using is inappropriate and this could explain the features depicted in Figure 2. The calibration factor is varying from -80.0 to -79.1 μGal/volt.

According to the error bar on the final value, we can say that we need four days of measurement to reach this value. It is also the time needed to achieve a 0.2% uncertainty. In a recent study, Francis (1997) found a similar result. Moreover the offset is varying from 426.8 to 426.4 which is small.

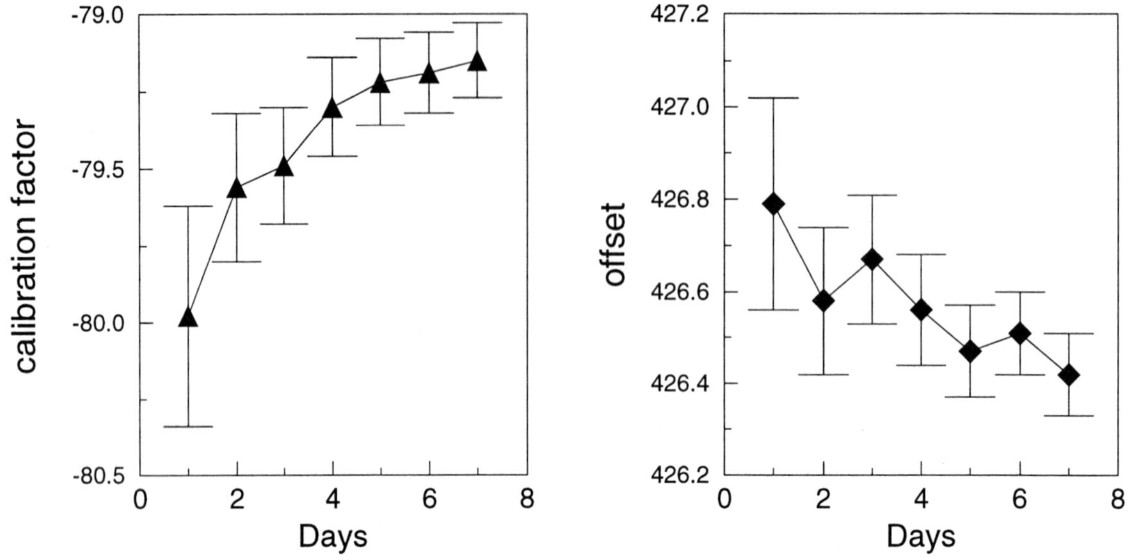

Figure 2 - Evolution with increasing time of the calibration factor (in µGal/volt) and of the offset (in µGal).

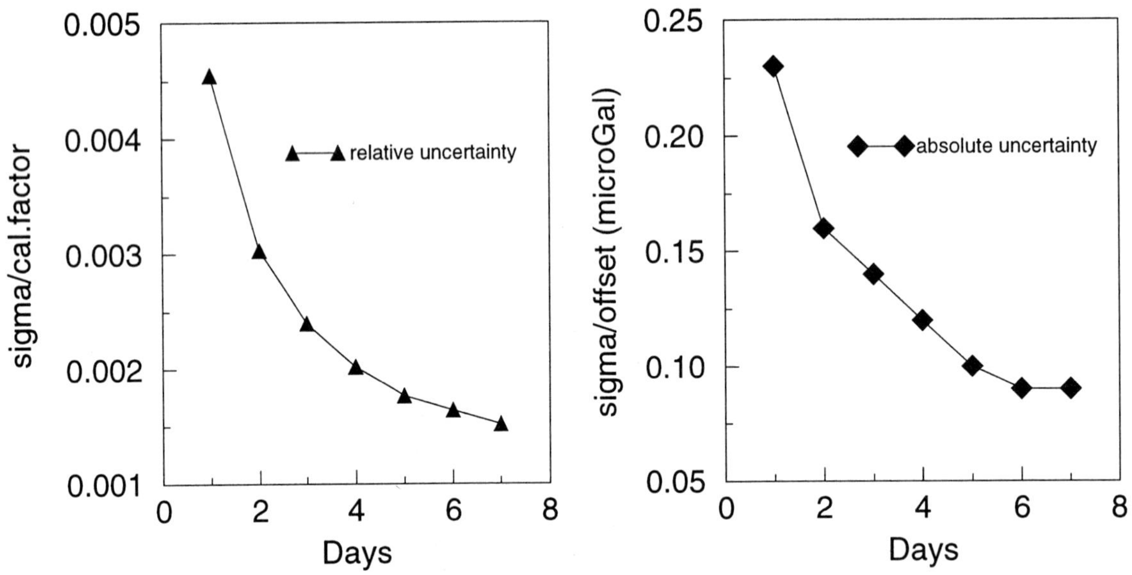

Figure 3 - Evolution of the uncertainty on the calibration factor and on the offset, with increasing time - march 1997

As can be seen on Figure 3, we have also studied the uncertainty on these two quantities. We show the relative uncertainty of the calibration factor b and the absolute uncertainty of the offset a. The two quantities are decreasing with time, which was expected since they are related to the number of measurements. From Figure 2, we see that the calibration factor indicates a 0.6% variation between 2 and 6.5 days. If one dismisses the May value, the variation seen in Figure 1 is almost similar and could reflect the effect of the duration of experiments.

We notice that the uncertainty for b is close to reach 0.1%. This is the value which is now needed in geophysics to study global geodynamics problems like Earth tides, ocean and atmospheric loading (e.g. Hinderer, 1996).

Conclusions

The main results of this study are the following:

Specifications: There is a good agreement between the observed values and the instrumental specifications given by the manufacturer. We have emphasised the importance of the site, whereas differences between two FG5s operating side by side are small.

Monthly calibration: Except for a bad experimental value (April), we find a good agreement of the values of the calibration factor over a 6-month time span, and also of the corresponding offset.

Daily calibration: To reach a 0.2% uncertainty in the calibration factor, at least 4 days of continuous AG measurements are needed in a medium quality site like J9.

Future: The repeated series of measurement will be very useful to check the stability (or the unstability) of the value of the mean gravity at J9 and Welschbruch. This is a very important result for the knowledge of the dynamics of the Rhine graben: is the graben going down or are the Vosges mountains going up?

Instrumental limitations: More work has to be done in order to check definitely if the uncertainty of 0.1% for the calibration factor, which is the value expected in present geophysical and geodynamical problems can be reached by parallel AG/SG comparisons.

Acknowledgements

We thank B. Luck for his kind technical support and the anonymous reviewer B. Richter for his constructive reading of the manuscript.

References

Francis, O., 1997. Calibration of the CO21 superconducting gravimeter in Membach (Belgium) using absolute gravity measurements, Proc. of the 13th International Symposium on Earth Tides, Brussels, submitted.

Freybourger M., Hinderer J., Trampert J., 1997. Comparative study of superconducting gravimeters and broadband seismometers STS-1/Z in seismic and subseismic frequency bands, Phys. Earth Planet. Int., **101**, 203-217.

Hinderer J., 1996. Gravity and the Earth's global structure and dynamics, Acta Geod. Geoph. Hung., **31** (3-4), 305-327.

Hinderer J., Florsch N., Mäkinen J., Legros H., Faller J.E., 1991 On the calibration of a superconducting gravimeter using absolute gravity measurements. Geophys. J. I., **106**, 491-497.

Hinderer, J., Florsch, N., and Mäkinen, J., 1995. Calibration of a superconducting gravimeter from repeated absolute gravity measurements, paper presented at the XXI IUGG Gen. Assembly, Boulder, USA.

Hinderer J., Amalvict M., Francis O., 1997, On the calibration of superconducting gravimeters with the help of absolute gravity measurements, Proc. of the 13th International Symposium on Earth Tides, Brussels, submitted.

Niebauer T.M., Sasagawa G.S., Faller J.E., Hilt R., Klopping F., 1995. A new generation of absolute gravimeters, Metrologia, **32**, 159-180.

Okubo S., Yoshida S., Sato T., Tamura Y., 1997. Verifying precision of a new generation absolute gravimeter FG5 - comparison with superconducting gravimeters and detection oceanic loading tide, Geoph. Res. Letters, in press.

An Airborne Geoid Mapping System for regional sea-surface topography: application to the Skagerrak and Azores areas

L. Bastos, S. Cunha
Observatório Astronómico, Monte da Virgem, P-4430 V. N. Gaia, Portugal
R. Forsberg, A. Olesen
KMS, Rentemestervej 8, DK-2400 Copenhagen NV, Denmark
A. Gidskehaug
Institute of Solid Earth Physics, Alle´gaten 41, N-5007 Bergen, Norway
U. Meyer, T. Boebel
Alfred Wegener Institut, Columbusstrasse, D-27568 Bremerhaven, Germany
L. Timmen, G. Xu, M. Nesemann
GeoForschungsZentrum,Telegrafenberg A17, D-14473 Potsdam, Germany
K. Hehl
Technische Fachhochschule Berlin, Luxemburgstrasse, D-13353 Berlin, Germany

Abstract

An airborne geoid mapping system is in development within the AGMASCO Project. The goal is the establishment of a system to be used mainly for regional, near coastal sea-surface topography determination by combining airborne gravity, airborne altimetry, DGPS and INS. Such a system can complimentary solve the problems of geoid modelling and sea-surface topography determination, bridging the gap between land areas and open ocean. It is therefore well suited for coastal and island areas. The reliability of the system is proven by using quasi-simultaneously ground truth shipborne gravity, ADCP and CTD measurements, and comparisons with satellite altimetry data.

The AGMASCO system, consisting of a laser and radar altimeter, a gravimeter, two GPS receivers, an INS and a data logging unit, was used in September 1996 to survey the Skagerrak sea area. Quasi simultaneous shipborne measurements were carried out by M/S Haakon Mosby, the research vessel owned by the University of Bergen.

In the paper we describe the system used in this campaign, present the methodology adopted for data processing and discuss the results achieved. The campaign in the Azores region is scheduled for October 1997. Details of the installation of the system, in a CASA C212 aircraft of the Portuguese Air Force, for this campaign will also be presented.

Introduction

AGMASCO (**A**irborne **G**eoid **MA**pping **S**ystem for **C**oastal **O**ceanography) is a project funded by the European Commission within the frame of the MAST III (**MA**rine **S**cience and **T**echnology) program.

The main goal is to develop a combined airborne gravity/airborne altimetry system which can contribute for geoid modelling and determination of the sea-surface topography (SST). The system is intended to be especially suited for application to coastal regions where the problem of geoid modelling and SST determination is more critical by the use of satellite altimetry methods.

The basic idea is to compute geoid heights from gravity measurements, ellipsoidal heights by DGPS and height above sea level by altimeter. These measurements can be related with the sea-surface topography.

The system can be a useful tool for oceanography. Using the obtained SST the ocean currents can be determined based on the geostrophic assumptions.

The feasibility and accuracy of such an airborne system will be demonstrated by comparisons with results from satellite altimetry and quasi-simultaneous ground truth ship gravity measurements. These gravity measurements combined with CTD (Conductivity Temperature Depth) and ADCP (Acoustic Doppler Current Profiler), also allow the determination of current velocities.

Within the AGMASCO two primary areas of interest have been defined:

- The Skagerrak sea area, between Denmark and Norway, was chosen because there are well known localised gravity anomalies and because the oceanographic features are significant and also quite well known. The area is therefore well suited for testing and validation of the system. It was surveyed in September 1996.
- The Azores region, in the North Atlantic, is a very interesting study area both from the geodynamic (Azores triple junction) and oceanographic (Azores current) point of view. It will be surveyed in October 1997.

For the airborne measurements two aeroplanes are available: a Dornier 228 from the AWI (Alfred Wegener Institut), Bremerhaven, was used for the test flights in the Braunschweig area and for the Skagerrak campaign; a CASA Aviocar C212, from the Geophysical Squadron of the Portuguese Airforce, will be used for the Azores campaign. The oceanographic measurements are carried out by M/S Haakon Mosby, the research vessel owned by the University of Bergen.

The AGMASCO system

The system involves the development of hardware components and dedicated software within the AGMASCO group. It was conceived in order to allow the determination of several parameters: the aeroplane position; the height above the sea surface; the disturbing accelerations; the gravity field. These parameters are used to obtain all the quantities of interest.

System Hardware. The hardware system has special prototype solutions: a radar altimeter, developed by the University of Stuttgart, Germany; a data logging unit, developed within an industry contract of GFZ (GeoForschungsZentrum); a strapdown accelerometer system. This sensor block consists of three QA-3000-030 accelerometers, produced by AlliedSignal Aerospace, USA, which are mounted orthogonally in an aluminium block. The mounting device including the electronics has been developed at Bayerische Kommission fue die Internationale Erdmessung, Munich, under Dr. G. Boedecker. The gravity sensor is a LaCoste & Romberg S99 air/sea gravity meter specially upgraded by ZLS Corp. for airborne use. The navigation system is based on GPS and INS market products. An Optech 501 SX laser altimeter, is also part of the system.

The system was adapted for use in the Dornier 228 and is now being installed in the CASA C212 . In addition to the operational LaCoste&Romberg gravimeter system a strapdown accelerometer system (SAGS) for gravity field determination will be on board of the Aviocar during the Azores campaign. A third GPS receiver was installed in this plane.

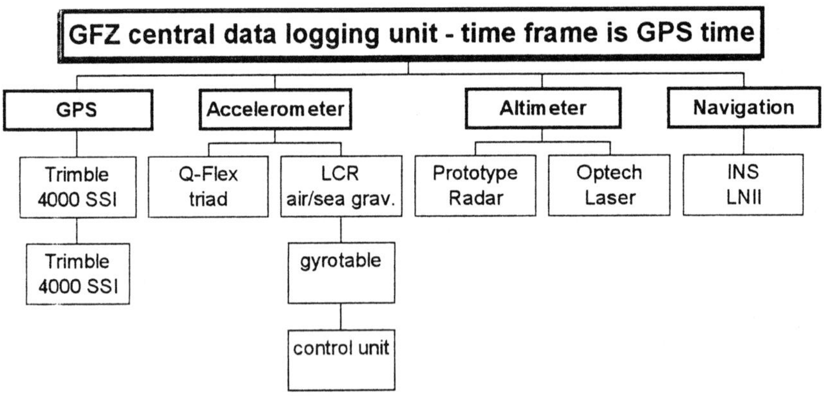

Figure 1 – AGMASCO system architecture.

System Requirements. In order to meet the established goals we have strict demands concerning the accuracy to be achieved by the different sensors. For a resolution of the gravity field of 1mgal the acceleration of the gravimeter has to be determined with an accuracy of the order of 0.001cm/s^2. Using the DGPS technology it is possible to determine vertical aeroplane accelerations with that accuracy. The major limitation is now the gravity sensor. We expect to improve the results by using the QA-3000-030 high accuracy accelerometers.

The altimeters show an accuracy in the 10 cm range. This is the position requirement. Processing of the INS data must give an attitude with an error less than 0.1 degree. This should be accomplished by a proper integration of the INS and GPS data.

Operational conditions. The AGMASCO system should allow a flexible solution for airborne gravity/altimetry. It works under velocities of the order of 200 to 250 km/h and at altitudes ranging from 400 to 700 m above sea level. The smoothness of the flight is a key issue to get valuable gravity data. The Dornier 228 is very stable and thus well suited for this experiment.

Validation of the system and data processing

Test flights. The validation of the system was accomplished with test flights made in the Braunschweig area (Germany, June, 1996) and a campaign in the Skagerrak region (September, 1996). We flew some profiles over land and over the North Sea for testing the sensors at different altitudes. For checking the gravity meter behaviour the aircraft passed over well-known gravity anomalies. The capability of the sensor to detect those anomalies was confirmed already by in flight inspection of the data.

One of the flights was used to calibrate the altimeters by passing the runway of the Braunschweig airport at very low altitude. A pre-flight survey of the runway heights by means of kinematic DGPS provided ground truth.

The first test of the strapdown accelerometer, to ensure the performance of the system has just been done along the border of the Alps in Germany.

Data processing. For the processing of the gravity data standard methods of physical geodesy (collocation and spherical FFT) are used.

GPS data is being processed using the kinematic software developed at GFZ (Xu, 1996) and compared to results from a commercial software (Xu et. al. ,1997).

Procedures to process the altimeter and INS data are still under development.

The validation of the results is done by direct comparison of the data from the different sensors as well as comparison with shipborne gravity measurements and satellite altimetry. The airborne data is evaluated in comparison with ground truth data upward continued to 1200ft elevation by using FFT methods after griding the available ship and land data.

A check was also made by comparing the tide-gauge reading at the Hanstolm harbour (North Denmark) with the ellipsoidal height of the altimeter (computed from DGPS) and the measured height above sea level (obtained with the altimeter). A difference of 8 cm was found which can be attributed to measurement errors.

A comparison of GPS heights with laser/radar altimeter heights is provided in Figure 2.

An arbitrary offset was introduced into the vertical to show the three lines. Altimeter data have been averaged to comply with the 1Hz rate of GPS.

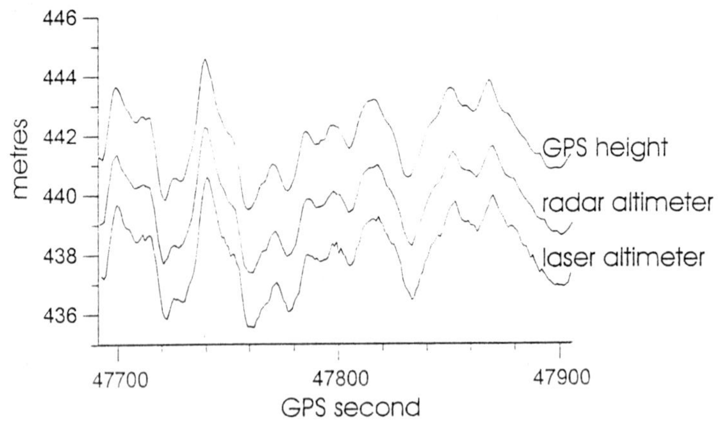

Figure 2 - Comparison of GPS heights with laser/radar altimeter heights.

Results from the Skagerrak campaign

The Skagerrak campaign was realized from 13 to 23 of September 1996 and involved airborne and shipborne measurements. Figure 3 , shows some of the profiles observed.

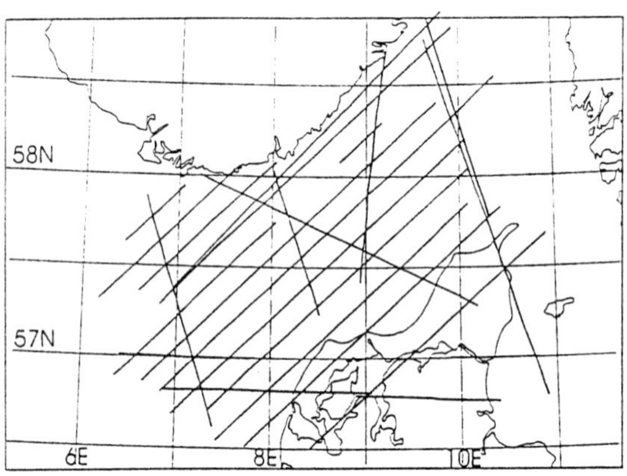

Figure 3 – Location of profiles, with processed gravity, in the Skagerrrak region.

Results from a track with a major gravity anomaly from a suspected buried volcano off Kristianssand, Norway, are shown in Figure 4. There is a very good agreement in the magnitude of the anomaly between the airborne gravity and the upward continued surface data. In some areas the surface data prediction suffers from gaps in the marine data. This was confirmed by recomputing the gravity data using a better set of marine data (Gidskehaug, 1997 - Private communication).

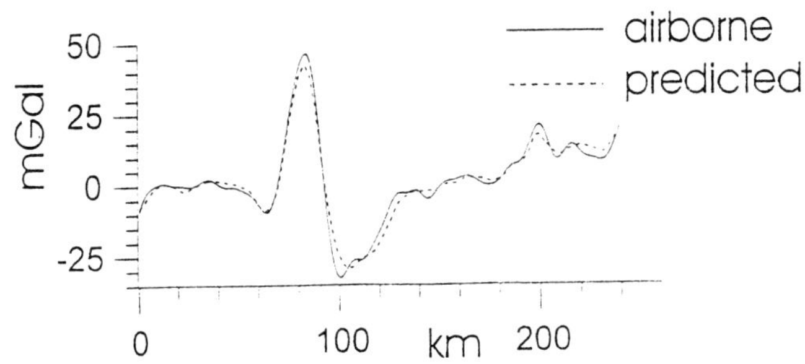

Figure 4-Anomaly from buried ‘ volcano’ in Skagerrak

An example of a more local anomaly (Mors salt dome, nothern Jutland) is shown in the Figure 5. The salt dome anomaly has a half-width/half-maximum measure of approximately 5 km, and is shown to be nicely resolved in the aerogravity (Olesen et al., 1997).

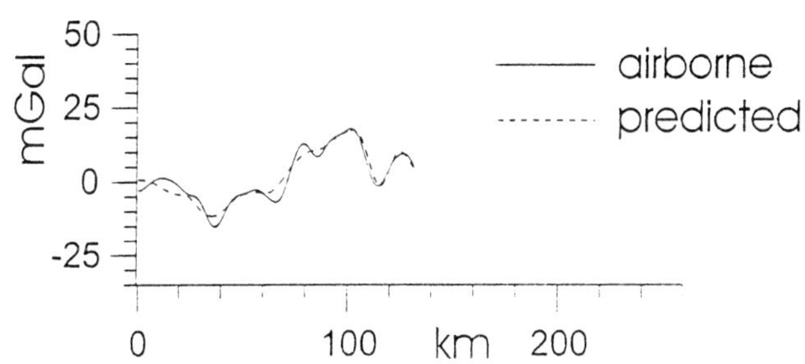

Figure 5 – Anomaly from Mors salt dome.

These anomalies are well detected in the shipborne data and confirm the results from aerogravity as can be seen.

The r.m.s. deviation from ground truth were around 3.5 mGal. This is mainly due to insufficient ground truth data. The magnitude of the r.m.s. cross-over errors of the airborne gravity are summarised in Table 1 and indicate the quality of the aerogravity. This was around 2 mGal over 7 km.

Unit: mgal	**Standard processing**	**Applying linear error model**
Before adjustment	3.1	2.8
After adjustment	2.4	2.2

Table 1 – R.m.s. cross-over from 55 crossings.

Conclusions and future plans

Both the analyses of the system test flights and the Skagerrak campaign show that the overall system performs well. Anomalies from air and sea fitted very well especially in the long wavelengths.

Crossover analyses as well as ground truth comparisons allow to say that our system has an accuracy of about 2 mGal over wavelengths of 5 to 7 Km. These results are quite promising and indicate the impact that the system can have for geoid determination as well as for geophysical exploration.

Concerning the oceanography part of the AGMASCO project we can already say that the use of ADCP with DGPS shows very good results. Further investigations have to go on regarding some small scale features of the circulation.

Next campaign is scheduled from 3 to 18 of October 1997 in the Azores region. The hardware system is just now being installed in a CASA C212 of the Portuguese Airforce. The impact of using a third GPS antenna and a strapdown sensor will now be evaluated.

Acknowledgements. This project is funded by the European Commission under the contract number MAS3-CT95-0014. This support is greatfully acknowledged.

The Portuguese participation in AGMASCO is further funded by the Portuguese government.

References

Boedecker, G., Leismüller, F., Spohnholtz, T., Cuno, J., Neumeyer K.H. (1994): Accelerometer / GPS Integration for Strapdown Airborne Gravimetry: First Test Results. Proc. IAG. Symposia 113 "Gravity and Geoid", 177-186, Springer.

Forsberg, R., Hehl, K., Bastos, L., Gidskehaug, A., Meyer, U. (1996): Development of an Airborne Geoid Mapping System for Coastal Oceanography (AGMASCO). Proc. International Symposium on Gravity, Geoid and Marine Geodesy, Tokyo, 1996.

Olesen, A., Forsberg, R., (1997): Airborne Gravimetry using the LaCoste and Romberg Gravimeter - An error analysis. Proc. International Symposium on Kinematic Systems in Geodesy, Geomatics and Navigation, KIS 97, Banff, Canada, June 3-7, 1997.

Xu, G. (1996): Kinematische Positionierung mit GPS in der Fluggravimetrie – Software, Entwicklung und Ergebnisse. Paper presented at the National Geodetic Workshop Geodatische Woche, October 7-11, 1996.

Xu, G. , Bastos, L., Timmen, L. (1997): GPS Kinematic Positioning in AGMASCO Campaigns - Strategic Goals and Numerical Results. Paper accepted for presentation at the ION 97 meeting , Kansas City, September 1997.

The German Gravity Base Net 1994 (DSGN 1994)

B. Richter, H. Wilmes, A. Franke, E. Reinhart
Bundesamt für Kartographie und Geodäsie (former: Institut für Angewandte Geodäsie)
Richard-Strauss-Allee 11, D-60598 FRANKFURT / MAIN, Germany
corresponding e-mail: richter@ifag.de
W. Torge (Institut für Erdmessung, HANNOVER, Germany)

Abstract

On behalf of the German Geodetic Commission the Bundesamt für Kartographie und Geodäsie re-measured the German Gravity Base Net established in 1976. After the reunification of Germany this network has been extended to the new German Federal States so that in total the network comprises 30 stations with a mean distance of 200 to 250 km. Each station consists of a main point and two sub-centres. The observations were performed in 1994 / 95. All centres of the 30 stations have been measured with the absolute gravimeter FG5-101 of the Bundesamt für Kartographie und Geodäsie. At six selected stations the Institut für Erdmessung, Universität Hannover carried out absolute measurements with its absolute gravimeter JILAG-3.

Whereas in the old part of the network comparisons could be made to the adjustment results of 1976, relative gravity measurements were carried out in the new part of the network as an independent control. The relative gravity ties were measured with 3 LCR gravimeters with SRW feedback and one Scintrex gravimeter CGM3. To minimise instrumental effects the LCR gravimeters were installed into temperature controlled aluminium boxes. For the transport, station cars were used with special spring-suspended containers. The instrumental accuracy for the absolute measurements can be estimated with 10-30 nm/s^2 (1-3 μGal) mainly depending on the seismic noise (microseismisity or antropogenic noise sources) at the stations. For the relative measurements an accuracy of 30-60 nm/s^2 (3-6 μGal) can be stated.

Introduction

In the mid of the 70's the German Geodetic Commission agreed to set up the German Gravity Base Net 76 (Deutsches Schweregrundnetz 76 (DSGN76)). The reasons for establishing this gravity network which covered the older part of the Federal Republic of Germany with a station spacing of 200-250 km was the great loss of gravity points in older networks, the availability of new gravimeter types (LCR), the possibility of absolute gravity measurements and the need of the state surveying offices to use a base-net as frame for more dense gravity networks. The station selection in geologically stable areas, the monumentation inside public buildings or churches and the achieved overall network accuracy of ± 100 nm/s² (10 μGal) should guarantee a long term base for the observation of secular gravity variations. A re-observation of the network was planned after 10 to 15 years.

Therefore in the beginning of the 90's the German Geodetic Commission decided to re-measure the gravity base net, established a working group as scientific board under the leadership of Prof. W. Torge, University Hannover and asked the Institut für Angewandte Geodäsie (now Bundesamt für Kartographie und Geodäsie) to carry out the organisation, measurement and adjustment. It was quite natural to extend the network to the new german states in Eastern Germany. Again the state of the art in gravimetric measurement techniques is taken into account to set up a new gravity base net with the highest standard. The survey of the gravity network is based mainly upon absolute gravity measurements which have been carried out on the centre location of each point group.

General Concept of Design and Realisation:

The stations in the old federal states of Germany are mainly identical with the old gravity network established in 1976 [Deutsches Schweregrundnetz 1976]. To achieve a homogeneous network covering the total area of Germany 10 new stations have been installed in the new German states which fulfil the same criteria as in the old network.

To supply a network with good long term stability and to enable reliable absolute gravity measurements the stations have been selected with regard to the following criteria:

- all stations of the network consist of a centre and two eccentric locations as control stations which are located in a distance of 500 m to 5 km to the centre and which have minimum gravity difference (< 50 μm/s² or < 5 mGal).;
- stations shall be in geologically stable areas (bedrock is optimal); areas with a high microseismic noise level and effects by large mass changes (reservoirs, rivers) have been avoided;
- stations are located inside solid public buildings (more than 10 years old, no constructive changes expected), either on a pier or the stable fundament of the building;

Additional conditions for the station centre caused by the absolute gravimeter sensor were:

- room size: > 2 x 3 m^2, room height 2.0 m; door width: > 80 cm; space for sensor: 1 x 1m^2;
- access to the room should be possible at every time of the day during the measurements;
- room temperature between 10° and 25°C, changes < 1°C/h and < 5°C / day.

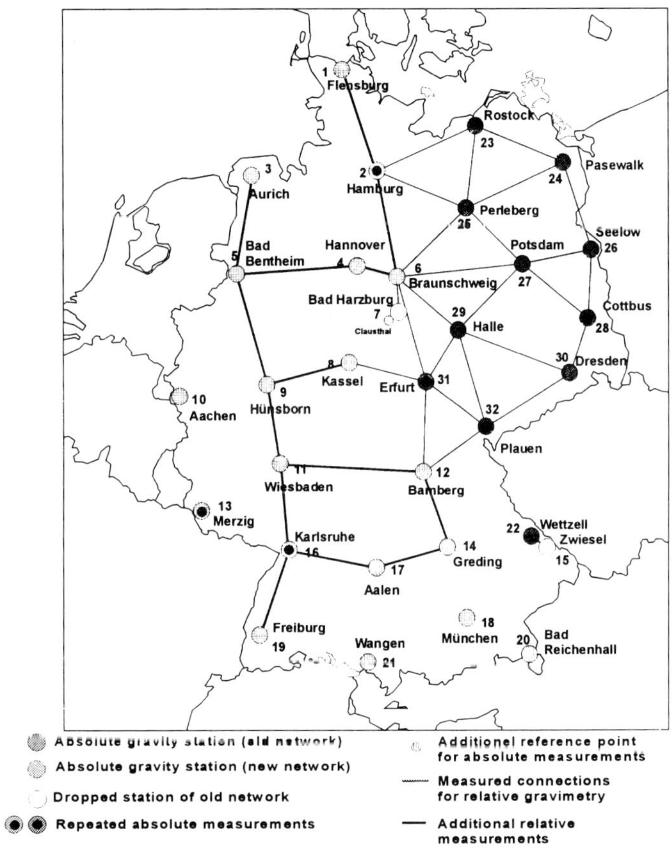

Fig. 1 The German Gravity Base Net 94 - station location and measured relative gravity connection.

In some of the stations of the DSGN76 the centre had to be moved to one of the former backup locations to fulfil the requirements posed by the absolute gravimeter. As far as possible and necessary, ground water gauges have been prepared in the vicinity of the absolute gravity sites in areas with high ground water level or expected ground water level changes. Measurements of the magnetic field intensity and direction have been performed on every station to prevent error influences upon the gravity results.

Besides the absolute measurements carried out at the centre station relative gravity measurements were used to determine gravity differences between the stations of the new network part, for the ties between station centre and eccentric stations at all stations of the network and for the determination of the vertical gravity gradient required for the evaluation of absolute gravity measurements. The design of the network and the measured relative ties are given in fig.1.

Control of possible gross errors was made in two ways, (1) by comparison with relative gravity measurements in the new part of the network (new federal states) and (2) by comparison with the former DSGN76 results. In case of differences bigger than 10-20 μGal additional relative gravity connections have been observed.

Instrumentation and Measurement

Absolute gravity measurements.

Absolute gravity measurements with the AXIS FG5-101 instrument have been carried out at all 30 station centres with 3600 drop samples each in an occupation time of 24 hours. To prevent or detect gross instrumental errors the measurement campaign was organised in „loops" of 7 to 12 stations with repeated occupations on selected sites (fig. 2).

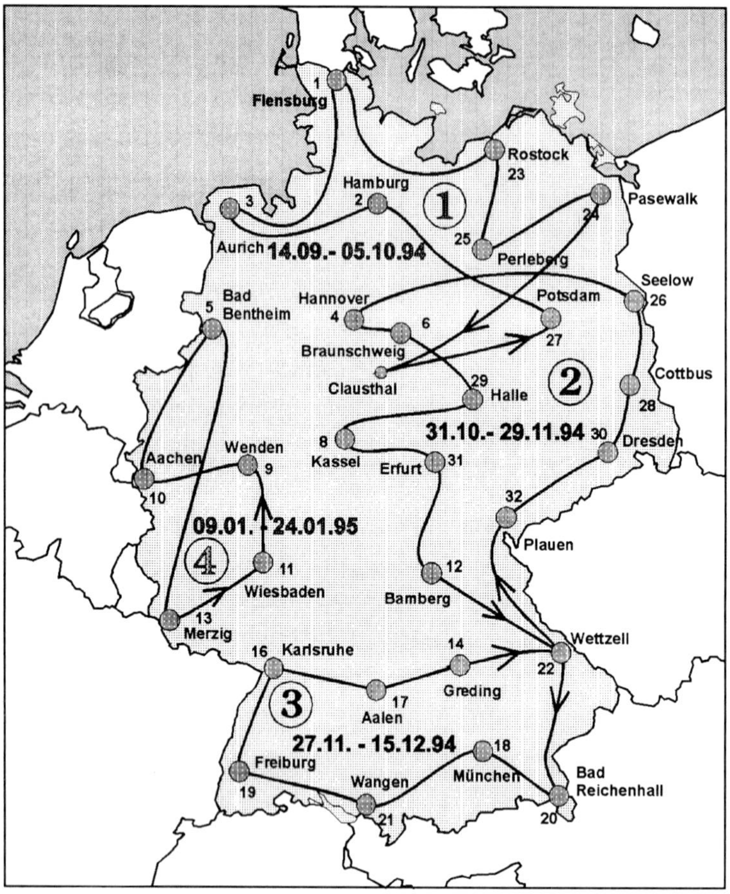

Fig. 2 Gravity ties measured with the absolute gravimeter FG5 - 101.

The major corrections are explained in the following:

- Earth tide corrections: The effect of lunar and solar mass attraction is computed and applied to the observations (each drop). The earth tide and tidal loading computation is based on a development by Timmen and Wenzel (1995).
- Gradient height correction: The observed gravity for each drop is referenced to a position at the top of drop. This is done by fitting the gravity value to a modified parabolic trajectory with a cubic and quartic terms.
- Vertical transfer correction: The observed gravity for each drop can be transferred to a specified height (datum height). The transfer is calculated by adjusting the gravity value using the difference between the instrument and the datum height multiplied by the gradient value.
- Local barometric pressure correction: The observed gravity is reduced to a nominal atmospheric pressure at each site by applying a correction based on the observed atmospheric pressure during the observations. A barometric admittance factor of -3 nm/s^2 / hPa is used. The nominal air pressure is computed from the height of the station in accordance to the standard atmosphere. This pressure correction is applied for each drop.

- Polar motion correction: This correction compensates for changes in centrifugal acceleration due to variations of the distance of the earth's rotational axis from the gravity station. This correction is computed using pole positions as published in the International Earth Rotation Service (IERS).

In addition at 6 selected gravity stations absolute measurements have been performed with the JILAG-3 absolute gravimeter of the Institut für Erdmessung, Universität Hannover.

Relative gravity measurements.
For the relative gravity measurements 3 LaCoste & Romberg Type G gravimeters (LCR) with feedback systems (conversion made by Institut für Erdmessung, Hannover) and 1 SCINTREX, Model CG3M have been selected out of a total of 10 instruments after extensive instrumental testing of the air pressure, air temperature and magnetic sensitivities. The 3 LCR gravimeters have been equipped with an additional thermostatised shielding and intermediate power supply. The gravimeters were transported in a conventional station car. A special open, spring suspended box has been constructed for each pair of the gravimeters placed in the back of the car.

In the network observations the measuring systems of the relative gravimeters were set up at the height of 1.25 m corresponding to the reference height of the absolute gravimeter in order to avoid error propagation from the vertical gradients into the adjustment of the network. Intensive measurements on the Hannover calibration line were used to determine a preliminary scale factor as well as for the determination of periodic spindle errors for the relative gravimeters.

The measuring scheme for the relative gravimeters was for the net gravity differences A-B-B-A, for the measurements within a station centre - Ecc1- Ecc2 - C. - Ecc2 - Ecc1 - c. and for the gradients 5 times the upper and lower position. Two gravimeters observed the above described scheme, the other two in a reversed way.

Statistics.
The relative gravity measurements took place from April to August 1994 and the absolute measurements from September 1994 to January 1995. In 1995 some absolute as well as relative re-measurements were performed especially the relative measurements in the old part of the base net.

Absolute measurements		
gravimeters	FG5-101 (BKG) JILAG 3 (IfE)	30 stations 5 stations
1 team	83 measuring days, incl. driving, 14 days for repeatitions	
	2 -3 person / team,	
	1 mini truck; 8000 km	

Relative measurements (net, gradient, backup, control measurements; 170 connections in total)		
gravimeters:	3 LCR gravimeters (G-528, G-563, G-1023), 1 SCINTREX M-202	
2 teams:	170 measuring days / team; 10-12 h / d	
	2 person / team	
	1 van / team; 40000 km	

Tab. 1 Summary of measuring days, personnel needed, cars and kilometres for the measurement of the DSGN94.

Results

All observations were treated in a very conventional combined adjustment with optimal weighting. The relative readings were reduced by station dependent Earth tides parameters, air pressure differences to a standard atmosphere, magnetic effects, spindle errors and the scale factor for the gravimeter feedback. The absolute observations are reduced by all effects discussed in the previous chapter.

$v_{ijk} = g_j - g^r_{jik} *(1 - m_i) + d_{ik} * t_k + n_{ik}$ relative gravity measurements

$v_{jl} = g_j - g^a_{jl}$ absolute gravity measurements

with
- g_j adjusted gravity value
- g^r_{ijk} observed gravity value at station j instrument i, day k; relative gravimeter
- g^a_{jl} observed gravity value at station j with Instrument l; absolute gravimeter
- m_i scale factor for relative gravimeter i
- d_{ik} daily (k) drift of relative gravimeter i
- n_{ik} daily (k) offset for relative gravimeter i

Optimal weighting function:			
Instrument	σ_0 [nm/s^2]	weight	number of observations
LCR G-528	71	1.1	163
LCR G-563	57	1.8	198
LCR G-1023	74	1.0	177
Scintrex CG3 M 202	53	2.1	224
FG5-101	37	4.3	28
JILAG-3	47	2.6	5

Table2: Weighting function derived in the combined analysis of absolute and relative gravity measurements.

The comparison between the DSGN 94 and DSGN 76 is shown in figure 3. Both networks are based on absolute gravity measurements. It seems that besides an offset there is a linear North-South trend which would mean in the sense of gravity a scale factor between the two measurement epochs.

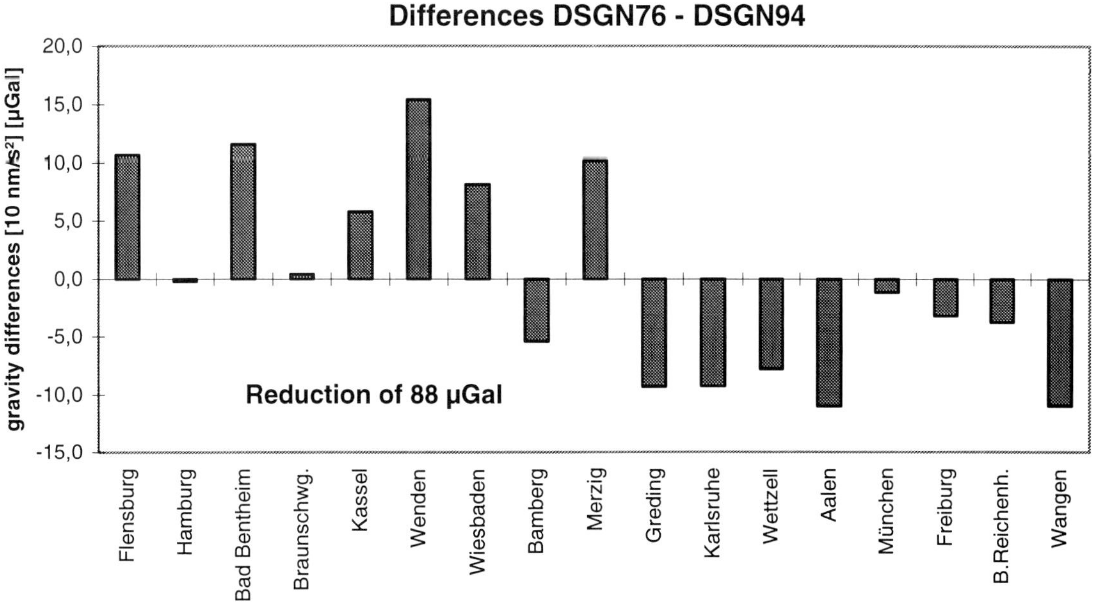

Fig. 3 Differences between the gravity base net DSGN76 and DSGN94.

References

Carter, W.E., G. Peter, G.S. Sasagawa, K.F. Klopping, K.A. Berstis, R.L. Hilt, P. Nelson, G.L. Christy, T.M. Niebauer, W. Hollander, H. Seeger, B. Richter, H. Wilmes, A. Lothhammer: New Gravity Meter Improves Measurements, EOS, Transactions, AGU, Vol. 75, No. 08, 1994, pp 90-92

Marson, I.; J.E. Faller; G. Cerutti; P. De Maria; J.-M. Chartier; L. Robertsson; L. Vitushkin, J. Friederich; K. Krautbluth; D. Stizza; J. Liard; C. Gagnon; A. Lothhammer, H. Wilmes; J. Mäkinen; M. Murakami; F. Rehren; M. Schnüll; D. Ruess; G.S. Sasagawa (1995): Fourth International Comparison of Absolute Gravimeters. Metrologia (32), pp. 137-144.

Niebauer, T.M.; G. Sasagawa; J.E. Faller; R. Hilt; F. Klopping (1995): A new generation of absolute gravity meters. Metrologia 32, 159-180.

Schnüll, M.; R.H. Röder and H.-G. Wenzel (1984) : An improved Electronic feedback for LaCoste&Romberg Gravity meters. Bureau Gravimetrique International, Toulouse, Bulletin d'Informations, No. 55, pp 27-36.

Timmen, L, H-G Wenzel: Worldwide Synthetic Gravity Tide Parameters. International Association of Geodesy Symposia, No:113 (Eds: H.Sünkel and I.Marson), 1995, pp 92-101. Springer.

SAR Data Acquisitions at Syowa Station in Antarctica

Koichiro Doi*, Kazuo Shibuya*, Taku Ozawa*, Hiroyuki Nakagawa**
* National Institute of Polar Research(NIPR)
9-10, Kaga 1-chome, Itabashi-ku, Tokyo 173 Japan
e-mail: doi@nipr.ac.jp
fax: +81-3-3962-4914
** Geographical Survey Institute(GSI)
1- Kitasato, Tsukuba, Ibaraki 305 Japan
fax: +81-298-64-1802

Abstract

We can acquire JERS-1 and ERS-1/2 SAR data at Syowa Station, Antarctica by an S/X-band parabola antenna whose diameter is 11 meters. The recorded SAR data tapes at Syowa Station are transported to NIPR every April by the icebreaker Shirase. The data are validated at NASDA/EOC and data with good quality are catalogued and made available to users by NASDA/EOC or via NIPR under collaborative research programs.

One of the purposes of the acquisition is to generate digital elevation model around Syowa station by means of interferometric SAR technique. An interferogram for the Amundsen Bay area is obtained from JERS-1 SAR data pair observed on November 26,1996 and January 9, 1997.

1. Introduction

Various geodetic observations are conducted at the Japanese Antarctic Station Syowa by the Japanese Antarctic Research Expedition(JARE), such as geodetic surveying, GPS measurements, absolute gravity meter measurements, continuous observations with a superconducting gravimeter. Geodetic VLBI experiments will also start from February, 1998.

In addition to the above geodetic programs, we have been acquiring several kinds of data from remote sensing satellites, for example, JERS-1(launched by NASDA) OPS(Optical Sensor) and SAR(Synthetic Aperture

- GPS positioning, both kinematic ('on-the-fly') and static
- Laser altimetry
- Radar altimetry (only in the interior parts of the ice sheet)
- Synthetic Aperture Radar (SAR) interferometry, both air- and spaceborne

Where SAR Interferometry (InSAR) is particularly useful due to the inherent high spatial resolution.

Test Area

The region around Scoresby Sund (see Figure 1) was chosen as a primary test area for SAR interferometry. The Danish airborne EMISAR has achieved images over the region in 1995 and 1997 and European Remote Sensing satellites (ERS-1/2) SAR images from 1996 were processed.

Jameson Land north of the sound is dominated by canyons. The Geikie Plateau south west of the sound is covered by a small ice cap with several outlet glaciers surrounded by steep mountains. The two regions represent major challenges to the applied techniques due to the steep slopes.

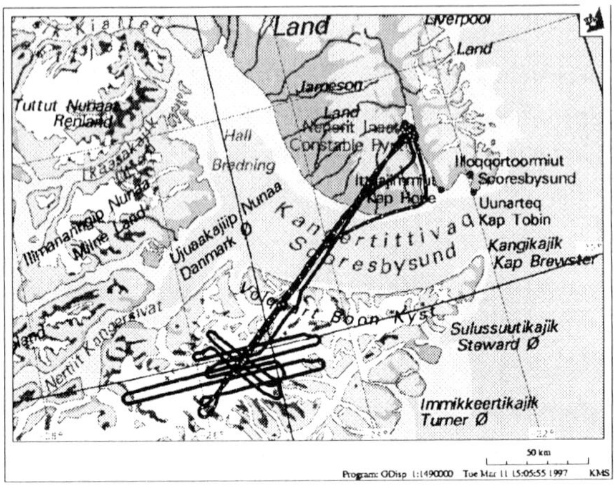

Figure 1: Flight tracks from July 16 and 17/96. The ice cap on the Geikie Plateau is covered by tracks in several directions.

InSAR DEM formation

4 SAR images acquired January/February 1996 by ERS-1/2 are used to create two interferograms (one is shown in Figure 2) covering the Geikie Plateau. In the process of creating the interferograms precise orbit data are used. By the use of two interferograms the temporal changes of the surface during the 1 day period between passes can be determined.

There are no reference points on solid ground available due to unwrapping difficulties in the mountainous regions. Instead areas with surface slopes oriented parallel with the satellite tracks are located (only ice flow perpendicular to the satellite tracks can be observed). Reference elevations are assigned to these region and the interferometric phase is inverted to absolute 3-D positions utilizing these values in calibrating the corresponding baseline and removing the absolute phase.

The SAR DEM covering the Jameson Land was created from a single interferogram formed from a ERS-1/2 tandem SAR pair from February 96 utilizing precise orbits from ESA as well. 8 geodetic control points across the region are used in the inversion to absolute positions.

Figure 2: Interferogram and amplitude image created from an ERS-1 scene (January 07/96, orbit 23430, frame 2187) and an ERS-2 scene (January 08/96, orbit 3757, frame 2187). One color cycle represents a phase change of 2π [Nielsen, 1996].

Laser Altimetry and GPS positioning

Airborne laser altimeter profiles were obtained in 1996 as a part of the Elevation Changes Of the Greenland Ice Sheet (ECOGIS) project. Both survey tracks are seen on the map (Figure 1).

An Optech laser altimeter was mounted onboard a Twin-Otter in combination with a Trimble GPS receiver and a dual-axis accelerometer. The maximum range of the laser is approximately 500 m and the size of the foot print increases with range so it is necessary to keep the aircraft in an altitude as low as possible.

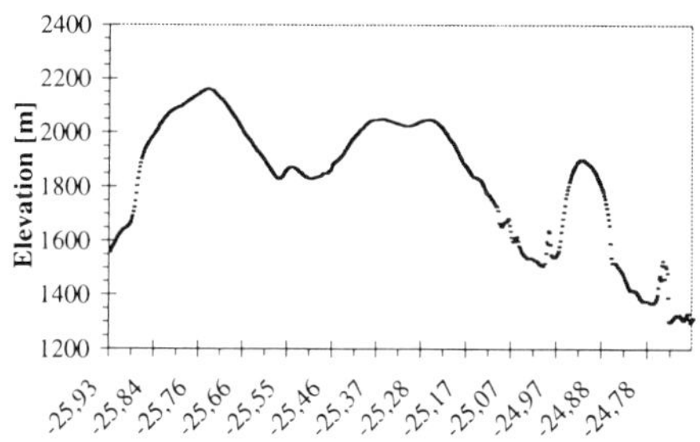

Figure 3: Example of surface elevations derived from laser altimetry.

The attitude (only pitch and roll) of the aircraft is estimated by the use of dual-axis accelerometer and GPS data. This provides the orientation of the laser beam and this combined with the laser altimeter range and the GPS positioning of the aircraft gives the position of the foot print in space. A derived elevation profile is shown in Figure 3. It is clear that the elevation is highly variable even in the ice covered regions.

Data Merging

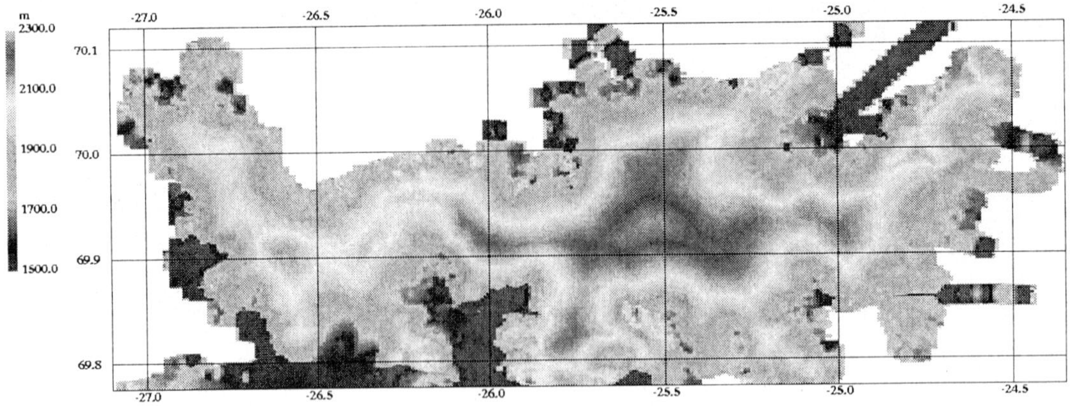

Figure 4: DEM covering the Geikie Plateau. Data from Static GPS, airborne laser altimetry and spaceborne SAR interferometry.

The DEM covering the Geikie Plateau (see Figure 4.) is based upon data from static GPS, airborne laser altimetry and SAR interferometry. The DEM has a grid spacing of 100 x 100 m. In some areas large jumps in elevation can be seen. These are caused by the unwrapping procedure not being able to determine the absolute phase in this region of the interferogram.

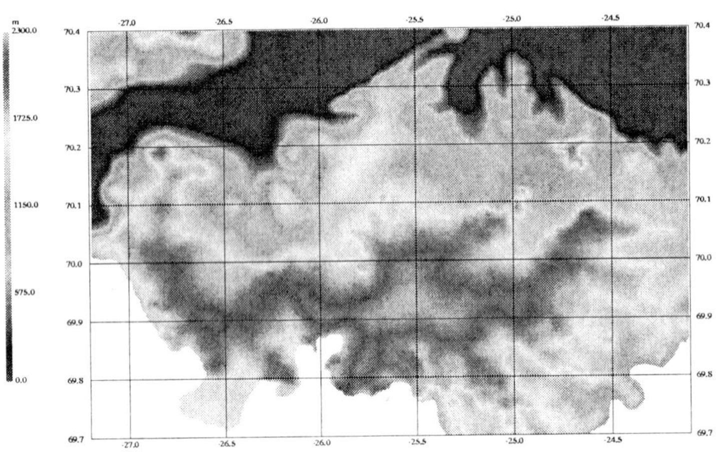

Figure 5: Plot of DEM covering the Geikie Plateau.

Another DEM (see Figure 5) covering a larger area consists of a mosaic of these data and older data from geodetic point measurements, DMA Digital Terrain Elevation Data (DTED) average data, fotogrammetry and manually scanned maps for the areas outside the smaller DEM data coverage. This DEM has a grid spacing of 1 x 1 km due to the highly varying resolution of the data.

These DEM's provides significant improvements because this region is generally poorly mapped. Older maps have elevation errors as large as several 100 m.

The complete DEM covering Jameson Land consists of a mosaic of data from SAR interferometry and older data such as DMA Digital Terrain Elevation Data (DTED) average data and manually scanned maps for the areas outside the SAR DEM data coverage. The grid spacing is 100 x 100 m. See Figure 6.

The DEM incorporating SAR data covering the Jameson Land has a much higher spatial resolution (in latitude, longitude and elevation) than the previous one and the accuracy is improved by a magnitude of 100 m [Nielsen et al., 1997].

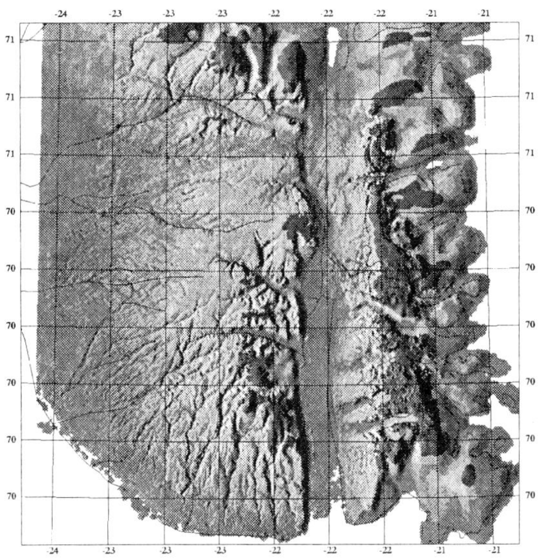

Figure 6: Shaded relief plot of complete DEM. Greytones are separated with 200 m equidistance. The darkest tone corresponds to elevations larger than 600 m [Nielsen et al., 1997].

Terrain Effects Computed by SAR DEM's

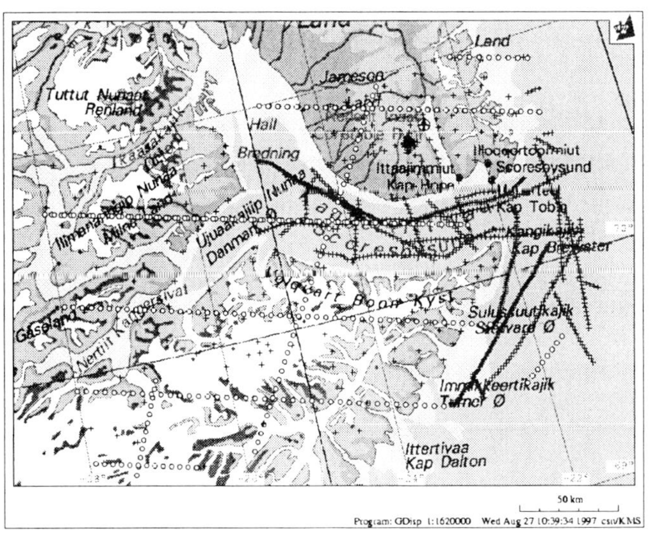

Figure 7: Plot of available gravity data. Markers: + indicates surface measurements and **o** indicates airborne measurements.

Existing gravity data in central East Greenland include land and marine gravity data surveyed 1982-96 (mostly by KMS), as well as airborne gravimetry of the Greenland Aerogeophysics Project 1991-92 [Brozena, 1992]. The location of the gravity points is shown in Figure 7.

Up to now existing DEM's of the region has been based on scanned or digitized maps, and have been quite poor. The most recent DEM has been compiled by [Ekholm 1996], and is given on a 2 km-resolution grid.

The new SAR-derived DEM's of Jameson Land and the Geikie Plateau were used to provide updated terrain effects for gravity in the two regions, and compared to similar computations based solely on the old, non-SAR DEM. Both classical terrain corrections and residual terrain model (RTM) effects relative to a 100 km-resolution mean surface was compared. Computations were done by prism integration using the TC program [Forsberg, 1984].

Results are given in Table 1. From the table it is seen that the impact of the SAR-derived DEM is very large, especially in the rugged Geikie area, with a bias of more than 18 mGal.

Table 1: Terrain effects computed by existing DEM and SAR-derived DEM's, East Greenland (unit: mGal or 10^{-5} ms^{-2})

Area	Type of terrain effect	No. of points	Terrain effect itself			Difference: New (SAR) - Old		
			mean	std.dev.	max abs.	mean	std.dev.	max abs.
Jameson Land	Classical t.c.	139	1.3	2.1	9.3	0.7	1.1	7.9
	RTM, surface data	139	2.0	34.3	77.7	-1.3	3.0	19.3
	RTM, airb. data	61	10.4	13.6	42.0	2.7	2.4	9.6
Geikie Plateau	Classical t.c.	35	8.2	2.7	14.2	2.6	2.5	8.4
	RTM, surface data	35	40.3	78.6	133.0	-0.4	5.1	12.2
	RTM, airb. data	67	42.4	40.1	119.2	18.1	12.3	49.6

Conclusions

It is possible to merge SAR interferometric data succesfully with other types of elevation data in order to compute terrain corrections for surface and airborne gravity surveys. The high resolution (in both latitude, longitude and elevation) of the interferograms significantly improves the general quality of the resulting DEM. By the use of other data sources it is also possible to fill data gaps in the interferograms.

The impact of SAR-derived DEM's on the terrain corrections is very large, especially in the previously poorly mapped Geikie area. This illustrates the importance of having good DEM's for use in connection with downward continuation of airborne gravimetry.

References

Brozena, J. M., 1992: The Greenland Aerogeophysics Project: Airborne Gravity, Topographic and Magnetic Mapping of an Entire Continent, *In: O. Colombo (Ed.): From Mars to Greenland, Proc. IAG Symposia 110*, Vienna 1991, Springer Verlag, pp. 203-214.

Ekholm, S., 1996: A Full Coverage, High-resolution, Topographic Model of Greenland Computed from a Variety of Digital Elevation Data, *J. of Geophysical Research*, Vol. 101, No. B10, 21961-21972.

Forsberg, R.: A study of terrain reductions, density anomalies and geophysical inversion methods in gravity field modelling. Rep. 355, Dept. of Geodetic Science, Ohio State University, 1984.

Nielsen, C.S., 1996: ERS-1/2 SAR Interferometri Anvendt til Estimering af Geikie Plateau Gletscherens Topografi, *M.Sc. thesis in geodesy (in Danish)*, Geophysical Dept., University of Copenhagen.

Nielsen, C.S.; R. Forsberg; S. Ekholm & J.J. Mohr, 1997: Merging of Elevations from SAR Interferometry, Satellite Altimetry, GPS and Laser Altimetry in Greenland, *ESA-SP-394, Proc. 3rd ERS Scientific Symposium*, in press.

The National Gravimetric Network of Uruguay

Walter H. Subiza
Servicio Geografico Militar, 8 de Octubre 3255
Montevideo 11600, Uruguay

Wolfgang Torge, Ludger Timmen
Institut für Erdmessung, University of Hannover, Schneiderberg 50
D-30167 Hannover, Germany

1. Introduction

A systematic survey of the Republica Oriental del Uruguay started in 1962, with a first international connection to Buenos Aires, transferring the absolute gravity datum from the Argentinian fundamental station to Montevideo. The Servicio Geografico Militar de Uruguay (SGM) then established a fundamental gravity network along the main roads between 1967 and 1968 including 924 stations. This relative type network consists of 25 loops, connected to the reference station in Montevideo. Densification surveys were mainly performed from 1984 to 1988 and increased the number of gravity stations to 2376 (average station distance 5 to 10 km).

In addition, new international relative ties to Brazil and Argentina were performed in 1987, 1989 and 1993, relating the gravity network to the International Gravity Standardization Net 1971 (IGSN71), and the Latin American Gravity Standardization Network 1977 (LAGSN77). As a part of the IfE Absolute Gravity Program "South America" 1989 - 1991 (IfE: Institut für Erdmessung, University of Hannover, Germany), three absolute gravity stations were established in Uruguay in 1989 and 1991, and connected to the fundamental network, as well as the IfE absolute station Buenos Aires. JILAG-3 gravimeter, a transportable free-fall apparatus was used (Torge, W., Timmen, L., Röder, R.H., Schnüll, M.: The IFE Absolute Gravity Program "South America" 1988-1991. Veröff. Deutsche Geodätische Kommission, Reihe B, Nr. 299, München 1994).

From these recent references to the international gravity standard (IGSN71 as well as absolute gravity standard provided by absolute gravimeters), and from an analysis of the network calculation procedures applied previously it was decided to perform a new calculation of the network, using all available absolute and relative data, and employing homogeneous and most advanced solution, reduction and adjustment procedures. SGM and IFE agreed to jointly carry out the task, the results of this cooperation project are presented here. The presentation describes the present state of the network, the surveys performed, the evaluation and reduction of the data as well as the adjustment, mainly by tables of results.

2. Network Adjustment and Results

The final gravity network adjustment was performed by introducing 4 stations with absolute gravity observations, serving for level and scale definition, and 2376 relative ones (standard deviations apriori: $s_{JILAG3} = \pm\ 0.08\ \mu ms^{-2}$, $s_{rel.grav.} = \pm\ 0.50\ \mu ms^{-2}$). For the absolute station Toledo, the 1989 and 1991 measurements were averaged. Table 1 gives the adjustment statistics, and figure 1 shows the distribution of the gravity stations. Altogether 5447 relative gravity observations with 8 different LaCoste-Romberg gravimeters have been used, with 3 instruments delivering 98 % of the data, only 0,3 % of the observations were rejected (Tab.

Fig. 1. Stations of the Uruguayan Gravimetric Network

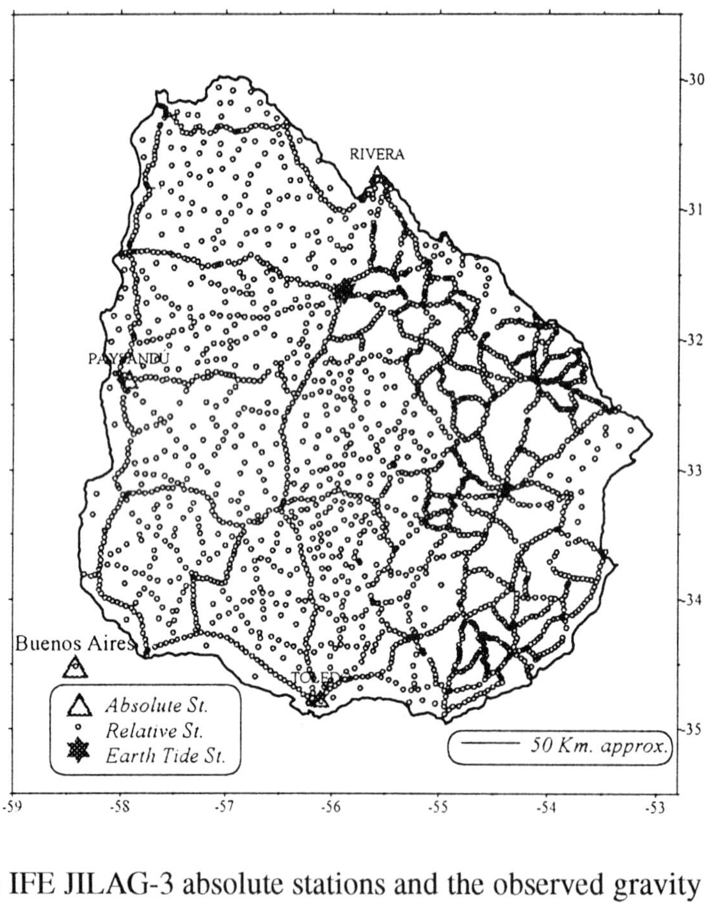

2). The IFE JILAG-3 absolute stations and the observed gravity values are given in Tab. 3, and shown in Fig. 2. From the adjustment, standard deviations between ± 0.17 and ± 0.50 μms^{-2} resulted for one gravity difference, and the calibration factors are uncertain between ± 4×10^{-5} and ± 18×10^{-3}. For the LCR gravimeters G013, G061 and G703 which practically determine the network, these figures are around ± 0.2 μms^{-2}, and ± 4...7 x 10^{-5} respectively

(Tab. 4). The mean standard deviation of the adjusted gravity values is ± 0.26 μms^{-2} with extreme values of ± 0.06 and ± 0.56 μms^{-2}. Comparing this result with the uncertainties of the three dominating gravimeters, an excellent performance of those instruments over 26 years observation time (1967 - 1993) may be stated. The absolute stations obtained only adjustment residuals of a few 0.01 μms^{-2}, and standard deviations of ± 0.06 to ± 0.08 μms^{-2}, which corresponds to the a priori error estimate (Tab. 5). Within the network frame, a special gravimeter calibration line in north-south direction was established and adjusted, based on the absolute stations Toledo and Rivera (Fig. 2). It covers a gravity range of 3751 μms^{-2} by 7 stations well determined with an average standard deviation of ± 0.1 μms^{-2} (Tab. 6). This calibration line can be used for a uniform standardization of all relative gravimeters used in the country.

Fig. 2. Relative Gravimeter Calibration Line

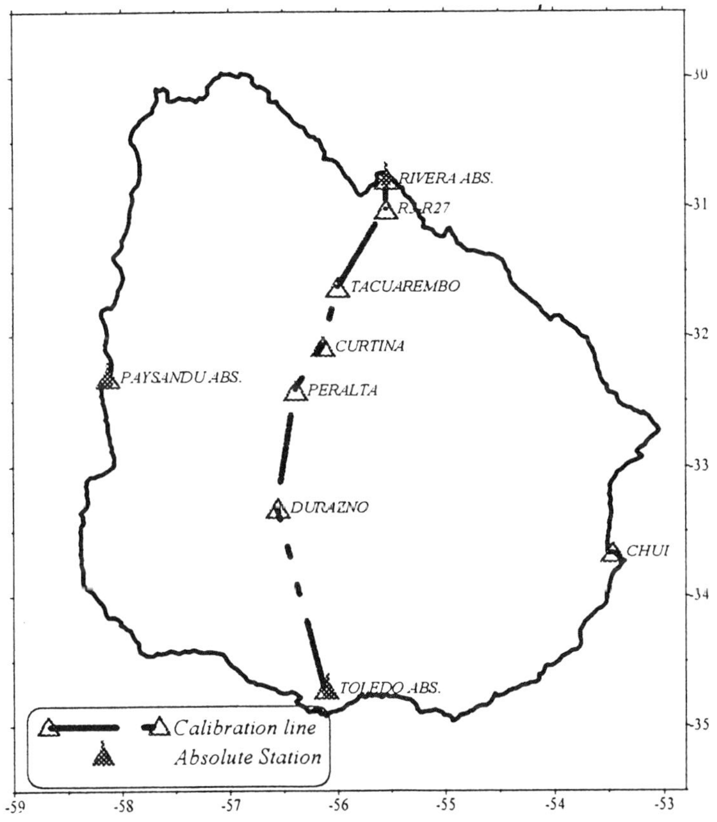

By including gravimetric ties to Argentine and Brazil into the calculation, the gravity standard of those countries, based on LAGSN77 and IGSN71 resp., could be compared with the the new gravity datum of Uruguay, revealing gravity differences of several 0.1 up to 1.2 μms^{-2} (Tab. 7). A direct comparison between JILAG-3 results and previous datum definitions (LAGSN77, IGSN71) was possible in two stations, and showed similar discrepancies (Tab. 8).

The software package GRAV (version 1.5, author: Prof. Dr.-Ing. H.G. Wenzel, University of Karlsruhe, Germany) was used to process all gravity measurements in the network. First, a preprocessing was performed to obtain a priori calibrated readings with applied earth tides. In a second step, the whole gravity network was processed using least squares adjustment, combining absolute and relative measurements.

3. Discussion and Conclusions

The main features of the new National Gravimetric Network of Uruguay may be summarized as follows:

- The absolute gravity datum is now accurate to ± 0.05 μms^{-2} ... ± 0.10 μms^{-2},
- The network scale is provided with an accuracy of ± 3 x 10^{-5}, and a calibration line (0.0037 μms^{-2}) realizes this standard,
- The average accuracy (std. deviation) of the stations is ± 0.3 μms^{-2}, and varies between ± 0.1 and ± 0.6 μms^{-2} depending on the number of observations involved,
- The IGSN71 standard in Uruguay is changed by - 0.5 μms^{-2}, while the scale factor changed at the order of 1 x 10^{-3}. This demonstrates the high quality of IGSN71 but also the improvement by absolute gravimetry,
- The gravity standard (IGSN71, LAGSN77) of Brazil and Argentina differs between -0.5 and -1.0 μms^{-2}, from the new gravity standard of Uruguay.

Tab. 1. Adjustment statistics

No. of absolute gravity stations	4
No. of relative gravity stations	2376
No. of absolute gravimeters	1
No. of relative gravimeters	8
No. of observed gravity differences	5447
No. of adjustment unknowns	2384

Tab. 2. Number of used and rejected gravity differences

LCR Gravimeter	Observations used	percentage	Rejected observations	percentage
G 013 (SGM)	2152	39.5 %	7	0.3 %
G 061 (SGM)	2314	42.5 %	4	0.2 %
G 703 (SGM)	881	16.2 %	2	0.2 %
G 945 (UT)	6	0.1 %	-	-
G 622 (ON)	32	0.6 %	-	-
G 602 (ON)	32	0.6 %	-	-
G 454 (USP)	18	0.3 %	-	-
G 679 (IGM)	12	0.2 %	2	14.3 %
Total	5447	100.0 %	15	0.3 %

Tab. 3 JILAG-3 absolute gravity results in Uruguay and Buenos Aires 1989/91

Station, IFE No.	Epoch	ϕ [deg]	λ [deg]	H [m]	dg/dH [$\mu ms^{-2}/m$]	g_{floor} [μms^{-2}]
Rivera, 212	3/89	- 30.90	- 55.54	213	- 3.08	9793443.77
Toledo, 222	3/89	- 34.74	- 56.09	65	- 3.07	9797158.55
Toledo, 222	12/91	- 34.74	- 56.09	65	- 3.07	9797158.56
Paysandú,232	12/91	- 32.38	- 58.03	61	- 3.04	9795235.26
B.Aires, 313	11/91	- 34.57	- 58.52	13	- 2.48	9796891.41

Tab. 4. Standard deviations of the individual LCR gravimeters (for one gravity difference measurement) and the adjusted calibration factors

LCR gravimeter	Standard deviation of one gravity difference [μms^{-2}]	adjusted calibration factor
G 013	± 0.24	0.99871 ± 0.00004
G 061	± 0.25	1.00033 ± 0.00007

G 703	± 0.17	1.00049 ± 0.00004
G 945	± 0.50	1.00081 ± 0.00016
G 622	± 0.38	1.00043 ± 0.00008
G 602	± 0.36	1.00029 ± 0.00007
G 454	± 0.28	1.00023 ± 0.00007
G 679	± 0.42	1.00934 ± 0.01803

Tab. 5. Network adjustment results for the absolute gravity stations and number of relative connections to other stations

Station, IFE No.	Adjustment residual [μms^{-2}]	Adjusted absolute gravity value [μms^{-2}]	No. of relative gravity ties
Rivera, 212	- 0.02	9793443.75 ± 0.08	22
Toledo, 222	- 0.01	9797158.55 ± 0.07	18
Paysandú, 232	+ 0.05	9795235.31 ± 0.06	12
Buenos Aires, 313	- 0.03	9796891.38 ± 0.07	2

Tab. 6. Gravimeter Calibration Line

Station	SGM Nr.	Gravity [μms^{-2}]	Ties
Toledo Absolute (IFE 222)	0003	9797158.55 ± 0.07	18
Nodal Durazno	0017	9795969.03 ± 0.06	87
Nodal Peralta	0027	9795051.66 ± 0.11	20
Nodal Curtina	0015	9794625.98 ± 0.11	17
Nodal Tacuarembó	0038	9794219.66 ± 0.07	90
Nodal Empalme R-5 R-27	0019	9793407.92 ± 0.11	46
Rivera Absolute (IFE 212)	0004	9793443.75 ± 0.08	22

Tab. 7. Comparison of the gravity datum with neighbouring countries

Gravity Differences at connected stations in Brazil and Argentine				
Station name	ϕ [deg]	λ [deg]	H [m]	Diff. (Urug.-Brazil/Arg.) [μms^{-2}]
Uruguaina, Brazil	- 29.74	- 57.09	60.34	- 1.2
Livramento, Brazil	- 30.90	- 55.50	183.58	- 0.7
Bagé, Brazil	- 31.30	- 54.10	202.63	- 0.9
Pelotas, Brazil	- 31.80	- 52.30	10.99	- 0.5
Chui, Brazil	- 33.70	- 53.50	13.39	- 0.8
Monte Caseros, Argentina	- 30.26	- 57.63	52.97	- 0.5
Parana, Argentina	- 31.78	- 60.47	59.98	+ 0.3

Tab. 8. Comparison between JILAG-3 and previous datum definitions

Station, IFE No.	$g_{IFE\ (89/91)}$ [μms^{-2}]	$g_{LAGSN77}$ [μms^{-2}]	g_{IGSN71} [μms^{-2}]	Diff.(IFE-Net) [μms^{-2}]
Montevideo (ecc.) 223	9797324.14	... 4.6	/	- 0.5
Buenos Aires, 311 (*)	9796900.71	/	... 0.3	+ 0.4

(*) gravity datum station IGSN71 "Buenos Aires" in Miguelete

ACKNOWLEDGMENT

The absolute gravity stations in Uruguay were established within the frame of the "IFE Absolute Gravity Program South America 1988 - 1991", supported by the German Research Society under grants To46/30-1 and 30-2. The Federal Ministry for Economic Cooperation and Development (BMWZ) supported the visit of one of the authors (W. Subiza) to IFE Hannover, for the common evaluation of the gravity data and the network adjustment. The software package GRAV used for data processing and adjustment was kindly provided by Professor H.-G. Wenzel, Geodetic Institute, University of Karlsruhe. All this support is gratefully acknowledged.

PAST AND FUTURE DEVELOPMENTS IN GEOPOTENTIAL MODELING

Richard H. Rapp
Dept. of Civil & Environmental Engineering & Geodetic Science
The Ohio State University
Columbus, OH 43210-1275
e-mail: rapp.1@osu.edu

ABSTRACT

This paper reviews the development and estimation of geopotential models over the past 96 years, starting from simple ellipsoidal normal gravity models to complex high degree (360/500) spherical harmonic expansions. The paper is written to show the evolutionary changes that have taken place in the mathematical models and data used. The discussion considers geopotential models from surface gravity data, satellite tracking data, and combination solutions that incorporate numerous data types including satellite altimeter data. A number of questions are posed that relate to future modeling efforts.

INTRODUCTION

A geopotential model is one that represents the Earth's gravity or gravitational potential and/or gravity values. The estimation of a variety of different models, from the simple to the complex, has evolved significantly in this century. Vetter (1994) discusses various geopotential models used in astrodynamics starting from the very early times (1950's) of space programs. Nerem, Jekeli and Kaula (1995) provide a review of the gravity field determination process, including a review of global scale modeling. Bouman (1997) provides a comprehensive discussion and analyses of geopotential estimation procedures and results starting from the 1970's. This current paper starts earlier than the previous reviews, going back to the time when estimates of normal gravity formula parameters were being determined.

GRAVITY FORMULA PARAMETERS

Theoretical studies done by many investigators (Clairaut, deSitter, Helmert, Pizzetti, Somigliana, Lambert, etc.) showed that variations of gravity on the surface of a rotating surface would take the form:

$$\gamma = \gamma_E \left(1+\beta_1 \sin^2\phi + \beta_2 \sin^2 2\phi + \cdots\right) \tag{1}$$

where γ is the "normal" value of gravity, γ_E is equatorial gravity, β_1, β_2 are constants to be determined or calculated from assumptions on the figure upon which the gravity variations were to be described, and ϕ is geodetic latitude. Based on estimates of the mass of the Earth, its equational radius, and flattening, the parameters of the gravity formula could be determined. The inverse aspect was to determine the parameters of the gravity formula using gravity measurements distributed as widely as possible. One of the first determinations in this century was by Helmert in 1901, who used gravity measurements at 1603 stations. As additional gravity measurements were acquired, various estimates of the parameters were made by Bowie (1917), Heiskanen (1928, 1957), Jeffreys (1937, 1941, 1948), Zhongolovich (1952), Uotila (1957), Kaula (1958), etc. The improved estimates not only included surface gravity data, but also information on the lunar orbit.

The gravity variations described by (1) are latitude dependent only. An extension of the assumption to the case where the equator of the Earth could be an ellipsoid led to a gravity formula that had a longitude dependent term:

$$\gamma = \gamma_E\left(1+\beta_1 \sin^2\phi+\beta_2 \sin^2 2\phi+\beta_3 \cos^2\phi \cos 2(\lambda-\lambda_0)\right) \tag{2}$$

where λ_0 is the longitude of the major axis of the equatorial ellipse. Parameters of the formula were determined by Helmert in 1915, Heiskanen in 1924 and 1928, Niskanen in 1945, Jeffreys in 1948, Uotila in 1957 and 1962, etc. The analyses evolved with the continuing improvement in the availability of surface gravity data.

The early parameter estimates were obtained from point gravity measurements where values were reduced to a zero elevation. As more data became available, gravity anomalies were used. To prevent the over weighting of areas in which dense data was available mean values in cells of various size (e.g. 1°x1°) were estimated. In the Heiskanen 1938 solution 1,802 1°x1° cells were used, while Uotila, in 1962, had 11,294 1° cells. The type of anomaly to use in the analyses was also pertinent to the estimates, with some groups favoring isostatic anomalies and others free-air anomalies. In addition, the magnitude of the adopted value of gravity at Potsdam, based on a measurement by Kühnen and Furtwanger in 1906, later turned out to be too large by about 14 mgal.

EARLY SPHERICAL HARMONIC EXPANSIONS OF TERRESTRIAL ANOMALIES

We next consider a spherical harmonic expansion of gravity anomalies, $\Delta g(\theta,\lambda)$, where θ is the co-geocentric latitude:

$$\Delta g(\theta,\lambda)=\sum_{n=0}^{M}\sum_{m=0}^{n}(a_{nm}\cos m\lambda+b_{nm}\sin m\lambda)P_{nm}(\cos\theta) \tag{3}$$

Dubovskii carried out two expansions to degree 6 in 1937. One expansion assumed zero value for the anomalies in unsurveyed areas while the second version assumed that Pratt isostatic anomalies were zero. In 1941-1943 Jeffreys used free-air gravity anomalies in 10° cells to determine—through a least squares adjustment—coefficients to degree 4. Based on error estimates, Jeffreys felt that no degree 4 terms were reliably determined nor were several degree 3 terms. Jeffreys emphasized the need for error analysis in the estimation of the coefficients. In a 1943 paper Jeffreys used the coefficients of the almost degree 3 expansions to calculate geoid undulations.

In 1952 Zhongolovich described expansions to degree 8 based on the analyses of 10°x10° anomalies estimated from approximately 26,000 point anomalies. A description of the two solutions (first variant, second variant) may also be found in Molodenskii, Eremeev, and Yurkina (1960). The data distribution was such that the data in the southern hemisphere was only 3% of that known in the northern hemisphere. The geoid undulations implied by one of the variants is given in Brovar, Magnitsky, and Shimbirer (1961). Uotila was to later note (1957) that the undulations implied by the Zhongolovich analyses were much larger (-160 to 180m) than seen in his analysis. Kaula (1959) described the estimation of 1°x1°, 5°x5°, 10°x10°, and 30°x30° anomalies. The 1° anomalies were estimated taking into account gravity and elevation data. A total of 8,000 1° values were estimated within 860 5°x5° cells. Ten degree anomalies were also formed from the 1°x1° cells, both extrapolated and interpolated. The 30° anomalies were adjusted to be consistent with one constraint based on satellite estimates of the 2, 4, and 6 zonal harmonies and five constraints from inadmissible harmonics. The corrections to the anomalies were used to obtain revised 10° cells which were then used to determine spherical harmonic coefficients to degree 8 using the orthogonality relationships. The coefficients were used to construct global geoid undulation maps. This solution was one of the first to introduce a constraint from satellite gravity field information into the analysis of terrestrial data.

In 1962 Uotila described the analysis of 1°x1° anomalies to determine, using a least squares adjustment, spherical harmonic coefficients to degree 4. Two solutions were made, one fixing the second, third, and fourth degree zonal harmonics from satellite data. Geoid undulation maps were also produced. Rapp (1969a) described the analyses of a global set of 5° anomalies (55% based on terrestrial gravity estimates) for harmonic coefficients to degree 8. Solutions for the coefficients were made using a least squares adjustment and a quadrature procedure. Undulation differences between the two solutions reached 20m, emphasizing the difference between the two types of solutions.

Except for special situations, the era of calculating harmonic coefficients from terrestrial data alone became less important because the terrestrial data coverage remained poor and information from the analyses of satellite data was increasing—both in accuracy and in maximum expansion degree possible.

EARLY SATELLITE AND COMBINATION GEOPOTENTIAL MODELS

With the launch of Sputnik in 1958, and the start of the Vanguard satellites, procedures for the analysis of the tracking data rapidly evolved. Initial results gave a few zonal harmonic coefficients and then a few tesseral coefficients (e.g. Cook, Izsak, Jacchia, King-Hele, O'Keefe, Kozai, Smith, D.E, etc.). For example, Kaula (1963) describes the use of Baker-Nunn camera observations of three satellites to obtain estimates of 35 selected coefficients with a maximum complete degree to 4 and selected coefficients to degree 7. In 1966 the Smithsonian Astrophysical Observatory documented (Lundquist and Veis, 1966) the first of three SAO Standard Earth models. The SE I was based solely on satellite observations and was complete to degree 8 with selected additional resonance and zonal terms. The SAO SE II was described by Gaposchkin and Lambeck in 1970 and SAO SE III in 1973 (Gaposchkin). Both SE II and III determined potential coefficient models that combined satellite information with terrestrial gravity data. SE II was complete to degree 16, while SE III was complete to degree 18.

The Applied Physics Laboratory (APL) of the Johns Hopkins University produced a number of geopotential models using Doppler data from the Transit navigation satellites. The APL 1.0 model (1963) was complete to degree 4, while the APL 3.5 model (1965) was complete to degree 8. Other models (e.g. APL 4.5 and 5.0) were also produced in this series. Guier and Newton (1965) describe a solution where estimates of odd degree zonal harmonies were combined with estimates of non-zonal harmonies and even degree zonal harmonies derived by King-Hele to produce a geopotential model complete to degree 8.

The Naval Weapons Laboratory (NWL) also developed geopotential models using the Doppler data from the Transit system. Anderle (1996) described the NWL-5E-6 solution that was complete to degree 7. The NWL-8 geodetic parameter sets (8B, 8C, 8D) (Anderle and Smith, 1967) included geopotential models that were complete to degree 12 with additional resonance terms. The preferred model was NWL-8D.

During the 1960's interest in combining satellite and terrestrial data in a joint estimation process increased. Kaula (1966) implemented a procedure that used harmonic coefficient estimation from gravity anomalies using the orthogonality relationships and those coefficients (estimated from satellite data) to obtain an adjusted set of potential coefficients and gravity anomalies. The procedure described by Kaula for this joint adjustment has been used for more than 30 years (with revisions) to obtain some of the high degree models we consider today (1997). In the Kaula 1966 study, 5° equal area anomalies were used with the mean of several sets of satellite-derived potential coefficients to estimate a geopotential model, complete to degree 12.

Results from an alternative way to combine terrestrial gravity data with satellite determinations of the geopotential were presented at the IUGG meeting in Lucerne, Switzerland in 1967. Rapp (1969b) described the estimation of 2,261 5° anomalies from terrestrial data and their combination in a least squares adjustment, with the degree 8 SAO SE I model, to obtain a set of potential coefficients to degree 14. In this solution, normal equations were formed using equation (3) as a basis where $\Delta\overline{g}$ (and its accuracy) were used to form normal equations in which the potential coefficients were parameters to be estimated. A similar solution was carried out by Köhnlein (1967) to degree 15.

At this point there were two different techniques that could be used for the combination of terrestrial and satellite data: one procedure required a global set of anomalies and involved an adjustment of satellite derived from the terrestrial data through orthogonality relationships; the other combined the satellite and terrestrial data through normal equations with the potential coefficients as parameters. Rapp (1969c) compared the two model approaches theoretically and carried out solutions with both methods using common data sets.

DEVELOPMENTS IN THE 1970's

With the basic techniques for determination of the spherical harmonic expansion of the gravitational field established, various groups turned to the processing of additional data and new data types (e.g. satellite laser ranging). The first model (GEM-1) in the GEM series was developed at NASA/Goddard Space Flight Center (Lerch *et al.*, 1972). This model was based on satellite tracking data and was complete to degree 12. The NASA analyses used the numerical integration of the satellite equations of motion, and variational equations, in the model estimation process, in contrast to the SAO studies that used analytic orbit theory. A companion model, GEM-2, was estimated to degree 16 using available 5° terrestrial gravity anomalies. Other models from Goddard quickly evolved as more data became available and computer resources increased. The combination of the satellite and terrestrial data was done through the merger of normal equations where equation (3) was the basis of the terrestrial data observation equation. A list of the GEM models produced is given in Table 1. The GEM-9 and 10 models (Lerch *et al.*, 1979) were complete to 20 (GEM-9) and to 22 (GEM-10), with selected terms to degree 30. The surface gravity data were 1,654 5° mean anomalies used by Rapp (1977) in an analysis where potential coefficients were estimated from the terrestrial data to degree 52 using a quadrature procedure. Two additional models, GEM-10B and GEM-10C, were estimated (Lerch *et al.*, 1981) (the models were presented at meetings in 1978 and formally published in 1981) using the same tracking data as used for the GEM-9 estimation. For GEM-10B (complete to degree 36) 1,654 5° equal area terrestrial anomalies were used as well as estimates of ocean geoid undulations obtained from 200 passes of GEOS-3 data. The GEM-10C model was complete to degree 180. The model was developed by first calculating a global set of 1°x1° geoid undulations using the terrestrial anomalies and anomalies derived (Rapp, 1979a) from GEOS-3 altimeter data and the GEM-10B model. In the oceans, the derived undulations were replaced by estimates from the GEOS-3 altimeter. The global set of undulations was then used in a quadrature procedure to determine potential coefficients to degree 180. The GEM-10C model was then formed from the GEM-10B model (2 to 36) with the coefficients 37 to 180 from the quadrature results.

The World Geodetic System 1972 (WGS72) was described by Sepplin (1974). Part of this system was a geopotential model which was based on the combination of the NWL10E geopotential model (based on Doppler tracking data) with other satellite tracking data (e.g. optical and secor) with terrestrial and astro-geodetic data. The adopted model was complete to degree 19.

Estimation of geopotential models in Europe was now also underway, with the development of the GRIM-1 and GRIM-2 models, the latter being a combination model (satellite and terrestrial data) (Balmino, Reigber, Maynot, 1978).

Rapp (1978) described the formation of a global 1°x1° anomaly field that was used with the GEM-9 potential coefficient model to determine a set of adjusted coefficients to degree 12 and an adjusted anomaly file that was used to determine, through quadrature procedures, a model to degree 60. Higher degree adjustments and model estimations were not possible because of the extensive computer resources needed in the formation and inversion of several matrices. In 1979 software developed by Rizos (1979) became available that was extremely efficient in the calculation of harmonic coefficients given a global data set (analysis) and the calculation of functional values (e.g. anomalies or geoid undulations) from a set of potential coefficients. What could not be done in 1978 because of inefficient code became possible in 1979 with the procedures developed by Rizos. It was then a natural step to take the adjusted 1° anomalies from the Rapp (1978) combination solution and determine the potential coefficients to degree 180 with the procedures and results described in Rapp (1979b).

During this time period several issues were considered dealing with the estimation of potential coefficients from terrestrial data. One item had been identified by Pellinen (1966), who described a smoothing parameter, β_n, that would yield mean cap values of a function defined by a set of harmonic coefficients. Since harmonic coefficients of the anomalies are derived from mean anomalies, the implied smoothing was to be considered. If $\Delta\bar{g}$ represents mean anomalies in a cell $d\sigma$, the quadrature procedure for estimating potential coefficients would be

$$\left\{ \begin{matrix} C_{nm} \\ S_{nm} \end{matrix} \right\} = \frac{1}{4\pi\gamma\beta_n(n-1)} \iint_\sigma \Delta\bar{g}\, P_{nm}(\cos\theta) \left\{ \begin{matrix} \cos m\lambda \\ \sin m\lambda \end{matrix} \right\} d\sigma \tag{4}$$

The calculation of β_n depends on the cap size of a circular cap that has an area equivalent to the cell size in which the anomaly is given. Closed expressions for β_n were given by Pellinen, as described by Rapp (1977) and studied further by Katsambalos (1979). The use of equations such as (4) enabled the determination of expansions to a degree higher than that implied by the Nyquist guideline that the highest frequency to be measured from a cell of size $\theta°$ would be $180°/\theta°$. For example, solutions were made to degree 52 from 5° anomalies (Rapp, 1977).

A second area of concern related to the reduction of the gravity anomalies to be used in the combination solution to the surface (Is it the geoid ?) on which the quadrature procedure was to be applied. Tests were carried out (Rapp, 1977) approximating a gradient reduction term by terrain corrections. Magnitudes were calculated for the correction to potential coefficients if terrain corrections were applied to the data. The results pointed out the need for such corrections. Similar computations were carried out for the effect of atmospheric corrections on anomaly data to potential coefficient estimations. Again, it became clear that such correction terms were needed.

Another area related to the proper formation of the boundary value condition in relating anomalies to potential coefficients. Most combination solutions (quadrature or direct) used spherical approximations. However, as higher degree solutions were being sought, a more rigorous approach was needed. Lelgemann (1973) provided equations (to eccentricity squared terms) to correct coefficients estimated with spherical approximation equation to those that would be found with a more rigorous formulation. The correction terms were evaluated in Rapp (1997) and found to have a magnitude on the order of 0.3% of the coefficients. Although small, such terms could cause systematic effects and it was thought desirable to consider such effects in future combination solutions.

The 1970's also saw the development of techniques to combine satellite and terrestrial gravity data where the gravitational field was modeled by surface density values (Koch, 1971, 1974; Koch and Witte, 1970; and Chovitz and Koch, 1979) or by discrete gravity anomaly values (Rapp, 1974). The advantage of such methods was the simple incorporation of terrestrial data in the solution. Once the discrete values (surface density or gravity anomalies) were estimated, they could be converted to a geopotential model whose maximum degree was consistent with the size of the cell used in the solution. The surface density solution used 104/192 density values leading to geopotential models of degree 11/15. The Rapp (1974) solution used 184 15° equal area cells leading to potential coefficient models of degree 12. Although models developed from discrete estimates of gravity field signal were conceptually of interest, the actual application in orbit analyses for high resolution geopotential models was not an efficient process (as compared to the use of potential coefficients) and interest in such procedures declined.

DEVELOPMENTS IN THE 1980's

The geopotential modeling effort actively continued in the 1980's, driven by more precise tracking data (primarily satellite laser tracking), a larger number of satellites with a variety of different inclinations and heights, satellite altimeter data (GEOS-3 and SEASAT), and improving terrestrial gravity anomaly data sets. In addition, the new data was making it possible to study new areas (e.g. plate motion and dynamic ocean topography) of research leading to a need for increasingly accurate geopotential models.

The efforts at NASA/GSFC now incorporated SLR data from Lageos and satellites that had laser reflectors. The GEM-L2 model (Lerch, Klosko, Patel, 1982) utilized GEM-9 data with 2 1/2 years of Lageos (launched in 1976) data. This model was complete to degree 20, although improvements over the GEM-9 model (due to the Lageos data) were primarily below degree 8. Data (laser, S band, and altimeter) from SEASAT was used with the GEM-9 model to produce several models tailored to SEASAT. The PGS-S4 model included the normal equations from GEM-10B, surface gravity, normal equations based on 5° data, and SEASAT altimeter data. This model was complete to degree 22 with additional terms.

With the approval of the TOPEX/POSEIDON (T/P) project, an effort to improve the Earth's geopotential was started with the goal of providing improved orbit determination for use with the T/P altimeter data. One of the first pre-launch models was developed at NASA/GSFC and designated GEM-T1 (Marsh *et al.*, 1988). This model was complete to degree 36 and based solely on satellite tracking data from 27 satellites.

The development of the GRIM models continued in Germany (DGFI) and France (GRGS). The GRIM-3 model (Reigber *et al.*, 1983) used optical, laser, and Doppler measurements and 1°x1° gravity anomalies from terrestrial and altimeter (GEOS-3) derived sources. The solution was complete to degree 36. An updated version (GRIM3-L1) of this model incorporated 16 months of Lageos SLR data, 1°x1° anomalies from the analysis of SEASAT data (Rapp, 1983a), an updated set of 1°x1° terrestrial anomalies (Rapp, 1983b), and several constraint equations from resonance effects. The anomaly data entered into the solution through normal equations where the anomalies where the observed values and the potential coefficients were the parameters. The maximum degree of the GRIM3-L1 model was 36.

The group at the Center for Space Research at the University of Texas at Austin started the estimation of geopotential models with the TEG-1S and TEG-1 model (Tapley *et al.*, 1988/1989). The TEG-1 model incorporated a variety of tracking data from 14 satellites, 1° surface gravity data, and direct SEASAT and GEOSAT altimeter data, and was complete to degree 50. The TEG-1S model excluded the direct altimeter data from the solution, although crossover information was included.

Numerous high degree expansions using different techniques and the improved terrestrial and altimeter derived anomalies were made during the 1980's. An important aspect in some solutions was the research described by Colombo (1981), who developed techniques to analyze and synthesize values on a grid on a sphere using FFT techniques. The quantities of interest could be given/computed as point or mean values. Colombo also showed how a least squares collocation procedure could be used to obtain high degree solution that would minimize the standard deviations of the adjusted coefficients. He also studied the Pellinen smoothing factors to determine improved values that would take into account noise in the anomaly data as well as cell size in determining optimal de-smoothing ($1/\beta_n$) factors. His suggestion was to use new β_n factors (q_n) in the discrete form of (2) using mean anomalies. The revised q_n values were: β_n^2 for $0 \le n \le N/3; \beta_n, N/3 < n \le N;$and 1 for n>N, where N=180°/θ°, and θ is the cell size in which the anomalies were given. A problem with the factors was the discontinuity that would take place when n=N/3 and at n=N. Numerical tests with these factors are described in Colombo (1981) and in Rapp (1986).

Rapp (1981) described a geopotential model (OSU81) that was complete to degree 180. This model used an updated 1° anomaly file (42,585 terrestrial-based values, 37,905 values derived from SEASAT altimeter data), and a set of a priori potential coefficients to degree 36 that were based on several models including some coefficients (e.g. zonal, resonant) determined in coefficient specific solutions. The 1° anomaly values lacking an estimate were computed from the a priori model and assigned an accuracy of ±30 mgals. The OSU81 model used optimum quadrature weights suggested by Colombo. The standard deviations of the non-adjusted coefficients were estimated using a propagated noise effect, with uniform anomaly accuracy and a component for the sampling error.

In 1985 Wenzel started procedures to carry out a sub-optimum least squares solution for geopotential coefficients combining an a priori geopotential model, terrestrial 1° anomalies, and 1° geoid undulations in the ocean areas derived from satellite altimeter data. A sub-optimum procedure was needed because of the large number of coefficients to be estimated (e.g. 32,761 for a degree 180 solution) in a rigorous least squares adjustment. The solution carried out by Wenzel assumed that the off diagonal elements of the normal equation of the combination solution were zero so that the inversion of the very large matrix became simply the inversion of a diagonal element. The advantage of the procedure (and others like it) was that a global anomaly field was not needed; different types of data (e.g. anomalies and undulations) could be combined and accuracy estimates for the adjusted coefficients could be determined. The model (GPM-2) developed by Wenzel was complete to degree 200.

Hajela (1985) described a solution to degree 250 using a least squares collaboration procedure suggested by Colombo (1981). The solution was made from 1° anomalies and was not a combination model as prior solutions. The power contained at the higher degrees was considerably smaller than other solutions, indicating too much smoothing could be taking place.

Rapp and Cruz (1986a) described several solutions using 1° data and GEM-L2 as the a priori model This solution used a downward continuation procedure to reduce surface anomalies to the ellipsoid and the ellipsoidal correction terms (to anomalies and to potential coefficients) developed by Cruz (1986). A usual quadrature type combination using the Colombo modified weights was used to determine an adjusted coefficient set to degree 20 and an adjusted 1° anomaly set. This global set of anomalies was expanded to degree 250 using the optimal estimation procedure implemented by Hajela. To reduce the power reductions at high degrees all 1° anomaly standard deviations were set to ±1 mgal. Two models (OSU86C,D) were released that merged the adjusted coefficients with the coefficients from the optimum estimation procedure. In these models the anomaly standard deviations in the combination solutions were restricted to a certain range to reduce the impact of large anomaly residuals in regions of high standard deviation.

Rapp and Cruz (1986b) described the first potential coefficient solution to degree 360. The first steps in the creation of the model were to create a nearly global 30'x30' anomaly file. This required the formation of a terrestrial 30' anomaly file that was merged with an altimeter derived from a GEOS-3/SEASAT data anomaly set. Of the 259,200 possible 30' values, 38% were based on adjusted 1° data with the rest coming from the terrestrial and altimeter data. The global 30' data was used in the quadrature procedure to determine the potential coefficients to degree 360. The ellipsoidal corrections and anomaly reduction to the ellipsoid were used. Two solutions (OSU86E,F) were generated using two different anomaly data sets. The final models were a merger of the coefficients from the adjustment of the GEM-L2 model and 1°x1° data with the coefficients from the quadrature procedure.

One should also note the development of the WGS84 geopotential model (White, 1986; DMA, 1987), which is complete to degree 180. This model was initially developed using satellite tracking data (Doppler, laser, GPS), surface gravity data, and geoid height information derived from SEASAT altimeter data. The merger of the normal equations and their solution gave a model complete to degree 41. This model was removed from a global 1°x1° anomaly set and the resulting residuals developed, with a quadrature procedure, to degree 180. The final WGS84 model was a merger of the degree 41 combination solution and the coefficients from degree 42 to 180 from the quadrature solution. The coefficients were initially available only to degree 18 and then later to degree 180.

In 1988 Weber and Zomorrodian suggested a procedure to create a tailored high degree geopotential model. The basic idea was to improve an existing model by the incorporation of new anomaly data in a region where improved geoid undulation computations (e.g.) were desired. The initial work described an improved model valid in Iran that was complete to degree 180 with GPM2 as the starting model. Based on the success of this work, others were developed for different areas. The IFE 88E2 model, based on the OSU86F model, was complete to degree 360 and tailored to the improved gravity data available in Europe (Basic *et al*, 1989). A discussion on the advantages and disadvantages of tailored geopotential models is found in Kearsley and Forsberg (1990).

As part of the pre-launch studies for the Topex/Poseidon mission, Tapley *et al.* (1989) described the estimation of a combination model to degree 50 (TEG-1) and a corresponding model (TEG-1S) , where direct altimeter data was withheld from the solution. Shum *et al.* (1989) described the estimation of the PTGF-4 and PTGF-4A geopotential models using data from 15 satellite and normal equations from 1°x1° terrestrial data created by Pavlis (1988). The two models were complete to degree 50 with the PTGF-4A including SEASAT and GEOSAT sea surface height data, which was not included in the PTGF-4 model. The PTGF-4A solution also estimated harmonic coefficients of the expansion, to degree 10, of dynamic ocean topography (the separation between the ocean surface and the geoid).

Improved models for the high degree combination solutions were made possible by the introduction of ellipsoidal harmonics and the use of conversion equations to transform ellipsoidal harmonics coefficients to spherical harmonic coefficients. The discussion of the transformation can be found in Gleason (1988) and Jekeli (1988). The introduction of these techniques into the quadrature based combination solution is described in Rapp and Pavlis (1990). The use of the almost closed expressions eliminated the need for the less than satisfactory series expressions used for correction terms. In addition, rigorous procedures were described in Pavlis (1988) to properly define the boundary value problem that was being carried and the development of appropriate correction terms to the mean gravity anomalies being used in the combination solution.

Rapp and Pavlis (1990) describe the development of two degree 360 geopotential models (OSU89A,B) that differ only in the treatment of anomaly estimates in which no directly observed gravity anomaly data was available. The OSU89B model incorporated anomalies for "empty" cells calculated from elevation data based on spherical harmonic models incorporating a topographic/isostatic hypothesis (Pavlis and Rapp, 1990). The combination model was based on the GEM-T2 model (Marsh *et al.*, 1989), which was complete to degree and order 36, with an additional 616 coefficients up to degree 50 and order 43. Improved 30' anomaly data sets were available for both terrestrial data and altimeter derived anomalies from GEOS-3/SEASAT data (Hwang, 1989) and a limited set of Geosat data in the oceans near Antarctica.

The developments in the 1980's were made possible, in part, by the introduction of vector processing computers, the increased speed of the computers, and the increased memory and storage capabilities. The new supercomputers required, for efficient use of the system, the rewriting of much software that had been previously used by sequential processors. The supercomputer provided the computer resources to process large amounts of new tracking data and enabled the extension of spherical harmonic models to higher degrees. The need for faster computers continues in the 1990's, where significant computer resources are required for the estimation of global geopotential models and the parameters associated with it.

DEVELOPMENTS IN THE 1990's

Geopotential models evolved rapidly starting early in the 1990's. The GEM-T3 model from NASA/GSFC was described by Lerch *et al.* (1992). This model was complete to degree 50, using tracking data from 31 satellites plus satellite altimeter data from GEOS-3, SEASAT, and GEOSAT, with improved surface gravity data. (A model based only on satellite tracking data was designated GEM-T3S and was also complete to degree 50.) The model estimation included three models (to degree 15) of the dynamic ocean topography for each of the altimeter missions as described by Nerem *et al* (1994b). The GEM-T3 combination model was found to give orbits more accurate than the GEM-T3S model. In the development of these and earlier models (e.g. GEM-T2) the optimal data weighting technique described by Lerch (1991) was used.

The TEG-2B model (Tapley *et al.*, 1991) was a combination model based on satellite tracking data, surface gravity data, and altimeter data and was complete to degree 50. An update of the GRIM models was started with the cooperative effort of DGFI/GRGS. The first in the GRIM4 series were the satellite-alone model (S1) and the combination model (C1) as described by Schwintzer *et al.* (1991). The C1 model added normal equations formed from 1° data sets with high frequency components (degree 51 to 360) removed using the OSU86F degree 360 model to reduce the aliening effects when the degree 50 model was estimated. The GRIM-4 model development continued with the S2/C2 model and the S3/C3 model described by Schwintzer *et al.* (1992). The S3 model was complete to degree 50, with coefficients to degree 20, plus additional zonal and resonant coefficients, reliably determined. The C3 solution (complete to degree 60) incorporated surface gravity data and direct altimeter data with a dynamic ocean topography model (to degree 10) also solved for. The most recent GRIM models are designated GRIM4-S4 and GRIM4-C4 (Schwintzer *et al.*, 1997). The satellite model is complete to degree 60 plus several resonance terms while the combination model is complete to degree 72 and included additional surface gravity data not previously used in GRIM model estimation. Mean (1° cells) geoid undulation values in ocean areas, based on a mean sea surface corrected by the Levitus dynamic ocean topography model, were introduced in the solution. The new tracking data included the GPS tracking of T/P.

Geopotential model development for the Topex/Poseidon project continued with the estimation of the three JGM models. These models were developed by NASA/GSFC and the University of Texas at Austin. The JGM-1 model was a pre-launch model while JGM -2 was a post-launch model that included T/P laser and DORIS tracking data (Nerem *et al.*, 1994a). Both models are combination models using OSU (1991 vintage) 1° surface data and are complete to degree 70. Direct altimeter data from GEOS-3, SEASAT and GEOSAT were also used in the solution. The JGM -3 model (Tapley *et al.*, 1996), released in 1994, was estimated starting from the JGM -1 model with the addition of SLR data to several satellites, DORIS tracking of T/P and SPOT2, and GPS tracking of T/P. The model, complete to degree 70, includes the surface gravity data used in previous JGM models.

The above discussion has been restricted to the low degree satellite and combination models where normal equations are formed to solve the coefficients of the maximum feasible degree, which has been degree 70 for the models described. Significant progress has been made in the development of high degree geopotential models and this will now be discussed.

Rapp, Wang, and Pavlis (1991) described the OSU91A model, which is complete to degree 360. This was a blended model where the coefficients to degree 50 were based on a combination solution starting from the GEM T2 model, incorporating 1° surface data and GEOSAT altimeter data, with coefficients from degree 51 to 360 from a quadrature type of combination solution as used in the OSU89 model estimation. The 30' anomalies used in the quadrature solution were based on a merger of terrestrial data, altimeter derived anomalies, and the topographic/isostatic anomalies for cells in which no other estimate was available.

During this time period, alternative techniques to the quadrature combination solution were developed that attempted to estimate a high degree model using a least squares adjustment technique with certain approximations that would allow high degree (360) models to be estimated. Bosch (1987) described an adjustment method that ordered the unknown coefficients in such a way that the normal equation became quite patterned, enabling a high degree model to be estimated with the neglect of certain off-diagonal terms. Gruber and Bosch (1992a) describe the degree 360 DGFI92A model that started from the normal equations of the GRIM4-S2 model adding 30' and 1° terrestrial anomaly information as well as a sea surface height model based on GEOSAT altimeter data. The least squares solution was carried out using the patterned structure approach. In 1992 Gruber and Bosch describe the OGE12 model which was then used to fill in gaps in the surface anomaly and sea surface height data sets. Gruber and Anzenhofer (1993) describe the GFZ93a and GFZ93b degree 360 geopotential model that are based on the GRIM4-C3 models and estimated using the block diagonal structure. The GFZ93b model used the Basic/Rapp (1992) mean sea surface while the GFZ93a model used a 9 month mean sea source derived from ERS-1 fast delivery data.

Gruber, Anzenhofer and Rentsch (1995) report on the development of the degree 360 GFZ95A model whose a priori basis was GRIM4-C4B. Improved terrestrial anomaly files and geoid height data (derived from a 2-year ERS-1 mean sea surface with dynamic ocean topography correction) were used in an iterative block-diagonal solution. The authors suggest an improved implementation of the procedure detailed by Bosch (1993), where by the full satellite only normal equations can be used with normal equations from the non-global anomaly and undulation data sets. A re-ordering leaves a full normal system in the upper left part of the normals with block-diagonal parts and sparse matrices for off diagonal components at higher degrees. This procedure was implemented (Gruber *et al.*, 1996) using new terrestrial data, a 3-year ERS-1 mean sea surface, altimeter derived gravity anomalies, with the normal equations and potential coefficients of the GRIM4-S4 model as the a priori model. Five different solutions (GFZ96) to degree 350 were computed testing different weighting procedures and data set combinations.

Several tailored geopotential models have also been developed in the 1990's. Li and Sideris (1994) describe the calculation of a degree 500 model using 5'x5' anomalies in Canada using the tailored model approach (Weber and Zomorrodian, 1988) starting from the OSU91A model. The new model (OSU91AT) was tested using geoid undulation information from GPS measurements as benchmarks. The tailored model to degree 360 gave slightly better results than the tailored model to degree 500. Motao *et al.* (1996) discuss the formation of a tailored (to degree 360) model (DQM94) using an improved set of gravity anomalies for China. The starting model was OSU91A. The tailored approach was implemented using ellipsoidal corrections for the anomalies, downward continuation and ellipsoidal harmonics. The residual coefficients were added to the original OSU91A model that left the OSU91A model unchanged to degree 19 and modified the residual by a weighting factor varying from zero at degree 20 to 1 at degree 180 and above. Anomaly and undulation tests were carried out to test the improvement of the DQM94 model over the OSU91A model.

Another type of tailored geopotential model is the kind designed to provide optimum orbit accuracy for a specific satellite. One example is the tuned model (PGS-S4) for SEASAT. A recent example is the DGM-E04 model (Scharroo, Visser, and Mets, 1998) which is a tuned version of JGM -3 to provide improved orbits for the ERS satellites. The tuning is carried out by adjusting a selected set (1100) of JGM-3 coefficients using crossover differences as the observations and including a scaled JGM-3 normal matrix to help stabilize the solution.

The final two geopotential models to be described for this paper were presented at the Gravity, Geoid, and Marine Geodesy meeting held in October 1996 in Tokyo. The TEG-3 model is described by Tapley *et al.* (1997). This model is a combination model to degree 70 with significant amounts of new, over JGM-3, satellite tracking data incorporated into the solution. The surface gravity data used for TEG-3 was the same as used in the JGM-3 model development.

Lemoine *et al.* (1997) describes the development of the EGM96 model which is complete to degree 360. This model incorporates new (over that used in JGM-3) satellite tracking data and terrestrial gravity data (1° and 30' cells) as well as 30' anomalies derived from GEOSAT (primarily) and ERS-1 altimeter data. The coefficients are a blend of three computational procedures. From degree 2 to 70 the coefficients are based on a least squares adjustment and terrestrial gravity data (1° and 30' cells) as well as 30' anomalies derived from GEOSAT (primarily) and ERS-1 altimeter data. The coefficients are a blend of three computational procedures. From degree 2 to 70 the coefficients are based on a least squares adjustment involving satellite tracking data, 1° terrestrial data, direct altimeter data, and fill-in anomalies in areas lacking data. The coefficients from degree 71 to 359 are taken from a block diagonal combination solution (Pavlis *et al.*, 1996) using normal equations derived from the satellite tracking data as a priori values. The coefficients at degree 360 are taken from a quadrature combination solution using the a priori satellite model and a global 30' anomaly set. (The adjusted anomalies from this solution were expanded to degree 500 but found useful to degree 460.)

The 1990's to date have been very productive in the estimation of improved geopotential models. The models have become increasingly accurate due, primarily, to more and new (e.g. GPS, TDRSS, and DORIS) tracking data, significant improvements in our terrestrial gravity data base, and in creating the mathematical model to unite the observations and parameters. The model improvement has been a significant challenge in the era of increased data accuracy. Areas in which the model development was critical included: Tides (both ocean and satellite effects); atmospheric drag; direct solar radiation pressure; Earth radiation pressure; spacecraft attitude effects; reference frame definitions and realization, etc. In addition, we have seen a significant effort to provide realistic error estimates for geopotential models (and related parameters). Such error estimates are needed so that propagated error estimates for quantities depending of the geopotential (e.g. satellite orbit error, geoid undulation, gravity anomaly, etc.) can be reliably determined.

TEMPORAL VARIATIONS OF THE GEOPOTENTIAL

The previous discussion has considered the geopotential to be constant in time. In fact, evidence has accumulated over the past ten years that demonstrates time changes can be estimated for selected zonal coefficients (Eanes and Bettadpur, 1996, Cazenave *et al.*, 1996, and Nerem and Klosko, 1996). Estimates of zonal rates to degree 5 have been made with the most reliable rate being for the degree 2 zonal coefficient. For the EGM96 geopotential model, the time rate of the zonal degree 2 term was fixed from a previous analysis (Nerem and Klosko, 1996) with the epoch for this term defined as 1986.0. All other coefficients (except for degree 2,1 terms which depend on polar motion) refer to a mean time span of the data used in the solution.

Geopotential changes are to be expected because of the changing mass distribution in the land, oceans, and atmosphere (NRC, 1997). The detection of the very small changes will be possible through the analysis of data from selected satellite geopotential mapping missions, as noted in the next section.

A LOOK TO THE FUTURE

In the near term (2-3 years) we can expect additional geopotential models that build on the increasing information available to the model community. New types of information could be incorporated into geopotential estimation. Suggestions have previously been made that dynamic ocean topography (DOT) models based on ocean circulation models might be used to correct altimeter derived sea surface heights to geoid undulations that are incorporated in the least squares estimation instead of solving for a DOT model. Questions exist as to the accuracy of a DOT model and the use of a static or time dependent model. The resolution, both in geographic sense and in a temporal sense, must be clearly defined.

Models higher than degree 360 models have been developed for special tailored models as well as developed from mean anomaly sets whose cell size would suggest that the maximum reasonable degree is $180°/\theta°$, although actual tests reveal expansions carried out above this maximum degree contain useful information. One would expect that more reliable high degree solutions could be obtained if cell sizes smaller than 30' could be estimated. The next logical size to work with would be 15' cells. Estimates of 15'x15' sea surface heights and gravity anomalies for much of the world are now possible. Combination solutions using the quadrature adjustment approach would not be a problem with the block diagonal approach being somewhat more complicated. With some additional effort, potential coefficient models to degree 720 are possible using 15' cell size for the anomaly estimates.

An alternative to higher degree geopotential models may be a combination of a high degree spherical harmonic model with harmonic wavelets. The wavelets can be introduced in a joint model in a way to be consistent with the data density available for a solution. The combined model can be estimated from both satellite data and terrestrial data. The resultant model can be used to calculate various gravimetric quantities such as geoid undulations, deflections of the vertical, etc. The theory for the development of the combined model procedure can be found in Freeden and Windheuser (1997) and Freeden and Schneider (1997).

Before developing such models, it may be appropriate to consider if a rectangular cell size should continue to be the way in which mean values are estimated. Jekeli (1996) notes the aliasing impact if the function averaging is done over constant angular blocks, as is the usual case. He suggests such errors can be reduced by forming spherical cap averages. To implement this procedure one would need to define the location of the spherical caps as well as carry out a new estimation of the mean values which is a substantial effort. Additional discussion on the recovery of harmonic coefficients from mean values on a sphere is given by Albertella and Sacerdote (1995).

As one goes to the higher degrees improved procedures will be needed to reduce the surface gravity anomalies to the ellipsoid. Downward contribution techniques need to be studied to determine the data needed and method to use for the reduction.

The most accurate high degree models are currently blended models where the low degree component comes from the combination of satellite tracking data, direct altimeter data, and terrestrial data. As part of the solution a model for DOT needs to be estimated. Could the block diagonal approach be expanded so that altimeter data and DOT estimates be incorporated in the model determination?

Improved data sets would also be of help in forming new models. Terrestrial gravity data in some selected land areas (parts of South and Central America, Asia, and Africa); in ocean areas (primarily southern oceans); and in the polar regions are still quite poor. Altimeter data is plentiful but not widely used in the shallow water areas. Improved shallow water tide models are needed so that data now edited from altimeter analysis is kept in the low degree combination model or in the recovery of gravity anomalies from altimeter data in shallow water or coastal water regions. Higher resolution and more consistent elevation data on a global basis would be helpful in the reduction of surface gravity data.

Much remains to be done to improve the data and analysis techniques now being used for the estimation of geopotential models. Steps can be taken as an evolutionary effort or one may wish to make significant changes in the way gravity anomaly data is represented and in the model to be estimated for the Earth's gravitational potential. Reliable accuracy estimates for any geopotential model are an important aspect of any modeling effort.

In a few years data will become available from satellite missions designed to determine a more precise knowledge of the Earth's gravitational potential. The missions currently underway are CHAMP (1999), GRACE (2001), and GOCE (2004?). The analysis of the data from each mission will provide a considerable challenge. Numerous questions need to be considered in such analyses. For example, what is the ideal representation of the geopotential for use in these missions? How can existing geopotential models be used in the data analysis? How can existing terrestrial gravity anomaly and satellite altimeter data be incorporated in the modeling effort? How will time changes in the gravity field be modeled? A variety of approaches to the analyses can be done prior to the acquisition of data from the new missions.

CONCLUSION

This paper has reviewed the estimation of the Earth's geopotential-from parameters of an ellipsoid gravity formula to geopotential models complete to spherical harmonic degree 500. Much can be done in the next few years to improve the accuracy of our existing models and to extend them to higher degrees is such high degree model will be useful. Alternative modeling techniques also provide additional opportunities for creative research. Much has been done in the past 96 years; much remains to be done.

ACKNOWLEDGMENT

The preparation of this report was supported, in part, by NASA (Contract NAS5-32352) through Hughes STX. Comments from the reviewers were helpful in improving the original text.

REFERENCES

Albertella A and F Sacerdote, Spectral analyses of block averaged data in geopotential global model determination, J Geodesy, 70:166-175, 1995.

Anderle R, Geodetic parameter set NWL-5E-6 based on Doppler satellite observations, in Proc. of 2nd Int Sym Geodetic Use of Satellites, G. Veis (ed), 179-220, Athens, Greece, 1966.

Anderle R and S Smith, NWL-8 Geodetic parameters based on Doppler satellite observations, NWL Technical Report No. 2106, Dahlgren, Virginia, July 1967.

Balmino G, C Reigber, B Maynot, Le modèle de potential gravitationel terrestre GRIM2: Determination, evaluation, Ann Geophys 34:55-78, 1978.

Basic T et al, A new geopotential model tailored to gravity data in Europe, in Gravity, Gradiometry, and Gravimetry (eds. Rummel and Hipkin), IAG Sym 103, Springer, Berlin, 1989.

Basic T and R Rapp, Oceanwide prediction of gravity anomalies and sea surface heights using Geos-3, Seasat, and Geosat altimeter data and ETOPO5U bathymetric data, Rpt 416, Dept of Geodetic Science and Surveying, Ohio State University, Columbus, OH, 94p, 1992.

Bosch W, High degree spherical harmonic analyses by least squares, in Proc. of IAG Sym XIX IUGG General Assembly, 194-205, 1987.

Bosch W, A rigorous least squares combination of low and high degree spherical harmonics, presented at IAG General Meeting, Session 11, Part D, Beijing, China, 1993.

Bouman J, A survey of global gravity models, Delft Institute for Earth-Oriented Space Research, Report No. 97.1, Delft University of Technology, Netherlands, July 1997.

Brovar V, V Magnitsky, and B Shimbirer, The Theory of the Figure of the Earth, Moscow 1961, translation, Foreign Technology Div., Air Force Systems Command, Wright-Patterson AFB, Ohio, FTD-TT-64-930, 1964.

Cazenave AA, P Gegout, G Ferhat and R Biancale, Temporal variations of the gravity field from Lageos 1 and Lageos 2 Observations, in Proc. Global Gravity Field and Its Temporal Variations, Rapp, Cazenave, Nerem (eds), IAG Symposium 16, Springer Berlin 1996.

Chovitz B and K-R Koch, Combination solution for gravity field including altimetry, J Geophys Res, 84(B8), 4041-4044, 1979.

Colombo O, Numerical methods for harmonic analysis on the sphere, Rpt 310, Dept of Geodetic Science and Surveying, Ohio State University, Columbus, OH, 140pp, 1981.

Cruz JY, Ellipsoidal corrections to potential coefficients obtained from gravity anomaly data on the ellipsoid, Rpt 367, Dept of Geodetic Science and Surveying, Ohio State University, Columbus, OH, 38pp, 1986.

DMA, Supplement to the Department of Defense World Geodetic System 1984, Tech Rpt Part I, Methods, Techniques and Data Used in WGS84 Development, DMA TR 8350, 2-A, 412p, Washington, DC, 1987.

Eanes RJ and SV Bettadpur, Temporal variations of Earth's gravitational field from satellite laser ranging, in Proc. Global Gravity Field and Its Temporal Variations, Rapp, Cazenave, Nerem (eds), IAG Symposium 16, Springer Berlin 1996.

Freeden W and F Schneider, An integrated wavelet concept of physical geodesy, No. 173, Univ of Kaiserslautern, Kaiserslautern, Germany, 27p, 1997.

Freeden W, U Windheuser Combined spherical harmonic and wavelet expansion-a future concept in Earth's gravitational determination, Applied and Computational Harmonic Analysis 4, 1-37, 1997.

Gaposchkin EM and K Lambeck, 1969 Smithsonian Standard Earth (II), Smithsonian Astrophys Obs, Spec Rep 315, 93 pp, 1970.

Gaposchkin EM(ed), 1973 Smithsonian Standard Earth (III), Smithsonian Astrophys Obs, Spec Rep 353, 388 pp, 1973.

Gaposchkin M, Earth's gravity field to the eighteenth degree and geocentric coordinates for 104 stations from satellite and terrestrial data, J Geophys Res, 79, 5377-5411, 1974.
Gleason D, Comparing ellipsoidal corrections to the transformation between the geopotential's spherical and ellipsoidal spectrum, manuscripta geod 13, 114-129, 1988.
Gruber T and W Bosch, A new 360 gravity field model, presented at the European Geophysical Soc meeting, Edinburgh, 1992a.
Gruber T and W Bosch, OGE12, A new 360 gravity field model, in Proc. (Montag and Reigber (eds)) of IAG Sym 112, Geodesy and Physics of the Earth, Springer, Berlin, 1992b.
Gruber T and M Anzenhofer, The GFZ 360 gravity field model, in Proc. (Forsberg and Denker (eds)), The European Geoid Determination, European Geophysical Soc XVIII General Assembly, Wiesbaden, Germany, published by Geodetic Div, Kort-og Matrikelstyrelsen, Copenhagen, 13-18, 1993.
Gruber T, M Anzenhofer and M Rentsch, The 1995 GFZ high resolution gravity model, in (Rapp, Cazenave, Nerem, eds), Global Gravity Field and Its Temporal Variations, IAG Sym 116, 61-70, Springer, Berlin, 1996.
Gruber T et al, Improvements in high resolution gravity field modeling at GFZ, in (Segawa, J, Fujimato, H, Okubo, S, eds) Proc. Gravity, Geoid, Marine Geodesy, Springer, Berlin, 1997.
Guier WH and RR Newton, The Earth's gravity field deduced from the Doppler tracking of five satellites, J. Geophys. Res, 70, 4613-4626, 1965.
Hwang C, High precision gravity anomaly and sea surface height estimation from GEOS-3/Seasat satellite altimeter data, Rpt 399, Dept of Geodetic Science and Surveying, Ohio State University, Columbus, OH, 155p, 1989.
Jeffreys H, The determination of the Earth's gravitational field, Mon. Not. R. Astron. Soc., Geophys. Suppl., 5, 1, 1-22, 1941.
Jeffreys H, The determination of the Earth's gravitational field, Mon. Not. R. Astron. Soc, Geophys. Suppl., 5, 3, 55-66, 1943.
Jeffreys H, The figures of the Earth and moon, Mon. Not. R. Astron. Soc, Geophys. Suppl., 5, 7, 219-247, 1948.
Jekeli C, The exact transformation between ellipsoidal and spherical harmonic expansion, manuscript geod 13, 106-113, 1988.
Jekeli C, Methods to reduce aliasing in spherical harmonic analyses, in (Rapp, Nerem, Cazenave, eds) Global Gravity Field and Temporal Variations IAG Sym 115, Springer, Berlin, 121-130, 1996.
Kaula WM, Statistical and harmonic analyses of gravity, J Geophys Res, 64(12), 2401-2421, 1959.
Kaula WM, Tests and combinations of satellite determinations of the gravity field with gravimetry, J Geophys Res, 71, 5303-5314, 1966.
Kearsley WH and R Forsberg, Tailored geopotential models-applications, and short comings, manuscripta geod, 15, 151-158, 1990.
Koch K-R, Surface density values for the Earth from satellite and gravity observations, Geophys. J.R. Astr. Soc., 21, 1-15, 1970.
Koch K-R, Earth's gravity field and station coordinates from Doppler data, satellite triangulation, and gravity anomalies, NOAA Technical Report NOS 62, 1974.
Koch K-R and B Witte, Earth's gravity field represented by a simple-layer potential from Doppler tracking of satellites, J. Geophys. Res., 76(35), 8471-8479, 1971.
Köhnlein W, The Earth's gravitational field as derived from a combination of satellite data with gravity data, presented at 14th General Assembly, IUGG, Lucerne, Switzerland, 1967.
Lelgemann D, Spherical approximation and the combination of gravimetric and satellite data, Bulletino di Geodesia e Scienze Affini, XXXII, M4, 1973.
Lerch FJ et al, Gravitational field models for the Earth (GEM 1&2), NASA/Goddard Space Flight Center Report X-553-72-146, May 1972.

Lerch FJ et al, Gravity model improvement using Geos 3 (GEM9 and 10), J Geophys Res, 84(B8), 3897-3916, 1979.
Lerch FJ et al, Goddard Earth models for oceanographic applications (GEM 10B and 10C), Marine Geodesy, Vol. 5(2), 145-187, 1981.
Lerch FJ et al., Gravity model improvement for SEASAT, J Geophys Res, 87(C5), 3281-3296, 1982.
Lerch FJ, S Klosko, G Patel, A refined gravity model from Lageos (GEM-L2), Geophys Res, Letrs 9(11), 1263-1266, 1982.
Lerch FJ et al, Geopotential models of the Earth from satellite tracking, altimeter and surface gravity observations, GEM-T3 and GEM-T3S, NASA Tech Mem 104555, Jan 1992.
Li Y and M Sideris, Minimization and estimation of geoid undulation errors, Bulletin Geod, 68, 201-219, 1994.
Lundquist CA and G Veis (eds), Geodetic Parameters for a 1966 Smithsonian Institution Standard Earth, 3 vols., Smithsonian Astrophys Obs, Spec Rep 200, 1966.
Marsh JG et al, A new gravitational model for the Earth from satellite tracking data: GEM-T1, J Geophys Res, 93(B6), 6169-6215, 1988.
Marsh J, et al, The GEM-T2 gravitational model, NASA Tech Memo, TM 100746, 52pp, 1989.
Motao H et al, On the improvement of geopotential model using gravity data in China, in (Rapp, Cazenave, Nerem, eds), Global Gravity Field and Its Temporal Variation, IAG Sym 116, 180-197, Springer, Berlin, 1996.
Molodensky, MS, VF Eremeev and MI Yurkina, Methods for the Study of the External Gravitational Field and Figure of the Earth, translated from Russian Israel Program for Scientific Translation, (OTS61-31207), 1962.
NRC, Satellite gravity and the geosphere, National Academy Press, Washington DC, 1997.
Nerem RS et al, Gravity model development for TOPEX/POSEIDON: Joint Gravity Models 1 and 2, J Geophys Res, 99(C12), 1994a.
Nerem RS et al, Ocean dynamic topography from satellite altimetry based on the GEM-T3 geopotential model, manuscripta geod 19, 346-366, 1994b.
Nerem RS and SM Klosko, Secular variations of the zonal harmonies and polar motion as geophysical constraints, in Proc. Global Gravity Field and Its Temporal Variations, Rapp, Cazenave, Nerem (eds), IAG Symposium 16, Springer Berlin 1996.
Pavlis N, Modeling and estimation of a low degree geopotential model from terrestrial gravity data, Rpt 386, Dept of Geodetic Science and Surveying, Ohio State University, Columbus, OH, 478p, 1988.
Pavlis NK and RH Rapp, The development of an isostatic gravitational model to degree 360 and its use in global gravity modeling, Geophys J Int, 100, 369-378, 1990.
Pavlis NK, JC Chan, and F Lerch, Alternative estimation techniques for global high-degree modeling, in (Rapp, Nerem, Cazenave, eds) Global Gravity Field and Temporal Variations IAG Sym 115, Springer, Berlin, 111-120, 1996.
Pellinen L, A method for expanding the gravity potential of the Earth in spherical functions, Trans. of the Central Scientific Res Inst of Geodesy, Aerial Survey and Cartography, Issue 171, 92 p, Nedra, Moscow, 1966 (translation ACIC-TC-1282).
Perasanz F, JC Marty, G Balmino, Dynamic orbit determination and gravity field model improvement from GPS, DORIS and laser measurements on TOPEX/POSEIDON satellite, J. Geodesy, 71:160-170, 1997.
Rapp RH, Gravitational potential coefficients from gravity data alone, Allgemeine Vermessungs-Nachrichten, 76, 228-233, 1969a.
Rapp RH, The geopotential to (14,14) from a combination of satellite and gravimetric data, Bulletin Geodesique, No. 91, 47-80, 1969b.
Rapp RH, Analytical and numerical differences between two methods for the combination of gravimetric and satellite data, Bolletino di Geofisica Theorica ed Applicata, 11(41), 108-118, 1969c.

Rapp RH, Procedures and results related to the determination of gravity anomalies from satellite and terrestrial gravity data, Rept 211 Dept of Geod Sci and Surv, Ohio State University, Columbus, OH, 1974.
Rapp RH, Determination of potential coefficients to degree 52 from 5° mean gravity anomalies, Bulletin Geodesique, 51, 301-323, 1977.
Rapp RH, Geos 3 processing for the recovery of geoid undulations and gravity anomalies, J Geophys Res, 84(B8), 3784-3792, 1979a.
Rapp RH, A review of anomaly and undulations determination from Geos-3 altimeter data, presentation, 1979 Spring Mt, American Geophysical Union, Washington, May 1979b.
Rapp RH, The Earth's gravity field to degree and order 180 using Seasat altimeter data, terrestrial gravity data, and other data, Rpt 322, Dept of Geodetic Science and Surveying, Ohio State University, Columbus, OH, 58pp, 1981.
Rapp RH, The determination of geoid undulation and gravity anomalies from Seasat altimeter data, J Geophys Res, 88, 1552-1562, 1983a.
Rapp RH, The development of the January 1983 1°x1° mean free-air anomaly data tape, internal report, Dept of Geod Sci and Surv, Ohio State University, Columbus, OH, 1983b.
Rapp RH Global geopotential solutions, in Mathematical and Numerical Techniques in Physical Geodesy, Lecture Notes in Earth Sciences, Vol. 7, 365-415, Springer, Berlin, 1986.
Rapp RH and J Cruz, The representation of the Earth's gravitational potential in a spherical harmonic expansion to degree 250, Rpt 372, Dept of Geodetic Science and Surveying, Ohio State University, Columbus, OH, 70pp, 1986a.
Rapp RH and J Cruz, Spherical harmonic expansions of the Earth's gravitational potential to degree 360 using 30' means anomalies, Rpt 376, Dept of Geodetic Science and Surveying, Ohio State University, Columbus, OH, 27pp, 1986b.
Rapp RH and NK Pavlis, The development and analysis of geopotential models to spherical harmonic degree 360, J Geophys Res, 95(B13), 21,885-21,911, 1990.
Reigber C, A Balmino, B Moynot, and H Miiler, The GRIM3 Earth gravity field model, manuscripta geod, 8(2), 93-138, 1983.
Reigber C et al., GRIM Gravity model improvement using LAGEOS (GRIM3-L1), J Geophys Res, 90(B11), 9285-9299, 1985.
Rizos C, An efficient computer technique for the evaluation of geopotential from spherical harmonic models, Australian J of Geodesy, Photogrammetry, and Cartography, No. 31, 161-170, 1979.
Rummel R and R Rapp, The influence of the atmosphere in detailed geoid computations and in potential coefficient determinations with gravity data, J Geophys Res, 81(32), 5639-5642, 1976.
Scharroo R, P Visser, G Mets, Precise orbit determinations and gravity field improvement for the ERS satellites, J Geophys Res (Oceans), ERS Spec Sect, in press, 1998.
Schwintzer P et al, A new Earth gravity model in support of ERS-1 and SPOT-2: GRIM4-S1/C1, final Rpt to DARA and CNES, DGFI/GRGS, Münich/Toulouse, 1991.
Schwintzer P et al, Improvement of GRIM4 Earth gravity models using Geosat altimeter and SPOT-2 and ERS-1 tracking data, in Proc. (Montag and Reigber, eds) Geodesy and Physics of the Earth, IAG Sym 112, Potsdam, Springer, Berlin, 1992.
Schwintzer P et al, Long-wave length global gravity field models: GRIM4-S4, GRIM4-C4, J Geodesy, 71:189-208, 1997.
Seppelin T, The Department of Defense World Geodetic System 1972, The Canadian Surveyor, 28(5), 496-506, 1974.
Shum C-K et al, An improved model for the Earth's gravity field, in Gravity, Gradiometry, Gradiometry, and Gravimetry (eds. Rummel and Hipkin), IAG Sym 103, 97-108, Springer, Berlin, 1989.

Tapley B et al, An improved model for the Earth's gravity field, in Progress in the Determination of the Earth's Gravity Field, Rpt 397, Dept of Geodetic Science and Surveying, Ohio State University, Columbus, OH, 1989.

Tapley B et al, The University of Texas Earth gravity model, presented at the IUGG General Assembly, IAG Sym, G3, Gravity Field Determination from Space and Airborne Measurements, Vienna, Austria, Aug 1991.

Tapley B et al, The Joint Gravity Model 3, J Geophys Res, 101(B12), 28,029-28,049, 1996.

Tapley B et al, The TEG-3 Earth gravity field model, in (Segawa, J, Fujimato, H, Okubo, S, eds), Proc. Gravity, Geoid, Marine Geodesy, Springer, Berlin, 1997.

Uotila UA, Determination of the shape of the geoid, Pub. Inst. Geod. Photogr. Cart. No. 7, 90-97, The Ohio State University, Columbus, OH, 1957.

Uotila UA, Harmonic analysis of worldwide gravity material, Annalis Academiae Fennicae, Ser. A, 111, Geologica-Geographica., 60, 1-17, 1962.

Weber G, H Zomorrodian, Regional geopotential model improvement for the Iranian geoid determination, Bulletin Geodesique, 62, 125-141, 1988.

Wenzel H-G, Hochaufloesende Kugelkenktionsmodelle fuer das Gravitationspotential der Erde, Wiss Arb Fachrichtung Vermessungswesen der Universitaet Hannover, 1985.

White H, The World Geodetic System 1984 Earth Gravitational Model, in Proc. 4th Int. Geodetic Symp. on Satellite Positioning, The University of Texas at Austin, Applied Res. Lab, Vol. 1, 93-116, 1986.

Vetter J, The evolution of Earth gravitational models used in astrodynamics, Johns Hopkins APL Technical Digest, 15(4): 319-335, 1994.

Zhongolovich ID, The external gravitational field of the Earth and the fundamental constants related to it, Acad. Sci. Publ. Inst. Teor. Astron., Leningrad, 1952.

Table 1. Geopotential Models Developed in the 1970's

Model	Type*	Date	Max. Complete Degree
GEM-1	S	1972	8
GEM-2	C	1972	16
GEM-3	S	1972	12
GEM-4	C	1972	16
WGS-72	C	1974	19
GEM-5	S	1974	12
GEM-6	C	1974	16
GEM-7	S	1976	16
GEM-8	C	1976	25
GEM-9	S	1979	20
GEM-10	C	1979	22
GEM-10B	C	1978	36
GEM-10C	C	1978	180
SAO SEIII	C	1973	18
GRIM1	S	1976	10
GRIM2	C	1978	23
OSU78	C	1978	60
OSU79	C	1979	180

* S = satellite tracking data; C = combination solution

Table 2. Geopotential Models Developed in the 1980's

Model	Type*	Date	Max. Complete Degree
GEM-L2	S	1982	20
PGS-S4	C	1982	22
GRIM3	C	1983	36
GRIM3B	C	1984	36
GRIM3-L1	C	1985	36
GEM-T1	S	1988	36
GEM-T2	S	1989	36
TEG-1	C	1988	50
PTGF-4/4A	C	1989	50
OSU81	C	1981	180
Hajela	T	1984	250
GPM2	C	1985	200
OSU86C,D	C	1986	250
OSU86E,F	C	1986	360
WGS84	C	1987	180
IFE88E2	t	1989	360
OSU89A/B	C	1989	360

* S = satellite tracking data; C = combination solution; T = terrestrial data only; t = tailored model

Table 3. Geopotential Models Developed in the 1990's

Model	Type*	Date	Max. Complete Degree
GEM-T3	C	1992	50
GEM-T3S	S	1992	50
TEG-2B	C	1991	50
GRIM4-S1	S	1991	50
GRIM4-C1	C	1991	50
GRIM4-S3	S	1992	50
GRIM4-C3	S	1992	50
JGM-1	C	1994	60
JGM-2	C	1994	70
JGM-3	C	1994	70
DGM-E04	t	1997	70
GRIM4-S4	S	1997	60
GRIM4-C4	C	1997	72
OSU91A	C	1991	360
DGFI92A	C	1992	360
OGE12	C	1992	360
GFZ93	C	1993	360
GFZ95A	C	1995	360
GFZ96	C	1996	359
Li/Sideris	t	1994	500
DQM94	t	1996	360
TEG-3	C	1997	70
EGM96	C	1997	360

* S = satellite tracking data; C = combination solution; t = tailored model

THE WORLD OF GRAVITY ACCORDING TO RAPP

Christopher Jekeli
Department of Civil and Environmental Engineering
and Geodetic Science
Ohio State University
2070 Neil Ave.
Columbus OH 43210
e-mail: jekeli.1@osu.edu

Abstract

A former advisee, student, apprentice, and now colleague gives a historical account of Richard H. Rapp's contributions to our knowledge of the Earth's gravity field. Clearly, his most prominent products are the spherical harmonic degree models that have emphasized theoretical rigor, wide geodetic applicability, and computational economy, and that are used by scientists and professionals throughout the world. Though the study and development of these models occupied the center of Richard Rapp's attention over most of his career, he contributed much to related endeavors, including gravity estimation techniques, geoid determination, satellite altimetry data processing, ellipsoidal parameter determination, world height datum definition, sea surface topography, and ocean circulation models. Richard Rapp's forte and credo are the numerics of data, always with a keen understanding of the underlying theory that they ought to support. His proven methods, organization, and enthusiasm were inculcated in many of his students who now also practice what he (still) preaches.

Introduction

A short examination of one of the problems of gravimetric geodesy, the determination of the geoid, serves as backdrop for a subsequent review of the principal areas where Richard H. Rapp has spent a career in working and contributing to refining the concept, definition, and most importantly, the calculation of the geoid. From classic considerations to modern concerns and questions for the future, none of this is new to many geodesists; but this paper is meant to be a brief tribute to Richard Rapp's work not only as it influenced the geodetic community as a whole, but also in the profound impact it had on his students and the Department of Geodetic Science at Ohio State University. It is impossible to use this vehicle to give credit to the many other geodetic activities Richard Rapp engaged in, though one may easily say they are numerous and of consistently outstanding quality. Nor can this paper properly review the current state of physical geodesy and the reader should not be offended if his or her work is not mentioned.

The Problem of Gravimetric Geodesy

The basic problem of gravimetric geodesy is the determination of the *geoid: an equipotential surface of the Earth's gravity field that closely approximates the level of the oceans*. It cannot be exactly sea level because the oceans, even in an average sense (mean sea level), are not in hydrostatic equilibrium implying that the mean sea surface has topography with respect to a surface of equal potential, such as the geoid; this *sea surface topography* is on the order of several decimeters, up to two meters.

Identifying precisely what surface should be the geoid has proved difficult since it cannot be *defined* independently as a uniquely and directly measurable or accessible fixed entity. In the words of Heck and Rummel (1991) it can be defined only *operationally*, that is, within the confines of model developments. Nevertheless, one must start somewhere and that place is the classic definition where the finer points raised by Heck and Rummel can safely be ignored from a certain conceptual viewpoint or if one accepts the limitations in accuracy or realizability that such a definition represents. That definition is given in the first sentence of this section.

The gravimetric determination of the geoid with respect to a given ellipsoid is well known to geodesists, being based on a solution to the disturbing potential, T, as given by (Heiskanen and Moritz, 1967):

$$T(\theta,\lambda,r) = \frac{R}{4\pi} \iint_{\sigma} \Delta g(\theta',\lambda')\, S(\psi,r)\, d\sigma \tag{1}$$

where Δg is the gravity anomaly on the geoid; $S(\psi,r)$ is the generalized Stokes function; R is a mean radius for the Earth; (θ,λ,r) are spherical polar coordinates; ψ is the central angle between (θ,λ) and (θ',λ'); and σ is the unit sphere. An equivalent solution is given by an infinite series of spherical harmonic functions:

$$T(\theta,\lambda,r) = \frac{kM}{R} \sum_{n=2}^{\infty} \left(\frac{R}{r}\right)^{n+1} \sum_{m=-n}^{n} C_{n,m}\, \overline{Y}_{n,m}(\theta,\lambda) \tag{2}$$

where kM is the product of Newton's gravitational constant and Earth's total mass; $\overline{Y}_{n,m}$ is a fully normalized surface spherical harmonic function of degree n and order m; and the (unitless) coefficients of the infinite series are given by

$$C_{n,m} = \frac{R^2}{4\pi\, kM\, (n-1)} \iint_{\sigma} \Delta g(\theta,\lambda)\, \overline{Y}_{n,m}(\theta,\lambda)\, d\sigma \tag{3}$$

These coefficients, thus, also depend on a global knowledge of Δg. But, they can be obtained to a limited degree also from the observation of satellite orbit perturbations.

Though analytically equivalent, the two solutions (1) and (2), by virtue of necessary further implementational approximations, represent alternatives in terms of accuracy and resolution, where the integral applies to local determinations and the series better describes the global field. The two solutions can be combined, again, yielding an equivalent form:

$$T(\theta,\lambda,r) = \frac{R}{4\pi} \iint_{\sigma_c} \left(\Delta g - \Delta g_s\right) S(\psi,r)\, d\sigma + \frac{kM}{R} \sum_{n=2}^{n_{max}} \left(\frac{R}{r}\right)^{n+1} \sum_{m=-n}^{n} C_{n,m} \overline{Y}_{n,m}$$
$$+ \frac{kM}{2R} \sum_{n=n_{max}+1}^{\infty} \left(\frac{R}{r}\right)^{n+1} (n-1)\, Q_n \sum_{m=-n}^{n} C_{n,m} \overline{Y}_{n,m} \tag{4}$$

where Δg_s is the gravity anomaly implied by the coefficients up to some finite degree, n_{max}; σ_c represents the area of a spherical cap centered at the computation point; and Q_n is a so-called Molodensky truncation coefficient, being well defined analytically.

These solutions are derived under the spherical approximation and assume that the disturbing potential is harmonic $\left(\nabla^2 T = 0\right)$ above the geoid and that the gravity anomalies reside on the geoid. Furthermore, the mass of the normal ellipsoid is assumed to be equal to Earth's total mass (including atmosphere). This is the reason for the absence of the zero-degree harmonic in the solution (2) (and (1)!). Finally, the origin of the coordinate system (and, therefore, the ellipsoid center) is assumed to be at Earth's center of mass. Thus, there are no first-degree harmonics in (2) (nor in (1)!).

Once a solution for the disturbing potential is found, the geoid undulation is given by:

$$N(\theta,\lambda) = \frac{T(\theta,\lambda,r_{geoid})}{\gamma(\theta)} - \frac{W_0 - U_0}{\gamma} = N'(\theta,\lambda) + N_0 \tag{5}$$

where $\gamma(\theta)$ is normal gravity evaluated on the ellipsoid; r_{geoid} is the radial distance to the geoid which may be approximated by the ellipsoid, and where an average value of normal gravity, γ, is adequate in the second term. The constant

$$N_0 = -\frac{W_0 - U_0}{\gamma} \tag{6}$$

accounts for the difference in global *geometric scales* between the geoid and ellipsoid. If the zero-degree harmonic of T is not zero an additional constant term, $k\delta M/(\gamma R)$, is included in N_0. The geoid component, N', in (5) thus refers to a "best-fitting" ellipsoid in terms of its geometric scale, that is, a best-fitting equatorial radius; since without N_0 the mean separation between these two surfaces is null.

We note the perhaps surprising fact that the potential anomaly, $\Delta W_0 = W_0 - U_0$, does not appear in the zero-degree harmonic of the disturbing potential, T, which identifies only differences in total mass. Therefore, this anomaly is intimately related, through (6), to the geometric, rather than physical, scale of the normal ellipsoid relative to the geoid. For a given normal ellipsoid, a determination of N_0 or a definition of ΔW_0 implies directly the difference in equatorial radius between that of the given ellipsoid and one that best fits the geoid in scale (but not flattening).

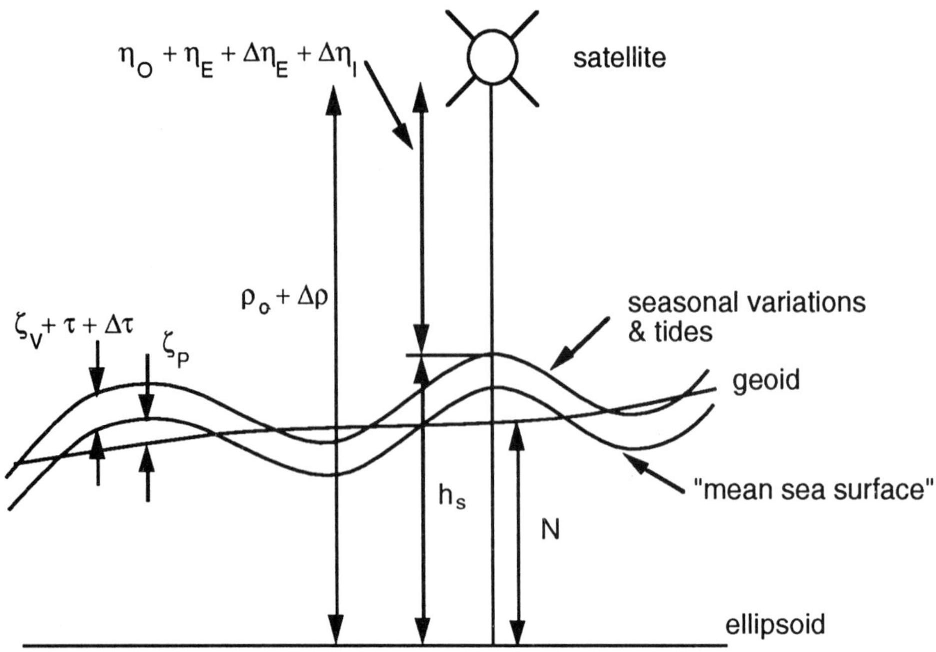

Figure 1: Relationships between geoid, mean sea surface, and altimeter measurement.

A more direct determination of the geoid undulation in ocean areas is possible using satellite altimetry. Figure 1 shows the basic relationship between N and the oceanographic variables on the one side and the satellite orbit / altimeter quantities on the other side. In equation form, we have

$$N + \zeta_P + \zeta_V + \tau + \Delta\tau = (\rho_c + \Delta\rho) - (\eta_O + \eta_E + \Delta\eta_E + \Delta\eta_I) \tag{7}$$

where ζ_P is the "permanent" or mean dynamic sea surface topography; ζ_V is the seasonal variation of ζ_P; τ is the tidal correction (including the solid Earth deformation) and atmospheric loading modeled with total error $\Delta\tau$; ρ_c is the satellite height above the given ellipsoid (obtained from the orbit) computed with error $\Delta\rho$; η_O is the observed altimeter range with corrections η_E for troposphere, ionosphere, and wave height effects and $\Delta\eta_E$ being the error in these; and $\Delta\eta_I$ is the altimeter instrument error. To an accuracy of 2 m, all terms (except N) on the left side may be neglected, resulting in a direct determination of the the geoid height subject to measurement and orbit error.

The variable sea surface component can be averaged away, thus obtaining a *mean sea surface*, by locally combining altimetry data from tracks over longer periods of time, even using data from subsequent generations of altimeter satellites. Radial orbit error can be reduced on a global scale by fitting the measured sea surface heights (possibly corrected for sea surface topography) to a reference geoid implied by a set of spherical harmonic coefficients using a low-degree polynomial. Locally, remaining orbit error can be reduced by removing a bias (and perhaps a trend) for each track using track crossing discrepancies.

Several possibilities exist to use altimeter-derived geoid undulations in the solution for a global geopotential model, and they fall under the altimetry-gravimetry boundary-value problem. One method transforms the undulation data to gravity anomaly estimates using an inverse Stokes formula, or the operational equivalent, least-squares collocation:

$$\Delta g = C_{\Delta g,N}\left(C_{N,N} + D_{N,N}\right)^{-1} N \tag{8}$$

where $C_{\cdot,\cdot}$ represents the covariance matrix for the indicated quantities and D is the error covariance matrix for N. Another possibility derives from the relationship of N to spherical harmonic coefficients (in spherical approximation):

$$N(\theta,\lambda) = R \sum_{n=2}^{\infty} \sum_{m=-n}^{n} C_{n,m} \overline{Y}_{n,m}(\theta,\lambda) \tag{9}$$

which serves as a model relating the observations, N, to the sought parameters, $C_{n,m}$.

Because the oceans cover over 70% of the Earth's surface, satellite altimetry lends itself to determining a best estimate of the zero-degree undulation and consequently a best ellipsoidal equatorial radius. Figure 2 yields the observation equation:

$$N_{grav} + v = N_{sat} - da \tag{10}$$

where N_{grav} is the gravimetric geoid undulation (the observation) given, for example, by (9) from a set of spherical harmonic coefficients *without* the zero-degree term. Therefore, it refers by definition to a best equatorial radius. N_{sat} is the geoid undulation obtained from satellite altimetry, as in (7), referring to a prior adopted ellipsoid. The residual v absorbs model, observation, and altimeter errors in a least squares adjustment for the equatorial radius correction, da.

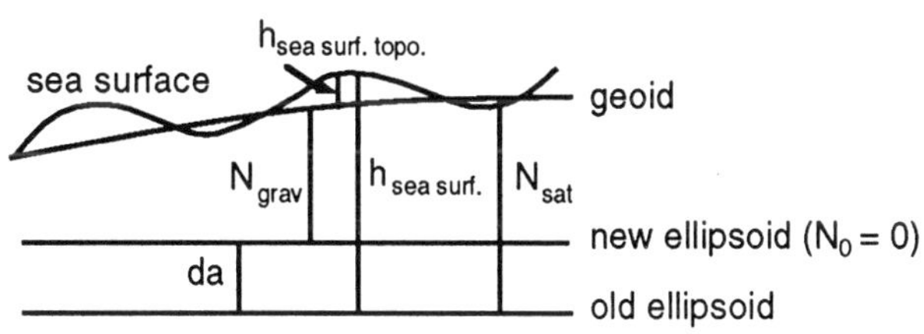

Figure 2: Method to determine best-fitting equatorial radius.

This brief presentation of the geoid determination problem provides the essential theory behind much of Richard Rapp's work in geodesy at Ohio State University, which is detailed in the following three sections.

Spherical Harmonic Models

Richard Rapp trained his primary research focus on the spherical harmonic modeling of the disturbing potential and the geoid, or the longer-wavelength geopotential signal, that is, the solution (2). This specialization has not only geodetic importance, but plays directly into geophysical and oceanographic interests, the latter, in particular, being quite active and relevant in the broader arena of climate change studies. The key element of any global solution such as (2) is, of course, the availability of global data. Before satellites, land data had accumulated over many (politically and geographically) accessible land areas, but the oceans were barren except for some measurements along ship tracks. The advent of satellites brought a new form of gravitational information - the observations of orbital perturbations. But the real boon in gravity data came from the satellite altimeter (GEOS-3 in 1975 was the first such dedicated satellite). Satellite altimetry yielded not only the geoid over the oceans, but also the means to estimate gravity anomalies (see (8)) over most of the ocean areas, thus bringing the worldwide data distribution needed in the computation of harmonic coefficients, equation (3), very close to reality.

The early spherical harmonic developments date back to around 1940; e.g., Jeffreys (1943) computed the first complete set of coefficients (to degree 3). Later, Kaula (1959) was first to combine satellite and terrestrial data in a least squares sense. Many other groups and investigators developed such models (see the review by Nerem et al., 1995), usually to low degree, and mostly on the basis of satellite data. Rapp's models begin around 1968 and concentrated on using global gravity data, but also in combination with satellite data. His degree-and-order (14,14) field (Rapp, 1968) was derived from a careful and meticulous culling of global terrestrial data as well as satellite-derived coefficients, ensuring consistent use of normal ellipsoid parameters for each of the data sources. The terrestrial anomalies were estimated over 5° equiangular blocks using available gravity data and topographic/isostatic models where such data were lacking; some blocks were set to zero if no data were available. Combined with the 1966 Smithsonian Astrophysical Observatory model of satellite-derived harmonics to degree 8, the resulting geoid model represented the state-of-the-art in data evaluation. Comparison of this model to other, then current models showed root-mean-square (rms) differences of 7 - 8 m; while comparisons to geoid undulations computed in the U.S. using the combined solution (4) and more detailed local gravity data showed agreement at the 3 m level (rms).

Subsequent developments through the 1970's considered improved reductions of gravity anomalies to sea level, atmospheric corrections, and ellipsoidal terms to account for the spherical approximation. These refinements in the computations offered verification of theoretical enhancements to the solution of the geodetic boundary-value problem using actual data (e.g, Rapp, 1975, 1981a; Rummel and Rapp, 1976). The next model (Rapp, 1977) to degree 52 was based on 5° *equal-area* averages of gravity anomalies (but no satellite-derived coefficients) and incorporated many of these refinements.

The confluence of three important events led to a major breakthrough in spherical harmonic development in the late 1970's. Electronic computing began its new, more interactive and powerful exponential growth swing, the first dedicated satellite altimeter (GEOS-3) was in orbit, and Oscar Colombo was at Ohio State University. Using least squares collocation, Rapp made use of the satellite-altimeter derived ocean geoid (having an accuracy of 1.9 m (Rummel and Rapp, 1977), sea surface topography was not an issue) to compute 28,176 1°x1° mean gravity anomalies, thus filling huge gaps in the global data base. With 22,474 1°x1° mean anomalies derived from surface measurements and 14,150 zero values in areas where no data were available, Rapp (1978) combined these data and a low-degree satellite-derived spherical harmonic model (up to degree 12 from GEM9) to obtain an adjusted, global set of 64,800 1°x1° mean anomalies. Using an efficient algorithm, developed by C. Rizos, to integrate this set of anomalies numerically according to (3), the first high-degree solution of the geopotential, complete to degree and order 180, was determined.

Colombo's (1981) systematic development of harmonic analysis on the sphere using the fast Fourier transform algorithm paved the way for routine and extremely efficient calculations of high-degree spherical harmonic models. The 1978 solution complete to degree 180, was quickly followed by another 180-degree model using updated terrestrial data, newly available SEASAT altimetry data, and Colombo's technique (Rapp, 1981b). In addition, the satellite-derived harmonics for this new combination model were complete to degree 36 and comprised a collection of the best available from a number of solutions; also, the fill-in anomalies were derived from this 36-degree satellite solution. The agreement with the previous model was 1.9 m (rms), while later comparisons to Doppler-derived geoid undulations (i.e., geoid heights from Doppler positioning and orthometric heights) showed agreement for both at 1.8 m (rms).

These early high-degree solutions represent milestones in spherical harmonic development. But compromises had to be made; for example, the quadratures techniques based on equation (3) (which is known as Method B, Rapp, 1969a) in its primitive form does not incorporate minimization of errors. Though Colombo soon also developed the spherical harmonic analysis using least squares collocation, for practical applications the data error models required very simple and unrealistic forms (which made the solution equivalent to a form of quadratures). Also, computer limitations prevented (they still do) a full, rigorous error covariance propagation. Yet, the data are the important ingredient; and new satellite-only geopotential harmonic models, increases and improvements in terrestrial data bases, new satellite altimeter missions with higher resolution and accuracy, and theoretical enhancements, including the use of Lagrangian perturbation theory to improve radial orbit accuracy, led to subsequent solutions complete to degree and order 360.

The OSU86C/D models, complete to degree 250, represent an intermediate step leading to the first 360-degree spherical harmonic models, OSU86E/F (Rapp and Cruz, 1986). This high-degree development was justified by the high resolution of the SEASAT data, yielding 30' mean gravity anomalies, even though many terrestrial anomalies had a resolution of only 1°. The C/D/E/F models used the GEM-L2 satellite-only solution up to degree 20 plus several individual harmonics of higher degree. The alternate versions of the model primarily examined the effect of geophysically predicting anomalies in data-poor areas. A comparison with Doppler-derived geoid undulations yielded rms differences of ±1.7 m for the 250-degree model and ±1.6 m for the 360-degree model. One should note that the Doppler-derived geoid undulations themselves had accuracies no better than ±1 m.

The next generation of models came as a result of the GEOSAT altimetry data, as well as new satellite-only solutions. However, the maximum degree was kept at 360. OSU89 was based on GEM-T2 harmonic coefficients complete to degree 36 plus 616 individual harmonics up to degree 50. 27,043 of the 30' mean anomalies were interpolated from 1° mean values and the fill-in values were derived from GEM-T2 (36-degree expansion) plus topographic/isostatic predictions (Rapp and Pavlis, 1990). OSU91 is similar to OSU89, with the exception of additional gravity data made available for China and the fill-in values spectrally relied more on the topographic/isostatic computations (Rapp et al., 1991). OSU89 undulations were compared to GEOSAT undulations and showed agreement at ±0.59 m (rms). A comparison of OSU91 undulations with accurate GPS-derived undulations (GPS minus orthometric heights) gave an rms difference of 0.56 m (Rapp, 1997). Even better agreement (0.34 cm) is found in ocean areas when comparing to altimeter-derived undulations. The latter derivation was made possible by solving for a degree-10 spherical harmonic model of the sea surface topography (Denker and Rapp, 1990). Table 1 gives a summary of the OSU models and their characteristics.

Table 1: OSU models developed by R.H. Rapp. See text for further details.

Name	n_{max} (total model)	ℓ_{max} (sat. mod.)	Δg resolution	Number of. Terr. Δg	Number of Alt. Δg	Number of Fill-ins
(14,14) 1968	14	8	5°	1,426	-	1,166
(52,52) 1977	52	-	5°	1,507	-	147
OSU78	180	12	1°	22,474	28,176	14,150
OSU81	180	36	1°	22,856	33,905	8,049
OSU86	250	20+	1°	18,255	32,274	14,238
OSU86	360	20+	30'	21,739	139,946	97,515
OSU89	360	36+	30'	72,209	136,270	50,721
OSU91	360	36+	30'	75,353	136,270	47,577

Ever conscious of the underlying theory, Rapp's models of the geoid referred to the enigmatic ellipsoid with undefined equatorial radius. In other words, no attempt was made to include a determination of the N_0 term, which would require an actual determination of the best-fitting ellipsoid.

Satellite Altimetry

Initial interest in satellite altimetry was motivated by the potential of substantially increasing the gravity data base for spherical harmonic developments. Rapp attacked the processing of altimetry data with the same care applied to the mean anomaly estimation over land areas. The most time consuming effort lay in editing through the data to eliminate egregiously bad data and detecting inaccurate corrections supplied with the data (such as tide models and tropospheric effects). The second major data processing effort entails the elimination of radial orbit error as discussed earlier. Finally, some interpolation scheme must be developed to create a regular grid of sea surface heights from the irregularly distributed altimeter measurements. The result of these preliminary correction and adjustment procedures is a *mean sea surface* which serves as basis for anomaly prediction. This mean sea surface, being adjusted to a reference geoid (and sea surface topography), by definition should have no zero-degree harmonic term; i.e., it refers to a best-fitting equatorial radius.

The GEOS-3 satellite altimetry data provided the first opportunity to create a mean sea surface (Rummel and Rapp, 1977). The reference, in this case, was the geoid implied by the GEM-7 geopotential field. Comparison of this mean sea surface to other gravimetric geoid determinations showed agreement at ±1.9 m (rms). Subsequent satellite altimetry data sets could be compared to this surface or used as a frame for combination and improvement of a mean sea surface. Rapp (1979) computed 1°x1° mean gravity anomalies from GEOS-3 data yielding ±12 mgal (rms) agreement with terrestrial (ship) anomalies, where most of the difference was deemed attributable to the latter.

The next global estimation of gravity anomalies from satellite altimetry came with SEASAT data (Rapp, 1983). Also ignoring the sea surface topography, this analysis yielded a 1° grid of mean anomalies and geoid undulations. The predicted standard error of the anomalies was about ±5 mgal, while comparison to the GEOS-3 anomalies showed a difference of ±8 mgal (global rms). A denser grid (0.125° x 0.125°) was created (Rapp, 1986) from the combination of GEOS-3 and SEASAT data. Derived 1° mean anomalies had estimated rms errors similar as before.

GEOSAT altimetry data provided the next framework for improved mean sea surface determination because of its much higher accuracy (altimeter and radial orbit) and its one-year span of 17-day-long repeat track data (ERM - Exact Repeat Mission). Accounting for sea surface topography (degree-10 spherical harmonic expansion) and ocean-bottom topography, gravity anomalies and geoid undulation were predicted on a 0.125° grid. Averaged to 1° cells, the gravity anomalies agreed with terrestrial anomalies at a global rms of ±9.2 mgal (Basic and Rapp, 1992).

The mean sea surface computed by Basic and Rapp (1992) from a combination of GEOS-3, SEASAT, and GEOSAT data was included in the initial TOPEX GDR (Geophysical Data Record). This OSU mean sea surface was evaluated against mean sea surfaces computed by other investigators and was found to be superior in terms of its gradients as compared to new data from ERS-1 altimetry, TOPEX altimetry, and GEOSAT altimetry (see Rapp et al., 1994a). Y. Yi (Rapp's Ph.D. student) developed a new mean sea surface (Yi, 1995) on a 0.0625° grid from TOPEX, GEOSAT, and ERS-1 altimetry data. This mean sea surface was placed on the later TOPEX GDR, as well as on the ERS-1 GDR.

Rapp and Yi (1997) refined this by removing some residual track-induced signatures from the surface using least-squares collocation, which was also used to predict 30' mean anomalies. Compared to ship data in various regions, these predictions agreed at about ±8 mgal (standard deviation).

Equatorial Radius

A very important by-product of satellite altimetry in conjunction with the development of geoid models is the equatorial radius of the ellipsoid that best fits the geoid (at least over the oceans) in geometric scale. Prior to satellite altimetry, however, Rapp was already engaged in the determination of improved parameters for the mean Earth ellipsoid on the basis of global collections of gravity anomalies. With his analysis that yielded the (14,14) solution of 1968, he found a new value for the equatorial normal gravity, derived from the non-zero mean of the terrestrial anomalies (and a revised flattening) that referred to the International Gravity Formula (1930). With this new equatorial gravity and a new value of kM for the Earth, Rapp (1969b) found a better value of the equatorial radius of the normal ellipsoid: a = 6378143 m. This was 17 m shorter than the just adopted value for GRS67, but only 6 m larger than the GRS80 value.

Analogous to the method based on satellite altimetry (equation (10)), a comparison of the gravimetric geoid (sans N_0) to geometrically determined geoid undulations (Doppler ellipsoidal minus orthometric heights) also yields a correction to the a priori ellipsoid radius, although the adjustment is limited to relatively few points on land areas. In this way, using the GEM-6 and the SAO models, Rapp (1974) found a new value: a = 6378139. ± 2.5 m. Again by the same method, with the OSU86 gravimetric geoid, he found: a = 6378136.2 ± (0.5 to 1.) m (Rapp, 1987).

The equatorial radius improvements derived from satellite altimetry began with a global averaging of the SEASAT data with respect to the GEOS-3 mean sea surface (Rapp, 1983). The average difference of 1.4 m between the two implied a better equatorial radius, equal to a = 6378135.6 m, compared to the GRS80 value of 6378137 to which the SEASAT heights referred. This method was used again later when TOPEX data were compared to the OSU mean sea surface created from GEOS-3, SEASAT, and GEOSAT altimetry fitted to the OSU91A gravimetric geoid, as well as a degree-10 sea surface topography model. The TOPEX data, referred to a = 6378136.3 m, had a global difference of 25.3 cm with respect to the mean sea surface, thus implying a new equatorial radius of 6378136.55 m (Rapp et al., 1994a); and with further TOPEX data, according to Rapp (1995): a = 6378136.59 ± 0.1 m. At this level of accuracy, specification of the tide system must be made (Ekman, 1989); the above value is in the *mean tide system.* Table 2 summarizes Rapp's determinations of the best equatorial radius of the normal ellipsoid.

Table 2: Best-fitting equatorial radii computed by R.H. Rapp.

Reference	value [m]	Std. Error [m]	Reference	value [m]	Std. Error [m]
GRS67	6378160	defined constant	Rapp (1983)	6378135.6	not given
Rapp (1969)	6378143	no given	Rapp (1987)	6378136.2	± 0.5 to 1.
Rapp (1974)	6378139.	± 2.5	Rapp et al. (1994)	6378136.55	not given
GRS80	6378137	defined constant	Rapp (1995)	6378136.59	± 0.1

Epilogue

The focus of Richard H. Rapp's work at Ohio State University was the computation of the global geoid in the form of a spherical harmonic expansion, and the creation, via estimations from satellite altimetry and terrestrial data, of a global set of mean gravity anomalies needed to do this computation. Pushing the accuracy and resolution of the geoid model from the early values of several meters over 1000 km to present-day values of a few tens of centimeters over some tens of kilometers was enabled not only by increases in data densification (primarily through altimetry) and accuracy, but also by Rapp's persistent and timely application of advancements in theory and computing power. A stickler for detail and rigor, he achieved models that have world renown for being the most accurate for geodetic purposes.

Foremost, in celebrating these achievements one should note that his work bestowed the obvious benefits to his students and the Department of Geodetic Science at Ohio State University in the forms of expert education, the maintenance of an active research program, and the extended visits of many prominent geodesists. Conversely, one should also note that his students shared in the development of the state-of-the-art in gravity field modeling and certainly contributed much of their time and effort (often seemingly in excess of 25 hours a day) to its fulfillment. Table 3 is just one indicator of the breadth of Richard

Rapp's influence on the department and on the world through his students that hail from New Zealand to Canada and from Nigeria to Korea.

Table 3: Rapp's Students and Reports of the Department of Geodetic Science, OSU.

M.S. Students	59
M.S.+Ph.D. Students	13
Ph.D. Students	12
Departmental Reports	62 (out of 430)

Second, though focused on geoid modeling, Rapp's work encompassed a variety of by-products, the best-fitting equatorial radius and the mean sea surface being just two prominent examples. Others include, but are not limited to, geopotential covariance models, such as the well known Tscherning/Rapp model (Tscherning and Rapp, 1974), world vertical datum studies (e.g., Rapp, 1994), and ocean circulation computations (e.g., Rapp and Smith, 1994b). Despite their cursory mention here, these achievements can easily be further expounded (lack of space precludes this here) and are of significant worth in their own right.

Finally, the current state of the art in global geoid modeling that Richard Rapp perfected has brought with it new issues for pragmatists concerning the definition of the geoid, as well as the normal ellipsoid. This is exemplified in two recent developments. The first is GEOID96, a detailed model of the geoid in the U.S., created by D. Milbert of the National Geodetic Survey (NGS). This model is deliberately biased (up to 1.6 m in height) with respect to a geocentric, reference ellipsoid in order to allow users to convert GPS heights based on the NAD83 (non-geocentric) horizontal datum coordinates directly into orthometric heights in the NAVD88 vertical datum (Milbert and Smith, 1996). The second example is the Earth Gravity Model 1996 (EGM96), a new degree-360 geopotential model recently completed by NIMA and NASA, with the cooperation and guidance of Richard Rapp (Lemoine et al., 1996). This model represents an evolutionary improvement to the WGS84 gravity model provided by NIMA, but the geometric parameters of the reference ellipsoid to which it refers retain their WGS84 values. This means that the EGM96 model for WGS84 includes a N_0 term, equal to -53 cm.

These biases in the geoid model result not from limitations in theory or data accuracy, but from a practical need to maintain longevity in datum definition. It is an interesting return to the classic datum definition problems, but only for the sake of practical continuity, not theoretical inability to do it differently. Yet, for the scientist this operational pragmatism, at a minimum, lacks aesthetics and is certainly fodder for future discussions and studies.

References

Basic, T. and R.H. Rapp, 1992: Oceanwide prediction of gravity anomalies and sea surface heights using Geos-3, Seasat, and Geosat altimeter data and ETOPO5U bathymetric data. Rep.416, Dep. Geod. Sci. & Surv., Ohio State Univ., Columbus, OH.

Denker, H. and R.H. Rapp, 1990: Geodetic and oceanographic results from the analysis of one year of Geosat data. *J. Geophys. Res.*, **95**(C8), 13,151-13,168.

Ekman, M., 1989: Impacts of geodynamic phenomena on systems for height and gravity. *Bulletin Geodesique*, **63**, 281-296.

Heck, B. and R. Rummel, 1991: Strategies for solving the vertical datum problem using terrestrial and satellite geodetic data. IAG Symposium 104 on Sea Surface Topography and the Geoid, H. Sünkel, T. Baker (eds.), pp.116-127, Springer-Verlag, Berlin.

Heiskanen, W.A. and H. Moritz, 1967: *Physical Geodesy,* W. H. Freeman, New York.

Jeffreys, H., 1943: The determination of the Earth's gravitational field (second paper), *Monthly Notices Roy. Ast. Soc.*, Geo. Supp. **5**, 55-66.

Kaula, W.M., 1959: Statistical and harmonic analysis of gravity, *J. Geophys. Res.*, **64**, 2401-2421.

Lemoine, F.G. et al., 1996: The development of the NASA GSFC and NIMA Joint Geopotential Model. Proc. International Symposium on Gravity, Geoid, and Marine Geodesy (GRAGEOMAR 1996), Univ. Tokyo, 30 Sep. - 5 Oct., 1996.

Nerem, R.S., C. Jekeli, and W.M. Kaula, 1995: Gravity field determination and characteristics: retrospective and prospective, *J. Geophys. Res.*, **100**(B8), 15,053-15,074.

Rapp, R.H., 1968: The gravitational potential of the Earth from a combination of satellite, observed, and model anomalies. *J. Geophys. Res.*, **73**(20), 6555-6562.

Rapp, R.H., 1969a: Analytical and numerical differences between two methods for the combination of gravimetric and satellite data. *Boll. Geof. Theo. Appl.*, **11**(41),108-118.

Rapp, R.H., 1969b: The geopotential to (14, 14) from a combination of satellite and gravimetric data, *Bulletin Geodesique*, **91**, 47-80.

Rapp, R.H., 1974: Current estimates of mean earth ellipsoid parameters. *Geophys. Res. Lett.*, **1**, 35-38.

Rapp, R.H., 1975: Effects of certain anomaly corrections terms on potential coefficient determinations of the Earth's gravitational field. *Bulletin Geodesique*, **115**, 57-63.

Rapp, R.H., 1977: Determination of potential coefficients to degree 52 from 5 degree mean anomalies. *Bulletin Geodesique*, **51**, 301-323.

Rapp, R.H., 1978: A global 1° x 1° anomaly field combining satellite, Geos-3 altimetry and terrestrial anomaly data. Report No. 278, Dep. Geod. Sci., Ohio State Univ., Columbus OH.

Rapp, R.H., 1979: Global anomaly and undulation recovery using Geos-3 altimeter data. Report No. 285, Department of Geodetic Science, Ohio State University, Columbus, OH.

Rapp, R.H., 1981a: Ellipsoidal Corrections for Geoid Undulation Computations Using Gravity Anomalies in a Cap. *J. of Geophys. Res.*, **86**(B1), 10843-10848.

Rapp, R.H., 1981b: The Earth's gravity field to degree and order 180 using Seasat altimeter data, terrestrial gravity data, and other data. Rep.322, Dep. Geod. Sci., Ohio State Univ., Columbus OH.

Rapp, R.H., 1983: The determination of geoid undulations and gravity anomalies from Seasat altimeter data. *J. Geophys. Res.*, **88**(C3), 1552-1562.

Rapp, R.H., 1986: Gravity anomalies and sea surface heights derived from a combined Geos3/Seasat altimeter data set. *J. Geophys. Res.*, **91**(B5), 4867-4876.

Rapp, R.H., 1987: An estimate of equatorial gravity from terrestrial and satellite data. *Geophys. Res. Lett.*, **14**, 730-732.

Rapp, R.H., 1994: Separation between reference surfaces of selected vertical datums, *Bulletin Geodesique*, **69**, 26-31.

Rapp, R.H., 1995: Equatorial radius estimates from Topex altimeter data. In: Festschrift to E. Groten; M. Becker, G. Hein, and R. Rummel (eds.), Institute of Geodesy and Navigation, Univ. FAF, Munich.

Rapp, R.H., 1997: Use of potential coefficient models for geoid undulation determinations using a spherical harmonic representation of the height anomaly/geoid undulation difference. *J. Geodesy*, **71**(5), 282-289.

Rapp, R.H. and J.Y. Cruz, 1986: Spherical harmonic expansion of the Earth's gravitational potential to degree 360 using 30' mean anomalies. Rep.376, Dep. Geod. Sci. & Surv., Ohio State Univ., Columbus, OH.

Rapp, R.H. and N.K. Pavlis, 1990: The development and analysis of geopotential coefficient models to spherical harmonic degree 360. *J. Geophys. Res.*, **95**(B13), 21,885-21,911.

Rapp, R.H., Y.M. Wang, and N.K. Pavlis, 1991: The Ohio State 1991 geopotential and sea surface topography harmonic coefficient models. Rep.410, Dep. Geod. Sci. & Surv., Ohio State Univ., Columbus, OH.

Rapp, R.H., Y. Yi, and Y.M. Wang, 1994a: Mean sea surface and geoid gradient comparisons with Topex altimeter data. *J. Geophys. Res.*, **99**(C12), 24,657-24,667.

Rapp, R.H. and D. Smith, 1994b: Preliminary estimates of gulf stream characteristics from Topex data and a precise gravimetric geoid. *J. Geophys. Res.*, **99**(C12), 24,707-24,723.

Rapp, R.H. and Y. Yi, 1997: Role of ocean variability and dynamic ocean topography in the recovery of the mean sea surface and gravity anomalies from satellite altimeter data. *J. Geodesy*, **71**(10), 617-629.

Rummel, R. and R.H. Rapp, 1976: The influence of the atmosphere on geoid and potential coefficient determinations from gravity data. *J. Geophys. Res.*, **81**, 5639-5642.

Rummel, R. and R.H. Rapp, 1977: Undulation and anomaly estimation using Geos-3 altimeter data without precise orbits. *Bulletin Geodesique*, **51**, 73-88.

Tscherning, C.C. and R.H. Rapp, 1974: Closed Covariance Expressions for Gravity Anomalies, Geoid Undulations and Deflections of the Vertical Implied by Anomaly Degree Variance Models. Rep.208, Dep. Geod. Sci., Ohio State Univ., Columbus OH.

Yi, Y., 1995: Determination of gridded mean sea surface from Topex, ERS-1, and Geosat altimeter data. Rep.434, Dep. Geod. Sci. & Surv, Ohio State Univ., Columbus, OH.

Gravity Field Improvement Activities at NASA GSFC

F. G. Lemoine
Laboratory for Terrestrial Physics, NASA GSFC, Greenbelt MD 20771, U.S.A.

E. C. Pavlis[1]
Joint Center for Earth System Technology, University of Maryland, Baltimore, MD 21250, U.S.A.

N. K. Pavlis, C. M. Cox, D. S. Chinn, M. H. Torrence, R. G. Williamson, and Y. M. Wang
Hughes STX Corp., Greenbelt, MD 20770, U.S.A.

Abstract

Since the completion of the EGM96 geopotential model, additional satellite tracking data has been added to the satellite-only geopotential model solution. The new data include, TRANET Doppler tracking data from the GEOSAT Geodetic Mission, TDRSS tracking of the Gamma Ray Observatory (GRO), the X-Ray Timing Explorer (XTE), the Earth Radiation Budget Satellite (ERBS), as well as additional data from the Extreme Ultraviolet Explorer (EUVE). The new data from the TDRSS tracked satellites make an important contribution to the satellite-only geopotential solution. Comparisons with independent 30'x30' altimeter derived anomalies from the GEOSAT Geodetic Mission provided by NIMA show that the ERBS data contribute by reducing the residual at degree 70 by 0.47 $mGal^2$. The data from XTE are valuable because of their unique inclination of 23°. Data from low inclination satellites is sparse in most satellite-only derived geopotential models, although substantial amounts of tracking data from EUVE (at an inclination of 28.5°) were included in EGM96. The performance of permutations and subsets of the EGM96 model are also shown to highlight different aspects of the model's performance.

Introduction

The EGM96 geopotential solution, completed in 1996 (Lemoine et al., 1997), was the product of a three year collaboration between the NASA Goddard Space Flight Center, the National Imagery and Mapping Agency (NIMA; formerly the Defense Mapping Agency or DMA) and The Ohio State University to produce an improved degree 360 spherical harmonic model representing the Earth's gravitational potential. The final model, EGM96, is a blended solution. It consists of a combination solution to degree and order 70, a block diagonal solution from degree 71 to 359, and a quadrature solution at degree 360. The combination model is based on satellite tracking data to over 20 satellites, including those tracked by Satellite Laser Ranging (SLR), the Global Positioning System (GPS), the Tracking Data Relay Satellite System (TDRSS) and TRANET Doppler, direct altimetry from TOPEX, ERS-1, and GEOSAT, as well as the normal equations of the 1°x1° surface gravity data (excluding the altimeter derived anomalies) to degree and order 70. The quadrature solution is based on the satellite only counterpart of EGM96 (EGM96S), as well as the surface gravity data and altimeter-derived anomalies from the GEOSAT Geodetic Mission processed by NIMA, as well as additional altimeter-derived anomalies from ERS-1 provided by Kort-og Matrikelstyrelsen (KMS) and by Schoene (1996). The unique aspect of the EGM96 project was the release by NIMA of new 30'x30' terrestrial data over many

---

[1] Also at Laboratory for Terrestrial Physics, NASA Goddard Space Flight Center.

International Association of Geodesy Symposia, Vol. 119
Forsberg et al. (eds.), Geodesy on the Move

regions of the globe including the former Soviet Union, Eastern Europe, South America, and other regions, as well as the acquisition of 30'x30' airborne gravity data, primarily over Greenland and adjacent regions of the Arctic, as reviewed in Kenyon and Pavlis (1996). The combination solution to degree 70 is a simultaneous solution of over 12,000 parameters including the geopotential coefficients, tides (112 constituents), pole position, station coordinates, and spherical harmonic representations of the dynamic ocean topography (DOT) for both TOPEX/ERS-1 and GEOSAT.

Testing of EGM96 Variants

After the creation of the EGM96 combination model, a number of variants were created in order to understand the performance sensitivity of the final combination solution. These permutations are listed in Table 1. The GPS/leveling tests with these final permutations are listed in Table 2, where the high degree solution used beyond degree 70, was the EGM96 block diagonal solution, HDM190. The pgs7337 solution verified that adjusting the once per revolution empirical acceleration parameters on GEOSAT, caused a loss in geopotential signal. The U.S. National Geodetic Survey (NGS) and British Columbia tests are neutral, howcvcr the 5 area average test shows a substantial improvement from 26.72 to 25.75 cm in EGM96. Three of the constituents (Europe, Canada, and Tennessee) of the five areas test are reduced by 0.9 to 1.9 cm in the standard deviation when the once per revolution parameters are not adjusted.

Table 1 Permutations of the EGM96 combination solution

Model	Description
pgs7337b	EGM96
pgs7337	pgs7337b but adjust empirical once per revolution accelerations on GEOSAT.
pgs7337d	pgs7337b with no EUVE data.
pgs7338	pgs7337 with altimetry downweighted by 50%.
pgs7339	pgs7337b without GEOSAT altimetry.

Table 2 Comparison of geopotential model-derived geoid undulations with values obtained from GPS/leveling

	Standard deviation of the difference between GPS/leveling and model derived geoid undulations (σ in cm)[1,2]		
Model	5 areas average	BC	US (NGS)
EGM96	25.75	51.66	52.59
pgs7337	26.72	51.57	52.71
pgs7337d	25.65	52.19	55.55
pgs7338	26.51	52.21	52.60
pgs7339	26.11	52.14	53.00

1 High degree field (L> 70) from HDM190 block diagonal solution.

2 The BC data include 298 points in British Columbia (Véronneau, 1995). The NGS data consist of 1889 measurements in the continental United States (Milbert, 1995). The five areas consist of 38 points in Australia, 63 points in Canada, 60 points in Europe. 46 points in Scandinavia and 49 points in Tennessee and are described in Rapp and Pavlis (1990). Each traverse was adjusted individually, and the average of all five standard deviations is presented.

The strength of EUVE in the solution is illustrated in pgs7337d by the degradation of the US (NGS) test from a standard deviation of 52.69 cm to 55.55 cm. Downweighting of the altimetry has little effect on the US/NGS GPS/leveling test, but degrades the GPS/leveling comparisons over British Columbia and over the five areas by 0.6 and 0.75 cm in the standard deviation.

The TOPEX/ERS-1 dynamic ocean topography (DOT) solutions from the EGM96 derivative fields were compared with a 24x24 spherical harmonic representation of the POCM-4B Ocean Circulation Model (Stammer et al., 1996), derived by Rapp (private communication, 1996). These comparisons involve computing the differences between the sea surfaces defined by the OCM and combination model solution for each harmonic degree over the ocean areas on a 1°x1° grid. The ocean domain is defined as those 1°x1° blocks whose depth was greater than 1000 meters. Only regions from 66°S to 65°N were included in the comparison. Removal of the GEOSAT altimeter data as in pgs7339 is undesirable since both the GPS/leveling comparisons (see Table 2) as well as the DOT versus POCM-4B comparisons for TOPEX are degraded (see Table 3). Downweighting of the altimeter data in pgs7338 has a negligible effect on the TOPEX DOT comparisons. However, the RMS geoid errors to 70x70 from the error covariance of pgs7338 increase slightly (19.1 cm RMS over the globe compared to 18.1 cm in EGM96; 28.7 cm RMS over land areas compared to 28.0 cm in EGM96; 13.5 cm over ocean areas compared to 11.8 cm in EGM96). The fifty percent reduction in the altimeter data weights has only a small impact on the predicted geoid errors to 70x70 because of the large quantity of TOPEX altimeter data included in the solution. EGM96 includes 2,892,900 altimeter normal points from TOPEX, 542,417 altimeter normal points from ERS-1, and 274, 812 altimeter normal points from GEOSAT.

Table 3 TOPEX/ERS-1 Dynamic ocean topography differences between a 24x24 spherical harmonic fit to POCM-4B, and DOT models from combination solutions

Model	Cumulative DOT Differences to degree n, (RMS, cm)	
	n=14	n=20
pgs7337b	12.50	13.26
pgs7337	12.54	13.32
pgs7337d	13.15	13.91
pgs7338	12.53	13.26
pgs7339	12.66	13.46

The DOT comparisons in Table 3 once again demonstrate the strength of the EUVE data. The TOPEX DOT comparisons at degree 14 degrade from 12.50 cm to 13.15 cm with pgs7337d. Without EUVE, the DOT versus POCM-4B differences increase substantially in the regions underneath the EUVE track, most especially in the Indian Ocean and the waters of Indonesia.

Testing of New TDRSS tracking data

Only data from TDRSS tracking of the EUVE satellite was included in EGM96. These data included 151,426 observations (Doppler and range) spanning the period from July 29 to September 16, 1994, corresponding to TOPEX/Poseidon (T/P) cycles 69 to 73. The noise on the TDRSS data tracking of EUVE is approximately 1.0 m for the range, and 0.1 cm/s for the Doppler. The technique utilized is discussed in Marshall et al.(1996). A separate "macro" model was developed to model the non conservative forces acting on the

TDRSS satellites and was applied in the TDRSS orbit determination (Luthcke et al., 1997). Tracking from four user satellites is available: the Gamma Ray Observatory (GRO), The Earth Radiation Budget Satellite (ERBS), the X-Ray Timing Explorer (XTE), as well as additional data from the Extreme Ultraviolet Explorer (EUVE). The orbit characteristics of these TDRSS user satellites are summarized in Table 4. These satellites fill important voids in the altitude and inclination distribution of satellites whose tracking data have been used to develop geopotential solutions. EGM96 marked the first time that a significant amount of high quality data was included from a low inclination satellite (EUVE).

Table 4 Orbital characteristics of TDRSS-tracked satellites

Spacecraft	GRO	ERBS	EUVE	XTE
Altitude (km)	380	585	510	579
Inclination	28.5°	57°	28.4°	23°

The data that are available span three time periods: November 1 to Dec. 1, 1992 (corresponding to cycles 5 through 7 of TOPEX/Poseidon), July 29 to Sept. 16, 1994 (cycles 69 to 73); and, January 5 to February 5, 1996 (corresponding to cycles 122 to 124 of TOPEX). During T/P cycles 5 to 7, TDRSS tracking data of EUVE, ERBS, and GRO were processed. Since these cycles had excellent T/P-TDRSS data coverage, superior TDRSS orbit accuracies were obtained. The TDRSS orbits were held fixed in the user-spacecraft solutions. During this period the TDRSS 4 was located at 41°W and TDRSS 5 at 171°W. During cycles 69 to 73 and 122 to 124, TDRSS 5 was located at 174.3°W, so the additional EUVE data from cycle 5 to 7 samples some regions not accessible with the cycle 69 to 73 EUVE data.

During the second time period, July 29 to Sept. 16, 1994, limited TDRSS tracking of the T/P spacecraft was available. TDRSS data of EUVE (cycles 69 to 73) and ERBS (T/P cycles 69 to 72) were processed in this time frame. The ERBS and EUVE user satellite data were used in a simultaneous solution for the TDRSS orbits. The arcs were 6 to 10 days in length, where the start and stop times were determined by the T/P cycle boundaries, and the times of TDRSS maneuvers.

The third time period from January 5 to February 5, 1996 involved only XTE, which had just been launched. Intensive TDRSS tracking of T/P had been requested to support the TDRSS orbit determination of the Space Shuttle mission STS-72.

All the new data were added to a satellite only model, pgs7501, that included a number of changes compared to the previous satellite only model, EGM96S. These changes included the following:

1. The TOPEX SLR and DORIS data were recalibrated and the weights of 1.82 m for the SLR, and 3.64 cm/s for the Doppler data were used (compared to 2.50 m and 1.58 cm/s in EGM96). This new weighting preserved the relative weighting of the SLR and DORIS data used in the second generation TOPEX precision orbit computations (Marshall et al., 1995).

2. The once-per-revolution empirical accelerations were adjusted on the TOPEX/GPS arcs. It had been observed in the development of EGM96, that when these ten day arcs, which relied on the first generation TOPEX precise orbit parameterization (Marshall et al., 1995), were included in the solution, the orbit fits on the 30-day arcs of LAGEOS data degraded. The cause of the deterioration in the fits, which was not observed with the 10-day arcs, was determined to be the solution for long period tides. The TOPEX/GPS

normal equations consisted of data from between T/P cycles 10 to 19, and so could not sample the long period tides (for instance Sa, and Ssa) properly. In EGM96 and EGM96S, the TOPEX/GPS tide contribution was detrended to alleviate this effect. In the current satellite-only model, pgs7501, the TOPEX/GPS tide partials were allowed to contribute to the general solution, with no detrending being implemented. Instead, the TOPEX/GPS empirical once-per-revolution parameters were adjusted, and the orbit fits to the 30-day arcs of LAGEOS data improved.

3. The Doppler tracking data from the GEOSAT Geodetic Mission were added to the solution, in order to strengthen the reference frame determination for GEOSAT. In JGM-1, JGM-2 (Nerem et al., 1994), and EGM96, only GEOSAT data from the Exact Repeat Mission were included. In EGM96S, these GEOSAT data included thirteen arcs from November 1986 to January 1987 and a total of 651,062 Doppler observations at a weight of 1.63 cm/s. Doppler data from March 30, 1985 through October 1, 1986 were added in the new satellite only model, pgs7501. The Geodetic Mission data included 109 arcs and 2,136,330 observations. Since so much more GEOSAT data had been added to the solution, a recalibration was necessary, and the new weight for these data was determined to be 3.13 cm/s. It was verified in a combination model that a reasonable estimate for the C_{10} of the GEOSAT dynamic ocean topography (DOT) was obtained, once the new Doppler data were added to the solution. In EGM96, the C_{10} of the GEOSAT dynamic ocean topography (DOT) was 2.39 ± 4.53 cm, whereas in the test combination solution (pgs7462k) the GEOSAT DOT C_{10} value was 12.50 ± 4.67 cm. In addition, the Z-bias in the coordinates of the GEOSAT Doppler stations, that had been observed in pgs7337b (EGM96) was eliminated.

The satellite-only model, pgs7501, included tracking of EUVE using GPS (Olson, 1996). The TDRSS tracking data of the user satellites are summarized in Table 5. The number of observations for each user satellite includes Doppler and range data collected through all the satellites of the TDRSS constellation.

Table 5 Summary of TDRSS data

Satellite	EUVE		ERBS		GRO	XTE
No. of Obs.	42,352	156,373‡	24,848	33,872	22,186	26,575
T/P Cycles	5-7	69-73	5-7	69-72	5-7	122-124

‡ EUVE data from cycles 69-73 included in EGM96.

The data weight calibration factors (Lerch, 1991; Nerem et al., 1994) represent the aggregate change in the geopotential coefficients divided by the aggregate change in the coefficient standard deviations between a master solution, and a subset solution excluding the data whose weight is to be calibrated. A calibration factor greater than unity indicates overoptimistic error estimates and that the data should be downweighted in the solution. The Lerch (1991) method of data weighting and calibration ignores the correlations in the error covariance. Since the information matrix for the continuous tracking types (TDRSS, GPS) has a less diagonally dominant structure than information matrices from satellites tracked by Satellite Laser Ranging (SLR), the calibration procedure needed to be modified in order to avoid overweighting the tracking data in a geopotential solution. In EGM96, calibration factors of 0.8 to 0.9 (rather than unity) were sought, and this goal was retained for the addition of the new TDRSS data.

The weights for the TDRSS data were first determined by calibrating each set of data with a subset of the base model pgs7501 that excluded the EUVE/GPS and EUVE/TDRSS tracking from cycles 69 to 73. Then all the data were added to pgs7501. Since the EUVE/GPS data represent tracking in an orbit identical to the EUVE/TDRSS data, both the

TDRSS and GPS data must be calibrated separately against a subset with no EUVE data (for instance pgs7530), and then again as an ensemble (EUVE/TDRSS data from cycles 5-7, and cycles 69-73, as well as the EUVE/GPS data). A synergism was found between the XTE data, and the data from EUVE. To avoid inflating the error covariance, these data were downweighted, even though the calibrations indicated the data were underweighted (see Table 6).

Table 6 Calibration of TDRSS satellite data

Master Field (M)	Satellite Data Calibrated	Subset Model (S)	Average Calibration Factor per degree	5°x5° mean altimeter derived anomalies ($mGal^2$)
pgs7529	Master	----------	------------------------	9.26 (M)
pgs7529	EUVE (all)	pgs7530	0.872	10.71 (S)
pgs7535	EUVE (TDRSS)	pgs7530	0.916	9.50 (M)
pgs7534	EUVE (GPS)	pgs7530	0.857	9.77 (M)
pgs7529	ERBS	pgs7531	0.919	9.73 (S)
pgs7529	GRO	pgs7232	0.807	9.35 (S)
pgs7529	XTE	pgs7533	0.529	9.40 (S)

Altimeter-derived anomalies provide an independent and accurate source of information that can be used to test and evaluate geopotential models derived solely from satellite tracking data (such as Doppler, SLR, GPS, and TDRSS). These mean 5°x5° anomalies (provided by NIMA using altimeter data from the GEOSAT Geodetic Mission), can be compared to degree 70 with corresponding anomalies from the satellite-only geopotential model. The high-degree contribution (beyond degree 70) is filtered out of the altimeter-derived anomalies using a high degree quadrature model. For the new satellite-only model, pgs7529, the comparison at degree 70 reaches 9.26 $mGal^2$ compared to 10.13 $mGal^2$ with EGM96S, and 16.35 $mGal^2$ with JGM-2S. The strength of a set of tracking data may be measured by the change in the altimeter-derived anomalies comparison as the data are removed from the pgs7529 solution. In this respect, the EUVE data (as an ensemble) and the ERBS data are the strongest datasets. Comparison of the altimeter-derived anomaly residuals by degree for EGM96S, and the new satellite-only solution reveals that the new TDRSS data improve the anomaly comparison by about 0.9 $mGal^2$ between degrees 35 and 45.

Summary

A new satellite-only model, pgs7529, has been developed that includes additional TDRSS tracking data from ERBS, GRO, and XTE. The new TDRSS data improve the solution primarily in the band between degree 35 and 45. The data from ERBS show unusual strength, however their utility is limited by the short (generally nine to ten minute) tracking passes that are available with these data. In comparison, for EUVE TDRSS, tracking passes are 21 and 30 minutes in length. Future investigations will focus on development of test solutions for selected terms beyond degree 70 to which these data are sensitive, and the incorporation of these data into a combination solution.

References

Kenyon, S. C., and N. K. Pavlis (1996), The development of a global surface gravity data base to be used in the joint DMA/GSFC geopotential model, In: Rapp. R, Nerem, R.S., and A. Cazenave (eds), "Global gravity field and its temporal variations," pp. 82-91, Springer Verlag, Berlin .

Lemoine, F. G., et al. (1997), The development of the NASA GSFC and NIMA joint geopotential model, In: Segawa J., Fujimoto H., and S. Okubo (eds), "Gravity, geoid and marine geodesy," pp. 461-469, Springer Verlag, Berlin.

Lerch, F. J. (1991), Optimum data weighting and error calibration for estimation of gravitational parameters, Bull. Geod., 65, 44-52.

Luthcke, S. B., et al. (1998), Enhanced radiative force modeling of the tracking and data relay satellites, in press J. Astro. Sci .

Marshall, J. A., et al. (1995), The temporal and spatial characteristics of TOPEX/ Poseidon radial orbit error, J. Geophys. Res., 100, 25331-25352.

Marshall, J. A., et al. (1996), An assessment of TDRSS for precision orbit determination, J. Astro. Sci, 44,1, 115-127.

Milbert, D. (1995), Improvement of a high resolution geoid height model in the United States by GPS height on NAVD88 benchmarks, in New Geoid in the World, IGeS Bulletin, No. 4, 13-36.

Nerem, R. S. et al. (1994), Gravity model Development for TOPEX/Poseidon: Joint Gravity Models 1 and 2, J. Geophys. Res., 99, 24421-24447.

Olson, T. R. (1996), Geopotential improvement from Explorer Platform single-frequency GPS tracking, Ph.D. Thesis, The Univ. of Colorado at Boulder, Boulder, Colorado, U.S.A.

Rapp, R. H. (1996), The Ohio State University, Ohio, United States, private communication.

Rapp, R. H., and N. K. Pavlis, The development and analysis of geopotential coefficient models to spherical harmonic degree 360, J. Geophys. Res., 95, 21885-21911, 1990.

Schoene, T. (1996), Alfred Wegner Institute, Bremerhaven .

Stammer, D., et al. (1996), How well does a 1/4° global circulation model simulate large-scale oceanic circulation? J. Geophys. Res., 101,25779-25812.

Véronneau, M. (1995), Geodetic Survey Division, Canada, private communications.

The Development by the National Imagery and Mapping Agency of a Global Surface Gravity Anomaly Database for the EGM96 Geopotential Model and Future Applications

S. C. Kenyon
National Imagery and Mapping Agency, Geodesy and Geophysics Dept.
3200 South Second Street, St. Louis, MO 63118, USA.
e-mail: kenyonS@nima.mil

ABSTRACT

The National Imagery and Mapping Agency (NIMA) gravity contribution to worldwide geodetic modeling is substantial. NIMA along with NASA/Goddard Space Flight Center (GSFC) have produced a global set of 30'x30' and 1°x1° surface mean free-air gravity anomalies in support of the EGM96 geopotential model development. The extensive data archive at NIMA is the main source of gravity information from which the 30' and 1° mean values were estimated. New terrestrial sources of gravity data used in the model include the former Soviet Union, Greenland, South America, Africa, Southeast Asia, Antarctica, and the Arctic. Improvements to current terrestrial sources have been made in North America, Europe, and Australia. Substantial effort was made to ensure accurate and consistent processing of the point gravity anomaly data used in the estimation of the 30' and 1° mean values. The mean values and their accuracies are estimated using least-squares collocation (LSC) and remove-restore techniques implemented at NIMA. For the reduction process, consistent 1' and 5' elevation databases were compiled by NIMA and GSFC using the best worldwide elevation sources currently available. In the future, NIMA will continue its worldwide data acquisition attempts to improve its point gravity anomaly database. These improvements, along with enhancements in gravity modeling techniques, will benefit worldwide geoid models, navigation systems, and mapping capabilities.

NIMA GRAVITY ANOMALY FILES

Surface Gravity Anomalies Over Land Areas Used in EGM96

The surface gravity data used in the EGM96 project has come predominately from data held in NIMA's Point Gravity Anomaly (PGA) file. This file contains in excess of 30 million point values collected and processed by NIMA during the last three decades through its independent collection efforts, reciprocal data arrangements, and cooperative agreements with foreign governments, academic institutions, and private concerns. The PGA file is the primary source of gravity anomalies used in statistical techniques that directly estimate the 30' mean terrestrial gravity anomalies.

Major terrestrial gravity acquisitions since 1990 include aerogravity over Greenland and parts of the Arctic and Antarctica, surveyed by the US Naval Research Lab (NRL), and cooperative gravity collection projects, several of which were undertaken in conjunction with the University of Leeds, England. These collection efforts have improved and densified data holdings over many of the world's land areas. Some of the notable geographic regions include Alaska, Canada, parts of South America and Africa, Southeast Asia, Eastern Europe and the former Soviet Union. In addition to the above gravity collections, there have been major efforts to improve NIMA's existing 30' mean anomaly database by mean anomaly contributions over various countries in Asia. There have also been 30' mean anomaly contributions by KMS (National Survey and Cadastre, Denmark) over the Gulf of Bothnia and the Baltic, and A.N. Marchenko (internal communication, 1996) contributed anomalies over the former Soviet Union.

Important steps in NIMA's gravity anomaly pre-processing algorithm include:

1. Gravity anomalies adjusted to the International Gravity Standardization Net 1971 (IGSN 71) system.
2. A major effort to reference all point gravity anomalies to the WGS 84 horizontal datum.
3. Compute Molodensky free-air gravity anomalies defined on the Earth's surface. The formula used to compute these anomalies is given by (Heiskanen and Moritz, 1967, eq. 8-9):

$$\Delta g = g - \gamma_{ell}\left[1 - 2\left(1 + f + m - 2f\sin^2\phi\right)\frac{H^*}{a} + 3\left(\frac{H^*}{a}\right)^2\right] \tag{1}$$

where g is the observed value of gravity on the Earth's surface and γ_{ell} is the value of normal gravity on the surface of the reference ellipsoid at WGS 84 geodetic latitude ϕ. The normal height H* of the gravity station is generally unavailable, so the orthometric height H is used instead. For the definitions of the quantities appearing in Eq. (1) see Heiskanen and Moritz (1967).

4. The geometry and the gravitational potential of the reference ellipsoid adopted for this project were defined as follows:

The fully normalized second degree zonal coefficient of the JGM-2 model (Nerem et al.,1994) as:

$$(\bar{C}_{2,0})^{\text{tide-free}}_{\text{JGM-2}} = -484.1654767 \times 10^{-6} \tag{2}$$

and the transformation:

$$(J_2)^{\text{zero}}_{\text{NASA/DMA}} = -(\bar{C}_{2,0})^{\text{tide-free}}_{\text{JGM-2}} \cdot \sqrt{5} - (-3.11080 \times 10^{-8} \cdot 0.3) \tag{3}$$

yields the 'zero' (permanent tide) J_2 value adopted for this project. This value, along with the Semi-major axis: a = 6378136.3 m, the Geocentric gravitational constant (including the mass of the atmosphere): GM = 3986004.415x10^8 $m^3\ s^{-2}$, and the Mean-

Earth spin rate: $\omega = 7292115 \times 10^{-11}$ *rad* s^{-1} uniquely define the 'zero' reference ellipsoid used in this project. All derived geometric and physical constants of the reference ellipsoid were computed in accordance to the recommendations of Moritz (1984). The normal gravity transformation from WGS 84 to the formula implied by the above constants was performed on the NIMA PGA database by differencing precise equations for normal gravity for the two reference systems. The steps (1-4) are important in reducing long-wavelength systematic errors present when calculating gravity anomalies (Heck, 1990).

The 5' worldwide elevation file, JGP95E, was developed to support the 30' surface gravity computations for the EGM96 project. A coordination between NASA/GSFC and NIMA led to the merger of all the different elevation sources available. JGP95E contains information from sources produced from NIMA's Digital Terrain Elevation Data (DTED), NGDC's TerrainBase 5-Minute Global Model (Row, et al.,1995), and orthometric heights obtained at NASA/GSFC from satellite altimetry (over ice) minus thc JGM-2/OSU91A gravity model. A global elevation file of orthometric heights (H) is important in this project for the following computations:

1. Terrain corrections applied to point Bouguer gravity anomalies.
2. Residual Terrain Model (RTM) effects applied to point free-air gravity anomalies.
3. Spherical harmonic representation of the elevation is used to create spherical harmonic Bouguer anomalies used in the remove-restore process of LSC
4. Restore 30' free-air anomaly from 30' Bouguer anomaly using 30' elevation file.
5. g_1 analytical continuation terms (Rapp and Pavlis,1990)
6. Topographic/Isostatic anomalies.
7. Development of correction terms that are required to convert height anomalies to geoid undulations.

Computational Methodology For Surface 30' x 30' Mean Gravity Anomalies

The computation of 30' mean free-air anomalies by NIMA is based on LSC (Moritz, 1980). NIMA has applied LSC using the Forsberg (1987) covariance model to estimate the 30' mean gravity anomalies and their associated accuracies directly using the PGA file.

The computational process at NIMA is to select the most accurate gravity data at appropriate spacing from the PGA file and then reduce the anomaly data for the effects of terrain (high frequency effects) and long wavelength effects using JGM-2/OSU91A. After these reductions, analytical covariance functions based on the three Forsberg model parameters are closely fitted to empirical functions that describe the covariance relationship between the reduced point anomaly field. The local Forsberg covariance parameters (D, T, C0) are then used in a LSC algorithm that utilizes closed auto- and cross-covariance expressions for gravimetric quantities. The LSC algorithm uses integral formulas for the mean representation of the gravimetric quantities and Cholesky decomposition to efficiently and accurately calculate the mean gravity anomalies from available PGA data in a specified cell.

There are two techniques to estimate 30' free-air gravity anomalies: the Bouguer anomaly methodology and the use of free-air anomalies in the computations (Kenyon and Pavlis, 1996). For all interior continental areas and islands the Bouguer anomaly methodology was used in the computations. The Molodensky free-air gravity anomalies were used for Greenland and the coastlines of all continental areas. The main difference in the Bouguer and free-air methodologies relates to the terrain reductions performed.

The 30' free-air gravity anomaly predictions were performed at the 30' mean elevation of the cell (from JGP95E) using LSC. The 30' Bouguer predictions were referred to the geoid (H=0) and then restored to the 30' mean elevation of the cell using the Bouguer reduction of 0.1119* 30' H, where H was determined from JGP95E.

For this project, the use of the JGM-2 (to degree 70) model (Nerem et al., 1994) augmented by the OSU91A (degree 71 to 360) model (Rapp et al.,1991) was selected as the most accurate geopotential model then available. This was used in the reduction of the point free-air and Bouguer anomalies and restoration of the 30' mean predicted anomalies.

There were 97,250 NIMA Terrestrial 30'x30' mean gravity anomalies computed from NIMA's PGA database and other various mean anomaly sources.

Surface 1° x 1° Mean Gravity Anomalies over Ocean Areas

The surface gravity data for ocean areas was compiled at a 1° resolution using two main sources of information. The two sources consist of the 1° Ohio State University (OSU) mean anomaly set that was used for the OSU91A geopotential model (Rapp, et al., 1991) and ocean gravity sources collected by NIMA. Accuracy estimates reflecting the number of point anomalies within 1° cells and comparisons with altimeter-derived 1° values from the GEOSAT Geodetic Mission (GM) data were used to determine the most representative surface values. The method of computation for the 1° surface gravity anomalies over the ocean areas consisted of using a technique called the "modified average free-air" procedure (Uotila, 1967).

There were 32,113 1°x1° mean terrestrial gravity anomalies computed over ocean areas using the two 1° gravity sources described in this section.

Final Statistics of Surface 30' x 30' Mean Gravity Anomalies For EGM96

The statistics of a global merged 30' mean free-air gravity anomaly file are shown in Table 1. The NIMA altimetric (abbreviated Altim.) 30' anomalies from predominately the GEOSAT mission are also shown in Table 1, together with the NIMA and OSU 30' terrestrial anomalies. 96.9 % of the Earth's surface is covered by high quality 30' mean anomalies from either actual surface gravity measurements or from satellite altimetry. The remainder are from the NIMA 1° database or Topographic/Isostatic fill-in anomalies (Pavlis and Rapp, 1990), which are used when no gravity information is available.

The development of the final global surface 30'x 30' mean gravity database for EGM96 has paralleled the computation of the ocean wide altimetry-derived anomaly file used in this project. The formation of these databases represent a major 3 year computational effort by NIMA. The final EGM96 model, completed in September 1996, employed these data has demonstrated significant improvements over pre-existing models in areas such as the former Soviet Union, Greenland, Canada, Scandinavia, Africa, China, and South America. The 30' x 30' NIMA gravity data and other NIMA/NASA EGM96 geopotential model related information can be accessed on the World Wide Web at http: //cddis.gsfc.nasa.gov/926/egm96/egm96.html.

Table 1. Statistics of global 30' NIMA/GSFC mean free-air anomaly merged file

Type	NIMA Terr. 30'	NIMA Altim. 30'	OSU Terr. 30'	NIMA Terr. 1°	Topo/Iso Fill-in 30'
Number	86740	146042	1064	6500	18854
% of Area	30.7	66.1	0.1	0.8	2.3
RMS Δg (mgal)	35.2	25.6	56.7	49.1	28.0
RMS σ (mgal)	5.4	1.7	16.9	35.7	36.0

FUTURE APPLICATIONS

NIMA is continuing its support of worldwide gravity acquisitions. The Anglo-Brazilian Gravity Project is filling in voids in the gravity coverage of Brazil, Argentina, and Chile and improving the South American Digital Terrain Model. The Arctic is continuing to be surveyed with aerogravity missions by NRL and the Airborne Geoid Mapping System for Coastal Oceanography (AGMASCO) project.

NIMA is also continuing its efforts in satisfying worldwide gravity requirements by participating in the refurbishment and upgrade of the Bell Gravity Gradiometer System (GSS) . The GSS is capable of highly accurate airborne gradient surveying with gravity anomaly recovery of 0.6-1.2 mgals. This complements the scalar airborne surveys that have been shown in Greenland and the Arctic to approach 2-4 mgals at varying resolutions making a combined scalar/gradient program very attractive.

A major current effort at NIMA is the definition of the worldwide gravity spacing to support the Advanced Inertial Navigation System. This problem is being solved through the use of stochastic models associated with LSC. The desired accuracy of the deflection of the vertical (DOV) along with covariance characteristics (variance and correlation lengths) of point gravity anomalies and PGA accuracies in local areas are used in analytical models to develop the desired PGA spacing. The NIMA PGA database is then queried to determine if sufficient data exists to meet these modeled gravity spacings and achieve the required DOV accuracy.

NIMA is contributing DOV information for different navigation systems including the Advanced Integrated Mapping System (AIMS) being developed by the Ohio State University Center for Mapping. The Ohio State AIMS will utilize DOVs in an Inertial Navigation System/GPS integrated system with laser and radar technology that will provide advanced mapping capabilities.

NIMA is also assisting in the development of the North American Geoid Project with the National Geodetic Survey (NGS) and Geodetic Survey Division, Canada, and has recently participated with the NGS in the CARIB97 geoid model for the Caribbean. The use of NIMA PGA data for continental wide geoid models will assist in the future realization of a unified world height system and vertical datum.

REFERENCES

Forsberg, R.,1987, A New Covariance Model for Inertia, Gravimetry and Gradiometry, *Journal of Geophysical Research*, 92(B2), 1305-1310.

Heck, B.,1990, An evaluation of some systematic error sources affecting terrestrial gravity anomalies, *Bulletin Géodésique*, 64: 88-108.

Heiskanen, W.A. and H. Moritz, 1967, *Physical Geodesy*, W.H. Freeman, San Francisco.

Kenyon S. C., and N. K. Pavlis, 1996, The development of a global surface gravity data base to be used in the joint DMA/GSFC geopotential model, *in Global Gravity Field and Its Temporal Variations*, R. H. Rapp, R. S. Nerem, and A. A. Cazenave (editors), IAG Symposia, Vol. 116, Springer Verlag, Heidelberg, pp. 82-91

Moritz, H.,1980, *Advanced Physical Geodesy*, Herbert Wichmann Verlag, Karlsruhe.

Moritz, H., 1984, Geodetic Reference System 1980, *Bulletin Géodésique*, Vol. 58(3)

Nerem, R.S., F.J. Lerch, J.A. Marshall, E.C. Pavlis, B.H. Putney, B.D. Tapley, R.J. Eanes, J.C. Ries, B.E. Schutz, C.K. Shum, M.M. Watkins, S.M. Klosko, J.C. Chan, S.B. Luthcke, G.B. Patel, N.K. Pavlis, R.G. Williamson, R.H. Rapp, R. Biancale, and F. Nouel,1994, Gravity model development for TOPEX/Poseidon: Joint Gravity Models 1 and 2, *Journal of Geophysical Research*, 99, 24421-24447.

Pavlis, N.K. and R.H. Rapp,1990, The development of an isostatic gravitational model to degree 360 and its use in global gravity modeling, *Geophys. J. Int.*, 100: 369-378.

Rapp, R.H. and N.K. Pavlis,1990, The development and analysis of geopotential coefficient models to spherical harmonic degree 360, *Journal of Geophysical Research,* 95: 21885- 21911.

Rapp, R.H., Y.M. Wang and N.K. Pavlis, 1991, The Ohio State 1991 geopotential and sea surface topography harmonic coefficient models, *Rep. 410*, Dept. of Geodetic Science and Surveying, Ohio State University, Columbus.

Row, L.W., D.A. Hastings and P.K. Dunbar, April 1995, TerrainBase Worldwide Digital Terrain Data, Documentation Manual, CD-ROM Release 1.0, National Geophysical Data Center, Boulder, Colorado.

Uotila, U.A., 1967, Computation of Mean Anomalies of 1° x 1° Blocks, *Rep. 95*, Dept. of Geodetic Science and Surveying, Ohio State University, Columbus.

ON THE POSSIBILITY TO ESTIMATE OCEAN BOTTOM TOPOGRAPHY FROM MARINE GRAVITY AND SATELLITE ALTIMETER DATA USING COLLOCATION

D. Arabelos

Department of Geodesy and Surveying, Aristotle University of Thessaloniki, GR-540 06 Thessaloniki

Abstract

The contribution of some global bottom topography models in smoothing quantities related to the gravity field such as gravity anomalies and sea surface heights is discussed. Observed gravity anomalies or gravity anomalies derived from altimeter data and a-priori statistical characteristics of depths are used in a least squares collocation procedure in order to produce new depths in Central Mediterranean and in northern Atlantic. Different approximations of the problem are discussed, based on one- or two layer models. The estimated bathymetry is compared with existing global and regional data bases. Many features of the sea floor described by the DTMs are also shown in the estimated bathymetry. The effect of the estimated bathymetry in smoothing gravity anomalies and sea surface heights is studied using both the original and the new depths through a residual terrain modelling (RTM)-reduction.

1. Introduction

The depth of the ocean plays a very important role in a variety of marine activities. Marine biologic processes depend on the ocean depth. The bathymetry information is necessary in many marine studies. The locations of the strong currents depend on the bathymetry (see, e.g., Knudsen and Andersen, 1996). In geodetic studies the effect of the bathymetry to various quantities related to the gravity field is of great importance for a wide field of oceanic applications. The effect of the bathymetry can be taken into account through the different kinds of reductions such as the topographic reduction, the isostatic reduction, the residual terrain modelling reduction (RTM), etc. The reductions are primarily aiming in smoothing the quantities related to the gravity field after subtracting the effect of the bathymetry. The smoothing of the data reaches the level of 50%, when high quality depths are available.

Ocean depth information is obtained from echo soundings during ship surveys. Unfortunately, the quality of the available ocean depth information is good mainly in shallow waters since these areas are important for many activities. Deeper waters are poorly surveyed regions. Recently, alternative to ship surveying methods for the estimation of the ocean depth were described by several authors, based on the inversion of surface gravity data (Knudsen, 1993; Smith and Sandwell, 1994; Tscherning et al., 1994). In ocean areas where gravity anomalies are not available, satellite altimetry with its very dense distribution and high accuracy may be converted to gravity with an error of the order of a few mGals.

In this paper the method of estimating the bottom topography by least squares collocation is tested using sea gravity information, or gravity data derived from satellite altimetry, and a-priori statistical characteristics of depths. The method has been described in detail in a number of earlier papers (e.g. Barzaghi et al., 1992; Knudsen, 1993; Tscherning et al. 1994, Arabelos and Tziavos, 1997). In the present paper the method is tested in extended regions of the Central Mediterranean and northern Atlantic.

2. Data

Our numerical investigations were carried out in two different test areas: (i) in a region of the Central Mediterranean bounded by the limits $33^\circ \le \phi \le 37^\circ, 16^\circ \le \lambda \le 20^\circ$ and (ii) in a region of the northern Atlantic bounded by the limits $72^\circ \le \phi \le 76^\circ, -8^\circ \le \lambda \le 12^\circ$.
The following data were used in the area (i):

(a) Moho depths at the 12'× 20' resolution from seismic investigation in the Mediterranean (Geiss, 1987). The area covered by this data set is bounded by the limits $35^\circ \le \phi \le 50^\circ, -5^\circ \le \lambda \le 25^\circ$, so that only the upper half of the test area is covered by Moho depths.

(b) The JGP95E 5'× 5' global elevation database, provided by Kenyon and Pavlis (private communication, 1996). In Fig. 1 the bottom topography in central Mediterranean is shown, based on JGP95E.

(c) A 5'× 7.5' bottom topography data set covering the area of the Mediterranean Sea. This regional DTM was digitized by the author from bathymetric maps of the Mediterranean (Morelli et all., 1975).

(d) 2,401 5'× 5' free-air gravity anomalies referred to GRS-80.

(e) 6,108 altimeter heights derived from a common adjustment of ERS-1/ERM, ERS-1/GM I, and TOPEX data. The adjustment procedure of these data is described by Arabelos and Vermeer (1996).

In the test area (ii) the following data were used:

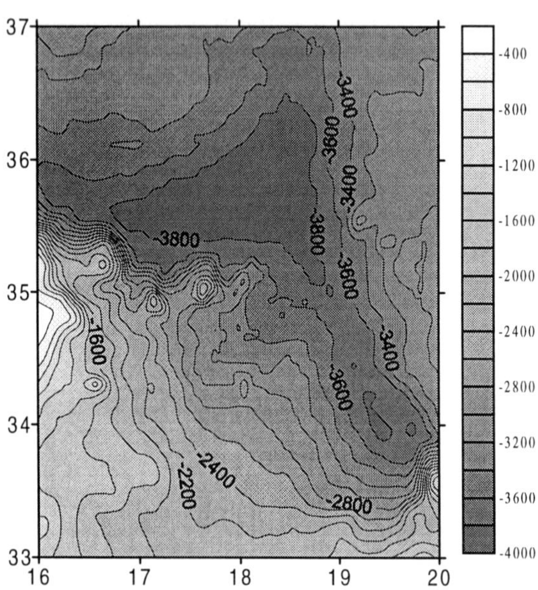

Fig. 1. Bathymetry (m) in Central Mediterranean based on JGP95E

(a) Topography data sets from the global 5'× 5' ETOPO5 (NGDC, 1988) and Terrain Base (Row et al., 1995) data bases extracted via NETSCAPE from the NGDC. In Fig. 2 the bottom topography in north North Atlantic is shown, based on TBASE.

(b) 29,760 point gravity anomalies extracted from GEODAS CD-ROM (GEODAS -Marine Geological and Geophysical Data from NGDC, published by NOAA). The area covered by this gravity data set is bounded by the limits $70^\circ \le \phi \le 83^\circ, -10^\circ \le \lambda \le 20^\circ$. The EGM96 model (Lemoine et al., 1996) was used for the experiments in this test area. From the observed gravity data set only the observations close to the nodes of a 5'× 20' grid were selected.

(c) Fully corrected altimetric sea heights, based on ERS-2 Ocean Product

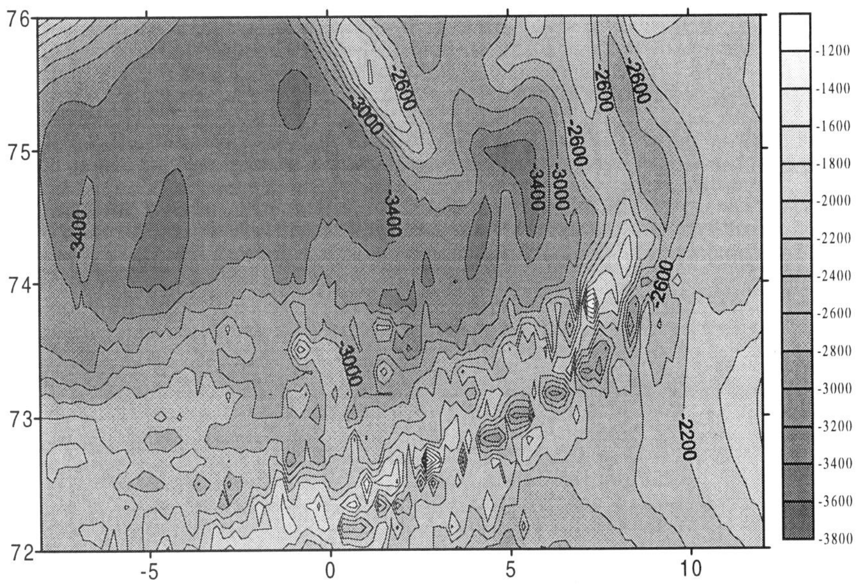

Fig. 2. Bathymetry (m) in northern Atlantic based on TerrainBase

(OPR) Version surface heights were included in 17 cycles (the period of each cycle is approximately one month). For the adjustment of these altimeter heights the same procedure as in Arabelos and Vermeer, 1996 was applied. The mean value and the std deviation of the reduced to EGM96 SSH are 0 and 0.43 m respectively.

3. Comparison of DTMs in Central Mediterranean and northern Atlantic

Central Mediterranean

In the present analysis the JGP95E 5'× 5' model was compared with Morelli's 5'× 7.5' DTM. Although both models present similar statistics and patterns in the test area, the comparison between them showed considerable differences ranging from -1.479 to 0.653 km with a std equal to 0.136 km.

In gravity field modelling, it is very important to attempt to smooth field related quantities, using bathymetry as additional information. Residual Terrain Model (RTM) reduction is a good way to do this since topographic or isostatic reduction could introduce large biases in the data. However, there is a number of parameters affecting the quality of the reduction such as the accuracy of the bottom topography model, the assumption for the density contrast and the geologic structures below the sea floor.

In order to examine the quality of the DTMs used in this test area we conducted the following experiment: RTM corrections for the gravity and altimeter data described in section 2 were computed using both models (JGP95E and Morelli's) with different density values. Better results in terms of std. deviation of the reduced quantities were found for density of the oceanic lithosphere equal to 2,000 kg m^{-3}. The statistics of the RTM reduced data is shown in Table 1. The std. deviation of the unreduced gravity anomalies and altimeter heights was 13.42 mGal and 0.35 m respectively.

The individual differences in the statistics of the reduced data appeared in Table 1 and the fact that the std. deviation of the reduced quantities still remains high in comparison to the std of the original quantities (78% and 85% for gravity and altimeter data respectively) could be attributed to the poor quality of the DTMs or to other unknown geological peculiarities.

Table 1. Statistics of the RTM reduced gravity anomalies (2,401 points) and SSH (6,108 points) in the Central Mediterranean. Unit for gravity is mGal and for SSH m.

	JGP95E (5'× 5')				Morelli (5'× 7. 5')			
	Mean	Std.dev.	Min.	Max.	Mean	Std.dev.	Min.	Max.
Gravity	0.11	11.42	-56.46	85.25	0.23	10.48	-46.17	66.64
Altimetry	-0.02	0.30	-0.99	0.83	0.03	0.31	-0.99	0.81

Northern Atlantic

Taking into account the drawback of LSC that one needs to solve a system of normal equations involving as many unknowns as the number of data, we restricted ourselves in a smaller area bounded by $72^\circ \le \phi \le 76^\circ, -8^\circ \le \lambda \le 12^\circ$. Another reason for this restriction is that our test area must be an oceanic area (without land parts).

The comparison between ETOPO5 and TerrainBase DTMs in the area $70^\circ \le \phi \le 80^\circ$, $-10^\circ \le \lambda \le 20^\circ$ showed differences ranging from –1,070 to 762 m. But in the test sub-area $72^\circ \le \phi \le 76^\circ, -8^\circ \le \lambda \le 12^\circ$ the two DTMs are identical.

An attempt to smooth the gravity and altimeter data in the test sub-area using RTM reductions based on ETOPO5 and TerrainBase in a wider area ($70^\circ \le \phi \le 78^\circ$, $-10^\circ \le \lambda \le 14^\circ$) gave better results for density of the oceanic lithosphere equal to 2,000 kg m^{-3}. These results are shown in Table 2. The std. deviation of the unreduced gravity anomalies and altimeter heights was 16.50 mGal and 0.26 m respectively.

Table 2. Statistics of the RTM reduced gravity anomalies (1,040 points) and SSH (2,471 points) in the northern Atlantic. Unit for gravity is mGal and for SSH m.

	ETOPO5 (5'× 5')				TerrainBase (5'× 5')			
	Mean	Std. dev.	Min.	Max.	Mean	Std. dev.	Min.	Max.
Gravity	3.95	15.67	-47.99	84.21	4.15	17.69	-49.46	77.69
Altimetry	-0.06	0.35	-0.95	1.18	-0.07	0.60	-1.07	2.21

Table 2 shows more disappointing results than those of Table 1, concerning the RTM reductions in the test area (ii) based on ETOPO5 or TerrainBase DTMs. Concerning the gravity anomalies, the std. deviation of the reduced data remains slightly lower than the std. deviation of the unreduced data only in the case of using ETOPO5 and density values lower than to 2,300 kg m^{-3}. For the SSH we have an increase instead of decrease of the std. deviation for both DTMs and for all density values. It is doubtful that these bad results are due to unknown geophysical signal. To author's opinion the main reason is the poor quality of the global DTMs.

4. Estimation of the model of bottom topography in the test areas

Central Mediterranean

The statistical characteristics of the Moho depths and bottom topography were used to construct a two-layer model for the test area. The density contrast of the first layer with respect to the sea water was taken equal to 1,600 kg m^{-3}, while of the second layer with respect to the first one was taken equal to 600 kg m^{-3}.

Using the parameters shown in Table 3 it was possible to get a very good agreement between the covariance function of the observed gravity anomalies and the covariance function of the gravity response of the two layers.

Table 3. A-priori statistical characteristics in the case of two-layer model

Layer No.	Mean depth (km)	Std.deviation (km)	Correl. length (km)	Density contrast (kg m^{-3})	Correlation to previous layer
1	-3	0.78	50.0	1,600	1.00
2	-18	2.00	70.0	600	-.97

After four iterations the mean value and the std. deviation of the differences (observations-model response) was 0 and 0.74 mGal respectively. Taking into account the accuracy of the gravity anomalies used (2...5 mGal) these results are considered satisfactory.

The statistical characteristics of the estimated bottom topography and Moho interface are shown in Table 4. The comparison of the mean depth and std. deviation of Tables 3 and 4 shows that there are differences between the a-priori given std. deviation and the corresponding std. deviation of the estimated layers. Also differences exist in the maximum and minimum depth of the observed and estimated layers.

Table 4. Statistical characteristics of the estimated layers from gravity anomalies. Unit is km.

Layer No.	Mean depth	Std. deviation	Min.	Max.
1	-2.89	0.69	-4.81	-0.55
2	-18.52	1.42	-22.13	-14.77

In Fig. 3 the estimated bottom topography is shown. The comparison of Figs. 1 and 3 shows large discrepancies between the JGP95E and the estimated bottom topography from gravity anomalies. The larger differences exist in the northern part of the test area. On the other hand, characteristic features in the centre as well as in the Southeast part are very well described in the estimated model.

The estimated bottom topography from gravity was consequently used for the computation of RTM reductions of gravity and altimeter data in the test area. For the density of the oceanic lithosphere the value 1,800 km m^{-3} was used. Table 5 shows the results of the RTM reduction. Concerning the gravity anomalies, the results in terms of std. deviation of the reduced values are very close to the corresponding results based on JGP95E (see Table 1). Unfortunately, the same is not valid for the altimeter data. In this case the std. deviation of the RTM reduced data is higher than the std of the observations.

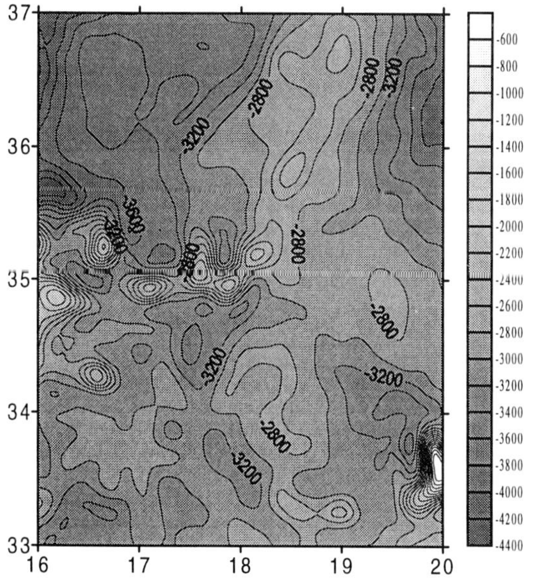

Fig. 3. Bathymetry (m) in Central Mediterranean estimated from observed gravity anomalies

The previously described experiments were repeated using gravity anomalies derived from satellite altimeter data, instead of observed marine gravity data. The altimeter data used for this purpose and the method applied have been described in (Arabelos and Vermeer, 1996). The std. deviation of the differences between observed and predicted data was 5.5 mGal. The same a-priori statistical characteristics for the two layers as in Table 4 were used also in the new experiment. The convergence in this case was not so fast as in the previous experiment: after four iterations the std. deviation of the differences (gravity anomalies-gravity response of the layers) remains over 4 mGal.

The statistical characteristics of the estimated layers are shown in Table 6.

Table 5. Results of the RTM reduction of gravity and altimeter data using the estimated bottom topography from observed gravity data.

	No.	Mean	Std. deviation	Min.	Max.
gravity (mGal)	2401	2.33	11.55	-56.48	54.25
altimeter data (m)	6108	0.12	0.37	-0.70	0.98

The comparison of the results of Tables 4 and 6 shows that the statistical characteristics of the layers estimated from gravity predicted from altimeter data are closer to the a-priori values than the corresponding of the layers estimated from marine gravity anomalies.

Table 6. Statistical characteristics of the estimated layers using gravity predicted from satellite altimeter data. Unit is km.

Layer No.	Mean depth	Std. deviation	Min.	Max.
1	-2.96	0.74	-4.79	-0.64
2	-18.35	1.60	-22.25	-14.58

The results of the RTM reduction of gravity and altimeter data based on the last estimated bottom topography are shown in Table 7. No significant differences exist between the results of Tables 5 and 7.

Table 7. Results of the RTM reduction of gravity and altimeter data using the estimated bottom topography from gravity anomalies predicted from altimeter data.

	No.	Mean	Std. deviation	Min.	Max.
gravity (mGal)	2401	1.85	11.94	-60.56	81.89
altimeter data (m)	6108	.12	.37	-.70	.98

From the experiments described up to now the conclusion is that marine gravity or gravity derived from satellite altimetry gives consistent results when used for the estimation of bottom topography at least in the present test area. The large discrepancies between "observed" and estimated bottom topography -in spite of using a more or less realistic model -is difficult to explain.

Finally it should be noted that the several models of bottom topography estimated by inverting the observed gravity anomalies or the gravity anomalies derived from altimeter data -apart of the individual differences -present the same pattern. This pattern does not fully agree with the bottom topography from JGP95E or Morelli's DTMs.

Northern Atlantic

The available observed surface gravity anomalies in this test area are not evenly distributed. For this reason the estimation of the bottom topography in this area was carried out using gravity anomalies derived from ERS-2 altimeter data. From the adjusted sea surface heights described in section 2 gravity anomalies were predicted on a 5'× 20' grid (~10.2 × 9.2 km) using the FFT method. The mean value and the std. deviation of the predicted gravity anomalies was 0.00 and 15.32 mGal respectively. A comparison between predicted and observed gravity anomalies at 1,000 points gave the results showed in Table 8.

Table 8. Statistics of the differences between observed and predicted from ERS-2 altimeter data gravity anomalies in the north North Atlantic. Unit is mGal.

	Observed	Predicted	Difference
mean	3.96	–0.46	4.41
std. deviation	16.10	13.31	7.66

The predicted gravity anomalies were used for the estimation of a new model of bottom topography. Since information about the Moho interface was not available, in the experiments a single layer model was used. The best agreement between the empirical covariance function and the covariance function estimated from the gravity response of the layer was achieved by changing considerably the statistical characteristics of the TerrainBase model. The final parameters used in this experiment were: mean depth = -2.9 km, std. deviation =0.43 km (instead of 0.497) and correlation length =8 km (instead of 30 km). As density contrast the value of 1,300 kg m^{-3} was adopted. After five iterations the convergence was satisfactory: the mean value and the std. deviation between observed gravity anomalies and gravimetric response of the layer were equal to 0.004 and 0.75 mGal respectively. The statistical characteristics of the estimated model are: mean depth = –2.9 km, std = 0.418 km, min depth =–5.33 km, max. depth =–1.19 km. The estimated bottom topography in this way is shown in Fig. 4. Comparing Figs. 2 and 4 we see that the characteristic deep structure directed from Southwest to Northeast in the TerrainBase is also well described by the estimated model. But the extended basin in the Northwest part of TerrainBase has been substituted by another more complicated structure.

Using the bathymetric model estimated from gravity anomalies predicted from altimeter data a new attempt to smooth the gravity anomalies in the test area (ii) by RTM reductions was made. Better results correspond to a density equal to 2,300 kg m^{-3}. The results of this attempt are shown in Table 9

Table 9. Results of the RTM reduction of gravity and altimeter data. The RTM reduction is based on the estimated bottom topography using gravity from altimeter data.

	No.	Mean	Std. deviation	Min.	Max.
gravity (mGal)	1,040	3.90	7.74	-30.51	47.34
altimeter data (m)	2,471	-0.06	0.12	-0.43	0.82

The smoothing effect of the estimated bottom topography on sea surface heights was more or less expected, since altimeter heights were used for the prediction of gravity anomalies. However, the smoothing effect on the observed gravity data is an extremely satisfactory result, since observed gravity and altimeter data are coming from independent data sources. From Table 9 it is shown that the smoothing of both kinds of data exceeds the level of 50%.

With the RTM reduced ERS-2 sea surface heights we tried again to predict gravity anomalies using the remove-restore technique. Unfortunately, the results of this last experiment were not better than the corresponding results of Table 8, when non reduced SSH were used to predict gravity anomalies: the mean value and std. deviation of the differences in this case are equal to 4.07 and 7.82 mGal respectively. A possible reason could be the fact that the gravity information included in the SSH has been exhausted through the RTM reduction. Nevertheless, the estimated bathymetry is still useful for geodetic applications: The smoothed quantities could be used for interpolation.

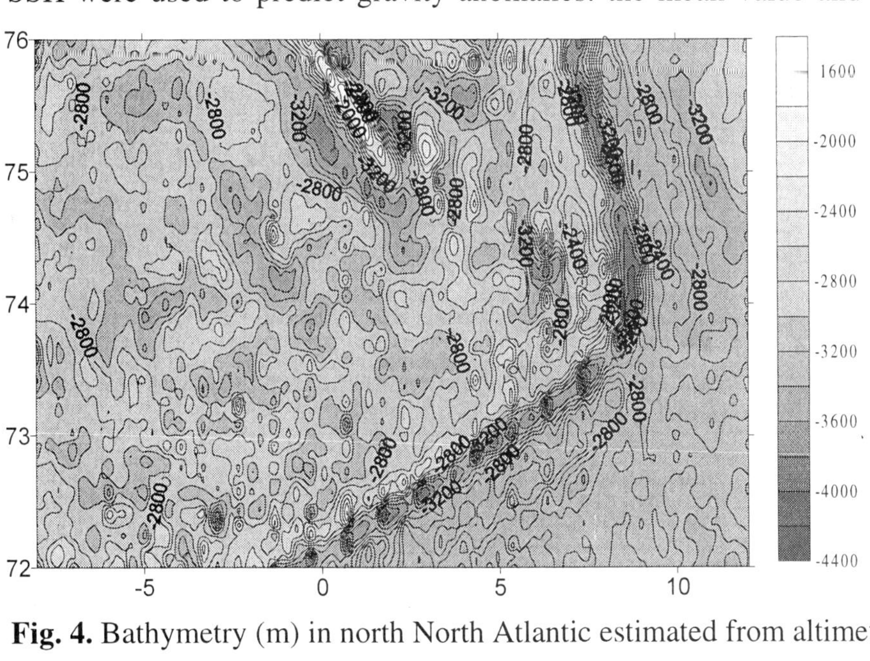

Fig. 4. Bathymetry (m) in north North Atlantic estimated from altimeter data.

5. Conclusion

The comparison between a global and a regional DTM in the Central Mediterranean showed differences ranging from -1.5 to 0.7 km. Similar problems were found comparing ETOPO5 and TerrainBase in the northern Atlantic. In the test areas the smoothing effect of these models on quantities related to the gravity field was very poor.

Observed gravity anomalies as well as gravity anomalies derived from altimeter data were used to estimate the bottom topography in the test areas using inversion of the gravity data through a least squares collocation procedure. The estimated topography in both test areas describes very well many characteristic features of the bathymetry. However, there are considerable discrepancies between "observed" and estimated bathymetry.

The RTM reduction of the gravity or satellite altimeter data using the estimated bottom topography in Central Mediterranean gives results very close to the corresponding RTM reduction using JGP95E or Morelli's model. Using the estimated bottom topography in the northern Atlantic the smoothing of gravity anomalies as well as of SSH exceeds the level of 50%.

The estimated bottom topography could be used in gravity field modelling and in other applications in areas where no high quality observed depths of the sea floor are available.

REFERENCES

Arabelos, D., and I.N. Tziavos, 1997. Gravity-field improvement in the Mediterranean Sea by estimating the bottom topography using collocation. Accepted for publication in Journal of Geodesy.

Arabelos, D., M Vermeer, 1996. Gravity field Mapping from a combination of satellite altimetry and sea gravimetry in the Mediterranean Sea. Reports of the Finnish Geodetic Institute, Rep. No. 96:1, Masala.

Barzaghi, R., A. Gandino, F. Sanso and C. Zenucchini, 1992. The collocation approach to the inversion of gravity data. Geophysical prospecting, Vol. 40, pp. 429-452, 1992.

Geiss, E, 1987. A new compilation of crustal thickness data for the Mediterranean area. Ann. Geophysicae, 5B, (6), 623-630.

Knudsen, P., 1993. Integrated inversion of gravity data. Final Report Norsk Hydro R&D Project, Kort og Matrikelstyrelsen, Geodetic Division, Technical Report No. 7, Kopenhagen.

Knudsen, P. and O.B. Andersen, 1996. Ocean Bottom Topography from ERS-1 Altimeter Data. Earth Observation Quaterly (EOQ), 51, ERSIN, 16-18, 1996.

Lemoine, F.G. et al., 1996. The Developement of the NASA GSFC and NIMA Joint Geopotential Model. Proceedings GRAGEOMAR, Tokyo, Sept.30 -Oct. 5.

Morelli, C, G. Cantar, M. Pisani, 1975. Bathymetry, Gravity and Magnetism in the Strait of Sicily and in the Ionian Sea. Bollettino di Geofisica Teorica ed Applicata, Vol. XVII, N. 65, 39-58.

NGDC, ETOPO5U, Digital Relief of the Surface of the Earth, National Geophysical Data Center Data Announcement 88-MGG-02, Boulder, CO, 1988.

Row, L. W. III, D.A. Hastings and P.K. Dunbar. 1995. TerrainBase Worldwide Digital Terrain Data. NGDC Key to Geophysical Records Documentation No. 30.

Smith, W.H.F., and D.T. Sandwell, 1994. Bathymetry prediction from dense satellite altimetry and sparse shipborne bathymetry. J. Geophys. Res., 99, 21803-21824, 1994.

Tscherning, C.C., Forsberg, R., and Knudsen, P., 1994. First experiments with improvement of depth information using gravity anomalies in the Mediterranean Sea. In Proc. of *Mare Nostrum - Geomed Report 4,* (eds. D. Arabelos and I.N. Tziavos), pp. 133-148, Thessaloniki, Greece.

Generating Orthometric Heights From The ERS-1 Altimeter Geodetic Mission Dataset: Results From An Expert Systems Approach

Berry, P.A.M., Sanders, R.F., Leenmans, C. & Bron, E.
Geomatics Unit, Faculty of Computing Sciences & Engineering,
De Montfort University, Leicester LE1 9BH, UK
Tel: +44 116 257 7497 email: pamb@dmu.ac.uk

Abstract

The ERS-1 mission resulted in the acquisition of large volumes of echoes over non-ocean surfaces. The dense data coverage generated from the geodetic mission offers the possibility of generating millions of height points over the earth's land surface. However, highly complex altimeter echoes are observed over much of the terrain not sampled by previous altimeters due to loss of lock. An expert system has therefore been developed to retrack individual waveforms using a range of retrackers specifically configured to interpret echoes over land. Data from the entire geodetic mission has been processed to generate a Regional Altimeter Result database (RAR) and this paper presents results illustrating the potential of this technique to generate terrain related information over remote areas. The results confirm the value of an expert system approach to land altimetry retracking, and demonstrate that additional terrain related information may also be derived.

1. Introduction

The ERS-1 mission resulted in the acquisition of large volumes of echoes over non-ocean surfaces, offering the possibility of generating many millions of height points over the earth's land surface if the complex echoes resulting from topographic surfaces can be interpreted. An expert system has therefore been developed to retrack individual waveforms using a range of retrackers specifically configured to interpret echoes over land. Data from the entire geodetic mission has been reprocessed and results have been analysed on a global scale.

2. Expert System Development

Using test areas in Zimbabwe and the central U.S.A., several retrackers specifically designed to deal with land waveforms were tested (1,3) and a new experimental retracker was developed to deal with waveforms identified as returning from moderately sloping surfaces (1). The entire geodetic mission altimeter dataset of Altimeter Waveform Product

(WAP) data (6) was then reprocessed using an updated preprocessing chain (Figure 2.1) and subset by region into 15° squares prior to ingestion into the expert system.
The existing version of the expert system was then further tuned using regions selected from the reprocessed dataset, and a further experimental retracker was added to form the Epsilon implementation of the expert system.

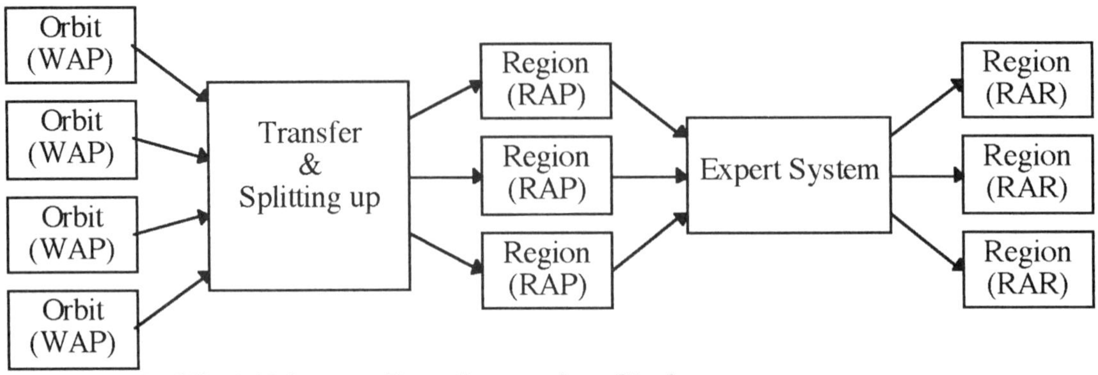

Figure 2.1 Modified Altimeter Data Processing Chain

3. Waveform Results

The entire global dataset was retracked to generate orthometric heights using the Epsilon-1 version of the expert system. In this section, a brief description of the various land specific retrackers is given. Results from the spatial analysis of retracker usage of the global results, and regional analyses using an enhanced system are discussed. All statistics were generated over 15 ° regions and combined to form summary tables of Northern and Southern hemisphere figures, with the exception of extreme South (Antarctica) and extreme North (Greenland) figures which are presented separately.

3.1 Spline Retracker

This retracker was specifically developed to deal more effectively with pre-peaked waveforms by fitting bicubic spline functions to the leading edge. Global statistics for the usage of this retracker are given in Figure 3.1.1. The Northern hemisphere statistics show a peak at 11% for this retracker with a broad distribution to about 25%. Over the smaller Southern hemisphere land mass a similar result is obtained. However over Antarctica a much higher proportion of Spline selected waveforms is observed.

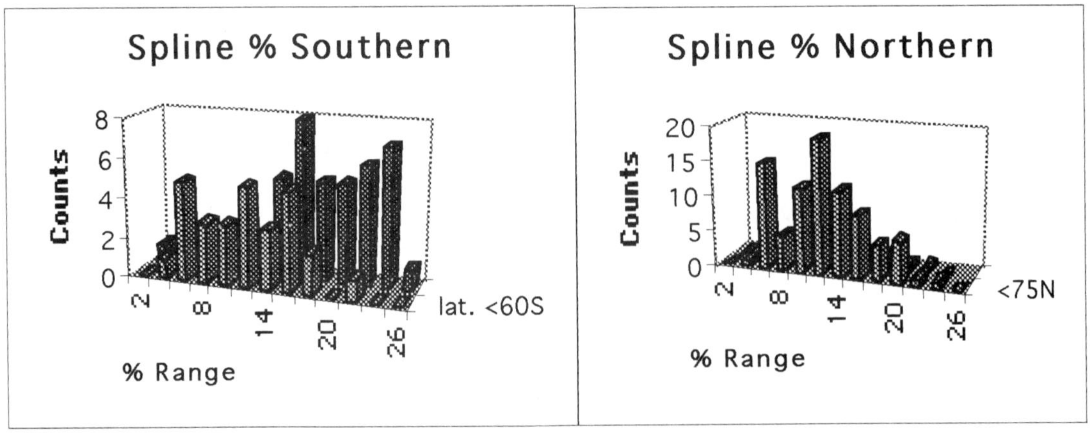

Figure 3.1.1 Global usage of Spline retracker

3.2 Narrow waveforms

This retracker analyses narrow waveforms, from high power returns indicative of good reflectors such as inland water or salars, to low power returns from predominately flat patches. The global usage is shown in Figure 3.2.1. This distribution peaks at just under 40%, in a wide distribution. The Antarctic statistics (not shown) have all counts at less than 2%, of which most are less than 1%.

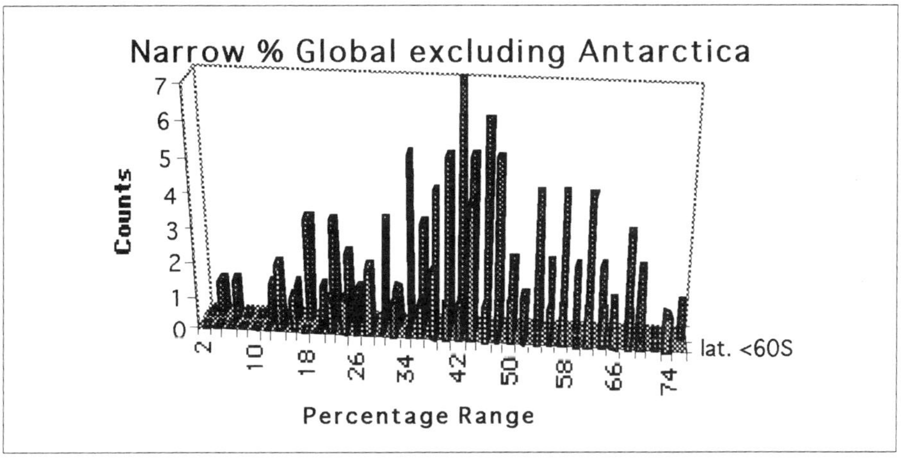

Figure 3.2.1 Global Narrow retracker usage

3.3 Bor Retracker

This retracker was designed specifically to deal with a class of waveforms for which slope information is encoded within the waveform shape in a form amenable to automatic analysis, with the aim of enabling the retracker to retrack to the nadir return. The slope component is identified and removed prior to retracking. The global usage statistics are presented in Figure 3.3.1. A fairly compact distribution is observed, peaking at about 9% in the Northern hemisphere. Antarctic waveforms trigger this retracker in 4 to 6% of cases.

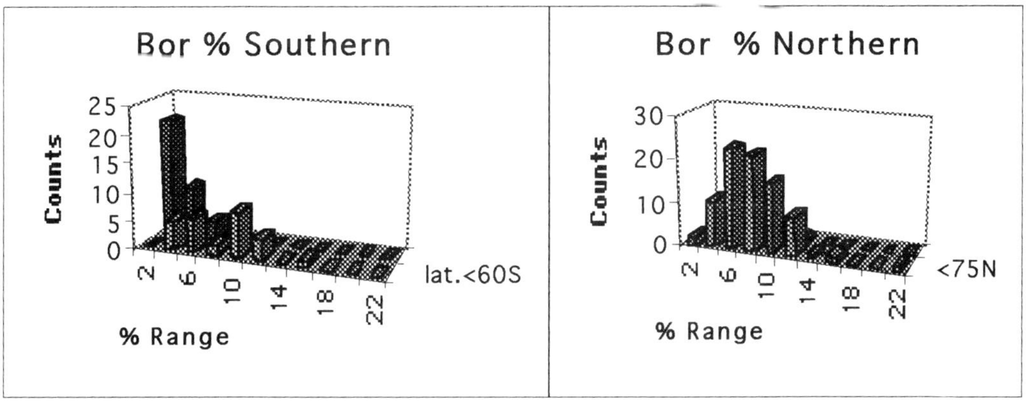

Figure 3.3.1 Bor global usage

3.4 Bog retracker

This experimental retracker was tested for a specific subgroup of Bor-type waveforms. However, less than 1% of waveforms from any region were selected for this retracker; during processing it became evident that the decision module criteria were set too high, and this retracker is therefore excluded from further consideration in this paper. (The criteria have now been reviewed).

3.5 Default Retracker

The system also amassed statistics detailing the proportion of waveforms which would have been reprocessed by the default (threshold) retracker if only the fully operational and validated retrackers (Bor, Spline and Narrow) had been switched on. These statistics are summarised in Figure 3.5.1. For land data, the default retracker is currently still generally used for around 25 - 38% of the waveforms, the percentages decreasing as each new retracker is added to the expert system. Over Antarctic waveforms, note that far higher percentages still trigger the default retracker.

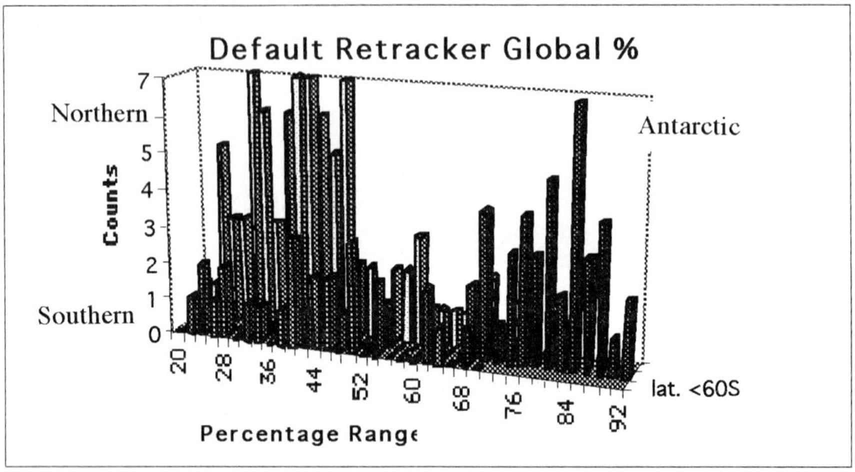

Figure 3.5.1 Global default retracker usage

4. **Spatial Correlation Results**

In this section, the detailed regional distribution of the land specific retrackers is considered and example results are presented over a 15 degree square of Africa. Included in this discussion is a new experimental retracker which has been switched on over selected regions with excellent results.

4.1 Bor results

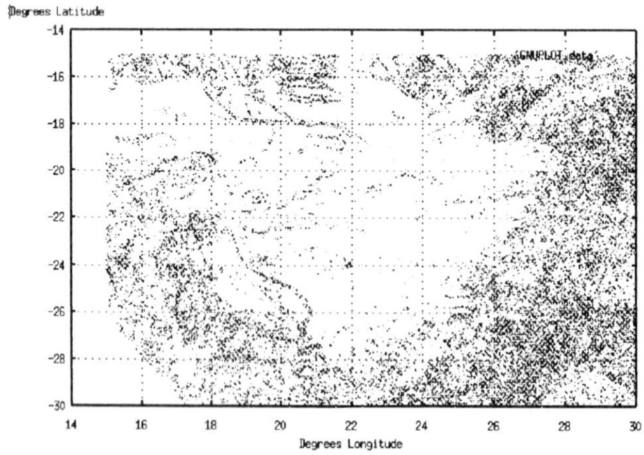

Figure 4.1.1 Bor returns over Africa: 15 to 30E, 30 to 15S.

Figure 4.1.1 shows the spatial distribution of returns retracked using the Bor retracker over the example region. Note the relative absence of Bor returns over the central region of the image (which includes the Kalahari desert), and the rather diffuse identification of the Zambesi and other neighbouring rivers in the upper left hand corner. The majority of the returns correspond to the significant terrain relief within the region.

4.2 Narrow waveform results

The distribution of Narrow waveforms is shown in Figure 4.2.1 as previously. The clear river identification, and the general high response over inland water and salars is as expected, as quasi-specular returns have been known for many years to occur in these circumstances (7). However, Figure 4.2.1 clearly indicates that many returns analysed by this retracker occur over rougher terrain, in fact the great majority of these data correspond to low power returns from flat patches in areas of significant terrain relief. Further analysis over a number of individual 15 degree areas has confirmed that this waveform group would benefit from additional segmentation. The expert system is now being redesigned to deal with low power returns separately using an additional retracker to optimise determination of the nadir return point.

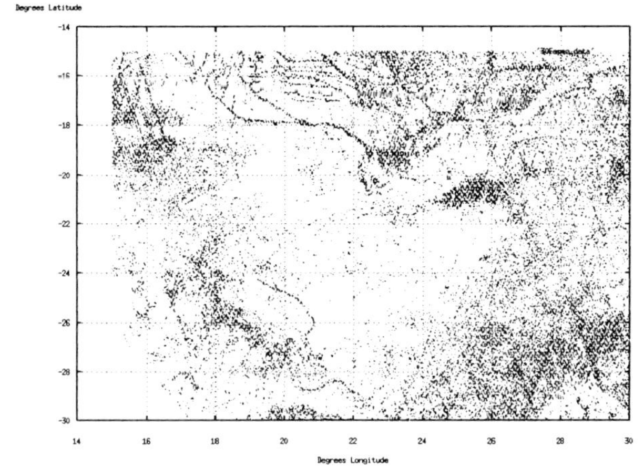

Figure 4.2.1 Narrow waveform returns over Africa: 15 to 30E, 30 to 15S

4.3 Experimental Retracker 7 (ER-7) Results

An additional experimental retracker (ER-7) designed to analyse a complex waveform group observed over extreme terrain, was included in the regional analysis for evaluation (Figure 4.3.1). A significant proportion of returns were selected by this retracker, and such strong spatial correlations with terrain were observed that in many areas the plots could be interpreted directly in terms of regional terrain type variation.

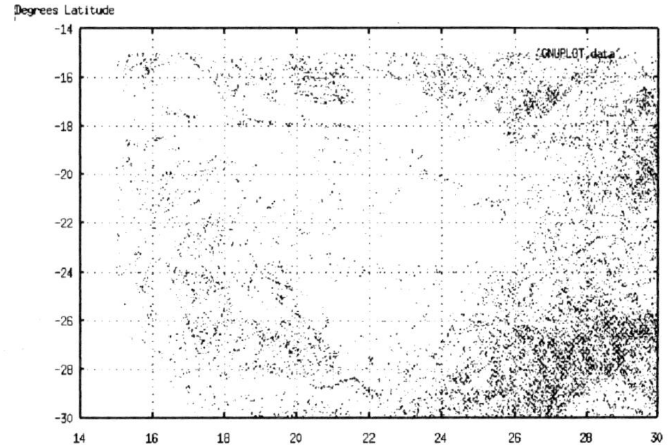

Figure 4.3.1 Experimental retracker 7 results over Africa: 15 to 30E, 30 to 15S

5. Discussion

The inclusion of the ice mode on ERS-1 has enabled recovery of altimeter echoes over the majority of the earth's land surface sampled during the ERS-1 geodetic mission. The resulting waveforms show complex and diverse structures. However, after an initial sorting of waveforms into a number of primary groups for retracking, a large number of identifiable patterns have emerged in the waveform shapes being processed by specific retrackers. Strong correlations with terrain features are evident in the retracker usage. Both the Bor retracker and ER-7 specifically switch on over areas of significant terrain relief, the Bor retracker preferentially selecting moderate slopes, and the experimental retracker 7 identifying more extreme terrain. This is a very encouraging finding, and demonstrates that the theoretical basis for these retrackers is well founded. Other work with orthometric height data generated by the expert system, presented elsewhere in this volume (2,4) confirms the excellent performance of the RAR dataset both in crossover difference analysis and use for validation of the GLOBE dataset.

All results to date confirm the initial finding that altimeter derived heights over land can be derived far more accurately than had previously been expected, if a sufficiently sophisticated system is used to interpret the complex returns from areas of significant terrain relief. The statistical results also demonstrate clearly the fundamental difference in waveform shape distributions obtained over ice and land surfaces and the consequent implications for retracking.

Work is now progressing on identification and interpretation of environmental information also found to be present in the altimeter land returns. With the increased understanding of the charateristics of land waveforms encoded into the expert system, data are now being examined to identify and interpret the additional environmental signals.

Acknowledgements

The authors thank the European Space Agency for supply of Altimeter Waveform Product data for this research, and the reviewers for helpful comments.

References

1. Berry, P.A.M., Bracke, H. & Jasper, A., 1997. 'Retracking ERS-1 Altimeter Waveforms over Land for Topographic Height Determination: an Expert Systems Approach'. Proceedings, Third ERS Symposium, Florence, March 1997.
2. Berry, P.A.M., Bron E.I., Leenmans, C. & Sanders, R.F., 1997a. 'Use of ERS-1 Land Altimetry to Validate the GLOBE Global Digital Elevation Model', this issue.
3. Bracke, H., Berry, P.A.M. & Dowson, M. 1996. 'Geodetic Applications of ERS-1 Altimeter Data Over Land'. American Geophysical Union Fall Meeting, December 1996.
4. Berry, P.A.M., Sanders, R.F., Bron E.I. & Leenmans, C., 1997c. 'Generating Orthometric Heights From The ERS-1 Altimeter Geodetic Mission Dataset: Results From An Expert Systems Approach', this issue.
5. 'GLOBE Dataset Version 0.5', 1997, National Geophysical Data Center, Boulder, Colorado
6. Wilson, H.K., 1994. 'WAP Compact User Guide', issue 1.4. National Remote Sensing Centre Report: PF-UG-NRL-AL-0001.
7. Guzkowska, M.A., Rapley, C.G., Ridley, J.K., Cudlip, W., Birkett, C.M. & Scott, R.F., 1990. 'Developments in inland water and land altimetry', ESA CR-7839/88/F/FL.

Use Of ERS-1 Land Altimetry to Validate the GLOBE Global Digital Elevation Model

Berry, P.A.M., Bron, E., Sanders, R.F. & Leenmans, C.
Geomatics Unit, Faculty of Computing Sciences & Engineering,
De Montfort University, Leicester LE1 9BH, UK.
Tel: +44 116 257 7497 email: pamb@dmu.ac.uk

Abstract

The requirement for a global 1km Digital Elevation Model is currently being addressed by the generation of a number of global digital elevation models such as the GLOBE dataset. However, the availability and quality of input ground truth varies widely over the earth's land surface. There is thus an urgent need for additional independent information both to validate GDEMs and to augment the existing ground truth. At this spatial scale, altimetry presents an attractive option, with the combination of near polar orbit and close track spacing during the ERS-1 geodetic mission providing data at a spatial scale of several km. However, to interpret these data effectively over land requires sophisticated post-processing.

This paper presents the results of a global comparison between altimeter derived heights and the GLOBE GDEM, using orthometric heights derived from ERS-1 Altimeter Waveform Product data retracked using an expert system. The results demonstrate the limitations in the current GLOBE dataset in coverage and accuracy both locally, with mismatches between individual tiles, and on a regional scale, with errors in GLOBE input datasets clearly apparent. The findings also demonstrate the unique ability of the ERS-1 altimeter derived dataset to detect these errors, and clearly show the potential of a correctly retracked altimeter dataset as a tool to assess and correct errors in regional DEMs.

1. Introduction

The geodetic phase of the ERS-1 mission resulted in the acquisition of large volumes of altimeter returns obtained over the earth's land surface. However, the echo structure is complex and thus difficult to interpret unambiguously, particularly over areas of significant topographic relief. This paper presents results derived using a reprocessed altimeter dataset from the ERS-1 geodetic mission; a rule based expert system (2) has been used which characterises individual echoes, sorts them by shape, then uses one of a range of 'retracking' algorithms to generate orthometric heights. These data have been used to carry out a global comparison with the GLOBE 1km Global Digital Elevation Model (GDEM) (5) both to assess the accuracy of the GLOBE dataset and to illustrate the power of altimeter derived orthometric heights to evaluate and correct errors in ground based DEMs.

2. **Datasets**

2.1 Altimeter Data.

Altimeter Waveform Product data (6) from the ERS-1 geodetic mission have been used for this research. These data comprise about 160Gbytes of altimeter data over both land and ocean surfaces, and are supplied on 120 Exabyte tapes. The data were preprocessed and filtered to remove erroneous records, divided regionally into 15 degree squares and archived. Each individual return shape or 'waveform' was then piped through a rule based expert system equipped with a range of algorithms specifically configured to analyse altimeter echoes over land. After analysing the waveform shape, the most appropriate algorithm for each waveform was selected by the expert system to retrack the echo to obtain an estimate of the range to the surface. Details of the processing are given elsewhere (1,3). The Epsilon-1 version of the expert system, containing 6 retrackers developed for a range of waveform characteristic shapes, was used for generation of all altimeter derived orthometric heights used in this study. The resulting Regional Altimeter Result (RAR) height dataset was referenced to WGS 84 using the OSU 91A geoid model and used for this comparison work. Note that only data obtained in 'ice mode' have been used in this study for consistency, so some coastline areas and countries where the altimeter was not switched into ice mode have consequently been omitted.

2.2 GLOBE.

The GLOBE GDEM is currently being developed by the National Geophysical Data Center (NGDC) with the aim of providing a global model at 1km resolution. The prototype version 0.5, used for this work, covers about 60% of the earth's land surface, using a 30 arc-second latitude-longitude grid (giving a spacing somewhat finer than 1km at the equator). Data from a variety of public domain datasets have been combined, referenced to the WGS'84 ellipsoid.

GLOBE version 0.5 has been stored on three CDs. Volume 1 contains 12056 averaged height files for individual 1 degree by 1 degree tiles (these files have a .avg extension). Volume 2 contains 12056 files of maximum (.max) and minimum (.min) height values for the tiles. Volume 3 contains 30-arc second discrete elevation values.

A number of problems in data handling were encountered when processing the GLOBE data. Several errors are documented on the internet (URL: http//www.ngdc.noaa.gov /seg/fliers/se-1106b.html). A validation test on the GLOBE dataset also revealed that many of the .min files contain heights greater than the .avg height (Figure 2.2.1).

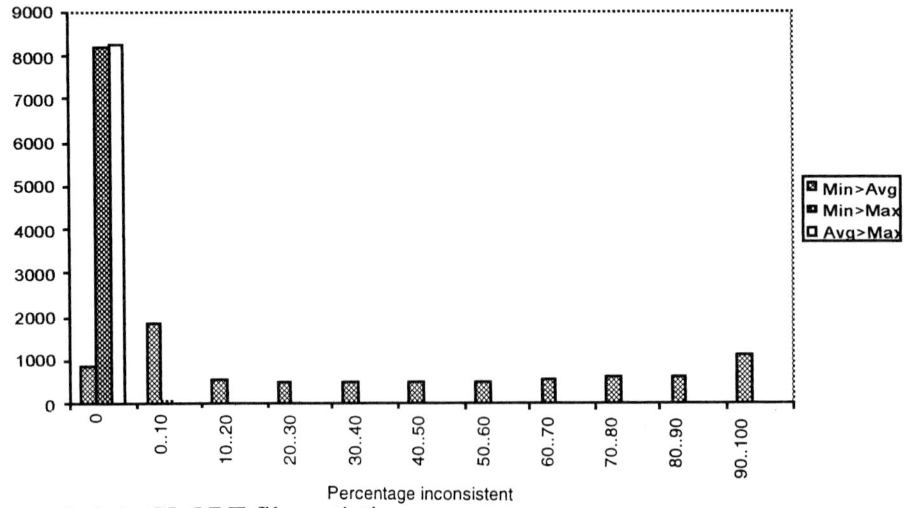

Figure 2.2.1 GLOBE file statistics

3. **Comparison**

3.1 Coverage.

The altimeter has near global sampling between 81.5N and 81.5S, with an along track spacing of data points of about 350m, but an across track spacing of several km. This varies according to latitude, being greatest at the equator.

The GLOBE dataset covers only 60% of the land surface, but individual one degree tiles have internally continuous coverage. Thus regions exist for which altimeter data are available but GLOBE data are not; conversely the altimeter subsamples the GLOBE dataset across track. The altimeter coverage is in general extremely good with the exception of extreme terrain (comparison data over the Andes and Himalayas are virtually non-existent, and coverage over other mountain ranges is poor). As the altimeter data have been retracked using a range of new retrackers specifically designed to interpret complex echoes over land (3,4), all heights derived from the altimeter have been retained in this comparison even when generated over extreme terrain.

3.2 Methodology.

Because detailed regional analysis of the retracked altimeter data has demonstrated that the new retrackers can identify the nadir return from the surface by interpreting the slope information encoded in the individual waveform shapes (enabling derivation of an accurate height directly below the altimeter (3,4)) the altimeter often subsamples the height within a single GLOBE pixel. Thus for this comparison, the altimeter data were not averaged; for each altimeter point a comparison was made with the GLOBE 30" cell within which the altimeter point was located. A global dataset was assembled, and differences computed, mapped and statistics generated for 15 degree regions globally.

3.3 Results.

Over the U.S.A., good agreement was generally found between the two datasets (a sample histogram is given in Figure 3.3.1). However, analysis of the difference plots showed that even in this well mapped region, the edges of individual GLOBE tiles were sometimes clearly visible, indicating some mismatch of tiles when mosaiced together. There is zero mean difference between the two datasets.

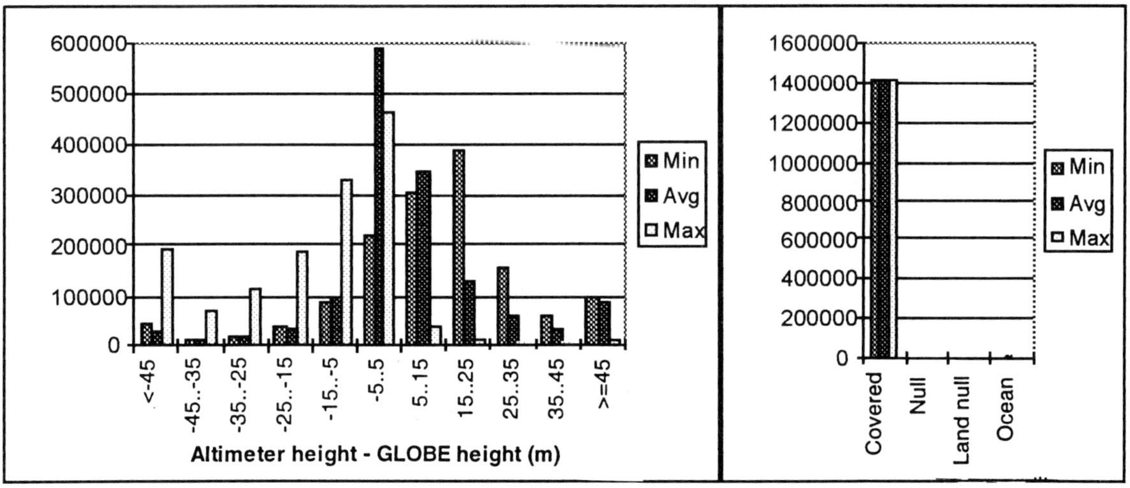

Figure 3.3.1 Sample comparison histogram over the U.S.A: 30-45N, 90 - 75W

Over one hundred separate comparisons over 15° regions were made to provide a global evaluation. One finding over most of the world other than the U.S.A. is a mean height difference between the two datasets of 5 - 15m, with the altimeter usually ranging higher than the GLOBE dataset. Other comparisons with DEM data (2) have shown no mean bias in the altimeter heights except over extreme topography. As the GLOBE data over the U.S.A. show no bias, this finding is attributed to problems within the GLOBE dataset. It is clearly impossible to present all the results in one paper; in adddition, greyscale limits the representation of difference maps. However, sample maps in this section have been included to illustrate some of the typical findings. Pixels where no altimeter data were available appear as white.

Figure 3.3.2 Subsection around 25N, 85E.

Figure 3.3.2 shows a small section around 25N, 85E illustrating the near global finding of discontinuities at boundaries of individual tiles, visible in greyscale as a shade difference running in bands across and down the plot. The diagonal lines show the altimeter coverage. These discontinuities, typically of the order of 5 - 15m vertical displacement, are often not apparent when height contour maps or DEM images are produced. They appear in the GLOBE dataset only when illuminated.

Figure 3.3.3 Central Siberian Plateau, 66 - 75N, 105 - 120 E

Figure 3.3.3 shows several large areas characterised by a consistent large vertical offset in the difference dataset, visible as dark grey. This is attributed to a datum problem in the source data used for this region, of at least 35m. Large scale errors of this type were observed in a number of locations.
Figure 3.3.4 illustrates one of a number of regions where an interesting effect due to the altimeter sampling is observed. Over these regions of high topographic variation, the altimeter remains in lock only in the valleys, producing characteristic undulating patterns in the difference plot. A further discussion of this terrain mapping effect is presented elsewhere (4).

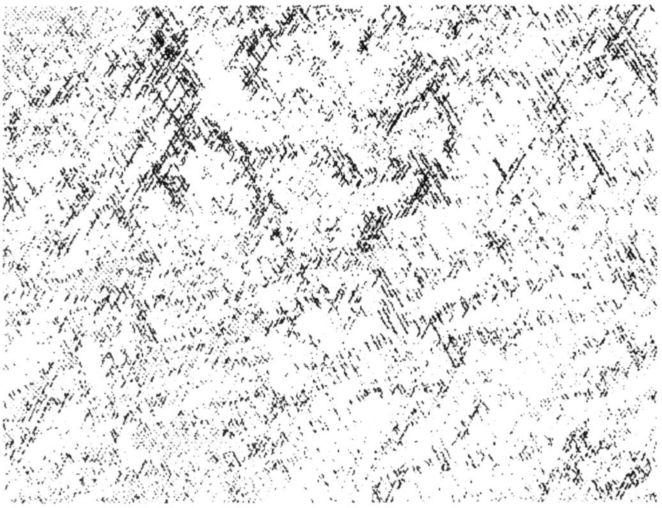

Figure 3.3.4 Southern Siberia, 52 - 60N, 112 - 120E

4. Discussion

The comparison between GLOBE and the RAR altimeter dataset has demonstrated the ability of this altimeter derived dataset to identify errors in the GLOBE data arising from different reference datums and mismatches between individual GLOBE tiles. The comparison has also illustrated the poor coverage of GLOBE data over some areas, particularly over South America and Africa. The most disturbing finding for users of the current issue of the GLOBE dataset is the prevalence of discontinuities at tile boundaries; in many areas individual tiles appear to have been lifted or dropped relative to the reference datum. These errors can be in excess of 90m in extreme cases, but are more generally between 5 and 15m. At this scale the errors are insidious, often not evident in height maps of a region. It is strongly recommended that users of the GLOBE dataset check data for this effect either by illuminating the data to provide a shadowmap, or by validation with another dataset over the area of interest.

This comparison has demonstrated globally the ability of the RAR altimeter dataset to identify errors in existing digital elevation models. These comparison data can clearly be used to correct DEM datum errors, and if required can reorient individual tiles of data to remove discrepancies, and bring disparate datasets into a single global datum: the potential also exists to map directly areas of the world where existing ground truth is poor or absent.

Acknowledgements
The authors would like to thank the European Space Agency for their supply of WAP data from the ERS-1 geodetic mission, Rutherford Appleton Laboratories for the loan of equipment to facilitate this research, and the reviewers for helpful comments.

References

1. Berry, P.A.M., Jasper, A. & Bracke, H., 1997b, 'Retracking ERS-1 Altimeter Waveforms over Land for Topographic Height Determination: An Expert Systems Approach', Third ERS Symposium, European Space Agency, Florence, March 1997.
2. Berry, P.A.M. & Leigh, M., 1997, 'Towards Validation and Correction of Global Digital Elevation Models with ERS-1 Altimetry', Third ERS Symposium, European Space Agency, Florence, March 1997.
3. Berry, P.A.M., Sanders, R.F., Bron E.I. & Leenmans, C., 1997c, 'Generating Orthometric Heights From The ERS-1 Altimeter Geodetic Mission Dataset: Results From An Expert Systems Approach', IAG Symposium, Rio de Janeiro, September 1997.
4. Dowson, M. & Berry, P.A.M., 1997. 'Geodetic Application Of ERS-1 Altimeter Data Over Land'. This issue.
5. 'GLOBE Dataset Version 0.5', 1997, National Geophysical Data Center, Boulder, Colorado
6. Wilson, H.K., 1995, 'Altimeter Waveform Product ALT.WAP Compact User Guide', National Remote Sensing Centre Limited, Farnborough, Hants.

Geodetic Application Of ERS-1 Altimeter Data Over Land

Dowson, M. & Berry, P.A.M.
Geomatics Unit, Faculty of Computing Sciences and Engineering,
De Montfort University, Leicester LE1 9BH, UK
Tel: +44 116 257 7462 email: mzd@dmu.ac.uk

Abstract

The high number of valid returns obtained over land from the ERS-1 altimeter during the geodetic mission presents a unique opportunity for orthometric height determination. In order to utilise these data fully, the precision to which altimeter derived heights can be generated over land surfaces must first be quantified. In this paper, retracked Altimeter Waveform Product (WAP) (7) data from the Geodetic Mission of ERS-1 are evaluated. The dataset upon which this research is based has been derived from the WAP dataset via the use of an Expert System which 'retracks' each altimeter waveform return, selecting one of a set of retracking algorithms designed to optimise the determination of the individual height. The resulting data at a number of test sites over four continents have been used to investigate the extent to which retracked altimeter data can be used to generate Ground Control Points for geodetic purposes.

The results obtained over these test areas demonstrate that altimeter heights over land can be determined with very high internal consistency, with between 20% and 50% of results in a 15° square agreeing to within 1m, and filtering for specific retracker types the percentage can be as high as 60%. However, interpretation of the complex echoes returned over areas of significant topographic relief necessitates the use of a sophisticated processing chain in order to generate optimal height values. This paper presents sample results and discussion drawing on the results from the test areas to infer the potential for generation of geodetic control points on a global scale.

1. Introduction

1.1 Context

The orthometric height information derived by retracking ERS-1 land altimetry data from the geodetic mission has many potential applications. One possible use of such data is to generate ground control points. However, this can only be done if it can be demonstrated that under certain conditions the heights obtained are extremely accurate. In order to use altimeter data obtained over a complex topographic surface, sophisticated post-processing techniques are required to recover orthometric height data. As part of an investigation into the applications of land altimetry data, retracked altimeter data from the ERS-1 geodetic mission (1,3) have been transformed to orthometric heights (referenced to the WGS84 ellipsoid using the OSU'91a geoid model) to yield a global Regional Altimeter Result (RAR) database containing over one hundred million data points. Detailed discussion of the techniques used in producing this dataset can be seen elsewhere (2,3). As part of the assessment and validation of this dataset, crossover analyses have been undertaken to examine the internal consistency of the RAR dataset. One advantage of this approach is that errors in the geoid model used to convert to orthometric heights do not contribute significantly to the height differences. The error budget for the ERS-1 altimeter is widely discussed elsewhere but for clarity, the most significant contributions are summarised here. Radial orbit error is quoted as less than 30cms rms, of which only a proportion is visible at crossovers. Including instrument and atmospheric error terms gives an error budget of less than 50cms excluding the contribution from waveform retracking. This is one reason for choosing 1m as the critical measurement for crossover pair height differences.

1.2 Expert System

A detailed description of the expert system used to derive altimeter heights is given elsewhere (1,3). A brief rationale is given here for clarity. Over land it has been found that the shape of the returned echo from the surface changes with terrain type; these echoes can be highly complex. In order to interpret these echoes an expert system approach has been taken. Here individual returns are classified according to their shape and one of a series of different retracking algorithms is selected and is used to generate an estimate of orthometric height.

1.3 Methodology

The assumption that the heights from ascending and descending tracks should be the same when interpolated to a crossover point is clearly invalid over land, due to short wavelength topographic differences in the terrain. Thus, interpolation of the data to an actual crossover point is not performed. Instead, crossover pairs are used, identifying pairs of points from different arcs that result from echoes from the surface lying in close spatial proximity. For the geodetic mission data, the density of crossover points means that a data driven approach rather than a global solution is more appropriate for crossover point detection. For this research, theoretical crossover point locations are calculated for each test area, and all data points whose geodetic latitudes and longitudes place them within a certain radius of these locations are identified. The spatial separation of each pair of points is then calculated, as is the height difference. Because of the inherent topographic variation of the underlying surface, a severe criterion was adopted to screen possible pairs of points; only those pairs whose nominal separations were less than 200m were included in this analysis. Note that only 'ice mode' data have been used for this study to maintain a consistent sampling of the earth's land surface.

2. Regional Analysis

In this section, a selection of regional results from the crossover analysis is presented. Note that several of the assessment areas discussed here have been chosen on the basis of the results from a parallel study comparing the RAR dataset with the GLOBE global digital elevation model (2).

2.1 USA

The region 30°N - 45°N, 105°W - 90°W in the USA was chosen for analysis because of the generally good agreement between altimeter and GLOBE heights in the region. Spatial plots indicate significant terrain related correlations in the locations of 'good' crossover pairs, with few qualifying pairs over the more severe terrain. Figure 2.1.1 shows the statistical distribution of crossover pair heights over the 15° region, indicating that 37% agree to within 1 metre.

2.2 Siberia

Two regions of Siberia were chosen for inclusion on the basis of the GLOBE comparison.

2.2.1 Northern region - N Siberia

The region 60°N 105°E - 75°N 120°E in N.Siberia was selected because a large regional offset was observed in the GLOBE difference analysis (2). This region was therefore split horizontally into two parts for analysis, along the latitude line 68°N. The results are summarised in Figure 2.2.1.1. The upper part of this region in fact shows slightly better crossover results than the lower, confirming the internal consistency of the RAR dataset over this region.

2.2.2 Southern region - S Siberia

Over the region 45°N 105°E - 60°N 120°E in S.Siberia altimeter data is obtained only intermittently as the instrument is unable to remain in lock over the severe terrain and regains the surface only briefly over the valleys (2). This area therefore represents a good test case for examining altimeter derived heights obtained from returns obtained at the limit of the ERS-1 altimeter's ability to sample the earth's land surface. The crossover results summarised in Figure 2.2.2.1 show an encouraging result with almost 29% of the crossover pairs agreeing to within 1m.

2.3 Zimbabwe

The area 24°E 24°S - 34°E 16°S encompassing Zimbabwe has been used as a test area for this work on several occasions (4,5), chosen primarily because the region is characterised by extensive areas of low slope. The spatial distribution of crossover pairs agreeing to within 1m vertically is given in Figure 2.3.1 and when compared to the contour map shown in Figure 2.3.2 shows clear correlation with terrain type, with predominantly flat and low slope terrain selected. The crossover results (Figure 2.3.3) show the superb internal agreement of these data, with over 49% of the 2249 crossover pairs agreeing to within 1m.

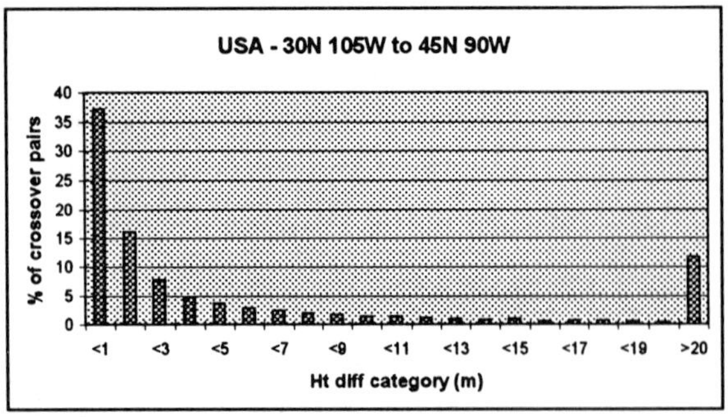

Figure 2.1.1 USA statistics

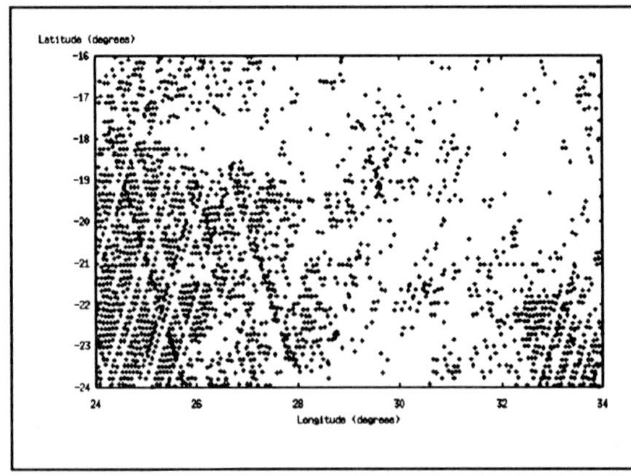

Figure 2.3.1 Zimbabwe crossover pair locations (Htdiff < 1m)

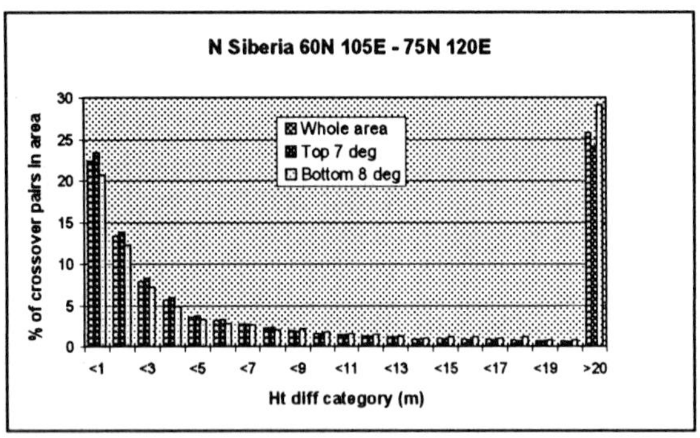

Figure 2.2.1.1 N.Siberia statistics

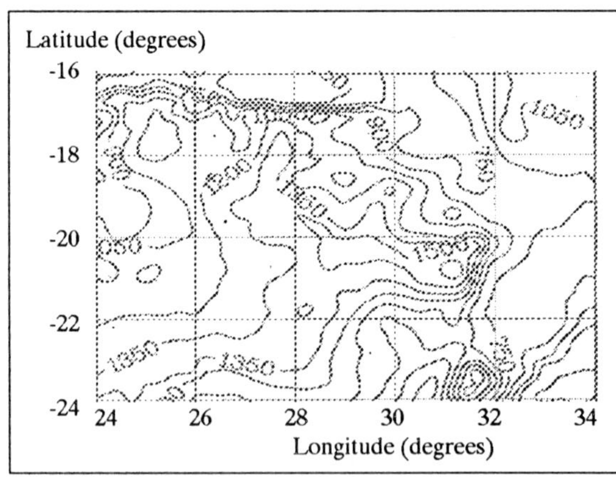

Figure 2.3.2 Zimbabwe contour map

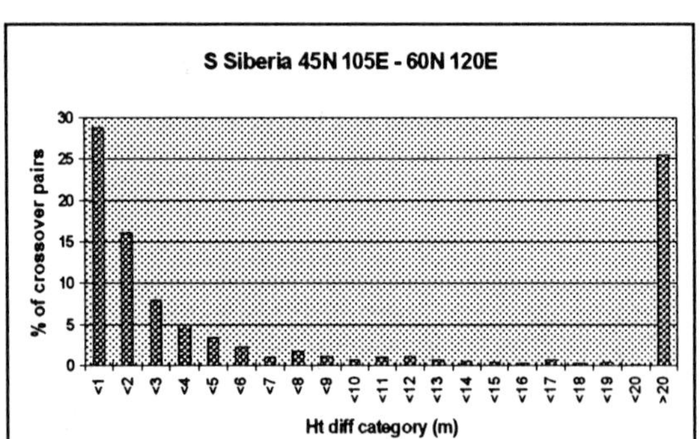

Figure 2.2.2.1 S.Siberia statistics

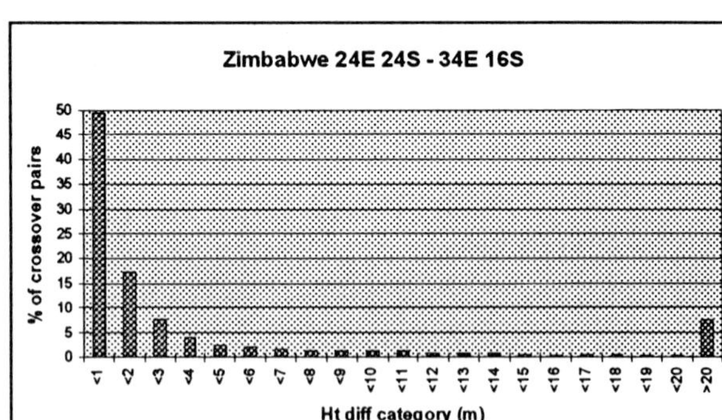

Figure 2.3.3 Zimbabwe statistics

2.4 Brazil & Paraguay - C.Brazil and S.Brazil

Two adjoining regions were considered. For ease of reference these are termed C.Brazil: 15°S 60°W - 0°S 45°W and S.Brazil: 30°S 60°W - 15°S 45°W. The statistics for both regions are given in Figures 2.4.1 and 2.4.2, with around 40% of the crossover pairs agreeing to within 1m.

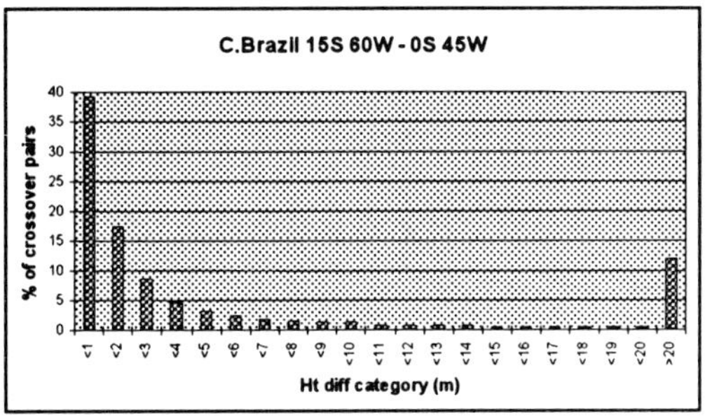

Figure 2.4.1 C.Brazil statistics

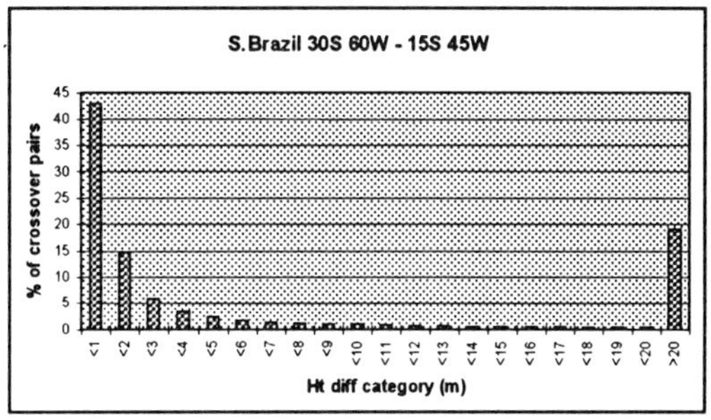

Figure 2.4.2 S.Brazil statistics

3. Retracker Analysis

In this section, consideration is given to the different retrackers used to derive orthometric heights in the RAR dataset. A primary reason for doing this is that retracker type correlates with terrain (1). Thus breaking down the statistics in this way allows more detailed analysis of crossover height difference distribution by effectively filtering crossovers on terrain type. These retrackers are discussed more fully elsewhere (3) but a brief summary is given here:

- Threshold — Default retracker (offset centre of gravity (6))
- Bor — Associated with areas of moderate slope: slope information encoded within waveform
- Spline — Double-peaked return strongly correlated with low slope
- Narrow — Quasi-specular return from flat surface

3.1 Brazil & Paraguay regional analysis

A regional study of retracker type at crossover pairs was carried out over both test areas. The results are shown in Table 3.1.1, and the correlation with retracker type is clear. The Narrow retracker gives the best results, with 54% and 63% of pairs where both members are of this type agreeing to within 1m. Results for the other retrackers show much poorer agreement, although even with the Bor retracker, 28% and 19% of data show close agreement over the two regions. However, as this retracker is specifically triggered by waveform shapes returned from sloping terrain (3) the extent to which crossover analysis is an appropriate tool for assessing the performance of this retracker is debatable.

3.2 Overall Statistics

Table 3.2.1 gives the summary statistics for retracker usage for all data over the test areas

C.Brazil: 15°S 60°W - 0°S 45°W				
Retracker type	Threshold	Bor	Spline	Narrow
Threshold	35% (484)	4% (16)	19% (130)	27% (322)
Bor		28% (40)	3% (4)	17% (52)
Spline			27% (28)	16% (46)
Narrow				54% (2479)
S.Brazil: 30°S 60°W - 15°S 45°W				
Retracker type	Threshold	Bor	Spline	Narrow
Threshold	32% (303)	2% (11)	22% (83)	29% (440)
Bor		19% (27)	1% (1)	8% (32)
Spline			24% (13)	15% (40)
Narrow				63% (2870)

Table 3.1.1 Retracker combinations where crossover height difference < 1m

	N.Siberia	S.Siberia	USA	Zimbabwe	C.Brazil	S.Brazil
Threshold	45.4%	40.4%	29.2%	53.7%	26.1%	28.4%
Bor	7.9%	9.0%	5.4%	8.2%	8.5%	6.8%
Spline	17.6%	12.5%	6.7%	16.5%	5.2%	7.5%
Narrow	29.1%	38.1%	58.7%	21.6%	60.2%	57.3%

Table 3.2.1 Percentage of retracker usage by region

detailed in Section 2. A wide variation in retracker choice is evident, with the default retracker followed by the Narrow retracker dominating the returns from both Siberia and Zimbabwe. This is an expected finding since this retracker tends to associate with areas of low slope. The Narrow retracker is dominant over the USA and South America test areas with low involvement from the other retrackers. Although crossover pairs where both members are retracked using the Narrow retracker are shown to give by far the best statistical performance, these results indicate that retracker usage (and therefore echo shape) does not simply translate into the crossover performance statistics.

4. Discussion

The crossover results show that between 20% and 50% of all crossover pairs within 15 degree regions show a height difference of less than one metre. This agreement is strongly terrain correlated, best over comparatively flat and dry terrain and worst over mountainous terrain with intermittent altimeter coverage. Although this is in accordance with the results of previous research (6), since where terrain changes rapidly, considerable height variation occurs on a short spatial scale, what is most encouraging is the high proportion of pairs which consistently agree to within 1 metre. In particular, where the Narrow retracker is used, results are extremely good, a natural consequence of the fact that these types of echoes result from flat patches and permit very precise retracking to be performed.

These results support the conclusion that the precision of orthometric height determination can be very high under certain conditions. It is also clear that orthometric height data to which a high confidence level may be attached can be selected on the basis of the retracker used and the terrain characteristics. However, for evaluating the RAR precision over terrain where considerable topographic variation occurs on a short spatial scale, a more sophisticated technique is required, since significant crossover pair height differences over such terrain do not necessarily imply data error. Having established the internal consistency of these height measurements using crossover analysis, this unique dataset is now being evaluated by comparison with detailed regional ground truth.

Acknowledgements

The authors wish to thank the European Space Agency for supplying Altimeter Waveform Product data for this research, and Mrs H.K. Wilson and the reviewers for helpful comments.

References

1. Berry, P.A.M., Bracke, H. & Jasper, A., 1997. 'Retracking ERS-1 Altimeter Waveforms over Land for Topographic Height Determination: an Expert System Approach'. Proceedings, Third ERS Symposium, Florence, March 1998.
2. Berry, P.A.M., Bron, E.I., Leenmans, C. & Sanders, R.F., 1997a, 'Use of ERS-1 Land Altimetry to Validate the GLOBE Global Digital Elevation Model', this issue.
3. Berry, P.A.M., Sanders, R.F., Bron, E.I. & Leenmans, C., 1997b. 'Generating Orthometric Heights from the ERS-1 Altimeter Geodetic Mission Dataset: Results from an Expert Systems Approach', this issue.
4. Dowson, M. & Berry, P.A.M., 1996, 'Potential of ERS-1 Derived Orthometric Heights to Generate Ground Control Points: First Results'. European Conference on Satellite Altimetry III, Oporto, October 1996.
5. Dowson, M. & Berry, P.A.M., 1997, 'Potential of ERS-1 Derived Orthometric Heights to Generate Ground Control Points'. Proceedings, Third ERS Symposium, Florence, March 1997.
6. Guzkowska, M.A., Rapley, C.G., Ridley, J.K., Cudlip, W., Birkett, C.M. & Scott, R.F., 1990. 'Developments in inland water and land altimetry', ESA CR-7839/88/F/FL.
7. Wilson, H.K., 1994. 'WAP Compact User Guide', issue 1.4. National Remote Sensing Centre Report: PF-UG-NRL-AL-0001.

Global Marine Gravity and Mean Sea Surface from Multi Mission Satellite Altimetry

Per Knudsen and Ole B. Andersen
Kort & Matrikelstyrelsen
DK-2400 Copenhagen NV

Satellite altimetry from the GEOSAT and the ERS-1 geodetic missions provide altimeter data with a very dense coverage. Hence, the gravity field may be recovered very detailed. Satellite altimetry from the 35 days repeat cycle mission of the ERS satellites and, especially, from the 10 days repeat cycle TOPEX/POSEIDON satellite mission provide accurate mean sea surface heights along the ground tracks of those missions. In this study the global marine gravity field has been derived using the altimetry from the geodetic missions. This was done in small cells covering the worlds ocean using advanced techniques for crossover adjustment, gridding by local collocation, and conversion of sea surface heights into gravity by FFT. Then averaged sea surface heights of the repeat missions were used to construct an accurate mean sea surface. To enhance the resolution of the mean sea surface the altimetry from the geodetic missions was utilized. Marine geoid heights derived from the altimetric gravity was used for this task. The advantage of such a procedure is that inconsistences along the edges of the small cells may be avoided.

Introduction

Satellite altimetry from the geodetic missions of the GEOSAT and the ERS-1 provides the opportunity for geodesists to make very detailed mapping of the marine gravity field. The newly released GEOSAT Geodetic Mission altimetry contain data from the first 18 months of the GEOSAT satellite mission (1985-1986) covering marine regions at latitudes between -72° and 72°. The satellite was operated in an non-repeating orbit yielding a very dense, though not completely homogeneous coverage of observations. The ERS-1 satellite completed its 336 days Geodetic Mission in March 1995. The altimetry from this mission cover regions at latitudes between -82° and 82° and provides a very dense and homogeneous coverage. At the equator the spacing between the satellites ground tracks of the ERS-1 Geodetic Mission is 8.3 km.

The first 18 cycles of the ERS-1 35 days repeat mission covered the period of about 20 months prior to the ERS-1 geodetic mission. Through each cycle 501 revolutions were completed which yields a track spacing of 80 km at the equator. Along those tracks averaged sea surface heights were obtained from the 18 ERS-1 cycles. From the TOPEX/POSEIDON mission the first 150 cycles were used (September 1992 to September 1996). Each cycle of 10 days consist of 127 revolution having a spacing of 315 km.

Mapping of the Gravity Field

Mapping of the Earth gravity fields from different data sources has previously been presented by, e.g. Balmino (1987), Sandwell (1992), Knudsen (1991), Knudsen et al. (1992), and Andersen et al. (1996). In this study the enhanced method described in Knudsen and Andersen (1997) and in Andersen and Knudsen (1997) was applied.

The estimation procedure for the marine gravity field was chosen, in order to derive a global map of the gravity field within a reasonable computational effort. Due to the very large amount of observations, rigorous methods such as least squares collocation are not feasible. Hence, a less rigorous, but very efficient approximation method based on the Fast Fourier Transform (FFT) was chosen, though it requires that the data are distributed in a regular grid. The result may be sensitive to cross-track gradients caused by sea surface variability arising as the distance between parallel tracks becomes very small in the geodetic mission altimetry. Such effects may be reduced by using altimetric slopes or second order derivatives (e.g. Hwang and Parsons, 1995; Rummel and Haagmans, 1990). However, it was decided to stay with the heights as the observations and elaborate existing software in the GRAVSOFT package (Tscherning et al., 1992) for gridding and FFT manipulations.

The determination of the gravity field was carried out in cells of the size of 2° latitude by 10° longitude. However, observations were extracted in cells extended by a border zone of 0.5° latitude by 1° longitude to a size of 3° latitude by 12° longitude. The selection of such small sub-areas was essential to the modelling of orbit errors and sea surface variability. A choice of larger cells caused problems in the removal of the sea surface variability, while smaller cells may corrupt parts of the regional geoid signal.

Edited and crossover adjusted altimeter data fitted to the EGM96 geoid model (Lemoine et al., 1997) were interpolated onto a regular grid using a local collocation technique. However, to filter out remaining sea surface variability that may cause erroneous cross-track gradients between parallel tracks, a modified covariance function was introduced. For data located on the same tracks an error covariance function was applied, hereby, assuming the error to be temporally uncorrelated. That is

$$c(r) = C_0(1 + \frac{r}{\alpha})e^{(-r/\alpha)} + D_0(1 + \frac{r}{\beta})e^{(-r/\beta)} \quad (1)$$

The parameters D_0 and β were empirically determined to a variance of $(0.1\ \mathrm{m})^2$ and a correlation length of 100 km respectively. For observations on different tracks D_0 was set to zero. In both cases the C_0 was fixed at $(0.2\ \mathrm{m})^2$ and the parameter α was fixed so that the correlation length was 15 km.

To enhance the filtering of the sea surface variability the estimation was carried out relative to a local weighted average of the nearest 100 observations. A weight function $1/(r^2+r_0^2)$ with r_0=0.1° is applied, so the local weighted average is a smooth representation of the surface heights that is less affected by the sea surface variability. Furthermore, in the collocation estimation, the 48 nearest observations are used to secure redundant geoid information at crossing tracks. The result of the gridding is a 1/16° by 1/16° grid covering an area extended by a border zone of 1° latitude by 3° longitude to a size of 4° latitude by 16° longitude. This grid of sea surface heights will be treated as geoid undulations through the following steps!

The gravity anomalies, Δg, were derived from the geoid undulations, N, using FFT techniques (Schwarz et al., 1990). In the frequency domain (u,v) that is

$$\Delta\tilde{g}(u,v) \approx \omega\,\gamma\,\tilde{N}(u,v)\,F(\omega) \tag{2}$$

where $\omega^2=u^2+v^2$ and γ is the normal gravity. Such a transformation from geoid undulations into free air gravity anomalies is a differentiation which enhances the high frequencies. Hence, it is sensitive to noise. Therefore a Wiener filtering function, $F(\omega)$, was introduced in Equation (2). This filter function is equivalent to a collocation filter that assumes Kaula's rule to be valid and uncorrelated noise (details in Forsberg and Solheim, 1988). That is

$$F(\omega) = \frac{\omega_c^4}{\omega^4 + \omega_c^4} \tag{3}$$

In this case the "cut-off" frequency, ω_c, where the filter is 0.5, was empirically determined to a wavelength of 12 km which roughly corresponds to about 10 cycles per degree or harmonic degree 3600. Finally, the contribution of the EGM96 gravity field was restored to obtain the free air gravity anomalies and isolated within the small 2° by 10° cell.

Determination of the Mean Sea Surface

For an estimation of the mean sea surface the gridded sea surface height from the computations of the gravity field may be used. However, as described above, very small cells were required to filter out the ocean variability properly. Hence, only the short wavelength parts of the mean sea surface are recovered accurately. Errors in the wavelengths of the same extension as the cells may still affect the gridded sea surface heights which may result in problems when the surfaces of the different cells are combined. In order to avoid inconsistencies along the edges of neighbouring cells it was decided to use the gravity field to recover the details of the mean sea surface.

The determination of the marine geoid was carried out relative to the EGM96 geopotential model from the global set of altimetric gravity anomalies using FFT (Equation 2 without the filter). Three parallels at 0°, 40°, and 60° were used through this task. Then this marine geoid was removed from the averaged ERS-1 and TOPEX/POSEIDON repeat mission sea surface heights.

The differences between the two sets of mean sea surface heights and the marine geoid were assumed to contain errors in the marine geoid, the sea surface topography, and systematic errors in the two altimeter data sets. The mean values of the ERS-1 and the TOPEX/POSEIDON data were 0.82 m and 0.38 m respectively. To avoid problems that such systematic differences may cause the respective mean values were subtracted from the altimeter data.

The mean sea surfaces associated with the two periods that are covered by the two repeat missions are not supposed to coincide. Long term sea level changes such as seasonal and interseasonal ocean variability may cause such differences. It was decided to fit the mean sea surface to the TOPEX/POSEIDON altimetry, since these data have the lowest errors and span the longest period of time. Hence, the ERS-1 mean sea surface heights were fitted to the TOPEX/POSEIDON mean sea surface heights. This was carried out in two steps: First a spherical harmonic expansion to degree and order 20 was fitted to the TOPEX/POSEIDON data. Then a second spherical harmonic expansion to degree and order 20 was fitted to the differences between the ERS-1 data and the first harmonic expansion located between latitudes -66° and 66° and zero at lower and higher latitudes respectively. Subsequently, the sea surface of the second harmonic expansion was removed from the ERS-1 mean sea surface heights.

Then a spherical harmonic expansion to degree and order 20 was fitted to the merged set of mean sea surface heights in order to obtain a determination of the long wavelength parts of the differences between the mean sea surface and the marine geoid. After this surface was removed from the mean sea surface heights the residuals are assumed to contain wavelengths between 2000 km which roughly corresponds to harmonic degree 20, and 100 km. Those residuals were recovered using the local collocation technique.

Finally, the mean sea surface was obtained by adding the marine geoid, the long wavelength harmonic expansion, and the medium wavelength residuals. Also the 0.38 m difference between the TOPEX/POSEIDON altimetry and the marine geoid was included to obtain consistency with the TOPEX/POSEIDON reference frame.

Results

A description and the results of an evaluation of the global marine gravity field can be found in Andersen and Knudsen (1997). In this section the determination of the mean sea surface is evaluated.

For each set of mean sea surface heights through the process Root-Mean-Square (RMS) values were computed. For subsets of seven consecutive data along track, the fourth height, the mean of the seven heights, and their difference were evaluated. Also the slopes were analysed in a similar manner. The mean taken over seven heights represents roughly a mean over 50 km along track. Hence, the RMS of this mean height reflects the magnitude of wavelengths longer than about 100-200 km and the difference. The RMS of the difference between the fourth height and the mean height reflects the magnitude of the shorter wavelength parts.

The results for the TOPEX/POSEIDON altimetry are shown in Table 1 and the results for the ERS-1 altimetry are shown in Table 2. Especially, the removal of the EGM96 geoid

	h	h mean	Δh	s	s mean	Δs
	m			mm/km		
data	30.825	30.805	0.777	100.3	72.0	50.8
-EGM96	0.776	0.773	0.082	8.2	7.7	4.2
-Geoid	0.711	0.710	0.026	3.6	2.5	1.6
-Harm	0.102	0.096	0.024	3.2	2.3	1.4
Residuals	0.051	0.043	0.019	2.5	1.8	1.0

Table 1. RMS values of heights and slopes of TOPEX/POSEIDON altimeter data, with the EGM96 removed and with the geoid and the harmonic expansion removed, and the residuals with respect to the final mean sea surface.

model reduces all RMS. The removal of the marine geoid has an effect on the height differences and the slopes. Hence, the marine geoid resolve the short wavelengths of the mean sea surface. The removal of the spherical harmonic expansion (to degree 20) reduces the heights as well as the mean heights.

The final mean sea surface fits the TOPEX/POSEIDON mean sea surface heights with 5 cm. The heights relative to the mean heights are fitted by 2 cm and the slopes fit within a few mm per km. For the ERS-1 mean sea surface heights the results are quite similar. However, the ERS-1 values are slightly higher which may be caused by slightly higher error in the data and in the orbit. An orbit error adjustment might improve this result.

	h	h mean	Δh	s	s mean	Δs
	m			mm/km		
data	30.501	30.459	1.102	149.7	102.6	72.0
-EGM96	0.702	0.695	0.071	9.2	7.5	3.4
-Geoid	0.646	0.645	0.031	4.1	2.7	2.0
fitted	0.640	0.639	0.031	4.1	2.7	2.0
-Harm	0.103	0.097	0.027	3.5	2.3	1.7
Residuals	0.066	0.060	0.022	2.8	1.8	1.4

Table 2. RMS values of heights and slopes of ERS-1 altimeter data, with the EGM96 and the geoid removed, fitted to TOPEX/POSEIDON, with the harmonic expansion removed, and the residuals with respect to the final mean sea surface.

Acknowledgment. This research is supported by the Danish Space Board. The GEOSAT data were kindly provided by NOAA. The ERS-1 data were obtained from ESA to which DUT/DEOS kindly provided the JGM-3 orbits. The repeat mission ERS-1 and TOPEX/POSEIDON data were obtained from the NASA Pathfinder programme.

References

Andersen, O. B., P. Knudsen & C.C. Tscherning, Investigation of methods for global gravity field recovery from dense ERS-1 geodetic mission altimetry. In: Global Gravity Field and its temporal variations. Ed.s: Rapp, Cazenave & Nerem, IAG Symposia no. 116, p. 218 - 226, Springer Verlag, 1996.

Andersen, O. B., and P. Knudsen, Global Marine Gravity Field from the ERS-1 and GEOSAT Geodetic Mission Altimetry, J. Geoph. Res., in press, 1997.

Balmino, G., B. Moynot, M. Sarrailh and N. Vales, Free air gravity anomalies over the oceans from Seasat and Geos3 altimeter data, EOS, 68, (2), 17-18, 1987

Forsberg R. and D. Solheim, Performance of FFT methods in local gravity field modelling. Proc. Chapman conference on progress in the determination of the earths gravity field, Edited by R. H. Rapp, Ft. Lauderdale, Florida, 100-103,1988.

Hwang, C., and B. Parsons, Gravity Anomalies Derived from Seasat, Geosat, ERS-1, and TOPEX/POSEIDON Altimetry and Ship Gravity: A Case Study over the Reykjanes Ridge, Geophys. J. Int., 122, 551-568, 1995.

Knudsen, P., Simultaneous Estimation of the Gravity Field and Sea Surface Topography From Satellite Altimeter Data by Least Squares Collocation. Geophys. J. Int.l, Vol. 104, No. 2, 307-317, 1991.

Knudsen, P., and O. B. Andersen, Improved Recovery of the Global Marine Gravity Field from the GEOSAT and the ERS-1 Geodetic Mission Altimetry, in J. Segawa et al. (Eds.), Gravity, Geoid and Marine Geodesy, IAG symposia, Springer-Verlag, Berlin, 429-436, 1997.

Knudsen, P., O. B. Andersen, and C.C. Tscherning, Altimetric Gravity Anomalies in the Norwegian-Greenland Sea - Preliminary Results from the ERS-1 35 days Repeat Mission. Geophys. Res. Lett., 19(17), 1795-1798, 1992.

Lemoine, F.G., D.E. Smith, R. Smith, L. Kunz, E.C. Pavlis, S.M. Klosko, D.S. Chinn, M.H. Torrence, R.G. Williamson, C.M. Cox, K.E. Rachlin, Y.M. Wang, S.C. Kenyon, R. Salman, R. Trimmer, R.H. Rapp, and R.S. Nerem, The development of the NASA GSFC and DMA joint geopotential model, in J. Segawa et al. (Eds.), Gravity, Geoid and Marine Geodesy, IAG symposia, Springer-Verlag, Berlin, 461-469, 1997.

Rummel, R., and R.H.N. Haagmans, Gravity Gradients fran Satellite Altimetry, Marine Geodesy, Marine Geodesy, 14, 1-12, 1990.

Sandwell, D. T., Antarctic marine gravity field from high-density satellite altimetry, Geophys. J. Int., 109, 437-448, 1992.

Schwarz, K.-P., M. G. Sideris, R. Forsberg: Use of FFT methods in Physical Geodesy. Geophysical Journal International, Vol. 100, 485-514, 1990.

Tscherning, C.C., R. Forsberg, P. Knudsen, The GRAVSOFT package for geoid determination, Proc. First workshop on the European Geoid, Prague, May, 1992.

Sea Level Variations Measured by TOPEX/POSEIDON and by Tide Gauges: A Validation Study in the Southern Baltic Sea

G. Liebsch and R. Dietrich
Technische Universität Dresden
Institut für Planetare Geodäsie
01062 Dresden

Abstract

To compare sea level heights resulting from satellite altimeter observations and from tide gauge measurements it is necessary to know the tides, the sea surface topography and the geoid in the area under investigation. The paper discusses the characteristics of these components for a test area in the southern Baltic Sea. It is shown that the Baltic Sea is a proper area for such investigations. The method used for the comparison of the sea level heights as well as the results for a possible altimeter drift are presented.

1 Introduction

Recent satellite altimetry missions, especially the TOPEX/POSEIDON mission, have reached an accuracy which in principle enables the determination of global long-term sea level changes (Nerem, 1995). For the correct interpretation of these changes a careful evaluation of the drift stability of the altimeter oscillator is necessary. One possibility to control this sensor is the comparison of the altimetric sea level heights with in-situ tide gauge measurements. Examples for such comparisons are reported (Christensen et al., 1994), (Menard et al., 1994), (White et al., 1994), (Morris and Gill, 1994) and (Mitchum, 1997). In this investigation we present a specific and rather direct approach in order to compare sea surface heights from tide gauges with those from satellite altimetry. The test area (see figure 1) is located in the southern Baltic Sea. This region with an area of about 50 km by 50 km is covered by 8 permanent tide gauges on the one hand, on the other hand it is crossed by the ascending TOPEX/POSEIDON altimeter track No. 111.

If both types of observations (tide gauges and altimeter) are not available at exactly the same location, one has to interpolate considering relative changes of

- tides,
- geoid and
- sea surface topography

Models or data needed for this purpose may be very simple, or the effect can be neglected, if there are some special characteristics of the test area. In our case this is valid to a great extend.

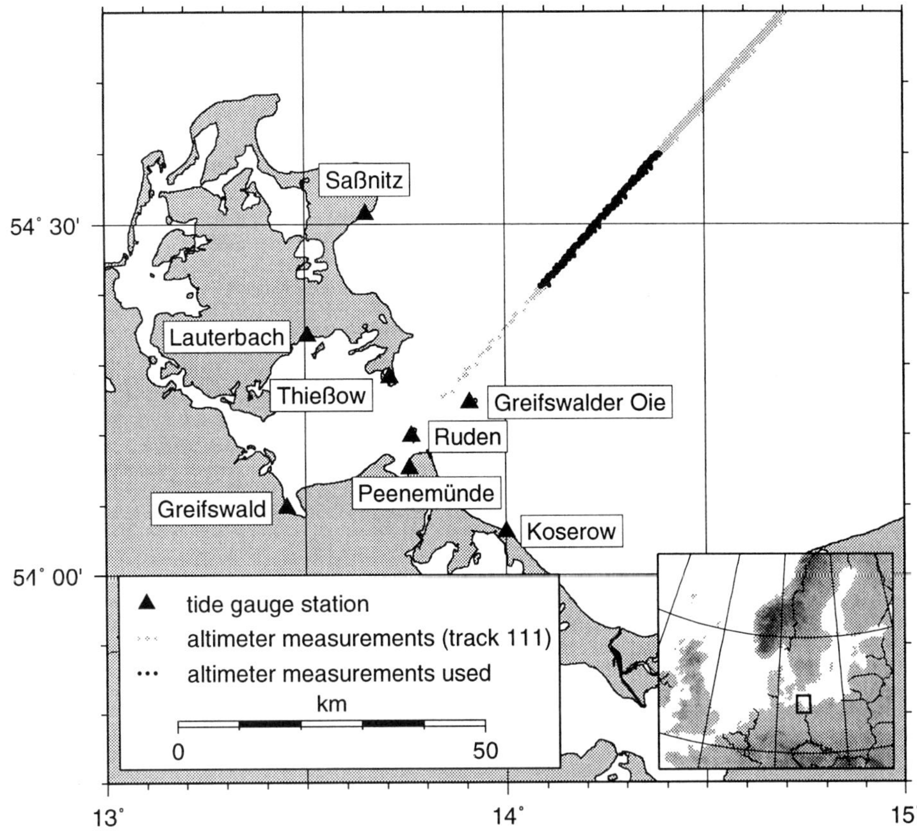

Figure 1: Test area

2 Data Description

The analysis carried out covers a 4-years time span (1992 – 1996). The available tide gauge data are hourly readings of 8 stations shown in figure 1. The zero points of all tide gauges are related to a common height system (called HN76), realized by spirit levellings. That means, all tide gauge readings are related to one and the same equipotential surface.

Looking at the tide gauge data (figure 2) one can state:

- A small sea level variability (±25cm, extrems within a 2.5m band).
- Very small tidal signals in that part of the Baltic Sea.
- A high coherence between the signals at different tide gauges (correlation coefficient above 0.92), see figure 2, bottom part.

Concerning the altimeter data in this investigation we used the Corrected Sea Surface Heights (CORSSH) altimeter product (AVISO, 1996). This product provides directly the altimetric sea surface heights above the reference ellipsoid, where all geophysical and media corrections are already applied. For comparisons with our in-situ tide gauge data we had to remove the correction for the inverse barometer effect, which was already applied to the altimetric sea surface heights. In the analysis data from cycles 2 to 149 were used (figure 1).

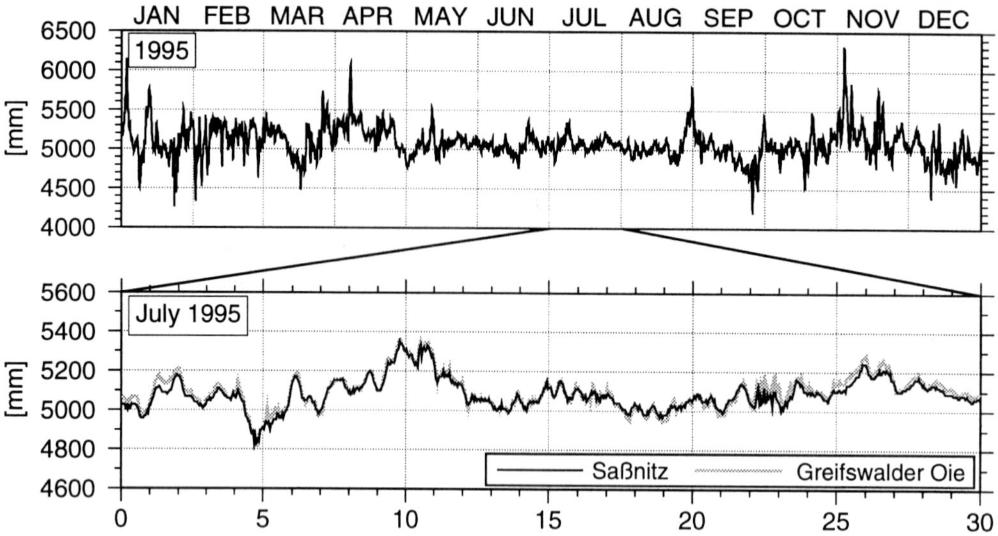

Figure 2: Tide gauge measurements

top — Hourly sea level measurements of the tide gauge Saßnitz for one year of observations

bottom — Comparison of hourly sea level measurements of the tide gauges Saßnitz and Greifswalder Oie for July 1995

3 Computation of the instantaneous sea surface topography at altimeter footprint from tide gauge data

Due to the high correlation between the tide gauge records in the area it is possible to approximate the instantaneous sea surface by a linear function of the form

$$z = a + bx + cy \tag{1}$$

with x, y ... coordinates in east and north direction
z ... estimated sea level at the location x,y
a, b, c ... parameters of the surface (sea level height at the origin of the coordinate system; linear trend of the sea surface in east and north direction)

Because

- the tide gauges are well distributed in space and
- the tide gauge measurements can be considered as heights z_i with respect to a common equipotential surface

it is possible to estimate the parameters a, b, c from the tide gauge measurements for each altimeter cycle by a least squares adjustment. After fitting of (1) the standard deviations of the residuals of the tide gauge measurements are typically in the order of 2...3cm.

The parameters a, b, c determined in the described way for each altimetric observation epoch were used to compute the instantaneous sea surface height at the altimeter foot prints. The altimeter observations were limited to an area between the latitudes 54.4^o and 54.6^o. The reasons for this limitation were

- the small number of observations close to the shore line and

- the suitability of the linear model, which was considered to be valid up to a distance of about 50km from the shore line, corresponding to the dimension of the tide gauge network.

This method was first applied by (Weise, 1985) in the same area to link the islands to the height system HN76.

4 Comparison

The comparison of the altimetric sea surface heights with the sea surface heights predicted from the tide gauges was carried out in several steps. The result of every step is shown in figure 3. The diagram on the top (a) shows the pure altimetric sea surface heights. Every grey dot symbolizes one single measurement. The geoid undulation is the most dominant term in the altimetric sea surface heights. It amounts to about 35m and varies by about 0.5m along the altimeter track. For comparison the geoid undulation along the altimeter track using the model EGM96 (Lemoine et.al., 1996) is presented in the figure (black dots).

In figure 3b this geoid undulation is removed. The remaining variability is in the order of 1m. Eliminating the predicted sea surface heights from the tide gauge measurements by formula (1) this variability can clearly be decreased to a level of about 20...40cm (figure 3c). Smaller values are reached in that part of the altimeter track, which is closer to the tide gauges (smaller latitudes). This effect is caused by the decreasing prediction accuracy of formula (1) with increasing distance from the tide gauges. Altogether one can conclude that the predicted sea surface heights and the altimetric sea surface heights are very well correlated. Therefore, the instantaneous ocean signal can largely be canceled in the differences.

The remaining differences have a offset of about 0.7m, which varies by about 0.1m (figure 3c). This offset may be caused by

- errors in the geoid model,
- an offset between the zero surface of the geoid model and the height system used to determine the zero point of the tide gauges,
- a systematic deviation of the stationary sea surface topography from the simple lineare model and
- a possible constant altimeter bias.

One can assume, that the differences are dominated by the first two factors.

In the present study we decided to estimate all mentioned effects by an empirical function. For this purpose we used a second order polynom. The coefficients were determined from the differences by least squares as a function of the latitude φ. The resulting polynom is shown in figure 3c in black colour. After subtracting the polynom the remaining residuals have a standard deviation of about 7cm. They are displayed in figure 3d.

Note that up to now all operations were performed in the space domain. Therefore, a possible long-term variation of the altimeter bias is not effected by these operations. Therefore we can rearrange all residuals as a function of time. These residuals in the time domain can be used in order to determine this kind of bias. We estimated a linear drift (black colour), which is 0.7mm/year with a standard deviation of 2.6mm/year (see figure 3e).

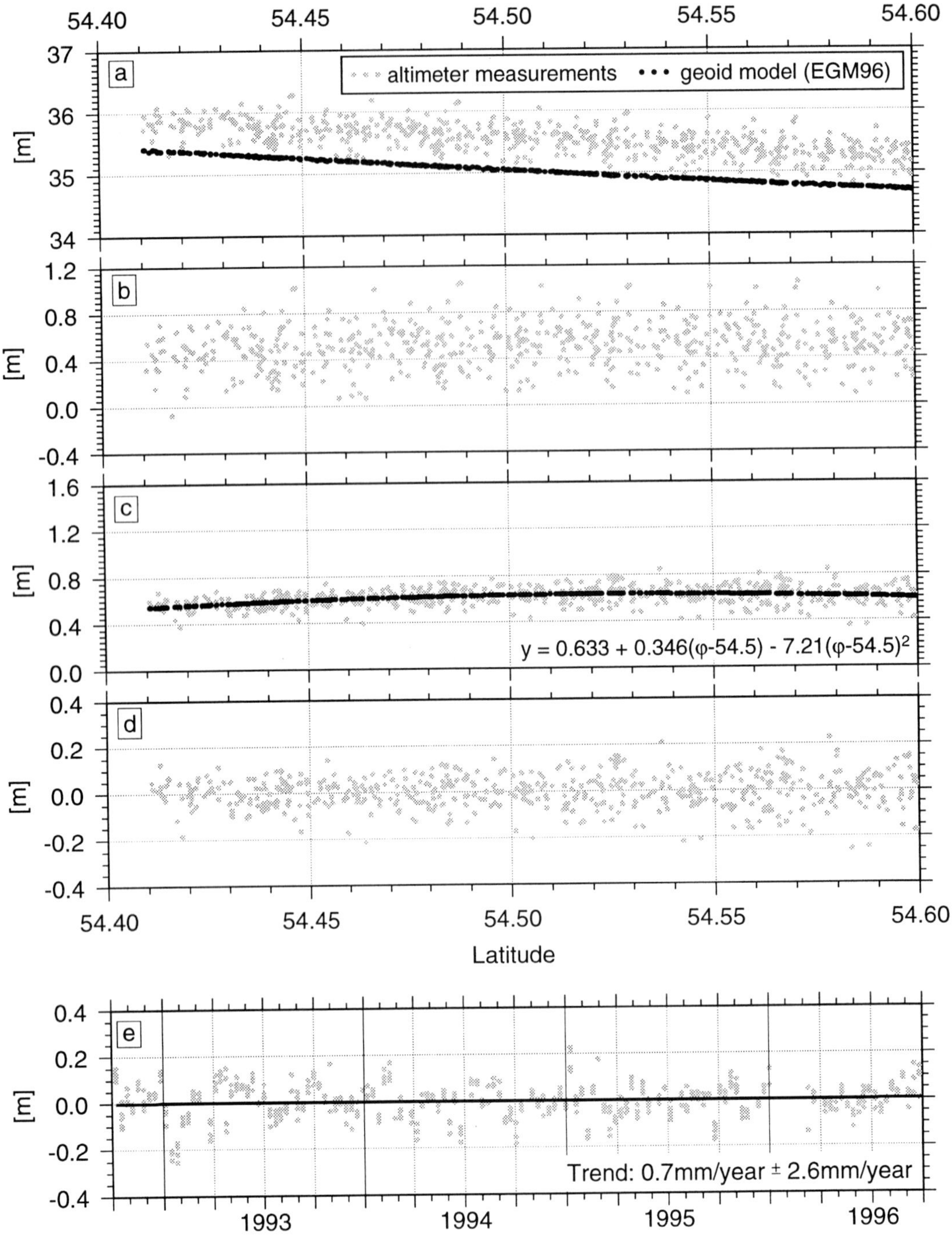

Figure 3: Comparison of altimetric sea surface heights with the sea surface heights predicted from the tide gauges (The abscissa of diagrams a – d represents the latitude of the altimeter measurements. In the last diagram the abscissa is subdivided into days with respect to the first TOPEX/POSEIDON cycle.)
a) Altimetric sea level heights above ellipsoid
b) Geoid undulation removed (model EGM96)
c) Instantaneous sea surface topography removed
d) Residuals after polynomial fit of 2nd order
e) Residuals after polynomial fit of 2nd order rearranged as a function of time

5 Conclusions and Outlook

We have presented an approach to compare altimetric sea surface heights with local tide gauge data. The resulting random noise level is about ±7cm per altimeter point. This value represents the accuracy for the comparison of one single altimetric measurement with the corresponding tide gauge measurement. It corresponds approximately to the expected accuracy taking into account the error budget of the altimetric sea surface heights in the order of 4.7cm (Fu et al., 1994) on the one hand, and of the tide gauge measurements including sea surface extrapolation in the order of about 3cm on the other hand. Further improvements may be reached using more tide gauges (better geometry of the model).

A possible long term drift of the altimeter bias can be estimated with a standard deviation of about 3mm/year for a 4 year data interval. An absolute calibration will be possible if GPS observations, related to the zero equipotential surface, are available. GPS observations at the tide gauges are also necessary for the separation of possible vertical crustal movements from tide gauge records. Furthermore, we conclude, that altimeter calibration results from different groups can be compared in a better way if there is a closer cooperation, e.g. agreements on standards in altimeter data handling and error estimation.

References

AVISO (1996). *AVISO Handbook for Merged TOPEX/POSEIDON Products*. AVI-NT-02-101-CN, 3 edition.

Christensen, E., Haines, B., Keihm, S., Morris, C., Norman, R., Purcell, G., Williams, B., Wilson, B., Born, G., Parke, M., Gill, S., Shum, C., Tapley, B., Kolenkiewicz, R., and Nerem, R. (1994). Calibration of TOPEX/POSEIDON at platform harvest. *JGR*, 99(C12):24465 – 24485.

Fu, L., Christensen, E., Yamarone, C., Lefebvre, M., Menard, Y., Dorrer, M., and Escudier, P. (1994). TOPEX/POSEIDON mission overview. *JGR*, 99(C12):24369 – 24381.

Lemoine, F., Smith, D., Kunz, L., Smith, R., Pavlis, E., Pavlis, N., Klosko, S., Chinn, D., Torrence, M., Wiliamson, R., Cox, C., Rachlin, K., Wang, Y., Kenyon, S., Salman, R., Trimmer, R., Rapp, R., and Nerem, R. (1996). The development of the NASA GSFC and NIMA joint geopotential model. In *Proc. Of the Int. Symp. On Gravity, Geoid and Marine Geodesy (GRAGEOMAR 1996), Tokyo, Sep. 30 – Oct. 5. 1996.*

Menard, Y., Jeansou, E., and Vincent, P. (1994). Calibration of the TOPEX/POSEIDON altimeters at Lampedusa: Additional results at Harvest. *JGR*, 99(C12):24487 – 24504.

Mitchum, G. (1997). Altimeter drift from tide gauges. *AVISO newsletter*, (5):8 – 11.

Morris, C. and Gill, S. (1994). Evaluation of the TOPEX/POSEIDON altimeter system over the Great Lakes. *JGR*, 99(C12):24527 – 24539.

Nerem, R. (1995). Measuring global mean sea level variations using topex/poseidon altimeter data. *JGR*, 100(C12):25135–25151.

Weise, H. (1985). Hydrokinematisches Nivellement. *Vermessungstechnik*, 33:406–409.

White, N., Coleman, R., Curch, J., Morgan, P., and Walker, S. (1994). A southern hemisphere verification for the TOPEX/POSEIDON satellite altimeter mission. *JGR*, 99(C12):24505 – 24516.

OBSERVED INCONSISTENCIES BETWEEN SATELLITE-ONLY AND SURFACE GRAVITY-ONLY GEOPOTENTIAL MODELS

Nikolaos K. Pavlis
Hughes STX Corporation
7701 Greenbelt Road, Suite 400
Greenbelt, MD 20770, USA
e-mail: npavlis@geodesy2.gsfc.nasa.gov

Abstract

The long wavelength inconsistencies observed between satellite-only and terrestrial-only gravitational solutions were re-examined, in view of the recent release by NIMA of an updated 1° mean gravity anomaly file (which was used in the development of the EGM96 geopotential model). The differences between the satellite-only model EGM96S and corresponding solutions obtained from 1°x1° terrestrial mean Δg were examined both spectrally and geographically. Up to $N_{\max} = 20$, the global RMS geoid undulation difference (δN) between these models was ±3.7 m, for the NIMA 1° Δg data. This is an improvement over the ±4.4 m, obtained when the older OSU 1° Δg data were used. In some geographic regions however, the NIMA Δg data produce larger δN values with EGM96S, than the corresponding OSU anomalies. When the marine Δg data were replaced by altimetric values, the RMS δN dropped to ±1.6 m, indicating that more than 50% of the observed differences at long wavelengths is due to the poor quality of the available marine Δg. The spectrum of δN (EGM96S minus a terrestrial-only solution) exceeds by more than an order of magnitude the undulation spectra predicted by some postulated models of vertical datum inconsistencies. The spectrum of undulation effects implied by the approximation $H^* \approx H$, is quite similar to that predicted by one of the vertical datum inconsistency models postulated by *Laskowski* [1983].

Motivation

Systematic errors associated with terrestrial gravity anomaly data have been identified in several investigations (e.g., *Kahn et al.* [1982], *Rapp and Cruz* [1986], *Pavlis* [1988]). Such errors, and the lack of accurate gravity data over extended areas of the Earth (e.g., Antarctica, south Pacific ocean), are considered to be responsible for the low accuracy of the long wavelength gravitational information, that can be extracted from terrestrial data. Appropriate treatment of these errors is particularly important when the terrestrial gravity data are combined with satellite tracking and altimeter data, in the development of

International Association of Geodesy Symposia, Vol. 119
Forsberg et al. (eds.), Geodesy on the Move

combination geopotential models. Several effects are suspected to be responsible for these systematic errors, but the specific origin of such errors in existing global data bases is difficult to assess. *Heck* [1990] reviewed the entire measurement and computation process, necessary to define a Δg value, and quantified the effects of some systematic errors affecting the Δg data. He concluded that there is little chance for the analyst to come through the "jungle of systematic effects present in the available gravity anomaly data", and recommended setting up a new global anomaly data bank, based on new measurements and consistent, precise evaluation procedures. This may be the ideal solution to the problem; however, the cost and time required to accomplish such a task, make it rather unlikely to happen in the foreseeable future.

Provided that the systematic errors in the existing terrestrial gravity data affect mostly low degree coefficients, a more feasible approach may be to develop techniques that would separate these systematics from the gravitational signal present in Δg, in least squares adjustment procedures where terrestrial and satellite data are combined. In such a scenario, one aims to filter out of the Δg data systematic errors, in some parametric form (e.g., regional biases or low degree polynomials, surface spherical harmonic coefficients), without necessarily attempting to characterize the specific cause(s) of the errors. This approach becomes rather appealing in view of upcoming geopotential mapping missions like CHAMP, GRACE and GOCE. If such an approach proves successful, it may alleviate the need to downweight the Δg data in combination solutions (more or less arbitrarily), and to exclude their contribution from the very low degree ($n \leq 5$) portion of combined models, as it was done in both the EGM96 [*Lemoine et al.*, 1997] and the GRIM4-C4 [*Schwintzer et al.*, 1997] models. A reasonable starting point towards the development and testing of such bias recovery techniques, is to quantify and examine the current status of these inconsistencies, in view of the recent releases of updated terrestrial 1° data, and the availability of improved satellite-only models, such as EGM96S.

Numerical Experiments and Results

Assessment of long wavelength systematic errors in Δg data may be made by comparing a gravitational solution obtained on the basis of the anomaly data alone, to a satellite-only model, such as EGM96S which was used here. Three global 1° mean anomaly files were formed and used to develop three corresponding "terrestrial" solutions (to degree 70), which were then compared to the EGM96S model. Table 1 lists the number of anomalies by data type for each of the three files. The high frequency (n>70) anomaly contribution to the data of all 3 files was removed, prior to the "terrestrial" solution development.

File A: Terrestrial 1° Δg from the OSU October 1990 data base [*Yi and Rapp*, 1991]. Surface gravity normal equations ($N_{\max} = 70$) formed and used in JGM-1, 2 and 3, were developed based on this file.

File B: NIMA (Sept. 1996) terrestrial 1° Δg file. This is the file used to form the ($N_{\max} = 70$) normal equations used in the development of EGM96.

File C: Same as file B, but employing altimetry-derived 1° mean Δg over (most of) the ocean. Altimetric values here are 1° averages of the 30' values computed at NIMA [*Trimmer and Manning*, 1996], and of 30' values provided by KMS and by T. Schoene.

Table 1. 1°x1° mean gravity anomaly (Δg) counts by data type.

Type of Δg	File A	File B	File C
Land	13919	18596	18319
Marine	32013	31039	1842
Airborne	0	2636	2108
Altimetric	0	0	37850
Fill-in	18868	12529	4681

Because the available 1° terrestrial anomaly data do not cover completely the Earth, one needs to fill-in the empty areas, in order to obtain a terrestrial-only solution, without severe spectral distortions [*Pavlis*, 1988, sect. 5.3.4]. The fill-in values were computed here based on the EGM96S coefficients ($n \leq 40$), augmented by topographic/isostatic potential coefficients ($40 < n \leq 70$). Two points have to be kept in mind when one examines the differences between EGM96S and the "terrestrial" solutions: (a) due to the use of the EGM96S coefficients in the computation of fill-in values, the results are optimistic (especially over areas occupied by fill-in values), and, (b) comparisons beyond $n \approx 20$ do not necessarily indicate problems with the Δg data, since the accuracy of the satellite-only model degrades at higher degrees, due to the attenuation of the gravitational signal with altitude.

The Δg differences of files B minus C were computed over ocean areas, not occupied by fill-in values. Over 33481 1° cells occupying 60.1% of the Earth's area, the mean and stand. deviation $d\Delta g$ were 0.2 and ±9.1 mGal, respectively, with $\min(d\Delta g) = -201$ mGal and $\max(d\Delta g) = 125$ mGal. Careful re-examination of the anomaly files is needed to identify the causes of the large discrepancies and correct possible erroneous values.

Results from comparisons in terms of 1° mean Δg and N values from EGM96S and the 3 terrestrial solutions are summarized in Table 2, while Fig. 1 displays geographically the δN (EGM96S minus solution B), to degree 20. Similar figures (not shown here) were made for the δN with solutions A and C. Comparative examination of these figures, and the results of Table 2, indicate that although the NIMA 1996 terrestrial 1° data resulted in a definite overall improvement with respect to the OSU 1990 data, little progress was made with regard to the marine anomalies. Also, in certain areas, such as the north Atlantic ocean, the NIMA data produced larger N differences with EGM96S, than the OSU data. A large tilt in the δN values over the Pacific ocean (from NW to SE) was also amplified with the use of the NIMA data. The spectra of undulation differences between EGM96S and the three terrestrial solutions are shown in Fig. 2.

Table 2. Gravity anomaly and geoid undulation differences in terms of 1° mean values. EGM96S minus three terrestrial solutions (A, B, C). $N_{max} = 20$.

Solution	Δg Difference (mGal)			N Difference (m)		
Difference	Min.	Max.	S. Dev.	Min.	Max.	S. Dev.
EGM96S - A	-17.2	16.0	3.3	-16.7	9.0	4.4
EGM96S - B	-11.3	11.0	2.2	-13.0	11.0	3.7
EGM96S - C	-11.4	8.0	1.5	-6.7	6.2	1.6

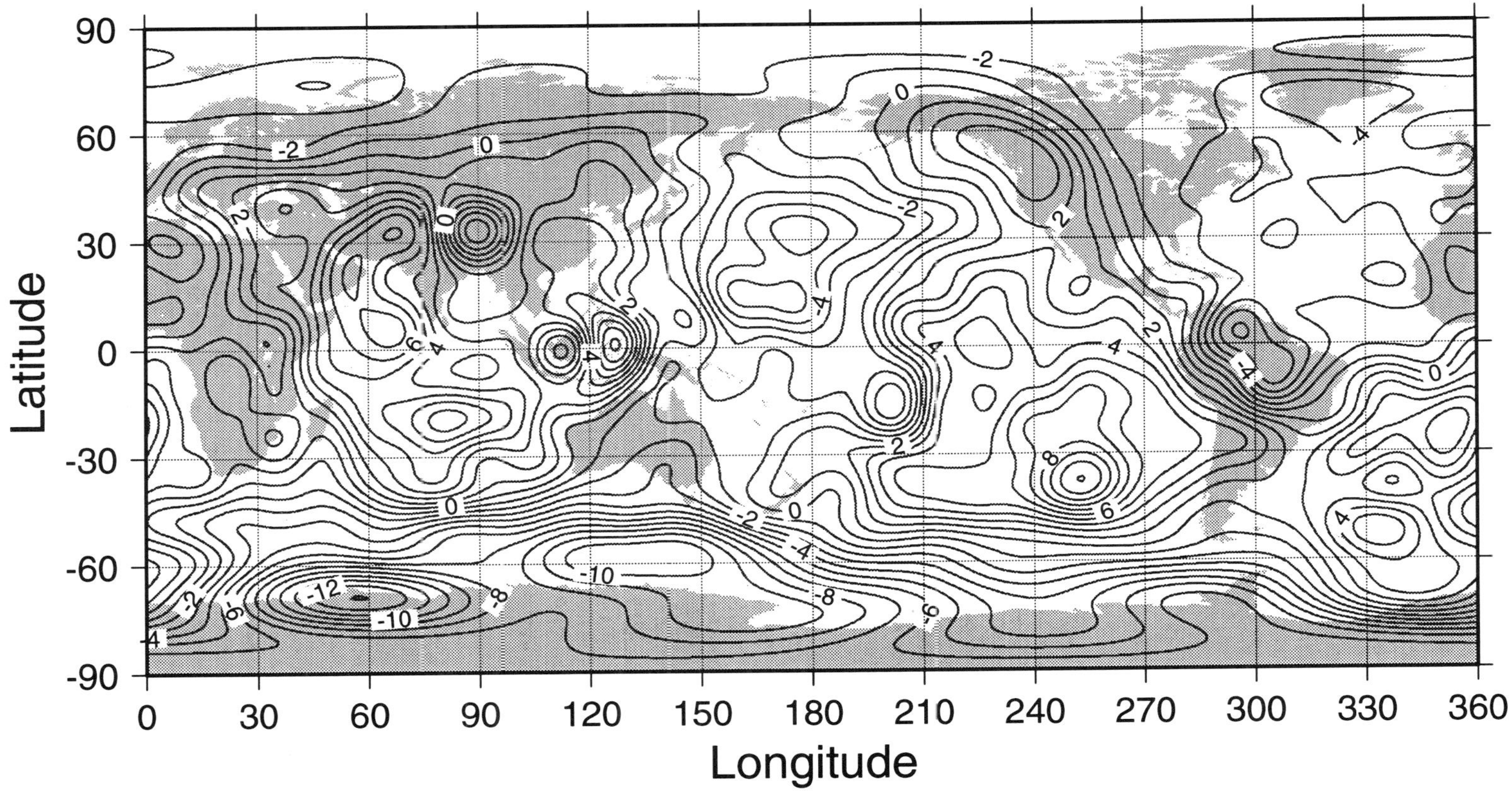

Figure 1. Geoid undulation differences: EGM96S minus Surface Gravity Solution B. $N_{max} = 20$.
Contour interval is 1 m.

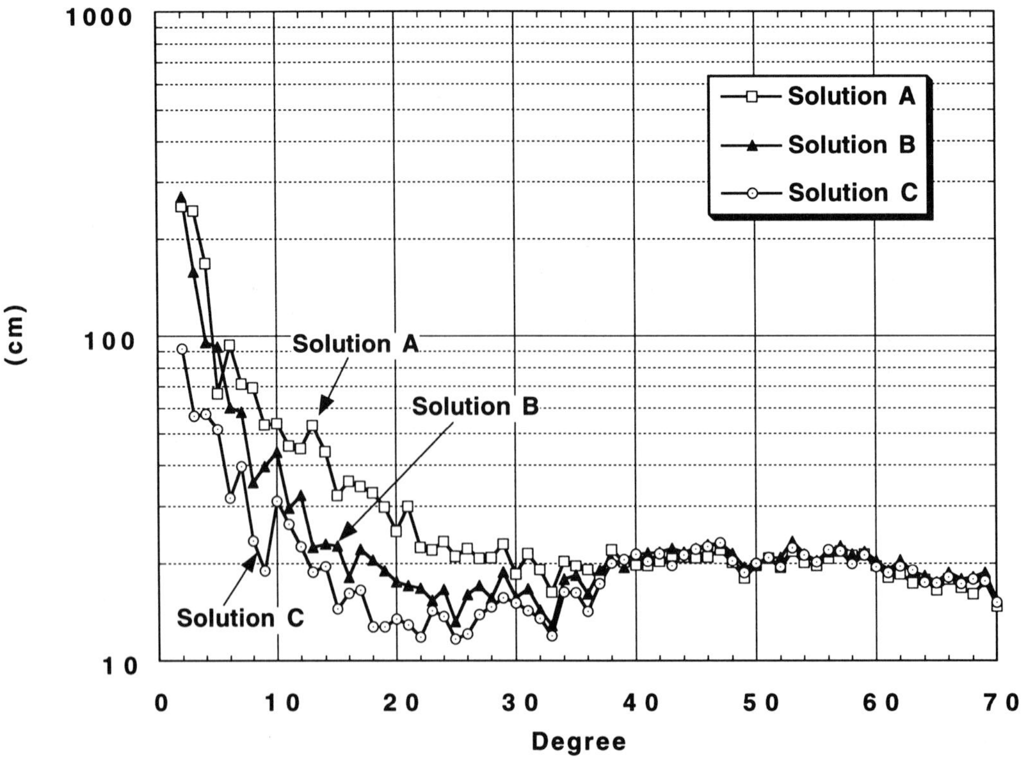

Figure 2. Global RMS per degree geoid undulation difference. EGM96S minus three terrestrial gravity solutions (see text).

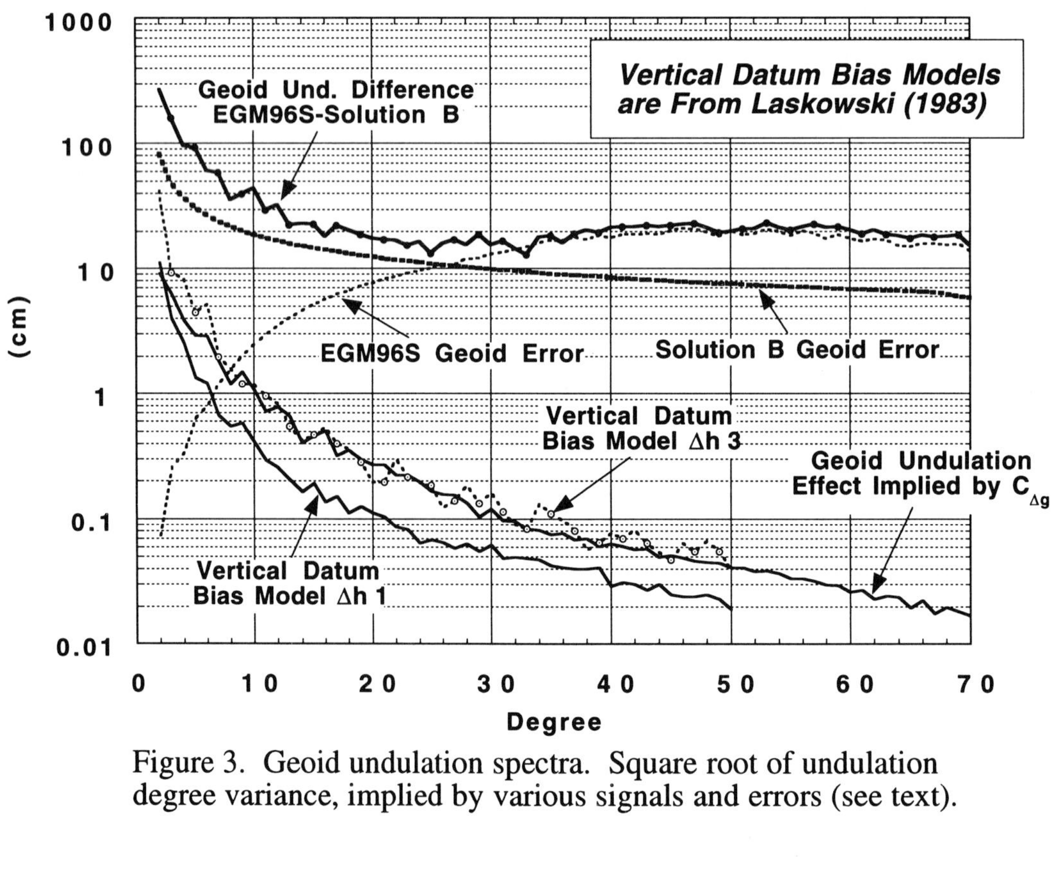

Figure 3. Geoid undulation spectra. Square root of undulation degree variance, implied by various signals and errors (see text).

The effect of approximating normal by orthometric heights, when defining a Molodensky free-air anomaly was also examined. Assuming that heights associated with gravity records are indeed orthometric, existing gravity anomalies may be corrected (approximately) for this effect by adding to them the term:

$$C(\Delta g) = -\left(\left(\partial\gamma/\partial h\right) + \left(\partial^2\gamma/\partial^2 h\right)\cdot H\right)\cdot(H^* - H) \qquad (1)$$

The $(H^* - H)$ term may be computed from Bouguer anomaly and elevation estimates. Using expansions to degree 360, $C(\Delta g)$ was found to have an RMS of ±70 μGal, and its implied undulation effect an RMS of ±10 cm, with an extreme value of -86 cm. In Fig. 3 the undulation spectrum of this effect is shown, along with the spectra of undulation effects predicted by two models ($\Delta h1$ and $\Delta h3$) of vertical datum inconsistencies, postulated by *Laskowski* [1983]. The effect of $C(\Delta g)$ has very similar spectrum as the bias model $\Delta h3$ (the two effects though are not necessarily correlated geographically). The undulation effects predicted by $\Delta h1$ or $\Delta h3$ are much smaller than the observed δN, implying that other errors affect existing Δg, probably more than vertical datum biases.

References

Heck, B., An evaluation of some systematic error sources affecting terrestrial gravity anomalies, *Bull. Geod.*, 64, 88-108, 1990.

Kahn, W.D., S.M. Klosko and W.T. Wells, Mean gravity anomalies from a combination of Apollo/ATS6 and GEOS3/ATS6 SST tracking campaigns, *J. Geophys. Res., 87*, B4, 2904-2981, 1982.

Laskowski, P., The effect of vertical datum inconsistencies on the determination of gravity related quantities, *Rep. 349*, Dep. of Geod. Sci. and Surv., Ohio State Univ., Columbus, 1983.

Lemoine, F.G., et al., The development of the NASA GSFC and NIMA joint geopotential model, in: Gravity, Geoid and Marine Geodesy, J. Segawa, H. Fujimoto and S. Okubo (eds.), IAG Symposia, Vol. 117, Springer-Verlag, Berlin, Heidelberg, 1997.

Pavlis, N.K., Modeling and estimation of a low degree geopotential model from terrestrial gravity data, *Rep. 386*, Dep. of Geod. Sci. and Surv., Ohio State Univ., Columbus, 1988.

Rapp, R.H. and J.Y. Cruz, The representation of the Earth's gravitational potential in a spherical harmonic expansion to degree 250, *Rep. 372*, Dep. of Geod. Sci. and Surv., Ohio State Univ., Columbus, 1986.

Schwintzer, P., et al., Long-wavelength global gravity field models: GRIM4-S4, GRIM4-C4, *J. Geod.*, *71*, 189-208, 1997.

Trimmer, R. and D.M. Manning, The altimetry derived gravity anomalies to be used in computing the joint DMA/NASA Earth Gravity Model, in: Global Gravity Field and Its Temporal Variations, R.H. Rapp, R.S. Nerem and A.A. Cazenave (eds.), IAG Symposia, Vol. 116, Springer-Verlag, Berlin, Heidelberg, 1996.

Yi, Y. and R.H. Rapp, The October 1990 1°x1° mean anomaly file including an analysis of gravity information from China, *Internal Rep.,* Dept. of Geod. Sci. and Surv., Ohio State Univ., Columbus, 1991.

A COMPARISON OF SATELLITE ALTIMETRY METHODS FOR THE RECOVERY OF GRAVITY FIELD QUANTITIES

I.N. Tziavos[1], R. Forsberg[2], M.G. Sideris[3], V.D. Andritsanos[1]

1 Department of Geodesy and Surveying, Aristotle University of Thessaloniki, 540 06 Thessaloniki, Greece

2 National Survey and Cadastre (KMS), Rentemestervej 8, DK-2400 Copenhagen NV Denmark

3 Department of Geomatics Engineering, The University of Calgary, 2500 Univ. Drive N.W., Calgary, Alberta T2N 1N4, Canada

Abstract

The recovery of quantities related to the gravity field (e.g., geoid heights, gravity anomalies, deflections of the vertical) is realized in the off Newfoundland sea area on the East coast of Canada using reliable marine gravity data and altimeter data from ERS-1 Geodetic Mission (GM). The methods tested include three basic methods, i.e., the Input-Output System Theory (IOST) and the Least Squares Adjustment in the Frequency Domain (LSAFD) methods used to compute geoid heights and gravity anomalies and the Inverse Stokes (IS) method used to compute deflections of the vertical. Evaluations are done in the spectral domain utilizing planar and spherical Fast Fourier Transform (FFT) algorithms. The input to these methods consists of (almost) directly observable data such as altimetric sea surface heights (geoid heights) and shipborne gravity anomalies, and derived data such as marine geoid heights, gravity anomalies and deflections of the vertical. The estimated results are intercompared in order to assess the agreement of the predicted quantities derived by different methods and with different input data and data errors. The geoid heights are additionally compared to TOPEX altimeter data.

Introduction

The present study is primarily aiming at the computation of geoid heights and gravity anomalies in the off Newfoundland sea area by combining adjusted altimetric sea surface heights (SSHs) from ERS-1-GM with real marine gravity anomaly observations by the reliable and efficient spectral IOST and LSAFD methods, employing non-isotropic information. The non-isotropic information corresponds to the computation of the Power Spectral Density functions (PSDs) directly from the data. Moreover, geoid heights are predicted in the same test area using only marine gravity data and the 1-D Fast Fourier Transform (1DFFT) procedure. From the comparisons between the gravimetric geoid heights and the geoid heights from the combined solution with TOPEX SSHs is shown how significant is the use of altimetry for the accurate gravity field modeling in sea areas.

In a last numerical test deflections of the vertical are computed (a) from marine gravity data using the Vening-Meinesz formula and employing the planar FFT technique and (b) from ERS-1-GM SSHs using the IS method in the frequency domain. The results derived are compared in order to assess how close results could be obtained by these two different methodologies.

Mathematical background

Only some basic information and formulas will be given in this chapter concerning the mathematical models used in the numerical tests of this study, since all these models have been discussed extensively in the literature of physical geodesy during the last two decades.

Geoid and Gravity by Input-Output System Theory (IOST). The two input, single output IOST general formula is written in the frequency domain as

$$\hat{Y} = B_1 Z + B_2 \Delta G \quad , \tag{1}$$

where $\hat{Y}$ is the spectrum of the output quantity (geoid or gravity) and Z and ΔG are the PSDs of the input SSHs (ζ) and gravity anomalies (Δg). The factors B_1 and B_2 are computed from the spectra of observations and the PSDs of the input signals and noises (see, e.g., Tziavos et al., 1996; Tziavos et al., 1997; Sideris, 1996; Li and Sideris, 1997; Li, 1996). It is worth mentioning here that for the noise PSDs a simulation technique was used. In order to simulate the noises of observations, Gaussian noises were generated and subtracted from the observations to estimate signals. Moreover, the PSDs of the signals and noises were computed directly from the data and not from the data covariance functions, and for this reason present non-isotropic characteristics. As it has been shown in previous studies (Li, 1996; Tziavos et al., 1996; Tziavos et al., 1997) the non-isotropic information is superior to the isotropic one, where the PSDs are computed from the modeled auto- and cross-covariance functions of the different observables (gravity, SSHs), .

Geoid and Gravity by Least Squares Adjustment in the Frequency Domain (LSAFD). The observation equation of this method is written in the frequency domain as

$$\begin{Bmatrix} Z \\ \Delta G \end{Bmatrix} = \begin{Bmatrix} B_1^{-1} \\ B_2^{-2} \end{Bmatrix} \hat{Y} + \begin{Bmatrix} N_\zeta \\ N_{\Delta g} \end{Bmatrix} \quad , \tag{2}$$

where again $\hat{Y}$ is the spectrum of the predicted quantity (geoid or gravity) and the quantities by capital letters represent the PSDs of signals and noises. Everything mentioned before for the noisy simulation and the computation of PSDs in the IOST method is also valid for the present method. Additional details can be found in Barzaghi et al. (1993).

Gravimetric geoid determination by the 1-D Fast Fourier Transform (1DFFT). The evaluation of the true discrete Stokes integral without approximation, parallel by parallel is given by means of the 1DFFT formula (Haagmans et al., 1993)

$$\zeta(\phi_P,\lambda_P) = \frac{R\Delta\phi\Delta\lambda}{4\pi\gamma}\mathbf{F}^{-1}\left\{\sum_{\phi_Q=\phi_1}^{\phi_M} \mathbf{F}[S(\psi_{PQ})]\mathbf{F}[\Delta g(\phi_Q,\lambda_Q)\cos\phi_Q]\right\} \quad , \tag{3}$$

with ϕ_P fixed and $S(\psi_{PQ})$ representing the Stokes function. $\mathbf{F}$ and $\mathbf{F}^{-1}$ denote the direct and the inverse 1DFFT.

Deflections of the vertical. The gravimetric deflections of the vertical in planar approximation were computed by the following formula evaluated also by FFT

$$\begin{Bmatrix} \xi \\ \eta \end{Bmatrix} = -\frac{1}{\gamma}\mathbf{F}^{-1}\begin{Bmatrix} \Delta G(u,v) \quad jv/(u^2+v^2)^{1/2} \\ \Delta G(u,v) \quad ju/(u^2+v^2)^{1/2} \end{Bmatrix} \quad , \tag{4}$$

where γ is a mean value of normal gravity. The deflections of the vertical can be computed from SSHs using the following formula by means of FFT (Sideris and Tziavos, 1988)

$$\begin{Bmatrix} \xi \\ \eta \end{Bmatrix} = -2\pi \mathbf{F}^{-1} \left\{ \begin{pmatrix} jv \\ ju \end{pmatrix} \; Z(u,v) \right\} , \tag{5}$$

where Z(u,v) is the spectrum of the SSHs (ζ).

Data Processing

Marine gravity data. The original shipborne free-air gravity data covering the off Newfoundland sea test area in the East coast of Canada belong to the Gravity Data Centre of Canada (responsible person Marc Veronneau). A number of 84169 point free-air gravity anomalies was originally selected in the region. Noticing that a gap existing in the north-east part of the area was filled by 3'x3' gravity data extracted from the KMS worldwide data bank and another gap in the south-west part of the area was filled by gravity anomalies computed from the EGM96 geopotential model. From this gravity data set a 3'x6' (~ 5.56 km x 7.15 km) grid was produced in the area $45^o \le \phi \le 55^o$, $-55^o \le \lambda \le -45^o$ (200x100 grid, 20000 points). This was finally our test area, where all the numerical tests described below were carried out. The 20000 gridded free-air anomalies are referred to the GRS80 reference system and are related to the EGM96 geopotential solution. The statistics of gravity anomalies before and after the subtraction of the contribution of the geopotential model is given in Table 1. The atmospheric correction, equal to 0.87 mGal has been also applied to these gravity anomalies. The residual gravity anomalies are still biased (- 2.15 mGal).

Table 1. Statistics of the 20000 gridded free-air gravity anomalies. Unit: [mGal].

	mean	min.	max.	sd	rms
original Δg	12.61	- 53.47	132.40	± 24.33	27.40
Δg from EGM96	14.76	- 39.88	107.57	± 22.11	26.58
Δg reduced to EGM96	- 2.15	- 54.97	88.12	± 10.97	11.18

Altimeter data. The ERS-1 radar altimeter data of the GM were used in our test area. To the selected altimeter data (39335 subsatellite points from 366 tracks) a simple bias crossover adjustment was applied after the removal of a number of erroneous observations. The number of the detected 11742 crossovers had before adjustment a mean value equal to -0.10 m and a rms value equal to 0.21 m. After the bias adjustment these values reached the level of 0.00 m and 0.09 m, respectively. Additional numerical tests in a larger area by applying a 3-parameters (bias and two tilts) adjustment did not improve the results in terms of the mean and rms values. From the adjusted altimeter points geoid heights were computed on the same 3'x6' gravity grid (see previous section). For these gridded altimetric geoid heights the EGM96 geopotential model is used as a reference surface as well. The statistical results of the 20000 gridded altimeter heights are listed in Table 2.

We used also a number of 817 TOPEX altimeter heights in the inner part of our test area. These values were used only as control values in order to assess the accuracy of the predicted geoid heights.

Table 2. ERS-1-GM altimetry statistics before and after crossover adjustment (200x100 grid). Unit: (m).

	mean	min.	max.	sd	rms
observed SSHs	20.34	- 1.39	38.11	± 8.48	22.03
SSHs reduced to EGM96 (before)	- 0.87	- 2.80	0.55	± 0.29	0.92
SSHs reduced to EGM96 (after)	- 0.01	- 0.62	1.27	± 0.25	0.25

Results and discussion

Geoid height predictions. The first numerical test deals with the computation of residual geoid heights in our test area on the previously defined 3'x6' grid. Using as input data the produced gravity and altimetry 3'x6' grids geoid heights were computed by the spectral IOST and LSAFD methods (combined geoid solutions). Using only the gravity grid a pure gravimetric geoid solution was computed. In the geoid computations 100% zero-padding was appended around the data and the kernel functions in order to eliminate edge effects. The above mentioned geoid solutions are intercompared, and compared also with the ERS-1-GM altimetric heights (see Table 2) and with the 817 TOPEX heights as well. The statistical results are tabulated in Table 3. It is noticed that the input noise was 0.1m for the altimetry and 3 mGal for the gravity, with corresponding variances $0.01 m^2$ and $9 mGal^2$.

The choice of the altimetry noise was based on previous studies (Li, 1996; Tziavos et al., 1996) and the gravity noise on available information related to observation errors. The PSDs used in IOST and LSAFD are non-isotropic, since they represent more accurately the detailed signal than the isotropic PSDs, which are based on the assumption of an isotropical structure of the covariance functions. Moreover, in order to produce signals we simulated the noises by a Gaussian model. Further investigation is necessary in this direction by testing additional models (uniform distributed noises, white noises).

Table 3. Geoid height predictions and geoid height differences between different methods. Unit: [m]

GEOID HEIGHTS: (1) IOST, (2) LSAFD, (3)1DFFT, (4) ERS-1-GM, (5) TOPEX					
Input errors: Altimetry: 0.10 m, Gravity: 3 mGal					
	mean	min	max	sd	rms
(1)	0.00	- 0.61	1.23	± 0.25	0.25
(2)	- 0.01	- 0.59	1.27	± 0.23	0.23
(3)	- 1.07	- 2.26	0.45	± 0.34	1.12
(4)	- 0.01	- 0.62	1.27	± 0.25	0.25
(5)	- 0.63	- 1.16	0.10	± 0.22	0.67
(1) - (2)	0.00	- 0.62	0.56	± 0.08	0.08
(1) - (3)	1.07	0.02	2.41	± 0.39	1.14
(3) - (4)	- 1.07	- 0.05	- 2.45	± 0.39	1.14
(1) - (5)	- 0.58 (0.00)	- 0.72 (- 0.12)	- 0.46 (0.14)	± 0.04 (± 0.04)	0.58 (0.04)
(3) - (5)	0.55 (0.00)	- 0.17 (- 0.26)	- 1.03 (0.25)	± 0.24 (± 0.08)	0.27 (0.08)

In order to absorb long-wavelengths errors between the different geoid comparisons, a 4-parameters datum shift was applied to the computed geoid heights. The number in parentheses in Table 3 refer to the results after removing the datum differences. From the results of Table 1 is obvious that the spectral IOST and LSAFD techniques gives close results at the level of 8 cm. The gravimetric geoid heights (solution 3) are biased due to the biased input gravity data (see Table 1). The excellent agreement of the combined solutions (1,2) with the TOPEX data at the level of 4 cm in terms of sd before the removal of systematic differences means that the contribution of satellite altimetry is very critical in these combined gravity field models. In this study the improvement in terms of sd reaches the level of 83% (the sd of differences in the gravimetric solution from 24 cm decrease to 4 cm in the combined one). The still remaining bias in the combined solutions can be also attributed to the gravity data, since the ERS1-GM heights are unbiased (mean value 1 cm).

The internal estimation errors were found to be equal to 4 cm. These errors depend on the input signal-to-noise levels of the gravity anomalies and altimeter heights. Detailed discussion about it can be found in Li (1996) and Tziavos et al. (1997).

For easier visualization purpose the residual geoid heights from the prediction by the IOST method are illustrated in Figure 1.

Gravity anomaly predictions. Gravity anomalies were predicted by IOST and LSAFD methods using ERS-1-GM altimetry and shipborne gravity data as in the case of geoid height prediction. Since there are no available control gravity values from an independent source, it is not possible to have an external estimation accuracy for gravity predictions. Therefore, the gravity anomaly prediction results from IOST and LSAFD and their differences are listed in Table 4. Regarding the internal estimation error, this was found close to 6mGal for both methods. From the results of Table 1 and Table 4 it is obvious that the observed and predicted gravity statistical characteristics are very close. The residuals from the gravity anomaly predictions by IOST are given in Figure 2.

Table 4. Gravity anomaly predictions and their differences. Unit: [mGal]

GRAVITY ANOMALIES: (1) IOST, (2) LSAFD					
Input errors: Altimetry: 0.10 m, Gravity: 3 mGal					
	mean	min	max	sd	rms
(1)	0.08	- 49.62	89.55	± 10.65	10.65
(2)	0.01	- 44.98	86.10	± 10.66	10.66
(1) - (2)	0.08	- 9.03	11.33	± 2.20	2.20

Deflection of the vertical predictions. In a third numerical tests components of the deflections of the vertical were predicted (a) from the 3'x6' gravity data and (b) from the 3'x6' ERS-1-GM SSHs. Both gridded evaluated by FFT using analytically defined spectra and 100% zero-padding. The prediction results and their differences are listed in Table 5. From these results more compatible are the predictions for the meridian component. However, these results can be considered as preliminary, since more research is necessary to this direction. Mentioning that the comparisons refer to the inner zone of the test area.

Table 5. Deflection of the vertical predictions and their differences. Unit: [arcsec].

Deflections of the vertical from: (1) Δg and (2) ERS1-GM SSHs										
	mean		min		max		sd		rms	
	ξ	η	ξ	η	ξ	η	ξ	η	ξ	η
(1)	- 0.03	- 0.23	- 5.47	- 4.87	4.92	5.38	± 1.15	± 1.54	1.15	1.56
(2)	0.10	0.07	- 4.14	- 5.52	4.59	7.45	± 1.01	± 2.00	1.01	2.00
(1) - (2)	0.13	0.30	- 3.98	- 3.19	3.80	4.74	± 0.65	± 1.08	0.66	1.12

Conclusions and recommendations.

The geoid prediction results from shipborne gravity and ERS-1-GM altimetry present a very good agreement with TOPEX SSHs in our test area. The sd of differences reaches the level of 4 cm. The detected bias of 58 cm is attributed to the biased input gravity anomalies. From these results is shown the importance of satellite altimetry in combined geoid models. The improvement of the prediction results reaches in our case the level of 83%, when altimetry is combined with sea gravimetry. On the whole, the spectral IOST and LSAFD methods are very efficient gravity field modeling by combining heterogeneous data over oceanic areas.

The gravimetric deflections of the vertical and those predicted from altimetry agree at the level of 0.65 and 1.08 arcsec for the meridian and prime vertical component, respectively. This accuracy can be further considerably improved by transforming the available software in spherical mode and using discrete spectra for the kernel functions.

Acknowledgments. *The financial support provided by grant number CRG 951041 to the first three authors under the NATO Collaborative Research Grants Program is gratefully acknowledged. Many thanks are due to Marc Veronneau who provided the gravity data and Jinwen Li for making available to us several computer routines.*

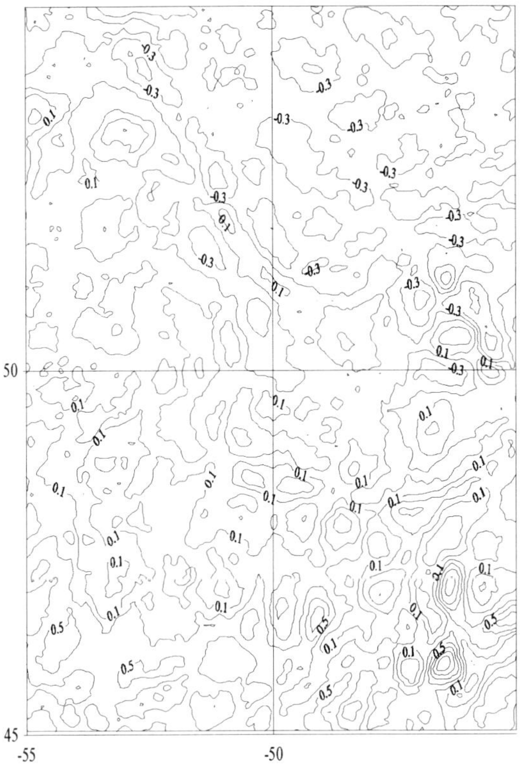

Fig. 1 Residual geoid from altimetry and shipborne gravimetry by IOST. Contour interval: 0.2 m.

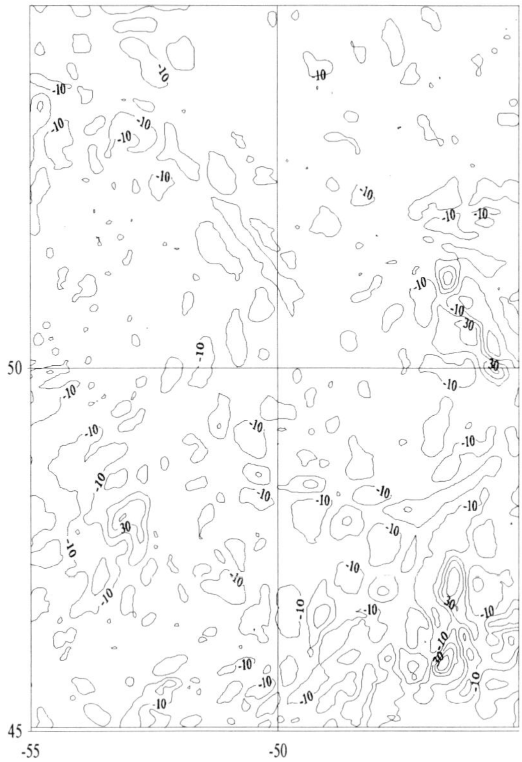

Fig. 2 Residual gravity from altimetry and shipborne gravimetry by IOST. Contour interval: 20 mGal.

References

Barzaghi, R., Fermi, A., Tarantola, S. and Sansò, F. (1993). Spectral techniques in inverse Stokes and overdetermined problems. *Surveys in Geophysics*, **14**, 461-475.

Haagmans, R., de Min, E. and van Gelderen, M., (1993). Fast evaluation of convolution integrals on the sphere using 1-D FFT, and a comparison with existing methods for Stokes' integral. *Manuscr. Geod.*, **18**, 227-241.

Li, J. (1996). Detailed marine gravity field determination by combination of heterogeneous data. UCSE Report No 20102, Dept. of Geomatics Engg., The Univ. of Calgary, Calgary, Alberta, Canada.

Li, J. and Sideris, M.G. (1997). Marine gravity and geoid determination by optimal combination of satellite altimetry and shipborne gravimetry data. *Journal of Geodesy,* **71** (4), 209-216.

Sideris, M.G. (1996). On the use of heterogeneous noisy data in spectral gravity field modeling methods. *Journal of Geodesy,* **70** (8), 470-479.

Sideris, M.G. and Tziavos, I.N. (1988). FFT-evaluation and applications of gravity-field convolution integrals with mean and point data. Bull. Géod., **62**, 521-540.

Tziavos, I.N., Li, J. and Sideris, M.G. (1997). Marine gravity field modeling using non-isotropic a-priori information. Presented at the *GraGeoMar96, Tokyo, Japan, Sept. 30 - Oct. 5, 1996,* to appear in Proceedings by Springer Verlag, 1997.

Tziavos, I.N., Sideris, M.G. and Li, J. (1996). Optimal spectral combination of satellite altimetry and marine gravity data. Presented at the *XXI EGS General Assembly, The Hague, Netherlands, 6-10 May, 1996, in Proceedings (Tziavos and Vermeer eds), "Techniques for local geoid determination", pp. 41-56,* Masala, 1996.

Gravity determination in ice covered regions by altimetry

Tilo Schöne and Hans Werner Schenke
Alfred Wegener Institute for Polar and Marine Research
27568 Bremerhaven, Columbusstrasse, Germany
TSchoene@AWI-Bremerhaven.DE

Abstract

Gravity field determination in ice covered regions of the polar marine areas is possible by using remote sensing techniques, such as satellite altimetry. An approach using standard altimetry products (ocean products, OPR) was developed and successfully applied in the Weddell Sea region, Antarctica. Comparisons to marine gravity data clearly demonstrate good agreement between both data sets.

Introduction

Marine gravity data provide a substantial insight into the Earth's structure. Geophysical features such as plate boundaries, fracture zones and bathymetric structures can be detected using marine gravity data. Due to technical and operational limitations of shipborne and airborne gravimetry measurements, satellite altimetry has become an important tool for geophysical studies in remote oceanic regions, like the ice covered Weddell Sea or parts of the Arctic.

From satellite altimetry the parameters of the marine gravity field may be derived; a dense coverage of data in both time and space is achieved with this technique. Based on experiences with GEOS-3 and SEASAT altimeter data, new and more sophisticated techniques were developed for the determination of a high resolution marine gravity field. More precise and extended data sets are available from the U.S. Navy Satellite GEOSAT for the period between 1986 and 1991. However, due to GEOSAT's inclination of 108° only areas between ±72° latitude have been covered.

In 1991, the European Remote-Sensing Satellite ERS-1 was launched carrying an improved type of a radar altimeter. Using ERS-1 altimetry, it is possible for the first time to survey large regions of Antarctica south of 72°S. Due to the permanent sea ice coverage in large areas of the Southern Ocean, only a small number of sub-satellite tracks can be used for gravity anomaly recovery studies.

Gravity field determination

The southern part of the Atlantic Ocean, i.e. the Weddell Sea, is permanently covered by sea ice. Sea ice produces a complex radar return echo. The accuracy and quality of the altimeter return signal is degraded by surface roughness and physical properties of the sea ice and cannot be determined with standard models. Thus, the onboard tracker of the satellite cannot detect the exact travel time of the pulse. As a result, outliers and noisy height measurements are included in the distributed altimeter products.

The processing and analyzing of satellite altimetry over open ocean areas is usually based on the on-board tracked data. The retracking techniques, commonly used for the precise determination of the surface heights of land ice or shelf ice, uses a function which is fitted to the shape of the return signal. The position of the leading edge of the return signal defines the exact travel time of the radar signal. Over sea ice the complex radar return signal cannot be processed using the same models and techniques. A special retracking procedure for sea ice altimeter data was developed by *Laxon & McAdoo* (1994) for gravity field determination in Arctic regions. The technique was developed by Laxon (1994). Recently *McAdoo & Laxon* (1996) have published results of gravity field computations in the Weddell Sea, using the ERS-1 waveform data with this techniques.

A new algorithm for the determination of mean altimetry height profiles was developed (*Schöne*, 1997) using the standard "Ocean Products" (OPR), which are continuously processed and distributed by the F-PAF (*Dumont & Stum*, 1994).

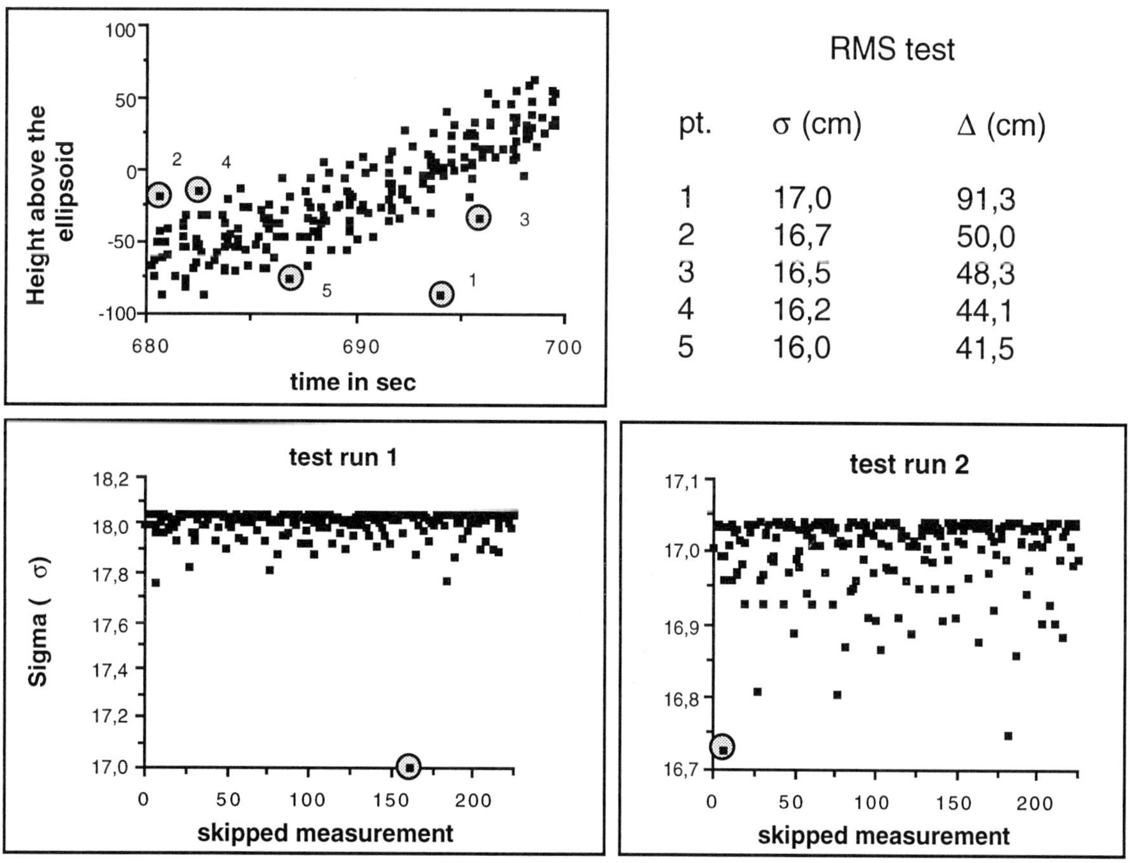

RMS test

pt.	σ (cm)	Δ (cm)
1	17,0	91,3
2	16,7	50,0
3	16,5	48,3
4	16,2	44,1
5	16,0	41,5

Fig. 1 rms test criterion. For details see the text above.

A two-phase adjustment was utilised for data editing and in the computation process. In order to achieve a good signal-to-noise ratio, measurements from repeat orbits were used to determine mean height profiles over sea ice. In the open ocean, good results were obtained by using a collinear analysis with a least-squares adjustment based on a three parameter error model for the reduction of orbit errors with 1 cyc/rev (*van Gysen et al.*, 1992). The general adjustment procedure required observations with a Gaussian error distribution. Since the use of identical data-editing criteria over sea-ice and on the open ocean would lead to the rejection of most over ice observations, it is required to relax the editing criteria. As a result, many outliers and errors remain in the data. In order to minimize their influence on the final results, the collinear analysis is used to adjust only measurements over the open ocean.

In the second phase, all data over ice (e.g. south of 60°S) are adjusted using a more robust criterion. A polynomial curve is fitted through all altimeter heights within a variable search radius around a location within the subtrack. The rms is then computed for this polynomial. Heights are then removed in an iterative process until no values lie outside of 2.58 σ_0. For each iteration, the height leading to the highest rms value is removed and the polynomial is recalculated (Fig. 1). The test run for the example shown in Fig. 1 was stopped after five iterations.

This two-step adjustment procedure has proven to give optimal results in the ice-covered regions. For the Weddell Sea region, *Schöne* (1997) has shown that the effects of sea-ice can be neglected for gravity anomaly determination. The resulting sea-surface heights lead to a significant improvement of the computed marine gravity field. Our computation procedure permits the determination of anomalies with wavelengths down to 30 km. The marine gravity field (Fig. 2) was determined by applying the same formulas which are generally used in ice-free ocean areas (*McAdoo & Marks*, 1992). This algorithm is based on fast Fourier techniques. A flat earth approximation can be used, dividing the whole area into smaller regions. Only data from GEOSAT and ERS-1 (3 days and 35 days orbit) were used for the computation.

Comparisons to marine gravity data

The calculated gravity anomaly field was verified by comparing it to marine gravity data measured with the "Bodenseegravimeter KSS 31" on RV "Polarstern" (*Meyer,* 1996). For this purpose, gravity anomaly data were extracted from the satellite gravity field along existing marine gravity profiles using a sampling interval of 6 km. See profile 1 (70°35' S/ 8°8' W - 42° S/ 12°1' E) (Fig. 3) and profile 2 (74° S/ 22°12' W - 76°57' S/ 49°43' W) (Fig. 4).

After bias removal, the data from the new altimeter gravity field show good agreement with the marine gravity measurements from profile no. 1. The standard deviation was computed to be ±11mgal. *Marks* (1996) has shown that marine gravity and altimeter derived gravity normally agree at a 5 mgal rms level. For comparison, gravity anomalies were extracted from *Sandwell and Smith* (1996) for each point of our profile no. 1. After bias removal, the resulting rms is ±10.7mgal. Since *Sandwell & Smith* have also included the ERS-1 168-day orbit data, we assume, that either the resolution of the present altimeter derived gravity field is generally low in this area, or there is an unknown mismatch between the different data sets. Profile no. 2, which lies south of 72° S has a standard deviation of ±22mgal. This difference may be explained as being a result of the filtering techniques used in our altimeter data processing, and by the low number of useful altimetry measurements in this region.

Looking at the calculated gravity model of the Weddell Sea as a whole, the decrease of accuracy is caused by the larger cross-track distance of the orbits. Since the measurements from the 168-day orbit over ice-covered regions in the southern latitudes are of limited use (due to the non-repeated orbits), only the data from the 35-day orbit are significant. To

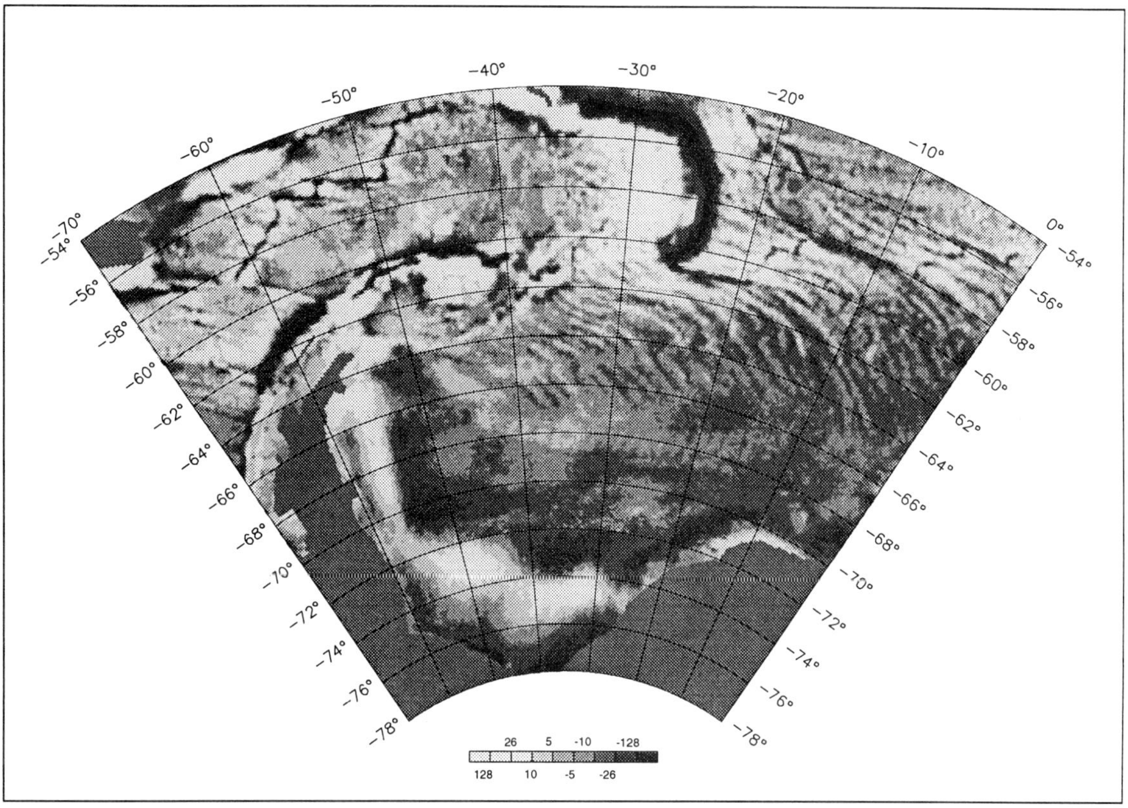

Fig. 2 Derived gravity anomalies in the area of the Weddell Sea (in mgal)

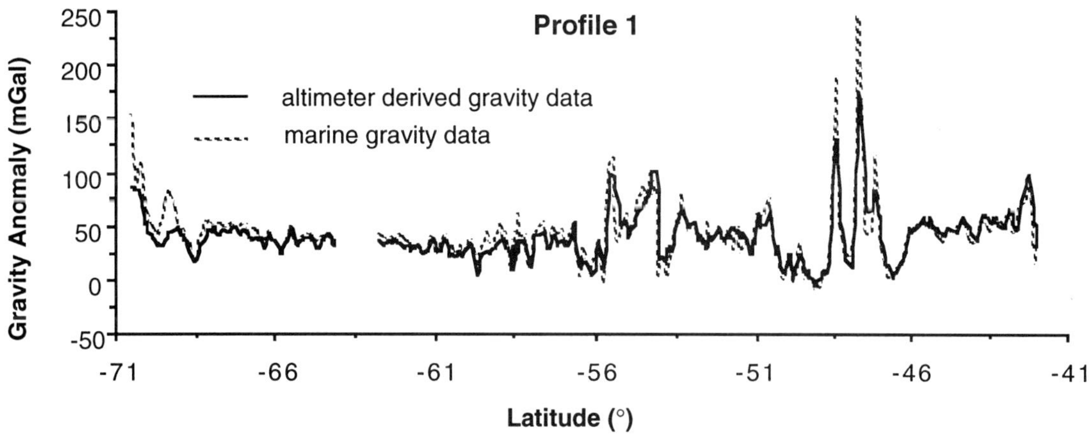

Fig. 3 Comparison of marine gravity and altimeter derived gravity anomalies north of 72° S (profile 1). This comparison shows the capability of satellite altimetry to resolve the gravity field.

investigate this effect, the marine gravity measurements (*Meyer*, 1996) were filtered using a low pass filter (Fig. 4). Wavelengths smaller than 150 km are suppressed. After applying this filter, the resulting rms is only ±8.5 mGal. Thus, the general resolution of the gravity field is limited by the large distances between subtracks in this region. In both test data sets (altimetry and marine), however, the general trends agree well.

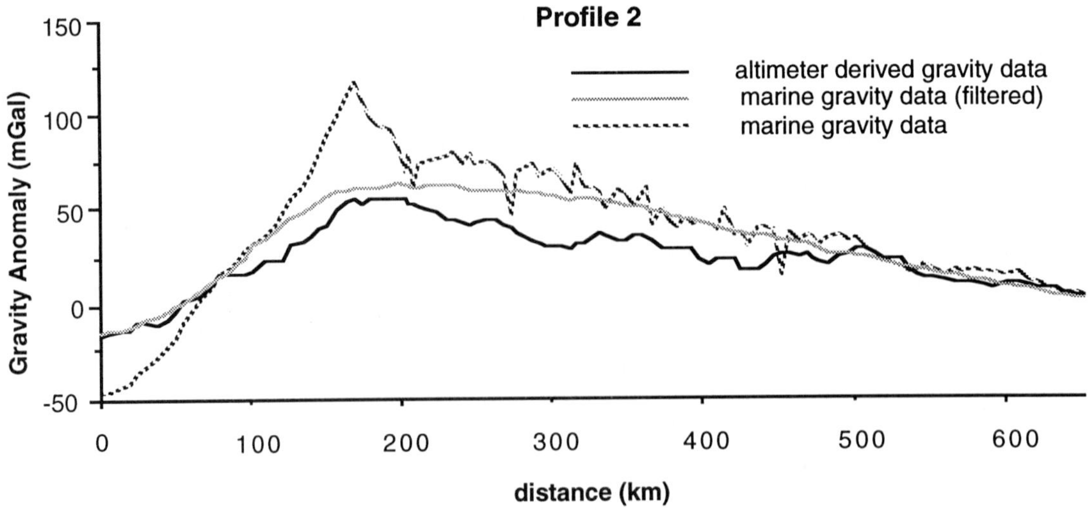

Fig. 4 Comparison of marine gravity and altimeter derived gravity anomalies south of 72° S (profile 2). Due to the larger cross-track distance of the subsatellite tracks, the estimated amplitude of the gravity field is too low.

Sea Surface Height and Geopotential Models

The filtered altimetry data were also used for the determination of an improved sea surface height model (SSH). The SSH is referenced to as well as influenced by the orbit model used in the OPR data set. The orbits refer to the GRS80 ellipsoid. Due to inconsistencies and

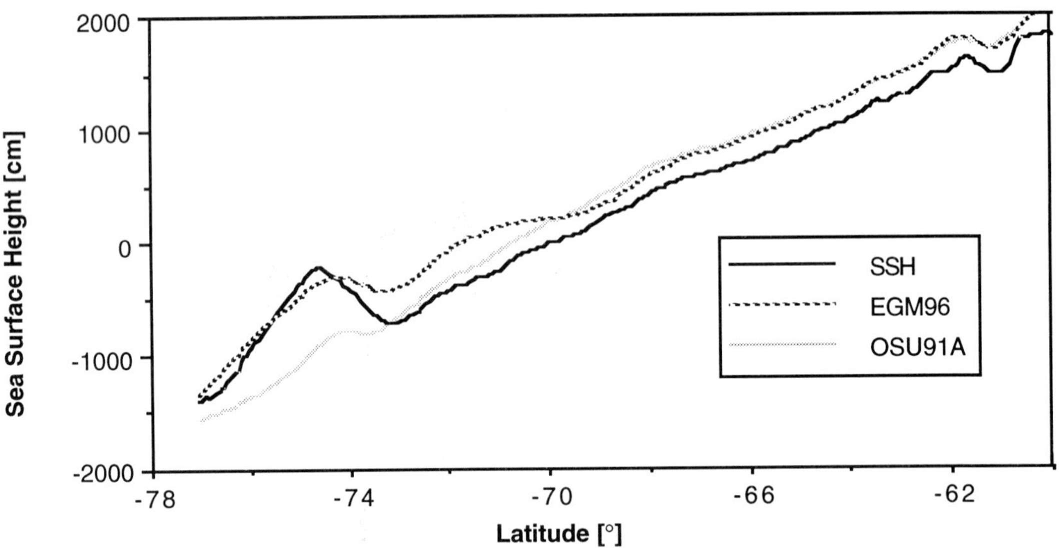

Fig. 5 Selected altimeter profile crossing the central Weddell Sea (60° S / 21°50' W - 77°5' S / 48°46' W) showing the computed sea surface height and geoid heights of the OSU91A and EGM96.

orbit errors of the OPR's, small aliasing errors may still remain in thederived SSH. After a crossover adjustment, the heights were gridded, using standard procedures. The resulting digital surface model (Fig. 5, selected altimeter subtrack) was compared to the OSU91A

(*Rapp et al.*, 1991) and to the most recent geopotential model EGM96 (*Lemoine et al.*, 1996). The latter has incorporated the derived gravity anomalies described above.

In the northern part of the Weddell Sea, the difference in the geopotential models reveals small deviations in the range of -1 m and -2 m due to the effects of the sea surface topography (SST). The deviations in the OSU91A in the southern part increase up to +8 m (Fig. 6). This discrepancy is beyond the typical size of SST in this region. We explain this difference as the effect of sparse gravity data. The discrepancy in the EGM96 model is smaller in the central Weddell Sea (-4 m to +2 m) (Fig. 6). A medium wavelengh oscillation, striking NW-SE, can be seen in the central and southern Weddell Sea. Using the new altimeter derived gravity data may lead to inhomogeneities in some parts of the Weddell Sea. Further studies of this effect are required.

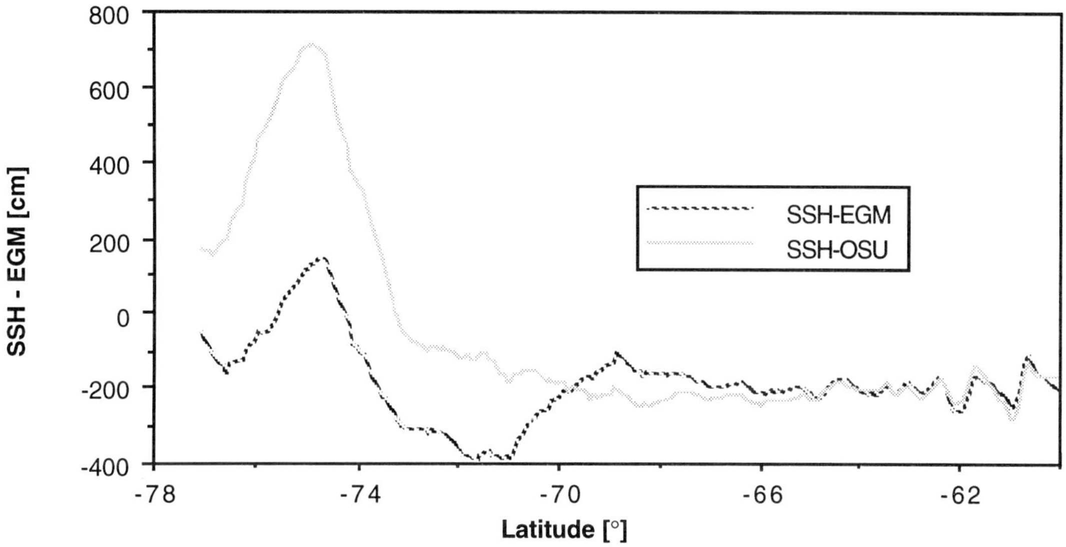

Fig. 6 Differences between SSH and geoid models. The difference in the northern part clearly reveals the sea surface topography (approx. -1.8 m in this area), in the southern part larger differences occur.

This is AWI contribution Nr. 1365

References

Dumont, J.P & J. Stum, 1994. Altimeter Products User Manual. CERSAT, C1-EX-MUT-A21-01-CN, CLS-ARGOS

Laxon, S. & D. McAdoo, 1994. Arctic Ocean Gravity Field Derived From ERS-1 Satellite Altimetry, Science, Vol.265, p.621-624

Laxon, S.W. 1994. Sea ice altimeter processing scheme at the EODC, International Journal of Remote Sensing, Vol.15, p.915-924

Lemoine, F.G. et al., 1996. The Development of the NASA GSFC and NIMA Joint Geopotential Model. Proceedings paper for the International Symposium on Gravity, Geoid, and Marine Geodesy, The University of Tokyo, Tokyo, Japan, Sep. 30 - Oct. 5, 1996

Marks, K.M., 1996. Resolution of the Scripps/NOAA marine gravity field from satellite altimetry. Geophysical Research Letters, Vol.23, No.16, p.2069-2072

McAdoo, D.C. & K.M.Marks, 1992. Gravity Field of the Southern Ocean From Geosat Data. Journal of Geophysical Research, Vol.97, p.3247-3260
McAdoo, D.C. & S.W. Laxon 1996. Marine gravity from GEOSAT and ERS-1 altimetry in the Weddell Sea, in: Geological Society Special Publication, Nr. 108, p. 155-164
Meyer, U., 1996. Untersuchungen des südöstlichen Weddellmeeres auf der Basis mariner Potentialfelddaten. Berichte zur Polarforschung (BzP) (in press)
Rapp, R.H. et al., 1991. The Ohio State 1991 Geopotential and Sea Surface Topography Harmonic Coefficient Models. OSU Report No. 410
Sandwell, D.T. & W.H.F. Smith, 1996. Marine Gravity Anomaly from Geosat and ERS-1 Satellite Altimetry. Journal of Geophysical Research, Vol.102, p. 10039-10054
Schöne, T., 1997. Ein Beitrag zum Schwerefeld im Bereich des Weddellmeeres, Antarktis, Nutzung von Altimetermessungen des GEOSAT und ERS-1. BzP 220
van Gysen, H., et al., 1992. Analysis of Collinear Passes of Satellite Altimeter Data. Journal of Geophysical Research, Vol.97, p. 2265-2277

APRIORI ESTIMATES OF THE SOLUTION OF THE GEODETIC BOUNDARY VALUE PROBLEM

Petr Holota
Research Institute of Geodesy, Topography and Cartography
250 66 Zdiby 98, Praha-východ, Czech Republic

Abstract. The purpose of this paper is to discuss an apriori estimate (bound) for the disturbing potential and its part representing higher degree harmonic components. The disturbing potential is treated as a solution of the geodetic boundary value problem and the estimates are considered in terms of the norm assigned to Lebesgue's space of square integrable functions defined on the boundary of the solution domain.

Introduction

Geodetic computations mostly have a quantitative character and a numerical realization of various quantities or functions represents a target to be achieved. Nevertheless any computation and also the computation of the (quasi-)geoid has also its functional-analytic aspect bearing a qualitative nature.

For illustration recall Stokes' problem in the exterior S_R of a sphere of radius R, i.e. in $S_R \equiv \{\boldsymbol{x} \in \boldsymbol{R}^3; |\boldsymbol{x}| > R\}$. Here $|\boldsymbol{x}| = (\sum_{i=1}^3 x_i^2)^{1/2}$ and x_i, $i = 1,2,3$, are rectangular Cartesian coordinates in Euclidean three-dimensional space $\boldsymbol{R}^3$. The solution of the problem is sought as a function T which is harmonic in S_R and such that

$$\langle \boldsymbol{x}, \boldsymbol{grad}\, T(\boldsymbol{x})\rangle + 2T(\boldsymbol{x}) = -R\,\Delta g(\boldsymbol{x}) \quad \text{for} \quad |\boldsymbol{x}| = R\,, \tag{1}$$

where $\langle\, , \rangle$ means the scalar product. If we expand T and Δg in spherical harmonics, i.e., $T(\boldsymbol{x}) = \sum_{n=0}^{\infty}(R/|\boldsymbol{x}|)^{n+1} T_n(\boldsymbol{x}/|\boldsymbol{x}|)$ and $\Delta g = \sum_{n=0}^{\infty} \Delta g_n$, where T_n and Δg_n are Laplace's surface harmonics of degree n, it follows then that for $|\boldsymbol{x}| = R$ we have

$$T(\boldsymbol{x}) = T_1\Big(\frac{\boldsymbol{x}}{|\boldsymbol{x}|}\Big) + \sum_{n=0, n\neq 1}^{\infty} \frac{R}{n-1}\,\Delta g_n\Big(\frac{\boldsymbol{x}}{|\boldsymbol{x}|}\Big)\,, \tag{2}$$

provided that $\Delta g_1 = 0$. Thus T is determined apart from the first degree harmonic component T_1. Note, however, that for the well known (physical) reasons associated with the position of center of the coordinate system we usually put $T_1 = 0$.

Let now ∂S_R be the boundary of S_R. Under this notation we easily deduce that

$$\int_{\partial S_R} T^2\, dS \leq R^2 \sum_{n=0, n\neq 1}^{\infty} \int_{\partial S_R} (\Delta g_n)^2\, dS = R^2 \int_{\partial S_R} (\Delta g)^2\, dS \tag{3}$$

which implies

$$\|T\|_{L_2(\partial S_R)} \leq R\|\Delta g\|_{L_2(\partial S_R)}\,, \tag{4}$$

where $\|.\|_{L_2(\partial S_R)}$ means the norm of a function from $L_2(\partial S_R)$, i.e. from the space of square integrable functions defined on ∂S_R. (In the title we have taken the liberty of replacing the latin a priori with the single apriori for estimates of this kind.)

In case that a part of the solution represented by low degree harmonics is known, we can split T as $T = T' + T''$, where

$$T'(\boldsymbol{x}) = \sum_{n=0}^{N}\left(\frac{R}{|\boldsymbol{x}|}\right)^{n+1} T_n\left(\frac{\boldsymbol{x}}{|\boldsymbol{x}|}\right)\,,\quad T''(\boldsymbol{x}) = \sum_{n=N+1}^{\infty}\left(\frac{R}{|\boldsymbol{x}|}\right)^{n+1}\frac{R}{n-1}\Delta g_n\left(\frac{\boldsymbol{x}}{|\boldsymbol{x}|}\right) \tag{5}$$

and $N > 1$. Hence

$$\int_{\partial S_R}(T'')^2\,dS \leq \left(\frac{R}{N}\right)^2 \sum_{n=N+1}^{\infty}\int_{\partial S_R}(\Delta g_n)^2\,dS \leq \left(\frac{R}{N}\right)^2\int_{\partial S_R}(\Delta g)^2\,dS \tag{6}$$

and it is obvious that

$$\|T''\|_{L_2(\partial S_R)} \leq (R/N)\,\|\Delta g\|_{L_2(\partial S_R)} \tag{7}$$

which in contrast to (4) is a more favourable estimate. It gives also a possibility to estimate how T'' may change in dependence on an incremental change of Δg. Indeed, due to the linearity of Stokes' problem we have $\|\delta T''\|_{L_2(\partial S_R)} \leq (R/N)\|\delta(\Delta g)\|_{L_2(\partial S_R)}$, where $\delta T''$ and $\delta(\Delta g)$ are the incremental changes of T'' and Δg, respectively. Recall, however, that similarly as above we have to suppose that $\delta(\Delta g_1) = 0$.

Finally, putting $\mathcal{E}[\,.\,] = \omega^{-1/2}\|\,.\,\|_{L_2(\partial S_R)}$, where $\omega = 4\pi R^2$ means the surface of the sphere ∂S_R, we also have $\mathcal{E}[\,\delta T''\,] \leq (R/N)\,\mathcal{E}[\,\delta(\Delta g)\,]$.

Example 1. The preceeding inequality contains some parameters. Numerically, taking e.g. $\mathcal{E}[\,\delta(\Delta g)\,] = 10^{-5}ms^{-2}$ (i.e. $1\,mgal$), $R = 6.3\cdot 10^6 m$ and $N = 20$, we obtain that $\mathcal{E}[\,\delta T''\,] \leq 3.2\,m^2s^{-2}$. Moreover, dividing this inequality by $g = 9.8\,ms^{-2}$, we arrive at $\mathcal{E}[\,\delta T''\,]/g \leq 0.3\,m$, which enables us to interpret our apriori estimate of $\mathcal{E}[\,\delta T''\,]$ in a more telling way, i.e. in terms of geoid undulations.

In reality (Molodensky's problem) the solution domain does not coincide with the exterior of the sphere and in the boundary condition the direction of the derivative is not perpendicular to the boundary. In consequence the orthogonal decomposition of T (i.e., $T = T' + T''$) and of the right hand side Δg as used in (5) does not hold anymore. A more general approach is needed.

Let U stand for the potential of the normal gravity of the Earth and ΔW and $\Delta\boldsymbol{g}$ be the potential and the (vectorial) gravity anomaly, respectively. These anomalies are related to U and to an adopted model of the Earth surface. We will denote the exterior of this model by Ω and recall that for the boundary $\partial\Omega$ of Ω we usually use the term "telluroid". Moreover, we will assume that $\Omega' = \boldsymbol{R}^3 - \bar{\Omega}$ (where $\bar{\Omega}$ means the closure of Ω) is a starshaped domain at the origin with the Lipschitz boundary such that $\langle\boldsymbol{x},\boldsymbol{n}\rangle > 0$ for almost all $\boldsymbol{x}\in\partial\Omega$ and $\boldsymbol{n}$ denoting the outer unit normal of $\partial\Omega$.

Now we are ready to write that in general the problem is to find T which is harmonic in Ω and such that

$$\langle\boldsymbol{h},\boldsymbol{grad}T\rangle + T = f \quad \text{on} \quad \partial\Omega\,, \tag{8}$$

where $f = \Delta W + \langle \boldsymbol{h}, \Delta \boldsymbol{g}\rangle$. The vector $\boldsymbol{h} = -[M_{ij}]^{-1}\boldsymbol{grad}\,U$, provided that for $[M_{ij}] = [\partial^2 U/\partial x_i \partial x_j]$ and $\boldsymbol{x} \in \partial\Omega$ the Hessian: $det\,[M_{ij}(\boldsymbol{x})] \neq 0$. Note that for the standard choice of U the vector $\boldsymbol{h}$ is close to $\boldsymbol{x}/2$.

As is well-known for an arbitrary constant vector $\boldsymbol{c}$ the function $u = \langle \boldsymbol{c}, \boldsymbol{grad}\,U\rangle$ is harmonic in Ω and represents a non-trivial solution of (8) in case of $f = 0$. Recall, however, that this solution is ruled out by the following asymptotic condition

$$T(\boldsymbol{x}) = c/|\boldsymbol{x}| + O(|\boldsymbol{x}|^{-3}) \quad \text{for} \quad \boldsymbol{x} \to \infty\,, \tag{9}$$

It completes the formulation of the problem. Here c is a constant and O means the usual Landau symbol. For reasons associated with our direct approach, we still put $\boldsymbol{\sigma} = \boldsymbol{h}\langle \boldsymbol{h}, \boldsymbol{n}\rangle^{-1}$, $\chi = \langle \boldsymbol{h}, \boldsymbol{n}\rangle^{-1}$, $g = -f\langle \boldsymbol{h}, \boldsymbol{n}\rangle^{-1}$, provided that $\langle \boldsymbol{h}, \boldsymbol{n}\rangle^{-1} \neq 0$ on $\partial\Omega$, cf. (Holota, 1996, 1997a). Under this notation (8) turns into

$$\langle \boldsymbol{\sigma}, \boldsymbol{grad}\,T\rangle + \chi T = -g \quad \text{on} \quad \partial\Omega\,. \tag{8$'$}$$

A Non-Spherical Case

Let T be a harmonic function in Ω such that it satisfies the boundary condition (8)$'$. Suppose that $S_R \equiv \{\boldsymbol{x} \in \boldsymbol{R}^3; |\boldsymbol{x}| > R\}$ is contained in Ω. Now in analogy to the spherical case, we split T into two parts: T', represented by a finite sum of low degree harmonics and T'' which is harmonic in Ω and such that asymptotically

$$T''(\boldsymbol{x}) = O(|\boldsymbol{x}|^{-N-2}) \quad \text{for} \quad \boldsymbol{x} \to \infty\,. \tag{10}$$

Thus $T = T' + T''$. Assuming now that T' is known, we try to estimate T'' in terms of the norm $\|\,.\,\|_{L_2(\partial\Omega)}$ of the space $L_2(\partial\Omega)$, i.e., our aim is to find an upper bound for

$$\|T''\|_{L_2(\partial\Omega)} = \Big[\int_{\partial\Omega} (T'')^2\, dS\Big]^{1/2}. \tag{11}$$

For this purpose we multiply equation (8)$'$ by T'' and integrate it over $\partial\Omega$. We obtain

$$((T'', T'')) = \int_{\partial\Omega} \bar{g}\, T''\, dS\,, \tag{12}$$

where

$$((T'', T'')) = -\int_{\partial\Omega} T''\langle \boldsymbol{\sigma}, \boldsymbol{grad}\,T''\rangle\, dS - \int_{\partial\Omega} \chi \cdot (T'')^2\, dS \tag{13}$$

and $\bar{g} = g + \langle \boldsymbol{\sigma}, \boldsymbol{grad}\,T'\rangle + \chi T'$, which in terms of (8)$'$ represents a reduced right hand side. Moreover, following (Holota, 1996,1997b), we can write for the first integral on the right hand side of (13) that

$$-\int_{\partial\Omega} T''\langle \boldsymbol{\sigma}, \boldsymbol{grad}\,T''\rangle\, dS = \int_{\Omega} |\boldsymbol{grad}\,T''|^2\, d\boldsymbol{x} - \frac{1}{2}\int_{\partial\Omega} L \cdot (T'')^2\, dS\,, \tag{14}$$

where $L = \langle \boldsymbol{curl}(\boldsymbol{n} \times \boldsymbol{\sigma}), \boldsymbol{n}\rangle$.

Remark 1. For the harmonic T'' (14) represents a generalized Green's identity. Indeed, for $\boldsymbol{\sigma} = \boldsymbol{n}$ we have $L = 0$ and (14) turns into the usual version of Green's identity. In (Holota, 1996, 1997b) we can also find a geometrical interpretation of L.

For $((T'', T''))$ we now have

$$((T'', T'')) = \int_{\Omega} |\boldsymbol{grad}\,T''|^2\, d\boldsymbol{x} + \int_{\partial\Omega} K \cdot (T'')^2\, dS\,, \tag{15}$$

where $K = -\chi - L/2$. In the sequel we will need a lower estimate of $((T'', T''))$ and an upper estimate for the right hand side in (12).

A Lower Estimate

Putting $D = \Omega - \bar{S}_R$, we can write

$$((T'',T'')) \geq \int_{S_R} |\boldsymbol{grad}\, T''|^2 \, d\boldsymbol{x} + \int_D |\boldsymbol{grad}\, T''|^2 \, d\boldsymbol{x} - k \int_{\partial\Omega} \langle \boldsymbol{x}, \boldsymbol{n} \rangle \cdot (T'')^2 \, dS \,, \quad (16)$$

where

$$k = \operatorname*{supess}_{x \in \partial\Omega} \left[\langle \boldsymbol{x}, \boldsymbol{n}_x \rangle^{-1} |K(\boldsymbol{x})| \right] . \quad (17)$$

Here we suppose that the essential supreme value concerns a Lebesgue measurable function defined and bounded almost everywhere on $\partial\Omega$. (Loosely speaking, we can say that k does not reflect extremes achieved on sets of a zero Lebesgue measure.) In addition from Green's identity, we deduce that

$$\int_{\partial\Omega} \langle \boldsymbol{x}, \boldsymbol{n} \rangle \cdot (T'')^2 \, dS = R \int_{\partial S_R} (T'')^2 \, dS - 3 \int_D (T'')^2 \, d\boldsymbol{x} - \int_D \langle \boldsymbol{x}, \boldsymbol{grad}\,(T'')^2 \rangle \, d\boldsymbol{x} \,, \quad (18)$$

where the third integral on the right hand side can be estimated by means of the inequality $|ab| \leq \varepsilon a^2/2 + b^2/2\varepsilon$ with an arbitrary $\varepsilon > 0$. Hence, using still $\boldsymbol{grad}\,(T'')^2 = 2T''\boldsymbol{grad}\, T''$ and recalling that $|\boldsymbol{x}| \leq R$ on D, we have

$$\int_{\partial\Omega} \langle \boldsymbol{x}, \boldsymbol{n} \rangle \cdot (T'')^2 \, dS \leq R \int_{\partial S_R} (T'')^2 \, dS + (\varepsilon - 3) \int_D (T'')^2 \, d\boldsymbol{x} + \frac{R^2}{\varepsilon} \int_D |\boldsymbol{grad}\, T''|^2 \, d\boldsymbol{x} \quad (19)$$

and is clear that for $\varepsilon = 3$ eq. (19) results in

$$\int_{\partial\Omega} \langle \boldsymbol{x}, \boldsymbol{n} \rangle \cdot (T'')^2 \, dS \leq R \int_{\partial S_R} (T'')^2 \, dS + \frac{R^2}{3} \int_D |\boldsymbol{grad}\, T''|^2 \, d\boldsymbol{x} \,. \quad (20)$$

This immediately yields

$$((T'',T'')) \geq (((T'',T''))) + \left(1 - \frac{kR^2}{3}\right) \int_D |\boldsymbol{grad}\, T''|^2 \, d\boldsymbol{x} \quad (21)$$

with

$$(((T'',T''))) = \int_{S_R} |\boldsymbol{grad}\, T''|^2 \, d\boldsymbol{x} - kR \int_{\partial S_R} (T'')^2 \, dS \,. \quad (22)$$

Our problem now is to examine $(((T'',T'')))$. In S_R we expand T'' in spherical harmonics, i.e., $T''(\boldsymbol{x}) = \sum_{n=N+1}^{\infty} (R/|\boldsymbol{x}|)^{n+1} T''_n(\boldsymbol{x}/|\boldsymbol{x}|)$, where the summation starts with $n = N+1$ in view of (10). We know that

$$\int_{\partial S_R} (T'')^2 \, dS = \sum_{n=N+1}^{\infty} \int_{\partial S_R} (T''_n)^2 \, dS \,. \quad (23)$$

Simultaneously, using Green's identity, we easily deduce

$$(((T'',T'''))) = \frac{1}{R} \sum_{N+1}^{\infty} (n + 1 - kR^2) \int_{\partial S_R} (T''_n)^2 \, dS \,. \quad (24)$$

Hence, assuming that $N+2-kR^2>0$ and putting $C(N)=(N+2-kR^2)/R$, we arrive at

$$(((T'',T''))) \geq C(N) \int_{\partial S_R} (T'')^2 \, dS \,. \tag{25}$$

Remark 2. In (Holota, 1996) the existence, uniqueness and stability of the solution of our oblique derivative boundary value problem have been proved for $3-kR^2>0$. Thus $N+2-kR^2>0$ for $N>1$ is even a less restrictive assumption.

The last inequality together with (21) now yields

$$((T'',T'')) \geq C(N) \int_{\partial S_R} (T'')^2 \, dS + \left(1-\frac{kR^2}{3}\right) \int_D |\boldsymbol{grad}\, T''|^2 \, d\boldsymbol{x} \geq$$

$$\geq \frac{3-kR^2}{3R} \left[(N+2) \int_{\partial S_R} (T'')^2 \, dS + R \int_D |\boldsymbol{grad}\, T''|^2 \, d\boldsymbol{x} \right] . \tag{26}$$

and it is obvious that

$$((T'',T'')) \geq \frac{3-kR^2}{R^2} \int_{\partial\Omega} \langle \boldsymbol{x}, \boldsymbol{n} \rangle \cdot (T'')^2 \, dS \tag{27}$$

in view of (20). Finally, recalling that $\langle \boldsymbol{x}, \boldsymbol{n}_x \rangle$ is positive for almost all $\boldsymbol{x} \in \partial\Omega$ (according to the assumption made in the introduction), we can conclude that

$$((T'',T'')) \geq c \, \|T''\|^2_{L_2(\partial\Omega)} \quad \text{with} \quad c = \frac{3-kR^2}{R^2} \inf_{x\in\partial\Omega} [\langle \boldsymbol{x}, \boldsymbol{n}_x \rangle] \,. \tag{28}$$

An Estimate of the Solution

As already mentioned we need an upper estimate of the right hand side of (12). It follows immediately from Hölder's inequality. Indeed,

$$\int_{\partial\Omega} \bar{g}\, T'' \, dS \leq \|\bar{g}\|_{L_2(\partial\Omega)} \|T''\|_{L_2(\partial\Omega)} \,. \tag{29}$$

Now combining (12), (28) and (29), we obtain

$$c\, \|T''\|^2_{L_2(\partial\Omega)} \leq ((T'',T'')) = \int_{\partial\Omega} \bar{g}\, T'' \, dS \leq \|\bar{g}\|_{L_2(\partial\Omega)} \|T''\|_{L_2(\partial\Omega)} \,. \tag{30}$$

In consequence

$$\|T''\|_{L_2(\partial\Omega)} \leq c^{-1} \, \|\bar{g}\|_{L_2(\partial\Omega)} \tag{31}$$

and naturally also $\|\delta T''\|_{L_2(\partial\Omega)} \leq c^{-1} \, \|\delta\bar{g}\|_{L_2(\partial\Omega)}$. Note that for $\partial\Omega$ of mild slopes and curvatures and for a standard choice of the normal potential the coefficient k is close to $2/R^2$ and $\langle \boldsymbol{x}, \boldsymbol{n}_x \rangle$ does not differ from R strongly. Thus with some approximation c^{-1} is close to R. This, however, yields an estimate on the level of efficiency as in (4). Naturally, recalling (7), we expected more, i.e. a smaller factor in front of the norm on the right hand side. We attempt to approach this problem in the next section.

An Improved Estimate of an Approximate Solution in a Finite Dimensional Space

Inspecting our computations, we can see that in (27) we have not made full use of the favourable value of the factor $C(N)$. Indeed, the coefficients in (20) are not well balanced with $C(N)$ and $(1-kR^2)/3$ in (26). Therefore, our aim is to modify (20).

In (19) we first try to estimate $\int_D (T'')^2\, d\boldsymbol{x}$ by means of $\int_{\partial S_R} (T'')^2\, dS$. This, however, is associated with the non-stability of the downward continuation problem. The estimate maybe obtained, but for T'' smoothed up to a certain degree only. In order to smooth T'' we will confine ourselves to a finite range of its spherical harmonic components, i.e., in D we will approximate T'' by

$$T^*(\boldsymbol{x}) = \sum_{n=N+1}^{n_{max}} \left(\frac{R}{|\boldsymbol{x}|}\right)^{n+1} T''_n\left(\frac{\boldsymbol{x}}{|\boldsymbol{x}|}\right) \tag{32}$$

and will hope that n_{max} can be sufficiently high. Moreover, to improve conditions for the computation of the desired estimate, we put $T^*_n = (R/R_0)^{n+1} T''_n$, where R_0 is a greatest radius such that D is contained in $S_{R_0} \equiv \{\boldsymbol{x} \in \boldsymbol{R}^3; |\boldsymbol{x}| > R_0\}$. Thus,

$$T^*(\boldsymbol{x}) = \sum_{n=N+1}^{n_{max}} \left(\frac{R_0}{|\boldsymbol{x}|}\right)^{n+1} T^*_n\left(\frac{\boldsymbol{x}}{|\boldsymbol{x}|}\right). \tag{33}$$

We evidently have

$$\int_{\partial S_R} (T^*)^2\, dS = R^2 \sum_{n=N+1}^{n_{max}} q^{n+1} \int_{\partial S_1} (T^*_n)^2\, dS\,, \tag{34}$$

where $q = R_0/R$ and by direct computation we obtain

$$\int_D (T^*)^2\, d\boldsymbol{x} \leq \int_{R \leq |\boldsymbol{x}| \leq R_0} (T^*)^2\, d\boldsymbol{x} = R_0^3 \sum_{n=N+1}^{n_{max}} \frac{1-q^{2n-1}}{2n-1} \int_{\partial S_1} (T^*_n)^2\, dS\,. \tag{35}$$

Comparing now (34) and (35), we see clearly that we cannot deduce the desired estimate for a full spectrum of harmonics since $q^{n+1} \to 0$ for $n \to \infty$. Therefore, we will rather look for a smallest positive constant c^*, such that

$$R_0^3 \frac{1-q^{2n-1}}{2n-1} \leq c^* R^2 q^{n+1} \quad \text{for} \quad n = N+1, N+2, \ldots, n_{max}\,. \tag{36}$$

In other words, putting (after some arrangement)

$$f(q,n) = \frac{1-q^{2n-1}}{(2n-1)q^{n-2}}\,, \tag{37}$$

we are looking for the smallest c^*, such that

$$R f(q,n) \leq c^* \quad \text{for} \quad n = N+1, N+2, \ldots, n_{max}\,. \tag{38}$$

Note that $f(1,n) = 0$ for all n. Thus for $q = 1$ we can put $c^* = 0$.

In order to make the situation more transparent we give below several values of $f(q,n)$ for $R = 6378\,km$ and $R_0 = 6356\,km$, i.e. for $q = 0.9965506$:

$$f(q,0) = 0.0034 \qquad f(q,1000) = 0.0157 \qquad f(q,2000) = 0.2490$$
$$f(q,500) = 0.0054 \qquad f(q,1500) = 0.0590 \qquad f(q,2500) = 1.1213$$

Inspecting the table of $f(q,n)$ quickly and taking e.g. $n_{max} = 1000$, we can put $c^* = 0.016\,R$. Hence for the mentioned parameters

$$\int_D (T^*)^2\,d\boldsymbol{x} \leq 0.016\,R \int_{\partial S_R} (T^*)^2\,dS \tag{39}$$

which enables to modify (19) as follows:

$$\int_{\partial\Omega} \langle \boldsymbol{x}, \boldsymbol{n}\rangle \cdot (T^*)^2\,dS \leq R\,[\,1 + 0.016(\varepsilon - 3)\,] \int_{\partial S_R} (T^*)^2\,dS + \frac{R^2}{\varepsilon} \int_D |\boldsymbol{grad}\,T^*|^2\,d\boldsymbol{x}\,. \tag{40}$$

Moreover, putting e.g. $\varepsilon = 22$, we obtain

$$\int_{\partial\Omega} \langle \boldsymbol{x}, \boldsymbol{n}\rangle \cdot (T^*)^2\,dS \leq 1.30R \int_{\partial S_R} (T^*)^2\,dS + \frac{R^2}{22} \int_D |\boldsymbol{grad}\,T^*|^2\,d\boldsymbol{x}\,. \tag{41}$$

Going now back to (26) and interpreting it for T^* and $N = 20$, we obtain

$$((T^*, T^*)) \geq \frac{22\,(3 - kR^2)}{3R} \left[\int_{\partial S_R} (T^*)^2\,dS + \frac{R}{22} \int_D |\boldsymbol{grad}\,T^*|^2\,d\boldsymbol{x} \right], \tag{42}$$

which in combination with (41) yields

$$((T^*, T^*)) \geq \frac{22\,(3 - kR^2)}{3 \cdot 1.30\,R^2} \inf_{x \in \partial\Omega} [\,\langle \boldsymbol{x}, \boldsymbol{n}_x \rangle\,]\; \|T^*\|^2_{L_2(\partial\Omega)}\,. \tag{43}$$

Finally, recalling (29) which holds for T^* as well, we can conclude that

$$\|T^*\|_{L_2(\partial\Omega)} \leq 0.17\,c^{-1} \|\bar{g}\|_{L_2(\partial\Omega)}\,, \tag{44}$$

provided that $N = 20$, $n_{max} = 1000$, $R = 6378\,km$ and $R_0 = 6356\,km$. This estimate is somewhat weaker than (7) in the spherical case, but it shows again the desired effect associated with the use of geopotential models.

Acknowledgements. The work on this paper has been supported by the Grant Agency of the Czech Republic through Grant No. 205/96/0956. This support is gratefully acknowledged.

References

Holota P. (1996). Variational methods for quasigeoid determination. In: Tziavos, I.N. and Vermeer, M. (eds.), Techniques for Local Geoid Determination, Proc. Session G7, EGS XXI Gen. Assembly, The Hague, The Netherlands, 6-10 May, 1996, Reports of the Finnish Geodetic Inst. No. 96:2, Masala, 1996, pp 9-21.

Holota P. (1997a). Variational methods for geodetic boundary value problems. In: Sansò, F. and Rummel R. (eds.), Geodetic Boundary Value Problems in View of the One Centimeter Geoid, Lecture Notes in Earth Sciences, Springer, Berlin etc., pp 468-481.

Holota P. (1997b). Coerciveness of the linear gravimetric boundary-value problem and a geometrical interpretation. Journal of Geodesy 71: pp 640-651.

APPLICATION OF MULTIRESOLUTION FILTERING IN SPECTRAL GEOID DETERMINATION

Christopher Kotsakis and Michael G. Sideris
The University of Calgary
Department of Geomatics Engineering
Calgary, Alberta, Canada

Abstract

The Wavelet Transform as a new tool for spectral gravity field modeling is introduced. Its very useful localization properties are explored and some comparisons with classic Fourier transform are made in order to realize the superiority of the Wavelet transform in approximating and analyzing general, non-stationary gravity field signals. Finally, some important issues for further research are outlined.

Introduction

Spectral methods for analyzing and synthesizing gravity field signals have become a standard procedure in every application of physical geodesy. Spherical harmonic expansions and Fourier transforms (FT) are used for global and local modeling, respectively, of the Earth's gravity field, and one could say that every possible advantage that these two methods offer has been extensively explored. The Wavelet transform (WT) provides an alternative method for spectral decomposition and transformation of signals and has already been considered as a substitute for the present techniques. In the following, we discuss and compare the WT over the classical FT and we try to demonstrate some of the advantages that will arise from a possible incorporation of the WT into local gravity field modeling applications.

Essentially, Wavelet and Fourier methods are both the same thing in the following sense:

1. they both pose the same restrictions on the analyzing signals (square integrability);
2. they both offer a unique and invertible transformation of the signal to/from the corresponding spectral domain; and
3. they both can provide a power spectrum that gives the distribution of the signal's power in the corresponding spectral domain.

On the other hand, complex exponentials are eigenfunctions of several important operators in physical geodesy, whereas wavelets do not in general diagonalize commonly used operators such as convolution. Is there, therefore, any real need to switch our current practice and attention into wavelet methods? We will see that there is, as long as the non-stationary behavior of the gravity field is a major concern. Due to their ability to localize both space and frequency characteristics, wavelets offer a much more consistent way of approximating irregular functions. This is in contrast with Fourier analysis, where a local disturbance in the original function (e.g. gravity anomalies) will produce distortions in its smoother parts even for small wavelength approximation.

Overview of Wavelet transform theory

Just like in the Fourier transform case, many different versions of the wavelet transform exist and in the following we will be dealing with the most general case, the Continuous Wavelet Transform (CWT). We are also going to restrict ourselves to 1D signals in order to keep the notation simple and the visualization possible. Extensions into higher dimensions, although not straightforward in some cases, can be made.

Since we are going to be brief, let us start by pointing out that more detailed treatments about the wavelet transform can be found in, e.g., Cohen and Kovacevic (1996), and Daubechies (1990). The CWT is defined through an operator $\boldsymbol{W}$ which maps a function f(x) in the space domain into a 2D distribution $W_f(\alpha,b)$ in the scale-translation or wavelet domain as follows:

$$W_f(\alpha,b) = \boldsymbol{W}\{f(x)\} = \alpha^{-1/2}\int f(x)\,\psi\big((x-b)/\alpha\big)\,dx \quad \text{with } \alpha, b \in R \text{ and } \alpha > 0 \quad (1)$$

The CWT is always taken with respect to a particular function $\psi(x)$ which is called the *mother wavelet* and will be considered real in all cases for this paper. We can reconstruct the original function through an inverse CWT operator $\boldsymbol{W}^{-1}$ as follows:

$$f(x) = \boldsymbol{W}^{-1}\{W_f(\alpha,b)\} = c_\psi^{-1}\iint W_f(\alpha,b)\,\alpha^{-1/2}\psi\big((x-b)/\alpha\big)\,\frac{db\,d\alpha}{\alpha^2} \quad (2)$$

where c_ψ is a positive constant that depends only on the used mother wavelet. The CWT satisfies also a power conservation property which is very much similar to Plancherel's identity from Fourier analysis. In complete analogy with Fourier theory we can define the square modulus of the CWT as the *wavelet spectrum*, which gives the power spectral density of the original signal distributed in the wavelet(α,b) domain.

The only requirement the mother wavelet has to satisfy is that $c_\psi<\infty$, which is equivalent to saying that $\psi(x)$ is a band-pass type of filter; see details in (Daubechies, 1990). In particular, this mild restriction allows us to use functions which have almost compact support in both the space and the frequency domain (figure 2), and provides the CWT with its very beneficial localization properties. This is best understood if we write equation (1) in the following convolution form:

$$W_f(\alpha,b) = \alpha^{-1/2}\, f(b) * \psi(b/\alpha) \quad (3)$$

where it has been assumed for simplicity that $\psi(x)$ is even. At each scale level α, the CWT is nothing more than a band-pass filtering of the original function, with the filter being the corresponding scaled version of the mother wavelet. As α becomes smaller, the bandwidth of $\psi(b/\alpha)$ is expanded and the filtering takes place mostly at the higher frequencies, while for large scale values the opposite happens. The situation is illustrated in figure 1, where the frequency ω_o corresponds to the peak of the FT of the mother wavelet. The space localization is achieved through a similar process. Due to the compact space support of $\psi(x)$, the convolution equation (3) means that each point (α,b) in the wavelet domain contains information for the behavior of the signal f(x) inside an interval whose width is specified in terms of the scale parameter α and its center is always b.

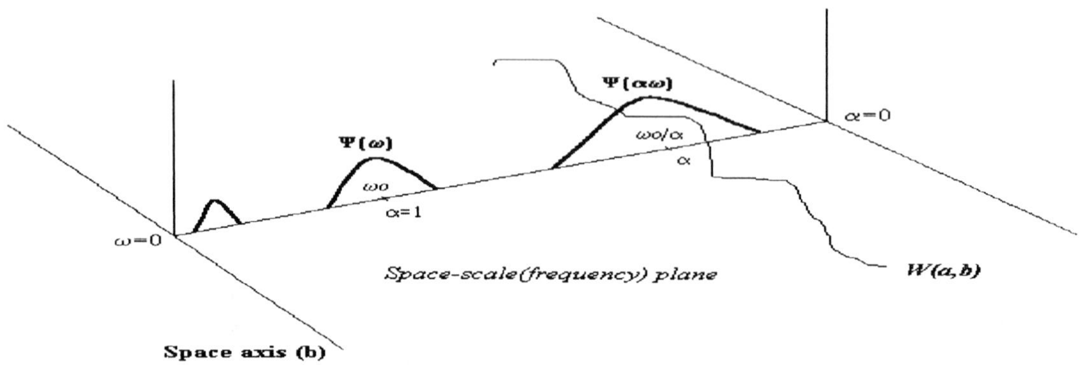

Figure 1. The CWT as a multiresolution filtering of the original signal

Although the whole procedure of the CWT seems quite complicated, it actually provides a much more *realistic* (physical) and *faster* framework for building irregular signals than the FT does. Instead of adding infinite sinusoids of increasing frequency, we use for each location varying scales to "measure" the similarity of our function with an accordingly scaled model (mother wavelet). This is exactly the multiresolution nature of wavelet methods, where the rather vague term *resolution* refers to the spatial scale we use to examine our signal. An excellent reference for the localization aspects of wavelets is Daubechies (1990).

Comparison between FT and WT for gravity field modeling

Let us assume that the value of the gravity anomaly signal Δg changes at a single location and try to see how Fourier and wavelet methods will respond to that change. A reconstructed delta function from its FT, up to a maximum frequency ω_{max}, is given by the equation:

$$\delta_F(x) = \omega_{max}\ sinc(\omega_{max} x) \tag{4}$$

In order to see what happens in the WT case, we will use the Morlet mother wavelet which is illustrated in figure 2, i.e.:

$$\psi(x) = k\,(1-x^2)\,e^{-x^2/2} \quad \text{and} \quad \Psi(\omega) = k\sqrt{2\pi}\,\omega^2\,e^{-\omega^2/2} \quad \text{with } k \text{ arbitrary const.} \tag{5}$$

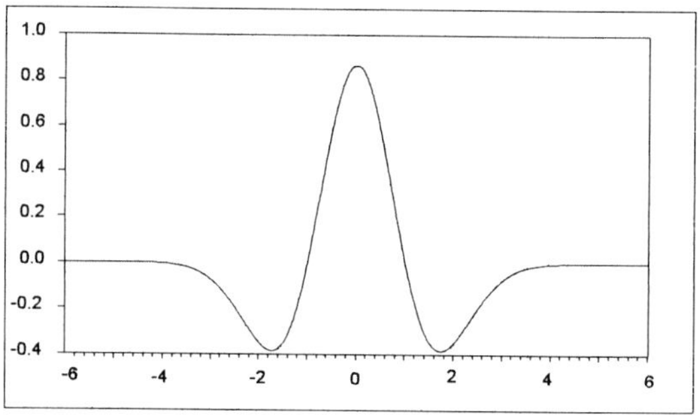

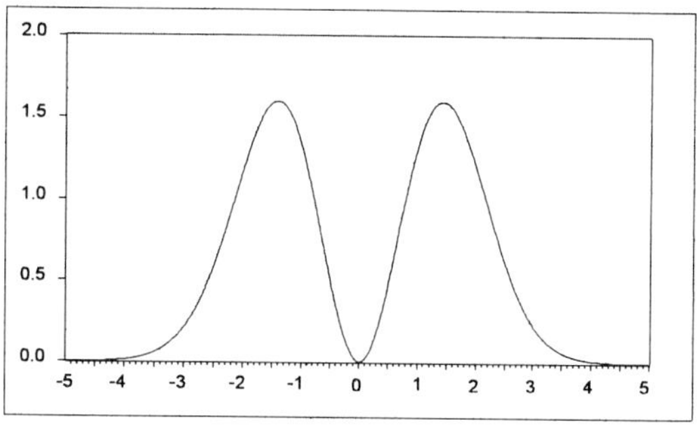

Figure 2. The Morlet mother Wavelet and its Fourier transform

The CWT of $\delta(x)$ with respect to $\psi(x)$ is easily found to be:

$$W_\delta(\alpha, b) = k\alpha^{-1/2}\left(1-(b/\alpha)^2\right)e^{-b^2/2\alpha^2} \tag{6}$$

and if we take the inverse CWT of the last equation, up to a minimum scale level α_{min}, we get the reconstructed delta function:

$$\delta_W(x) = \frac{k^2\sqrt{\pi}}{8c_\psi}\,\frac{6\alpha_{min}^2 x - x^3}{\alpha_{min}^3 x}\,e^{-x^2/4\alpha_{min}^2} \quad \text{and} \quad \delta_W(0) = \frac{k^2 3\sqrt{\pi}}{4c_\psi \alpha_{min}} \tag{7}$$

The comparison between eqs. (4) and (7) for various minimum recovery scale levels, using the interconnection formula $\omega=\omega_o/\alpha$, is shown in figure 3. Although this example is extreme and of theoretical interest only, similar approximation behavior is also obtained for other, more "realistic" irregular functions.

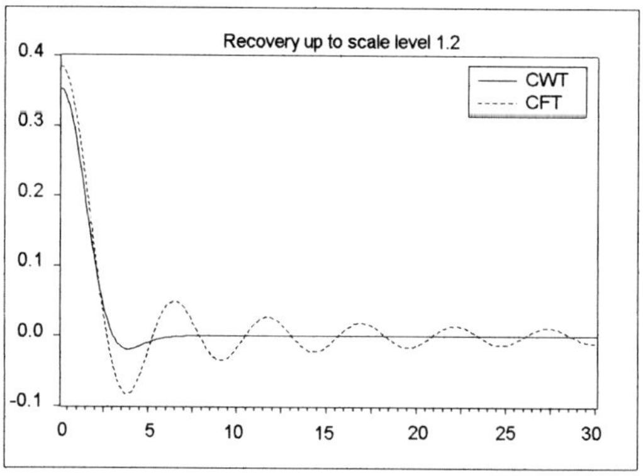

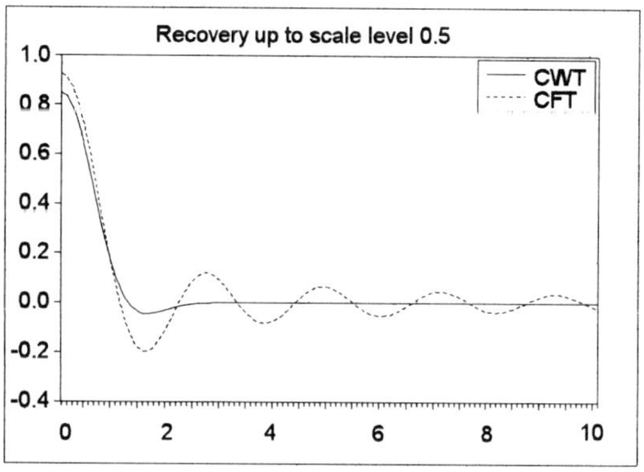

Figure 3. The Dirac delta function recovered through the CWT and CFT

We see therefore that, not only offer wavelets a more physical framework for approximation, they also provide much better convergence rates than Fourier methods. Since the recovered gravity anomaly signal up to a certain level, with respect to a specific spectral representation, is equal to the convolution of the original Δg signal with the delta function recovered up to the same level (proof is trivial for the CWT case), we can conclude from the last two graphs that multiresolution methods will do better in approximating general, non-stationary signals. This is not of course a rigorous treatment of the problem, but at least demonstrates the potential of wavelet methods.

Stokes' equation.

The previous example was aiming at demonstrating how wavelets can provide better spectral resolution and approximation convergence rates than Fourier methods for single, non-stationary gravity or height data sets. Similar advantages exist also when we have to compute, e.g., geoid undulations from gravity anomalies. Let us first see what happens to the Stokes formula when we work in the wavelet domain. If we apply eq.(3) for the case of geoid undulations N, we have:

$$W_N(\alpha, b) = \alpha^{-1/2}\ N(b) * \psi(b/\alpha) \tag{8}$$

and taking into account that $N(b) = S(b) * \Delta g(b)$, where S is the Stokes operator, we finally get:

$$W_N(\alpha, b) = S(b) * W_{\Delta g}(\alpha, b) \tag{9}$$

All the localization information that is contained in the wavelet transform of gravity anomalies is transformed, through the Stokes operator, into equivalent localized information in terms of geoid undulations. In this way, local irregularities in Δg data will be mapped in a more meaningful and useful way into geoid undulation spectral content, i.e.:

- FT - a single Δg influences mostly the low frequencies of the geoid
- WT - a single Δg influences mostly the low scales (high frequencies) of the geoid and only around that single location.

The above are best understood with the help of the CWT of the Dirac function, shown in figure 4.

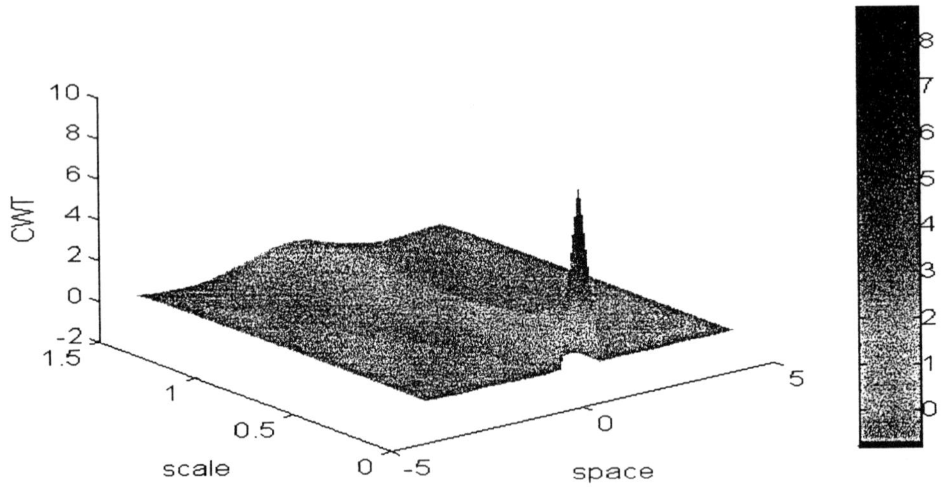

Figure 4. The CWT of the Dirac delta function

Noise filtering.

This is probably one of the most challenging applications of multiresolution methods. When we work with noisy data sets under a Fourier framework we basically apply the classic Wiener filter (collocation with the stationarity assumption required) to our gravity data, which oversmoothes possible irregular features in the signal; see (Sideris, 1995). Wavelets, on the other hand, offer space-variant noise filtering capabilities. Most of the work in this area, both theoretical and practical, has been done in the statistical community where thresholding methods have been developed with very satisfactory results. We still need, however, good theoretical models on how to distinguish non-stationary gravity field signals from noise.

Summary - Conclusions

Wavelets have definitely the potential to produce more useful and reliable results in gravity field modeling than Fourier methods. Since we are dealing with non-stationary signals in general, the ability to measure their power at different scales and different positions in their domain gives us a much more clear picture of both data and results. It also opens naturally the very important field of irregular sampling, since under a wavelet framework we need denser data only where the gravity signal power remains high at small scales.

The theoretical power of the CWT stems from the fact that they provide, in contrast to Fourier expansions, an "overcomplete" system of basis functions. Using functional analysis terminology, the CWT is a (tight) *frame* instead of a complete orthonormal expansion; see (Daubechies et al., 1986). This overcompleteness is exactly what gives the higher resolution and the sparse spectral representation of non-stationary functions through the CWT. In practice, where we use a discrete number of wavelet functions, this overcompleteness remains as long as we pose some additional restrictions on the mother wavelet. We can even obtain orthonormal wavelet expansions if we pose more restrictions on $\psi(x)$. Fortunately enough, these additional restrictions do not destroy the compact support of $\psi(x)$ and therefore the localization properties remain. The issue, however, of using overcomplete expansions for the representation of the gravity field is something that needs further justification.

A number of non-trivial problems need also to be solved before wavelets can become a rigorous tool for geoid determination. The combination of multiresolution methods for local gravity field approximation with spectral information obtained from global geopotential models, is probably the most difficult. Also, the type of the used mother wavelet and its influence in the results is something that should be studied in detail, as well as the extension of the theoretical results into higher dimension signals (2D,3D).

References

Cohen, A. and Kovacevic, J. (1996): Wavelets: The Mathematical Background. *IEEE Proceedings,* **84**(4), April 1996.

Daubechies, I. (1990): The Wavelet Transform, Time-Frequency Localization and Signal Analysis. *IEEE Transactions on Information Theory*, **36**(5), 961-1005.

Daubechies, I., Grossmann, A. and Meyer, Y. (1986): Painless non-orthogonal expansions. *Journal of Mathematical Physics*, **27**(5), 1271-1283.

Sideris, M.G. (1995): On the use of heterogeneous noisy data in spectral gravity field modeling methods. *Journal of Geodesy*, **70**, 470-479.

GEOID COMPUTATION BY COLLOCATION IN SCALING SPACES

W. Keller
University Stuttgart
Germany

1 Introduction

Since the 70's, collocation in reproducing kernel Hilbert spaces has been advocated by many authors: Tscherning and Rapp(1974),Sünkel(1986) . The only disadvantage of this method is that the number of equations which have to be solved equals the number of observations. Hence, for large data sets the computational power even of large computers is frequently exceeded.
Despite the fact that the number of equations cannot be reduced it would already be helpful if the matrix had a sparse structure. This is not the case for Hilbert spaces which are traditionally used, because the base functions of these Hilbert spaces have a global support. On the other hand it is impossible to span the space $L^2(R^2)$ by locally supported base functions. This would only be possible for subspaces which are restricted in their resolutions.
The decomposition of $L^2(R^2)$ into a nested sequence of subspaces with increasing resolution is exactly the idea of multiscale analysis in wavelet theory. Because the base functions of scaling spaces of such a multiscale analysis have a local support, the reproducing kernel is rapidly decreasing and produces a sparse system of collocation equations.
In the paper two very simple examples of multiscale annalysises of $L^2(R^2)$ are studied, showing the potential applicability of collocation in their scaling spaces.

2 Multiscale analysis

A multiscale analysis of $L^2(R^2)$ is a decomposition of this space which resembles a *Russian Doll*:

- Each space is contained in the preceeding space,
- the resolution decreases from space to space by the factor 2 and
- the sequence of subspaces spans the whole $L^2(R^2)$.

A mathematically more precise definition of a multiscale analysis is

Definition 1 *A multiscale analysis (MSA) of $L^2(R^2)$ is a sequence of spaces V_n with*

- $V_n \subset V_{n-1} \subset \ldots \subset L^2(R^2)$
- $\bigcap_{n \in Z} V_n = \{0\}$
- $\bigcup_{n \in Z} V_n = L^2(R^2)$
- $f(\bullet) \in V_n \leftrightarrow f(2 \cdot \bullet) \in V_{n-1}$
- *There is a function* φ *with* $V_0 = span\{\varphi(\bullet - n) \mid n \in Z^2\}$

The function φ is called scaling function.

If collocation is to be carried out in the scaling spaces V_n, one has to make sure that V_n is a reproducing kernel Hilbert space. The scaling function φ provides us with a criterion to decide wether a reproducing kernel exists.

Definition 2 *A scaling function φ is called orthogonal if the set*

$$\{\varphi(\bullet - k) \mid k \in Z^2\} \tag{1}$$

forms a complete orthonormal base of V_0.

If the MSA is build from an orthogonal scaling function then indeed the scaling spaces have reproducing kernels.

Lemma 1 *Let φ be the orthogonal scaling function of a MSA. Then every element V_n of the MSA is a reproducing kernel Hilbert space with the kernel*

$$K_n(\mathbf{x}, \mathbf{y}) = 2^{2n} \sum_{\mathbf{k} \in Z^2} \varphi(2^{-n}\mathbf{x} - \mathbf{k}) \cdot \varphi(2^{-n}\mathbf{y} - \mathbf{k}) \tag{2}$$

The proof can be found in [3].
Since the scaling function φ has a local support the reproducing kernel is rapidly decreasing and leads to a sparse collocation matrix.

Obviously, not every MSA will be of relevance for geodesy. The following basic conditions have to be fulfilled:

1. The elements of V_n must have a geodetic interpretation.
2. A closed expression of the series expansion of the reproducing kernel has to be found.
3. The dual space of V_n has to contain geodetic relevant functionals.

Hence, for every potential MSA prior to its application the above mentioned conditions have to be checked.

3 Examples

3.1 MSA based on the Shannon wavelet

The scaling function connected with the one-dimensional Shannon wavelet is given by

$$\bar{\varphi}(x) = \frac{\sin(\pi x)}{\pi x} \tag{3}$$

Its obvious extension to the two-dimensional case is

$$\varphi(\mathbf{x}) = \bar{\varphi}(x_1)\bar{\varphi}(x_2) \tag{4}$$

For later use the following compact notation is introduced

$$\varphi_{n,\mathbf{k}}(\mathbf{x}) := 2^{-n}\varphi(2^{-n}\mathbf{x} - \mathbf{k}) \tag{5}$$

As a direct consequence of the results given in [3] the following lemma is obtained

Lemma 2 *The spaces*

$$V_n := span\{\varphi_{n,\mathbf{k}}(\mathbf{x}) \mid \mathbf{k} \in Z^2\} \tag{6}$$

form a MSA of $L^2(R^2)$

For this MSA the above mentioned conditions have to be checked.

1. characterization of V_0
Every element $f \in V_0$ has the following expansion

$$f(\mathbf{x}) = \sum_{\mathbf{k} \in Z^2} f_{\mathbf{k}} \varphi(\mathbf{x} - \mathbf{k}) \tag{7}$$

Since $\varphi(\mathbf{k}) = \delta_{0,\mathbf{k}}$ holds the realtion $f(\mathbf{k}) = f_{\mathbf{k}}$ is true. Hence,

$$f(\mathbf{x}) = \sum_{\mathbf{k} \in Z^2} f(\mathbf{k}) \varphi(\mathbf{x} - \mathbf{k}) \tag{8}$$

Equation (8) is exactly Shannon's theorem. This theorem is true iff

$$\mathcal{F}\{f\}(\omega) = 0 \Leftrightarrow |\omega_i| > \pi \quad i = 1, 2 \tag{9}$$

Hence, the spaces V_n consist of band-limited functions and therefore are a useful model for geodetic data, because practically all geodetic measurements are band-limited.

2. Determination of the reproducing kernel
Unfortunately, it is impossible to find a closed expression for the reproducing kernel, if its arguments $\mathbf{x}, \mathbf{y}$ can vary arbitrarily in R^2. On the other hand, due to the band limitation of functions in V_n it is sufficient to know them on a grid.

Definition 3 *The set*

$$G_n := \{(2^n \mathbf{r}, 2^n \mathbf{s}) \mid \mathbf{r}, \mathbf{s} \in Z^2\} \tag{10}$$

is called a grid of order n.

If the arguments $\mathbf{x}, \mathbf{y}$ of the kernel are restricted to the grid G_n it gets a very simple structure:

Lemma 3 *For $\mathbf{x}, \mathbf{y} \in G_n$ the reproducing kernel K of V_n is*

$$K(\mathbf{x}, \mathbf{y}) = \begin{cases} 2^{-2n} & , \quad \mathbf{x} = \mathbf{y} \\ 0 & , \quad \textit{else} \end{cases} \tag{11}$$

The proof of this lemma is obvious.

The essence of the lemma above is that the collocation matrix gets a diagonal structure if both the data points and the prediction points lie on the grid G_n.

3. Identification of linear functionals
Identification of a linear functional means finding its Riesz-representer in V_n. First it will be done for a special example and later it can be concluded that the same reasoning is possible for all functionals of the same structure.
Let us assume the gravity anomalies Δg to be in V_n. For one special $\mathbf{x} = 2^n \mathbf{r} \in G_n$ the geoid undulations N is given by the following lemma:

Lemma 4 *The Riesz-representer of the linear functional $N^{\mathbf{r}}$ is given by*

$$\mathbf{R}N^{\mathbf{r}} = \sum_{\mathbf{k} \in Z^2} a_{\mathbf{k}}^{\mathbf{r}} \cdot \varphi_{n,\mathbf{k}} \tag{12}$$

with the coefficients $a_{\mathbf{k}}^{\mathbf{r}}$ given by

$$a_{\mathbf{k}}^{\mathbf{r}} := \frac{1}{2\pi\gamma} \int_{R^2} \int_{R^2} \frac{\varphi_{n,\mathbf{r}}(\mathbf{x}) \varphi_{n,\mathbf{k}}(\mathbf{y})}{|\mathbf{x} - \mathbf{y}|} d\mathbf{y} d\mathbf{x} \tag{13}$$

In a completely analog manner the reasoning can be repeated for all functionals which can be written as a convolution. Hence, almost all geodetic integral functionals can be treated in the framework of collocation in the scaling space V_n.
The collocation solution now gets very simple. In its standard form it is

$$\begin{aligned}\hat{N}^{\mathbf{r}} &= \left[< \mathbf{R}N^{\mathbf{r}}, K_n(\bullet, \mathbf{k}_1) >, \ldots, < \mathbf{R}N^{\mathbf{r}}, K_n(\bullet, \mathbf{k}_n) >\right] \times \\ &\times \begin{bmatrix} K_n(\mathbf{k}_1, \mathbf{k}_1) & \ldots & K_n(\mathbf{k}_1, \mathbf{k}_n) \\ \vdots & & \vdots \\ K_n(\mathbf{k}_n, \mathbf{k}_1) & \ldots & K_n(\mathbf{k}_n, \mathbf{k}_n) \end{bmatrix}^{-1} \begin{bmatrix} g_{\mathbf{k}_1} \\ \vdots \\ g_{\mathbf{k}_n} \end{bmatrix}\end{aligned}$$

After a short algebra it simplifies to

$$N^{\mathbf{r}} = \sum_{j=1}^{n} a^{\mathbf{r}}_{\mathbf{k}_j} \cdot g_{\mathbf{k}_j} \tag{14}$$

Hence, the collocation solution is reduced to a weighted mean of the given gravity anomalies $g_{\mathbf{k}_j}$ and no matrix inversion is necessary.

The last open question is the computation of the weights $a^{\mathbf{r}}_{\mathbf{k}}$. It would be useless if all the computational savings had to be reinvested in the computation of the weights. Fortunately, this is not the case. Due to the fact that the scaling functions are even: $\varphi_{n,\mathbf{k}}(\mathbf{x}) = \varphi_{n,\mathbf{k}}(-\mathbf{x})$ the definition (13) of the weights can be written as a double convolution

$$a^{\mathbf{r}}_{\mathbf{k}} = \varphi_{n,\mathbf{k}} * \varphi_{n,\mathbf{r}} * \frac{1}{|\mathbf{x}|} \tag{15}$$

Using the convolution theorem of Fourier transform the following efficient computation formula is obtained:

$$a^{\mathbf{r}}_{\mathbf{k}} = \frac{1}{\gamma}\mathcal{F}^{-1}\{\chi_{[-2^{-n}\pi, 2^{-n}\pi]^2} \frac{1}{|\omega|}\}(2^n\mathbf{r} - 2^n\mathbf{k}) \tag{16}$$

Here the function χ is the indicator function

$$\chi_{[-2^{-n}\pi, 2^{-n}\pi]^2}(\mathbf{x}) = \begin{cases} 1 & , \quad |x_i| < 2^{-n}\pi \quad i = 1, 2 \\ 0 & , \quad \text{else} \end{cases} \tag{17}$$

and (16) can be efficiently evaluated using FFT.

3.2 MSA based on the Haar Wavelet

Another simple wavelet is the Haar wavelet. All the investigations of the previous subsection can be repeated for MSA based on the Haar wavelet. The scaling function of the one-dimensional Haar wavelet is given by

$$\bar{\varphi}(x) := \begin{cases} 1 & , \quad 0 \le x < 1 \\ 0 & , \quad \text{else} \end{cases} \tag{18}$$

Its extension to the two-dimensional case follows exactly (4). The two-dimensional scaling functions generate a MSA:

Lemma 5 *The spaces*

$$V_n := span\{\varphi_{n,\mathbf{k}}(\mathbf{x}) \mid \mathbf{k} \in Z^2\} \tag{19}$$

form a MSA of $L^2(R^2)$

1. characterization of V_0
The functions $\varphi_{0,\mathbf{k}}$ are the characteristic functions of the squares $[k_1, k_1+1) \times [k_2, k_2+1)$, having their south-west corner in $\mathbf{k}$ and a side-length equal to 1. Hence, the elements of V_0 are block mean-value data – another important geodetic data type.

2. Determination of the reproducing kernel

Lemma 6 *Let $[\mathbf{x}]$ denote the vector of the largest integers smaller than $\mathbf{x}$. Then*

$$K_n(\mathbf{x}, \mathbf{y}) = \begin{cases} 2^{-2n} & , \; if\,[2^{-n}\mathbf{x}] = [2^{-n}\mathbf{y}] \\ 0 & , \; else \end{cases} \tag{20}$$

is the reproducing kernel of V_n.

This means the kernel equals 2^{-2n} if both $\mathbf{x}$ and $\mathbf{y}$ belong to the same block. Otherwise the kernel equals zero.

3. Identification of linear functionals
Assuming now that both the given gravity anomalies Δg and the unknown geoid undulations N are block mean-values, with a repetition of the same arguments as in the previous subsection the literally identical results can be obtained.

Lemma 7 *The Riesz-representer of the linear functional $N^{\mathbf{r}}$ is given by*

$$\mathbf{R}N^{\mathbf{r}} = \sum_{\mathbf{k} \in Z^2} a_{\mathbf{k}}^{\mathbf{r}} \cdot \varphi_{n,\mathbf{k}} \tag{21}$$

with the coefficients $a_{\mathbf{k}}^{\mathbf{r}}$ given by

$$a_{\mathbf{k}}^{\mathbf{r}} := \frac{1}{2\pi\gamma} \int_{R^2} \int_{R^2} \frac{\varphi_{n,\mathbf{r}}(\mathbf{x}) \varphi_{n,\mathbf{k}}(\mathbf{y})}{|\mathbf{x} - \mathbf{y}|} d\mathbf{y} d\mathbf{x} \tag{22}$$

If both the given gravity anomalies and the unknown geoid undulations are block mean-values, the collocation solution again gets very simple:

$$\hat{N}^{\mathbf{r}} = \sum_{j=1}^{n} a_{\mathbf{k}_j}^{\mathbf{r}} \cdot g_{\mathbf{k}_j} \tag{23}$$

It is again a weighted mean but now of block mean-values instead of point-values . A matrix inversion is obsolete again.
The computation of the weights (22) can also be efficiently achieved using FFT.

4 Numerical examples

Both the collocation solutions have been compared to the corresponding FFT solution for the conversion of gravity anomalies into geoid undulations. The data set consists of 16384 point gravity anomalies from the north part of the German state of *Baden-Württemberg*.
In the Shannon case 8129 point gravity anomalies were randomly chosen from the original 16384 values. From these 8129 gravity anomalies 16384 geoid undulations were computed by Shannon collocation and compared to the same number of geoid undulations obtained from FFT. The difference of both solutions is considered as the error of the Shannon collocation.
The results of this test are not satisfactory. They are summarized in the table 1. The situation changes if the full data set of 16384 values is used for the Shannon collocation. In this case the collocation solution and the FFT solution coincide completely. This result can be explained: For complete data grids both methods can be considered as filters. The filter function of the FFT is $\frac{1}{|\omega|}$ and those

Table 1: Accuracy of geoid undulations computed by Shannon collocation from randomly selected gravity anomalies

	Errors of collocation	Variation of geoid
rms	0.23 m	2.73 m
maximum	0.72 m	3.92 m
minimum	-0.81 m	-4.18 m

of the collocation is $\chi_{[-2^{-n}\pi,2^{-n}\pi]^2}\frac{1}{|\omega|}$. This means the collocation differs from FFT by an additional band-restriction. Due to the sampling the data is band-limited from the very beginning. Hence, the additional band-restriction of Shannon collocation has no effect and both methods coincide.
A similar test was carried out for the Haar collocation. In contrast to the Shannon case here the gravity anomalies were not selected randomly. The reduction of input data was achieved by the data compression feature of the Haar wavelet. The wavelet-compression generated a data reduction rate of about 15. The Haar collocation solution was then compared to the usual FFT solution. The difference between both solutions was considered as the error of the Haar collocation solution. The results are displayed in table 2. Despite the high data compression rate the accuracy is quite good.

Table 2: Accuracy of geoid undulations computed by Haar collocation from compressed gravity anomalies

	Errors of collocation	Variation of geoid
rms	0.16 m	2.73 m
maximum	0.83 m	3.92 m
minimum	-0.86 m	-4.18 m

If the coplete data-set is used the error does not vanish completely but becomes neglectable.

5 Conclusion

The aim of this paper was to investigate collocation in special reproducing kernel Hilbertspaces which lead to a diagonal collocation matrix. It was not intented to provide a general alternative to classical collocation. Classical collocation is unpreceeded in its flexibility and its *Plug-and-Play* use.
Only in some cases MSA based collocation can be taken into consideration for incomplete data grids or in cases where a data compression is needed.

Finally, one should not forget that there is an infinite variety of orthogonal scaling functions. Here only two very simple examples have been studied. In cases where very special data are given it might be worthwile to use collocation in spaces which are spanned by a scaling function which exactly reflects the properties of the data. For such kinds of computation further investigations are necessary.

References

[1] Sünkel H. (ed.), *Mathematical and Numerical Techniques in Physical Geodesy* , Lecture Notes in Earth Sciences, Vol. 7. , Springer-Verlag, New york 1986

[2] Tscherning C. C., Rapp R. H. , Closed Covariance Expressions for Gravity Anomalies, Geoid Undulations and Deflections of the Vertical Implied by Anomaly Degree Variance Models, *OSU-Rep. No. 208*, Columbus Ohio, 1974

[3] Walter G. G., *Wavelets and other orthogonal systems with applications.* CRC Press, Boca Raton Ann Arbor London Tokyo, 1994

AN EFFICIENT TECHNIQUE FOR THE COMPUTATION OF THE GRAVIMETRIC QUANTITIES FROM GEOPOTENTIAL EARTH MODELS

Hussein A. Abd-Elmotaal
Civil Engineering Department, Faculty of Engineering,
Minia University, Minia 61111, Egypt

Abstract

This paper presents an efficient computer technique for computing the gravimetric quantities from high degree global geopotential earth models. The developed technique is designed to compute the gravimetric quantities for an equally spaced grid in the longitude direction. It can be used to compute the gravimetric quantities either for a local region or for the whole globe. Computational timing tests have been carried out and the results prove that the developed technique saves significant computer time.

1 Introduction

Many geodetic applications use the global spherical harmonic expansions. Such applications include recovering the long wavelength part of the gravity field in case of the gravimetric geoid determination using Stokes and Vening Meinesz integrals, the determination of the orthometric heights without levelling and, in general, computing various gravimetric quantities. For these purposes, efficient computer techniques are always needed. The principal problem in performing such computations is the required computer time. This paper presents an efficient computer technique that can be used to perform the above mentioned computations.

The basic equations for the determination of the gravimetric quantities from the global geopotential earth models are given. The developed technique is described. The computational steps for the developed technique are then explained. Some computational timing tests are carried out and comparisons are made.

It should be pointed out that many alternative techniques were carried out by many authors, such as, Rapp (1982), Tscherning et al. (1983), Colombo (1981), Rizos (1979) and Tscherning et al. (1994).

2 Basic Equations for Gravimetric Quantities

The basic equations for determining the gravimetric quantities (namely, the gravity anomaly Δg, the height anomaly ζ and the components of the deflection of the vertical ξ and η) can be expressed as (Rapp, 1982, p. 5)

$$\Delta g(r,\theta,\lambda) = \frac{GM}{r^2}\sum_{n=2}^{\infty}(n-1)\left(\frac{a}{r}\right)^n \sum_{m=0}^{n}\left(\bar{C}^*_{nm}\cos m\lambda + \bar{S}_{nm}\sin m\lambda\right)\bar{P}_{nm}(\cos\theta)\,, \quad (1)$$

$$\zeta(r,\theta,\lambda) = \frac{GM}{\gamma r}\sum_{n=2}^{\infty}\left(\frac{a}{r}\right)^n \sum_{m=0}^{n}\left(\bar{C}^*_{nm}\cos m\lambda + \bar{S}_{nm}\sin m\lambda\right)\bar{P}_{nm}(\cos\theta)\,, \tag{2}$$

$$\xi(r,\theta,\lambda) = \frac{GM}{\gamma r^2}\sum_{n=2}^{\infty}\left(\frac{a}{r}\right)^n \sum_{m=0}^{n}\left(\bar{C}^*_{nm}\cos m\lambda + \bar{S}_{nm}\sin m\lambda\right)\frac{d\bar{P}_{nm}(\cos\theta)}{d\theta}\,, \tag{3}$$

$$\eta(r,\theta,\lambda) = \frac{GM}{\gamma r^2\sin\theta}\sum_{n=2}^{\infty}\left(\frac{a}{r}\right)^n \sum_{m=0}^{n} m\left(-\bar{C}^*_{nm}\sin m\lambda + \bar{S}_{nm}\cos m\lambda\right)\bar{P}_{nm}(\cos\theta)\,, \tag{4}$$

where $\bar{C}^*_{nm}$ is the difference between the actual coefficients $\bar{C}_{nm}$ and those implied by the reference equipotential ellipsoid $\bar{C}^u_{nm}$, given by

$$\begin{aligned} \bar{C}^*_{n0} &= \bar{C}_{n0} - \bar{C}^u_{n0} \qquad \text{if } m = 0\,,\\ \bar{C}^*_{nm} &= \bar{C}_{nm} \qquad\qquad \text{if } m \neq 0\,, \end{aligned} \tag{5}$$

$\bar{C}_{nm}$ and $\bar{S}_{nm}$ are the fully normalized potential coefficients, GM is the geocentric gravitational constant, r is the geocentric radius, θ is the polar distance, λ is the geodetic longitude, a stands for the equatorial radius of the mean earth's ellipsoid and $\bar{P}_{nm}$ denotes the fully normalized associated Legendre functions. The polar distance θ can be expressed in terms of the geocentric latitude ψ as:

$$\theta = 90^\circ - \psi\,, \tag{6}$$

where ψ is related to the geodetic latitude ϕ through the following expression (Torge, 1980, p. 50):

$$\tan\psi = (1-f)^2\tan\phi\,, \tag{7}$$

where f is the flattening of the earth's ellipsoid.

The radial component of the gravity disturbance δg can be given by (Heiskanen and Moritz, 1967, p. 246)

$$\delta g = \Delta g + \frac{2\gamma}{r}\zeta\,. \tag{8}$$

Hence we avoid a direct evaluation of δg and compute it after computing Δg and ζ.

3 The Developed Technique

Studying the basic equations of the gravimetric quantities given in section 2, one can see that the summation along m depends on the longitude λ. In this section, a technique that can be used to eliminate λ from the inner summation in these expressions will be given.

Assume that we would like to compute the gravimetric quantities on an equally spaced grid in the longitude direction. Thus, one may express λ as

$$\lambda = \lambda_\circ + k\Delta\lambda\,, \tag{9}$$

where $\Delta\lambda$ is the distance between each pair of the computational points in the longitude direction, k is ranging between 0 and l, and l is defined by

$$l = \frac{\lambda_r - \lambda_\circ}{\Delta\lambda}\,, \tag{10}$$

where λ_o and λ_r are the most left and most right longitudes of the region, respectively. Thus, the total number of points per row is thus $l + 1$.

Let us express $\cos m\lambda$ and $\sin m\lambda$ using the well known decomposition formulas as:

$$\cos m\lambda = \cos m(\lambda_o + k\Delta\lambda) = \cos m\lambda_o \cos mk\Delta\lambda - \sin m\lambda_o \sin mk\Delta\lambda \tag{11}$$

$$\sin m\lambda = \sin m(\lambda_o + k\Delta\lambda) = \sin m\lambda_o \cos mk\Delta\lambda + \cos m\lambda_o \sin mk\Delta\lambda \tag{12}$$

Insert (11) and (12) into (1), (2), (3) and (4), and truncate the summation along the degree n to the maximum available degree N_{max}, this gives immediately

$$\Delta g(k) = \frac{GM}{r^2} \sum_{m=0}^{N_{max}} \sum_{n=m}^{N_{max}} (n-1) \left(\frac{a}{r}\right)^n \left[a_{nm} \cos mk\Delta\lambda + b_{nm} \sin mk\Delta\lambda\right] \bar{P}_{nm}(\cos\theta), \tag{13}$$

$$\zeta(k) = \frac{GM}{\gamma r} \sum_{m=0}^{N_{max}} \sum_{n=m}^{N_{max}} \left(\frac{a}{r}\right)^n \left[a_{nm} \cos mk\Delta\lambda + b_{nm} \sin mk\Delta\lambda\right] \bar{P}_{nm}(\cos\theta), \tag{14}$$

$$\xi(k) = \frac{GM}{\gamma r^2} \sum_{m=0}^{N_{max}} \sum_{n=m}^{N_{max}} \left(\frac{a}{r}\right)^n \left[a_{nm} \cos mk\Delta\lambda + b_{nm} \sin mk\Delta\lambda\right] \frac{d\bar{P}_{nm}(\cos\theta)}{d\theta}, \tag{15}$$

$$\eta(k) = \frac{GM}{\gamma r^2 \sin\theta} \sum_{m=1}^{N_{max}} m \sum_{n=m}^{N_{max}} \left(\frac{a}{r}\right)^n \left[a_{nm} \sin mk\Delta\lambda - b_{nm} \cos mk\Delta\lambda\right] \bar{P}_{nm}(\cos\theta), \tag{16}$$

where a_{nm} and b_{nm} are given by

$$a_{nm} = \bar{C}^*_{nm} \cos m\lambda_o + \bar{S}_{nm} \sin m\lambda_o, \tag{17}$$

$$b_{nm} = -\bar{C}^*_{nm} \sin m\lambda_o + \bar{S}_{nm} \cos m\lambda_o. \tag{18}$$

The expressions (13) to (16) can be written in the following form:

$$\Delta g(k) = \frac{GM}{r^2} \sum_{m=0}^{N_{max}} (\alpha_m \cos mk\Delta\lambda + \beta_m \sin mk\Delta\lambda), \tag{19}$$

$$\zeta(k) = \frac{GM}{\gamma r} \sum_{m=0}^{N_{max}} (\mu_m \cos mk\Delta\lambda + \nu_m \sin mk\Delta\lambda), \tag{20}$$

$$\xi(k) = \frac{GM}{\gamma r^2} \sum_{m=0}^{N_{max}} (\epsilon_m \cos mk\Delta\lambda + \tau_m \sin mk\Delta\lambda), \tag{21}$$

$$\eta(k) = \frac{GM}{\gamma r^2 \sin\theta} \sum_{m=1}^{N_{max}} m (\mu_m \sin mk\Delta\lambda - \nu_m \cos mk\Delta\lambda), \tag{22}$$

where α_m, β_m, μ_m, ν_m, ϵ_m and τ_m are given by

$$\alpha_m = \sum_{n=m}^{N_{max}} (n-1) \left(\frac{a}{r}\right)^n a_{nm} \bar{P}_{nm}(\cos\theta), \tag{23}$$

$$\beta_m = \sum_{n=m}^{N_{max}} (n-1) \left(\frac{a}{r}\right)^n b_{nm} \bar{P}_{nm}(\cos\theta), \tag{24}$$

$$\mu_m = \sum_{n=m}^{N_{max}} \left(\frac{a}{r}\right)^n a_{nm}\bar{P}_{nm}(\cos\theta)\,, \tag{25}$$

$$\nu_m = \sum_{n=m}^{N_{max}} \left(\frac{a}{r}\right)^n b_{nm}\bar{P}_{nm}(\cos\theta)\,, \tag{26}$$

$$\epsilon_m = \sum_{n=m}^{N_{max}} \left(\frac{a}{r}\right)^n a_{nm}\frac{d\bar{P}_{nm}(\cos\theta)}{d\theta}\,, \tag{27}$$

$$\tau_m = \sum_{n=m}^{N_{max}} \left(\frac{a}{r}\right)^n b_{nm}\frac{d\bar{P}_{nm}(\cos\theta)}{d\theta}\,. \tag{28}$$

Let us introduce the following recursion formulas (Colombo, 1981, p. 20)

$$A\cos kx = 2A\cos x\cos(k-1)x - A\cos(k-2)x\,, \tag{29}$$

$$B\sin kx = 2B\cos x\sin(k-1)x - B\sin(k-2)x\,. \tag{30}$$

The addition (or subtraction) of these two expressions gives:

$$D(k,x) = 2\cos x\,D(k-1,x) - D(k-2,x)\,, \tag{31}$$

where, e.g., $D(k,x)$ is given by

$$D(k,x) = A\cos kx \pm B\sin kx\,.$$

Using this useful formula, final expressions for the gravimetric quantities Δg, ζ, ξ and η can be written as:

$$\Delta g(k) = \frac{GM}{r^2}\sum_{m=0}^{N_{max}} D_{\Delta g}(k)\,, \tag{32}$$

$$\zeta(k) = \frac{GM}{\gamma r}\sum_{m=0}^{N_{max}} D_{\zeta}(k)\,, \tag{33}$$

$$\xi(k) = \frac{GM}{\gamma r^2}\sum_{m=0}^{N_{max}} D_{\xi}(k)\,, \tag{34}$$

$$\eta(k) = \frac{GM}{\gamma r^2\sin\theta}\sum_{m=1}^{N_{max}} mD_{\eta}(k)\,, \tag{35}$$

where

$$D_i(k) = 2\cos(m\Delta\lambda)D_i(k-1) - D_i(k-2) \tag{36}$$

with the following starting elements:

$$D_{\Delta g}(0) = \alpha_m\,, \tag{37}$$

$$D_{\Delta g}(1) = \alpha_m\cos(m\Delta\lambda) + \beta_m\sin(m\Delta\lambda)\,, \tag{38}$$

$$D_{\zeta}(0) = \mu_m\,, \tag{39}$$

$$D_{\zeta}(1) = \mu_m\cos(m\Delta\lambda) + \nu_m\sin(m\Delta\lambda)\,, \tag{40}$$

$$D_{\xi}(0) = \epsilon_m\,, \tag{41}$$

$$D_{\xi}(1) = \epsilon_m\cos(m\Delta\lambda) + \tau_m\sin(m\Delta\lambda)\,, \tag{42}$$

$$D_{\eta}(0) = -\nu_m\,, \tag{43}$$

$$D_{\eta}(1) = \mu_m\sin(m\Delta\lambda) - \nu_m\cos(m\Delta\lambda)\,. \tag{44}$$

4 Computational Steps

Let us summarize the steps of the developed technique as follows:

1. Generate two arrays of $\cos m\lambda_\circ$ and $\sin m\lambda_\circ$ for $0 \leq m \leq N_{max}$. The dimension of these arrays is $N_{max}+1$. The computations of $\cos m\lambda_\circ$ and $\sin m\lambda_\circ$ can be speed up by using the following recursion formulas analogous to (29) and (30)

$$\sin m\lambda_\circ = 2\cos\lambda_\circ \sin(m-1)\lambda_\circ - \sin(m-2)\lambda_\circ\,, \quad (45)$$

$$\cos m\lambda_\circ = 2\cos\lambda_\circ \cos(m-1)\lambda_\circ - \cos(m-2)\lambda_\circ\,. \quad (46)$$

2. Generate the a_{nm} and b_{nm} arrays using (17) and (18). The dimension of these arrays is $(N_{max}+1)(N_{max}+2)/2-3$.
3. Working through row-by-row, first working at a single order m, calculate the fully normalized associated Legendre functions and their derivatives only for a certain order m complete from $n = m$ to $n = N_{max}$ (one can use the recursive algorithm described by Colombo (1981)). The required dimension for $\bar{P}_{nm}(\cos\theta)$ and $d\bar{P}_{nm}(\cos\theta)/d\theta$ is then only $N_{max}+1$.
4. Generate α_m, β_m, μ_m, ν_m, ϵ_m and τ_m using (23) to (28). Here α_m, β_m, μ_m, ν_m, ϵ_m and τ_m are single values (remember we are working at a certain order m).
5. Compute the contribution of the order m on the gravimetric quantities $\Delta g(k)$, $\zeta(k)$, $\xi(k)$ and $\eta(k)$ for each point of the row using (32) to (35) for $0 \leq k \leq l$.
6. Repeat steps 3 to 5 for $0 \leq m \leq N_{max}$ to get the contribution of all orders m on the gravimetric quantities $\Delta g(k)$, $\zeta(k)$, $\xi(k)$ and $\eta(k)$ for all points at the same row.
7. Compute the gravity disturbance $\delta g(k)$ using (8) (after the final computation of $\Delta g(k)$ and $\zeta(k)$) for all points at the same row.
8. Repeat steps 3 to 7 for all rows of the region.

5 Timing Tests

To obtain time comparisons, some tests were performed on an IBM-AT Pentium/200 computer. The used geopotential coefficient set is complete to degree and order 360. The listed computer times exclude the time for reading the geopotential coefficients and removing the reference field.

Two other techniques are used for the time comparison. The first is the Rapp (1982) technique which depends on the point-by-point basis of computing the gravimetric quantities and is a representative of the many techniques that use similar recursion procedures. The second is the GEOCOL program provided by GRAVSOFT package (Tscherning et al., 1994).

The first test is the computation of the gravity anomaly Δg at different sets of points. These different sets of points are, single point, 720 points at a $30'$ interval in the longitude direction on a given latitude and a global $30' \times 30'$ grid. The results are shown in Table 1. It shows that for computing Δg on a single point, Rapp's technique takes the shortest computer time. For computing Δg for a limited area or on a global grid, GEOCOL program saves significant computer time but the developed technique takes the shortest computer time.

The second test is the computation of Δg, ζ, ξ, η and δg at the same sets of points. Here only Rapp (1982) and the developed techniques are used. The results are given in Table 2. It shows that the developed technique dramatically saves the required computer time.

Table 1. CPU time for computing Δg at different sets of points.

Used Technique	CPU Time (sec.)		
	1 point	720 points on a latitude	global $30' \times 30'$ grid
Rapp (1982)	0.11	13.73	4942*
GEOCOL	0.17	1.10	368.7
Developed Technique	0.22	0.39	101.5

* extrapolated value by multiplying the time of 720 points by 360

Table 2. CPU time for computing Δg, ζ, ξ, η and δg at different sets of points.

Used Technique	CPU Time (sec.)		
	1 point	720 points on a latitude	global $30' \times 30'$ grid
Rapp (1982)	0.17	23.07	8305*
Developed Technique	0.27	0.60	178.8

* extrapolated value by multiplying the time of 720 points by 360

6 Conclusion

An efficient computer technique to compute the gravimetric quantities Δg, ζ, ξ, η and δg from high degree geopotential earth models is described in this paper. It proved to save significant computer time in most cases. If the computation of the gravimetric quantities is required only at a single point, the Rapp (1982) technique would be efficient. If the computation of these quantities is required either for a limited region or for the whole globe, the developed technique is most efficient.

References

Colombo, O. (1981) Numerical Methods for Harmonic Analysis on the Sphere, *Department of Geodetic Science, The Ohio State University, Columbus, Ohio,* **310**.

Heiskanen, W.A. and Moritz, H. (1967) *Physical Geodesy,* Freeman, San Francisco.

Rapp, R.H. (1982) A Fortran Program for the Computation of Gravimetric Quantities From High Degree Spherical Harmonic Expansions, *Department of Geodetic Science, The Ohio State University, Columbus, Ohio,* **334**.

Rizos, C. (1979) An Efficient Computer Technique for the Evaluation of Geopotential From Spherical Harmonic Models, *Aust. J. Geodesy, Photogrammetry and Surveying,* **31**, 161–169.

Torge, W. (1980) *Geodesy,* Walter de Gruyter, Berlin, New York.

Tscherning, C.C., Knudsen, P. and Forsberg, R. (1994) Description of the GRAVSOFT Package, *Geophysical Institute, University of Copenhagen, Technical Report.*

Tscherning, C.C., Rapp R.H. and Goad, C. (1983) A Comparison of Methods for Computing Gravimetric Quantities From High Degree Spherical Harmonic Expansions. *Manuscripta geodaetica,* **8**, 249–272.

MODIFIED KERNELS IN SPECTRAL GEOID DETERMINATION: FIRST RESULTS FROM WESTERN AUSTRALIA

W E Featherstone
School of Spatial Sciences, Curtin University of Technology,
GPO Box U1987, Perth WA 6845, Australia.

M G Sideris
Department of Geomatics Engineering, The University of Calgary,
University Drive NW, Calgary, Alberta, Canada.

Abstract

The deterministic kernel modifications proposed by Wong and Gore (1969), Meissl (1971), and Vanicek and Kleusberg (1987) have been used in the one-dimensional fast Fourier transform (1D-FFT) implementation of Stokes's integral. Geoid results over Western Australia are compared with Global Positioning System and Australian Height Datum data to illustrate improvements made upon the spherical Stokes kernel.

Introduction

In the mid-1980s, the fast Fourier transform (FFT) began to find wide-spread use in geoid determination because of its efficient evaluation of convolution integrals, when compared to classical numerical integration. For many years, the planar, two-dimensional FFT was used (eg. Schwarz *et al.*, 1990). Strang van Hees (1990) then introduced the spherical, two-dimensional FFT. However, each of these FFT implementations are subject to several approximation errors, the most notable of which is the simplification of Stokes's kernel. Therefore, Forsberg and Sideris (1993) proposed the spherical, multi-band FFT, which reduces the impact of the simplified kernel. Haagmans *et al.* (1993) refined this approach to give the spherical, one-dimensional FFT, which requires no simplification of Stokes's kernel. Ironically, the 1D-FFT is a combination of the FFT and numerical integration, but is indisputably faster than numerical integration alone.

The geoid can be computed from a combination of a global geopotential model (N_L) and residual gravity anomalies (Δg^L) in the area of interest via the 1D-FFT by

$$N = N_L + \frac{R\Delta\phi\Delta\lambda}{4\pi\gamma}\mathbf{F}_1^{-1}\left[\sum_{\phi}\mathbf{F}_1[S(\Delta\lambda)]\mathbf{F}_1\left[\Delta g^L \cos\phi\right]\right] , \tag{1}$$

with

$$N_L = \frac{GM}{r\gamma}\sum_{n=2}^{L}\left(\frac{a}{r}\right)^n \sum_{m=0}^{n}[\delta C_{nm}\cos m\lambda + S_n \sin m\lambda]P_{nm}(\cos\theta) , \tag{2}$$

and

$$\Delta g^L = \Delta g - \frac{GM}{r^2}\sum_{n=2}^{L}\left(\frac{a}{r}\right)^n (n-1)\sum_{m=0}^{n}[\delta C_{nm}\cos m\lambda + S_n \sin m\lambda]P_{nm}(\cos\theta) , \tag{3}$$

where R is the radius of a spherical Earth, γ is normal gravity on the reference ellipsoid, $S(\Delta\lambda)$ is Stokes's kernel function, $\Delta\phi$ and $\Delta\lambda$ are the latitudinal and longitudinal grid spacing of the residual gravity anomalies respectively, $\mathbf{F_1}$ and $\mathbf{F_1}^{-1}$ are the discrete one-dimensional fast Fourier transform operator and its inverse respectively, which are applied along the parallels, GM is the geocentric gravitational constant, (r,θ,λ) are the spherical polar coordinates of the computation point, a is the semi-major axis length of the reference ellipsoid, $P_{nm}(\cos\theta)$ are the fully normalised associated Legendre polynomials for degree n and order m, δC_{nm} and S_{nm} are the fully normalised spherical harmonic coefficients of the geopotential model, which have been reduced by the even zonal harmonics of the reference ellipsoid, and L is the maximum degree of their expansion.

Regional determinations of the geoid using the FFT often convolve the whole grid of gravity anomalies with Stokes's kernel. Conversely, geoid determinations using numerical integration use residual gravity anomalies over a spherical cap of radius (ψ_0) about each computation point. Whether the integration is performed over a spherical cap or the whole data area, this still represents an approximation of Stokes's integral. This results in a truncation error due to the neglect of residual gravity anomalies in the remote zones outside the integration domain. This truncation error is commonly neglected during practical geoid computations, particularly when using the fast Fourier transform approach (eg. Schwarz *et al.*, 1990). However, such an assumption is not strictly valid because neither the integration kernel nor the residual gravity anomalies are zero outside the integration domain.

Instead, the impact of this truncation error can be reduced through a modification to Stokes's integration kernel. Several authors have proposed reductions of the truncation error for a spherical cap using deterministic kernel modifications (eg. Molodensky *et al.*, 1962; Wong and Gore, 1969; Meissl, 1971; Vanicek and Kleusberg, 1987) or stochastic modifications (eg. Wenzel, 1982; Sjöberg, 1991; Vanicek and Sjoberg, 1991). Kernel modifications have also been suggested to reduce the truncation error for integration areas other than a spherical cap (eg. Neyman *et al.*, 1996; Zelin and Zoufa, 1992).

The deterministic kernel modifications

Wong and Gore (1969) and Vanicek and Kleusberg (1987) propose to remove the low-degree Legendre polynomials from Stokes's kernel, which also reduces the magnitude of the truncation error. This will be called the Wong and Gore kernel, and is given by

$$S^{M}(\Delta\lambda) = S(\Delta\lambda) - \sum_{n=2}^{M} \frac{2n+1}{n-1} P_n(\Delta\lambda) \quad \text{for} \quad 0 \le \Delta\lambda \le \psi_0 \ , \tag{4}$$

where $P(\Delta\lambda)$ are Legendre's polynomials along the parallel, and M ($\le L$) is the degree of spheroidal modification to the integration kernel.

If the Stokes integration is performed over the whole sphere using the Wong and Gore kernel, the orthogonality relations dictate that the result is identical to that achieved when the low frequencies ($2 \le n \le M$) are removed from the kernel, the gravity anomalies, or both. However, when the integration is performed over a limited area, as is the case in any regional geoid determination, the Wong and Gore kernel only acts as a partial high-pass filter. Therefore, if the values of M and L are different, different results can be expected.

Meissl's (1971) kernel modification simply subtracts the value of Stokes's kernel at the truncation radius from the kernel inside the spherical cap. Thus, the Meissl kernel is

$$S_{me}(\psi) = S(\psi) - S(\psi_0) \quad \text{for} \quad 0 \le \Delta\lambda \le \psi_0 \ . \tag{5}$$

This modification causes the Fourier series of the truncation error to converge to zero more rapidly, which can be proven by applying Green's second identity (eg. Jekeli, 1981).

Vanicek and Kleusberg (1987) and Vanicek and Sjöberg (1991) apply theory similar to that used by Molodensky *et al.* (1962) to Eq. (4). This modification scheme gives

$$S^{M*}(\Delta\lambda) = S^{M}(\Delta\lambda) - \sum_{n=2}^{M} \frac{2n+1}{2} t_n P_n(\Delta\lambda) \quad \text{for} \quad 0 \le \Delta\lambda \le \psi_0 \ , \tag{6}$$

where the coefficients t_n are analogous with the spherical truncation coefficients Q_n defined by Molodensky *et al.* (*ibid.*), and are given by Vanicek and Sjöberg (*ibid.*) as

$$\sum_{n=2}^{M} \frac{2n+1}{2} t_n e_{kn} = Q_k - \sum_{n=2}^{M} \frac{2n+1}{n-1} e_{kn} \ , \tag{7}$$

where the coefficients e_{kn} can be computed using the recursive relations of Paul (1973).

Comparison of Modified 1D-FFT Results with GPS at AHD Benchmarks

In this case study, the deterministic modifications outlined above were implemented in the 1D-FFT software developed at the University of Calgary (Sideris, 1994). Stochastic kernel modifications were not considered because accurate estimates of the errors in the terrestrial gravity data are not currently available. In order to prevent the whole grid of gravity anomalies being used during the 1D-FFT geoid computation, the kernel was set to zero outside the cap radius (ψ_0) before being transformed to the frequency domain. Similarly, each deterministically modified kernel was computed before its transform to the frequency domain. As is customary, the 1D-FFT geoid results were compared with Global Positioning System (GPS) and optical levelling data to determine if any improvements are made when using spherical integration caps and deterministically modified kernels.

All geoid computations were conducted on a 5′ by 5′ grid over Western Australia in an area bound by 37°S < ϕ < 13°S and 112°E < λ < 131°E. The terrestrial gravity data used comprise simple 5′ by 5′ means of the validated Australian gravity data-base (Featherstone *et al.*, 1997), supplemented with marine gravity anomalies derived from combined satellite altimeter missions (Sandwell *et al.*, 1995). These free-air gravity anomalies were reduced by the L = 360 expansion of EGM96 (Lemoine *et al.*, 1997), then gridded using splines (Smith and Wessel, 1990). Digital elevation models were not used in the computations because the direct and indirect effects of the terrain are known to be small over most of Western Australia (Zhang and Featherstone, 1997). The 1D-FFT geoid solutions, computed using various cap radii and modified kernels, were compared with 65 discrete geoid heights from Western Australian GPS networks (Stewart *et al.*, 1997) co-located with optically levelled benchmarks on the Australian Height Datum (AHD), which is a third-order datum. The results are summarised in Table 1.

Table 1. Statistical fit of the modified 1D-FFT free-air co-geoid models to 65 discrete GPS-AHD geoid heights in Western Australia (units in metres).

kernel	degree	ψ_0 (°)	max.	min.	mean	std	rms
EGM96 only	L=360	n/a	1.831	-0.255	1.024	0.345	1.080
Stokes	n/a	0.25	1.781	-0.831	1.054	0.376	1.119
"	"	0.50	1.711	-0.660	1.051	0.372	1.115
"	"	1.00	1.904	-0.632	1.044	0.411	1.122
"	"	1.50	2.127	-0.669	1.047	0.440	1.136
"	"	2.00	2.251	-0.564	1.045	0.459	1.141
"	"	30.00	2.291	-0.438	1.179	0.491	1.276
Wong and Gore	M=360	0.25	1.823	-0.533	1.032	0.343	1.087
"	"	0.50	1.851	-0.603	1.034	0.358	1.094
"	"	1.00	1.834	-0.626	1.035	0.362	1.097
"	"	1.50	1.826	-0.636	1.035	0.363	1.097
"	"	2.00	1.821	-0.629	1.036	0.363	1.098
"	"	30.00	1.823	-0.631	1.036	0.364	1.100
Wong and Gore	M=36	0.25	1.786	-0.796	1.051	0.371	1.114
"	"	0.50	1.733	-0.669	1.049	0.364	1.111
"	"	1.00	1.709	-0.643	1.045	0.374	1.110
"	"	1.50	1.770	-0.660	1.045	0.379	1.112
"	"	2.00	1.768	-0.658	1.045	0.379	1.111
"	"	30.00	1.771	-0.611	1.073	0.387	1.140
Meissl	n/a	0.25	1.811	-0.624	1.039	0.350	1.096
"	"	0.50	1.777	-0.685	1.046	0.375	1.105
"	"	1.00	1.732	-0.648	1.045	0.362	1.106
"	"	1.50	1.746	-0.659	1.045	0.377	1.111
"	"	2.00	1.849	-0.646	1.044	0.389	1.114
"	"	30.00	2.266	-0.464	1.153	0.492	1.125
Vanicek/Kleusberg	M=36	0.25	1.786	-0.798	1.051	0.371	1.115
"	"	0.50	1.730	-0.668	1.050	0.365	1.111
"	"	1.00	1.752	-0.641	1.045	0.381	1.112
"	"	1.50	1.873	-0.663	1.046	0.394	1.118
"	"	2.00	1.933	-0.627	1.044	0.400	1.119
"	"	30.00	2.146	-0.537	1.101	0.456	1.192

In Table 1, a bias of ~1m exists because the zero-degree term has been considered (Kirby and Featherstone, 1997), and there is a ~0.7m separation between the geoid (W_0=constant) and the AHD (Rapp, 1994). It is interesting to note that the inclusion of the terrestrial gravity data often degrades the standard deviations of the agreements with the 65 GPS-AHD control points. Similar results are seen in Forsberg and Featherstone (1998) for the whole Australian continent. This is probably due to a combination of errors in the terrestrial gravity data and global geopotential model, and errors in the control data through distortions in the AHD in Western Australia (Featherstone and Stewart, 1998). Nevertheless, as this investigation is only concerned with the relative merits of using spherical caps and modified kernels in the 1D-FFT, the exact origin or removal of these errors is immaterial.

The computations have utilised cap radii ranging from 0.25° (the Nyquist frequency of EGM96) to 2° (the minimum radius from the edge of the data area and a control point). A cap-size of 30° has also been specified, so as to allow the whole gravity data grid to be used by way of comparison. The most striking observation is that the use of a cap improves the standard deviation of the comparisons over those achieved when using the whole data area (eg. 0.376m for ψ_0=0.25° versus 0.491m for the whole grid). This clearly illustrates that it is preferable to implement the 1D-FFT with a limited spherical cap in Western Australia.

The results are improved further when the deterministically modified forms of Stokes's kernel are used. The modified kernels yield improved agreements over the spherical Stokes kernel for each cap-radius. This indicates that it is preferable to use a limited spherical cap together with a modified integration kernel in 1D-FFT geoid computations over Western Australia. Indeed, this approach makes the spectral and numerical integration approaches to geoid computation even more similar, but with the attraction of the improved computational efficiency offered by spectral methods.

Finally, the M=36 and M=360 spheroidal modifications have been used for the Wong and Gore kernel in order to illustrate the (partial) high-pass filtering effect of a high degree of spheroidal modification. The M=360 modification gives slightly better agreements than the M=36 kernel because it filters out most of the medium frequency errors from the residual gravity anomalies. Note also that the statistics of the differences in Table 1 are weakly dependent on the cap radius for the M=360 modification. This is because this modification is a more efficient high-pass filter than a limited spherical cap.

Conclusions

From these preliminary results, it is clear that spectral geoid computations should use a spherical cap of limited extent instead of the whole gravity grid over Western Australia. This causes the FFT approach to more closely mimic numerical integration. A limited spherical cap acts as a (partial) high-pass filter, and thus removes the detrimental effects of erroneous residual gravity data. Moreover, small but consistent improvements are observed when a deterministically modified Stokes kernel is also used. This is because the kernel modification reduces the truncation error associated with performing the convolution over a limited area. Therefore, it is recommended that limited integration caps and modified kernels are used in future 1D-FFT geoid determinations of Western Australia.

Acknowledgments: *We would like to thank the Western Australian Department of Land Administration, Australian Surveying and Land Information Group, Australian Geological Survey Organisation, US National Imagery and Mapping Authority, and US National Aeronautics and Space Administration for providing data.*

References

Featherstone WE, Kearsley AHW, Gilliland JR (1997) Data preparations for a new Australian gravimetric geoid, *Aust Surv*, 42(1): 33-44.

Featherstone WE, Stewart MP (1998) Possible evidence for systematic distortions in the Australian Height Datum in Western Australia, *Geom Res Aust*, (in press).

Forsberg R, Sideris MG (1993) Geoid computations by the multi-band spherical FFT approach, *Bull Geod*, 18: 82-90.

Forsberg R, Featherstone WE (1998) Geoids and cap sizes. *Proc IAG Conf*, Rio (this volume).

Haagmans RR, de Min E, van Gelderen M (1993) Fast evaluation of convolution integrals on the sphere using 1D-FFT, and a comparison with existing methods for Stokes's integral. *manuscr geod*, 18(5): 227-241.

Jekeli C (1981) Modifying Stokes's function to reduce the error of geoid undulation computations, *J Geophys Res*, 86(B8): 6985-6990.

Kirby JF, Featherstone, WE (1997) A study of zero- and first-degree terms in geopotential models over Australia, *Geom Res Aust*, 66: 93-108.

Lemoine FG, Smith DE, Smith R, Kunz L, Pavlis NK, Klosko SM, Chinn DS, Torrence MH, Williamson RG, Cox CM, Rachlin KE, Wang WM, Pavlis EC, Kenyon SC, Salman R, Trimmer R, Rapp RH, Nerem RS (1997) The development of the NASA, GSFC and DMA joint geopotential model, *Proc IAG Conf*, Tokyo.

Meissl P (1971) Preparations for the numerical evaluation of second-order Molodensky-type formulas. *Report 163*, Dept Geod Sci & Surv, Ohio State University, Columbus.

Molodensky MS, Eremeev VF, Yurkina MI (1962): *Methods for Study of the External Gravitational Field and Figure of the Earth*, Israeli Programme for the Translation of Scientific Publications, Jerusalem.

Neyman YM, Li J, Liu Q (1996) Modification of Stokes and Vening-Meinesz formulas for the inner zone of arbitrary shape by minimisation of upper bound truncation errors, *J Geod*, 70: 410-418.

Paul MK (1973) A method of evaluating the truncation error coefficients for geoidal height, *Bull Géod*, 47: 413-425.

Rapp RH (1994) Separation between reference surfaces of selected vertical datums. *Bull Géod*, 69: 26-31.

Sandwell DT, Yale MM, Smith WHF (1995) Gravity anomaly profiles from ERS-1, TOPEX and Geosat altimetry, *EOS Trans AGU*, 76(17), S89.

Schwarz KP, Sideris MG, Forsberg R (1990) The use of FFT techniques in physical geodesy, *Geophys J Int* 100: 485-514.

Sideris MG (1994) Geoid determination by FFT techniques. *Lecture Notes*, International School for the Determination and Use of the Geoid, DIIAR, Politecnico di Milano.

Sjöberg LE (1991) Refined least squares modification of Stokes's formula. *manuscr geod*, 16: 367-375.

Smith WHF, Wessel P (1990) Gridding with continuous curvature splines in tension, *Geophys*, 55(3): 293-305.

Stewart MP, Ding X, Tsakiri M, Featherstone WE (1997) The 1996 STATEFIX Project Final Report. *Contract Report*, School of Surveying and Land Information, Curtin University of Technology.

Strang van Hees GL (1990) Stokes's formula using fast Fourier techniques. *manuscr geod*, 15: 235-239.

Vanicek P, Kleusberg A (1987) The Canadian geoid - Stokesian approach, *manuscr geod*, 12: 86-98.

Vanicek P, Sjöberg LE (1991) Reformulation of Stokes's theory for higher than second-degree reference field and modification of integration kernels, *J Geophys Res*, 96(B4): 6529-6540.

Wenzel H-G (1982) Geoid computation by least squares spectral combination using integral kernels, *Proc IAG Conf*, Tokyo.

Wong L, Gore R (1969) Accuracy of geoid heights from modified Stokes kernels, *Geophys J R astr Soc*, 18: 81-91.

Zelin G, Zuofa L (1992) Modified Stokes's integral formulas using FFT. *manuscr geod*, 17: 227-232.

Zhang KF, Featherstone WE (1997) A preliminary evaluation of the terrain effects on gravimetric geoid determination in Australia, *Proc IAG Conf*, Tokyo.

GEOIDS AND CAP SIZES

Rene Forsberg
Geodynamics Dept., KMS, Rentemestervej 8,
DK-2400 Copenhagen NV, Denmark, E-mail: rf@kms.min.dk
Will Featherstone
School of Spatial Sciences, Curtin University of Technology,
Perth, WA 6845, Australia, E-mail: Featherstone_WE@cc.curtin.edu.au

ABSTRACT. This paper gives results of applying Stokes' function with different integration radii (cap sizes) for new FFT geoid solutions of Australia and Scandinavia. While Australian results show that a limited cap size or a modified Stokes kernel is necessary for optimal results, this is not true for Scandinavia, where large cap sizes are necessary. This must ultimately reflect problems in either the fit of the spherical harmonic model, or systematic errors in the terrestrial gravity data. The overall quality of the computed geoids are compared to GPS levelling, ranging from 30 cm in the Australian-wide GPS network, to 2-5 cm in national Scandinavian networks, reflecting mainly differences in quality of the available levelling data.

INTRODUCTION

The determination of continental-scale regional geoid models by FFT methods or Stokes integration invariably has an assumption - implicit or explicit - on integration cap size. The performance of various geoid models depend strongly on this cap size, and results can be seriously degraded if too large cap sizes are used. The degradation depends very much on reference field model quality, systematic errors in availaible gravity data, and is different from region to region. In the sequel, different spherical FFT geoid computations of two large areas - Australia and Scandinavia - illustrate these effects by comparisons to GPS-levelling results.

The geoid solutions have been obtained by the multiband spherical FFT method (Forsberg and Sideris, 1993), which like other spherical FFT methods directly implement Stokes integration

$$N = N_{ref} + \frac{R}{4\pi\gamma} \int_{\psi=0}^{\psi_0} \int_{\alpha=0}^{2\pi} (\Delta g - \Delta g_{ref}) \; S(\psi) \; \psi d\psi d\alpha \tag{1}$$

through a sequence of frequency-domain convolutions, and all carried out relative to a spherical harmonic reference field

$$T_{ref} = \frac{GM}{R} \sum_{n=0}^{N} \left(\frac{R}{r}\right)^n \sum_{m=0}^{n} (C_{nm}\cos m\lambda + S_{nm}\sin m\lambda) P_{nm}\sin(\phi) \tag{2}$$

In the present geoid solutions, EGM96 (Lemoine et al, 1997) has been used. It is clear that any systematic errors in Δg can propagate into serious geoid errors when the integration kernel cap size ψ_0 is large. To circumvent this problem the cap size can either be limited to rather small distances, or modified Stokes' kernels can be used, where the influence of the longest wavelengths are attenuated, e.g. by the Wong and Gore (1969) modification

$$N = N_{ref} + \frac{R}{4\pi\gamma} \int_{\psi=0}^{\psi_0} \int_{\alpha=0}^{2\pi} (\Delta g - \Delta g_{ref}) \left[S(\psi) - \sum_{n=2}^{n_{max}} \frac{2n+1}{n-1} P_n(\psi)\right] \psi d\psi d\alpha \tag{3}$$

This corresponds effectively to reducing the influence of the n_{max} lowest harmonics of Δg prior to Stokes' integration, thus limiting the influence of long-wavelength errors. The modified Stokes function (in the square brackets) are readily implemented by using recursive algorithms for the Legendre functions P_n.

CAP SIZES AND THE GEOID OF AUSTRALIA

The Australian continent provides a unique oppurtunity to test Stokes' integration over very large regions, and cap size effects here and elsewhere have been reported by numerous authors, see e.g. Kearsley (1988) or Zhang (1997). The Australian geoid computation was done in this study by gridding available free-air data, including marine gravimetry data around the coast, as provided by the Australian Geological Survey Organization (AGSO), onto a 6' grid by least-squares collocation, covering the region 48°-8°S, 108°-162°E. Gridding was done on gravity data reduced for EGM96 and atmospheric effects, the results of Table 1 show that Australian gravity field is very smooth, and that EGM96 provides a reasonably good fit to the data.

Table 1. Statistics of original and reduced 6' selected Australian gravity data

Unit. mgal	mean	std.dev.	min	max
Original data	-0.6	12.3	-95.2	184.3
EGM96-reduced	0.4	27.5	-223.9	275.7

A gravimetric geoid was computed by the multi-band spherical FFT method, using 5 bands and a 50% zero-padded grid of 600 x 810 points, with subsequent restoral of the EGM96 geoid. Several geoid solutions were computed, corresponding to different integration radii ψ_0 and use of theWong-Gore Stokes kernel modification. To be more strict and in the wording of classical geoid determination, the computations resulted in a free-air co-geoid, as the gravimetric terrain correction and corresponding Helmert indirect effects were not considered.

The geoid solutions were compared to GPS-levelling heights of the Australian National GPS Network ("ANN"), a data set of 59 points in ITRF92-epoch 1994 provided by the Australian Surveying and Land Information Group (AUSLIG), cf. Fig. 1. Additionally, a local Western Australian GPS-levelling data set "MERL" (21 GPS points on 3rd order levelling bench-marks) was used for a local comparison. Results are given in Table 2, and show clearly that the inclusion of larger and larger integration caps clearly degrades results, and optimal cap size is somewhere below 0.5°. This illustrates the very good performance of EGM96, and could point to minute systematic errors in the terrestrial gravity data. A test was additionally carried out with terrain corrections (with terrain corrections for all of Australia computed on a 0.01° grid by FFT), but the general trend persisted. When using the Wong-Gore modification, good results are obtained even for large cap sizes, showing the usefulness of the modified kernels.

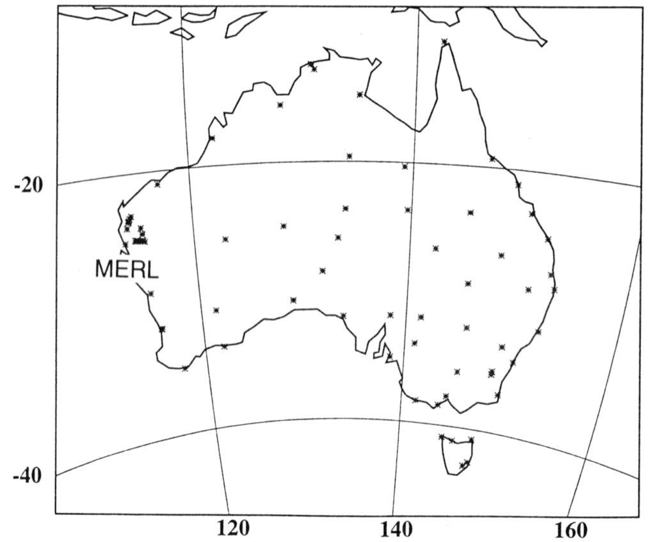

Fig. 1. Location of Australian National GPS Net (ANN), and local WA net (MERL).

When noting the relatively poor 30-cm performance of the gravimetric geoid when compared to the continental-scale ANN network, it should be borne in mind that the Australian Height Datum (AHD) does not correspond to an equipotential surface (the height datum has been defined by fixing 30 tide gauges to zero around the Australian coast), and a majority of the discrepancy probably is due to unaccounted sea-surface topography at the tide gauges.

Table 2. Difference of FFT geoids to GPS levelling in Australia (unit: m)

Cap size ψ_o	Geoid from national GPS net (ANN)		Geoid from local WA net (MERL)	
	mean	std.dev.	mean	std.dev.
Unmodified: 0.2°	.030	.324	-.119	.063
0.5°	.024	.309	-.154	.078
1.0°	.039	.311	-.217	.071
2.0°	.038	.367	-.267	.133
5.0°	-.008	.491	-.476	.131
10.0°	-.125	.786	-.807	.128
Wong-Gore: 10°, $n_{max} = 36$	.036	.318	-.127	.077
10°, $n_{max} = 180$	.030	.360	-.096	.075
10°, $n_{max} = 360$	.006	.367	-.110	.080

CAP SIZES AND GEOID OF SCANDINAVIA

Geoid computations in the Nordic and Baltic region have been carried out in a continuing project over a number of years, cf. (Forsberg, 1990). The most recent solution - NKG96 - is a 9-band spherical FFT solution, covering the area 52°-72°N, 1°-33°E, based on a zero-padded 1.5' x 3' gravity grid of 1600 x 1280 points. The NKG96 solution is based on significant new gravity data from the Baltic regions, as well as very dense gravity data over much of the Nordic region. Satellite altimetry has been draped upon existing marine gravity data in the Baltic Sea using least-squares collocation to fill data voids. In addition, detailed digital elevation models have been taken into account through an RTM terrain reduction of all gravity data; for details see Forsberg et al. (1996). Table 3 shows the statistics of the gravity data reductions in the Nordic Area. The EGM96 fits a little bit poorer in Scandinavia than in Australia, probably due to the combination of a rougher gravity field due to the topography, and a more inhomogeneous gravity data set. After RTM reduction (and atmospheric correction) the reduced field is indeed very smooth.

Table 3. Statistics of gravity data reductions in Nordic area 53-73°N, 1-33° E (unit: mgal)

Points: 182239	mean	std.dev.	min	max
Original data	-0.8	25.0	-141.2	193.2
Δg - EGM96 ref.	-1.5	15.5	-167.8	132.0
Δg - ref. - RTM	0.5	9.8	-68.8	89.1

For the present investigations the NKG96 geoid computation was repeated with the same data and procedures, the only change being that the integration cap size ψ_0 is varied (and that no quasigeoid/geoid conversion and fitting of the gravimetric geoid to local sea level has been done). The geoids computed with the different cap sizes are subsequently compared to various high-quality GPS-levelling data sets, all at first-order levelling benchmarks with GPS coordinates expressed in ITRF93. The GPS/levelling data sets used for comparison here are (cf. Fig. 2):

- The "IfE" line, 46 pts. on a N-S GPS traverse from the Danish/German border to Tromsø, Norway, carried out by the National Survey Agencies in cooperation with Institut für Erdmessund, Hannover,
- The "SWET" E-W GPS profile, 34 pts. from Bergen, Norway to the Finnish/Russian border, carried out by the Nordic Geodetic Surveys in support of geoid determination (Poutanen, 1994),
- The Swedish National GPS net (SWEPOS, 21 points, data provided by B.-G. Reit, Sweden),
- The Danish National GPS network (REFDK, 67 pts., data provided by B. Madsen).

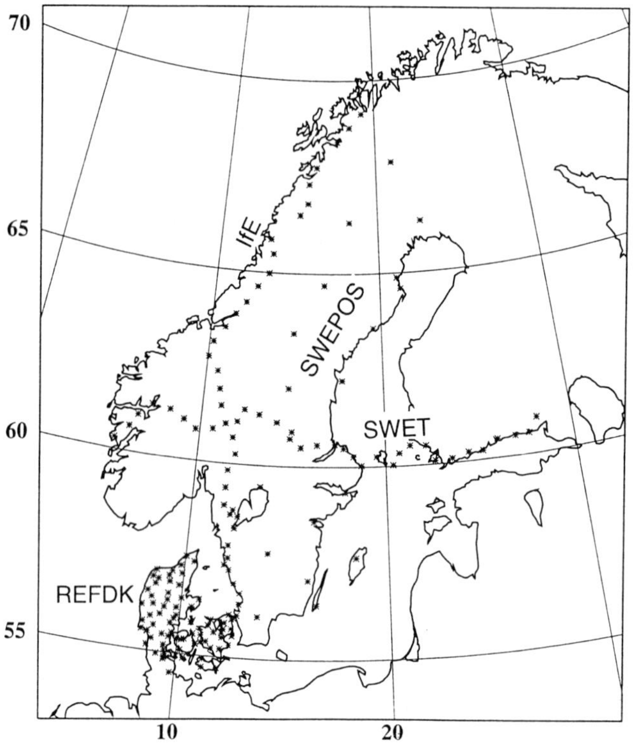

Fig. 2. Location of used Nordic area GPS-levelling points.

Table 4 shows the results of comparing the GPS data sets to geoid models computed with different cap sizes, as well as an example of using a modified kernel (n_{max} = 36). It is seen that the standard deviation of the geoid fits in Scandinavia become *better* with a larger cap size, opposed to the experience from Australia. This could of course indicate problems with the EGM96 model in Scandianavia, but given the small standard deviation of the reduced gravity data (Table 3) it might also in part be due to the relatively homogeneous Scandinavian gravity data base, which has been carefully checked and evaluated for systematic errors. The mean values change quite dramatically for changing cap sizes, but given the local sea-level reference for the height datum it is difficult to say which is the better mean value.When using a modified kernel results are not improved, showing that even the harmonics below 36 are improved by local gravity data in the Scandinavian geoid solution.

From Table 4, it is seen that generally excellent geoid fits are obtained to the GPS lines, with around 10 cm r.m.s. on the long lines and over Sweden, and 2 cm r.m.s. in Denmark. These numbers refer to data directly without any kind of fitting. If a 4-parameter detrending surface is used, the r.m.s. numbers are around 5 cm and 2 cm, respectively, showing probably that some systematic long-wavelength systematic geoid errors remain in the geoid solutions. This is also evident when studying the behavior of the individual data sets as a function of cap size: The decay of r.m.s. fit is not uniform, and some cap sizes (e.g. 5° for Sweden) may give worse results than smaller cap sizes.

Table 4. Fit of geoids of Nordic area to GPS for different cap sizes and kernels (unit: m)

Cap size and kernel	N-S GPS line (IfE)		E-W line (SWET)		Sweden (SWEPOS)		Denmark (REFDK)	
	mean	std.dev.	mean	std.dev.	mean	std.dev.	mean	std.dev.
0.2°	-.519	.209	-.487	.267	-.593	.111	-.701	.136
0.3°	-.539	.194	-.493	.236	-.590	.118	-.703	.133
0.5°	-.576	.144	-.524	.174	-.572	.099	-.709	.106
1.0°	-.625	.132	-.596	.161	-.564	.067	-.725	.067
2.0°	-.681	.137	-.716	.203	-.575	.108	-.741	.062
5.0°	-.802	.115	-.791	.120	-.765	.130	-.814	.028
∞	-1.044	.097	-1.061	.118	-1.001	.078	-1.078	.023
∞ $n_{max}=36$	-.565	.133	-.554	.163	-.506	.097	-.702	.070

CONCLUSIONS

Depending on the actual area in question, cap sizes may need to be limited, or modified Stokes kernels used, in order to obtain optimum geoid results. This is the case in Australia, but apparently not in Scandinavia. The differences may be due to systematic errors in the spherical harmonic reference models and/or in the terrestrial gravity data. The lesson is, therefore, that the best kernels and integration cap radii are area-specific, and no universal "optimum" geoid computation parameters exist.

REFERENCES

Forsberg, R., J. Kaminskis and D. Solheim: *Geoid of the Nordic and Baltic region from gravimetry and satellite altimetry*. Proc. Symp. on Gravity, Geoid and Marine Geodesy, Tokyo, 1997 (in print).

Forsberg, R.: *A new high-resolution geoid of the Nordic area*. In: Determination of the geoid. IAG symposium 106, Milano, Proc. published by Springer Verlag, 1990.

Forsberg, R., M. G. Sideris: *Geoid computations by the multi-band spherical FFT approach*. Manuscripta Geodaetica, 18, pp. 82-90, 1993.

Kearsley, A. H. W.: *Tests on the recovery of precise geoid height differences from gravimetry.* Journal of Geophysical Res., vol. 93, B6, pp. 6559-6570, 1988.

Lemoine, F.G., D. Smith, R. Smith, L. Kunz, E. Pavlis, N. Pavlis, S. Klosko, D. Chinn, M. Torrence, R. Williamson, C. Cox, K. Rachlin, Y. Wang, S. Kenyon, R. Salman. R. Trimmer, R. Rapp and S. Nerem: *The development of the NASA GSFC and DMA joint geopotential model.* Proc. Symp. on Gravity, Geoid and Marine Geodesy, Tokyo, 1997 (in print).

Poutanen, M, K. Gjerde, L. Jivall, F. Madsen: *Final results of the SWET-92 GPS campaign*. Proc. 12th General meeting of the Nordic Geodetic Commission, Ullensvang, Norway, 1994, pp.294-303.

Wong, L., R. Gore: *Accuracy of geoid heights from modified Stokes kernels*. Geophys. J. R. astr. Soc., vol. 18, pp. 81-91, 1969.

Zhang, K.: *An evaluation of geoid determination techniques and their application to Australian GPS heighting*. Ph.d. thesis, School of Surveying and Land Information, Curtin University of Technology, Perth, WA, 1997.

THE EFFECTS OF VARYING CAP SIZES ON GEOID COMPUTATIONS: EXPERIENCES WITH FFT AND RING INTEGRATION

M.B. Higgins
Queensland Department of Natural Resources
Locked Bag 40, Coorparoo Delivery Centre, Qld 4151 AUSTRALIA
R. Forsberg
KMS (National Survey and Cadastre),
Rentemestervej 8, DK-2400 Copenhagen NV DENMARK
A.H.W. Kearsley
School of Geomatic Engineering
University of New South Wales, Sydney, NSW 2052, AUSTRALIA

Abstract

Recent computations comparing geoid heights computed with varying cap sizes in both Australia and Scandinavia, using both FFT and Ring Integration techniques, show unambiguously that the quality of the gravimetric solution varies significantly with cap size. The Australian tests show that the two different approaches produce similar results, and that when compared against control of good quality and over all possible combinations of the points, the best comparison (for both techniques) occurred for an integration cap size between 0.2 and 0.4 degrees. Increasing the cap size beyond this often degraded the comparisons. These results have serious consequences for it reinforces the view that the short wavelength contribution to the geoid height varies with cap size, and that this is not an artifact of the Ring Integration technique. It is therefore important to modify either the cap size and/or the kernel function of the integration to achieve optimal results.

Introduction

In late 1996, the FFT based geoid height computation software of Forsberg (*SPFOUR*) was tested using data sets in Australia. In applications to date, (e.g. Forsberg *et al*, 1996), SPFOUR has not employed a kernel modification to account for the truncation error which arises in the so-called *remove restore* technique when using a global geopotential model (GGM) and local gravity. One of the first Australian tests, using a 100km GPS network across the state of Queensland, showed that SPFOUR gave geoid heights which did not fit the existing control as well as the GGM. For those calculations, SPFOUR used a cap radius of 5 degrees and it was suspected that the large cap size may have been responsible for the degraded results.

In the past, the Ring Integration technique (referred to as *RINT*, Kearsley, 1986) developed at the University of New South Wales has also used a kernel function which was not modified to account for truncation error. Recent joint research with Featherstone has investigated the effects of varying cap size and the use of various modified kernels in the RINT technique. As a step toward better understanding the application of kernel modification to both numerical integration and FFT based techniques, this paper examines the effect of varying cap size on the unmodified kernels of both SPFOUR and RINT. The examination uses a comparison process which has been used with RINT for some time.

The Comparison Process

In the past, in the absence of kernel modification, the choice of optimum cap size for RINT has known to be significant and has been decided empirically by comparing the resulting gravimetric geoid heights (N_{GRAV}) to geometric geoid heights deduced at GPS occupied level marks (N_{GEOM} = ellipsoidal height (h) - leveled height (H)). Comparisons are done using a point residual ($\delta N = N_{GEOM} - N_{GRAV}$) at each GPS station for a given cap size and taking the mean and RMS of all the residuals. For a given set of GPS stations, the computation of mean and RMS of δN are repeated for each cap size of interest. Given that there can be biases and trends between the height datums involved, it is also often useful to do comparisons using baselines. For each baseline, a residual ($\delta\Delta N$) is calculated between the change in geometric geoid height ($\Delta N_{GEOM} = \Delta h - \Delta H$) and the change in gravimetrically determined geoid height (ΔN_{GRAV}). As with point residuals, the mean and RMS of all the baseline residuals (i.e. of $\delta\Delta N=\Delta N_{GRAV}-\Delta N_{GEOM}$) are computed at each cap size.

Data used for the Tests

The data used for the geoid computations in these tests were as follows;

- The short wavelength components of the 1992 release of Australian Geological Survey Organisation (AGSO) gravity data base, at a nominal spacing of 11km.
- The long wavelength components were derived from the EGM96 global geopotential model to degree and order 360 (see Lemoine *et al.*, 1996).
- Terrain corrections to the gravity were computed with FFT software by Forsberg and using DEM data at a density of 30". The DEM was produced by Curtin University based on spot heights from the AGSO gravity data and from topographic data from the Australian Surveying and Land Information Group (AUSLIG).

RINT works on the original randomly spaced AGSO data with the DEM used to improve interpolation (as in Higgins *et al,* 1996). For the FFT based SPFOUR, the AGSO data were gridded at a density of 3' using Bouguer anomalies with the DEM used to reconstruct free air anomalies at the grid points.

Tests using the Queensland 100km GPS Network

The 100km network covers the Australian state of Queensland and was observed using GPS in sessions of at least 5 hours and processed using IGS precise ephemeris and seeded by connections to the ITRF (via the 500km Australian National Network, Morgan *et al*, 1996). The resulting ellipsoidal heights have a precision in ITRF92 of 10cm (95% confidence). Of the 220 stations in the network, 149 have leveled heights of 4th order quality or better and have been used in these tests. For these tests, both RINT and SPFOUR were run with the cap size (radius) varying, in steps of 0.1 degrees, from 0.1 to 1.0 degrees. Figure 1 shows statistics for point residuals (δN as outlined earlier) for all 149 stations and for each cap size. A zero cap size represents use of the GGM (EGM96) alone.

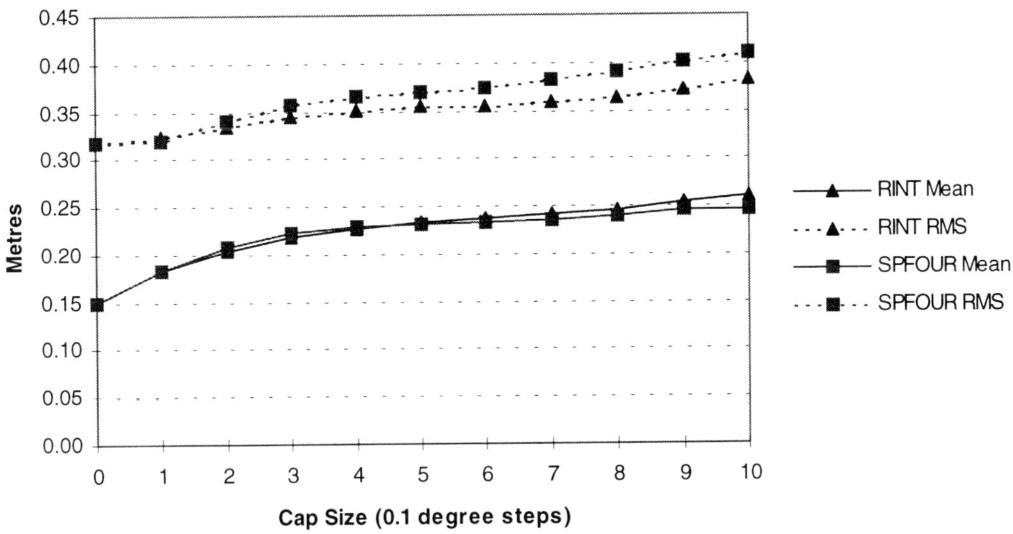

Figure 1: RINT vs SPFOUR - Queensland 100km Network - Point Residuals

From Figure 1 we see

- The strong similarity between the behaviour of the results from RINT and SPFOUR as the cap size is varied.
- Both techniques degrade from the GGM results as the cap size increases. We believe this is due to biases between the local gravity data and the GGM and/or biases between the height datum and the computed gravimetric geoid.

Baseline residuals (δΔN) were also analysed. Baseline combinations between all stations give a possible range from 2km to 2000km with a mean length around 600km. It was decided that such an overall statistic would be less useful than one from grouping the results according to baseline length. RMS statistics are shown in Figure 2 for varying baseline lengths (*100k* means lines from 50to 100km).

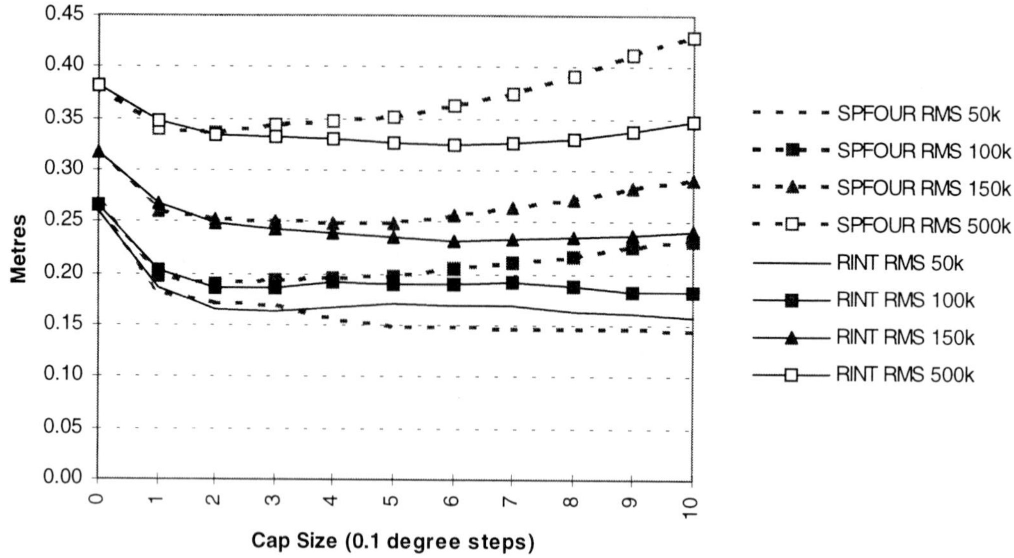

Figure 2: RINT vs SPFOUR - 100km Network - RMS of Baseline Residual - Varying Baseline Length

From Figure 2 we note:

- Results improve rapidly over the GGM with even small amounts of local gravity data.
- For longer baselines, the result degrades with increasing cap size; while for shorter baselines the solutions remain more stable over the range of cap sizes represented.
- The difference in variability between SPFOUR and RINT is also worth noting. We suspect that this is due to differences in the way the programs process the original gravity data. Such differences will be more pronounced further away from the point of computation.
- Overall a cap size between 0.2 and 0.5 degrees seems optimal. This seems reasonable given the underlying reference GGM is complete to degree and order 360.

Tests using the South East Queensland GPS Network

The network in South East Queensland has a total of 270 GPS stations at a density of 5km and with a precision of 5ppm (95% confidence). 149 points with leveled heights of 1st, 3rd or 4th order quality, have been used in these tests. The ellipsoidal heights are based on ITRF92, by adjusting the network constraining two 100km network stations.

As with the 100km network tests, both RINT and SPFOUR were run varying the cap size from 0.1 to 1.0 degree. Figure 3 shows statistics for point residuals (δN) for all 149 stations and for each cap size:

- As with the 100km network tests, the RMS results from both methods behave similarly. However the mean for both techniques do not. We suspect this is due to the differences in the treatment of the gravity data - the FFT takes the raw data and preprocesses it to form a grad; whereas RINT uses the raw data to estimate mean values for the compartments formed by the rings. However further investigation is needed to confirm this view.

- Once again, the improvement over the GGM is rapid with a very small optimal cap size of 0.2 degrees.

Baseline residuals (δΔN) and the effect of the levelling order was also investigated (see Figure 3.

For each order, baselines ranged from 1 to 90km with an average around 30km.

- Again, the shape of the RMS curves for both techniques is similar as cap size is varied.
- The variation with order is influenced by both the changing precision of the levelling (the higher the terrain the less precise the levelling and the gravity).
- In any case, the interesting point is that, once again, for all orders there is a very small optimal cap size of between 0.2 and 0.4 degrees.

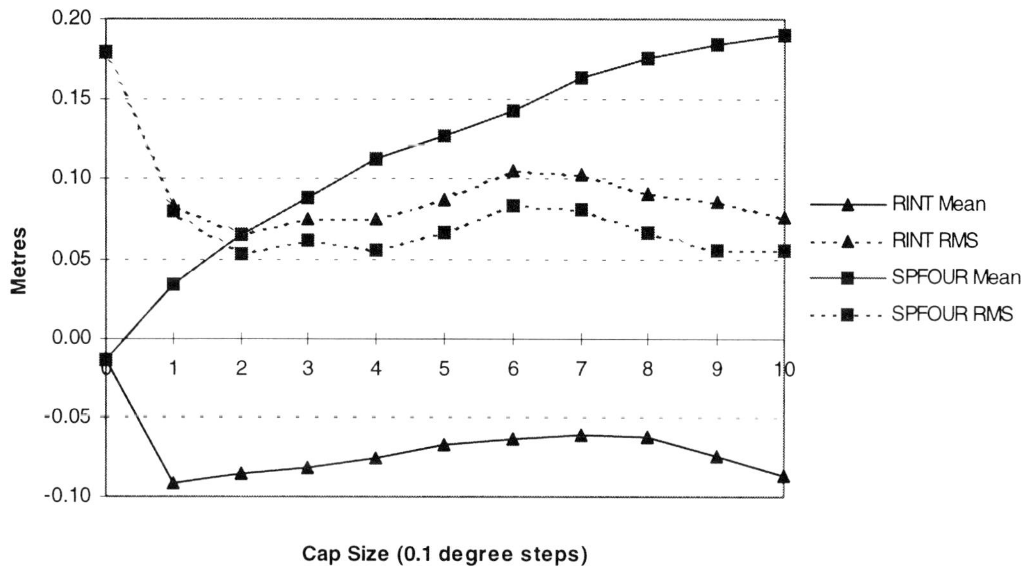

Figure 3: RINT vs SPFOUR - SE Queensland - Point Residuals

Conclusion and Future Research

The tests presented in this paper show that the results for the short wavelength components of the geoid height from RINT and behave in similar fashion. Point and baseline comparisons against control and over all possible combinations of points show that the best result occurs for a comparatively small integration cap size with a radius between 0.2 and 0.4 degrees. The comparisons also showed that increasing the cap size beyond this and up to 1.0 degree degrades the comparisons. This relatively small cap size for a first minimum has been noticed in RINT computations for some time. However, the results presented here show that this relationship between cap size and agreement with control is not limited to ring integration - the same features are also present when using an FFT based approach.

In future we will extend the tests to better understand the behaviour of point and baseline results (δN and δΔN) for cap radii beyond 1 degree. We also need to better understand the correlation between cap size variation and such aspects as baseline length, quality of control and terrain variability. As well as those empirical investigations, Featherstone and

Sideris (1997) have summarised possible kernel modifications for FFT evaluations, and we are working with Featherstone and Sideris on the application of these kernel modification functions to the RINT approach.

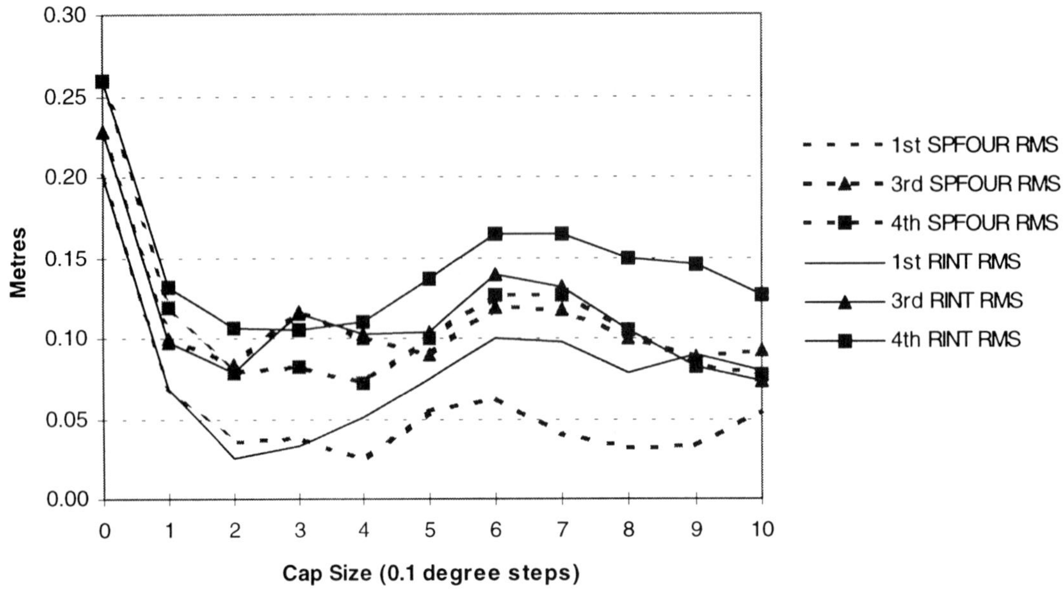

Figure 4: RINT vs SPFOUR - SE Queensland - RMS of Baseline Residual - Varying Order of Leveling

References

Featherstone, W. E. and Sideris, M. G., Modified kernels in spectral geoid determinations, proc. IAG Scientific Assembly, Rio de Janeiro, Sept., 1997 (in press)

Forsberg, R., J. Kaminskis, and D. Solheim, 1996, Geoid of the Nordic and Baltic Region from Gravimetry and Satellite Altimetry, Proceedings of the International Symposium on Gravity, Geoid and Marine Geodesy, Tokyo, Japan, 30 Sept - 5 Oct., pp. 540 - 547.

Higgins, M.B., A.H.W. Kearsley, and M.B. Pearse, 1996, Using Digital Elevation Models in the Computation of Geoid Heights, Geomatics Research Australasia, December.

Lemoine, F.G., D.E., Smith, R. Smith, L. Kunz, N. Pavlis, M. Klosko, D.S. Chinn, M.H. Torrence, R.G. Williamson, C.M. Cox, K.E. Rachlin, Y.M. Wang, E.C. Pavlis, S.C. Kenyon, R. Salman, R. Trimmer, R.H. Rapp and R.S. Nerem, 1996, The Development of the NASA GSFC and DMA Joint Geopotential Model, Proceedings of the International Symposium on Gravity, Geoid and Marine Geodesy, Tokyo, Japan, Sept 30 - Oct 5, pp. 461 - 469.

Kearsley, A.H.W. 1986, Data requirements for determining precise relative geoid heights from gravimetry. Journal of Geophysical Research, Vol. 91 B9 pp 9193-9201, August 10.

Morgan, P., Bock, Y., Coleman, R., Feng, P., Garrad, D., Johnston, G., Luton, G., McDowell, B., Pearse, M., Rizos, C., and Russell, T., 1996, A Zero-order GPS network for the Australian region, *Unisurv S46, Geomatic Eng., UNSW*, Sydney, 2052, Australia, 187 pp.

A COMPARISON OF TECHNIQUES FOR THE INTEGRATION OF SATELLITE ALTIMETER AND SURFACE GRAVITY DATA FOR GEOID DETERMINATION

J F Kirby
School of Surveying and Land Information,
Curtin University of Technology, GPO Box U1987,
Perth, WA 6845, Australia.

R Forsberg
Geodetic Section, KMS, Rentemestervej 8,
Copenhagen NV, Denmark DK 2400.

Abstract

Two methods are tested whereby satellite altimeter measurements of the geoid height are combined with surface measurements of the free-air gravity anomaly. The study area comprises the oceans around the Australian continent. The first method involves draping a grid of the free-air anomaly from satellite data onto a grid of the ship and land data. The second method utilises grids of the altimeter-derived geoid height, combining these with the surface data in an iterative superposition. Preliminary results show that the draping method yields a fit of 5.4 mgal between the satellite and marine data, while the iterative procedure returns 8.1 mgal. Further work can be done, however, to improve these results. The impact of the combined marine gravity datasets is illustrated by comparing the effects on an Australia-wide spherical-FFT geoid solution.

Introduction

Although there exists an almost global coverage of satellite altimetry data, with the newly released Geosat/GM data and ERS-1 data greatly improving resolution, there is also an extensive global marine gravity dataset. This dataset should not be ignored as it can quite easily be incorporated with satellite altimetry data to give a more accurate map of the geoid than presumably could ever be obtained from purely altimetric data.

A method extensively used to combine heterogeneous gravity data is least squares collocation (LSC) [Moritz (1980)]. However, LSC is notoriously costly in computer execution time. To combine M data points, a matrix of size $M \times M$ must be inverted; "inversions of this size obviously present time problems even on a supercomputer and results will suffer from round-off errors" [Schwarz et al. (1990)].

Other procedures to transform gravity anomalies to geoid heights are based directly upon the classical Stokes integral [Heiskanen and Moritz (1967)]. These include direct numerical integration of the anomalies through ring integration methods [*e.g.* Kearsley (1985)] or quadrature integration; Fourier-domain approximations of the Stokes integral, a number of different techniques being summarised in Tziavos (1996); and a method by Sideris (1995) to construct a geoid from irregularly spaced gravity data using Fourier techniques.

This paper tests two approaches to the integration of satellite and surface data: a straightforward "draping" of the marine gravity onto the altimeter field; and an iterative scheme whereby the two heterogeneous datasets are integrated through the fast Fourier transform (FFT).

The Data

The two methods were tested on data over and around the Australian continent. The gravity data were supplied by the Australian Geological Survey Organisation (AGSO), revalidated by Featherstone et al. (1997). They comprise 111,396 free-air anomalies at sea and 526,091 on land. Both datasets were gridded together at 6 minutes using the GMT software [Wessel and Smith (1995)].

The altimeter data were also gridded at 6 minutes from the global 2 minute grid of Sandwell et al. (1995). Both datasets had the EGM96 geopotential model to degree and order 360 removed.

The resulting geoid models were compared with GPS/levelling data at 34 stations of the Australian Fiducial and National Networks (AFN/ANN) around the coast of Australia, where a geometric control geoid height could be determined.

The Iterative FFT Method

The algorithm combining heterogeneous gravity datasets [Kirby (1996)] makes use of the wavenumber domain relationship between geoid height and free-air anomaly, and can thus employ the fast Fourier transform to carry out rapid conversions between these various forms of potential field data. The algorithm presented here has been given the name IFC, for Iterative Fourier Combination.

The procedure utilises the planar 2-D Fourier-domain representation of the boundary value problem of physical geodesy:

$$\mathcal{F}\{\Delta g\} = \gamma\, k\, \mathcal{F}\{N\} \tag{1}$$

after Hipkin (1988), where k is the two-dimensional wavenumber.

The IFC routine requires not only a grid of gravity values, but also a grid of weights for the dataset. This grid should reflect the relative influence which the gridnode corresponding to the data point has in the dataset combination. This depends upon the distance of the node from the actual observation points.

A provisional gravity field model, h, is created which is improved upon in successive iterations of a weighted superposition with grids of geoid or free-air anomaly measurements. Conversion is always performed through equation (1), while the superpositions take the forms:

$$h \rightarrow \frac{h + \omega_{terr}\Delta g_{terr}}{1 + \omega_{terr}} \tag{2}$$

for gravity anomaly combination, and

$$h \rightarrow \frac{h + \omega_{sat}N_{sat}}{1 + \omega_{sat}} \tag{3}$$

for geoid undulation combination. Here, ω represents the weight grids for the terrestrial ($terr$) and satellite (sat) datasets.

This procedure can be iterated any number of times until the provisional model, h, stabilises. Stabilisation occurs at the iteration count when the RMS difference between successive 'like' provisional models reaches a previously specified value, indicating convergence.

The Draping Method

The second method of combination tested involved "draping" the terrestrial gravity data onto the Sandwell altimeter grid. This was achieved by frequency filtering the satellite grid with a 200 km high-pass filter. The difference between this and the marine gravity was then gridded using the Geosoft collocation package, and added back to the filtered altimeter grid only where no marine data existed.

A geoid model was then produced from this combined gravity field using the 2-D multiband spherical FFT [Forsberg and Sideris (1993)], with equation:

$$\mathcal{F}\{N\} = \frac{R}{4\pi\gamma}\,\mathcal{F}\{\Delta g \cos\phi\}\;\mathcal{F}\{S(\psi)\} \tag{4}$$

where $S(\psi)$ is the Stokes function.

Results

To assess their accuracy, the models were compared with the original data by interpolation at the observation locations. Table 1 shows the statistics of the difference between the original data observations, and values interpolated from the indicated models at their locations.

The accuracy of the GMT gridding procedure was assessed by comparing the AGSO ship and land data with their gridded counterpart. Rows 1 and 5 in Table 1 show the accuracy for gridding the marine data is 4.341 mgal RMS , and 3.675 mgal RMS for the land data.

Points of interest in this table are the large RMS difference between the marine gravity and the altimeter data: 10.617 mgal, indicating that one of these datasets has gross errors.

data	model	max	min	mean	std dev	rms
	AGSO	74.015	-170.035	-0.050	4.341	4.341
AGSO marine	Altimeter	147.783	-239.579	-1.676	10.483	10.617
(mgal)	IFC	136.966	-233.471	-1.181	8.060	8.146
	Draped	95.536	-221.555	-0.103	5.447	5.448
AGSO land	AGSO	82.154	-62.403	-0.032	3.675	3.675
(mgal)	IFC	92.462	-69.082	-0.206	4.665	4.670
	Draped	97.617	-96.074	-1.113	6.795	6.885
Altimeter	IFC	98.650	-90.113	-0.211	2.357	2.357
(mgal)	Draped	159.166	-104.167	-0.084	6.158	6.159
AFN/ANN	IFC	1.133	-0.671	0.054	0.375	0.379
(metres)	Draped	1.196	-0.396	0.065	0.332	0.338

Table 1: Statistics of the difference between the data observations in the first column, and the values interpolated from the indicated model in the second column.

model	max	min	mean	std dev	rms
gravity (mgal)	159.514	-110.888	-0.388	7.085	7.095
geoid (metres)	4.384	-3.798	-0.023	0.303	0.304

Table 2: Statistics of the difference between the IFC and draped potential fields.

The draped free-air field shows a better accuracy at sea than on land, 5.448 mgal against 6.885 mgal; while the reverse is true for the IFC gravity field: 4.670 mgal on land versus 8.146 mgal at sea.

The comparisons between the combined models and the 6 minute altimeter grid are also shown in Table 1. The IFC field shows a marked improvement over the draped field, with an RMS difference of 2.357 mgal against 6.159 mgal.

Finally, the geoid models from the two methods were compared with the coastal AFN/ANN data. The improvement of the draped solution over the IFC solution is slight: 33.8 cm against 37.9 cm RMS .

Figures 1 and 2 show the resultant combined geoid models. Note that the IFC solution has more successfully integrated the ship track data with the altimeter field to produce a smoother grid. However, the draped gravity field shows better agreement with the AGSO marine data. This discrepancy is most probably due to errors in the marine data.

Table 2 shows the statistics of the difference between the IFC and draped gravity field and geoid models.

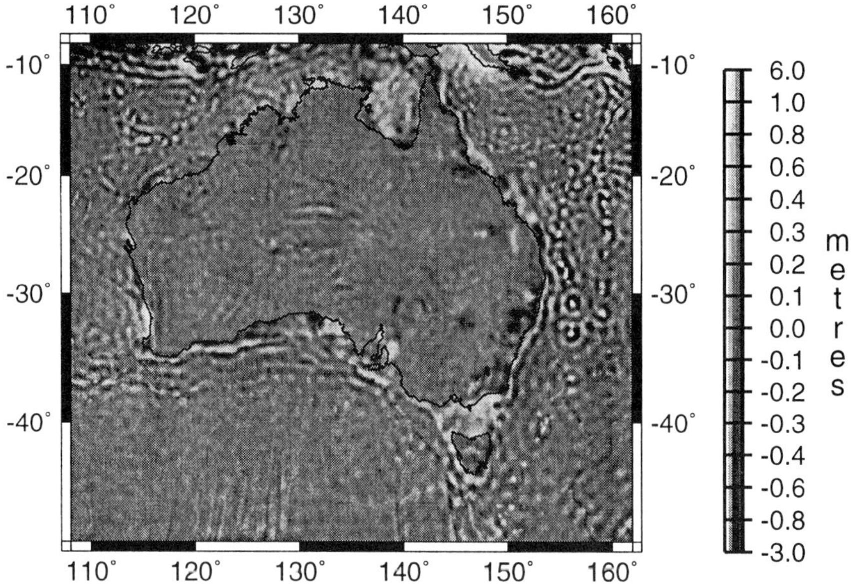

Figure 1: The geoid from the IFC method.

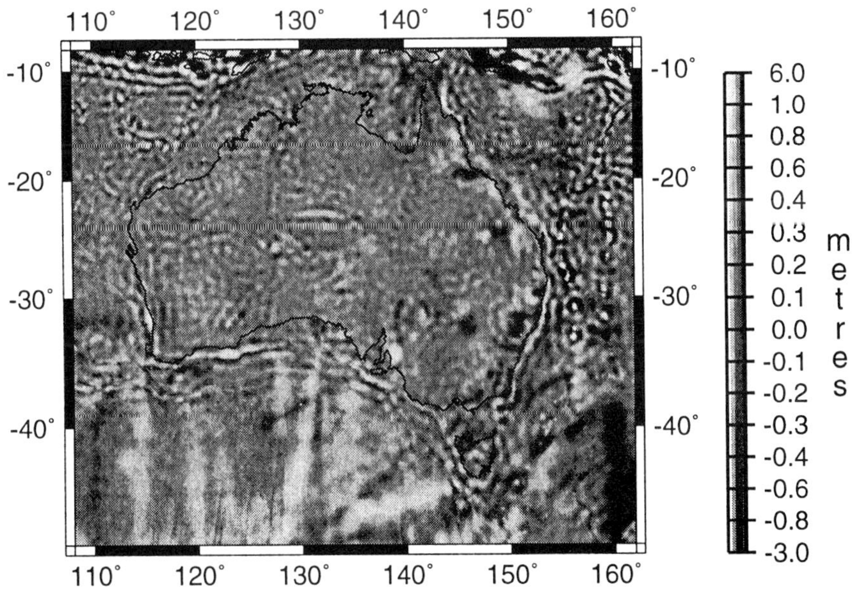

Figure 2: The geoid from the draping method.

Discussion and Conclusions

From Table 1, it can be seen that the majority of the differences occur at sea, where gravity data is usually ill constrained. Indeed, the ship data shows a very poor fit to the altimeter gravity field. However, both combination solutions have succeeded in blending the two datasets together, and, given accurate ship data, the benefits of this process would be multiple.

Further work should look at combining the benefits of both techniques, and this is indeed the subject of a proposal by the first author.

References

Featherstone, W. E., Kearsley, A. H. W. and Gilliland, J. R. (1997). Data preparations for a new Australian gravimetric geoid, *The Australian Surveyor* **42**: 33–44.

Forsberg, R. and Sideris, M. G. (1993). Geoid computations by the multi-band spherical FFT approach, *manuscripta geodætica* **18**: 82–90.

Heiskanen, W. A. and Moritz, H. (1967). *Physical Geodesy*, W. H. Freeman, San Francisco.

Hipkin, R. G. (1988). Bouguer anomalies and the geoid: a reassessment of Stokes' method, *Geophysical Journal* **92**: 53–66.

Kearsley, A. H. W. (1985). Towards the optimum evaluation of the inner zone contribution to geoidal heights, *Australian Journal of Geodesy, Photogrammetry and Surveying* **42**: 75–98.

Kirby, J. F. (1996). *The development and application of a new algorithm for ocean geoid recovery*, PhD thesis, The University of Edinburgh.

Moritz, H. (1980). *Advanced Physical Geodesy*, Wichmann, Karlsruhe.

Sandwell, D. T., Smith, W. H. F. and Yale, M. M. (1995). Gravity anomaly profiles from ERS-1, Topex and Geosat altimetry, *EOS, Transactions of the American Geophysical Union, Spring Meeting* **76**(17): S89.

Schwarz, K. P., Sideris, M. G. and Forsberg, R. (1990). The use of FFT techniques in physical geodesy, *Geophysical Journal International* **100**: 485–514.

Sideris, M. G. (1995). Fourier geoid determination with irregular data, *Journal of Geodesy* **70**: 2–12.

Tziavos, I. N. (1996). Comparisons of spectral techniques for geoid computations over large regions, *Journal of Geodesy* **70**: 357–373.

Wessel, P. and Smith, W. H. F. (1995). New version of the Generic Mapping Tools released, *EOS, Transactions of the American Geophysical Union* **72**: 441,445–446.

APPLICATION OF SPHERICAL WAVELETS FOR REGIONAL GRAVITY FIELD RECOVERY — A COMPARATIVE STUDY

J. Kusche, K.H. Ilk, S. Rudolph
Institute of Theoretical Geodesy, University of Bonn, Germany

M. Thalhammer
Institute of Astronomical and Physical Geodesy, University of Munich, Germany

Abstract. This paper is related to the application of spherical wavelets for gravity field recovery from SST/SGG–observations. The questions of geographical truncation and spatial resolution are studied, important for any regional gravity field recovery. A comparison with conventional regional parametrization functions reveals the basic pros and cons of a wavelet parametrization.

Analysis of SST/SGG data

The aim of a future SST/SGG mission is a global gravity field model of high resolution and precision. The computation of spherical harmonic coefficients as a global parametrization causes problems if an ideal data coverage is not available and if observations are not given at a well–defined boundary surface. The global support of spherical harmonics does not allow to adapt precision to areas of economical or geodynamical interest, and combination with inhomogeneous terrestrial data is not a trivial problem. A recovery strategy based on partial regional solutions has the potential to circumvent these problems. A global solution can be derived if individual regional solutions are merged subsequently to a global one. Finally, regional approaches allow to restrict the number of observations and unknowns to a managable size. Nethertheless, they have a few problems. Individual regional solutions are not able to represent the complete spectrum of the global gravity field and a proper handling of the geographical truncation error has to be found. This is a general problem in regional recovery applications not only restricted to the analysis of satellite–based gravity data.

The SST observation equation, when transformed onto the spectral domain, can be expressed by a combination of operators [1]

$$b_\nu = r_\nu^{obs} - r_\nu^{ref} = \mathcal{I}_\nu \circ \mathcal{P} \circ \mathcal{D}^\beta \circ \mathcal{K} \circ \chi = A_\nu \circ \chi \,, \tag{1}$$

with pseudo–observations from measured relative distances $r(t)$, velocities $\dot{r}(t)$ or accelerations $\ddot{r}(t)$, e.g.

$$r_\nu^{obs} = \frac{2T}{\nu\pi} \int\limits_0^1 \cos(\nu\pi\tau)\, \dot{r}(\tau)\, d\tau\,, \qquad \nu = 1, \ldots, N\,. \tag{2}$$

The (reduced) tidal force $\delta \boldsymbol{g} = \boldsymbol{g} - \boldsymbol{g}^{ref}$ is projected onto the intersatellite direction by $\mathcal{P}$ and then transformed into the spectral domain, represented by the operator $\mathcal{I}_\nu$. The tidal force can be constructed from an integral representation of the potential $T = V - V^{ref}$

$$\delta \boldsymbol{g} = \mathcal{D}^\beta \circ T = \mathcal{D}^\beta \circ \mathcal{K} \circ \chi = \int_\Omega \mathcal{D}^\beta\, \mathcal{K}(\cdot, Q)\, \chi(Q)\, d\Omega\,, \tag{3}$$

containing the unknown parametrization function $\chi(Q)$, the kernel $\mathcal{K}(P,Q)$ and a differential operator $\mathcal{D}^\beta$ with $\beta = 1$ (SST) or $\beta = 2$ (SGG). The inverse problem

$$\chi = A_\nu^+ \circ b_\nu \tag{4}$$

is improperly posed and therefore a regularization technique has to be applied. From (1) it is obvious that one has to find a suitable parametrization for the kernel and the unknowns fulfilling the requirements above. For more details the reader is referred to [1].

Spherical Wavelets as a multiresolution parametrization

The spherical scale–discrete wavelet–transform $L_2(\Omega) \to L_2(I\!N \times \Omega)$, as introduced by [2], is said to generate a multiresolution of the earth's gravity potential:

$$ST(T)(Q) = (T * \Phi)(Q) = \int_\Omega T(Q')\, \Phi(Q,Q')\, d\Omega\,, \tag{5}$$

$$WT^j(T)(Q) = (T * \Psi^j)(Q) = \int_\Omega T(Q')\, \Psi^j(Q,Q')\, d\Omega\,, \qquad j = 1, \ldots, J\,. \tag{6}$$

In wavelet terminology Φ is called the scaling function, which canonically generates wavelet scales Ψ^j. Reconstruction (e.g. of the potential itself) is done by

$$T(Q) = \int_\Omega ST(T)(Q')\, \Phi(Q,Q')\, d\Omega + \sum_{j=1}^{J} \int_\Omega WT^j(T)(Q')\, \tilde{\Psi}^j(Q,Q')\, d\Omega\,. \tag{7}$$

In [2] the Abel–Poisson–Kernel was proposed as a candidate for scaling function

$$\Phi(Q,Q') = \Phi(t) = \sum_{n=0}^{\infty} \frac{2n+1}{4\pi} \Phi_n P_n(t)\,, \qquad \Phi_n = e^{-n\rho_1}\,, \qquad \rho_j > 0,\ \lim_{j\to\infty} \rho_j = 0 \tag{8}$$

$$\Psi^j(Q,Q') = \Psi^j(t) = \sum_{n=0}^{\infty} \frac{2n+1}{4\pi} \Psi_n^j P_n(t)\,, \qquad \Psi_n^j = \tilde{\Psi}_n^j = \left(e^{-2n\rho_{j+1}} - e^{-2n\rho_j}\right)^{1/2}\,. \tag{9}$$

It is obvious from (7) that wavelet multiresolution may be considered as a kind of multi–layer parametrization with $ST(T) = \chi^0, WT^j(T) = \chi^j$. For notational aspects see [2].

Comparative investigations

Geographical truncation effect. In regional techniques one has to estimate the effect of geographical truncation on SST/SGG–functionals at altitude, see e.g. [3]. In case of multiresolution parametrization different truncation radii on each scale are possible. We calculated Molodensky–type coefficients for scale discrete wavelets and compared with a single layer representation. As it is well–known, extending the integral (3) merely over a spherical cap with radius ϕ yields an error in the three components

$$\delta T_x = \frac{1}{\sin\theta}\sum_{n=1}^{\infty}\sum_{m=-n}^{n}\tilde{Q}_n T_{nm}\frac{\partial Y_{nm}}{\partial\lambda}\,, \qquad \delta T_y = -\sum_{n=1}^{\infty}\sum_{m=-n}^{n}\tilde{Q}_n T_{nm}\frac{\partial Y_{nm}}{\partial\theta}\,,$$

$$\delta T_z = \sum_{n=0}^{\infty}\sum_{m=-n}^{n} Q'_n T_{nm} Y_{nm}\,. \tag{10}$$

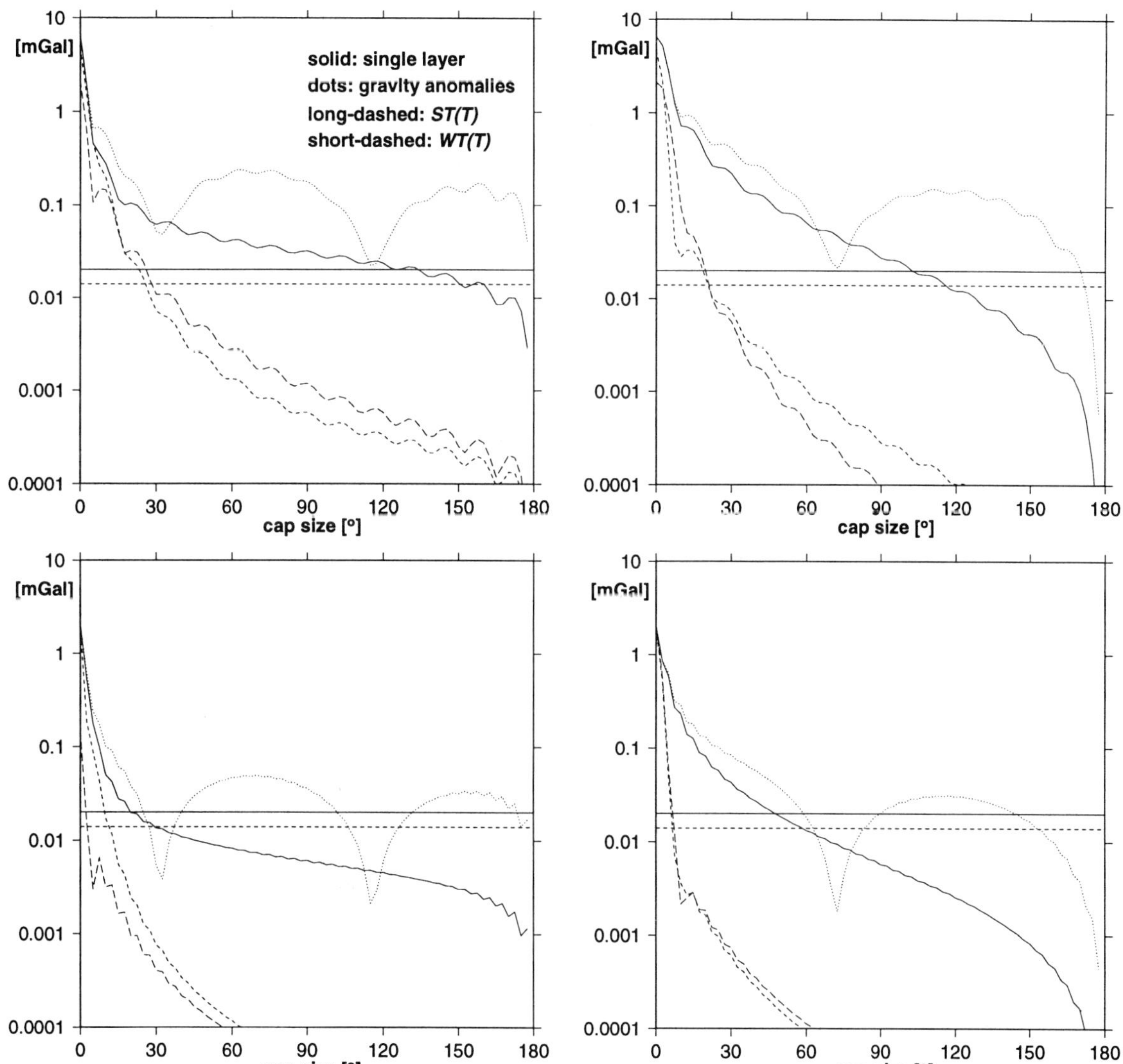

Fig. 1. Left column radial component $\sigma(\delta T_z)$, right column tangential component $(\sigma^2(\delta T_x)+\sigma^2(\delta T_y))^{1/2}$. Upper row n=13...1000 from degree variance model, lower row n=37...1000. Altitude h=200km.

In a multiresolution truncation on different scales contributes to Q'_n and $\tilde{Q}_n$

$$Q'_n = \sum_{j=0}^{J} Q'^j_n \,, \qquad \tilde{Q}_n = \sum_{j=0}^{J} \tilde{Q}^j_n \,. \tag{11}$$

The coefficients must be calculated by (remember that wavelet kernels may not have a closed expression)

$$Q'^j_n = -\sum_{k=0}^{\infty} \frac{(k+\frac{1}{2})(k+1)}{r} \left(\frac{R}{r}\right)^{k+1} R_{kn}(\phi_j)\, S^j_{kn} \,, \tag{12}$$

$$\tilde{Q}^j_n = \sum_{k=0}^{\infty} \frac{(k+\frac{1}{2})}{r} \left(\frac{R}{r}\right)^{k+1} \left\{ R_{kn}(\phi_j) + \frac{\sin\phi_j}{n(n+1)} P_k(\phi_j) P_{n1}(\phi_j) \right\} S^j_{kn} \,. \tag{13}$$

As auxiliary quantities the $R_{kn}(\phi_j) = \int_{\phi_j}^{\pi} P_k(\psi) P_n(\psi) \sin\psi \, d\psi$ can be calculated recursively. The key quantities are *i)* $S_{kn} = \frac{2n+1}{2k+1}$ *for single layer densities, ii)* $S_{kn} = \frac{n-1}{k-1}$ *(gravity anomalies)*, and *iii)* $S^0_{kn} = \Phi_k \Phi_n$, $S^j_{kn} = \tilde{\Psi}^j_k \Psi^j_n, j > 0$ *for scale discrete wavelets.* To illustrate our investigations we determined global means for truncation errors as a function of cap radius. We used a degree variance model excluding lower degree coefficients, which is certainly a simplification. We assume $J = 1$ for wavelet decomposition and scale parameters $\rho_1 = 0.03, \rho_2 = 0.1 \cdot 10^{-5}$. Fig. 1 reveals an appealing property of wavelet–type parametrization: Truncation radii are significantly smaller compared with gravity anomaly or single layer density representations. Note that the solid horizontal line marks the expected measurement accuracy [1].

Scale and space discretization. It is easy to see that wavelet scale discretization and space discretization — choice of block sizes — are intimately related. First we consider an arbitrary SST/SGG–observable reconstructed from wavelet parametrization of the gravity field, e.g.

$$T_z = -\sum_{n=0}^{\infty} \sum_{m=-n}^{n} \left\{ \Phi_n \Phi_n + \sum_{j=1}^{J} \tilde{\Psi}^j_n \Psi^j_n \right\} \frac{(n+1)}{r} \left(\frac{R}{r}\right)^{k+1} T_{nm} Y_{nm} \,.$$

Obviously $(\Phi_n)^2$ acts as a low–pass filter, and the $\tilde{\Psi}^j_n \Psi^j_n$ are band–pass filters [2]. In the following we restrict ourselves to $J = 1$: our single band–pass $\tilde{\Psi}^1_n \Psi^1_n =: \tilde{\Psi}_n \Psi_n$ essentially acts as a high–pass. As demonstrated in fig. 2 varying the scale parameter ρ_1 mutually shifts low–pass and high–pass.

Given ρ_1, we find the highest relevant degree n' from weighting the accumulated signal content against measurement accuracy ϵ

$$\left(\sum_{n=n'}^{\infty} (\Phi_n)^4(\rho_1)\, \sigma^2_n(T_z) \right)^{1/2} < \frac{\epsilon}{\sqrt{2}} \,, \qquad \left(\sum_{n=n'}^{\infty} (\tilde{\Psi}_n \Psi_n)^2(\rho_1)\, \sigma^2_n(T_z) \right)^{1/2} < \frac{\epsilon}{\sqrt{2}} \,.$$

This procedure finally yields appropriate block sizes Δ_Φ and Δ_Ψ from the rule of thumb $\Delta = \frac{\sqrt{180 \cdot 360}}{n'}$. A ratio $\Delta_\Phi / \Delta_\Psi = 2 \ldots 3$ can be achieved by selecting ρ_1 around 0.03. This means that the block size for the "coarse structure" of the gravity field may be $2 \ldots 3$ times less dense than for the "detail structure".

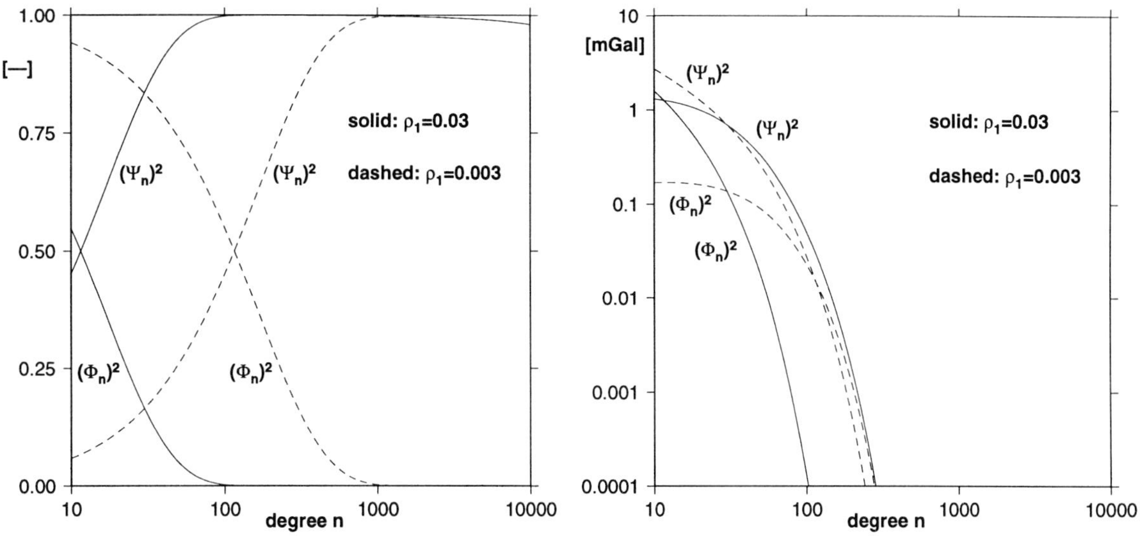

Fig. 2. Left side: Low–pass and high–pass for different ρ_1. Right side: Same filters applied to $\sigma_n(T_z)$

However, these considerations can only provide a rough estimate. A numerical simulation may give a deeper insight and allows comparison of different parametrizations under well–defined conditions. We proceeded in four steps: *i) Calculate true values:* SST/SGG–observables for a test area from an earth gravity model *ii) Compute exact parametrization:* single layer densities/gravity anomalies/wavelet coeffcients for a specified block size on whole sphere *iii) Reconstruct SST/SGG–observables* in test area from exact but discrete parametrization by spherical convolution. *iv) Judge the result* by means of an appropriate norm (L_2, L_1, L_∞) of the errors reconstructed vs. true values.

Parametrization & block size	δT_z $[10^{-2} mGal]$			δT_{zz} $[10^{-3} EU]$		
	rms	average	max	rms	average	max
gravity anomalies $\Delta = 0.5^o$	<0.1	<0.1	0.2	<0.1	<0.1	0.3
gravity anomalies $\Delta = 1^o$	1.1	0.8	4.1	3.5	2.5	14.1
wavelets $\Delta_\Phi = \Delta_\Psi = 0.5^o$	<0.1	<0.1	0.2	0.1	<0.1	0.2
wavelets $\Delta_\Phi = \Delta_\Psi = 1^o$	1.4	1.0	6.2	4.0	2.8	18.2
wavelets $\Delta_\Phi = 2^o, \Delta_\Psi = 0.5^o$	0.3	0.2	0.8	0.5	0.3	1.1
wavelets $\Delta_\Phi = 2^o, \Delta_\Psi = 1^o$	1.9	1.3	7.3			

Table 1: Differences from true values vs. reconstructed values

Fig. 3 shows gravity anomalies, wavelet scale coefficients $T * \Phi$ and wavelet coefficients $T * \Psi$ for the test area (although computed globally). We used degrees 37...180 from OSU91a with signal strength at altitude $h = 200km$ of $\pm 7mGal$ resp. $\pm 0.6EU$. A sample of the results of the simulation is given in table 1 for gravity anomalies (with negligible differences to single layer densities) and wavelets (with scale parameter $\rho_1 = 0.03$, which gives a good compromise). We found slight advantages for conventional parametrization techniques. This was confirmed by similar computations of $\delta T_x, \delta T_y$.

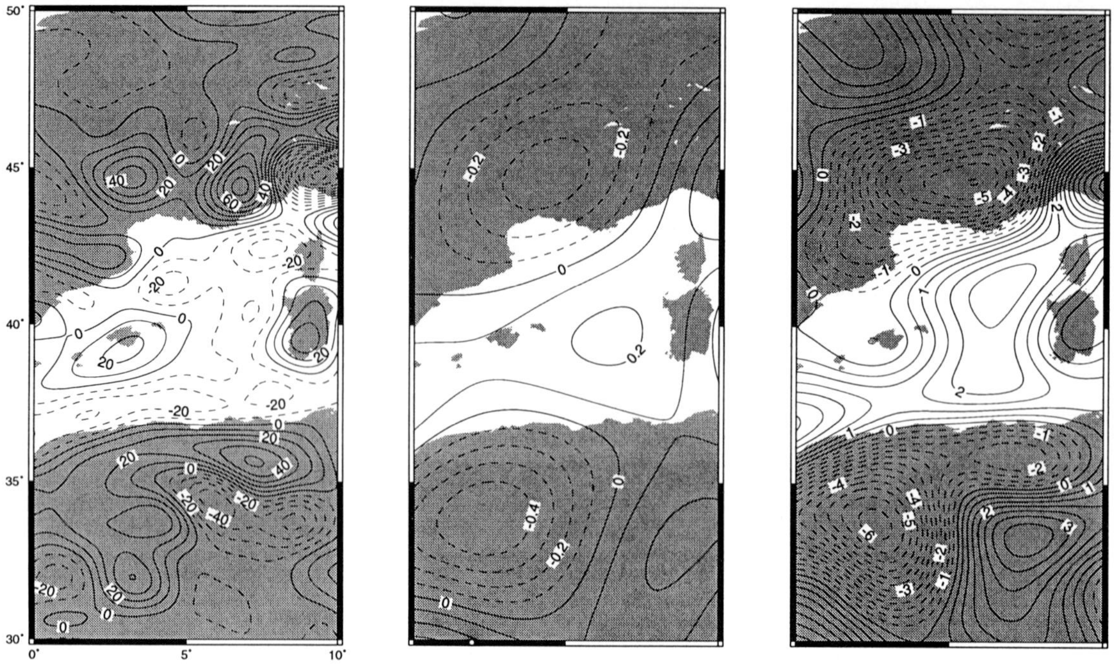

Fig. 3.: Gravity anomalies [mGal], wavelet scale coefficients $T * \Phi$ [$m^2 s^{-2}$], wavelet coefficients $T * \Psi$ [$m^2 s^{-2}$], test area western mediterranean sea

Conclusions and outlook

In general it is believed that wavelet techniques could play a very useful role in gravity field mapping, at least in combination with spherical harmonics describing a low–frequency reference part of the field. If compared to a representation by gravity anomalies and single layer densities we found that spherical wavelets in the medium frequency band show outstanding properties related to the geographical truncation effect and comparable results with respect to discretization in space. These facts recommend wavelets for the modelling of the medium wavelength part of the gravity field, especially in the context of regional spaceborn or airborn gravity data analysis techniques. However, open questions are related to the "right" choice of the scaling function. Furthermore an appropriate regularization technique has to be found to overcome the downward continuation. Finally the construction of wavelets on geophysically more relevant surfaces like ellipsoid, geoid or earth's surface is still an unsolved problem.

References

[1] CIGAR III/2 (1995): Study of the gravity field determination using gradiometry and GPS, Final Report. ESA contract No. 10713/93/F/FL, European Space Agency

[2] Freeden W., Windheuser U. (1997): Combined Spherical Harmonic and Wavelet Expansion — A Future Concept in Earth's Gravitational Determination, Appl. Comp. Harm. Anal. (ACHA), 4:1–37

[3] Thalhammer, M. (1994): The geographical truncation error in satellite gravity gradiometer measurements, manuscripta geodetica 19:45–54.

COMMENTS ON TWO DIMENSIONAL CONVOLUTIONS OF THE GEODETIC PROBLEMS IN PLANAR AND SPHERICAL COORDINATES

Jiancheng Li Jinsheng Ning Dingbo Chao
Wuhan Technical University of Surveying and Mapping
39 Luoyu Road, Wuhan 430070, China

Abstract

Almost all integral formulae in geodetic problems can be expressed in a convolution form, which makes it possible to perform a convolution evaluation by the fast Fourier transform or the fast Hartley one. The 2D convolution forms are usually in planar or spherical coordinates. Unfortunately, the accuracy of results from the evaluation of the 2D convolution in planar coordinates is lower than that in spherical coordinates and the numerical integration in the past. This conflict is caused by inadequately ignoring terms of the kernel function which follow the principal term. To satisfy the convolution theorem, it is necessary for the 2D convolutions in spherical coordinates to take the latitude approximation in their kernel function. Consequently, it leads to larger errors in results. The proposed new idea is that all terms of Stokes kernel functions can be taken into account by transforming all variables $\sin(\psi/2)$ of the functions to the l, a straight line length corresponding to the spherical distance ψ, and the functions could strictly be expressed in planar coordinates. In this case, any approximation will be avoided. Therefore, the results from the 2D convolution in planar coordinates can be obtained with much better accuracy than that in spherical coordinates, and they are very close to that from the 1D convolution or the numerical integration. Based on the above discussion, the evaluations of the Stokes formula are carried out using the 1D convolution and the 2D convolution in planar coordinates, and also that in spherical coordinates for the comparisons between them.

1. Introduction

Since the 80s, with the speed development of computer techniques, a fast Fourier transform (FFT) technique has been widely applied in convolution evaluations for almost all integral formulae in physical geodesy. Consequently, the computation efficiencies are enormously promoted. In the meantime, M.G. Sideris and R. Forsberg et al. made an initiative contribution to the application of FFT in physical geodesy (Sideris, 1985, 1987; Sideris and Schwarz, 1988; Forsberg, 1985; Schwarz et al., 1990, Sideris, 1996). At present, there are three types of convolution evaluation formulae in use, i. e. a planar convolution, a 2D spherical convolution and a 1D spherical convolution. The existing planar convolution expression suffers from planar approximate errors and truncation errors of the Stokes kernel function in which only the principal term of the function is taken into account.

To overcome the approximation errors of the planar convolution, a 2D spherical convolution expression was developed by Strang Van Hees (1990), but there is still an error due to an approximate mean latitude in the expression. Therefore, a multi-band spherical FFT method was proposed by Forsberg and Sideris (1993) for modifications of the Strang Van Hees spherical FFT approach. Furthermore, an approach of 1D spherical convolution for 2D evaluation was presented by (Haagmans et al, 1993) to avoid any approximate error in 2D spherical convolutions, but the cost of this method is a lower computation efficiency.

It is worth to studying the problem about how we can improve the strictness of 2D convolution evaluations, and keep their good computation efficiency in the same time. The intention of this paper is to make an effort in this direction. We have developed a more precise expression in planar Cartesian coordinates for the Stokes kernel function, which includes all terms of the function by use of the relation between the spherical variable $\sin(\psi/2)$ and the corresponding planar distance l. It is expected to improve the computation accuracy and efficiency of the related 2D convolution evaluation.

2. Mathematical background

The disturbing potential T at any point on Earth's surface can be expressed as follows (Heiskanen and Moritz, 1967):

$$T = \frac{R}{4\pi}\int_{\sigma} \Delta g S(\psi) d\sigma \tag{1}$$

According to Bruns' theorem, the geoidal height can be computed by

$$N = \frac{T}{\gamma} = \frac{R}{4\pi\gamma}\int_{\sigma} \Delta g S(\psi) d\sigma \tag{2}$$

where R is a mean radius of the Earth,

Δg is the free air gravity anomaly on a surface of the sphere of radius R,

γ is the mean normal gravity of the Earth,

ψ is the spherical distance between the computation point and the variable point of the integration,

$S(\psi)$ is the Stokes kernel function, given by

$$S(\psi)=\frac{1}{s}-6s-4+10s^2-3(1-2s^2)\ln(s+s^2) \tag{3}$$

where $s=\sin(\psi/2)$.

A local Cartesian coordinate system is defined, in which the origin is at the computation point P, and the X axis points to the North, the Y axis to the East. Therefore, to satisfy the convolution form, the Stokes function was approximated as (Schwarz et al., 1990)

$$S(\psi)\approx\frac{1}{s}=\frac{1}{\sin(\psi/2)}\approx\frac{2R}{r}=\frac{2R}{\sqrt{(x_P-x)^2+(y_P-y)^2}} \tag{4}$$

where r is the chord distance between P and the variable point of the integration. It equals to the distance $\frac{l}{\cos(\psi/2)}$ in the plane xy, and $r=l$ due to $\cos(\psi/2)\approx 1$.

The differential relation between the local Cartesian coordinates (x,y) and the spherical coordinates (φ,λ) reads,

$$dx=Rd\varphi \tag{5}$$

$$dy=R\cos\varphi d\lambda \tag{6}$$

Inserting surface element $d\sigma=\frac{1}{R^2}dxdy$ and (4) into (2), we see that the Stokes formula in planar approximation can be written as

$$N(x_P,y_P)=\frac{1}{2\pi\gamma}\iint_E \Delta g\frac{1}{\sqrt{(x_P-x)^2+(y_P-y)^2}}dxdy \tag{7}$$

cf. Schwarz et al., 1990; Sideris, 1996.

The above equation is permitted to use FFT/FHT techniques for convolution evaluation.

However, due to inadequately ignoring the terms of the kernel function which follow the principal term expressed by equation (4), it leads to larger errors of results obtained from (7). In fact, according to (4), s can be expressed as

$$s=\frac{l}{2R}=\frac{1}{2R}\sqrt{(x_P-x)^2+(y_P-y)^2} \tag{8}$$

Consequently, a more precise convolution expression of the Stokes formula in planar approximation has the form,

$$N(x_P,y_P)=\frac{1}{4\pi\gamma}\iint_E \Delta g S(x_P-x,y_P-y)dxdy \tag{9}$$

where the Stokes kernel function $S(x,y)$ is derived from (3) by transforming variable s

into l according to (8), which is equivalent to (3). Equation (9) satisfies the convolution theorem without any truncation error, and it can efficiently be evaluated by FFT/FHT.

The 2D spherical convolution representation of the Stokes formula is given in (Strang Van Hees, 1990) as

$$N(\varphi_P,\lambda_P)=\frac{R}{4\pi\gamma}\iint_{\sigma}\left[\Delta g\cos\varphi\right]S(\varphi,\lambda)\,d\varphi d\lambda \tag{10}$$

where the Stokes kernel function $S(\psi)$ is the same as in (3), but it is expressed by,

$$\sin(\psi/2)=\left[\sin^2\frac{1}{2}(\varphi_P-\varphi)+\sin^2\frac{1}{2}(\lambda_P-\lambda)\cdot\cos\varphi_P\cos\varphi\right]^{1/2} \tag{11a}$$

or

$$\sin(\psi/2)=\left\{\sin^2\frac{1}{2}(\varphi_P-\varphi)+\sin^2\frac{1}{2}(\lambda_P-\lambda)\cdot\left[\cos^2\varphi_m-\sin^2\frac{1}{2}(\varphi_P-\varphi)\right]\right\}^{1/2} \tag{11b}$$

where φ_m is the mean latitude of the computation area. Note that strict value should be $(\varphi_P+\varphi)/2$. This approximate treatment will cause a significant computational error in convolution (10) evaluated by FFT/FHT.

A strict spherical 1D convolution representation of equation (2) is given by Haagmans et al. (1993),

$$N(\varphi_i,\lambda)=\frac{R}{4\pi\gamma}F^{-1}\left\{\int_{\varphi}F\left[\Delta g(\varphi_i,\lambda)\cos(\varphi_i)\right]F\left[S(\varphi,\varphi_i,\lambda)\right]d\varphi\right\} \tag{12}$$

In practice, this technique performs a 1D Fourier transform in the longitude only and then takes a sum in latitude.

3. Numerical results

A large area of central China selected for the test is located in 31°N -41°N and 95°E-115°E. The 2 30 ×2 30 free air gravity anomaly was used for computing geoidal heights via 1D convolution, 2D planar and spherical convolution in this area. The global geopotential model EGM96 was used as a reference gravity field.

The remove/restore method was employed for the calculation of quasi-geoid height (Wang, 1993). The residual gravity anomalies, obtained by subtracting anomalies computed from the geopotential model EGM96, are used in equations (7), (9) and (10) as well as in (12). To the results thus obtained it should still be added the effect of geoid undulations from the EGM96 model. For comparison, the integration domain for evaluating the Stokes integral takes two spherical caps with the radiuses of 50 and 1°50 , respectively.

If we suppose that the results from the 1D convolution technique be the standard ones, the differences between the standard values and those from other convolution techniques discussed in this paper may indicate which performs better. The statistics of such a comparison are tabulated in the following tables 1-3.

Tab. 1: The statistics of the residual gravity anomalies (Unit: mGal)

Max	Min	Mean	RMS	Std
187.984	-209.781	-5.262	±30.511	±30.054

Tab. 2: The statistics of the comparison between results from the 1D convolution of the Stokes formula and those from the 2D planar and the spherical convolution with an integration cap of 50 (Unit: meter)

Methods	Max	Min	Mean	RMS	Std
new 2D planar convolution	0.094	-0.093	-0.001	±0.012	±0.012
old 2D planar convolution	0.073	-0.203	-0.017	±0.034	±0.029
2D spherical convolution	0.183	-0.092	0.001	±0.028	±0.026

Tab. 3: The statistics of the comparison of between results from the 1D convolution of the Stokes formula and those from the 2D planar and the spherical convolution with an integration cap of 1°50 (Unit: meter)

Methods	Max	Min	Mean	RMS	Std
new 2D planar convolution	0.082	-0.098	-0.001	±0.016	±0.016
old 2D planar convolution	0.092	-0.338	-0.036	±0.090	±0.064
2D spherical convolution	0.179	-0.116	0.013	±0.035	±0.033

4. Conclusions

The problem of 2D convolutions in planar coordinates in computations in physical geodesy was investigated in this paper. A more precise Stokes kernel function in planar coordinates without any truncation error was derived, which includes all the terms of the function. According to the numerical test for these new planar convolution formulae, we can give the following comments on the problem of the 2D convolution evaluations:

(1) For calculating geoidal heights, it shows much better accuracy to use the new planar convolution expressions than to use either the simple planar one (used in the past) or the spherical one.

(2) The proposed new planar convolution formulation can be used to evaluate any

convolution in planar coordinates in physical geodesy, such as even the Vening Meinesz integral and the inverse Stokes etc..

(3) The accuracy of numerical results from all 2D convolutions in planar coordinates in physical geodesy are generally better than those from the corresponding 2D convolutions in spherical coordinates.

We did only a preliminary work in this research area. Further investigations are needed for improving 2D convolution expressions in physical geodesy because they have higher computation efficiency than a 1D convolution evaluation process.

Acknowledgments

We wish to thank Professor P. Holota and an anonymous reviewer for the valuable comments and suggestions on an earlier version of this paper. The National Natural Science Foundation Council of China for Outstanding Young Scientists has supported this work under grant 49625408.

References

Forsberg R (1985) Gravity field terrain effect computations by FFT. *Bull. Geod.* 59, pp. 342-360.

Forsberg R and Sideris MG (1993) Geoid computations by the multi-band spherical FFT approach, *Manuscripta geodaetica* 18, pp. 82-90.

Haagmans R, E De Min, and M Van Gelderen (1993) Fast evaluation of convolution integrals on the sphere using 1D FFT, and a comparison with existing methods for Stokes'. *Manuscripta geodaetica* 18, pp. 227-241.

Heiskanen WA and H Moritz (1967) Physical Geodesy. Freeman & company, San Francisco-London.

Schwarz KP, MG Sideris, and R Forsberg (1990) The use of FFT techniques in physical geodesy, *Geophys.J.Int.*, 100, pp.485-514.

Sideris MG (1985) A fast Fourier transform method for computing terrain corrections, *Manuscripta geodaetica* 10, pp. 66-73.

Sideris MG (1987) Spectral methods for the numerical solution of Molodensky's problem, UCSE Report #20050, Department of Surveying Engineering, The University of Calgary, Calgary, Alberta.

Sideris MG and Schwarz KP (1988) Recent advances in the numerical solution of the lenear Molodensky problem. *Bull. Geod.* 62, pp. 521-540.

Sideris MG (1996) FFT geoid computations in Canada, New geoid in the words, BULLETIN D'INFORMATION N.77, IGES BULLETIN N. 4, Special Issue, pp.37-52.

Strang van Hees, G., Stokes' formulae using fast Fourier techniques, Manuscripta Geodaetica, Vol. 15, pp.235-239, 1990.

Wang YM (1993) Comments on proper use of the terrain correction for the computation of height anomalies. *Manuscripta geodaetica* 18 pp. 53-57.

OVERDETERMINED GRAVITY GRADIOMETRY BOUNDARY VALUE PROBLEM: THEORY AND SIMULATION

Z.C.Luo J.S.Ning D.B.Chao
School of Geoscience and Surveying Engineering
Wuhan Technical University of Surveying and Mapping
39 Luoyu Road, Wuhan 430070, P.R.China
E-mail: zcluo@hp05.wtusm.edu.cn

Abstract. In theory, the recovery of the Earth's gravity field using satellite gravity gradiometry data is formulated as a solution to a BVP of satellite gravity gradiometry, and in nature, it is an overdetermined BVP. In this paper, a quasi-solution to this BVP as well as its rigorous mathematical representation are presented on the basis of a general theory of quasi-solutions of overdetermined BVP's. Practical solution models are derived which lead to the solution of a system of linear equations in a zonal form, in particular, the geopotential coefficients for odd-degree terms and those for even ones are separated to be solved from two independent systems. The theory of quasi-solutions permits to combine all the components of the gravity gradient. Finally, the numerical example was performed and some conclusions derived.
Keywords. overdetermined gravity gradiometry boundary value problem, corresponding well-posed boundary value problem, quasi-solution, satellite gravity gradiometry.

1 Formulation and estimation principle

Within the framework of analytical theory the recovery of the Earth's gravity field from satellite gravity gradiometry (SGG) data is frequently treated as a boundary value problem(BVP) for the Laplace equation. Depending on the region under consideration (global or local) and the data source (one type of component or some combination of gravitational tensor) various approaches for the solution of the corresponding BVP have been already published(Sacerdote and Sansò, 1985; Holota, 1986; Rummel and Colombo, 1985; Rummel et al., 1989; Brovelli et al., 1991; Keller and Hirsch, 1994; Robbins, 1985; Tscherning et al., 1990; Luo et al., 1996, 1997.). This paper aims at the global gravity field determination from SGG data with a BVP approach based on a general theory of quasi-solutions of OBVP's presented by Zhu and Yu(1992).

Although all components or some combinations of the gravitational tensor can be obtained from satellite gravity gradiometry, only five independent components are to be considered in the following discussion. At first a geocentric spherical surface Σ of radius R_s representing a mean satellite orbit height is chosen as boundary surface. Supposing the boundary surface Σ is covered continuously by gradiometry data, then the overdetermined gravity gradiometry BVP can be formulated as follows:

$$\begin{cases} \nabla^2 T = 0, \text{outside } \Sigma \\ B_j T = f_j, \text{on } \Sigma, j = 1,2,\cdots,5 \\ T = 0(r^{-3}) \end{cases} \tag{1}$$

or

$$\{\nabla^2, B\} T = \{0, f\}, T = 0(r^{-3}) \tag{1'}$$

Here ∇^2 is the Laplace operator, B_j and f_j are boundary differential operators and boundary values, respectively, $B = \{B_j, j = 1, 2, \cdots, 5\}$, $f = \{f_j, j = 1, 2, \cdots, 5\}$. Furthermore, the boundary values f_j are defined as

$$f_1 = T_{zz}\,|_\Sigma, f_2 = \sin^2\theta T_{yy}\,|_\Sigma, f_3 = \sin\theta T_{yz}\,|_\Sigma$$
$$f_4 = \sin\theta\cos\theta T_{xz}\,|_\Sigma, f_5 = \sin^2\theta\cos\theta T_{xy}\,|_\Sigma$$

Where $T_{ij}(i, j = x, y, z)$ is the second order derivative of the disturbing potential T in a local Cartesian coordinate system.

Based on the definition of f_j, the corresponding expression of boundary differential operators B_j are

$$B_1 = \frac{\partial^2}{\partial r^2}, B_2 = \frac{\sin^2\theta}{r}\frac{\partial}{\partial r} + \frac{\sin\theta\cos\theta}{r^2}\frac{\partial}{\partial\theta} + \frac{1}{r^2}\frac{\partial^2}{\partial\lambda^2}$$
$$B_3 = \frac{1}{r^2}\frac{\partial}{\partial\lambda} - \frac{1}{r}\frac{\partial^2}{\partial r\partial\lambda}, B_4 = \frac{\sin\theta\cos\theta}{r^2}\frac{\partial}{\partial\theta} - \frac{\sin\theta\cos\theta}{r}\frac{\partial^2}{\partial r\partial\theta}$$
$$B_5 = \frac{\sin\theta\cos\theta}{r^2}\frac{\partial^2}{\partial\theta\partial\lambda} - \frac{\cos^2\theta}{r^2}\frac{\partial}{\partial\lambda}$$

where r, θ, λ are geocentric distance, geocentric co-latitude and geocentric longitude, respectively.

In general, problem (1) or (1′) has no solution in a strict sense, but it has unique solution if only one boundary condition is considered, that means the knowledge of f_j determines T partly. Now we consider a Sobolev space $H^2(\Omega)$, where Ω denotes the domain outside Σ. According to the trace theorem in Sobolev spaces (Adams 1975), $f_j \in H^{1/2}(\Sigma)$. Hence we can define a multiplication space:

$$J \equiv \{f = (f_1, f_2, f_3, f_4, f_5); f_j \in H^{1/2}(\Sigma), j = 1,2,\cdots,5\} \tag{2}$$

equipped with an inner product

$$\langle f, g\rangle_J = \sum_{j=1}^{5} \langle f_j, g_j\rangle_{H^{1/2}(\Sigma)}, \forall f, g \in J \tag{3}$$

Thus the norm in J is

$$\| f \|_J = \{\sum_{j=1}^{5} \| f_j \|^2_{H^{1/2}(\Sigma)}\}^{1/2}, \forall f \in J \tag{4}$$

moreover, construct a subspace $\underset{\sim}{J}$

$$\underset{\sim}{J} \equiv \{BT; T \in H^2(\Omega), \nabla^2 T = 0, T = 0(r^{-3})\} \tag{5}$$

It is clear that J is a Hilbert space, and it can be proved that $\underset{\sim}{J}$ is a closed subspace of J. According to the geometrical projection theorem in a Hilbert space, a unique projection $f^{(0)} \in \underset{\sim}{J}$ must exist, such that $f = f^{(0)} + \varepsilon$, where $f^{(0)} \in \underset{\sim}{J}$, $\varepsilon \in \underset{\sim}{J}^{\perp}$, and that the following condition

$$\| f - f^{(0)} \|_J = \inf_{g \in \underset{\sim}{J}} \| f - g \|_J \tag{6}$$

holds. From the definition of $\underset{\sim}{J}$, we know that a function $T \in H^2(\Omega)$ must exist, such that

$$\{\nabla^2; B\} T = \{0; f^{(0)}\}, T = 0(r^{-3}) \tag{7}$$

Problem (7) is called the corresponding well-posed boundary value problem related for problem (1), and its solution is called a quasi-solution in the paper.

Obviously, we have to determine $f^{(0)} \in \underset{\sim}{J}$ that satisfies condition (6) for solving problem (7). For this reason, the explicit form of the norm of $H^p(\Sigma)$ must be known, where p is a positive real number. For Σ being a spherical surface, Zhu and Yu (1992) have derived the explicit formula of the norm of $H^p(\Sigma)$ in terms of spherical harmonic expansion coefficients, i.e.

$$\| f \|_{H^p(\Sigma)} = [\sum_{n=0}^{\infty} \sum_{m=0}^{n} (1+n)^{2p} (\bar{a}_{nm}^2 + \bar{b}_{nm}^2)]^{1/2} \tag{8}$$

where $\bar{a}_{nm}$, $\bar{b}_{nm}$ are fully normalized spherical harmonic coefficients of function $f \in H^p(\Sigma)$, and Σ denotes the surface of the unit sphere.

Up to now we have given the definition of the quasi-solution to problem (1) and an estimation principle for its solution.

2 Quasi-solution model

On the basis of the expansion theorem in Hilbert spaces, the boundary value function f_j can be expanded into spherical harmonics

$$f_j(\theta, \lambda) = \frac{GM}{R_S} \sum_{n=0}^{\infty} \sum_{m=0}^{n} [\bar{a}_{nm}^{(j)} \cos m\lambda + \bar{b}_{nm}^{(j)} \sin m\lambda] \bar{P}_{nm}(\cos\theta)$$
$$j = 1, 2, \cdots, 5 \tag{9}$$

where

GM is the gravitational constant times the mass of the earth

$\bar{P}_{nm}(\cos\theta)$ denotes the normalized Legendre functions

$\bar{a}_{nm}^{(j)}, \bar{b}_{nm}^{(j)}$ are normalized spherical expansion coefficients of degree n and order m (which can be obtained by spherical harmonic analysis of the known boundary values f_j).

Let the disturbing potential $T \in H^2(\Omega)$, Its spherical harmonic expression is

$$T(r, \theta, \lambda) = \frac{GM}{r} \sum_{n=2}^{\infty} \left(\frac{R}{r}\right)^n \sum_{m=0}^{n} (\bar{C}_{nm} \cos m\lambda + \bar{S}_{nm} \sin m\lambda) \bar{P}_{nm}(\cos\theta) \tag{10}$$

where

R is a reference radius of the earth

$\bar{C}_{nm}, \bar{S}_{nm}$ are completely normalized potential coefficients of degree n and order m.

Now our aim is to determine all the geopotential coefficients $\bar{C}_{nm}$ and $\bar{S}_{nm}$ by means of the coefficients $\bar{a}_{nm}^{(j)}$ and $\bar{b}_{nm}^{(j)}$ of the corresponding boundary values f_j. In practice, we can compute the coefficients $\bar{C}_{nm}$ and $\bar{S}_{nm}$ up to some maximum degree N only. Omitting the complicated derivation, we arrive at the following mathematical formulas for the computation of $\bar{C}_{nm}$

$$
\left\{
\begin{aligned}
& K^n F_{nm}^{(3)} \bar{C}_{nm} + K^{n+2} F_{nm}^{(4)} \bar{C}_{n+2,m} + K^{n+4} F_{nm}^{(5)} \bar{C}_{n+4,m} = Q_{nm} \\
& K^n F_{n+2,m}^{(2)} \bar{C}_{nm} + K^{n+2} F_{n+2,m}^{(3)} \bar{C}_{n+2,m} + K^{n+4} F_{n+2,m}^{(4)} \bar{C}_{n+4,m} \\
& \qquad + K^{n+6} F_{n+2,m}^{(5)} \bar{C}_{n+6,m} = Q_{n+2,m} \\
& K^n F_{n+4,m}^{(1)} \bar{C}_{nm} + K^{n+2} F_{n+4,m}^{(2)} \bar{C}_{n+2,m} + K^{n+4} F_{n+4,m}^{(3)} \bar{C}_{n+4,m} \\
& + K^{n+6} F_{n+4,m}^{(4)} \bar{C}_{n+6,m} + K^{n+8} F_{n+4,m}^{(5)} \bar{C}_{n+8,m} = Q_{n+4,m} \\
& \cdots\cdots\cdots\cdots\cdots\cdots\cdots\cdots\cdots\cdots\cdots\cdots \\
& K^{N-8} F_{N-4,m}^{(1)} \bar{C}_{N-8,m} + K^{N-6} F_{N-4,m}^{(2)} \bar{C}_{N-6,m} + K^{N-4} F_{N-4,m}^{(3)} \bar{C}_{N-4,m} \\
& \qquad + K^{N-2} F_{N-4,m}^{(4)} \bar{C}_{N-2,m} + K^{N} F_{N-4,m}^{(5)} \bar{C}_{N,m} = Q_{N-4,m} \\
& K^{N-6} F_{N-2,m}^{(1)} \bar{C}_{N-6,m} + K^{N-4} F_{N-2,m}^{(2)} \bar{C}_{N-4,m} + K^{N-2} F_{N-2,m}^{(3)} \bar{C}_{N-2,m} \\
& \qquad + K^{N} F_{N-2,m}^{(4)} \bar{C}_{Nm} = Q_{N-2,m} \\
& K^{N-4} F_{Nm}^{(1)} \bar{C}_{N-4,m} + K^{N-2} F_{Nm}^{(2)} \bar{C}_{N-2,m} + K^{N} F_{Nm}^{(3)} \bar{C}_{Nm} = Q_{Nm} \\
& \qquad n = 2,3,\cdots,N,\ m = 0,1,\cdots,n
\end{aligned}
\right.
\tag{11}
$$

where

$$
\begin{aligned}
K &= R/R_s \\
Q_{nm} &= (n-1)[B_{nm}^{(1)} \bar{a}_{n-2,m}^{(4)} - C_{nm}^{(1)} \bar{b}_{n-2,m}^{(5)}] + (n+1)[(n+1)(n+2)\bar{a}_{nm}^{(1)} \\
& \quad - m(n+2)\bar{b}_{nm}^{(3)} + A_{nm}^{(2)} \bar{a}_{nm}^{(2)} + B_{nm}^{(2)} \bar{a}_{nm}^{(4)} - C_{nm}^{(2)} \bar{b}_{nm}^{(5)}] \\
& \quad + (n+3)[A_{nm}^{(3)} \bar{a}_{n+2,m}^{(2)} + B_{nm}^{(3)} \bar{a}_{n+2,m}^{(4)} - C_{nm}^{(3)} \bar{b}_{n+2,m}^{(5)}]
\end{aligned}
$$

and $F_{nm}^{(i)}, A_{nm}^{(i)}, B_{nm}^{(i)}, C_{nm}^{(i)}$ are the function of degree n and order m only (Luo, 1996), which can be easily computed.

Formula (11) is the case that $(N-n)$ is even, if $(N-n)$ is odd, then $N-1$ is used instead of N. Furthermore, formula (11) shows that its coefficient matrix is a band matrix with the maximum degree of $\left[\frac{N-n}{2}\right]+1$ and the band width of 5, where $[\cdot]$ means integering. For the computation of $\bar{S}_{nm}$, it is enough to substitute G_{nm} for Q_{nm} in formula (11).

$$
\begin{aligned}
G_{nm} &= (n-1)[B_{nm}^{(1)} \bar{b}_{n-2,m}^{(4)} + C_{nm}^{(1)} \bar{a}_{n-2,m}^{(5)}] \\
& \quad + (n+1)[(n+1)(n+2)\bar{b}_{nm}^{(1)} + m(n+2)\bar{a}_{nm}^{(3)} + A_{nm}^{(2)} \bar{b}_{nm}^{(2)} + B_{nm}^{(2)} \bar{b}_{m}^{(4)} \\
& \quad + C_{nm}^{(4)} \bar{a}_{nm}^{(5)}] + (n+3)[A_{nm}^{(3)} \bar{b}_{n+2,m}^{(2)} + B_{nm}^{(3)} \bar{b}_{n+2,m}^{(4)} + C_{nm}^{(3)} \bar{a}_{n+2,m}^{(5)}]
\end{aligned}
$$

3 Simulation results

In this section a small numerical experiment has been performed to check the correctness and effectiveness of the quasi-solution model presented in the paper. To this aim we have produced a $30' \times 30'$ grid of values of T_{zz} and T_{yy} on the sphere of $R_s = 6578$ km by using the WDM 94 model (Ning et al., 1994) from degree 51 to degree 360. We have then used formula (11) to compute the corresponding potential coefficients $\bar{\bar{C}}_{nm}$ and $\bar{\bar{S}}_{nm}$ from degree 51 to degree 260, with no other noise but the intrinsic numerical one. The comparison results are listed in Table 1 in terms of the relative differences defined as

$$d_c = |(\bar{\bar{C}}_{nm} - \bar{C}_{nm})/\bar{C}_{nm}|, d_s = |(\bar{\bar{S}}_{nm} - \bar{S}_{nm})/\bar{S}_{nm}|, \quad (12)$$

which seems very satisfactory.

Table 1 The comparison results of potential coefficients computed by the quasi-solution model in terms of the relative differences (Output degrees: 51～260)

The range of relative difference	No.		Percentage	
	$\bar{C}_{nm}$	$\bar{S}_{nm}$	d_c	d_s
$\leqslant 10^{-6}$	25861	27056	0.7869	0.8285
$10^{-6} \sim 10^{-5}$	4307	4472	0.1310	0.1369
$10^{-5} \sim 10^{-4}$	1179	824	0.0359	0.0252
$10^{-4} \sim 10^{-3}$	816	241	0.0248	0.0074
$10^{-3} \sim 10^{-2}$	645	47	0.0196	0.0014
$10^{-2} \sim 10^{-1}$	55	15	0.0017	0.0006
$10^{-1} \sim 10^{0}$	2	0	0.0001	0.0000
Σ	32865	32655	1.0000	1.0000

By adding a zero average white noise of uniform variance $10^{-2}E$ to the above simulation data, we also get the corresponding results listed in Table 2, which seems good indeed and shows that the quasi-solution model is not very sensitive to the noise. Moreover, the cumulative commission error to geoid undulation is not larger than 12cm.

Table 2 The comparison results of potential coefficients computed by the quasi-solution model in terms of the relative differences (Output degrees: 51～260)

The range of relative difference	No.		Percentage	
	$\bar{C}_{nm}$	$\bar{S}_{nm}$	d_c	d_s
$\leqslant 10^{-6}$	6459	6637	0.1965	0.2032
$10^{-6} \sim 10^{-5}$	9642	9735	0.2934	0.2981
$10^{-5} \sim 10^{-4}$	14964	14703	0.4553	0.4503
$10^{-4} \sim 10^{-3}$	782	766	0.0238	0.0235
$10^{-3} \sim 10^{-2}$	824	742	0.0251	0.0227
$10^{-2} \sim 10^{-1}$	182	69	0.0055	0.0021
$10^{-1} \sim 10^{0}$	12	3	0.0004	0.0001
Σ	32865	32655	1.0000	1.0000

Summarizing, we can state that the quasi-solution model to overdetermined gravity gradiometry boundary value problem is very efficient and practical, and it permits to combinations of all various components of the gravity gradient for refining the earth's gravity field. The solution is harmonic outside the earth and in the sense of norm $\|\cdot\|_J$ it fits optimally all the boundary conditions. Simulation results show that the results computed from T_{zz} and T_{yy} independently (Luo, 1996) can be more or less improved. Finally, we want to stress that the quasi-solution is a result of a spherical approximation of the disturbing potential. Furthermore, a nonlinear error should be taken into account in case that results of higher accuracy are required. The quasi-solution model given in the paper will be thoroughly verified in future.

Acknowledgments This work was supported by the National Natural Science Foundation of China and the Foundation of Laboratory for Modern Geodynamics, WTUSM. The authors would like to thank Dr. P. Holota and one anonymous reviewer for suggesting significant improvements in the manuscript.

References

Adams RA(1975) Sobolev's Space. Academic Press, New York.

Brovelli M, Migliaccio F, Sansò F (1991) A BVP Approach to the Reduction of Spaceborne Gradiometry: Theory and Simulations, IAG 110:169 – 180.

Holota P (1986) Boundary Value Problems and Invariants of the Gravitational Tensor in Satellite Gradiometry. Lecture Notes in Earth Sciences, Vol. 25, pp. 447 – 457.

Keller W, Hirsch M (1994) A boundary value approach to downward continuation. Manuscripta geodaetica, 19, pp. 101 – 118.

Luo ZC (1996) The Theory and Methodology for the Determination of the Earth's Gravity Field from Satellite Gravity Gradiometry Data. Dissertation (in Chinese), Wuhan Technical University of Surveying and Mapping.

Luo ZC, Ning JS, Chao DB (1996) Some Problems on Spherical Harmonic Analysis of Satellite Gravity Gradiometry. Geoinformatics, Vol. 1, pp289 – 294.

Luo ZC, Ning JS, Chao DB (1997) Spectral Method for the Downward Continuation of Satellite Gravity Gradiometry Data. ACTA GEODAETICA et CARTOGRAPHICA SINICA, 26, pp. 168 – 175.

Ning JS, Li JC, Chao DB (1994) The Research of the Earth's Gravity Field Model WDM94 Complete to Degree 360. Journal of Wuhan Technical University of Surveying and Mapping, 19, pp. 283 – 291.

Robbins J W (1985) Least Squares Collocation Applied to Local Gravimetric Solutions from Satellite Gravity Gradiometry Data. Report No. 368, Department of Geodetic Science and Surveying, The Ohio State University.

Rummel R, Colombo OL (1985) Gravity Field Determination from Satellite Gradiometry. Bulletin Geodesique 57, pp. 233 – 246.

Rummel R, Teunissen P, van Gelderen M (1989) Uniquely and Overdetermined Geodetic Boundary Value Problems by Least Squares. Bulletin Geodesique, 63, pp. 1 – 33.

Sacerdote F, Sansò F (1985) Overdetermined Boundary Value Problems in Physical Geodesy. Manuscripta geodaetica, 8, pp. 195 – 207.

Tscherning CC, Forsberg R, Vermeer M (1990) Methods for Regional Gravity Field Modelling from SST and SGG Data. Report 90: 2, Finnish Geodetic Institute, Helsinki.

Zhu ZW, Yu JH (1992) The Quasi-solution to overdetermined Geodetic Boundary Value Problems. Science in China (Series B), Vol. 35, pp. 103 – 112.

INVESTIGATIONS OF THE EARTH'S RELATIVISTIC GRAVITY FIELD

Wenbin Shen

Section of Physical Geodesy, Graz Technical University, A-8010 Graz, Austria

Abstract. The author investigated the Earth's relativistic gravity field and the coordinate transformation between the geocentric star-fixed system and the geocentric earth-fixed system. The standard measure rod and clock are defined as those which are at rest and at infinity with respect to the geocentric star-fixed system. Compared with gravity, gravitation is more fundamental. The definition of the *equifrequency geoid* is proposed. Based on this definition the geopotential difference as well as the height difference between two arbitrary stations (even if they are far away from each other) on the Earth's surface can directly be determined using the frequency shift method.

1. Introduction

One basic task of geodesy is to determine the gravity field of the Earth. Theoretically, if the gravity on the Earth's surface is measured and reduced into the geopotential value on the surface of the Earth, the gravity field in the external space of the Earth can uniquely be determined (in the frame of Newtonian mechanics). However, there is a drawback (cf. Jeffreys 1970). The errors are unavoidably introduced during the reduction from gravity to geopotential.

If one resort to the concept of the *equifrequency geoid* (see section 5), the above drawback can be overcome (cf. Shen et al. 1993, 1994; Shen 1997). Here, one can directly determine the geopotential on the Earth's surface using the *frequency shift method*, and, consequently, the Earth's external gravity field can be determined. The basic idea of the frequency shift method is as follows:

Suppose that there are two points A and B. An emitter at A emits a light signal with frequency f, and a receiver at B receives this signal with frequency f'. Then with the following formula

$$\Delta W_{AB} = W_B - W_A = \frac{f' - f}{f} \tag{1}$$

one can determine the geopotential difference between A and B. Or, we suppose that a signal source (a satellite source or a distant star) emits signals with the same frequency f_0 and two receivers at A and B receive these signals with frequencies f and f' at the same time. Then, using formula (1) we can also directly determine the geopotential difference between A and B (these two cases are denoted by *GPS frequency shift method* and *VLBI frequency shift method*, cf. Shen et al. 1993, 1994).

We note that we use the *light unit system* throughout this paper: $c = 1$.

2. Measure Standards

According to Newtonian theory, space and time are absolute, and, consequently, once a length measure standard and a time measure standard are chosen, they keep invariant. However, according to the theory of relativity, the length and time measure standards may vary with different positions and different reference systems. This implies that the observed values at points A and B are not absolute. Hence, we must establish a "common measure standard system", into which all observed values must be transformed.

Time measure is realized by the time keeping of a precise clock. In the viewpoint of relativity, the running rate of a clock depends not only on its velocity (the state of motion) but also on its position at which there exists a gravitational potential. Hence, generally, clocks located at two different points on (above or under) the Earth's surface run with different rates. However, we can assume that there exists an ideal clock at infinity at rest with respect to the *cosmic microwave background* (cf. Shen 1997). This clock is not influenced by the Earth's rotation, neither influenced by the gravitation nor by the gravitational potential. Under this assumption, we can establish the following relation:

$$T = T_0(1 - \frac{GM}{r} - \frac{1}{2}v^2), \tag{2}$$

where T_0 and T are the time "readings" of the *ideal clock* and the *ground clock* respectively, M is the Earth's mass, G is the gravitational constant, r is the distance from the mass center of the Earth to the ground clock, and v is the velocity of the ground clock with respect to the *geocentric star-fixed system* (GSS). Equation (2) shows that the ground clock runs slower than the ideal clock does.

Because of the existence of the gravitational field, the length of a measure rod depends on the directions along which the rod is located. Suppose that the gravitational field consists of two parts: the main part that is caused by a uniform sphere (with the Earth's mass), whose metric is the Schwarzschild metric $g_{\mu\nu}$ (cf. Weinberg 1972); and the perturbation part that is caused by irregular masses, whose metric is denoted by $h_{\mu\nu}$. According to the Schwarzschild metric, the length of the rod will not be influenced by the gravitational field when it is set along the transverse direction of a spherical field, denoted by L_t, and it will be influenced by gravitational field when it is set along the radial direction of a spherical field, denoted by L_r (cf. Lorentz et al. 1923). Suppose that a rod L_0 is at rest at infinity , which is neither influenced by the gravitation nor by the Earth's motion. Consequently it can be considered as an ideal length. Then we have (cf. Lorentz et al. 1923):

$$L_t = L_0, \quad L_r = (1 - \frac{GM}{r})L_0. \tag{3}$$

The Earth's rotation is not taken into account in the Schwarzschild metric. In fact, if we resort to the Kerr metric (i.e., the Earth's rotation is taken into account, cf. Weinberg 1972), we can still get the above result (cf. Shen 1997).

3. Gravitation and Inertia

In Newtonian theory, gravitation is related neither to the state of the particle's motion nor to the state of the gravitational source's motion. In general relativity however, gravitational effect is related to both the states of the particle's motion and the motion of the gravitational source. The nature of gravitation is still a great mystery. However, one fact is definite: gravitational fields originate from gravitational masses, just as electromagnetic fields originate from (electric) charges. Essentially, gravitation is a physical phenomenon arising from the mass sources.

When a body accelerates with respect to an inertial reference system, it "feels" a force which is opposite to the direction of the acceleration. Based on the well-known rotation experiment of the water bucket, Newton (1686) stated that inertia arises from the acceleration of the body relative to the absolute space. However, Mach (1883) stated that inertia arises from the acceleration relative to the total matter of the universe. Who is right?

Trying to answer this question, let us consider the following *Gedankenexperiment* (cf. Shen 1996): we assume that the universe is completely empty. If Mach is right, there will be no inertial force at all. However, if Newton is right, the inertial forces will exist. In fact, once an acceleration reference system is introduced, an *inertial force* followes; once a rotating reference system is introduced, the *inertial centrifugal force, Coriolis force* and *Euler force* are present. It seems that the inertial forces are not related to the total matter of the universe. Unfortunately, since there are no "empty universes", we cannot make a final judgement: Who is correct, Newton or Mach?

4. Gravity

Generally, gravity can be defined as follows (cf. Shen 1996):

Referring to a celestial body, at an arbitrary point P (either outside or inside the body), gravity is the superposition of the gravitation and the inertial forces caused by the celestial body, provided that P is at the moment rigidly fixed to the body.

Any inertial force is due to the use of a non-inertial reference system. However, the reference system does neither influence gravitational fields, nor the curvature of the spacetime. Reference system effects can be calculated theoretically, provided that a definite reference system is previously selected . Hence, once the Earth's gravitational field is determined, the gravity field is determined.

In practice, (generally) the observed quantities contain the reference system effects. Consequently, it is necessary to discuss the coordinate transformation between reference systems. Once a reference system is transformed, the corresponding metric $g_{\mu\nu}$ as well as the Christofel symbols $\Gamma^{\lambda}_{\mu\nu}$ will change. Since $g_{\mu\nu}$ and $\Gamma^{\lambda}_{\mu\nu}$ have the characters of the gravitational potential and the gravitation (cf. Weinberg 1972), and are related to the choices of the reference systems, they can be considered as having the characters of the geopotential and the gravity.

Suppose that there exists a gravitational field caused by a non-rotating uniform sphere. Then, the field is expressed by the *Schwarzschild metric*. In a spherical coordinate system S (e.g., the geocentric star-fixed system), the spacetime interval can be expressed as

$$d\tau^2 = (1 - \frac{2M}{r})dt^2 - (1 - \frac{2M}{r})^{-1}dr^2 - r^2 d\theta^2 - r^2 \sin^2\theta \varphi^2. \tag{4}$$

Now we consider the metric form expressed in another coordinate system S′, which is chosen in such a way that it rotates with a constant angular velocity $\vec{\omega}$ with respect to S (e.g., S′ is the *geocentric earth-fixed system* GES), and its origin and the third space axis always coincide with the origin and the third space axis of S. Suppose that point P has coordinates (t, r, θ, φ) in S and coordinates $(t', r', \theta', \varphi')$ in S′. Obviously, r' and θ' are equal to r and θ respectively, and φ' varies with time, given by $\varphi' = \varphi - \omega t$.

However, we will encounter a difficulty in considering the transformation between the time coordinates t and t' (cf. Shen 1996). Not going into details, it is reasonable to choose the transformation equation as (cf. Mϕller 1972; Shen 1996): $t' = t$. Then, we can write the coordinate transformation between S and S′ as follows:

$$t' = t, \;\; r' = r, \;\; \theta' = \theta, \;\; \varphi' = \varphi - \omega t. \tag{5}$$

Substituting (5) into (4) we get

$$\begin{aligned} d\tau'^2 &\equiv d\tau^2 \\ &= (1 - \frac{2MG}{r'})dt'^2 - (1 - \frac{2MG}{r'})^{-1}dr'^2 - r'^2 d\theta'^2 - r'^2 \sin^2\theta'(d\varphi' + \omega dt')^2 \\ &= (1 - \frac{2MG}{r'} - r'^2\omega^2\sin^2\theta')dt'^2 - (1 - \frac{2MG}{r'})^{-1}dr'^2 - r'^2 d\theta'^2 \\ &\quad -r'^2\sin^2\theta' d\varphi'^2 - 2r'^2\omega\sin^2\theta' d\phi' dt'. \end{aligned} \tag{6}$$

Observed in system S, $g_{\mu\nu}$ is given by (4), based on which $\Gamma^{\lambda}_{\mu\nu}$ has only the character of gravitation. However observed in system S′, $g_{\mu\nu}$ is given by (6), based on which

$\Gamma^{\lambda}_{\mu\nu}$ contains not only gravitational effects but also the inertial effects. We note that, if we further construct the Riemann tensor $R_{\mu\nu\alpha}{}^{\beta}$, it will be found that no matter which kind of reference system is chosen, the Riemann tensor $R_{\mu\nu\alpha}{}^{\beta}$ is not influenced (cf. Shen and Moritz 1996, Shen 1997). This implies that the choice of the reference system does not influence the nature of the spacetime (gravitation).

5. The definition of the Equifrequency Geoid

The *classical definition of the geoid* is not satisfactory. The main drawback is that, according to the classical definition, the geoid is "unobservable". However, based on the definition of the *equifrequency geoid* (introduced later), the geoid is "observable". This is the main reason why the concept of the equifrequency geoid is introduced (cf. Shen et al. 1993, 1994). To introduce this (new) concept, we first define the *equigeopotential surface* (cf. Shen et al. 1993, 1994):

The equigeopotential surface is such a closed curve surface on which there is no gravity frequency shift.

Based on this definition we can further define the *relativistic geoid* as follows (cf. Shen et al. 1993):

The relativistic geoid is the closed curve surface nearest to the mean sea level on which there is no gravity frequency shift.

Or, the relativistic geoid can be simply defined as the *equifrequency surface nearest to the mean sea level.* We call the relativistic geoid defined as above the *equifrequency geoid* (cf. Shen et al. 1993). Based on the concept of the equifrequency geoid we can determine the relativistic geoid by observing the frequency shifts. Similarly, based on the frequency shift observations the geopotential values on the whole surface of the Earth can be directly measured (the basic principle see section 1, or Shen et al. 1993, 1994) and consequently the Earth's external geopotential and gravity field can be determined by solving the boundary value problem (cf. Shen 1997).

6. Conclusions

Based on the classical definition of the geoid, the geopotential difference between two points A and B can only be determined by gravimetry combining with leveling, and the propagation error increases with the increase of the distance between the two points considered. However, based on the definition of the equifrequency geoid, the determination of the geopotential difference between any points A and B can be realized with the aid of the frequency shift method (cf. section 1, or Shen et al. 1993, 1994, Shen 1997). Especially, if the two points A and B are located in different continents, the advantage of the new method is obvious, because in this case it is still possible to directly determine the geopotential difference between these two points (using VLBI frequency shift method, see section 1, or Shen 1997 for details). This implies that, based on this new method, it is possible to realize the unification of the *world height system* (cf. Shen 1997).

REFERENCES

Jeffreys S H (1970), The Earth: Its origin, history and physical constitution (fifth ed., 1st ed. 1957). Cambridge University Press, Cambridge.

Lorentz H A, Einstein A, Minkowski H, Weyl H (1923), The Principle of Relativity: A Collection of Original Memories. Methuen, London.

Mach E (1883), Die Mechanik in Ihrer Entwicklung (6. Auflage 1908), F.A.Brockhaus, Leipzig; The Science of Mechanics (transl. by T.J.McCormack, 5th ed. 1942). Open Court, LaSalle, Ill.

Mϕller C (1972), The Theory of Relativity (2nd ed., 1st ed. publ. 1952). Oxford University Press, Oxford.

Newton I (1687), Philosophiae Naturalis Principia Mathematica (English transl. by Andrew Motte, revised and annotated by F.Cajori). University of California Press (1966), California.

Shen W (1996), On the Separability of Gravitation and Inertia According to General Relativity. Dissertation. Graz Technical University, Graz.

Shen W (1997), Relativistic Physical Geodesy. Habilitation. Institute of Theoretical Geodesy, Graz Technical University, Graz.

Shen W, Chao D, Jin B (1993), On the Relativistic Geoid. Bollettino di Geodesia e Scienze Affini, Vol. 52, pp. 207–216.

Shen W, Chao D, Jin B (1994), The Concept and Application of the Equifrequency Geoid (in Chinese). Journal of Wuhan Technical University of Surveying and Mapping, Vol. 19, pp. 232–238.

Shen W, Moritz H (1996), On the Separation of Gravitation and Inertia and the Determination of the Relativistic Gravity Field in the Case of Free Motion. The Journal of Geodesy, Vol. 70, pp. 633–644.

Weinberg S (1972), Gravitation and Cosmology. John Wiley & Sons, New York.

GEOID AND MASS DENSITY - WHY AND HOW?

Gabriel Strykowski, National Survey and Cadastre,
Rentemestervej 8, DK-2400 Copenhagen NV, Denmark

Abstract. *The concept of the geoid, originally introduced by Gauss and Listing, is inconsistent with the methods of geoid modelling used in Geodesy. The reason is, that the implicit/explicit choice of a simple mass density model used for the reduction of surface gravity data can deviate from the physical reality. The (possible) inconsistency is valid at all freqeuencies. Some ideas of how to handle these problems in modelling are put forward.*

1. Introduction

Newton's law for the gravitational potential V is (Heiskanen and Moritz, 1967):

$$V(P)=\iiint_{\Omega}\frac{G\rho(Q)}{|P-Q|}d\omega \tag{1}$$

where ρ is mass density, G is the gravitational constant, Ω is the mass support [i.e. Ω is the figure of the attracting body, e.g. the figure of the Earth], $Q \in \Omega$ and $P \in \mathbb{R}^3$ [i.e. P is a point *outside* or *inside* the (Earth's) masses].

The definition of the geoid by Gauss and Listing [after Grafarend, 1994] is: The *geoid* is the equipotential surface W(P)=const coinciding with the *mean sea level*, where

$$W(P)=V(P)+\Phi(P) \tag{2}$$

and where Φ is the centrifugal potential [$\Phi=\frac{1}{2}\omega^2 p^2$, where ω is a (constant) angular velocity of the Earth and *p* is the orthogonal distance from the rotational axis of the Earth].

2. Geodesy- from surface gravity to the geoid

In Geodesy, the *geoid* and the *quasigeoid* are two surfaces used for the definition of heights. The formal expresions are (Forsberg, 1996; Heiskanen and Moritz, 1967, chapter 8):

a. The classical geoid - "Stokes"

$$N=\frac{R}{4\pi\gamma}\iint_{\sigma}\Delta g S(\psi)d\sigma \tag{3}$$

where N are the geoidal heights, R is the radius of the spherical Earth, γ is the normal gravity, S is Stokes' function, ψ is the spherical distance, Δg are the gravity anomalies and σ is the surface of the spherical Earth.

b. The quasigeoid - "Molodensky" (here to the 1.order)

$$\zeta=\frac{R}{4\pi\gamma}\iint_{\sigma}(\Delta g+g_1)S(\psi)d\sigma \tag{4}$$

ζ is the height of the quasigeoid, g_1 is the first term of Molodensky's series expansion, Δg in Eq. (4) has a different meaning than in Eq. (3) (Heiskanen and Moritz, 1967, Eq.(8-7)).

Referring for details to the above references one can state that:

(i) ρ is not used explicitly in Eqs. (3)-(4). In Eq.(3) the surface gravity measurements are reduced to the gravity anomalies Δg on the *geoid*, cf. (Heiskanen and Moritz, 1967, sec. 8-2). This involves the use of simple mass density models, e.g. 2.67 g/cm^3 for the topography.

(ii) The *geoid* is modelled from measurements of gravity on or above the Earth's surface, i.e. outside the Earth's masses, and from height differences of the visible topography.

(iii) One can obtain N from ζ via the corresponding orthometric- and normal heights. The expression uses the same information as in (i)-(ii), cf. (Heiskanen and Moritz, 1967, Eq.(1-103)). Thus, (i)-(ii) cannot be circumvented by obtaining *geoid* from *quasigeoid.*

3. Discretizing the global mass density distribution

The physical mass density of the Earth ρ exists. Consider ρ being expressed mathematically in the usual geocentric spherical coordinates, $\rho\equiv\rho(r,\theta,\lambda)$, where r is radial distance, θ is the polar distance and λ is the geocentric longitude. Great simplification of formulas is obtained by regarding the free air as a part of ρ with a zero mass density value, see e.g. (Grafarend, 1994). For the mathematical convenience we will change the definition of Ω in Eq. (1) to a sphere of radius R_{max} enclosing the Earth's masses.

Consider next a (finite) number of disjoint, concentric spherical shells filling up Ω. A particular shell is identified by index *i*, *i*=1,...,*I*. The "thickness" of each shell is denoted by $\Delta r(i)$ and the centre radial distance by r_i, i.e. $r \in]\, r_i - \frac{1}{2}\Delta r(i),\ r_i + \frac{1}{2}\Delta r(i)]$. Mass distribution inside each shell can be condensed [by a technique similar to Helmert condensation, see (Heiskanen and Moritz, 1967, sec. 3.7)] to a surface mass density μ_i associated with a surface of a sphere with radius r_i, see e.g. (Sigl, 1985, sec. 1.4). [The use of the surface mass densities is not strictly necessary, e.g. one could parametrize directly the mass distribution inside each shell. However, the use of surface mass densities simplifies the formulas.]

Consider next the values of μ_i being expressed using fully normalized surface spherical harmonics (Heiskanen and Moritz, 1967) truncated to a maximal degree N(i). Formally, each μ_i can be expressed in $\mathbb{R}^3$ using the Dirac delta function δ:

$$\mu_i(r,\theta,\lambda)=[\sum_{n=0}^{N(i)}\sum_{m=0}^{n}\overline{a_{nm}}(i)\,\overline{R_{nm}}(\theta,\lambda)+\overline{b_{nm}}(i)\,\overline{S_{nm}}(\theta,\lambda)]\delta(r_i-r) \tag{5}$$

The notation in Eq.(5) follows (Heiskanen and Moritz, 1967). The coefficients of the expansion refer to a particular spherical shell (index *i*). Notice that the discretization of ρ is arbitrarily fine. The "vertical" and the "horizontal" resolutions are governed by the choices of $\Delta r(i)$ and N(i) respectively.

4. Can V be modelled without the explicit use of mass density?

The inner potential $V_i^{(i)}$ of μ_i, i.e. for $r \le r_i$:

$$V_i^{(i)}(r,\theta,\lambda)=4\pi G[\sum_{n=0}^{N(i)}\frac{r_i}{2n+1}(\frac{r}{r_i})^n\sum_{m=0}^{n}\overline{a_{nm}}(i)\,\overline{R_{nm}}(\theta,\lambda)+\overline{b_{nm}}(i)\,\overline{S_{nm}}(\theta,\lambda)] \tag{6}$$

The outer potential $V_i^{(e)}$ of μ_i, i.e. for $r \ge r_i$:

$$V_i^{(e)}(r,\theta,\lambda)=4\pi G[\sum_{n=0}^{N(i)}\frac{r_i}{2n+1}(\frac{r_i}{r})^{n+1}\sum_{m=0}^{n}\overline{a_{nm}}(i)\,\overline{R_{nm}}(\theta,\lambda)+\overline{b_{nm}}(i)\,\overline{S_{nm}}(\theta,\lambda)] \tag{7}$$

The derivation of Eqs.(6)-(7) is explained in details in Appendix A. To find the expression for the potential V inside/outside the masses one can start with finding the index J so that $r_J < r \le r_{J+1}$. Thus, the expression is:

$$V(r,\theta,\lambda)=\sum_{i=1}^{J}V_i^{(i)}(r,\theta,\lambda)+\sum_{i=J+1}^{I}V_i^{(e)}(r,\theta,\lambda) \tag{8}$$

Example: What is wrong with the "geodetic geoid"?
Referring to Eq. (5) consider anomalous mass distribution $\Delta\mu$, $\Delta\mu(r,\theta,\lambda) = \Delta\mu_i(r,\theta,\lambda) + \Delta\mu_j(r,\theta,\lambda)$, $i \neq j$ and $r_i < r_j$. Furthermore, assume that $\Delta\mu_i(r,\theta,\lambda)=C(i)\overline{F_{nm}}(\theta,\lambda)\delta(r-r_i)$ and $\Delta\mu_j(r,\theta,\lambda)=C(j)F_{nm}(\theta,\lambda)\delta(r-r_j)$, where (C, F_{nm}) denote either $(\overline{a_{nm}},\overline{R_{nm}})$ or $(\overline{b_{nm}},\overline{S_{nm}})$, i.e. any of the two possible kinds of fully normalized surface spherical harmonics and the associated coefficient. Consider next $C(j)= - (r_i/r_j)^{n+2}\,C(i)$. By inserting into Eq.(8) we get:

$$\Delta V(r,\theta,\lambda)=\begin{cases}0 & for\quad r\ge r_j\\ \dfrac{4\pi G}{2n+1}[r_i(\dfrac{r_i}{r})^{n+1}-r_j(\dfrac{r_i}{r_j})^{n+2}(\dfrac{r}{r_j})^n]C(i)F_{nm}(\theta,\lambda) & for\quad r_i\le r\le r_j\\ \dfrac{4\pi G}{2n+1}[r_i(\dfrac{r}{r_i})^n-r_j(\dfrac{r_i}{r_j})^{n+2}(\dfrac{r}{r_j})^n]C(i)F_{nm}(\theta,\lambda) & for\quad r\le r_i\end{cases} \tag{9}$$

Eq. (9) shows that ΔV, i.e. the contribution to the potential from $\Delta\mu$, is zero for $r \ge r_j$. Thus, the information about $\Delta\mu$ may not be present in the surface gravity data (from which the "geodetic *geoid*" is modelled, see sec. 2). However, $\Delta\mu$ affects V for $r<r_j$ and, thus, possibly the *geoid* of Gauss and Listing, see sec. 1. One should notice, that this possible in-consistency between the "geodetic geoid" and the geoid of Gauss and Listing is valid for an arbitrary frequency (index *nm*). Notice also that even for $r<r_j$ the contribution to the potential depends both on r_j and r_i. Thus, for example, it is not sufficient to know that the visible topography is compensated. One must also know (and for the arbitrary frequency *nm*) where the compensating masses are located in depth.

5. Geoid and mass density - Why?

In sec 4 it is demonstrated that there is an inconsistency between the original definition of the *geoid* by Gauss and Listing, see sec.1, and the "geodetic *geoid*" modelled from surface gravity data, see sec. 2. Is the stated (mathematical) inconsistency a physical reality, i.e. can the true mass density distribution of the Earth cause such inconsistency? Yes, the con-cept of compensating masses, which is a physical reality in many areas of the world, will cause such problems. Another example is the possible (negative) correlation of gravity signals generated by undulating geological layers. Thus, the stated problems can be real.

What is the "geodetic geoid" then? The "geodetic *geoid*" (i.e. a model obtained using Eqs. (3)-(4)) is associated with a simplified mass density model, the gravity response of which is consistent with the gravity measured on the Earth's surface. However, it is not a geoid in a sense of Gauss and Listing. Roughly speaking, the difference reflects the ambi-guity of a solution to the inverse gravimetric problem (Strykowski, 1995). One should also notice, that the formal assessment of the accuracy of the geoid, as it is done in Geodesy, cannot handle the stated problems. Such assessment is not absolute, i.e. the assessment of the accuracy of the "geodetic geoid" as compared to the geoid of Gauss and Listing. The reason is, that the usual assessment of the accuracy is conducted from the surface data.

6. Geoid and mass density - How?

From Eq. (1) the relationship between V and ρ is linear. Thus, $V(\rho+\Delta\rho) = V(\rho) + V(\Delta\rho)$. From sec. 5 the "geodetic *geoid*" is obtained using a simple mass density model ρ_G. The "true" mass density ρ_T is related to ρ_G by the correction term $\Delta\rho$, $\Delta\rho = \rho_T - \rho_G$. Thus, in principle, in order to correct the "geodetic *geoid*" towards Gauss and Listing *geoid*, it is only necessary to compute the correction term $\Delta V = V(\Delta\rho)$.

The correction can be partial, e.g. a more realistic mass density model for the topography. The correction can be conducted using Eq. (8) or similar, see e.g. (Strykowski, 1997a,b). By regarding the free air as a part of the mass density model the formulas are greatly simplified, cf. sec. 3.

Is it necessary to measure mass density? It is usefull (as a-priori model) to know some-thing about mass densities of local geological layers (Strykowski, 1996). However, mass density values can be estimated (Strykowski 1997a,b) using independent geometrical information (topography/seismograms).

7. Conclusions

In this paper it is shown explicitly that the geoid, as introduced by Gauss and Listing, is inconsistent, at all frequencies, with the methods of geoid modelling used in Geodesy. Some general ideas of how to correct for this inconsistency in practise are put forward.

References:

Forsberg, R. (1996). *Geoid Information and GPS - a Review and Nordic Status*. Lecture Notes for Nordiska Kommssionen för Geodesi courses, "Geodetic Applications of GPS", Båstad, Sweden, August, 1996.

Grafarend, E.W. (1994). *What is Geoid?* In: Geoid and its Geophysical Interpretation. Vaniček , P. and N.T. Christou (Eds.), CRC Press, pp. 3-32.

Heiskanen, W.A. and H. Moritz (1967). *Physical Geodesy*. W.H. Freeman and Co.

Sigl, R. (1985). *Introduction to Potential Theory*. Abacus Press in association with Herbert Wiechmann Verlag.

Strykowski, G. (1995). *Geodetic and Geophysical Inverse Gravimetric Problem, the Most Adequate Solution and the Information Content*. In: Sünkel, H. and I. Marson (Eds.): Gravity and Geoid, Proc. IAG Symposium 113, Graz, Austria, 11-17 sept, 1994, Springer Verlag, pp. 215-224.

Strykowski, G. (1996). *Borehole Data and Stochastic Gravimetric Inversion*. PhD-thesis, University of Copenhagen, Publ. of National Survey and Cadastre, series 4, vol. 3, ISBN 87-7866-013-0.

Strykowski, G. (1997a). *Formulation of the Mathematical Frame of the Joint Inverse Gravimetric-Seismic Modelling Problem Based on Analysis of Regionally Distributed Borehole Data*. Proc. Int. Symp. On Gravity, Geoid and Marine Geodesy, Sept. 1996, Tokyo, Japan, Springer Verlag, pp. 360-367.

Strykowski, G. (1997b). *Experiences with a Detailed Estimation of Mass Density Contrasts and the Regional Gravity Field using Geometrical Information from Seismograms*. Accepted for publ. Proc. Session EGS 1 of 22nd General Assembly of EGS, Vienna, Austria, 21-25, April, 1997.

Appendix A. The proof of Eq. (6) and Eq.(7)

In this short appendix Eqs. (6) - (7) are proved, as they are not commonly known. Referring to (Sigl, 1985, sec. 2.2), and for a continous mass density distribution μ_k given on a spherical surface Ω' (i.e. a surface of a sphere in $\mathbb{R}^3$) with radius r_k, we get the following expression for the outer potential:

$$V_i^{(e)}(P) = G\iint_{\Omega'} \frac{d\mu_k(Q)}{|P-Q|} \tag{A.1}$$

Eq. (A.1) is identical to the Eq. (II.42) in (Sigl, 1985), but transformed to the notation used in this paper, see Eq. (1). To prove Eq. (6) it is sufficient to prove it just for one component of the series expansion into fully normalized spherical harmonics. The full formula is a linear combination of such expressions. The chosen (arbitrary) component of the series expansion is called C_{nm} and $F_{nm}(\theta,\lambda)$, see Example in sec. 4.

In Eq.(A.1) the explicite integration is carried out with respect to r', θ' and λ'. As it can be seen from Eq.(5), the radial dependence is separable and expressed by the Dirac delta function. Thus, after the integration with respect to the radial dependence the radial distance of Q is fixed to r_k, and the remaining double integration is with respect to the spherical surface $\sigma=\sigma(\theta',\lambda')$, i.e.

$$d\mu_k(Q(r_k,\theta',\lambda'))=C_{nm}F_{nm}(\theta',\lambda')r_k^2 d\sigma(\theta',\lambda') \tag{A.2}$$

where $d\sigma(\theta',\lambda')=\sin\theta'\, d\theta'\, d\lambda'$. Thus, for $r\geq r_k$ we get:

$$V_i^{(e)}(r,\theta,\lambda)=GC_{nm}\iint_\sigma \frac{r_k^2 F_{nm}(\theta',\lambda')}{|P(r,\theta,\lambda)-Q(r_k,\theta',\lambda')|}d\sigma(\theta',\lambda') \tag{A.3}$$

The next step is to evaluate $V_i^{(e)}$ for $r=r_k$, i.e. for $P\in\sigma$. This is, however, straight forward using the series expansion of a distance |P-Q| into Legandre polynomials $P_n(\psi)$, where ψ is the spherical distance (Heiskanen and Moritz, 1967, Eq.(1-81)). This series expansion can be further expanded by decomposition formulas (Heiskanen and Moritz, 1967, Eqs. (1-82') - (1-83')) into products of fully normalized spherical harmonics, so that the dependence on (θ,λ) and (θ',λ') is separated. Finally the orthonormality of the fully normalized spherical harmonics can be used yielding:

$$V_i^{(e)}(r_k,\theta,\lambda))=\frac{4\pi r_k}{2n+1}GC_{nm}F_{nm}(\theta,\lambda) \tag{A.4}$$

The factor $4\pi r_k^2$ comes from dσ and from the normalization formula (Heiskanen and Moritz, 1967, Eq.(1-74)) and $1/[(2n+1)r_k]$ from the expansion of |P-Q|.

The potential of a surface mass density is continuous in $\mathbb{R}^3$ (Heiskanen and Moritz, 1967, sec. 1-2). It means, that for $r=r_k$ we have $V_i^{(e)}=V_i^{(i)}=V$. From here we can use the theorem described in (Heiskanen and Moritz, 1967, sec. 1-16) dealing with the expression for inner- and outer spherical harmonic continuation of a harmonic function given on a surface of a sphere. Utilizing that V in Eq.(A.4) is already expressed in terms of fully normalized surface spherical harmonics, we arrive the following final expressions:

$$V_i^{(e)}(r,\theta,\lambda)=[\frac{4\pi r_k}{2n+1}G](\frac{r_k}{r})^{n+1}C_{nm}F_{nm}(\theta,\lambda) \tag{A.5}$$

and

$$V_i^{(i)}(r,\theta,\lambda))=[\frac{4\pi r_k}{2n+1}G](\frac{r}{r_k})^{n}C_{nm}F_{nm}(\theta,\lambda) \tag{A.6}$$

Eqs. (6)-(7) are linear combinations of Eqs. (A.5)-(A.6). The correctness of Eqs. (A.5)-(A.6) can be tested independently, (Heiskanen and Moritz, 1967, Eq. (1-18)). ■

OPERATIONAL MERGING OF SATELLITE, AIRBORNE AND SURFACE GRAVITY DATA BY DRAPING TECHNIQUES

Gabriel Strykowski and René Forsberg, National Survey and Cadastre, Rentemestervej 8, DK-2400 Copenhagen NV, Denmark

Abstract. *In merging surface gravity data with airborne- and satellite-gravity data, a problem of inconsistency at long- and medium-wavelengths occurs. To deal with this problem, a draping technique can be used, which is an empirical technique by which the consistency is imposed upon airborne- and satellite-gravity data by appropriate corrections. The present paper describes some of the techniques used in a recent attempt to merge gravity data over parts of Greenland and the surrounding waters (59°-76^oN,15°-72^oW). The least squares collocation draping techniques were implemented in the GRAVSOFT software system.*

1. Introduction

National Survey and Cadastre, KMS (Kort- & Matrikelstyrelsen in Danish), has for years been collecting and processing the gravity data over Scandinavia, Greenland and surrounding areas. Of particular area of interest is Greenland, where KMS field campaigns over the years have improved the surface gravity data coverage over the coastal areas. All of Greenland, including the ice cap, was surveyed from the air in 1991-1992 in the Greenland Aerogeophysical Project (GAP) by the US Naval Research Laboratory, US Defence Mapping Agency and KMS (Forsberg and Brozena, 1993).

Before 1991, the offshore areas of Greenland were covered by few and often old marine gravity data. One of the problems was that for many gravity tracks, the depths to the sea bottom were not measured. Thus, it was not possible accurately to determine the Bouguer anomalies for the gravity stations. Between 1991 and 1995, the increased interest from oil companies gave rise to a number of seismic surveys along the shelf areas of Greenland. The survey ship (a Royal Danish Navy vessel "Thetis") was equipped with a LaCoste and Romberg marine gravimeter (from 1992 the gravimeter was upgraded to SEASYS). The use of P-code GPS navigation made it possible to model the fictitious forces caused by the ship movements very accurately. The new marine gravity data were processed by KMS and yielded a r.m.s. of less then 0.5 mgal at cross-over points (Strykowski *et al.*, 1996).

The data coverage offshore Greenland is further improved by the gravity anomalies obtained from satellite altimetry (Knudsen and Andersen, 1997). The altimetric data were converted to gravity anomalies on a regular 0.0625^o x 0.0625^o grid (which reflects the spatial resolution of the satellite tracks in Greenland waters quite well). Because of the permanent (during the lifetime of ERS-1) presence of the sea ice in north-eastern Greenland, only a few (and uncertain) altimetric gravity values exist. It is expected, that with better weather conditions in the future, the gravity data coverage will be improved.

2. Least Squares Collocation Draping Techniques

In the present context, the term "draping techniques" is used in plural to indicate that the draping is not confined to one type of data, and that one can think of a number of different ways of how to conduct draping. In general, the idea is to use the data of high quality as a reference for correcting the data of lower quality.

Denote a type of data of high quality by x_1 [e.g. x_1 are high quality free-air anomalies or high quality Bouguer anomalies]. The type of data of lower quality is denoted by x_2. Using technique of least squares collocation (Moritz, 1980), one can use a number of data of type x_1 [i.e. $x_1(\mathbf{l}_i)$ where $\mathbf{l}_i$, i=1,...,I denotes the location of the i'th data point] to predict a value of the (existing) data point of type x_2 at its location $\mathbf{l}_k$. The prediction is carried out only if $\mathbf{l}_k$ is in the vicinity of $\mathbf{l}_i$ for i=1,..,I. The predicted value is denoted $\overline{x_1}(\mathbf{l}_k)$.

In this way K residuals (k=1,...,K) $\epsilon_{12}(\mathbf{l}_k)$ are obtained, which are define as:

$$\epsilon_{12}(\mathbf{l}_k) \equiv x_2(\mathbf{l}_k)-\overline{x_1}(\mathbf{l}_k) \tag{1}$$

The residuals are then "softly" gridded (predicted by least squares collocation with a relatively large standard deviation) to a correction grid ϵ_{12} covering the whole area. Typically, a second-order Markov covariance model of the form:

$$C_{\epsilon\epsilon}(s)=C_0(1+\alpha\, s)e^{-\alpha s} \tag{2}$$

is used, where s is a distance, C_0 is the signal variance, and α is a parameter of the covariance model. The correlation length of the covariance function is chosen empirically based on the data distribution of x_2 and on the expected behaviour of the error signal. In practice, the correction value ϵ_{12} will tend to zero for locations at a distance greater than the correlation length from all $\mathbf{l}_k$. Finally, all data of type x_2 are corrected using the correction grid ϵ_{12} to predict the correction term.

3. Merging Gravity Data by Least Squares Collocation Draping Techniques - the Greenland Example

In the introduction, various types of gravity data in Greenland and in the adjacent sea areas are briefly described. The data were subdivided into 4 classes (denoted DATA1, DATA2, DATA3 and DATA4). This classification reflects, respectively, the hierarchy of gravity

Table 1. Classification of Greenland gravity data (see Introduction)

Name	Description
DATA1	New marine gravity data and land gravity data of high quality
DATA2	Old marine gravity data without bathymetric information and old "ice" data
DATA3	Gravity anomalies from satellite altimetry data
DATA4	Aerogravimetric data from the GAP

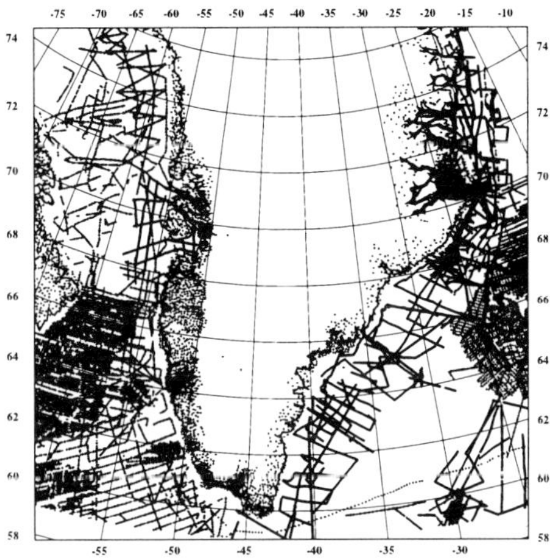

Fig. 1. Locations of DATA1 gravity stations. (see Table 1)

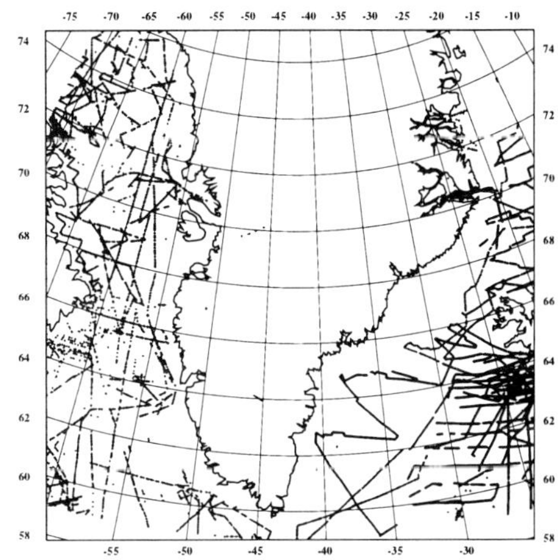

Fig. 2. Locations of DATA2 gravity stations. (see Table 1)

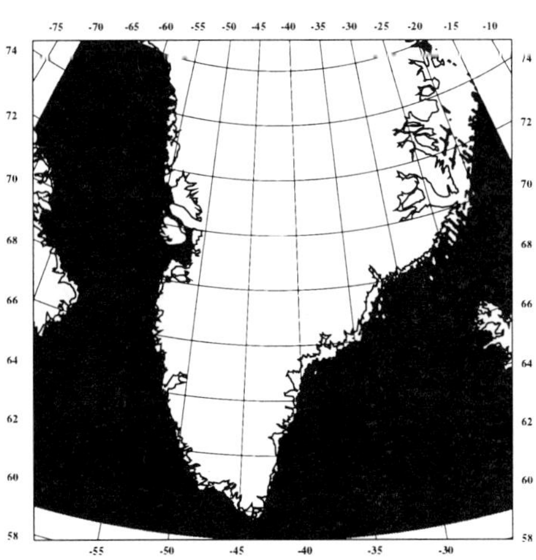

Fig. 3. Locations of DATA3 gravity stations. (see Table 1)

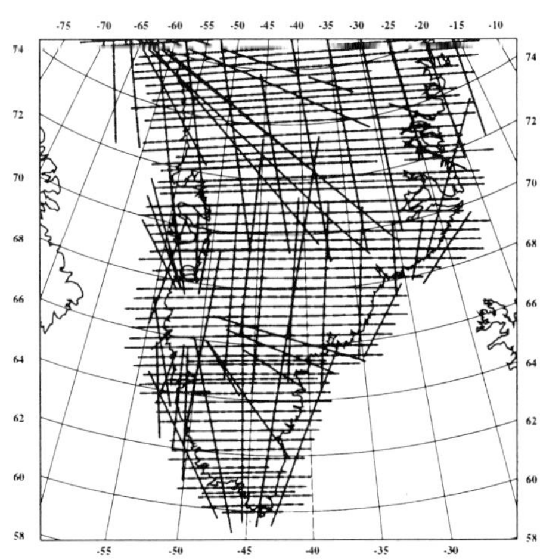

Fig. 4. Locations of DATA4 gravity stations. (see Table 1)

information (DATA1 are the gravity data of highest quality). Table 1 contains a short description related to this classification. Figures 1-4 show, respectively, the locations of the gravity data of these four types.

The first step in practical modelling was to drape DATA2 on DATA1. As seen from Table 1, the bathymetric information for DATA2 gravity stations was not available. It was decided to neglect the few existing land data on ice, see Fig. 2. The free-air gravity anomalies for these stations were already verified before DATA2 measurements were included into KMS's data base. This was verified once again by plotting the contoured free-air gravity anomalies for the joint data set: DATA1 and DATA2. Thus, the free-air gravity anomalies of DATA2 stations can be regarded as "errorless" (within the limits of the specified error variance, see below). Consequently, the main idea behind draping of DATA2 on DATA1 was to provide "the best possible" depths, so that the Bouguer anomalies of DATA2 stations were consistent with the high quality Bouguer anomalies of DATA1 stations.

The first estimate of depths to the sea floor was obtained by interpolating from the ETOPO5 global 5' x 5' bathymetric/topographic grid, which had been enhanced with the local bathymetric information from DATA1 stations. The criterion for choosing bathymetric information was the limitation on the maximum distance to the DATA2 station (less then 10 km). Subsequently, by using the ("errorless") free-air gravity values, the corresponding first estimates of the Bouguer anomalies for DATA2 stations were obtained.

Next followed the least squares collocation draping of the estimated DATA2 Bouguer anomalies on the corresponding values for the DATA1 stations, see sec. 2. The covariance model used was the second-order Markov model with a correlation length of 20 km. The final step of this procedure was to use the ("errorless") free-air gravity anomalies and the draped Bouguer anomalies (of DATA2 stations) to determine the corrected depths.

Before draping the DATA3 gravity values, a joint data set consisting of DATA1 and the draped DATA2 was formed. The joint data set is called DATA1+2. For the purpose of final interpolation, it was found necessary to assign different weights (expressed as error variances) to the two subsets of DATA1+2. It was decided that DATA1 gravity values were assigned a value of 1 $mgal^2$ while DATA2 gravity values were assigned a value of 3 $mgal^2$.

To create DATA3, the grid of satellite altimeter-derived free-air anomalies was converted to point values at the location of the grid points. The depths to the sea floor and the Bouguer anomalies are unknown. Again, these depths were interpolated from ETOPO5 enhanced with bathymetric information from DATA1 stations. Subsequently, the procedure described above was repeated to drape DATA3 on DATA1+2. Fig. 5 shows the draping corrections of DATA3 Bouguer gravity anomalies for Southern Greenland. DATA3 were assigned an error variance of 5 $mgal^2$ and were joined with DATA1+2 to form DATA1+2+3.

The treatment of DATA4 was very special. As seen from Fig. 4, the locations of most of DATA4 gravity stations is complementary to locations of DATA1+2+3 stations. (The idea behind GAP, see Introduction, was to cover the inaccessible inner parts of Greenland with gravity data). Furthermore, DATA4 gravity stations are located at a flight level of 4km above sea level. Thus, the first step in preparing DATA4 gravity values was to "downward continue" the measurements from the flight level to the level of the topography using the procedure in Forsberg and Kenyon (1994).

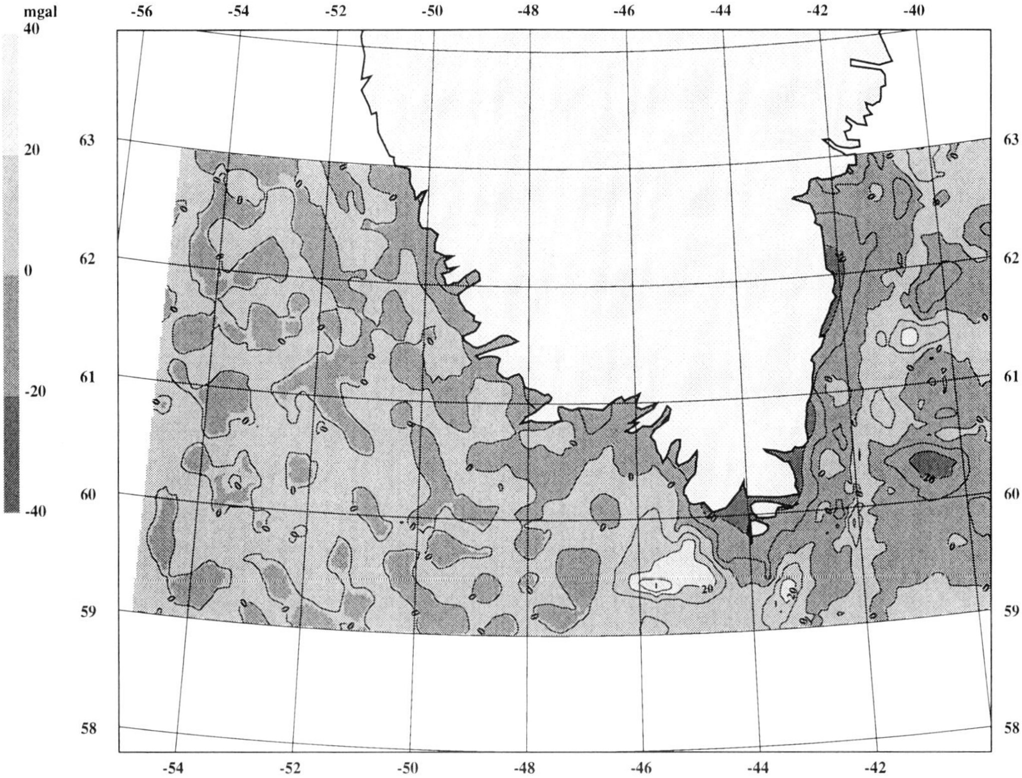

Fig. 5. Draping correction of DATA3 on DATA1+2 in Southern Greenland. (c.i. 10 mgal)

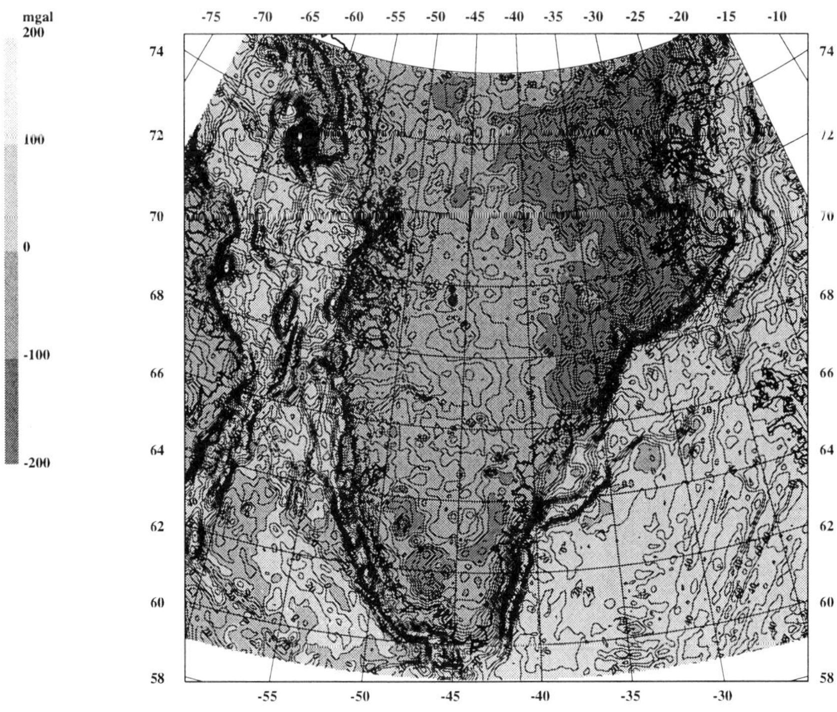

Fig. 6. Greenland gravity (sea: free-air anomalies, land: Bouguer anomalies). (c.i.10mgal).

Two different approaches of how to treat DATA4 were tested. In western Greenland, a formal draping was attempted. The DATA4 gravity values were assigned the error variance of 10 $mgal^2$. This large variance means, in practice, that DATA4 gravity values have al-most no influence in areas where there are other data. Thus, the most efficient way to treat DATA4 gravity values is to remove them (instead of draping them) whenever they are in the vicinity of other gravity stations. This was tried in eastern Greenland. All DATA4 gravity stations in a distance less than 10 km from the locations of DATA1+2+3 gravity stations were removed. Furthermore, all DATA4 gravity measurements over "ice-free" land areas were also removed (compare figures 1 and 4).

The main purpose of using the draping techniques described above was to prepare a consistent set of gravity data from which the final gravity map could be compiled. The final interpolation by weighted means makes use of the error variances assigned to the subsets of the final set of gravity data, DATA1+2+3+4. Fig. 6 shows the resulting gravity grid comprising free-air anomalies at sea and Bouguer anomalies on land. All computations (i.e. the draping and the final interpolation to a grid) were carried out using the GRAVSOFT software system (Tscherning *et al.*, 1992).

4. Conclusions

In this paper, the operational least squares draping techniques that can be implemented in GRAVSOFT software system are described. These techniques were used in a recent attempt to model the gravity field over Greenland and the adjacent sea areas by merging land gravity data, airborne gravity data and gravity anomalies from satellite altimetry.

References:

Forsberg, R., and J.M. Brozena (1993). *The Greenland Airborne Gravity Project - Comparison of Airborne and Terrestrial Gravity Data.* In: H. Montag and Ch. Reigberg (Eds.). Geodesy and the Physics of the Earth. IAG Symposia 112, Springer Verlag, pp. 171-175.

Forsberg, R., and S. Kenyon (1994). *Evaluation of Downward Continuation of Airborne Gravity Data - the Greenland Example.* In: M.E. Cannon and G. Lechapelle (Eds.). Proc. Int. Symp. On Kinematic Systems in Geodesy, Geomatics and Navigation. Banff, Canada, Aug. 30- Sep. 2, 1994, Publ. Univ. of Calgary, pp. 531-538.

Knudsen, P., and O. B. Andersen (1997). *Improved Recovery of the Global Marine Gravity Field from the GEOSAT and the ERS-1 Geodetic Mission Altimetry.* Proc. Int. Symp. on Gravity, Geoid and Marine Geodesy, Sept. 1996, Tokyo, Japan, Springer Verlag, pp. 429-436.

Moritz, H. (1980). *Advanced Physical Geodesy.* Herbert Wichmann Verlag.

Strykowski, G., R. Forsberg, and M.D. Larsen (1996). *Acquisition and Processing of High Precision Greenland Marine Gravity Data.* Phys. Chem. Earth, 21, pp. 353-356.

Tscherning, C.C., R.Forsberg, and P. Knudsen (1992). *The GRAVSOFT Package for Geoid Determination.* In: Holota P. and M. Vermeer (Eds.) Proc. IAG, First Continental Workshop on the Geoid in Europe. ISBN 80-901319-2-1, pp 327-334, Prague, Czechoslovakia, May 11-14, 1992.

THE EUROPEAN GRAVIMETRIC QUASIGEOID EGG97 - AN IAG SUPPORTED CONTINENTAL ENTERPRISE -

Heiner Denker, Wolfgang Torge
Institut für Erdmessung, University of Hannover, Schneiderberg 50
D-30167 Hannover, Federal Republic of Germany

ABSTRACT AND INTRODUCTION

In 1990, the Institut für Erdmessung (IfE), University of Hannover, started with the calculation of a gravimetric (quasi)geoid for Europe and the surrounding marine areas, operating as the computing center of the International Association of Geodesy (IAG) Geoid Commission. The requirements for accuracy and resolution were derived from the potential of GPS heighting and satellite altimetry, and defined at the "cm" to "dm" level over distances from a few km to some 1000 km, which requires a spatial resolution of a few km.

The data base established and continously extended and updated at IfE now includes several global gravity models, about 2.7 million (mainly point) gravity data, and about 700 million terrain heights. In some marine areas, gravity anomalies derived from ERS-1 satellite altimetry had to be included in the solution. All data sets were transformed to uniform standards in gravity, position and height.

Several quasigeoid solutions have been presented and discussed since 1990. The modeling strategy is based on the spectral combination technique in connection with the remove-restore procedure. The final solution combines the global Earth model EGM96 of NASA/NIMA with the high resolution gravity and terrain data stored in the IfE data base, including data from Russia and other Eastern European countries. The quasigeoid model is provided in a $1' \times 1.5'$ and a $10' \times 15'$ grid. In areas with a good data coverage and quality the accuracy estimates range from ± 1 ... 5 cm over 10 km to a few 100 km distance, and ± 5 ... 20 cm over a few 1000 km, respectively. Medium and long wavelength (larger than a few 100 km) errors have been found by comparisons with GPS/leveling control points, and are attributed mainly to errors of the global model, but also systematic errors in the gravity and GPS/leveling data are possible in some regions. The EGG97 quasigeoid model is now made available on a CD-ROM to users in geodesy, geophysics and engineering.

1. Computation Technique

The IfE gravity field modeling effort for Europe has concentrated on the calculation of height anomalies or quasigeoid undulations ζ. This has the advantage that only gravity field data observed at the Earth's surface and in its exterior enter into the calculations, while no assumptions about the gravity field in the Earth's interior are needed. Subsequently, a transformation from height anomalies ζ to geoid undulations N can be performed easily by introducing a density model.

The remove-restore technique is used to combine a high-degree spherical harmonic model and a digital terrain model (DTM) with terrestrial gravity field observations (point gravity data, etc.). For the field transformation from gravity to height anomalies, the least squares spectral combination technique is applied in order to reduce long-wavelength distortions, which may result from the use of Stokes's formula (see e.g. *Denker* et al. 1994). In the least squares spectral combination technique, instead of the Stokes kernel, a modified integral kernel

$$W(\psi) = \sum_{l=2}^{\infty} \frac{2l+1}{l-1} w_l \, P_l(\cos\psi) \ . \tag{1}$$

is used (see e.g. *Wenzel* 1982). In Eq. (1) l is the degree, P_l are the Legendre polynomials, ψ is the spherical distance, and w_l are the spectral weights with

$$w_l = \frac{\sigma_l^2(\varepsilon_1)}{\sigma_l^2(\varepsilon_1)+\sigma_l^2(\varepsilon_{\Delta g})} \ . \tag{2}$$

The w_l in Eq. (2) depend on the error degree variances of the potential coefficient model $\sigma_l^2(\varepsilon_1)$ and the gravity anomalies $\sigma_l^2(\varepsilon_{\Delta g})$, where the latter ones can be computed from the error covariance function of the terrestrial gravity data (see Eq. 3).

2. Data Description

In the course of the European Geoid Project, about 2.7 million gravity data and about 700 million topographical height data have been included in the project data base. The gravity data coverage for the computation area is depicted in Figure 1, showing that land gravity data with a resolution of at least 10 km were attained for all European countries, while for some sea areas the coverage with terrestrial gravity data is still insufficient. Therefore, the ship data were merged with altimetrically derived gravity anomalies from ERS-1 (*Andersen* et al. 1996) in most parts of the European seas.

The digital terrain models were regridded to a common block size of $7.5'' \times 7.5''$ (or multiples of this block size), with existing gaps being filled by values from ETOPO5. Figure 2 depicts the coverage of high resolution DTM's stored in the IfE data base. Prior to utilizing all the collected data, a transformation to uniform standards in gravity, position and height was performed whenever this was possible. Furthermore, all data were validated using batch and interactive procedures developed at IfE (see also *Denker* et al. 1994).

3. THE QUASIGEOID MODEL EGG97

In 1997, a new quasigeoid solution, EGG97, was computed for the whole of Europe. For the long wavelength gravity field information, the geopotential model EGM96 (*Lemoine* et al. 1997) was employed. The short wavelength gravity field components were modeled using the residual terrain model (RTM) reduction technique, where the reference topography was constructed from the DTM's using a $15' \times 15'$ moving average filter. The terrain reductions for the gravity observations were computed by numerical integration techniques, while the terrain effects for the height anomalies (restore part) were computed by spherical 1D FFT techniques.

At first, residual gravity anomalies were computed and gridded by a fast least squares prediction technique onto a $1.0' \times 1.5'$ grid covering the area from 25°N-77°N and 35°W-67.4°E. This yields $3{,}120 \times 4{,}096 = 12{,}779{,}520$ grid points. The field transformation from residual gravity to residual height anomalies was carried out using the spectral combination technique. The numerical evaluation of the integral formula was done by a 1D FFT technique in connection with a detailed/coarse grid approach to further speed up the computations. For the spectral combination technique the following error covariance function for the terrestrial gravity data was used:

$$\mathrm{cov}\left(\varepsilon_{\Delta g}, \varepsilon_{\Delta g}\right) = 4\ \left[\mathrm{mgal}^2\right]\ e^{-4\psi[^\circ]}\ . \tag{3}$$

This model uses correlated noise and was suggested and applied by *Weber* (1984). The spectral weights were derived on the basis of Eq. (2) using the above error covariance function for the terrestrial gravity data and the error degree variances from EGM96. It was decided to do the combination only up to degree 50, while between degrees 50 and 10000 (corresponding to the grid size used) the total weight was given to the terrestrial gravity data ($w_l = 1.0$). However, this does not imply that the global model EGM96 is completely disregarded above degree 50, as especially in areas with larger data gaps the high degree gravity information of the model is considered in the gridding process and thus practically taken over in the final quasigeoid model. A cosine tapering window was applied between degrees 10000 and 30000 in order to prevent oscillations of the integral kernel.

The major contribution to the final quasigeoid (internal name EGG97.03) comes from the spherical harmonic model EGM96 with values ranging from -43.3 m to $+67.9$ m and a standard deviation of ± 25.6 m. The standard deviations of the contributions from the DTM and the terrestrial gravity data are ± 0.03 m and ± 0.41 m, respectively. However, the maximum DTM effects are about 0.8 m, while the maximum effects of the terrestrial gravity data are 4.3 m. Furthermore, for the final EGG97 model, a zero-degree undulation correction of -0.5 m was considered, taking into account the results obtained by *Rapp* and *Balasubramania* (1992) as well as from comparisons with GPS/leveling data. A tidal correction (std.dev. ± 2.1 cm) was applied to refer the undulations to the zero tide system as recommended by IAG. The Molodensky correction terms have been neglected so far, which can obtain maximum values of 10 cm in the Alps and 1 cm in the highlands, respectively. Finally, geoid undulations were also derived based on a Bouguer plate model with constant density for the computation of the mean value of gravity. This corresponds to the so-called Helmert heights.

The spectral combination technique also permitted the derivation of error estimates for the resulting height anomalies. For the EGG97 model, $\sigma_{\Delta g} = \pm 2$ mgal gives standard deviations for height anomaly differences of ± 7.6 cm over 100 km and ± 12.7 cm over 1000 km, respectively. A more optimistic error estimate for the terrestrial gravity data of $\sigma_{\Delta g} = \pm 1$ mgal gives standard deviations of ± 3.9 cm over 100 km and ± 7.6 cm over 1000 km, respectively. The corresponding values for $\sigma_{\Delta g} = \pm 4$ mgal are ± 15.2 cm and ± 23.9 cm, respectively.

4. Evaluation of the Quasigeoid Model EGG97

The quasigeoid model EGG97 was verified by means of satellite altimeter data and GPS/leveling data. However, in this paper we will restrict the quality assessment of EGG97 to the comparison with three GPS/leveling data sets covering small (NDS92, Lower Saxony, 300 km extension), medium (RBF, France, 1000 km extension) and large scales (European GPS traverse, 3000 km length). Statistics of the discrepancies between the GPS/leveling data and some recent quasigeoid solutions are given in Table 1. The comparisons were always done using a bias fit as well as a bias and tilt fit in order to account for inaccuracies in the absolute positioning and for long wavelength errors of all data sets involved (GPS, leveling, quasigeoid).

From Table 1, it becomes clear that the new quasigeoid solution EGG97 is a significant improvement over the older models EGG1 (*Torge* et al. 1982) and EAGG1 (*Brennecke* et al. 1983), with the RMS discrepancies decreasing by a factor of 2 to 5. Moreover, in most cases one can observe a significant improvement for the bias and tilt fit versus the bias fit, thus indicating that small long wavelength discrepancies exist between the gravimetric and the GPS/leveling results (magnitude 0.1 to 0.5 ppm). The EGG96 (*Denker* et al. 1997) and EGG97 models give comparable results. The only difference between these models is the use of a different global model (EGM96 versus JGM3_OSU91A) and the use of one new gravity data source in EGG97.

References

Andersen, O.B., P. Knudsen, C.C. Tscherning (1996). Investigation of methods for global gravity field recovery from the dense ERS-1 geodetic mission altimetry. Proceed. IAG Symp. No. 116, Global Gravity Field and Its Temporal Variations, Boulder, Co., USA, July 12, 1995, 218-226, Springer-Verlag, Berlin, Heidelberg, New York.

Brennecke, J., D. Lelgemann, W. Torge, H.-G. Wenzel (1983). A European astro-gravimetric geoid. Deutsche Geod. Komm., Reihe B, Nr. 269, Frankfurt/Main.

Denker, H., D. Behrend, W. Torge (1994). European gravimetric geoid: Status report 1994. Proceed. IAG Symp. No. 113, Gravity and Geoid, Graz, Austria, Sept. 11-17, 1994, 423-433, Springer, Berlin, Heidelberg, New York.

Denker, H., D. Behrend, W. Torge (1997). The European gravimetric quasigeoid EGG96. Proceed. IAG Symp. No. 117, Gravity, Geoid and Maine Geodesy, Tokyo, Japan, Sept. 30 - Oct. 5, 1996, 532-539, Springer-Verlag, Berlin, Heidelberg, New York.

Lemoine, F., et al. (1997). The development of the NASA GSFC and DMA joint geopotential model. Proceed. IAG Symp. No. 117, Gravity, Geoid and Maine Geodesy, Tokyo, Japan, Sept. 30 - Oct. 5, 1996, 461-469, Springer-Verlag, Berlin, Heidelberg, New York.

Rapp, R.H., N. Balasubramania (1992). A conceptual formulation of a world height system. Dept. Geod. Sci. and Surv., Rep. 421, Columbus.

Torge, W., G. Weber, H.-G. Wenzel (1982). Computation of a high resolution European gravimetric geoid. Proceed. 2nd Internat. Symp. on the Geoid in Europe and Mediterranean Area, 437-460, Rome.

Weber, G. (1984). Hochauflösende Freiluftanomalien und gravimetrische Lotabweichungen für Europa. Wiss. Arb. Univ. Hannover, Nr. 135.

Wenzel, H.-G. (1982). Geoid computation by least squares spectral combination using integral kernels. Proceed. IAG General Meet., 438-453, Tokyo.

Table 1: Statistics of the differences from the comparison of recent quasigeoid solutions with different GPS/leveling data sets. Units are meters.

Quasigeoid Solution	Bias Fit		Bias + Tilt Fit	
	RMS	Max.	RMS	Max.
NDS92 (Lower Saxony; 41 stations)				
EGG1 (1982)	0.108	0.368	0.061	0.221
EAGG1 (1983)	0.069	0.171	0.063	0.162
OSU91A (1991)	0.195	0.912	0.143	0.651
EGM96 (1996)	0.148	0.462	0.136	0.567
EGG96 (1996)	0.039	0.090	0.015	0.032
EGG97 (1997)	0.038	0.090	0.013	0.033
RBF (France; 965 stations)				
EGG1 (1982)	0.664	2.937	0.381	2.079
EAGG1 (1983)	0.460	1.882	0.387	1.803
OSU91A (1991)	0.374	2.240	0.337	2.027
EGM96 (1996)	0.369	2.025	0.301	1.698
EGG96 (1996)	0.106	0.341	0.070	0.407
EGG97 (1997)	0.128	0.353	0.080	0.484
European GPS Traverse (67 stations)				
EGG1 (1982)	0.605	1.351	0.275	0.778
EAGG1 (1983)	0.240	0.607	0.175	0.501
OSU91A (1991)	0.319	0.926	0.260	0.795
EGM96 (1996)	0.304	0.875	0.251	1.114
EGG96 (1996)	0.307	0.851	0.157	0.436
EGG97 (1997)	0.294	0.793	0.175	0.470

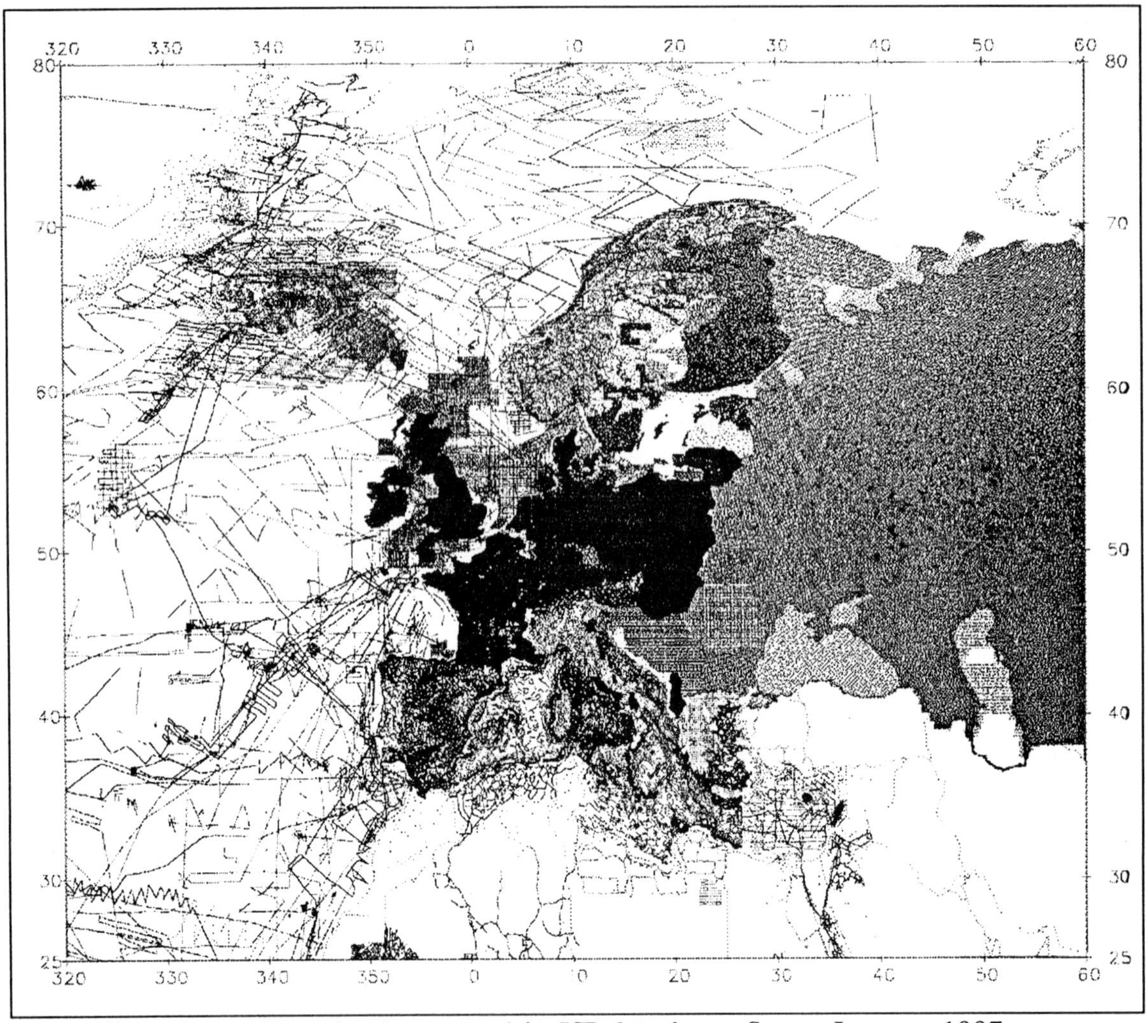

Fig. 1: Locations of gravity data stored in IfE data base. Status January 1997.

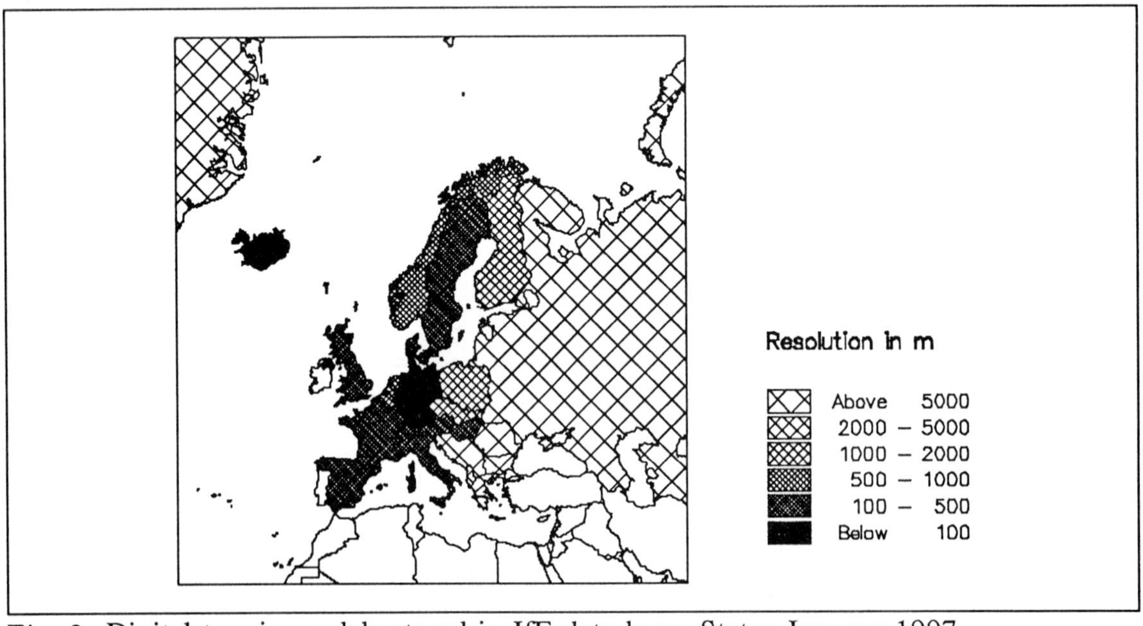

Fig. 2: Digital terrain models stored in IfE data base. Status January 1997.

A PRELIMINARY GEOID MODEL FOR ARGENTINA

Graciela Font
Facultad de Ciencias Astronómicas y Geofísicas
Paseo del Bosque s/n - 1900 - La Plata - Argentina
graciela@fcaglp.fcaglp.unlp.edu.ar

María Cristina Pacino
Instituto de Física de Rosario - Universidad Nacional de Rosario
Av. Pellegrini 250 - 2000 - Rosario - Argentina
mpacino@unrctu.edu.ar

Denizar Blitzkow
Escola Politécnica da Universidade de São Paulo - Brazil
dblitzko@usp.br

Claudia Tocho
Facultad de Ciencias Astronómicas y Geofísicas
Paseo del Bosque s/n - 1900 - La Plata - Argentina
ctocho@fcaglp.fcaglp.unlp.edu.ar

ABSTRACT

Instituto de Física de Rosario (IFIR), Facultad de Ciencias Astronómicas y Geofisicas de La Plata (FCAGLP) in a cooperation with Escola Politécnica of the University of São Paulo are involved in an effort for the determination of the geoid in Argentina. For this purpose, softwares have been set up at FCAGLP in order to perform the main tasks for geoid computations: processing of gravity data to derive mean gravity anomalies, dealing with geopotential models, performing the numerical integration of the modified Stokes's formula. The recent NASA/NIMA geopotential model, EGM96, is used as a reference field in the remove-restore technique. A preliminary model has been computed for Argentina with a resolution of 20' x 20'. A total of 55 GPS points established on the geometric leveling network have been used for a comparison with the undulations derived from gravimetric information. The activities related to the geoid in Argentina are carried out under the umbrella of the Sub-Commission for the Geoid in South America (SCGSA).

INTRODUCTION

A preliminary geoid model for Argentina with a resolution of 20' x 20' has been constructed using a set of more than 15,000 gravity stations measured by IGM (Instituto Geográfico Militar) and national universities.

All gravity values have been adjusted to IGSN71 and referred to the Fundamental Point of Miguelete (979690.03 mGal) of the Gravimetric Net for Argentina.

Gravity anomalies were referred to the WGS84 ellipsoid. To remove the atmospheric effect from the gravity anomalies, a correction has been added to the free air gravity anomalies (Blitzkow et al., 1996) and a curvature correction has been applied to the Bouguer anomalies (Green & Fairhead, 1991).

Terrain corrections were calculated for all gravity points until 166.7 km. To estimate this correction a 3'digital terrain model has been used.

DATA PROCESSING

Fortran software provided by Escola Politécnica of the University of São Paulo has been set up at FCAGLP (Facultad de Ciencias Astronómicas y Geofisicas of the University of La Plata) in order to perform the main tasks for the geoid computations.

The principal steps in the processing sequence will be briefly described in the following sections:

Sorting the Data Base

The data base available has been sorted in blocks of 1° x 1° (increasing in longitude from West to East and decreasing in latitude from North to South).

Each record has the following informations: latitude and longitude in decimal degree, observed gravity in mGal, free air and Bouguer anomalies in mGal, terrain correction in mGal and identification number.

Estimation of mean anomalies value for blocks of 5'x 5'

The ANOME program provided by Escola Politécnica was used to obtain a mean anomaly value for blocks 5'x 5'. This blocksize is in agreement with the density of gravity points. The arithmetic mean is used to estimate a mean anomaly for each block of 5' x 5' if at least two points exist in the cell. ANOME really estimates mean value for free air and Bouguer anomalies, height and terrain correction. The arithmetic mean value for empty cells of 5' x 5' was calculated using data from the nearest cells.

EGM96 geopotential model (Lemoine et al, 1996) has been used only up to n = 50, in order to solve the modified Stokes's integral. In this case, radius of integration around the computation point is 3.6°.

The data base for Argentina is delimited by latitudes 22°S and 55°S and longitudes 53°W and 72°W. The 3.6° data around this frame was provided by Escola Politécnica. In offshore areas, free air anomalies have been derived from satellite altimetry (Sandwell D.T. & Smith W.H., 1996).

From this complete file of 5'x 5' mean values, mean values of 20'x 20' have been derived.

The mean free air anomaly has been restored from the mean Bouguer anomaly and the mean height taken from the DTM. Finally, the 20' x 20' mean terrain correction has been added in order to obtain the Faye anomaly (Blitzkow et al., 1994).

The Modified Stokes's integral

The geoid height in terms of a series of spherical harmonic functions can be split in to two different spectral components, one of long wavelength which is easily obtained from a geopotential model and another of short wavelength which can be estimated using a modification of the Stokes's integral (Vanicek et al., 1987) and (Blitzkow et al., 1991), if gravity data is available in the region around the computation point.

The long wavelengh component from the 20' mean gravity anomalies has been removed using EGM96 up to degree and order 50. After the application of the modified Stokes's integral the corresponding geoid height component has been restored in a similar way from the same model. This is commonly called the "remove restore" technique.

The preliminary geoid model for Argentina with a resolution of 20' x 20' is shown in Figure 1.

GPS / Leveling

Unfortunately, there are only a few GPS / leveling points available in Argentina (Figure 2). A total of 55 GPS points established in the geometric leveling network have been used for a comparison with the undulations derived from gravitational information as is shown in Table I. They have been divided to two groups: the first one concerns 29 points in the Province of Buenos Aires where the quality of all elements can be guaranteed; the rest covers the whole country.

The first group shows a systematic difference in the origin of the orthometric heights of 0.6 m related to this realization of the geoid. Current work on national height origin will clarify on this discrepancy.

CONCLUSION

It is difficult to evaluate the error of the geoid height determined with the present data. It depends on: the quality and distribution of gravity data, the accuracy of the leveling network, and especially the consistency of the digital terrain model. As the amount of data used was limited, we hope to improve the results as soon as new gravity values from oil companies and other Universities as well as new digital terrain model can be incorporated.

The comparison with GPS derived geoid height gives an idea about the consistency of both determination.

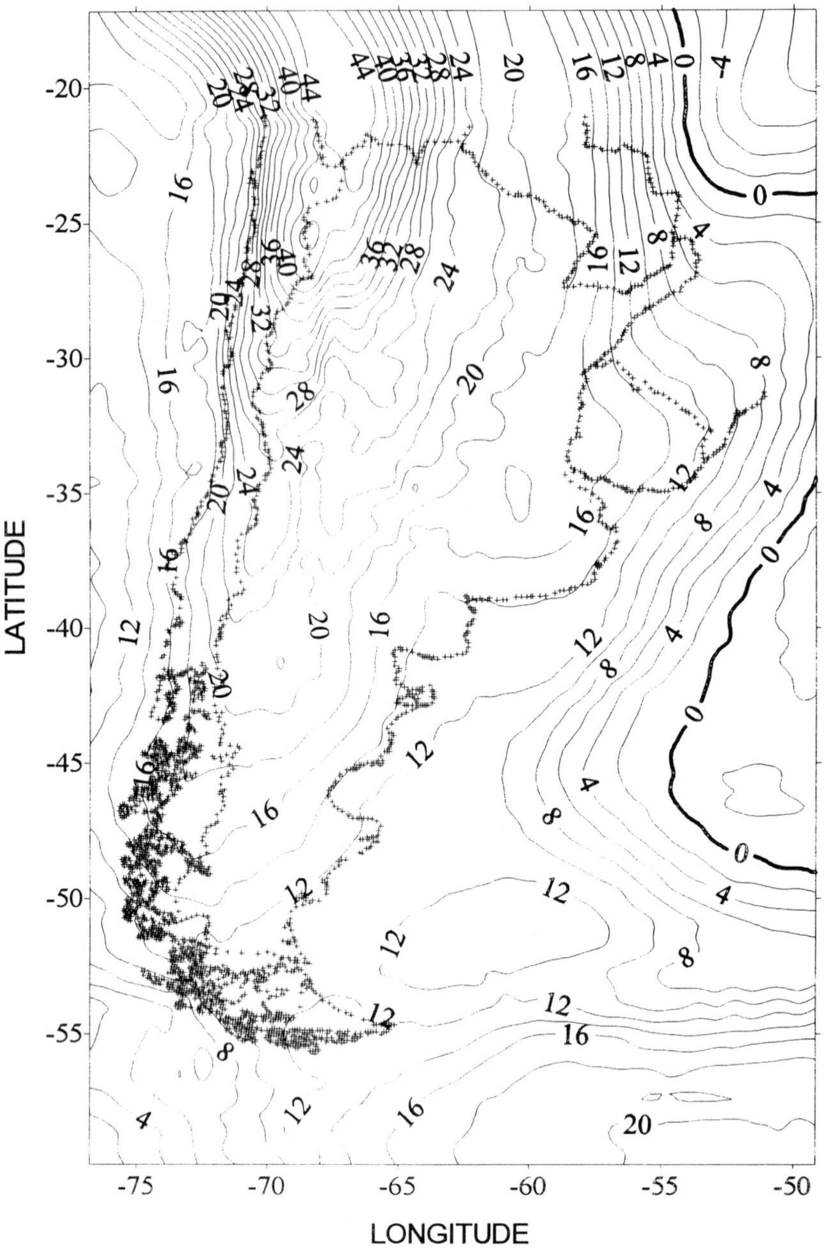

Figure 1
The preliminary gravimetric geoid model for Argentina (contours in meters above WGS84)

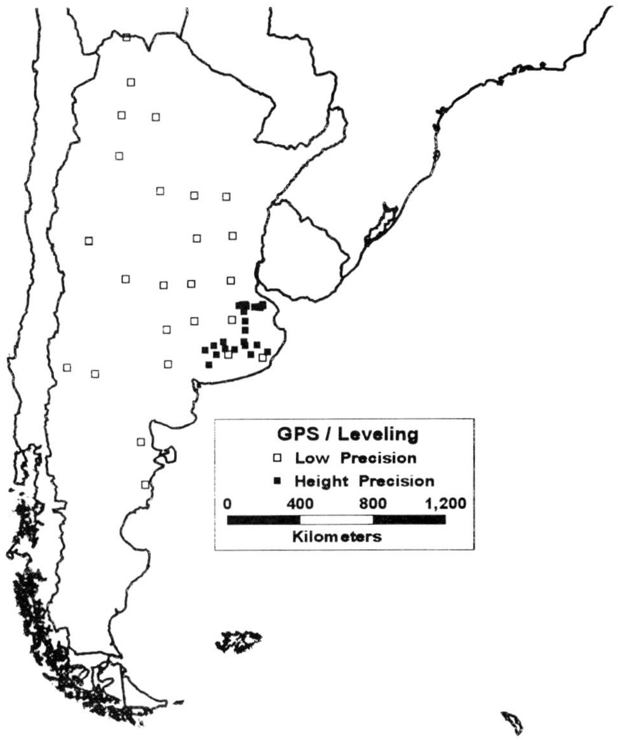

Figure 2

GPS / leveling points in Argentina

Latitude	Longitude	N (GPS / leveling)	N(Gravimetric)	Difference	
-35.1680	-59.1770	17.63	17.22	0.41	
-35.1730	-58.1139	16.62	16.42	0.20	
-35.1747	-59.2630	17.68	17.25	0.43	
-35.1810	-59.1246	17.67	17.19	0.48	
-35.1891	-58.0428	16.57	16.45	0.12	
-35.1914	-59.3176	17.74	17.26	0.48	
-35.2255	-58.0674	16.65	16.45	0.20	
-35.2324	-59.0811	17.78	17.24	0.54	
-35.2512	-59.5097	17.90	17.37	0.53	max: 1.36
-35.2573	-58.2153	16.80	16.56	0.24	min: 0.12
-35.2799	-59.0711	17.85	17.21	0.64	mean: 0.607
-35.2933	-58.2342	16.82	16.55	0.27	std: 0.320
-35.3023	-58.5921	17.35	16.82	0.53	
-35.3097	-58.2387	16.83	16.54	0.29	
-35.5111	-59.1933	18.28	17.30	0.98	
-35.9899	-59.1150	17.69	17.08	0.61	
-36.4193	-59.1430	17.23	16.91	0.32	
-36.9652	-59.1846	17.11	16.60	0.51	

-36.9782	-60.4473	17.73	17.06	0.67	
-37.1453	-58.3781	16.40	15.41	0.99	
-37.1646	-61.0341	17.26	16.41	0.85	
-37.1669	-59.1050	17.18	16.18	1.00	
-37.3241	-60.3335	17.15	16.25	0.90	
-37.3615	-59.7796	17.38	16.02	1.36	
-37.3730	-61.5848	16.88	16.10	0.78	
-37.4814	-57.7194	15.66	14.57	1.09	
-37.6087	-58.7528	16.48	15.26	1.22	
-37.6116	-60.8698	16.19	15.75	0.44	
-38.1102	-61.3573	15.35	14.84	0.51	
			Mean	0.60	All data
-22.1538	-65.4908	46.72	43.49	3.23	max: 5.44
-24.3069	-65.3366	35.86	36.55	-0.69	min: - 4.39
-25.8980	-65.9230	41.32	35.88	5.44	mean: 0.494
-26.0206	-64.0608	28.37	27.46	0.91	std.: 1.366
-27.8542	-66.1819	30.62	32.01	-1.39	
-29.6407	-63.9371	26.70	24.40	2.30	
-29.8684	-62.0212	22.04	20.34	1.70	
-29.9550	-60.2223	17.01	18.12	-1.11	
-31.8628	-59.8795	17.48	17.69	-0.21	
-31.8810	-68.1780	23.02	25.57	-2.55	
-31.8867	-68.1849	24.31	25.57	-1.26	
-31.9464	-61.9232	19.16	18.30	0.86	
-33.8272	-66.1535	23.83	24.46	-0.63	
-34.0136	-59.9812	17.81	17.86	-0.05	
-34.1511	-62.3229	18.73	18.03	0.70	
-34.2097	-63.9163	21.28	20.49	0.79	
-35.9090	-59.9144	16.52	17.30	-0.78	
-35.9708	-62.1708	18.61	17.10	1.51	
-36.3549	-63.8244	20.54	16.61	3.93	
-37.6109	-60.1188	15.47	15.64	-0.17	
-37.7527	-58.0092	14.65	14.68	-0.03	
-37.9403	-70.0816	24.85	24.01	0.84	
-38.0278	-63.8398	15.98	15.01	0.97	
-38.3383	-68.3790	20.16	20.87	-0.71	
-41.7990	-65.8025	11.32	15.71	-4.39	
-43.9094	-65.6785	14.09	14.32	-0.23	

Table I

Differences between GPS / leveling and gravimetric geoid heights over Argentina. The first 29 points refer to Buenos Aires area are "well - controlled points"

ACKNOWLEDGMENTS

We wish to thank all the organizations and people who provided the data and software for this work:
The Instituto Geográfico Militar (IGM) for the gravity data set.
Geophysical Exploration Technology (GETECH) - University of Leeds responsible for the collection and compilation of gravity data as well as the topographic data for Argentina.
National universities and oil companies for the gravity data set.

REFERENCES

Blitzkow, D., 1997. NASA GSFC NIMA Model Evaluation EGM96 Report.

Blitzkow, D., 1996. O Problema de Valor de Contorno da Geodésia. Resultados Práticos para a América do Sul, São Paulo, Brazil.

Blitzkow, D., & Fairhead, J. D., 1996. Data coverage improvement for geoid computation in South America. International Symposium on Gravity, Geoid and Marine Geodesy (Tokyo - Japan, 1996), pp 158.

Blitzkow, D., Fairhead, J. D. & Lobianco, M. C., 1994 . A preliminary gravimetric geoid for South America.. New Geoids in the World. IGeS Bulletin 4. International Association of Geodesy, pp 53.

Blitzkow, D., Cintra J.P., Fortes L.P.S. 1991. A contribution to the geoid determination . In: Recent geodetic and gravimetric research in Latin America. Springer Verlag, pp 65-81.

Green, C. M. & Fairhead, J. D., 1991. The South American Gravity Project. In: Recent Geoid and Gravimetric Research in Latin America. Springer Verlag,pp 82-95.

Lemoine F. G, et al, 1996 . The development of the NASA GSFC and NIMA Joint Geopotential Model., International Symposium on Gravity, Geoid and Marine Geodesy (Tokyo - Japan, 1996), pp 146.

Vanicek P., Kleusberg A., Chang R. G., Fashir H., Christou N., Hofman M. , Kling T., Arsenault T., 1987 . The Canadian Geoid, Technical Report 129. Dept. of Surveying Engineering, University of New Brunswick, Canada.

Sandwell D.T. & Smith W.H., 1996. Marine gravity anomalies from Geosat and ERS-1 satellite altimetry. Submitted to Journal of Geophysical Research.

THE GRAVIMETRIC GEOID IN ALGERIA : FIRST RESULTS

S. BEN AHMED DAHO, S. KAHLOUCHE
National Centre of Spatial Techniques
Po. Box 13 Arzew - 31200 - ALGERIA

Abstract

The first gravimetric determination of the geoid on a small zone in the north of Algeria has been made using Least Squares Collocation (**LSC**) method and the Gravsoft Software.
The gravimetric geoid has been computed on a limited area, covering a 8° x 3° zone in the north of Algeria, between the latitudes 34° and 37° N and longitudes 0° and 8° E. The solution of the geoid in the GRS80 system was based on the following data types : a) a global geopotential model **OSU91A**, b) a set of 2367 point free air gravity anomalies surrounding regions, and c) a digital terrain model ETOPO5 (10 km x 10 km). The terrain correction was applied to the data and the corresponding indirect effect was taken into account. The maximum value of the formal error estimated by LSC is about 81 cm in the area where gravity data is clearly missing.
Unfortunately, the data type GPS/levelling is not available in the north of the Algeria, this will help to make an independent verification of the precision of the geoid.

1. Introduction

The precise geoid model knowledge and its derivatives (gravity anomalies, vertical gradient of the gravity, deviation of the vertical, etc.) have various applications, notably in geodesy for the adequate exploitation of the measurements derived from tridimensional positioning such as GPS, and for the establishment of the altimetric networks, in oceanography for the determination of the topography of the ocean surface, and in geophysics for mining and oil prospecting.

This work constitutes the first test of calculation of the gravimetric geoid over a zone in the North of Algeria, although the results obtained are considered as preliminary. It has never been done before, on one hand because the need did not arise and on the other hand because the gravity measurements were not available.

2. Gravity Anomalies

The gravity anomalies used in this work were supplied by the Bureau Gravimétrique International (B.G.I.). These measurements with an initial precision of 5 mgals were expressed in the Geodetic Reference System GRS67 by the B.G.I.

2.1. Selection of measurements

In order to save time of calculation for this first test, we have used the gravity data between [33° - 38°] N in latitude and of [- 3° - 9°] E in longitude. The selected zone for this application contains 2367 measurements. The choice was emphasized by the relatively high density of the points and the high variation of the topography of the region. All the gravity measurements were transformed from the system GRS67 to the GRS80.

Finally, we applied an atmospheric correction recommended by the International Association of Geodesy (IAG, 1971) in order to eliminate the influences of the atmospheric masses. Figure I shows the free air gravity anomalies of the zone of survey and their distribution.

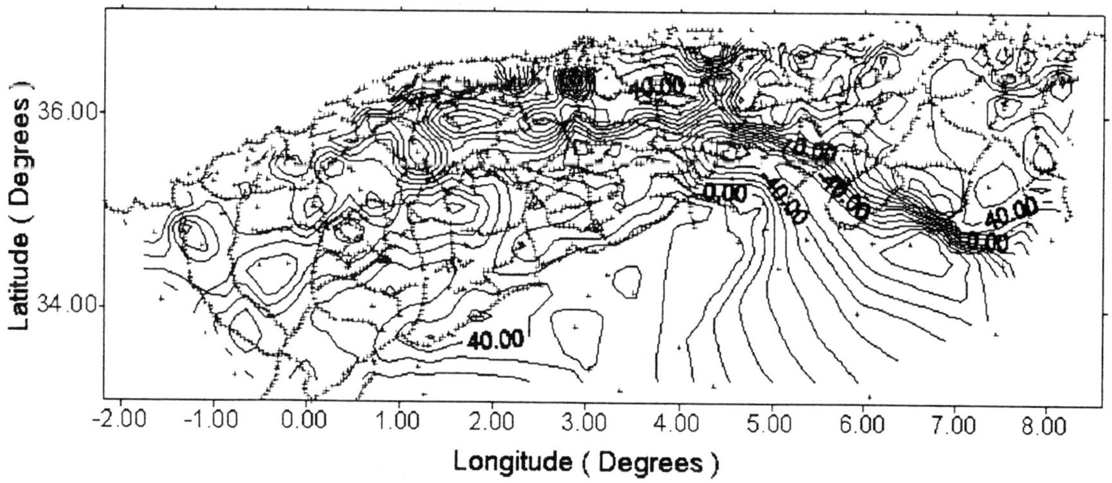

Figure I : Free air anomalies isolines contoured at 10 mgal intervals.

3. Geopotential model

The problem of choosing a geopotential model which fits best the gravity field in Algeria is not definitely solved. In the present work, the model OSU91A (Rapp & al., 1991) complete to a degree and an order of 360 was adopted as a reference model in order to eliminate the long wavelengths of gravity field. Figure II shows a map of the free air anomalies minus the contribution of the OSU91A model.

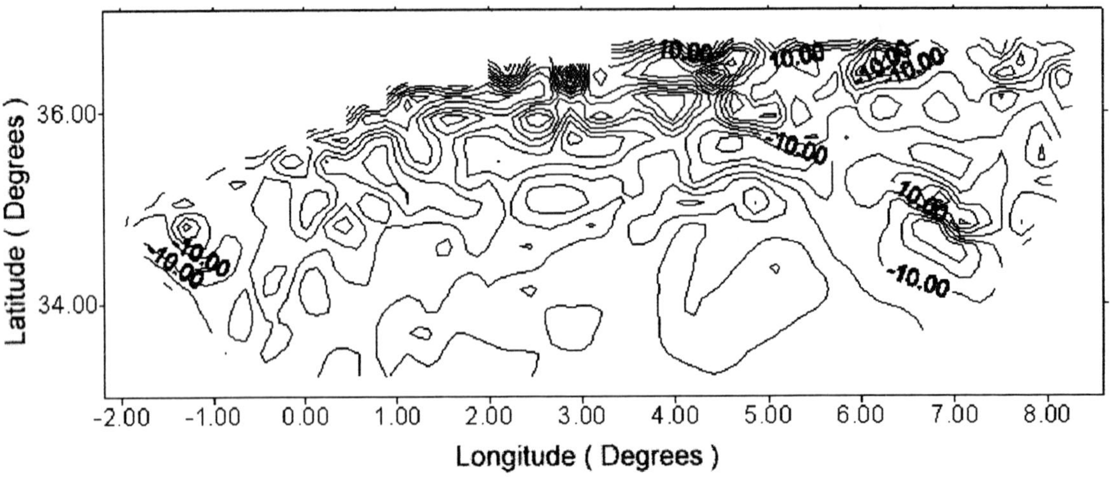

Figure II : Free air minus model gravity anomalies contoured at 10 mgal interval.

4. Gravity data validation

The validation is an extremely strict procedure which guarantees the quality and the integrity of the gravity data bank. It is applied systematically to all sets of data before integration into the bank. The principle consists of comparing the observed value and the predicted one, estimated by a powerful technique.

The validation procedure has been applied using least squares collocation. The block of data were partitioned into 11 rectangular zones which have been treated separately for numerical reasons. These zones are disjoint and only have been used by classification and validation purposes. The reduced data of each zone were divided into two parts A and B which have no common observations, but have the same distribution. From these anomalies, two empirical covariance functions were computed separately for all zones. These empirical values were used in order to estimate the parameters values of a local covariance function model (Tscherning, 1974).

The prediction of gravity anomalies have been done on all set B using data of set A, and compared with the observations in set B. However, if the difference between Δg_{obs} and Δg_{pred} was greater than 20 mgals, the observation was rejected. The same procedure was applied to set A with the data of set B.

By removing the suspected points in sets A and B, the procedure is repeated with prediction on all the points of sets A and B alternatively, and if the same point appears suspected then we conclude, that it has definitively a large error. The error ratio we have detected using this method was about 4.7 %.

5. Terrain correction

In order to smooth the anomalous gravity field of the test area, i.e. to remove the irregular variations of the topography, the residual-terrain-model (RTM) reduction is applied, which takes into account only the short-wavelengths of the topography. The method employs a smooth reference surface which may be derived either from a spherical harmonic expansion of the global topography or by averaging the local elevation data.

The computation of the effects of the topography according to the RTM reduction modelling method is based on a global topographic model ETOPO5 of 5' x 5' which were used up to a distance of 200 km. The reference surface of 15' x 15' needed for the RTM reduction is formed from the 5' x 5' blocks. This grid is smoothed further by taking moving averages over the 3 x 3 adjacent blocks. Table I shows the statistics in mgals of the gravity reductions for the gravity data, and map of the residual anomalies reduced of the terrain effect and the contribution of the global model OSU91A is shown on Figure III.

Anomalies	mean	std.dev.	min	max.
Δg_{obs}	29.78	29.99	- 82.59	165.37
$\Delta g_{obs} - \Delta g_{OSU91A}$	1.32	24.07	-70.99	125.56
$\Delta g_{obs} - \Delta g_{OSU91A} - \Delta g_{RTM}$	6.58	20.16	-32.48	109.61

Table I : Statistics of gravity reductions.

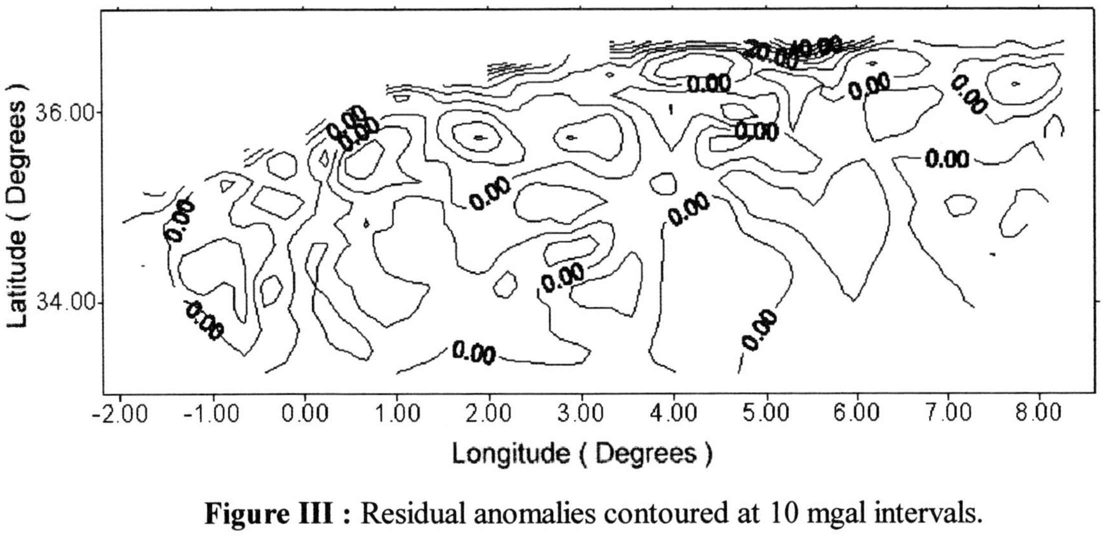

Figure III : Residual anomalies contoured at 10 mgal intervals.

6. Gravimetric geoid computation

The method used to determine the geoid heights by collocation is called a remove-restore technique. The procedure is summarised as follows :

- remove the contribution of the long wavelength components of the gravity field and the effects of terrain masses from the original data,
- estimate residual geoid signals by collocation method,
- restore the effects removed in the first step to obtain total geoid signals.

7. Estimation of covariance function

For the calculation of LSC, an analytic expression of the covariance function is more convenient. For this purpose, we adopt a classical analytic covariance function for the disturbing potential proposed by Tscherning & Rapp (1974), which is expressed by a sum of series of Legendre polynomials (Moritz, 1980) as :

$$K(\psi) = \alpha\left[\frac{GM}{R_E}\right]^2 \cdot \sum_{l=2}^{N} \varepsilon_l \left[\frac{R_E^2}{rr'}\right]^{l+1} P_l(\cos\psi) + \sum_{i=N+1}^{+\infty} \frac{A}{(i-1)(i-2)(i+4)}\left[\frac{R_B^2}{rr'}\right]^{i+1} P_i(\cos\psi)$$

where ε_l : error degree variances of the reference field,
N : the maximum degree of the reference field coefficients (here 360),
α : scale factor of errors associated with the reference field,
R_B : radius of a Bjerhammar sphere,
r and r' : radial distances of the points,
R_E : mean earth radius,
A : scale factor of the degree variances.

The actual use of this model as a local covariance function requires the estimation of three parameters: the radius of the sphere of Bjerhammar (R_B) and two scale factors α and A.

The results of the adjustment of the empirical covariance function on the Tscherning & Rapp model were obtained by the program COVFIT8 (Knudsen, 1987). Figure IV shows the empirical and analytical covariances of the residual anomalies.

8. Predictions

The LSC computations were done using software GRAVSOFT (Tscherning, 1994) package, developed at the National Survey and Cadastre and recommended by the IGeS.

Once all the gravity anomalies were validated, a geoid was computed in the zone [34° - 37° N] et [0° - 8° E]. Predictions were done on a regular geographical grid with sides of 5' both in N-S and E-W directions. Figure V is a map of the geoid surface contoured at 50 cm interval. The maximum value of the formal error estimated, which is defined by this covariance function and by the assigned error on the gravity anomalies (5 mgal in our case) is 81 cm on the area where there is no data, and 19 cm along the gravimetric survey, while the r.m.s. is 40 cm.

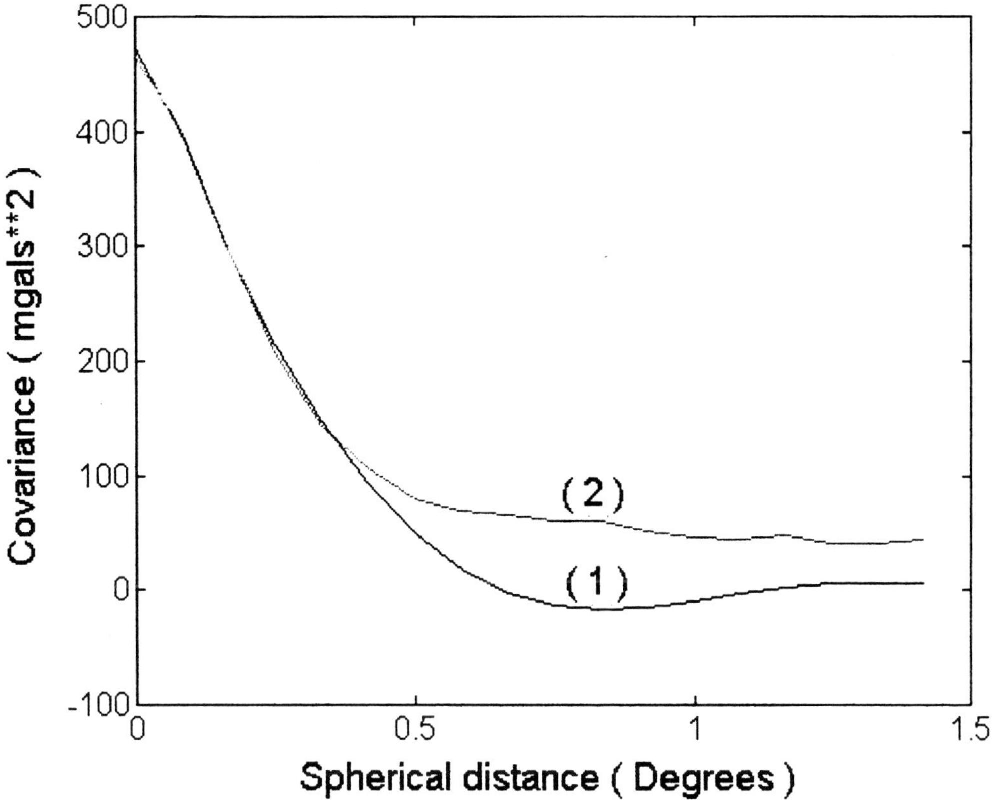

Figure IV : Covariance functions of residual gravity anomalies.
(1) : Analytical, **(2) :** Empirical

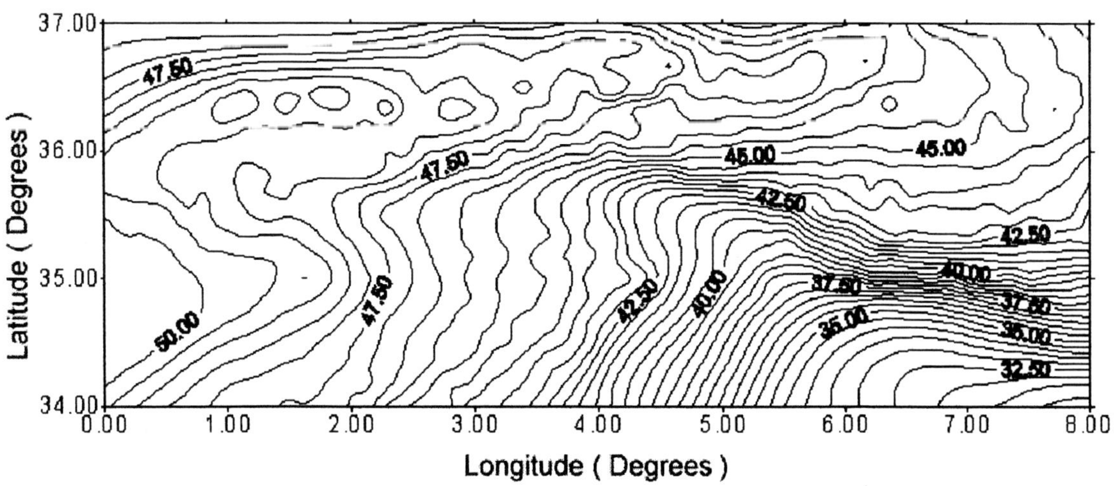

Figure V : Gravimetric geoid computed by collocation with 0.5 m. interval contour.

Conclusion

The method used in this work is not recent in its principle, but it is of a major interest for the modelling of local gravity field from heterogeneous data sources.

With the precision of the available gravity data, and its distribution, the geoid in the test zone were computed with an internal maximal error of 81 cm. Unfortunately, GPS/levelling data necessary for making an independent check of the geoid accuracy is not available in the north of Algeria.

Finally, in the perspective of the determination of a real geoid of Algeria, with a compatible precision to the one provided by the three-dimensional positioning GPS, one needs : 1) a dense and uniform gravimetric coverage on all the territory, 2) deviations of the vertical in order to adjust the errors of contours, 3) height anomalies to verify the quality of our predictions, and 4) a high resolution topographic model in order to guarantee an error of 1 mgal in the terrain corrections and a global geopotential model which fits the used data in an optimal sense.

Acknowledgements

The authors thank Prof. C. C. Tscherning for his support, and to have accepted to procure the GRAVSOFT software, and Prof. G. Balmino for providing gravity data on Algeria.

References

Forsberg R., 1985. Gravity field terrain effect computations by FFT. Bulletin Géodésique, vol. 59, N° 4.

Forsberg R., Tscherning C.C., 1981. The use of Height Data in Gravity Field Approximation by collocation. Journal of Geophysical Research, vol. 86, N° B9.

Knudsen P., 1987. Estimation and Modelling of the Local Empirical Covariance Function using gravity and satellite altimeter data. Bulletin Géodésique, vol. 61.

Moritz H., 1980. Advanced Physical Geodesy, H. Wichmann-Abazcus Press, Karlsruhe-Tundridge Wells.

Rapp R. H., Sanso F., 1995. Determination of the geoid present and future. International Association of Geodesy symposia, N° 106.

Benahmed Daho S., 1996. Détermination du géoïde gravimétrique par la méthode de collocation. Magister thesis, CNTS / Arzew.

Tscherning C.C., 1974. A FORTRAN IV program for the determination of the anomalous potential using stepwise least squares collocation. Reports of the Department of Geodetic Science, N° 212, Ohio State University.

Tscherning C.C., 1974. Closed Covariance Expressions for Gravity Anomalies, Geoid Undulations, and Deflections of the Vertical Implied by Anomaly Degree-Variance Models, Dept. of Geodetic Science, report 208, the Ohio state University.

Tscherning C.C., 1994. Geoid determination by Least Squares Collocation using GRAVSOFT. Lecture notes for the international school for the determination and use of the geoid, Milano, October.

GEOID STUDIES IN THE NORTH-EAST ATLANTIC (AZORES -PORTUGAL)

J.C.Catalao
Faculdade de Ciências de Lisboa, R. Escola Politécnica 58,
1200 Lisboa, Portugal
E-mail : mjoao@cc.fc.ul.pt

M. J. Sevilla
Instituto de Astronomia y Geodesia, Universidad Complutense de Madrid,
28040 Madrid, Spain
E-mail : maast01@sis.ucm.es

Abstract

During the past two years several studies concerning geoid computation have been done in the north-east Atlantic region between the Azores Archipelago and Portuguese mainland. The gravimetric geoid has been derived from gravity data, and uses the remove-restore procedure with the central step performed by least squares collocation method. Several comparisons between different geoid solutions were made considering two different DTMs (ETOPO5U and local bathymetric data) and different resolutions of the reference surface of the Residual Terrain Model computation. For the original ETOPO5U data the differences between altimeter and gravimetric geoid has a standard deviation of 0.24m with an amplitude of 1.26 m. It was found that ETOPO5U must be shifted 5' West in order to get better residual anomalies (lower variance). This new file was merged with local bathymetric data and edited to produce a new DTM. A second solution was obtained with this merged DTM, and the differences between altimeter and gravimetric geoid has a standard deviation of 0.18m and an amplitude of 1.25 m . The differences between this new gravimetric geoid and the altimetric surface approximate the Sea Surface Topography in this area very well.

1. Introduction

Topographic information has been widely used in combination with other data types for gravity field modelling. It is known that when using the least squares collocation method for precise gravimetric geoid determination, the topographic effects on gravity anomalies must be removed in order to get a smooth gravity field and better results in the estimated quantities (Tscherning, 1984). More particularly, the topographic effect on the covariance function has the ability to reduce the anisotropy indice and increase the correlation distance (Forsberg, 1984).

In the frame of GEOMED project, ETOPO5U bathymetric data was analysed by Arabelos (1993) and Furst *et al.* (1993) and it was found that a 5' West shift must be applied in order to get a better fit (in a least squares sense) with local data.

In this work, ETOPO5U was validated with local bathymetric data (the Centro Geofisica Universidade de Lisboa data bank) and we evaluated the effects of erroneous bathymetric data on the estimation of gravity field related quantities. In particular, this effect on gravity anomalies and on geoid undulation was analysed. The Residual Terrain Model (Forsberg *et al.* 1981) was also analysed as a tool for topographic effects computation. The external quality control, of geoid undulation, was done by comparison with ERS1 altimetric data.

2. Bathymetric data

The data sources containing bathymetric mean values were:

- ETOPO5U: a 5' x 5' digital elevation model covering the North Atlantic area in between the geographic limits : [$-34° < \lambda < -6°$; $33° < \phi < 45°$]
- CGUL: also a 5' x 5' digital elevation model with only bathymetric data obtained by a compilation and validation of several sources of bathymetric data (surveying and cartographic maps).
  The covered area of this bathymetric data is: [$-15° < \lambda < -8°$, $33° < \phi < 43°$]

ETOPO5U Model

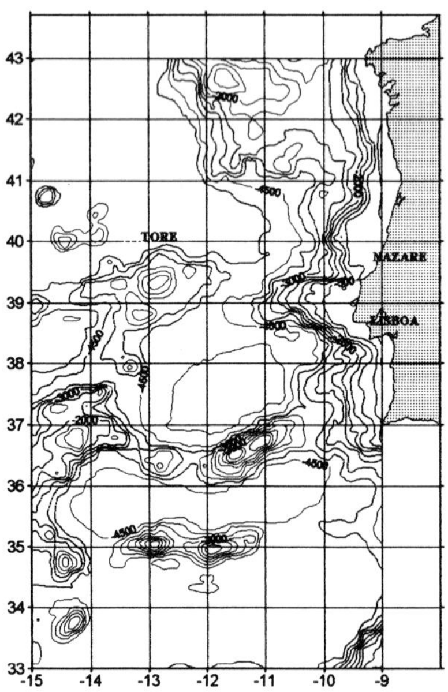

Figure 1. - ETOPO5U bathymetric data. contour interval 500m.

CGUL Model

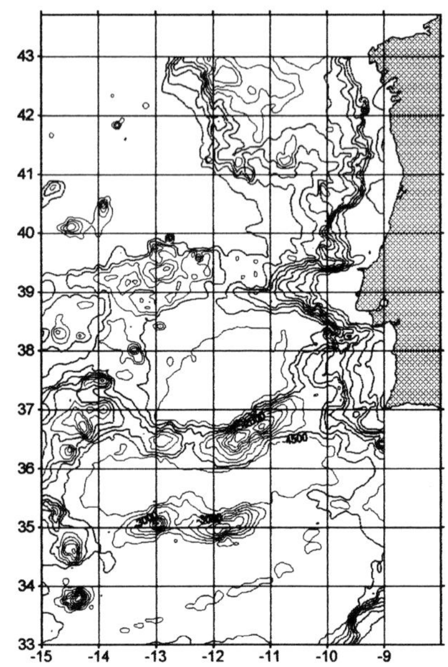

Figure 2. - CGUL bathymetric data. Contour interval 500m.

In Figure 1 the contour lines for ETOPO5U bathymetric data with a contour interval of 500 meters are plotted, and in Figure 2 the contour lines for CGUL bathymetric data with the same contour interval are plotted. The smoothness of ETOPO5U data particularly in the Tore bank (λ= -13°; ϕ = 39°) and in Nazare canyon (λ=-10; ϕ = 39.5) is noticeable. In Gorringe Bank it seems that both models have the same behaviour.

3. Data analysis

Assuming that CGUL data file is error free we proceed with the validation of ETOPO5U model in two steps.

- First we compare ETOPO5U in its original form and shift it 5' West to coincide with CGUL model. The differences were calculated on the basis of three grids with different resolutions: 5'x5', 10'x10' and 30'x30'. The choice of these grids is related to the RTM reductions in which we must consider an inner, outer and reference grid.
- Second, as our main interest on ETOPO5U model is its application to topographic reductions, we also compare the residual terrain model obtained from both models.

Both data files (ETOPO5U and CGUL) were compared in the common area delimited by [$-15° < \lambda < -8°$; $33° < \phi < 43°$]. The statistics of the differences are shown in Table 1.

	5' Grid		10' Grid		30' Grid	
	Original	Shift 5'	Original	Shift 5'	Original	Shift 5'
Min.	-2102.4	-1724.0	-1379.4	-1152.4	-510.0	-528.0
Max.	2643.0	2313.0	1658.2	1537.9	689.0	411.0
Mean	23.53	-21.48	21.64	-23.72	23.5	-23.8
Std. dev.	301.04	284.60	248.5	231.74	147.61	137.2

Table 1. Differences between ETOPO5U data (original and with a shift 5' W) against CGUL data (units - m)

The Residual Terrain Model was computed for each node of the 5'x5' grid as the difference between its height (negative height) and the height of a bathymetric reference model that we selected to be a grid of 30'x30'. The choice of this reference grid must relate to the geopotential model chosen as reference model. In Table 2 we can see that there are no significant differences between the two models.

	ETOPO5U	CGUL
Min.	-1593.5	-1803.0
Max.	3432.0	3711.0
Mean	-4.44	-12.79
Std .dev.	389.6	443.48

Table 2. Statistics of Residual Terrain Model (units -m)

4. Gravity reductions

In order to estimate the effect of erroneous bathymetric data on the gravity reductions we did some comparative tests with free-air anomalies in the test area. The free-air anomalies were first reduced to the OSU91A geopotential model and then the RTM effects in each gravity point data was computed (Sevilla *et al.* 1993). The RTM effects were computed with the "TC" program (Forsberg, 1984) with a reference grid of 30', an outer grid of 10' and a detailed grid of 5'. Finally the residual anomalies were computed as :

$$\Delta g_{res} = \Delta g_{FA} - \Delta g_{OSU91A} - \Delta g_{RTM}$$

	Free air anomalies	Original ETOPO5U	Shift 5' ETOPO5U	CGUL
Min.	-101.91	-80.65	-63.21	-57.47
Max.	345.85	99.28	83.89	69.65
Mean	8.05	1.63	1.67	1.70
Std. dev.	74.04	19.26	18.19	15.17

Table 3. Residual Anomalies using the original ETOPO5U data, the ETOPO5U data shifted 5' and the CGUL data (mGal)

From the results of Table 3 it is indicated that ETOPO5U model must be shifted 5' W in order to get a lower variance of the residual anomalies. However with this shift we only get an improvement of 1 mGal on the standard deviation. The real improvement is obtained with the new bathymetric model in which we gain 4 mGal in the standard deviation and 60 mGal in the signal amplitude compared with the original ETOPO5U. Conserning the empirical covariance function for the residual anomalies, the behaviour is very similar with only small differences in the correlation distance (greater for CGUL model) and in the location of the first zero crossing.

5. Geoid computation

The geoid undulation was computed in the area defined by $[-14° < \lambda < -10° ; 36° < \phi < 40°]$ by means of least squares collocation. The data used were: free-air gravimetric anomalies, the OSU91A geopotential model and both the ETOPO5U and CGUL bathymetric models. The Remove-Restore technique was applied with the central step performed by Least Squares Collocation (Catalao *et al.* 1994). Using the two data sets of residual anomalies, computed with the aforementioned two DTM's, the geoid undulation was obtained by:

$$N_{col} = C_{N\Delta g}\, C^{-1}\, \Delta g_{res} \ ; \quad N = N_{OSU91A} + N_{col} + N_{RTM}$$

The two solutions obtained were compared with ERS1 altimetric data processed by Fernandes *et al.* (1995). The results of the comparison (the residual geoid) are shown in Table 4.

	Original ETOPO5U	CGUL
Min.	-1.08	-0.99
Max.	0.18	0.26
Mean	-0.37	-0.27
Std. Dev.	0.24	0.18

Table 4. Differences between gravimetric geoid computed with the original ETOPO5U models and CGUL model and ERS1 altimetric data (in meters)

The results of Table 4 confirms that the errors of ETOPO5U has a small impact on the geoid determination using RTM and collocation. In a previous study with ETOPO5U heights in the Hellenic area, Arabelos (1993) reached quite similar results with a Std. dev. of 9 cm for the differences between geoid height using two different height models. Besides the evaluation of ETOPO5U these result of comparison of the geoid to the altimetric surface obtained by ERS1 seems to be very good for both models.

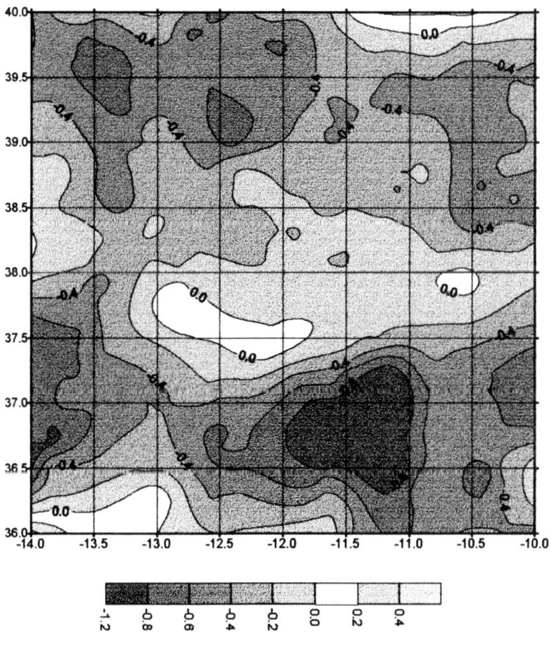

Figure 3 - Difference between Geoid undulation computed with ETOPO5U bathymetric model and ERS1 altimetric data. Contour interval 0.2 meters

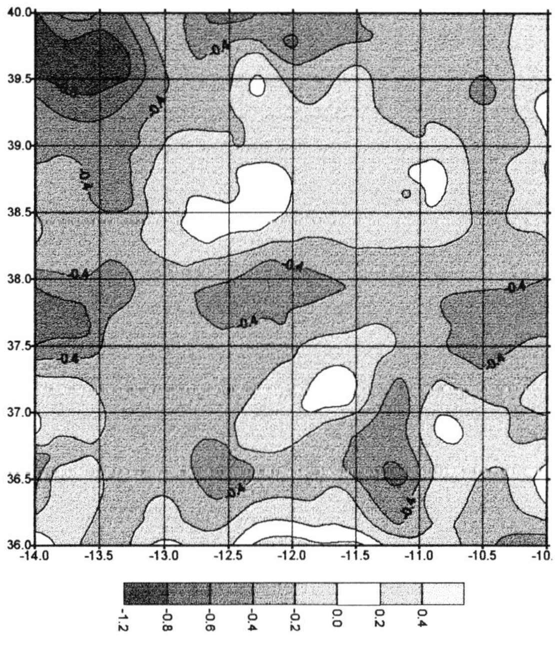

Figure 4 - Difference between Geoid undulation computed with CGUL bathymetric model and ERS1 altimetric data. Contour interval 0.2 meters

In Figure 3 and Figure 4 we present the differences between the geoid surface and ERS1 altimetric data for both the ETOPO5U model and the CGUL model. Although the similar statistical parameters of the residual geoid for both models are similar, it is clear that the geographic distribution of the residual geoid has a very different behaviour.

6. Conclusions

The 5' x 5' height model ETOPO5U was compared with a local bathymetric data CGUL in the ocean area around Portugal. It was confirmed that a 5' West shift must be applied to ETOPO5U data. Despite this shift, the resulting bathymetric data still has differences greater than 2000 meters and a Std. dev. of 284 meters. For mean heights computed on a grid of 30' x 30', the maximum differences between the two models decreases to 500 meters.

Despite the large absolute differences between models the resulting Residual Terrain Model are very similar with a Std. dev. of 389 m for ETOPO5U and a Std. Dev. of 443m for CGUL. The effect of the two models on the gravity reductions was analysed in a sub-area of 4° x 4°. The comparison between residual anomalies reveals a good agreement in their statistical parameters with a slightly lower Std dev for CGUL model. This fact confirms this model is of better quality than ETOPO5U.

Finally the geoid was computed for the same sub-area and compared with ERS1 altimeter data. The comparison shows that the less accurate bathymetric data of the original ETOPO5U has a small effect on the computed geoid, with a difference of 6 cm (Std. dev.) between the models.

7. References

Arabelos, D. (1993). Validation of ETOPO5U in the Hellenic area. In: Sanso F (Ed) Mare Nostrum . GEOMED Report n.3, DIIAR, Milan, pp 65-77.

Catalão, J. and Sevilla, M.J. (1994). Geoid Computation in the North-East Atlantic (Açores-Portugal). First Results. In Arabelos D. and Tziavos I. (Ed) Mare Nostrum. GEOMED Report n. 4, Dpt Geodesy and Surveying, Thessaloniki, pp. 77-90.

Fernandes, M.J., Bastos, L. and Catalão, J. (1995). Comparison between altimetric and gravimetric geoids in the Azores-Gibraltar Region. Presented at the IUGG XXI General Assembly, Symposium G10: Gravity Field Determination. 2-14 July, Boulder, CO.

Forsberg, R. and Tscherning, C.C.(1981). The use of height data in gravity field approximation by Collocation. JGR, Vol.86, No B9, September 10, pp 7843-7854.

Forsberg, R. (1984). A study of terrain reductions, density anomalies and geophysical inversion methods in gravity field modelling. Reports of the Dep. of Geodetic Science and Surveying No 355, The Ohio State University, Columbus.

Furst,W., Hausleiteiner,W., Hock, E., Schuh, W. Sunkel, H. (1993). Validation of the global digital height model ETOPO5U in a local area referring to the Italian height model. In: Sanso F (Ed) Mare Nostrum. GEOMED Report n.3, Milan, pp. 151-167.

Sevilla, M.J. and Catalão, J. (1993) Analysis and validation of B.G.I. gravimetric data in the North-East Atlantic (Açores-Portugal). In: Sanso, F. (Ed) Mare Nostrum nº 3, pp. 126-138. DIIAR, Milan.

Tscherning, C.C. (1984) Local Approximation of the gravity potential by Least Squares Collocation. In : K.P.Schwarz (Ed.) : Proceedings of the International Summer School on Local Gravity Field Appoximation, Beijing, China, Aug. 21-Sept. 4.

THE GRAVIMETRIC GEOID FOR LIBYA

A. Łyszkowicz
Space Research Centre, PAS, Department of Planetary Geodesy,
00-716 Warsaw, Bartycka 18A, Poland

Ahmida Ali Wahiba
Surveying Department of Libya

Abstract

In most positioning, mapping and exploration applications, conventional spirit levelling is being replaced by modern methods such as GPS and use of geoid information. The objective of this paper is to compute a precise geoid/quasigeoid model for Libya based on high resolution gravity and terrain data in combination with a global geopotential model.

This paper gives overview on the work connected with the computation of the first gravimetric geoid for the area of Libya. The data used are; (a) 5'x 5' gridded value of Bouguer anomalies (onshore) and free-air anomalies (offshore) based on the African Gravity Project, and released for geoid computation by Geophysical Exploration Technology University of Leeds (GETECH), (b) global (5'x5') Terrain Base global digital terrain model (c) the OSU91A and EGM96 geopotential models. Two geoid/quasigeoid models have been computed. First model is referred to OSU91A and second one to EGM96 global model.

Introduction

In geodetic applications, knowledge of the geoidal heights and deflections of the vertical are required for the many purposes e.g. in the reduction of distance measurements to a reference surface. To date, various approaches have been proposed by geodesists to determine the geoidal heights and deflection of the vertical. The primary approaches include the gravimetric method, the astro-geodetic method and astro-gravimetric method, satellite dynamic solution for potential coefficients, and direct determination from 3-D geocentric coordinates and orthometric heights.

Geoid determinations with real data have been performed in many areas of Europe,

North America and Australia. Recently several geoid computations for the northern part of Africa have been accomplished. For example Benaim et al.(1997) for Morocco, Bouziane (1996) for Algeria and Hanafy et al.(1995) for Egypt. This paper gives an overview on the work concerning computations of the first gravimetric geoid for territory of Libya.

Computational method

In this study generally adopted strategy for local geoid undulation (N) computation is composed of combination of the geopotential model (GM), mean free air gravity anomalies Δg_{FA} , heights (H) and is based on the following formulas:

$$N = N_{GM} + N_{\Delta g} + N_{H} \tag{1}$$

$$\Delta g = \Delta g_{FA} - \Delta g_{GM} - \Delta g_{H} - \delta\Delta g \tag{2}$$

The term N_{GM} gives the contribution of the GM coefficients, while the term $N_{\Delta g}$ gives the contribution of the reduced mean free air gravity anomalies with the effects of the global model (Δg_{GM}) terrain (Δg_{H}) and indirect effect on gravity ($\delta\Delta g$)[1] removed. N_H gives the indirect effect of the terrain reduction on N. The contributions of the GM to N and Δg can be found in many publications and will not be repeated here.

$N_{\Delta g}$ is computed by Stokes' integral which, in spherical over the area E, has the form

$$N_{\Delta g}(\varphi_p, \lambda_p) = \frac{R}{4\pi\gamma} \iint_E \Delta g(\varphi, \lambda)\, S(\psi)\, d\sigma \tag{3}$$

where R is the mean radius of the earth, $S(\psi)$ is the spherical Stokes kernel and ψ is the spherical distance. This convolution integral is evaluated in the frequency domain by the multi-band fast Fourier Transform approximation (Forsberg and Sideris,1993). Proper zero padding (100%) was applied to the gridded data to eliminate the effects of circular convolution.

Since Stokes' formula is valid for Δg on the geoid, all masses above it must be mathematically shifted inside the geoid via terrain reduction. Thus the term Δg_H in equation (1) is the classical terrain correction. In practice the free-air anomalies Δg_{FA} for the territory of Libya were obtained from terrain corrected Bouguer anomalies. Therefore in our case there were no need to compute term Δg_H.

The term N_H in equation (1) is called the indirect effect on the geoid, and accounts for the change of equipotential surface after a terrain reduction is applied to Δg. In our case

---

[1] Usually negligible ($= 0.3086 \times N_H$ in mGal)

the geoid term N_H will be computed from digital terrain model using Helmert's condensation method (Wichiencharoen,1982).

$$N_H = -\frac{\pi G \rho H^2}{\gamma} \quad (4)$$

where G is gravitational constant, ρ is the density of topographical masses, γ is the mean gravity and H is height.

The vertical control network of Libya was adjusted two times. First time as a network of normal heights and second time as the network of orthometric heights. In this way we have to vertical systems in Libya. Therefore in practical applications the knowledge of the geoid as well as the quasigeoid is required. To obtain the quasigeoidal model the computed geoid (N) is converted to the quasigeoid (ζ) by the relation:

$$\zeta = N - \frac{\Delta g_B}{\gamma} H \quad (5)$$

where Δg_B is simple Bouguer anomaly, γ is the mean normal gravity and H is normal height.

Used data

Geopotential models. For the determination of the long wavelength part of the earth's gravity field (N_{GM}) a global geopotential model is indispensable. Recent high resolution models with emphasis on accuracy arc based on combinations of terrestrial and satellite observations including satcllite altimetry. They were developed by the Ohio State University: OSU81 to degree and order 180, OSU86E,F to degree 360, OSU89A,B to degree 360 and OSU91A to degree 360. The latest geopotential model EGM96 is a spherical harmonic expansion of the Earth's gravitational potential to degree and order 360. It is the result of collaboration between the NASA Goddard Space Flight Centre, the National Imagery and Mapping Agency and the Ohio State University. The solution represents a major update to previous global models, such as OSU91A. In our computations both OSU91A and EGM96 models were used.

Gravity data. The gravity data used in our geoid computations are: (a) 5'x 5' gridded value of mean Bouguer anomalies (onshore) and mean free-air anomalies (offshore) based on African Gravity Project, and released for geoid computation by Geophysical Exploration Technology University of Leeds for the area $20^o < \varphi < 35^o$ and $9^o < \lambda < 25^o$. Computation of mean gravity anomalies is based on about 150 000 land and 3 000 marine point gravity data[2] which cover more or less uniformly the whole territory of Libya except south-east part of the country.

---

[2] no permission to publish diagram

Topographic data. For the computations of the indirect effect corrections and for the transformation of the geoid into quasigeoid, mean elevations for the area of Libya are necessary. This data set was taken away from the Terrain Base global digital terrain model consisting of a matrix of land elevations and depth values gridded at 5-minutes intervals for the whole world (Row et al.,1995).

Practical computations

Residual free-air anomalies computation. In the first step Bouguer anomalies were converted into the free-air anomalies by restoring the Bouguer plate with density $\rho = 2.67$ $g\ cm^{-3}$. In the next step two sets of residual anomalies were produced. From the free-air anomalies the contributions of the OSU91A and EGM96 model were subtracted. The statistics of the source and residual anomalies are given in Table1.

Table 1.

Anomalies	Mean	Std. dev.	Min.	Max.
source	4.10	30.01	-194.00	292.86
residual - OSU91A	-0.66	19.43	-99.84	190.28
residual - EGM96	-0.40	18.48	-103.51	184.97

The residual anomalies computed from the EGM96 model have better statistical characteristics which means that new geopotential model EGM96 better represents the gravity field of Libya than OSU91A.

Geoid undulation computation. This step comprises the main part of the geoid undulations computation. The input data in this case are the residual anomalies as computed from equation (2). The software used in this study is the original program SPFOUR written by Forsberg. Two sets of residual geoids were computed. One set is referred to OSU91A model and another to EGM96. The indirect effect on the geoid was computed using equation (3) with the 5' x 5' grid point elevation. By adding the reference undulations and indirect effect to the residual geoid undulations the final solutions were obtained. Statistics of the residual and final geoid solutions are given in table2.

The quasigeoid correction was computed using equation (5) with the 5'x 5' grid point elevations and free-air gravity anomalies. The results of computation vary between -10 cm and +32 cm. By adding quasigeoid corrections to geoid solutions one obtained final quasigeoid models: qgeoid97a and qgeoid97b.

The final results are presented in a form of the contour maps and also in a digital form by the computed geoid/quasigeoid heights for a regular grid of 5'x 5'. Totally, 34933 geoid/quasigeoid heights are stored in grid files, from which geoid/quasigeoid heights can be interpolated for any arbitrary point on the territory of Libya.

Achievable accuracy

It is not easy to give estimate of the geoid undulation accuracy. A number of different estimates can be found in current textbooks. With the development of GPS networks connected to the levelling lines it is possible to compare gravimetric derived geoid/quasigeoid undulations with the undulations obtained from GPS and levelling. In Libya the Institute Geographic Nationale established Doppler satellite supernet of 45 stations in 1976 -1978. In 1983-1985 a large horizontal Doppler network based on supernet was developed. Doppler satellite networks together with levelling data could serve to gravimetric geoid/quasigeoid accuracy estimations. In our study because of the lack of appropriate data we are forced to make a rough estimation of our computed geoid/quasigeoid models.

Table 2. Different geoid solutions (statistics in metres).

Geoid solution	Mean	Std. Dev.	Min.	Max.
model EGM96	25.99	5.65	8.79	37.80
residual geoid97b	-0.53	2.36	-4.29	9.39
indirect effect	-0.01	0.03	-0.65	0.00
geoid97b	25.45	4.71	8.30	37.75
quasigeoid corr.	-0.02	0.03	-0.08	0.32
qgeoid97b	25.42	4.71	8.30	37.71

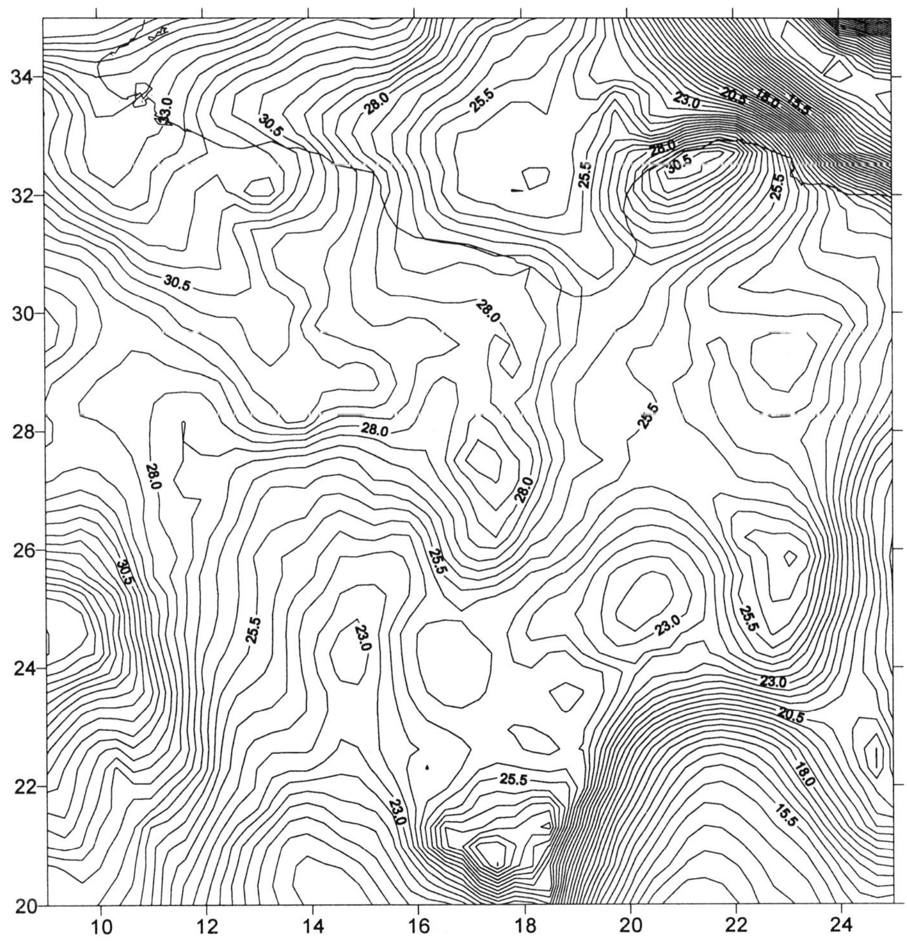

Fig. 1. Gravimetric geoid for the area of Libya

The accuracy of geoidal height difference depends on the accuracy of the three components: $\Delta N_{GM}, \Delta N_{\Delta g}, \Delta N_{H}$. Studies performed by many authors (e.g. Schwarz et al.,1987) have shown that average error, in ppm for contribution from geopotential model is below 2 ppm. The other two contributions are rather small and in our case should be below 1 ppm. Thus the total relative accuracy of our geoid/quasigeoid models is estimated on the level few ppm (about 2 ppm). To estimate the realistic accuracy of the computed geoid/quasigeoid models further investigations are needed.

Conclusion

The first gravimetric geoid/quasigeoid for the territory of Libya has been computed. Its accuracy is supposed to be few ppm which with combination with GPS measurements can provide vertical positioning for many sort of projects in a very fast and cost effective manner. To estimate the realistic accuracy of the computed geoid/quasigeoid models and to fit them correctly to the national vertical datum of Libya further investigations are needed.

Acknowledgements

The financial support provided by a Space Research Centre of Polish Academy of Sciences grant is gratefully acknowledged.

References

Benaim E.H., Takhchi M., Tikdirnine L., Sevilla M.J., 1997, *The first northern Moroccan gravimetric geoid*, paper presented at XXII General Assembly of European Geophysical Society, Vienna, Austria, 21-25 April 1997

Bouziane M., 1996, *The New Algerian Quasi-Geoid 96 Computed Using Spherical FFT*, International Geoid Service, Bulletin N. 5,pp.40-56,Milano, Italy

ForsbergR., Sideris M.G.,1993, *Geoid computations by the multi-band spherical FFT approach*, Manuscripta Geodaetica, 18, pp.82-90

Hanafy M.S.A, El Tokhey M.E.A., 1995, *Towards a High Precision Geoid for Egypt*, Gravity and Geoid, Symposium No. 113, Springer, pp.597-606

Row L.W., Hastings D.A., Dunbar P.K.,1995, *Terrain Base, Worldwilde Digital Terrain Data, CD.-ROM Release 1.0*, National Oceanic and Atmospheric Administration, Boulder, Colorado

Schwarz K.P., Sideris M.G., Forsberg R.,1987, *Orthometric Heights Without Levelling*, Journal of Surveying Engineering, vol. 113, No 1

Wichiencharoen C.,1982, *The Indirect Effects on the Computation of Geoid Undulations*, The Ohio State University, Department of Geodetic Science and Surveying, Report No. 336

COMPARISON BETWEEN ALTIMETRIC AND GRAVIMETRIC GEOID IN THE SOUTH - WEST MEDITERRANEAN BASIN

S. KAHLOUCHE, M.I KARICHE, S.A BENAHMED DAHO
National Center of Spatial Techniques
Geodetic Laboratory - BP 13 ARZEW- 31200- ALGERIA

Abstract

The processing of altimetric data collected by the Topex-Poseidon satellite on the Western Mediterranean sea, based on the Barrick-85 model, permitted the extraction of the instantaneous sea surface height. The analysis of 76 passes distributed over 8 orbital arcs, after correction of the disturbing effects (troposphere, ionosphere,...), the reduction of the instantaneous measurements and elimination of the weak periodic variable phenomena, allowed us to give a mean altimetric profile.

Thc proccssing of thc mcan arcs on thc crossing points by thc lcast squarcs mcthod allowed us to get a mean level on the western Mediterranean sea with a precision of a centimeter. The results confirm the dominance of an East-West current along the Algerian coast and the existence of localized eddies off the North African coasts (Bernard & al, 1993).

The sea surface obtained from Topex data were compared to the global geoid deducted from the OSU91A potential model. The metric difference obtained shows insufficient adequacy of the global solutions for geodetic applications. The comparison of the solution above with a gravimetric geoid provided by the collocation method (using GRAVSOFT software and a set of BGI free air gravity anomalies surrounding the region) near the Algerian coasts, improves the results appreciably and shows the importance and the influence of the inshore quality data.

The introduction of exhaustive and precise altimetric data, and of gravimetric measurements with sufficient density, will permit to define a more precise altimetric surface of the geoid by the integration of the existing tidegauge measurements.

1. Introduction

The analysis of altimetric measurements requires high accuracy in the determination of the different parameters (position of satellite, corrections) allowing calculation with sufficient accuracy of the sea height relative to the reference ellipsoid .

The detection of oceanographic phenomena (currents, tides, mean level) with the accuracy of the altimetric measurements (2~3 cm) requires the satellite orbit (reference of the measurement) to be known with at least an equivalent accuracy.

The present research takes mainly character of a study, because it was initiated on the basis of a set of experimental data of the satellite Topex on a limited time scale (three months) and with a low spatial homogeneity.

2. Data used

The used data represent 76 crossings of the Topex satellite on the Mediterranean through September and December 1992. This volume of data represents 8 arcs of the satellite; each arc being repeated on average 10 times (cycles); though some arcs don't count the totality of the ten cycles.

The data in our possession corresponds to the calibration- validation phase of the satellite, a phase during which the orbit of satellite is known with approximately 1 m. Generally, during this period, the tide and the effect of charge are not available; the electromagnetic bias and the speed of wind are computed with not validated algorithms (Bonnefond, 1994).

3. Data preprocessing

In the first place, the data are controlled in order to eliminate erroneous observations, i.e those corresponding to a footprint point not located in the sea, or collected under the unfavorable weather conditions, then rectified of disturbing effects : troposphere, ionosphere, electromagnetic bias, etc. The percentage of measurements eliminated and not included in the processing varies between 1% (arc 172) and 10% (arc 009).

The mean value of the tropospheric correction due to the dry air is in the order of 2 m up, while the one due to the humid air does not exceed 20 cm. The differences between the up and down relative path corrections are in a millimeter level for the dry correction and in a centimeter level for the humidity (Kariche, 1997).

The ionospheric correction applied is the average between the values provided by the altimeter bifrequency model of the NASA and the measurements of Doppler effect done by the DORIS beacons. The difference between the two processes is in the order of a centimeter for corrections less than 5 cm.

The whole of the instrumental corrections (shift of phase center of the antenna, disturbing effects, altimeter bias,...), are gathered in one corrective parameter. The effect of these disruptions translates the altimetric measurement by an error which could reach 5 m. The electromagnetic bias is a function of the state of the sea (height of the waves), which is deduced from the intensity and waveform of the reflected signal.

4. Mean arcs

The mean sea level is obtained by processing of the repetitive collinear profiles with statistical methods and by orbital error modeling. The adjustments are done by a polynomial representation of the errors for the processing of the collinear profiles and by least squares for the crossing points (Barzaghi & al, 1992).

The processing of the instantaneous heights of the sea surface are done by simple difference between the altitude of the satellite on the reference ellipsoid GRS80 and the altimetric measurements. The used orbital parameters are those of NASA, except for the cycles where only the data of CNES are available. However, about 30% of the deviation between the CNES and the NASA measurements presents a bias of +12.5 cm in satellite height.

The mean points thus obtained, apart from the measurement precision itself, has an accuracy which depends only on the number of averaged measurements.

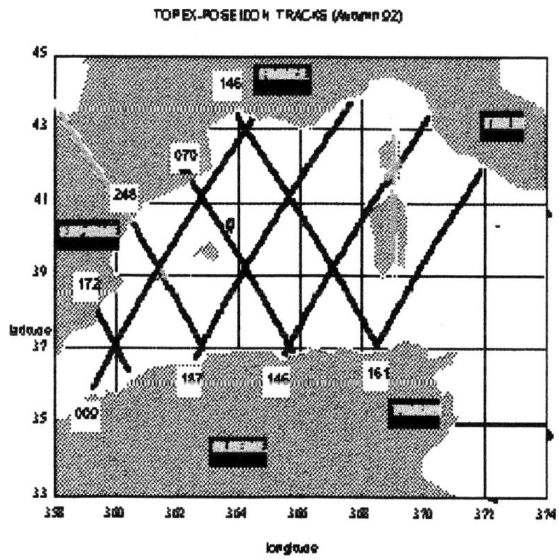

Arc	N° of points	Mean Variation	Standard deviation
009	1094	-2.4	9.8
070	931	+0.4	7.2
085	1397	+1.7	7.8
146	1319	+0.4	7.2
161	1435	-0.2	8.8
172	1442	-5.2	22.9
187	1398	+0.2	7.2
248	595	+15.3	34.3

Accuracy of the instantaneous heights (cm)

For the following processing, the mean points which do not contain at least 60% of cycles were not taken into account; this improves the accuracy in the determination of the mean profiles. Proceeding like this (averaging), we limited the influence of variable phenomena whose period is less than 3 months.

5. Processing the crossing points

On the intersection of the ascending and descending arcs, the differences of the sea heights are computed in the sense "ascending minus descending" mean profile, where the orbital errors were modeled by a bias and a drift parameter.

The height differences on the crossing points are not homogeneous, they are obtained by a linear interpolation from a different number of measurements from a point of crossing to an other: this is caused by the quality of our data.

From our results, the great standard deviations obtained on the crossings arcs could be interpreted like this (Kariche, 1997):

- a strong variability which would indicate the presence of a very important circulation, but in the Mediterranean, this is less probable.
- a radial orbital error from 1 m to 1.5 m.
- a mistake during the data extraction from the AVISO file.

The orbit errors are then modeled solely by a bias, this is sufficient considering the length of the arcs and the weak gradient surface (Arent & al., 1992).

In order to eliminate the case of singularity from the parameters matrix, the arc 187 (supposed not containing orbital error) was fixed. The choice of this arc was justified by the fact that, according to the last processing, it has more stability and homogeneity, compared to the other arcs. For the other arcs, the radial orbit errors are between 9 cm (arc 85) and 70 cm (arc 248).

6. Processing the mean sea level

The mean profiles rectified from the orbital errors determined by the processing of the crossing points, represent the mean heights of the sea surface under the satellite tracks.

N° arc	N° of pts	standard deviation	standard error
009	151	9	6
070	99	7	2
085	144	7	2
146	137	7	2
161	150	8	4
172	165	19	5
187	159	7	2
248	153	7	3
Total	1158	9.7	3.6

Precision of the mean profiles (cm)

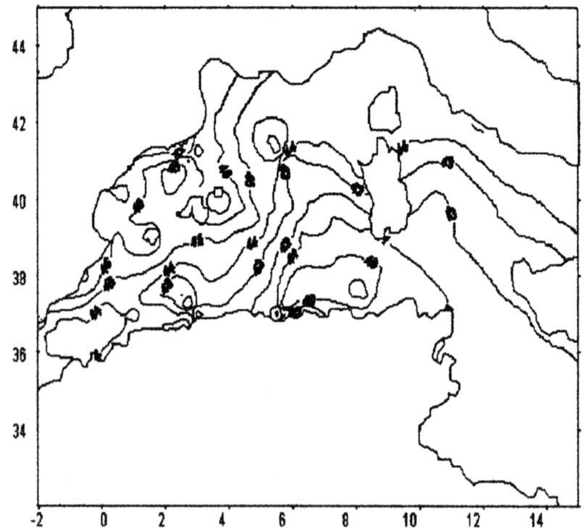

Mean altimetric level (equid.: 1 m)

The gridding of these mean profiles on a regular grid of 0°25x 0°25 (in latitude and longitude) allows to obtain a mean surface for the 3 months data period. This mean surface is computed by using the spline functions in order to interpolate the heights in the grid nodes. The mean level obtained is free from variable phenomena with a period less than 3 months. The processing was done on 1158 points, with an accuracy of 10 cm.

7. Comparison with a global geoid

The altimetric geoid heights obtained on the GRS80 ellipsoid, by interpolation on a regular grid of 0.5° by 0.5° are compared with the one deduced from a global geoid based on the potential model OSU91A.

The deviation between the two geoid surfaces has an average of 2 m. Some significant differences appear mainly in the East-Mediterranean (until 5 m) and in the surroundings of the Baleares islands (4 m). These differences could be attributed on one hand to the accuracy of the potential model used and on the other hand to the limited spatial and temporal coverage of altimetric data. Better results would be expected if data from dense satellite altimetry (ERS and GEOSAT) was utilised (Bernard & al., 1993).

Indeed, the processing of 3 months of Topex observations is not sufficient to determine exactly all the disturbing oceanographic phenomenon. But we estimate the accuracy of our processing in the order of a few decimeter (with orbital, tides and circulation errors).

The differences between the two surfaces are computed on the nodes of the same regular grid (0°25 x 0°25). However, it would be necessary to note that the processing were done on a rectangular area (located between the parallel 35°and 42 ° and the meridians 2° West, 13 ° East) which some parts are not covered by the altimetric data (on land or on the East part of the basin). This explains the high variations and the important difference deviation. The statistics of the deviations are:

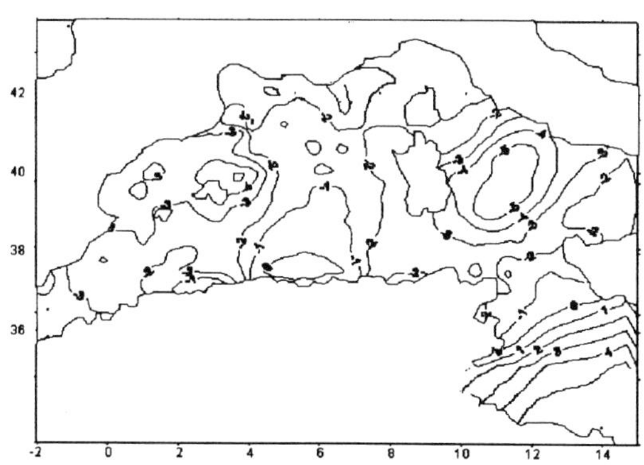

Height differences in m

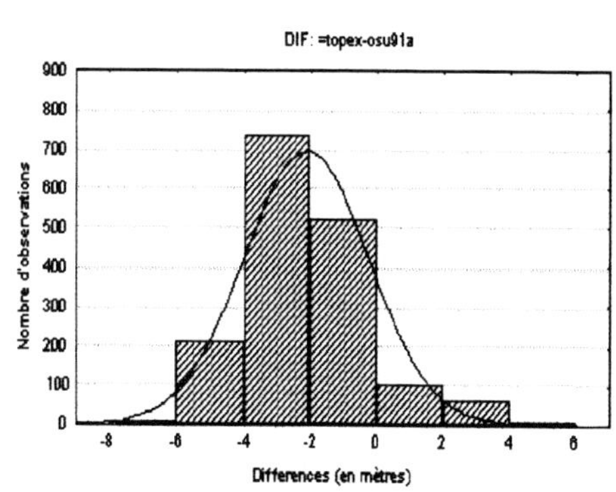

Distribution of the height differences

The differences greater than 4 m are located either on land, or in the parts not covered by the data. If we consider only the sea covered part, the separations between the two surfaces were reduced globally, and the separation of the whole of the differences reduces to 34 cm.

8. Position with regard to the local geoid

The local geoid computed with the GRAVSOFT software and the BGI gravimetric data from the North of Algeria were extrapolated by collocation in the areas located between (35°& 37°) North (latitude) and (0°& 8°) East (longitude). The altimetric geoid obtained was extrapolated on the emerged areas of North Algeria in order to provide the comparison of the results obtained by the different methods (Benahmed Daho, 1996).

The main results of the geoid differences are (in thick line: collocation geoid; in thin line : altimetric geoid):

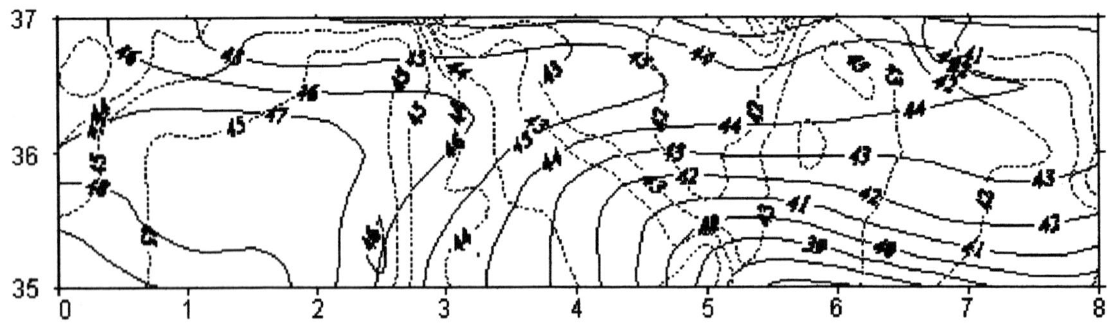

The differences between the two surfaces are :

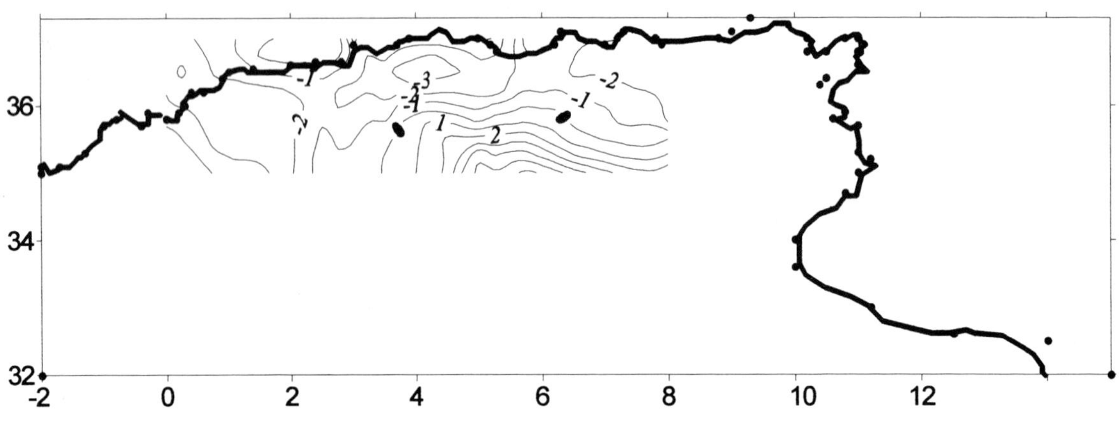

Note that the two grids have few common points. For the comparison, the altimetric heights were extrapolated on the places not covered by the altimetric measurements (land points).

Conclusion

In the first place, the objective is to process the Topex altimetric data and to compute an altimetric geoid from a set of data collected in the western Mediterranean.

The results obtained are far from being perfect considering their accuracy; this is due to the volume and the quality of the used data (relative to the period of satellite validation and calibration). But the mean surface, obtained with a standard deviation less than 10 cm, and containing the effects of the tides in the order of 30 cm (for the area of the Mediterranean studied), is therefore estimated with an submeter error.

The comparison surface used is a global geoid (from OSU91A), itself containing errors in its part covering the Mediterranean sea (deficiency of gravimetric data). The heights of this geoid on the same ellipsoid of reference is estimated with a metric accuracy. We may therefore say that the differences between these two surfaces which are in average of 1.5 meters are not significant. On the other hand, the more significant differences (3 to 4 m) were observed in the surrounding of islands of the Mediterranean: they are due to the fact that in our model, tidal effects and of the state of the sea were not taken into account the latter of which become very complex and sometimes significant around the islands and the coastal areas (Bernard & al., 1993).

Besides, the comparison with the results obtained by collocation do not allow to conclude on the quality of the data and on the methodology developed, considering the limited quantity and the nature of the used data.

The integration of exhaustive data (altimetry, gravimetry) with the existent tidegauge data would allow to further improve the geoid determination in the Mediterranean.

Bibliography

N.Arent, K.R.Koch, 1992. Method for obtaining geoid undulations from satellite altimetry data by a quasi geostrophic model of sea surface topography. Manuscripta Geodaetica 17.

R.Barzaghi, F.Sanso, 1992. Altimetry rank deficiency in crossover adjustment - Manuscripta Geodaetica 17.

S. Benahmed Daho, 1996. *Détermination du géoïde gravimétrique par la méthode de collocation.* Magister thesis, CNTS Arzew.

J.F.Bernard, F.Barlier, 1993. First Seasat altimeter data analysis on the western Mediterranean sea. JGR 88.

S.Bhaskaran, G.W.Rosborough, 1993. Computation of regional mean sea surface altimeter data - Manuscripta Geodaetica 18.

P.Bonnefond, 1994. *Méthode géométrique de trajectographie par arcs courts.* Doctoral thesis; CERGA Grasse.

M.I.Kariche, 1997. *Traitement des données du satellite Topex-Poseidon, Comparaison avec un modèle global de potentiel* - Magister thesis - CNTS Arzew.

Y.Ming Wang, R.H.Rapp, 1992. The determination of a one year mean sea surface height track from Geosat altimeter data and ocean variability implication. - Bull. Geod 66.

DATA PROCESSING FOR A GEOID MODEL IN ARGENTINA

María Cristina Pacino
Facultad de Ciencias Exactas, Ingeniería y Agrimensura (UNR) e IFIR
Av. Pellegrini 250 – 2000 – Rosario - ARGENTINA
mpacino@unrctu.edu.ar

Graciela Font
Facultad de Ciencias Astronómicas y Geofísicas (UNLP)
Paseo del Bosque s/n – 1900 – La Plata – ARGENTINA
graciela@fcaglp.fcaglp.unlp.edu.ar

Denizar Blitzkow
Escola Politécnica – Universidade de Sao Paulo – BRASIL
dblitzko@usp.br

ABSTRACT

Instituto de Física Rosario (IFIR), Facultad de Ciencias Astronómicas y Geofísicas de La Plata (FCAGLP), in a cooperation with the Politechnic School of the University of Sao Paulo (EPUSP), have been involved in the last few years in an effort for the determination of the Argentinian geoid. For this purpose as many gravity data as possible have been compiled and tests were undertaken in order to solve the problems concerning the quality of the data. The main gravity data set was basically compiled at IFIR. This data base consists of about 15.000 gravity stations, mainly from IGM (Instituto Geográfico Militar) and from National Universities. They were referred to different references, so that in a first step transformations were performed in order to have the gravity values homogeneously related to IGSN 71. The gravity anomalies have been computed and referred to WGS 84. Taking advantage of the South American Gravity Project (GETECH – University of Leeds), links have been established between IFIR, IGM, the Sub-Commission for the Geoid in South America (SCGSA) and GETECH. In this way new efforts have been carried out in the last few years to improve the gravity data coverage and the DTM. For this purpose topographic maps have been digitised, in particular in the Andes, and a 3'x3' grid generated for Argentina. All these efforts are described in the paper.

INTRODUCTION

In December of 1994, the Subcommission of Geodesy of the Comisión Nacional de la Unión Geodésica y Geofísica Internacional (CNUGGI – Argentina) created a working group, the Geoid Group (GG) to develop and implement a strategy to determine an enhanced geoid model for Argentina, which, in combination with GPS data, will provide an alternative method for orthometric height determination and many other applications, such as mapping, geodynamics or positioning.

Since then, efforts have been undertaken by the GG in the collection and compilation of gravimetric and topographic data from academic and non-academic organizations in Argentina.

Argentina consists in the East of one of the widest plains in the world and, in the West, the highest american peaks, reaching nearly up to 7000 m high. These features are correlated with Bouguer anomalies of near zero over the Pampean Plane dropping –400 mGal in the Andes, as expected from isostasy. Furthermore, the Chile trench, 200 to 300 km westward of the Andes presents maximum depths of about 8000 m below sea level and maximum Bouguer anomalies of about +350 mGal. In this way, a geoid model for the whole Argentinian territory requires the joint efforts of many institutions. The SCGSA is playing a leadership role by supporting and contributing to the research and development on our geoid model.

The intention of this paper is to give a brief overview on the status of the computations.

GEOID COMPUTATION METHOD

The computation of the geoid for Argentina combining a geopotential model (GM), mean free-air gravity anomalies Δg_F and heights H in a digital terrain model DTM was based on the following formula (Sideris, 1995, Lyszkowicz & Forsberg, 1995).

$$N = N_{GM} + N_{\Delta g} + N_{H}$$

where N_{GM} is the contribution of the spherical harmonic model, N_H is the indirect effect term and $N_{\Delta g}$ is the contribution of the terrestrial gravity field observations, i.e. free-air anomalies after removal the effect of the global geopotential model and the topography.

THE GEOPOTENTIAL MODEL

Nowadays, a very important source of information for gravity field approximation is a good reference field from a geopotential model, since the long wavelength contributions of the gravity field are provided by a set of spherical harmonic coefficients (GM)

In this way, several models are now available. We chose to use the EGM 96 NASA/NIMA geopotential model. The evaluation of the model has been carried out in South America under the auspices of the SCGSA, and with an important cooperation of GETECH – University of Leeds. We can certainly say that this new model in South America is better than any other previous model (Blitzkow, 1997).

GRAVITY DATA

Fig. 1A gives a graphical representation of the gravity data coverage in Argentina collected by SAGP. Besides land gravity measured points, the compilation includes gravity values calculated from digitized gravity anomalies maps.

For this preliminary geoid model we desregarded, in a first step, gravity data digitized from contour maps, until we can get a more detailed DTM of the areas.

Land gravity measured stations of Argentina are collected at IFIR. The Data Base is composed of more than 20.000 gravity values from the following institutions:

- Instituto Geográfico Militar (IGM)
- Instituto de Física Rosario (IFIR)
- Instituto de Geodesia – Universidad de Buenos Aires (IG – UBA)
- Instituto Argentino de Oceanografía (IADO)
- Centro de Investigaciones regionales geológicas (CIRGEO)
- Instituto Antártico Argentino (IA)
- Universidad Nacional de La Plata (UNLP)
- Universidad Nacional de Tucumán (UNT)
- Universidad Nacional de Salta (UNS)
- Universidad Nacional de San Juan (UNSJ)

The accuracy of these data ranges from 0,1 mGal to 10 mGal due to different sources used in the collection of initial data. At this moment we still cannot assure the accuracy of each gravity station. So, in this first step we only chose those from IGM, UNLP and IFIR, which guarantee homogeneous measuring criteria (Fig.1B).

As the different files referred to different datums, datum transformations were performed to standarize all the data. The standarized data are referred to Miguelete's fundamental value (979 690,03 mGal).

All data sets have been checked carefully to remove repeat points, and validated for gross errors by applying different procedures (Blitzkow et al., 1995, Guspí et al., 1995).

After the validation process was completed, terrain correction, free-air and Bouguer anomalies were computed. The calculation was done at GETECH – University of Leeds during one month stay of two of the authors.

The standarized data are referred to the Geodetic Reference System 1980 (Moritz, 1984) for theoretic gravity.

The free-air correction has been applied using the empirical equation:

$$FAC = (0{,}3083293357 + 0{,}0004397732 \cos^2 \phi)\, h - 7{,}2125\ 10^{-8}\ h^2$$

Which is a function of the latitude ϕ and the orthometric height h.

For the Bouguer correction an infinite slab expression has been used with a density of 2,67 g/cm^3 for land stations. A curvature correction has also been applied (Green & Fairhead, 1991).

Terrain corrections were calculated for all gravity points taking into consideration only the "outer zone" (from 5 km to 166,7 km). To estimate this correction the 3'x3' DTM has been used. The calculation was performed at GETECH. For more details and information refer again to Green & Fairhead, 1991.

To remove the atmospheric effect from the gravity anomalies, the following correction has been added to the free-air gravity anomalies (Pavlis, 1991):

$$\delta g_A = 0{,}8658 - 9{,}727\ 10^{-5}\ H + 3{,}482\ 10^{-9}\ H^2 \quad \text{(mGal)}$$

where H is the orthometric height of the gravity station in meters.

After computing gravity anomalies, the estimation of mean values and an interpolation process was carried out, according to Blitzkow et al., 1995. To remove as much predictable high-frequency content as possible, gridding was performed on terrain corrected Bouguer anomalies (Δg_B). The next step was to restore the Bouguer plate term to the Δg_B grid using the 3′x3′ DTM, yielding a grid of Faye anomalies.

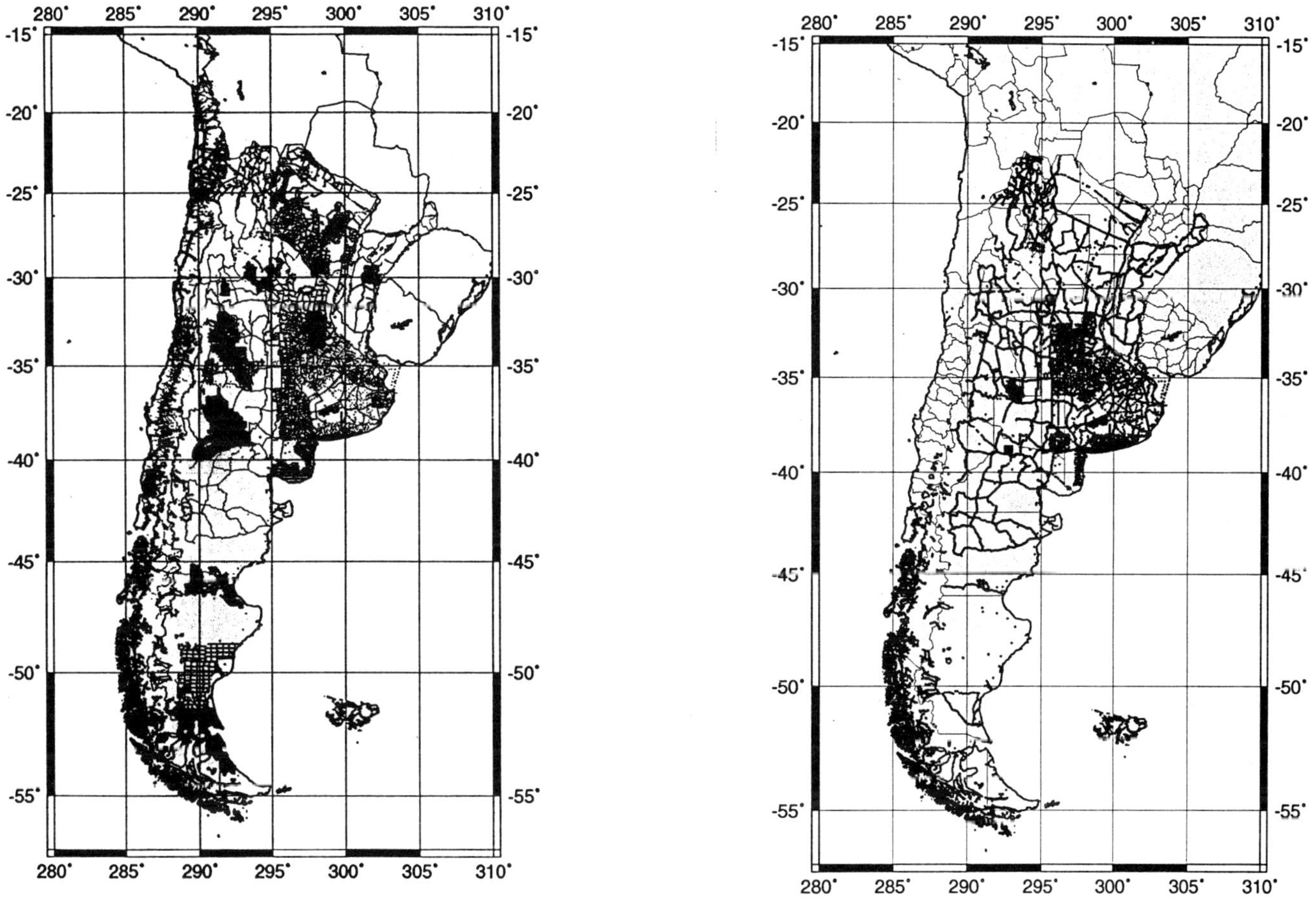

Fig. 1: (A) Gravity data coverage in Argentina collected by SAGP – (B) Gravity data used for the geoidal model of Argentina.

TOPOGRAPHIC DATA

A 3'x3' grid of topography was produced so that terrain corrections could be applied. Four types of elevation data have been used to construct the grid:

1- Worldwide topographic grid ETOPO5
2- Values picked from topographic maps
3- Height at gravity stations
4- Shore line locations

A detailed description of these sources and the way they have been used can be found in Green & Fairhead, 1991.

New efforts have been carried out in the last few years to improve the DTM. For this purpose 1:100.000 topographic maps have been digitized. The data points derived from these maps may be considered accurate between 20 and 50 meters.

Fig. 2 shows the status of digitising maps in Argentina.

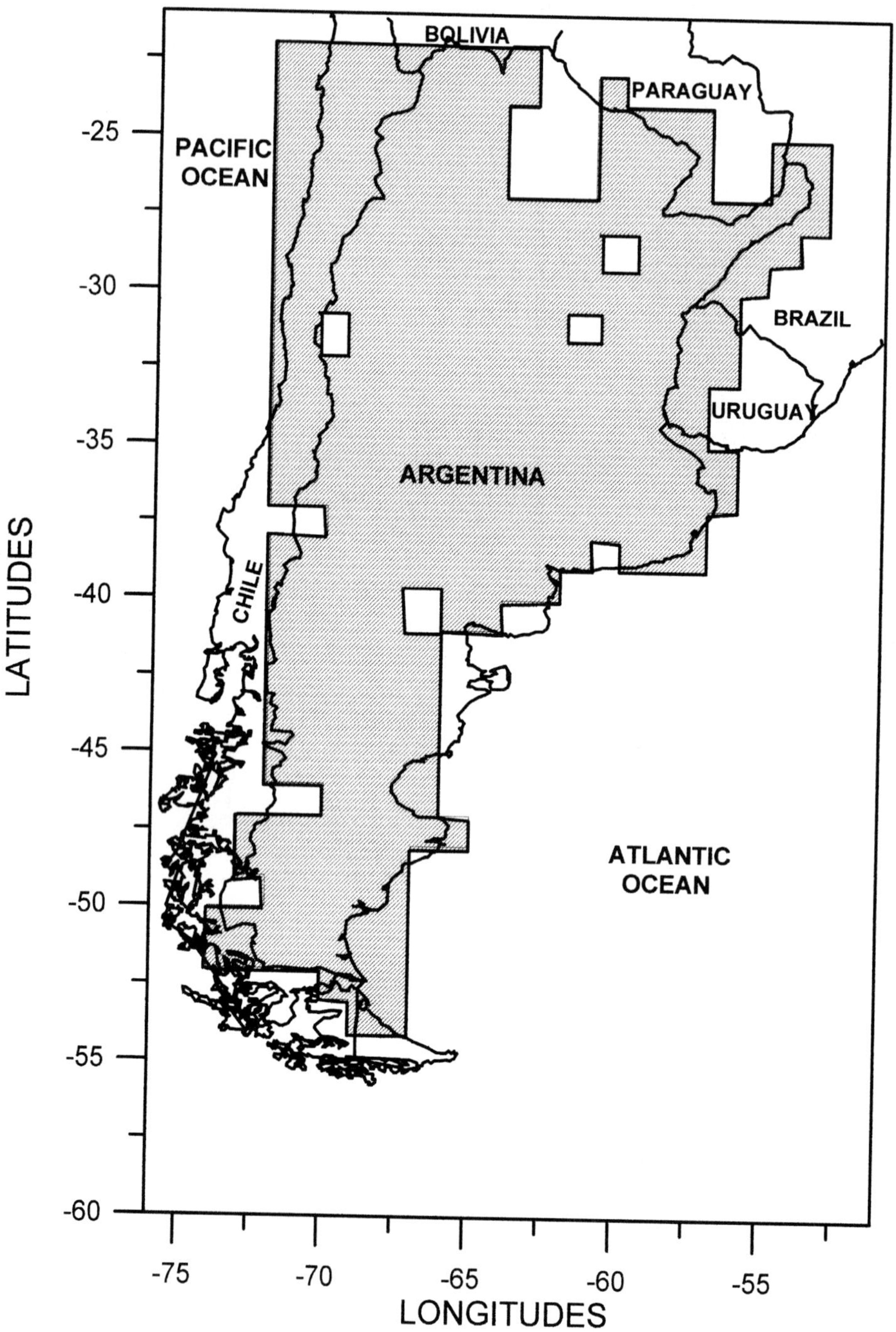

Fig.2: Status of digitising maps in Argentina.

RESULTS AND CONCLUSIONS

A complete file has been created with different 5'x5' mean values: free-air anomaly, Bouguer anomaly, height and terrain correction. From this file, mean values of 20'x20' have been derived. A mean Faye anomaly has been obtained by restoring the mean free-air anomaly from mean Bouguer anomaly and mean height.

The mean gravity values (Faye anomalies) will be used in the modified Stokes integral. The procedure will be described in an other paper of this section (Font et al., in press).

To conclude, significant progress was made since the initiation of the project three years ago regarding the collection of gravity and terrain data and GPS/levelling data for further evaluation of the results. It is expected an improvement in the geoid model for Argentina from the incorporation of more than 5000 new gravity values and nearly 100 GPS/levelling points.

ACKNOWLEDGEMENT

The authors are very grateful to Prof. J. D. Fairhead (GETECH – University of Leeds) for his support to the geoid related activities in South America. We also acknowledge all the institutions and National Universities in Argentina which supplied gravity data and specially the SCGSA for the support to this work.
This work was partially supported by PIA N° 6896 (CONICET – Argentina)

REFERENCES

Blitzkow, D., 1997. NASA (GSFC) / NIMA Model evaluation. International Geoid Service, Bull. N°6: 71-81.

Blitzkow, D., Fairhead, J. D. & Lobianco, M. C., 1995. A preliminary gravimetric geoid for South America. International Geoid Service, Bull. N°4: 53-61.

Font, G., Pacino, M. C., Blitzkow, D. & Tocho, C., 1997. A preliminary geoid model for Argentina. Books of Abstracts IAG 97, pp.170.

Guspí, F., Introcaso, A. & Pacino, M. C., 1995. Carta gravimétrica de Argentina 1995: su construcción. Actas I Congr. Arg. De Geociencias y Geotécnicas: 88-93.

Green, G. M. & Fairhead, J. D., 1991. The South American Gravity Project. In: Recent Geodetic and Gravimetric Research in Latin América. Edited by W. Torge. Springer Verlag, Berlín.

Lyszkowicz, A. & Forsberg, R., 1995. Gravimetric Geoid for Poland Area Using Spherical FFT. International Geoid Service, Bull. N°4: 153 – 161.

Moritz, H., 1984. Geodetic Reference System 1980. Bulletin Géodésique, 58: 388-398.

Pavlis, N. K., 1991. Estimation of geopotential differences over intercontinental locations using satellite and terrestrial measurements. Rep. N° 409 of the Dept. of Geodetic Science and Surveying. The Ohio State University. Columbus.

Sideris, M. G., 1995. FFT Geoid Computations in Canada. International Geoid Service, Bull. N°4: 37-52.

DECIMETRIC GEOID FOR BUENOS AIRES PROVINCE BASED ON GPS AND GLOBAL MODELS

R. Perdomo, C. Mondinalli, D. Del Cogliano
Facultad de Cs. Astronómicas y Geofísicas
Universidad Nacional de La Plata y CONICET
Paseo del Bosque (1900) La Plata - Argentina

Abstract

A precise GPS network was measured directly on the existing high precision leveling network of the Buenos Aires Province. The surveyed region is a North-South transect of about 400 km long, crossed at its southern border by an East-West transect 600 km long. Typical distances between adjacent points are 50 km excepting four densification areas where the distances are 5 to 10 km. The GPS network included 3 control points from the GFZ SAGA network. One of them is La Plata IGS permanent station. The observations were made with Ashtech Z12 and Trimble 4000 SSE receivers. IGS precise ephemeris were used. The baselines solutions were obtained with the GPPS software and the compensation of redundant observations and final adjustment to the SAGA datum (close to the ITRF94), with the Fillnet software. The estimated accuracy of the three components is about 1 to 2 cm. Similar accuracy is estimated for the orthometric heights, so the "observed" geoid undulations are accurate to only a few cm.

Comparisons of these results with the calculated undulations obtained with OSU91A global model show a mean difference of about -70 cm (in the sense observed - model) with standard deviation (SD) of 20 cm. A slightly bigger SD is obtained when using the GFZ96R global model but in this case the mean difference is only about -30 cm. In the case of the model EGM96, the SD is 25 cm and the mean difference, -20 cm. Several functions are being tested to adjust the global models to the observed undulations, however, with very simple interpolation functions, an accuracy of only a few centimeters is possible in the flat regions, while one to two decimeters accuracy is obtained in mountainous areas.

1. Introduction

The main objective of this work is to obtain a direct method to transform ellipsoidal heights, as obtained with GPS, into classical "mean sea level" heights as those obtained from the Buenos Aires leveling network (Nivelacion, 1962).

The best method to present the final results is still under discussion and depends strongly on the observed behavior of the two fundamental surfaces in the area.

In this presentation, the differences between the GPS ellipsoidal heights "h" and the published mean sea level heights "H", at about 50 points of the Buenos Aires province were obtained. These differences are called Nobs (= h - H). They represent a first approximation of discrete geoid undulations with respect to the GRS80 (ITRF) ellipsoid.

These values are compared with global geoid model undulations (Nmod) with two objectives: to estimate the accuracy of the models in the region, and to obtain a set of differences "Dif" easier to interpolate than the direct Nobs values:

Dif = Nobs - Nmod

2. The main characteristics of the surveyed region

The Buenos Aires province is very important from an economic point of view. It has a large surface area (about 600 km x 600 km) and it is the most populated among the Argentine provinces. The land is mainly devoted to agriculture and cattle raising.

It has a long coastline on La Plata River (north-east) and on the Atlantic Ocean (east). Most of the surface is flat excepting two orographic systems situated in the South (Tandilia) and southwest (Ventania). Tandilia is about 335 km long and 60 km wide. Its highest peak is of the order of 500 m. Ventania is smaller but higher (the highest peak is about 1200 m).

3. The GPS network

The GPS network was established coinciding with the existing official leveling network. A basic geodetic structure with baselines of about 50 km long was first developed. Subsequently, four densification areas with points at distances of 10 km or less were established. The whole network has its origin in the La Plata IGS station. Two control points are located, one in Tandilia and the other in Ventania. These three points are included in SAGA project (Klotz et al, 1997). They are very important to the study of the accuracy of the network and to realize an adequate reference frame.

The 50 point network was adjusted fixing the three SAGA points. IGS precise orbits were used. The estimated accuracy for the latitude and longitude was about 1 cm and about 1.5 cm for the ellipsoidal heights (ITRF coordinates and GRS80 ellipsoid).

The densificated areas were selected in such a way that similar information was obtained in quite different regions: two of them are located in the plains (Brandsen and Lobos) and the other two are close to the mountains (Tandilia and Ventania).

4. The global models used for comparison

Three global models were used to obtain Nmod in the surveyed region. The OSU91A (Rapp et al, 1991), the GFZ96R (Gruber et al, 1997), and the EGM96 (Lemoine et al, 1996). All these models are developed to degree and order 360.

The geoid undulations obtained with these models are called Nosu, Ngfz and Negm, respectively.

5. Results

The Table I shows the observed results Nobs. The units are degrees for Latitude and Longitude and meters for the ellipsoidal height and Nobs.

Table I

Longit.	Latitude	Ellip. height	Nobs	Longit.	Latitude	Ellip. height	Nobs
-59.18	-35.17	45.49	17.18	-58.22	-35.26	31.57	16.35
-58.11	-35.17	30.31	16.17	-59.07	-35.28	41.77	17.39
-59.26	-35.17	45.46	17.23	-58.23	-35.29	33.55	16.37
-58.24	-35.18	30.80	16.44	-58.59	-35.30	43.59	16.90
-59.12	-35.18	46.38	17.22	-58.24	-35.31	34.39	16.38
-58.04	-35.19	30.76	16.12	-62.20	-38.35	256.35	15.21
-59.32	-35.19	49.79	17.28	-59.19	-35.51	41.69	17.83
-58.07	-35.23	32.75	16.20	-59.12	-35.99	49.31	17.24
-59.08	-35.23	41.50	17.33	-59.14	-36.42	73.45	16.78
-59.51	-35.25	52.56	17.45	-59.18	-36.97	134.85	16.66

Table I (cont)

Longit.	Latitude	Ellip. height	Nobs	Longit.	Latitude	Ellip. height	Nobs
-60.45	-36.98	199.78	17.27	-59.08	-37.41	227.99	16.64
-58.38	-37.15	81.64	15.95	-59.01	-37.45	235.48	16.53
-61.03	-37.16	189.64	16.80	-57.72	-37.48	39.41	15.21
-59.10	-37.17	159.91	16.73	-58.94	-37.50	197.21	16.36
-58.73	-37.31	132.22	16.40	-58.89	-37.51	211.79	16.31
-60.33	-37.32	250.15	16.69	-58.81	-37.55	160.02	16.18
-58.99	-37.35	190.82	16.58	-62.46	-37.57	309.74	16.58
-61.81	-37.83	355.86	15.96	-62.79	-37.57	235.66	16.87
-59.78	-37.36	265.60	16.93	-61.90	-37.57	271.59	16.41
-61.58	-37.37	215.60	16.43	-58.75	-37.61	171.18	16.04
-60.87	-37.61	222.00	15.74	-61.69	-37.89	366.77	15.78
-61.90	-37.64	287.43	16.20	-61.54	-37.97	308.63	15.48
-61.90	-37.71	294.30	16.04	-61.36	-38.11	313.77	14.90
-61.86	-37.77	312.75	16.02	-62.22	-38.17	284.81	15.57
-62.33	-37.81	335.93	16.54	-62.02	-38.32	249.24	15.07

Figure 1

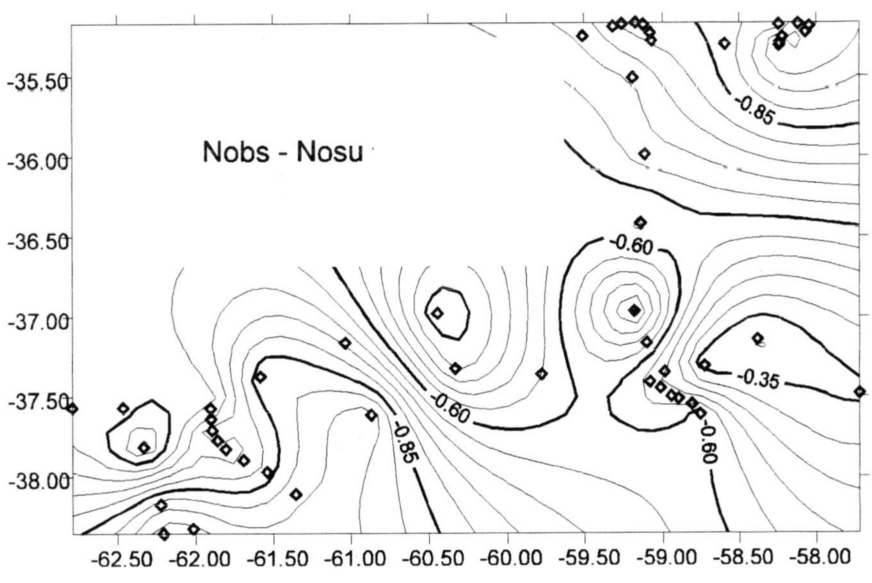

The figure 1 shows the GPS network and the contour map of the differences Nobs - Nosu. The two densificated areas in the north are in the plains. The other two, in the southern region, are close to the mountains).

Figure 2

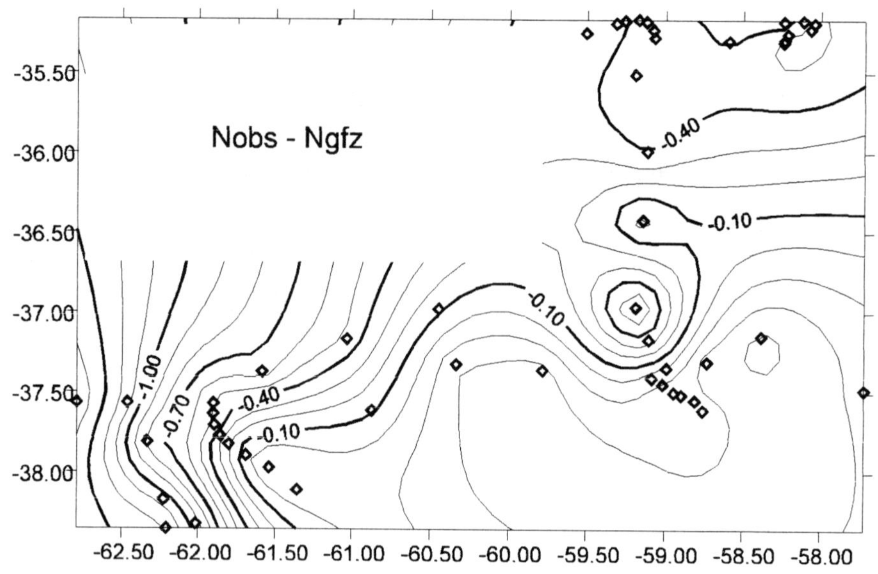

The figure 2 shows the differences Nobs - Ngfz. The variations of the differences are sligtly bigger than those in figure 1. The figure 3 shows the differences Nobs - Negm. It is quite clear that the differences vary smoothly from north to south and from east to west.

Figure 3

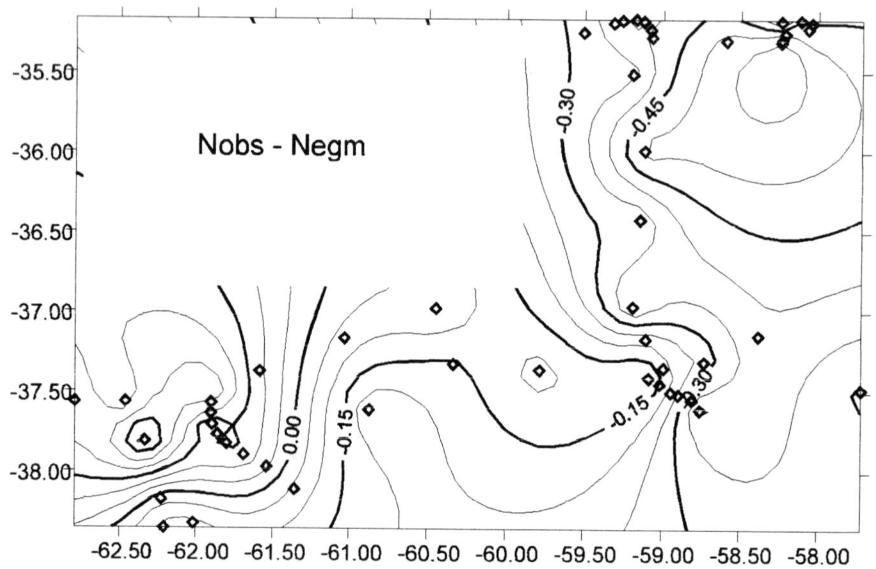

6. Conclusions and comments

The figures 1, 2 and 3 show that:

- There is an offset which averages about -0.70 m for Nobs-Nosu with a standard deviation (SD) of 0.23 m. This offset is about -0.30 m for Nobs-Ngfz with SD of 0.40 m. For Nobs-Negm, the mean difference is small, only - 0.15 m and the SD is about 0.25 m.

- If this mean difference is subtracted from the individual Dif, the three models are always within 0.50 m from the observed value. This means simply that when only sub meter accuracy is needed, these global models corrected only by a constant offset may be used to correct ellipsoidal heights (h) to obtain approximate orthometric heights (H).

- In the densificated areas of Brandsen (-35.2,-58.1) and Lobos (-35.2,-59.1), the observed variations in Dif are so smooth that they may be interpolated to the centimeter level. This suggests that for the plain region of the province, the development of a GPS network with baselines of about 50 km long, measured on the existing leveling network might be enough to get a grid of corrections to h to obtain H.

- In the mountains or near them, the observed Dif may reach several decimeters over distances of 50 km. Evidently it is necessary to add more information in these areas to get an accurate way to transform h in H at the centimetric level.

7. References

- Nivelación. Publicaciones generales de la Dirección de Geodesia de la Prov. de Buenos Aires, serie 1a. nro. 31, 1962.
- Klotz, J., Angermann D. & Reigber C. GPS observed crustal deformation in the central and southern Andes. Proceed. of IAG Scientific Assembly, 1997 (in press).
- Rapp. R., Wang, Y., Pavlis, N. The Ohio State University 1991 Geopotential and Sea Surface Topography Harmonics Coeff. Models. Rep. 410. Dep. of Geod. Sc. and Surv., The Ohio State University, 1991.
- Gruber, Th., Anzenhofer,M., Rentsch,M. Improvements in High Resolution Gravity Field Modeling at GFZ. Proceed. of IAG Symp. Gravity, Geoid and Marine Geodesy, Tokyo, 1996, 1997 (in press).
- Lemoine F. et al. The development of the NASA GSFC and DMA joint geopotential model IAG Symp. , 1996 (in press).

TOWARDS A REGIONAL GEOID DETERMINATION IN WESTERN VENEZUELA

E.Wildermann, M. Hoyer, G.Acuña
Laboratorio de Geodesia Física y Satelital
Escuela de Ingeniería Geodésica, Universidad del Zulia
Maracaibo, Venezuela; e-mail: ewilderm@luz.ve

ABSTRACT

An effort for regional geoid determination has been established at the Zulia University for the area between geographic latitude 7° to 12° North and longitude 68° to 73° West. Data collection, observation and validation is under progress. The creation of a digital height model, a combined geoid traverse crossing the Andes mountain chain, some first results of a GPS - levelling project around Maracaibo Lake and future plans will be considered.

INTRODUCTION

The determination of the vertical reference surface has been declared a priority programme by the Venezuelan National Cartographic Council after a workshop held at Caracas in 1993 (Hoyer, 1993). The increasing number of applications of geodetic space techniques, mainly GPS, in daily surveying work with an urgent need to transform the ellipsoidal heights to a physical height system has intensified the demand for a high precision geoid.

Unfortunately, however, numerous problems such as organisation, economics, logistics and comprehension have complicated the proposed project over recent years. Some natural reasons also exist, for instance, the lack of high quality data due to rugged topography combined with vigorous climate conditions.

Gravimetric data, as the main source of geoid information, is mainly acquired under geophysical exploration purposes, showing variations in accuracy and in uniform coverage over the study area. Quite often, the data available are considered to be confidential because of their possible use in the identification of new petroleum deposits.

Using a regional geoid determination strategy proposed for Venezuela and countries with similar conditions (Hoyer, Wildermann, in press), the Zulia University working group decided to begin a stepwise nation-wide geoid calculation starting with a zone between geographic latitude 7° to 12° North and longitude 68° to 73° West using the following steps:

- collection, revision and preparation of existing data sources,
- extension of the information base,
- concentration of measuring efforts in problematic areas,
- topographic corrections to gravity data,
- use of integration and/or collocation algorithms for data processing, and
- the present limitation to the region of the country which is easily accessible from the University at Maracaibo.

All these aspects shall be attained through cooperation with public and private national and/or international institutions. In general, an accuracy at dm-level is to be expected, for areas with good data coverage a reduction to the 10 cm level seems to be possible.

DESCRIPTION OF THE STUDY AREA

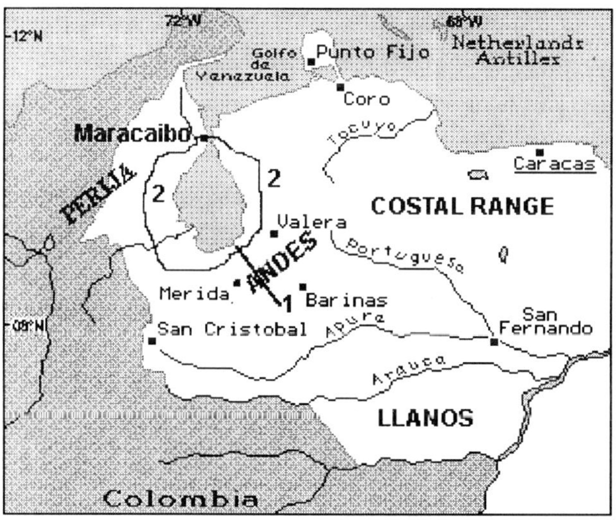

Fig. 1: Geographical situation at the Venezuelan study area (1= Andes traverse, 2= GPS-levelling)

Two high mountain chains intersect the study area. At the western border to Colombia, the Perijá mountains, with limited access possibilities by roads and little related infrastructure mean that there is a zone with no gravity measurements. The Venezuelan High Andes and its extension into the Coastal Range mountains can be accessed by some paved streets alongside the main valleys, where roughly all available data has been obtained. Most of the higher mountain parts are normally without easy access and consequently lack observation coverage.

In the north-west, the Maracaibo Lake area shows intensive oil industry activities. Only at some western shoreline parts access is difficult due to tropical vegetation swamps. Data from different sources can be found in the industrialised zones, but mostly, these are observed under distinct usage conditions.

In the segment to the east of the Maracaibo Lake, arid and lower lying ground can be found. The amount of data is small and normally restricted to crossing streets.

On the south side of the Andes and Coastal Range, the Venezuelan Llanos (lowlands) stretch without great topographic diversity to the Rio Apure-Orinoco river stream system. Access is limited part of the year due to some large inundation zones. The data coverage is sparse, mainly obtained under exploration purposes.

DATA SOURCE REVISION

At the 1993 workshop, all possible data contributors for geoid calculations were gathered to discuss the possibilities of cross-institutional cooperation. An analysis of the available data in the study area shows the following aspects:

- The data are very heterogeneous and of different accuracy (see Table 1).
- Only a few sources will fulfil all requirements of geoid calculations. For instance, these are gravity data over first-order levelling lines, gravimetric networks with geodynamical objectives, astrogravimetric data along the Andes traverse and GPS observations over levelling points around Maracaibo Lake.
- Gravimetric data are mainly obtained for geophysical exploration needs. Quality is reduced due to no measurement redundancy and lack of accurate point heights.
- An increasing number of GPS points over levelling marks can be used to test the astrogravimetric geoids and strengthen future solutions.
- A lack of digital height information specially in the mountainous areas.

Type of data	Institution	Accuracy	Use for geoid calculations
Astrogeodetic data:	- Dirección de Cartografía Nacional (DCN) - Zulia University	± 0.2" ±0.5-1.0"	Geoid gradients at Laplace stations and profile geoid calculations.
Gravimetric data:	- DCN - Zulia University	± 0.02 to ± 0.5mgal	High precision gravity networks, observations over levelling points.
	- University Simón Bolivar - Venezuelan Oil Industry - Dept. of Energy and Mining	up to mgal level	Gravimetric and aerogravimetric data, mainly for geophysical purposes.
Satellite data: Doppler	- DCN - Venezuelan Oil Industry - Dept. of Foreign Office - Private companies - Zulia University	1/2m to 1m level	Definition of geodetic reference system, frontiers and exploration. Only very few points with height information.
GPS	- DCN - Venezuelan Oil Industry - Dept. of Defense - Private companies - Zulia University	cm to dm level	Network definition. Oil exploration work. Increasing connection with levelling network. Some data used for local and regional geoid studies.

Table 1: Principal data sources for geoid calculations in Western Venezuela

Updated information on some of the main data sources is given in Table 1, which also shows accuracy and possible usage. Considering the problems with the gravimetric data and the increasing availability of satellite positions with orthometric height information, the latter is now considered as an important additional data source. The main advantage of these data is some data integration over large areas for undulation processing is not needed, as is the case of Stokes formula.

CURRENT PROGRESS AND FUTURE PLANS

Digital height information

The lack of digital height information for the project has been attacked in co-operation with the National Cartographic Administration, Caracas. Using 1:100,000 topographical maps as a base, height information will be digitised at first with a 2 km-interval separation. The choice of the 1:100,000-base has been made due to availability of this source over nearly all of the Venezuelan territory.

The Zulia State and adjacent zones were successfully completed in 1996. The model is currently being extended over the Occidental region of the country. From the 160 maps covering the proposed computation area, nearly 40% have been digitised, and the data is currently being tested. In mountainous areas, a 1km to 500m interpolation interval is also under preparation.

Combined Andes geoid traverse

A local geoid study under extreme topographic undulations, see Fig. 1 (No. 1), had been conducted using different data sources and processing algorithms in cooperation between the University of Hannover, Germany, and Maracaibo (Wildermann, 1988). An update, using newer, more detailed geopotential modelling (OSU91A) and connections to the GPS levelling project around Maracaibo Lake (described next) shows good coincidence between the astrogeodetic-gravimetric and GPS-benchmark results (Wildermann, 1996). The approximation of the gravity field by the newer geopotential models is better. However, local effects in the order of up to 3m still dominate the regional geoid structures, especially due to the large height variation of more than 4500m in less than 40km of horizontal distance.

GPS observations over levelling lines around Maracaibo Lake

A GPS project measured ellipsoid heights over 55 points of the first-order levelling network around Maracaibo Lake, see Fig. 1 (No. 2). This was performed through a collaboration between the National Cartographic Administration, the Venezuelan oil industry, some private companies, and the Zulia University. The GPS project was used to test commercial and scientific processing software. With the latter, a centimetre-accuracy level in all three coordinate directions could be obtained fixing the network through common stations with the CASA (Central and South American) GPS Geodynamical project (Drewes et al, 1995).

Discrepancies of 1m to 2m between the GPS observed geoid heights and the EGM96 geopotential model have been obtained. The resulting differences of some representative stations along the traverse is given in Figure 2. In the southern part between 450km and

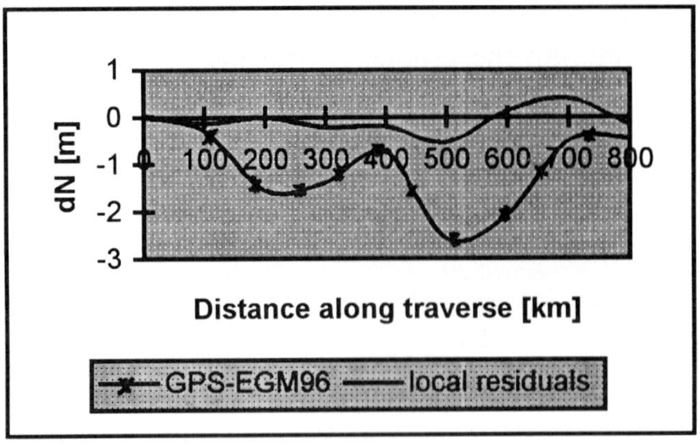

Figure 2: Residual undulations dN

650km, a high correlation with the Andes mountain block can be seen. In the north-east, between 150km and 300km of distance, a similar influence of the Perijá mountains is be found. A low-degree polynomial, fitted to the GPS minus model differences, reduces the remaining residuals below the 0.5m level. The Table 2 summarises the main characteristics for the observed GPS-levelling undulations, the differences after modelling with EGM96 and, finally, the residuals after fitting with local polynomials. A detailed analysis and interpretation of the project is under preparation.

Data characteristics	$N_{GPS\text{-}Lev}$ [m]	$N_{GPS\text{-}Lev} - N_{EGM96}$ [m]	after local fitting [m]
Range	18.385	3.639	1.764
Minimum value	-21.788	-3.208	-0.739
Maximum value	-3.403	0.431	1.025
Mean value	-10.302	-1.020	0
RMS	--	--	±0.347

Table 2: Statistics for 46 GPS points around Maracaibo Lake

Levelling of REGVEN and/or supplementation with points by GPS-levelling

The new REGVEN (Red GPS Venezuela, national GPS network), observed and processed by the National Cartographic Administration and the German Geodetic Research Institute, Munich, provides a great part of the Venezuelan country with a homogeneous, high precise GPS reference system, which will be established as the new national high-order network (Drewes et al., 1997). At present, levelling of these stations is under progress, thus permitting the establishment of a new high quality data source for geoid calculations with a resolution of approximately 100 km. Extensions of the network will incorporate more levelling points with GPS observations, giving an uniform data concentration over the study area with a higher resolution for geoid calculations.

GPS at high precision gravity network points

Other data sets with good orthometric height information are the gravimetric networks in the region, for instance, the Andes Geodynamic Precision Gravimetric Network covering the high mountain part (Drewes et al., 1991). The latter contains stations distributed with a

25 km separation along the Andes mountains between the Colombian border and the Caribbean Sea. From repeated measurements, gravity and orthometric heights are well known. Observing these stations now by satellite techniques permits the determination of their ellipsoidal heights with a high level of accuracy, if these measurements are connected to the REGVEN points in the region. An observation campaign is under preparation.

CONCLUDING REMARKS

A regional geoid determination in western Venezuela is prepared by combining mainly gravimetric and GPS-levelling data, geopotential modelling and use of a digitised heights. Various individual data sets will be merged to obtain geoid undulations at dm accuracy level, or even better in areas with good data coverage, - a demand stated by an increasing number of geodetic users and others.

Acknowledgements:
The work partly has been supported by CONDES (Consejo de Desarrollo Científico y Humanístico) of Zulia University, Maracaibo, Venezuela. The authors acknowledge the helpful hints of the reviewers for the final English version.

REFERENCES

H. Drewes, W. Torge, R. Röder, C. Badell, D. Bravo, O. Chourio, 1991: Absolute and relative gravimeter observations in Venezuela and monitoring recent gravity variations; IAG Proceedings "XX. General Assembly of the IUGG, Symposium G1", Vienna, Austria; Springer-Verlag, Heidelberg - New York

H. Drewes, K. Kaniuth, K. Stuber, H. Tremel, H.-G. Kahle, Ch. Straub, N. Hernandez, M. Hoyer, E. Wildermann, 1995: The CASA'93 campaign for crustal deformation research along the south Caribbean plate boundary; Journal of Geodynamics, Vol. 20, No. 2, 129-144

H. Drewes, H. Tremel, N. Hernandez, 1997: Adjustment of the new Venezuelan National GPS network within the SIRGAS reference frame; Scientific Assembly of the IAG, Symposium 1, Rio de Janeiro, Brasil

M. Hoyer (ed.), 1993: Memorias del Taller de Trabajo para la Formulación del Proyecto de Determinación del Geoide en Venezuela (Consejo Nacional de Cartografía); Junio 1993, Caracas

M. Hoyer, E. Wildermann, in press: Proposal of geoid calculations in Latin American countries; accepted for publication in Survey Review, Great Britain

E. Wildermann, 1988: Untersuchungen zur lokalen Schwerefeldbestimmung aus heterogenen Daten dargestellt am Beispiel der Geotraverse venezolanische Anden; Wissenschaftliche Arbeiten der Fachrichtung Vermessungswesen der Universität Hannover, No. 155; Hannover

E. Wildermann, 1996:Procesamientos de cálculos geoidales en áreas montañosas; Trabajo de Ascenso; Universidad del Zulia; Maracaibo

Improved Gravimetric Geoid AGG97 of Austria

Norbert Kühtreiber
Institute of Theoretical Geodesy, Section of Physical Geodesy
Technical University, Graz, Austria

Abstract

In 1995 at the IUGG-meeting in Boulder, the project of recomputation of the Austrian geoid under the title GEOID2000 was announced. The aim of this project is to get a geoid with an absolute accuracy of one centimeter. One of the studies within this project is dedicated to the computation of a pure gravimetric geoid and the accuracy which can be reached.

The present gravimetric geoid is computed by the remove-restore technique on the basis of the geopotential model EGM96 and a high resolution height model of Austria with a grid distance of 44 m $\times$ 49 m. Stokes' formula was evaluated by 2D FFT with the software package GRAVSOFT (Tscherning et al., 1991). An external check of the resulting geoid heights was done with the help of GPS reference points.

1 Introduction

Kühtreiber and Rautz (1996) presented a gravimetric geoid of Austria. This geoid computation was based on gravity points which were selected to get a coverage of approximately one point per 6 km $\times$ 6 km. The long wavelength part of the Earth's gravity field was modelled by the geopotential model OSU91A. The short wavelength topographic effect was computed on the basis of a 350 m $\times$ 390 m digital terrain model. The geoid solution was computed by using least squares collocation. The geoid was checked by GPS reference points with known orthometric heights. The geoid heights differed within a range of -11 cm to 19 cm with a standard deviation of $\pm$7 cm.

2 Source data

2.1 Gravity data

The present geoid computation is based on a data set of 28363 gravity anomalies. This data set includes irregular distributed point values and gridded mean values.

Within Austria a sample of 12743 point gravity anomalies out of all known gravity points was selected to get a coverage of approximately one point per 2 km × 2 km. Figure 1 shows the distribution of the gravity anomalies which is fairly good. Besides that, two things have to be mentioned. The white spots, areas with no measurements, will be filled for the final GEOID2000. The western part of Austria which is dominated by the eastern Alps shows a coverage less than one point per 2 km × 2 km. Measurements are still going on to densify the coverage.

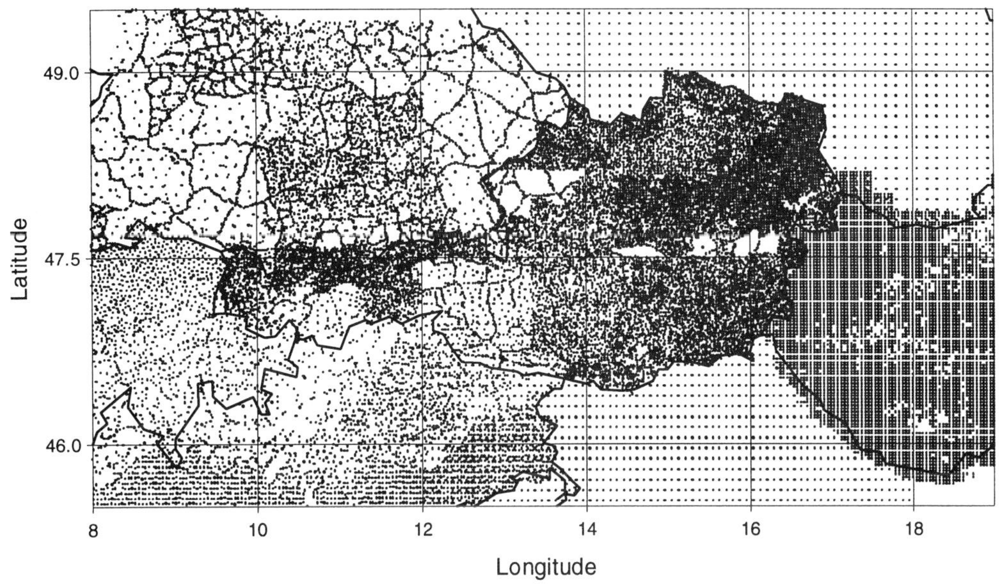

Fig. 1. Gravity anomalies used for the computation

Amount and kind of the data provided by the neighbouring countries of Austria differs. Germany, Switzerland and Italy provided point gravity values with changing density (see Figure 1). For Czech and Slovakia gridded mean values with a blocksize of $5' \times 7.5'$ were available. Hungary provided mean values with a blocksize of $1.5' \times 2.5'$.

Neglecting the different nature of point and mean gravity anomalies (mean values were handled as point values) free-air anomalies Δg_{FA} in a uniform reference system were derived.

2.2 High resolution height model of Austria

The Federal Office of Metrology and Surveying established a new height model for the project GEOID2000 based on grids with different resolutions determined photogrammetrically. The grid distance of these grids vary from 30 m × 30 m for rough topography to 160 m × 160 m in flat areas. From these base models, a digital terrain model was derived with the uniform resolution of $1''40625 \times 2''34375$, or approximately, 44 m × 49 m, see Graf (1996).

3 Gravity reduction

The geoid determination was done by the common remove-restore concept. Following the remove-restore procedure, first of all the free-air anomalies Δg_{FA} are reduced due to long wavelengths which are modelled by a geopotential model,

$$\Delta g_{FA-red} = \Delta g_{FA} - \Delta g_{EGM96}, \tag{1}$$

where Δg_{FA-red} are the reduced free-air anomalies and Δg_{EGM96} is the effect of the geopotential model EGM96 which approximates the long wavelength part. Within Austria the reduction using the geopotential model EGM96 gives a considerable improvement compared to the reduction using OSU91A (Kühtreiber, 1997).

In a second step the short wavelength topographic effect is subtracted by using the RTM-method (RTM is the abbreviation for residual terrain model):

$$\Delta g_{RTM} = \Delta g_{FA-red} + A_{RTM}, \tag{2}$$

where

$$A_{RTM} = -0.1119\,(H - H_{ref}) + A_t. \tag{3}$$

A_t is the classical terrain correction, H is the height of the measurement point. The RTM-reduced gravity anomalies Δg_{RTM} are referred to a height H_{ref} of a smoothed height model. The smoothing is obtained by applying a moving average operator with an integration area of $30' \times 30'$ to the detailed topographic model. The quality of the removing process is proofed through the statistics given in table 1. Whereas the standard deviation of Δg_{FA} is ± 41.3 mgal the standard deviation of Δg_{RTM} is only ± 11.1 mgal.

Table 1. Gravity reduction using the standard density of $2.67\,\mathrm{g/cm}^3$ and the geopotential model EGM96. Anomalies are given in mgal. Statistics based on 28363 points

	Min	Max	Mean	Std.dev.
Δg_{FA}	−239.5	200.2	3.8	±41.3
Δg_{FA-red}	−270.8	133.6	−12.0	±32.8
Δg_{RTM}	−117.5	102.1	−0.6	±11.1

4 Height anomalies and Geoid heights

Stokes' formula was evaluated by 2D FFT. Therefore a regular grid of 441×301 gravity anomalies for the area $45° \le \varphi \le 50°$ and $8° \le \lambda \le 19°$ had to be prepared. This grid with a spacing of $1' \times 1'5$ (approximately $2\,\mathrm{km} \times 2\,\mathrm{km}$) was computed by least squares prediction on the basis of Δg_{RTM}. For details concerning the prediction of gravity anomalies, see Kraiger (1987). All following considerations are refered to values defined in this grid.

In principle, the RTM-method provides height anomalies $\zeta_{RTM-red}$. In a final step, the long wavelength effect of the geopotential model N_{EGM96} and the indirect effect ζ_{RTM} are restored to $\zeta_{RTM-red}$ and we get

$$\zeta = \zeta_{RTM-red} + N_{EGM96} + \zeta_{RTM} . \tag{4}$$

For details on RTM-reduction refer to Mainville et al. (1995), Sideris and Forsberg (1991) and Forsberg (1984). In order to get geoid heights, the following approximate formula (Wang, 1993) was used:

$$N = \zeta + H \frac{\Delta g_B}{\gamma} \tag{5}$$

with Δg_B the Bouguer anomaly, γ an average normal gravity of 980000 mGal and H the elevation. The grid of Bouguer anomalies needed in eq. (5) was computed by least squares prediction on the basis of Bouguer anomalies computed for the data set of Section 2.1. The elevation H is derived from the high resolution height model of Austria.

Figure 2 shows the contour plot of the gravimetric geoid of Austria (AGG97).

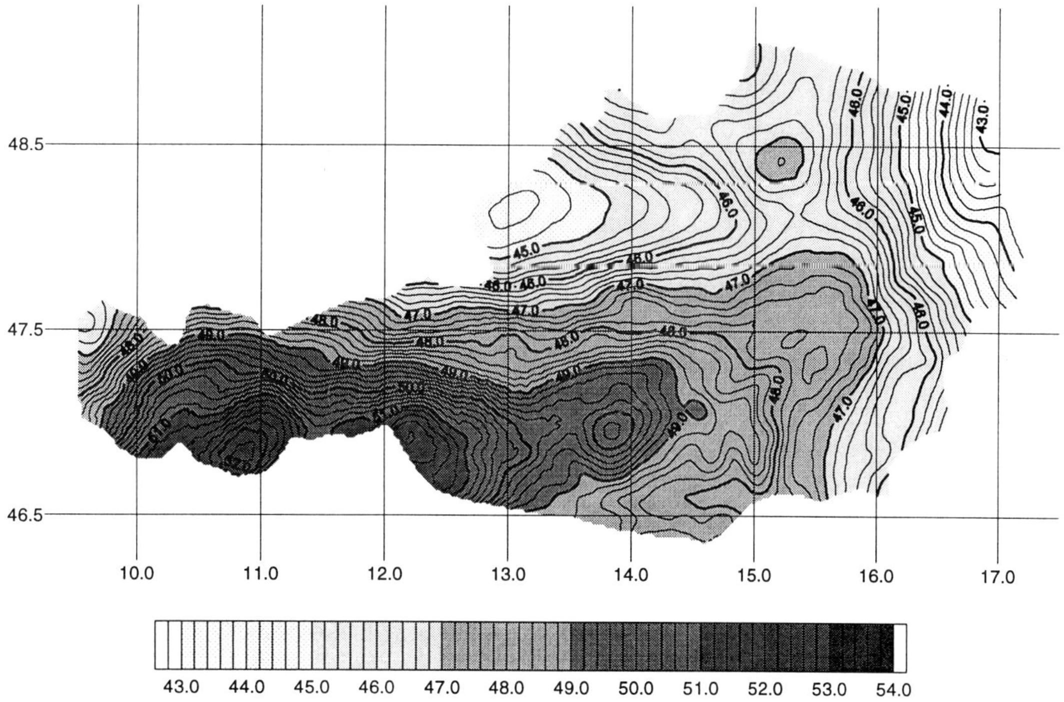

Fig. 2. Austrian Gravimetric Geoid - AGG97. Contour interval 20 cm

5 Accuracy

The quality of the gravimetric geoid was checked by 39 GPS reference points with known orthometric heights. Table 2 compares the accuracy of the gravimetric solution of 1997 N_{1997} to the gravimetric solution of 1996 N_{1996} (Kühtreiber and Rautz, 1996) and the astrogeodetic geoid of 1987 N_{1987} (Sünkel, 1987). The standard deviation of N_{1997} is ± 4.1 cm. The comparison of the two gravimetric solutions shows that the result improved more than 40% from N_{1996} to N_{1997}. Concerning the comparison of N_{1987} and N_{1997} an improvement of 20% can be recognized.

Table 2. Comparison of geoid heights at GPS reference stations. Values are given in cm. Statistics based on 39 points.

	Min	Max	Mean	Std.dev.
N_{1987}	−10.3	9.7	0.2	±5.1
N_{1996}	−11.0	19.0	3.0	±7.0
N_{1997}	−7.1	8.3	−2.6	±4.1

6 Conclusions

At the moment the gravimetric geoid 1997 of Austria is the most accurate geoid solution for Austria. Its accuracy is checked against 39 GPS reference points. The standard deviation of the differences between gravimetric geoid heights and GPS derived geoid heights is ± 4 cm. Further investigations will be done to increase the accuracy.

References

Forsberg, R., A study of terrain reductions, density anomalies and geophysical inversion methods in gravity field modelling, *Dept. of Geod. Sci. Rep., Rep. 355*, Ohio State University, 1984.

Graf, J., Das digitale Geländemodell für Geoidberechnungen und Schwerereduktionen in Österreich, Proceedings of the 7th International meeting on alpine gravimetry, *Österreichische Beiträge zu Meteorologie und Geophysik, Heft 14*, 121–136, 1996.

Kraiger, G., Untersuchungen zur Prädiktion nach kleinsten Quadraten mittels empirischer Kovarianzfunktionen unter besonderer Beachtung des Krümmungsparameters, *Mitteilungen der Geodätischen Institute der Technischen Universität Graz, Folge 53*, 1987.

Kühtreiber, N. and Rautz, K., Gravimetric geoid of Austria, Proceedings of the 7th International meeting on alpine gravimetry, *Österreichische Beiträge zu Meteorologie und Geophysik, Heft 14*, 103–112, 1996.

Kühtreiber, N., Precise geoid determination using a density variation model, accepted for publication in *Physics and Chemistry of the Earth*, 1997.

Mainville, A., Véronneau, M., Forsberg, R., Sideris, M.G., A comparison of geoid and quasigeoid modelling methods in rough topography, *Gravity and Geoid, IAG Symposium, No 113*, 491–500, 1995.

Sideris, M.G., Forsberg, R., Review of geoid prediction methods in mountainous regions, *Proceedings of the IAG symposia on the determination of the geoid - present and future*, 1991.

Sünkel, H., The Gravity Field in Austria, Geodätische Arbeiten Österreichs für die Internationale Erdmessung, Neue Folge, Band IV, 1987.

Tscherning, C.C., Knudsen, P., Forsberg, R., Description of GRAVSOFT package. *Geophysical Institute, University Copenhagen, Technical report*, 1991, 2. Ed. 1992, 3. Ed. 1993, 4. Ed., 1994.

Wang, Y.M., Comments on proper use of the terrain correction for the computation of height anomalies, *Manuscripta Geodaetica 18*, 53–57, 1993.

METROLOGY OF THE EARTH'S ROTATION

OVER 1988-1997

Daniel Gambis
IERS/CB, Paris Observatory
61 avenue de l'Observatoire, Paris, France

ABSTRACT

In this paper, we summarize the evolution concerning the Earth Orientation metrology over the first ten years of the IERS activity. This evolution is primarily highly linked to the techniques enhancement; secondly data analyses have improved, models of geophysical phenomena perturbing the earth orientation have been refined and smaller effects so far neglected are now taken into account. Precisions and temporal resolutions have increased in the quantities estimated. Finally electronic systems have developped more efficient means of data transmissions.

EVOLUTION OF THE EARTH ORIENTATION MONITORING OVER THE LAST DECADE

The main task of IERS is the maintenance of both a conventional terrestrial reference system based on observing stations, a conventional celestial reference system based on extragalactic radio sources and also the matrix product allowing the transformation between these two systems which takes into account precession, nutation, polar motion and Universal Time. The various improvements concerning the Earth Orientation metrology can be attributed to different causes; at first more interest (and money) have been put into the developpement of techniques in the frame of national and international projects. To date, various techniques allow the determination of all or a part of the Earth Orientation Parameters: Laser Ranging to the Moon (LLR) and to dedicated artificial satellites (SLR), Very Large Baseline Interferometry on extra-galactic sources (VLBI) and more recently GPS and DORIS introduced respectively in 1992 and 1995 in the IERS activities. The capabilities of these new techniques are permanently evolving, in particular the precision and the temporal resolution have significantly improved in 10 years. SLR precision has evolved from a few centimeters to a sub-centrimetric precision for a number of stations. VLBI technique has also evolved; for instance UT1 determination had a precision of about 100 μs in the mid 1980's (Robertson et al., 1985), the present precision is in the range of 20 μs. The full CORE program (Continuous Observation of the Earth) initiated by the VLBI community has a potential precision of 1 μs for daily a value of UT1 (GSFC report, 1997). New techniques have also recently appeared and developped: IERS has integrated GPS results in 1992. the polar motion combination series is to date mainly based on this technique which allows a daily estimation.

In parallel, the number of analysis centers implied in the Earth Orientation monitoring activity has progressed (table 1).

Table 1 - Evolution of the number of contributing analysis centers over 10 years.

	Operational	Annual Global Analysis
1988	2 VLBI 1 LLR 2 SLR	5 VLBI 4 LLR 5 SLR
total	5	14
1997	3 VLBI 1 LLR 5 SLR 7 GPS	7 VLBI 4 LLR 10 SLR 7 GPS 1 DORIS
total	16	29

Combination. The different Earth Orientation Parameters (EOP) series obtained by the individual techniques are not homogeneous in time length, quality, time resolution, which supports the concept of combined solutions benefitting from the various contributions (Gambis, 1996, 1997). In particular in the EOP combined series derived by the IERS/BC, the accuracies have evolved from 1 mas to about 0.2 mas for pole components, 0.05 to 0.02 ms for Universal Time and 0.5 to 0.2 mas for nutation offsets thanks to the improvement of the various techniques and also to the evolution in the analysis procedures. The temporal resolution has also increased over the decade from 3 days for 1 day for polar motion owing to GPS contribution. The restitution delay of the operational solution has also evolved and the discrepancy between the "frozen" Bulletin B derived every month and the operational C04 solution which is permanently updated has significantly decreased (figure 1) for all parameters. The agreements are over 1997 in the range of 0.1 mas for pole and 80 µs for Universal Time.

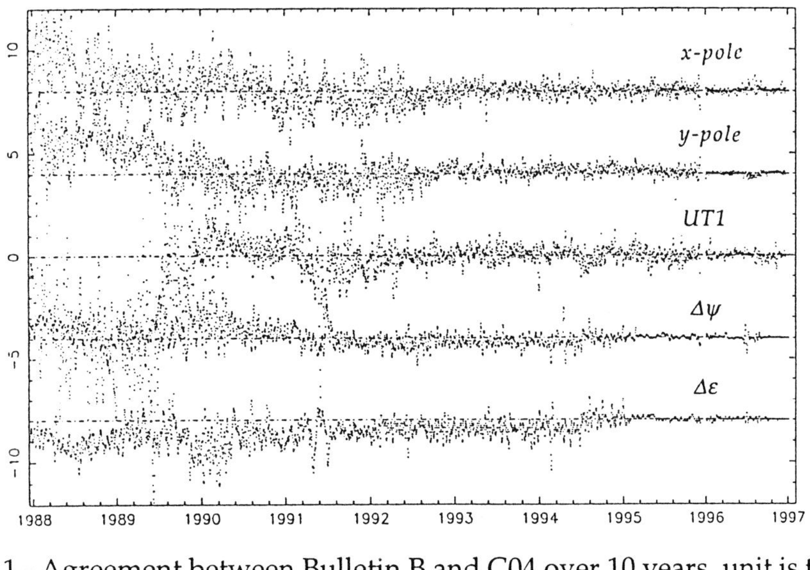

Figure 1 - Agreement between Bulletin B and C04 over 10 years, unit is the mas for polar motion and nutation offsets and 0.1 ms for UT1.

Another way of estimating the precision of EOP series is an external check; the comparison to another current solution like NEOS. The agreement has also improved, i.e 0.2 mas for pole components and 20 μs for UT1.

Table 2 - Agreement between the combined series NEOS and C04 in 1990 and 1997. Unit is the mas for pole components and 1 μs for Universal Time.

NEOS <=> IERS C04		
	1990	1997
x	0.60	17
y	0.40	17
UT1	0.40	22

PRESENT ACTIVITY, SOME ASPECTS

In this section we will select two main aspects on the activities of the Central Bureau which could seem in opposition but are nevertheless highly connected; the long-term maintenance of the Earth Orientation System realized through the EOP solutions and the use of GPS LOD estimates for the operational quasi-real determination of Universal Time.

1) Maintenance of the IERS EOP reference system; recomputation of the long-term IERS/BC EOP solutions

One task of the Central Bureau is to control and maintain the reference system defined by the EOP series in order to ensure the consistency between the terrestrial and celestial reference frames. In this objective, taking advantage of the re-analyses of long Earth Orientation Parameters series (EOP) by various centers, the IERS/CB combined solutions have been recomputed in July 1997. In the course of the analyses the determination of the systematic corrections of individual series entering the combination and the weighting procedure have been re-examined. Within each independant technique (SLR, VLBI and GPS), partial solutions are now previously made (Eisop and Gambis, 1997). This previous inter-technique combination are also now operationally made. This allows a better determination of the systematic errors. Figure 2 shows the difference of the 3 inter-technique solutions for x-pole coordines with respect to the combined C04 solution estimation of the RMS are given on table 3. In addition, smoothing characteristics have been modified to take into account the respective improvement of the individual series. This series EOP(IERS)97 C 04 is given in the 1997 system. It is also permanently updated on a near real-time basis. Others series, normal points solutions given at various time intervals are also proposed to users, i.e. EOP(IERS) 97 C01, C02, C03. These series are respectively consistent one to another.

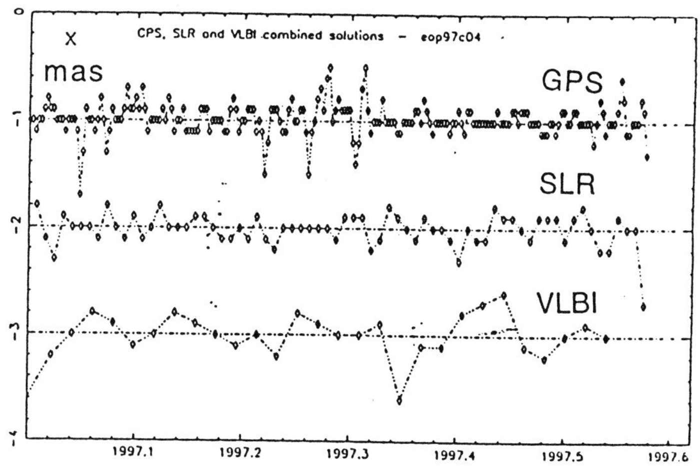

Figure - 2 Difference of the 3 inter-technique solutions for x-pole coordinates (mas) with respect to the combined C04 solution, biases are arbitrary.

Table 3- RMS agreement with EOP (IERS) C04

	x pole (mas)	y pole (mas)
EOP (IERS) 97 P 01	0.11	0.11
EOP (IERS) 97 L 01	0.11	0.12
EOP (IERS) 97 R 01	0.16	0.15

2) Use of UT1 GPS estimates for near real time applications

Due to the difficulty of determining the long-term behaviour of the non rotating system realized through the orbit orientation, Universal Time UT1 cannot be accurately derived from satellite techniques but only from inertial methods like VLBI. On the other hand, techniques like SLR and GPS can determine the length of day variations (LOD), derivative of Universal Time together with the orbital parameters; of course their spectrum show similar systematic errors than those of Universal Time directly estimated.

The various determinations which are made by the analysis of satellite data follow differents strategies: CSR is deriving UT estimates using the satellite ascending node residuals, some of the GPS analysis centers integrate their estimates of LOD to derive a "free-running" Universal Time series, some are constraining their determination using a-priori VLBI values in order to keep the consistency with the non-rotating inertial reference frame. Various studies (Gambis et al , 1993; Ray, 1997; Gambis, 1996) have shown that the high-frequency signal contained in the LOD estimates on time scales limited to a couple of months derived fron SLR and GPS can be used to densify the series obtained by the VLBI technique and also for near-real time earth orientation monitoring. For clarification it was felt (Ray, 1997) that the acronym UT1 should be reserved to Universal universal time derived from inertiel techniques (astrometry, VLBI). We shall adopt in the following the acronym UT for series partially constructed from various techniques.

Since december 1995, the Central Bureau of IERS is operationally publishing a mixed Universal Time solution based on a combined short-term GPS UT solution calibrated by the long-term VLBI UT1 series. The stategy has now evolved: since spring 1997 a combined GPS LOD solution is calculated using the 7 GPS Analysis Centers estimates and integrated to give an "internal free-running" UT solution which is finally calibrated by VLBI and labelled (IERS) 97 P 01. The availabilty of rapid LOD estimates derived from CODE and EMR allows to extrapolate Universal Time from the last VLBI value on a quasi-real time. This procedure takes into account a model to correct long-term errors in the GPS UT series (Gambis, 1996). This series after long-term corrections removed, is piped to the last available VLBI estimate or C04 solution. This extrapolation of UT is now calculated on a regular basis and enters our current analyses since january 1997. Figure 3 shows the performances of this strategy obtained in operational mode compared to the results obtained by the classical IERS/BC predictions based on an autoregressive process. Table 4 gives the statistical numbers derived from this procedure over 1997. The improvement is significant, about 1 order of magnitude for 1 to 3 weeks. The availability of these near-real time estimates also enable a better UT1 prediction.

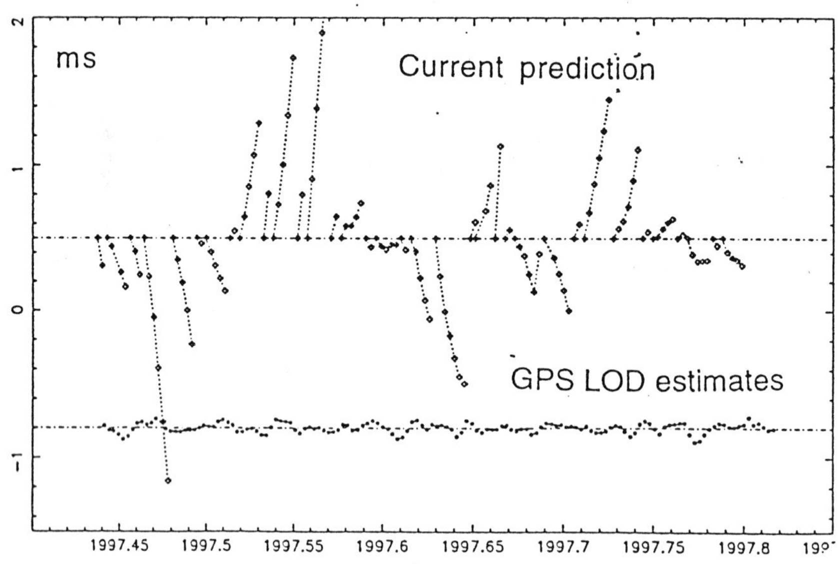

Figure 3 - Performances of GPS LOD estimates used for Universal Time extrapolation compared those obtained from classical IERS/BC predictions based on an autoregressive process.

Table 4.- RMS errors (in microseconde) of the Universal Time solution based on GPS and compared to the current prediction based on an auto-regressive process.

UNIT : 1 μs	1 week	2 weeks	3 weeks
Pure Prediction	1150	4000	7000
GPS estimates	150	250	400

CONCLUSIONS

The activity of Earth Orientation monitoring has dramatically increased over the last decade thanks to primarily the techniques improvement, analysis procedures and through the developpement of the electronic communications. These improvements make it necessary to more accurately maintain the Earth Orientation system to ensure the best possible consistency between the terrestrial and celestial reference frames. When possible, taking advantage of the re-computation of individual solutions, the combined IERS/BC solutions are recomputed. The operational users community have also requirements concerning the quality and the rapid availability of the EOP solutions. This and the "friendly" competition with other agencies like NEOS or the IGS, has lead us to improve our operational products using all available data. The illustration of this is the use of GPS LOD for Universal Time estimation on a quasi-real time basis.

The evolution in the respective contribution of the observing techniques and the increasing temporal resolution and accuracies obtained by the techniques require to maintain and constantly improve the combination procedures. New approaches using non-linear methods are investigated for modelization, filtering and prediction of the solutions (Frede and Mazzega, 1997). A global optimal combination taking into account the realisations of both celestial and terrestrial reference systems together with the Earth Orientation system will be in a near future a great challenge for IERS.

Acknowledgements. The author thanks E. Eisop from IERS/BC for his valuable work of collecting, analysing the Earth Orientation Parameters and realizing the various plots.

References

Eisop E. and D. Gambis, 1997, the combined solutions of the Central Bureau of IERS, Journées systèmes de référence, Prague, september 1997.

Frede V. and P. Mazzega, 1997, detectability of deterministic non-linear processes in Earth Rotation parameters time series, in prep.

Gambis D., Essaifi N., Eisop E. and M. Feissel, 1993, Universal time derived from VLBI, SLR and GPS, IERS technical note 16, Dickey and Feissel (eds), 15-20.

Gambis D. , 1996, Multi-technique EOP combinations, proceedings of the 1996 IGS Analysis Center workshop, Silver spring.MD, edited by P. Van Scoy and R.E. Neilan, Pasadena, CA, JPL, JPL Publication 96-23.

Gambis D., 1997, Monitoring Earth Rotation using various techniques, current results and future prospects, I.M. Wytrzyszczak, J.H. Lieske and R.A. Feldman (eds), Dynamics and astronomy of Natural and Artificial Celestial Bodies, 475.

GSFC report, 1997, CORE/IVN Geodetic VLBI for the new millenium.

Gray, J.E. and Allan, D.W. 1974: A method for estimating the frequency stability of an individual oscillator, Proc 8th Ann. Symp. on Frequency Control}, 2439, 277--287.

Ray J.R., 1996, Measurements of length of day using the Global Pösitioning system, JGR, 101, 20141.

Ray J., 1997, Proposal for the nomenclature and use of UT1-like measurements derived from satellite observations, personal communication.

Robertson D.S, W.E. Carter W, J.A. Campbell and H. Schuh, 1985, Nature, 316, 424-427,

Earth Orientation Parameter Analysis from VLBI Data

V. Thorandt[1), A. Reinhold[1), D. Ullrich[1), J. Ihde[1), J. Campbell[2), A. Nothnagel[2)
[1)Bundesamt für Kartographie und Geodäsie*, Außenstelle Leipzig;
Karl-Rothe-Straße 10-14, D-04105 Leipzig, Germany
[2)Geodätisches Institut der Universität Bonn (GIUB)
Nußallee 17, D-53115 Bonn

Figure 1: Fundamentalstation Wettzell, one of the three German VLBI-telescopes involved into the solution

General

Since more than twenty years the Bundesamt für Kartographie und Geodäsie (BKG)*, an institution for geodesy and cartography of the Federal Government, contributes successfully to space geodesy and to reference frames based on these techniques.

Closely cooperating nations and economies need highly accurate maps and geodetic information on a global scale. The methods and techniques of space geodesy as well as satellite borne navigation systems have developed solutions for global geodetic reference frames with even cm-precision for distances or coordinates in intercontinental and global dimensions.

---

* formerly Institut für Angewandte Geodäsie (IfAG)

The necessary precision and the non-rigid state and behaviour of the earth demand continuous observations of the space geodetic targets, the continuous processing of the data, keeping the observational tools up-to-date, developing new techniques and models. In a worldwide cooperation the International Earth Rotation Service at Paris (IERS) takes care of proper models and the standards on which the processing of space geodetic observations is based, IERS is further put in responsibility for the celestial and terrestrial reference systems and the transformation parameters between these systems, and last not least IERS is authorized to publish regularly the conventional coordinate frames, i.e. IERS Celestial Reference Frame (ICRF) and IERS Terrestrial Reference Frame (ITRS), as well as the connection between both, the Earth Orientation Parameters (EOP, this means the coordinates of the pole, the offsets of the earth rotation and the observed corrections to the conventional model of nutation).

The realization of the ITRS depends on the contribution of continuous observational space geodetic data over long timescales and the necessary equipment and observatories for receiving these data (IERS, 1995). Such contributions are performed by national scientific agencies or merely scientific institutions, and in this was BKG is playing its part since the emerging of the modern methods of space geodesy. In this frame the development of the Geodetic Fundamental Station at Wettzell and the Antarctic Station O'Higgins and the intensive observational activities have to be seen.

Analysis solution GIUB 97 R01

In close cooperation the VLBI group at the University of Bonn and the VLBI group at the Bundesamt für Kartographie und Geodäsie have analyzed 1052 sessions (January 1984 December 1996) of worldwide geodetic VLBI observations for earth orientation parameters (EOP), source positions, station coordinates and displacement vectors. Only data of fixed radio telescopes and sessions combining at least three stations were included in the analysis. The data were analysed with the CALC/SOLVE/GLOBL software system (Caprette et al., 1990) to the solution GIUB 97 R01.

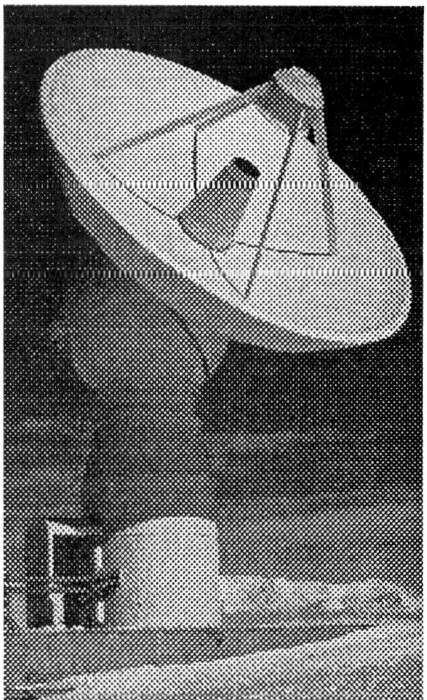

Figure 2: Antenna of the Geodetic Observatory O'Higgins, Antarctic Peninsula, managed by BKG

The orientation of the celestial reference frame was defined by using the ICRF95 Right Ascension Value

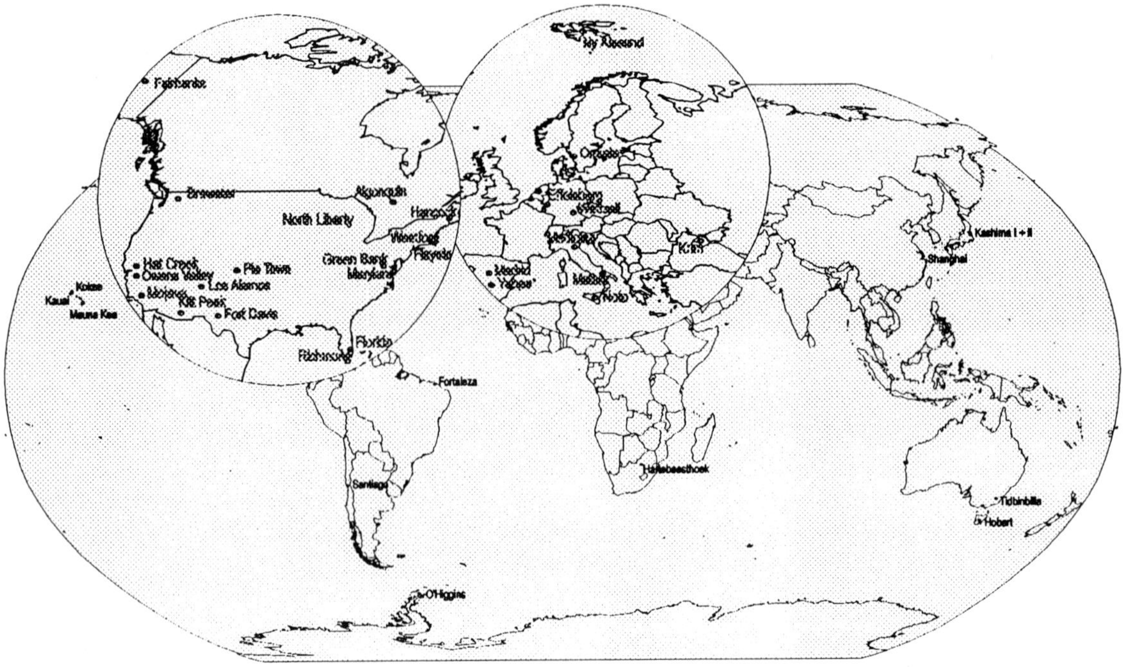

Figure 3: World-wide distributed VLBI/VLBA stations, included into the solution

of OJ287 (IERS, 1996) and fixing the nutation angels to the IAU model values on April 2, 1988. The initial orientation and translation of the terrestrial reference frame was defined by fixing the ITRF94 coordinates of Wettzell and the IERS Rapid Service EIP values on April 2, 1988. To avoid singularity of the displacement vector field another six parameters were modelled, i.e. the high changes of HartRAO were fixed to be zero and the motion of the Wettzell station, furthermore the orientation of the Wettzell — Westford baseline vector was modelled according to the NNR-NUVEL-1A model (De Mets et al., 1994). The displacements of all other stations were estimated either prerequisite, with loose constraints or with the restriction of no velocity components over time depending on their data history. The one sigma formal errors of the parameters are based on observation weights which are adjusted so that the Chi-square per degree of freedom ratio per baseline is close to unity. The hydrostatic component of atmospheric refraction was calibrated using the MTT model (Herring, 1992) based on surface meteorological data while the wet component was estimated with a 3-hour piecewise-linear function where surface meteorological data is available. Only if station logs indicate a rapidly changing climatic environment or meteorological data is not available shorter intervals were chosen. The station clocks were generally modelled with second order polynominals. Only where the residuals showed an abnormal behaviour a small number of additional piecewise linear clock parameters was introduced keeping the number of parameters as small as justifiable by the distribution of post fit residuals. the apriori EOP values for each observation were calculated by linearly interpolating the USNO Rapid Service Series. The reference epochs for UT1-UTC and polar motion are the midnight epochs of each measurement period while the nutation offsets are estimated at the beginning of each experiment.

Technical description summary of solution GIUB 97 R 01

1.	Software used:	CALC 8.1/SOLVE
2.	Data span	January 84 — December 96
3.	Celestial Reference Frame	RSC (GIUB) 96 R 01
	a) Nature:	extragalactic
	b) Definition of the orientation:	Fixing RA of OJ297 [ICRF (1995) WGRF] and IAU nutation for 1988 Apr 02
4.	Terrestrial Reference Frame	SSC (GIUB) 96 R 01 (Wettzell ITRF94 coordinates plus EOP fixed for 88APR02)
	a) Relativity scale:	Solar System Bary Center (SSB)
	b) Velocity of light:	299792458 m/s
	c) Gravitational constant:	not applicable
	d) Definition of the origin:	by fixing WETTZELL ITRF1994 coordinates
	e) Definition of the orientation:	by fixing USNO concrete EOP series for 1988 Apr 02
	f) Reference epoch:	1993.0
	g) Tectonic plate model:	NNR-NUVEL1A
	h) Constraint for time evolution:	The reference station WETTZELL and the direction of the baseline WESTFORD to WETTZELL move according to NNR-NUVEL1A, the height evolution of the station HARTRAO is constrained to be zero
5.	Earth orientation:	EOP (GIUB) 96 R 91
	a) A priori precession model:	IAU (1976)
	b) A priori nutation model:	IAU (1980)
	c) Short period tidal variations in x, y, UT1	None

Summary of results of the solution GIUB 97 R01

Some of the final results (Polar Motion, UT1-UTC and Nutation) of the adjustment of the solution GIUB 97 R01 are given in Figure 4 to 7. The calculated values of the solution for 43 sites, 1052 sessions and 293 sources are:

1. Station coordinates and velocities (X0, Y0, Z0, Xdot, Ydot, Zdot)
2. Right ascension and declination of radio sources
3. Earth orientation parameters (X, Y, UT1-UTC, $d\psi$, $d\varepsilon$), cf. Figures 1 to 4
4. Others: clock polynomials, wet zenith tropospheric delays
5. Reference epochs for UT1-UTC and polarmotion; midnight epochs of each meas
6. Nutation offsets were estimated at the beginning of each experiment

The results of the GIUB 97 R01 solution were sent to the Central Bureau of IERS, Paris. They will be a part of the Annual Report global analysis by the IERS Analysis Centres for 1996.

This analysis and its results are a first step towards a regular analysis of worldwide VLBI data by BKG. During the last years the amount of VLBI data gathered in Leipzig have increased steadily and the remaining gaps will be filled in near future.

Together with BKG participation in the IRIS series and the operation of the radio telescopes in Wettzell, Germany, and O'Higgins on the Antarctic Peninsula the regular VLBI data analysis will serve as part of the German contribution to the worldwide activities in geodetic VLBI for the determination of Earth rotation parameters.

References

DeMets, C., R. G. Gordon, D. F. Argus, S. Stein: Effect of recent revisions to the geomagnetic reversal time scale on estimates of current plate motions. Geophys. Res. Letters., 21 (1994) 20, p2191-2194, 1994

Caprette, D. S., C. Ma, J. W. Ryan: NASA Technical Memorandum 100765, A-1, NASA Goddard Space Flight Center, Greenbelt, MD, 1990

IERS1995 Annual Report, Observatoire de Paris, 1996

Herring, T. A.: Modelling Atmospheric Delays in the Analysis of the Geodetic Data. Symposium on Refraction of Transatmospheric Signals in Geodesy; Delft, May 19-22, 1992. - pp. 157-164 (Publications on Geodesy: New Series/Netherlands Geod. Comm. Delft; 36)

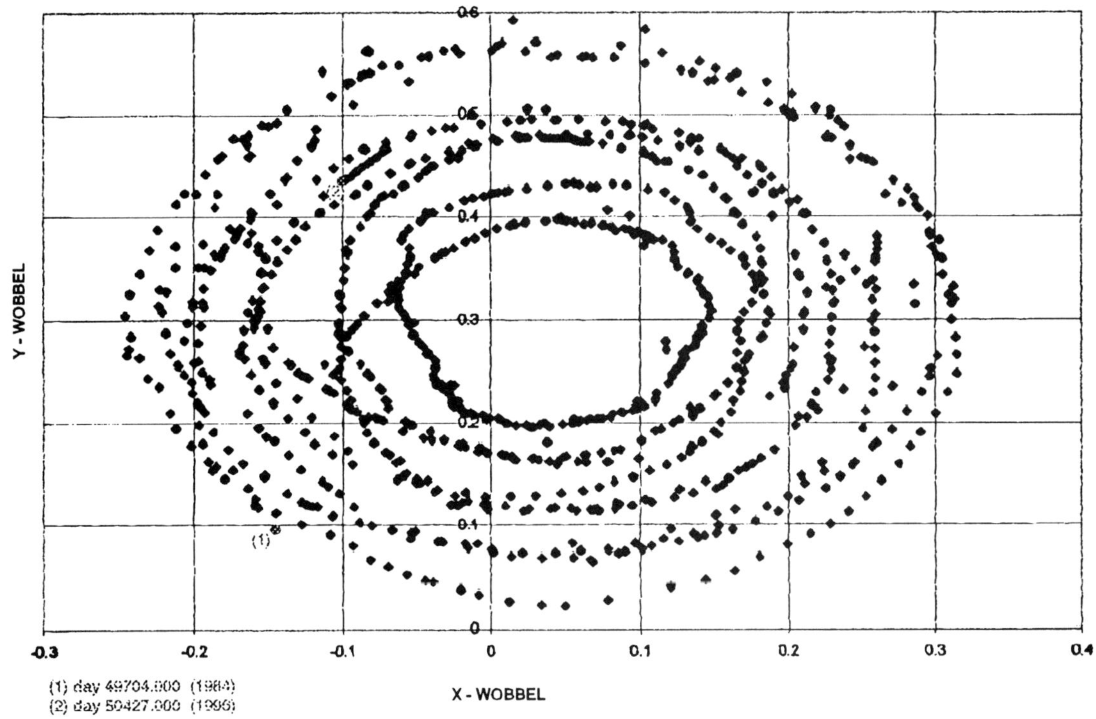

Figure 4: Polar Motion

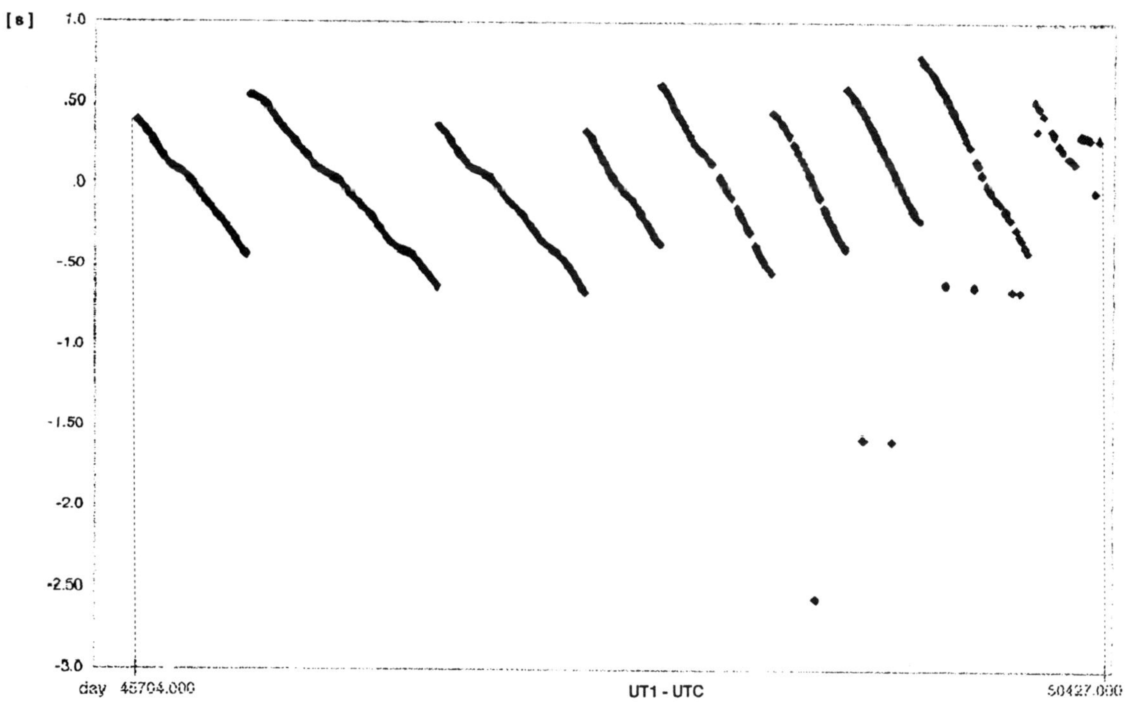

Figure 5: UT1 - UTC

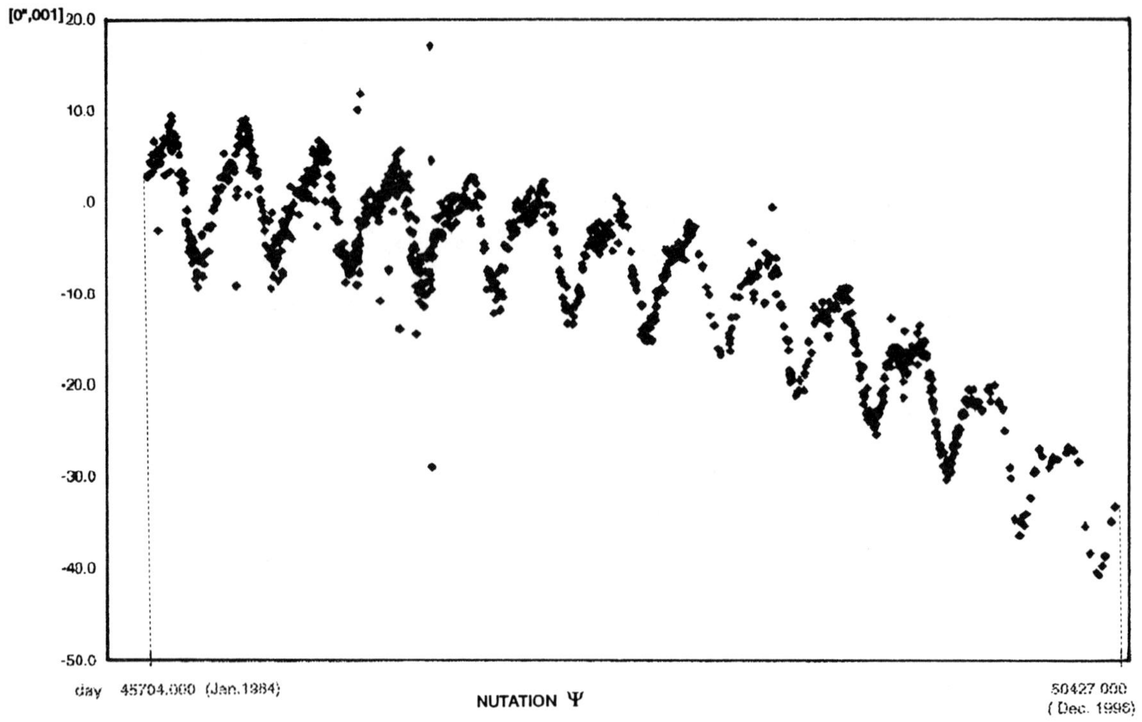

Figure 6: Nutation dΨ

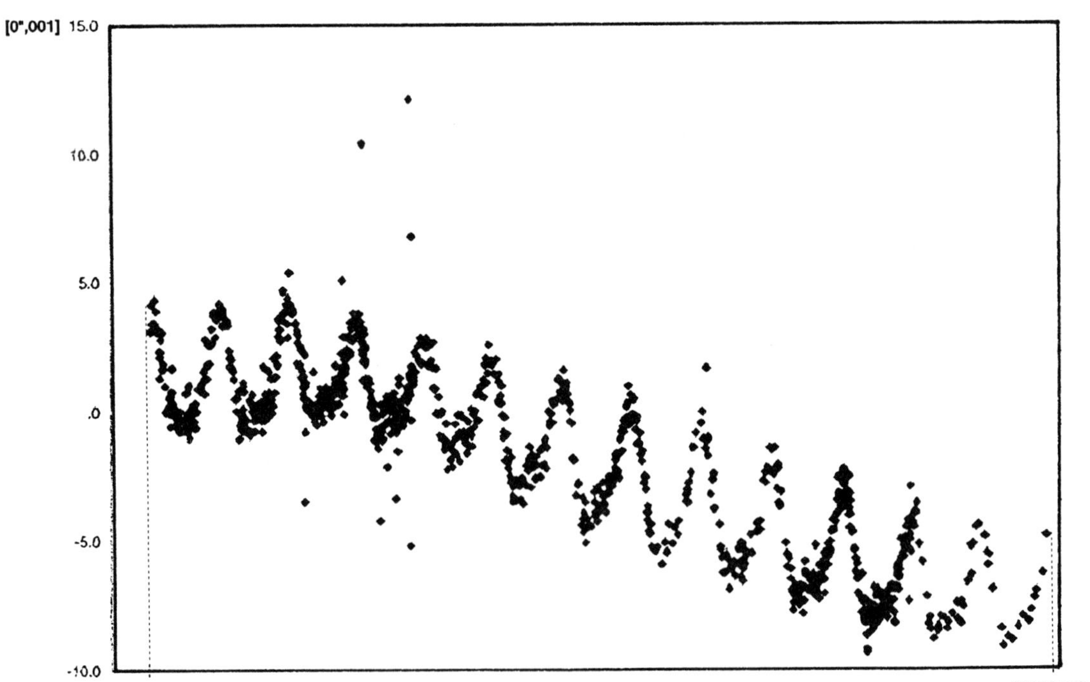

Figure 7: Nutation dε

TEMPO: DETERMINING, COMBINING, AND PREDICTING EARTH ORIENTATION PARAMETERS FOR SPACECRAFT NAVIGATION

R. S. Gross, D. H. Boggs, J. O. Dickey, S. H. Oliveau, T. F. Runge, and J. A. Steppe

Jet Propulsion Laboratory, California Institute of Technology,
4800 Oak Grove Drive, Pasadena, CA 91109, USA

In support of spacecraft tracking and navigation at the Jet Propulsion Laboratory (JPL), Earth orientation measurements are currently acquired, processed, and delivered twice-per-week to the JPL navigation teams by the Time and Earth Motion Precision Observations (TEMPO) project. Because the Earth's orientation changes rapidly and unpredictably, measurements must be acquired frequently and processed rapidly in order to meet the near-real-time Earth orientation calibration requirements of the navigation teams. These requirements are currently met by (1) conducting twice-per-week single baseline VLBI observing sessions; (2) rapidly processing the VLBI data to determine the baseline variation-of-latitude and UT0 components of the Earth's orientation; and (3) using the Kalman Earth Orientation Filter (KEOF) to combine the TEMPO VLBI measurements with other publicly available Earth orientation measurements, including determinations and forecasts of the axial component of the atmospheric angular momentum (AAM), in order to generate and deliver the Earth orientation calibrations (polar motion and UT1) required by the navigation teams.

In the near future, the Global Positioning System (GPS) will be used to provide daily determinations of polar motion and length-of-day within 24 hours of acquisition. TEMPO VLBI observing sessions will still be conducted, although less frequently and with the data processed less rapidly, in order to provide the benchmark universal time measurements between which the GPS length-of-day measurements will be integrated. The KEOF will be used to combine the GPS polar motion and length-of-day measurements with the TEMPO VLBI variation-of-latitude and UT0 measurements, along with other publicly available Earth orientation and AAM determinations, in order to generate and deliver the required polar motion and UT1 calibrations to the JPL navigation teams.

POSSIBLE TEMPORAL VARIATIONS OF THE FREE CORE NUTATION AND FORCED NUTATIONS

LI JINLING ZHENG DAWEI
Shanghai Observatory, Shanghai 200030, P. R. China
E-mail: jll@center.shao.ac.cn

ABSTRACT. The IAU 1980 nutation daily offsets in longitude ($\delta\psi$) and obliquity ($\delta\epsilon$) from VLBI observations are analyzed by wavelet and multistage filter methods. Results suggest that the temporal variations of the free core nutation(FCN) is more intense than those of the annual nutation. The FCN's magnitude is larger before 1990 than that afterwards and there exists phase variation around the middle of 1993. If such "frequent" variations are of geophysical origin, the intensity of strong electromagnetic or viscus coupling between the core and the mantle would be limited. Results from wavelet analyses indicate that near 500 days period there exists no counterpart of the prograde free core nutation (PFCN) associated with the motion of the solid inner core, but there exist discernible signals in quasi-biennial and quasi-quadrennial frequencies. Temporal variations of the forced nutation terms with biggest amplitudes can also be distinguishable.

1. Introduction

There are excellent theoretical reasons to believe that the existence of the earth's liquid core, besides shortening the period of the Chandler wobble, gives rise to a second free wobble mode, the earth's nearly diurnal free wobble. Its period is shorter than a sidereal day by a factor of the order of the ellipticity of the core-mantle boundary. Seen from space this nearly diurnal geographic motion of the pole appears as a retrograde nutation of the earth's rotation axis, the retrograde free core nutation (FCN), with a theoretical period about 460 days (Rochester et al. 1974; Sasao & Wahr 1981).

The detection of the FCN is among the most intriguing phenomena in geophysics, because the possibility of observing it holds the promise of yielding crucial information on the physical properties of the core, the nature of the core-mantle interaction, and the shape of the core-mantle boundary. Too, it is essential to the improving of the precision of the earth nutation theory (Mathews et al. 1991a,b).

Based on a few years Very Long Baseline Interferometry (VLBI) observations, Herring et al. (1986) revealed the errors at the milliarcsecond (mas) level in the IAU 1980 nutation theory (Seidelmann 1982), and set a bound on the FCN's magnitude, which is less than $1.0mas$ with 99.1% confidence interval. Since after, the discrepancies between the IAU 1980 nutation theory and observations have been studied to determine empirical corrections to individual terms of the nutation theory (Herring et al. 1991; McCarthy & Luzum 1991; Williams et al. 1991; Charlot et al. 1995; Souchay et al 1995; Walter & Sovers 1996; Souchay et al. 1996), and geophysical explanations of the discrepancies have been considered (Gwinn et al. 1986; Zhu & Groten 1989; Kinoshita & Souchay 1990; Mathews et al 1991a,b; Souchay & Kinoshita 1996).

The period of FCN is dependent on the core flattening and the amplitude seems to be fluctuating with time (Gwinn et al. 1986). The annual nutation is expected to receive excitations from the atmosphere and the oceans, and so its amplitude may also vary from year to year, in particular in the out-of-phase component (Souchay et al. 1995). Since the FCN and the annual nutation is very close in frequency, from VLBI observations it is easy to get an overall impression on the temporal variations of their composite, but it is difficult to evaluate the individual role to the variations of this composite.

In this paper, observations of the IAU 1980 nutation daily offsets in longitude ($\delta\psi$) and obliquity ($\delta\epsilon$) (in EOP(GSFC) 97 R 01, provided by Dr. Ma Chopo) from the CALC 8.1 solution of Mark III VLBI delay data from the beginning of 1983 to the end of 1996 are analyzed in order to determine the characteristics of the possible temporal variations of the FCN and the annual nutation. We will first introduce the MultiStage Filter (MSF), which was originally suggested by Zheng & Dong (1986). Then, with the application of the MSF to pole daily offsets, we show and discuss the results about the possible temporal variations of the FCN, the annual nutation, and other forced nutational terms with biggest magnitudes.

2. The temporal variations of the FCN

According to digital filter theory, the multistage filter is designed as follows.

$$W = c(I - V^n)^m$$

where c is real constant, n and m are positive integers, I is unity metrics and V is Vondrak filter (Vondrak 1977). The frequency response function for the multistage filter is characterized by small bandwidth of truncated frequency, so it can be treated as narrow band filtering. However, the multistage filter is of more serious edge effect than Vondrak filter. This effect can be largely reduced by extrapolating the data series with Leap-Step Auto-Regressive (LSAR) model (Dong & Zheng 1985).

In order to simplify our analyses, band filtering with period interval between 200 days and 1000 days is realized to the pole offsets. The resultant data is denoted by S_1. Results from wavelet analyses and least squares (LS) adjustment are shown in Fig.1, the left part is about the offsets in longitude and the right is in obliquity.

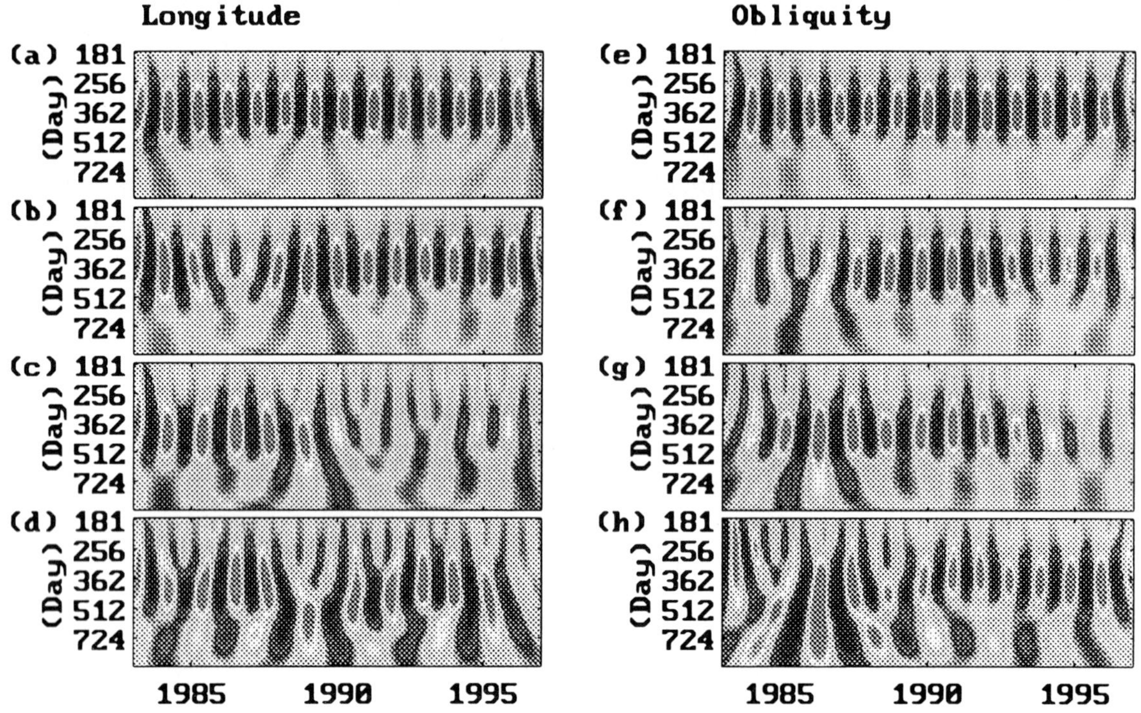

Figure 1. Results from MSF and wavelet analyses of pole offsets. The temporal variations of FCN are more intense than those of the annual nutation. There exists signal in quasi-biennial frequency.

Fig.1(a) and (e) are wavelet analyses of S_1. There are 14 peak-valley (PV) pairs and the data span is 14 years, so the demonstrated signal corresponds to the annual nutation. Signals in nearby frequencies are indistinguishable because of the contrast of intensity.

Following the LS solutions to the in-phase component of the annual nutation are removed from S_1, the resultant data series is denoted by S_2. Wavelet analyses of S_2 are shown in Fig.1(b) and (f), in which there exist still 14 PV pairs and so the out-of-phase component of the annual term becomes the strongest.

In the sense of a best fit to data, LS solutions to the out-of-phase component of annual term are removed from S_2, and wavelet analyses of the new data series (denoted by S_3) are shown by Fig.1(c) and (g). The FCN turns to be the dominator because there are 12 PV pair, corresponding to an average period 425 days. It is clear that the magnitude of FCN is larger before 1990 than afterwards.

After the removal of the LS solutions to FCN from S_3, wavelet analyses are demonstrated by Fig.1(d) and (h). The appearance of Fig.1(d) and (h) becomes complicated. Around 400 days, the intensities of peaks and valleys as well as the corresponding periods are all changing with time, but there still about 12 PV pairs. Therefore, the temporal variations of the FCN is more intense than those of the annual term. By comparing the locations of peaks and valleys in Fig.1(d) with (c) and those in Fig.1(h) with (g), from the middle of 1993 the peaks in Fig1.(d) and (h) correspond respectively to the valleys in Fig.1(c) and (g), which means there is phase variation in FCN at the middle of 1993. If the temporal variations in magnitude and phase of the FCN are of geophysical origin (rather than due to noises, or deficiencies associated with the data processing), the coupling between the core and the mantle would not be too strong, otherwise such "frequent" variations would be restricted. Around 700 days, there appears a discernible quasi-biennial signal, which will be discussed in the following section.

3. Possible temporal variations of forced nutations

The earth's nutational motions can be theoretically predicted from motions of the sun and moon, depending on properties of the earth. Since the relative motions and masses of the earth, moon, and sun are known to high accuracy, the properties of the earth, which determine the eigenfrequencies and eigenfunctions of its normal modes, make their imprint on the nutational responses to the lunisolar perturbations through the resonances associated with the normal modes. Famous normal modes to the computations of forced nutations are the Chandler wobble, the free core nutation and the tilt-over mode. By taking into consideration of the interactions of the motions of the solid inner core with the rest of the earth, Mathews et al (1991a;b) discussed thoroughly the roles of other two normal modes, the prograde free core nutation (PFCN) and the inner core wobble (ICW). The periods of the two normal modes are 475 days and 2409 days for PREM and 512 days and 1815 days for earth's model 1066A. The most important normal mode to the computation of the earth's nutation is the FCN. Gwinn et al. (1986) have pointed out that a small change in the value of the FCN frequency could produce the correction demanded by VLBI measurements of the amplitude of the retrograde annual nutation. As discussed in the previous section, the FCN is characterized by temporal variations in magnitude, phase, and even its period. If the temporal variations are of geophysical origin, this suggests the temporal variations of the earth's properties, and so the temporal variations of the forced nutations.

In order to detect possible temporal variations of the forced nutations, LS adjustment solutions to corrections of coefficients of forced nutations with largest magnitudes are solved for from the pole offsets. The results are given in Table 1, and are generally consistent with the 1996 Nutation Theory of the International Earth Rotation Service (McCarthy 1996). The FCN is solved for in the sense of the best fit to observations, and the solutions is as follows.

$$\begin{aligned}\delta\psi &= -0.10_{mas} sin(2\pi(t-51544.5)/425.8) + 0.53_{mas} cos(2\pi(t-51544.5)/425.8)\\ \delta\epsilon &= 0.24_{mas} sin(2\pi(t-51544.5)/415.3) + 0.04_{mas} cos(2\pi(t-51544.5)/415.3)\end{aligned}$$

where t is the date (MJD). The uncertainty on the amplitude is ± 0.03 mas in longitude and ± 0.01 mas in obliquity.

After the removal of the solutions to FCN and forced nutations from the pole offsets, wavelet analyses of the residuals are shown by Fig.2. First, if the corrections to forced nutations are normally parameterized and the signals are stable, the residuals should

Table 1. Corrections to coefficients of nutations in longitude and obliquity (mas).

Period (day)		This paper	McCarthy (1996)	Period (day)		This paper	McCarthy (1996)
Linear	$\Delta\epsilon/dt$	-0.28 ± 0.03	-0.24	121.75	$\Delta\epsilon_c$	0.03 ± 0.01	0.040
(mas/yr)	$\Delta\psi/dt$	-3.00 ± 0.03	-2.99		$\Delta\psi_s$	0.00 ± 0.02	0.013
					$\Delta\epsilon_s$	-0.01 ± 0.01	-0.018
					$\Delta\psi_c$	-0.06 ± 0.02	-0.054
-6798.38	$\Delta\epsilon_c$	2.81 ± 0.07	2.856	27.56	$\Delta\epsilon_c$	-0.03 ± 0.01	0.013
	$\Delta\psi_s$	-7.15 ± 0.17	-6.677		$\Delta\psi_s$	-0.12 ± 0.02	-0.002
	$\Delta\epsilon_s$	1.10 ± 0.17	1.553		$\Delta\epsilon_s$	0.08 ± 0.01	0.039
	$\Delta\psi_c$	3.56 ± 0.06	3.646		$\Delta\psi_c$	-0.09 ± 0.02	-0.094
-3399.19	$\Delta\epsilon_c$	-0.19 ± 0.05	-0.247	13.66	$\Delta\epsilon_c$	-0.07 ± 0.01	0.164
	$\Delta\psi_s$	1.27 ± 0.06	1.229		$\Delta\psi_s$	-0.01 ± 0.03	-0.320
	$\Delta\epsilon_s$	0.21 ± 0.06	-0.029		$\Delta\epsilon_s$	0.20 ± 0.01	0.136
	$\Delta\psi_c$	-0.14 ± 0.04	-0.071		$\Delta\psi_c$	0.40 ± 0.03	0.269
365.26	$\Delta\epsilon_c$	2.12 ± 0.01	1.988	13.63	$\Delta\epsilon_c$	0.02 ± 0.01	0.076
	$\Delta\psi_s$	5.11 ± 0.02	4.938		$\Delta\psi_s$	-0.09 ± 0.03	-0.152
	$\Delta\epsilon_s$	-0.28 ± 0.01	-0.198		$\Delta\epsilon_s$	0.07 ± 0.01	0.032
	$\Delta\psi_c$	1.01 ± 0.02	1.121		$\Delta\psi_c$	0.07 ± 0.03	0.034
182.62	$\Delta\epsilon_c$	-0.51 ± 0.01	-0.542	9.13	$\Delta\epsilon_c$	-0.08 ± 0.01	-0.004
	$\Delta\psi_s$	1.52 ± 0.02	1.686		$\Delta\psi_s$	0.02 ± 0.02	-0.037
	$\Delta\epsilon_s$	-0.54 ± 0.01	-0.464		$\Delta\epsilon_s$	0.02 ± 0.01	0.035
	$\Delta\psi_c$	-1.62 ± 0.02	-1.400		$\Delta\psi_c$	0.11 ± 0.02	0.077

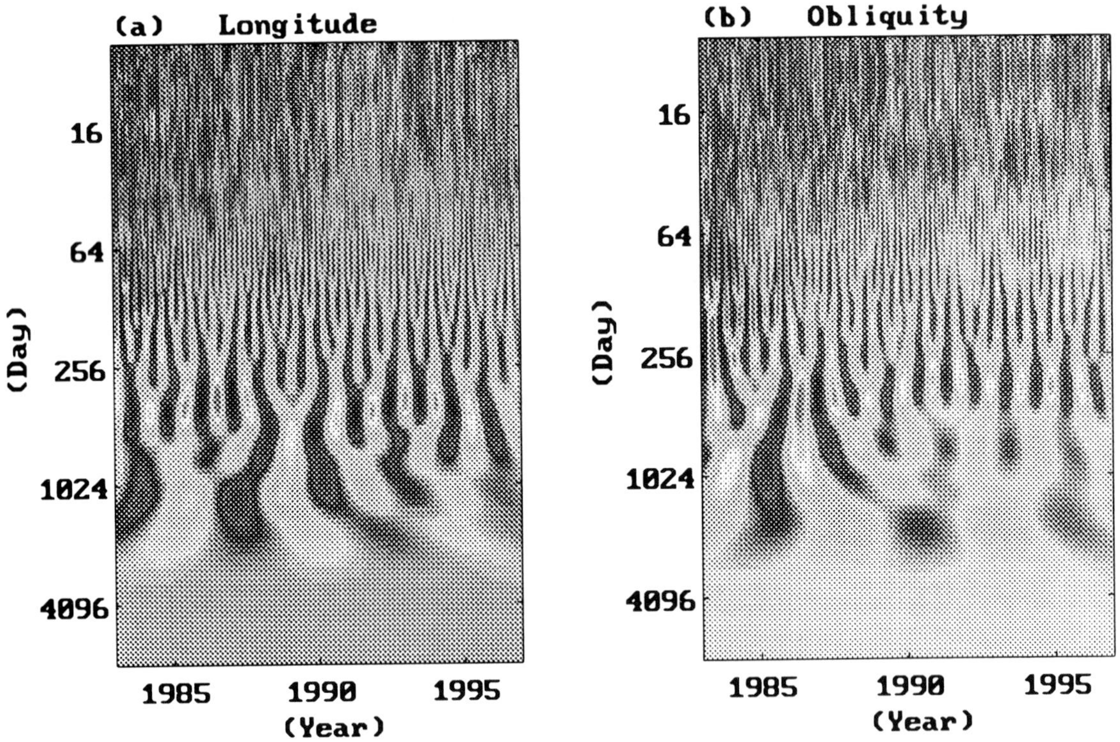

Figure 2. Results from LS adjustment and wavelet analyses of pole offsets. The temporal variations of the FCN and the forced nutations are demonstrated. There exist discernible signals in quasi-biennial and quasi-quadrennial frequencies.

be white noises. In Fig.2, signals in the fortnightly, monthly and semiannual frequencies are still discernible, which indicates the nature of the temporal variations of forced nutations. Second, as already mentioned, there are about 12 PV pairs near the 425 days period, the temporal variations of FCN is dominant. Third, the signals within the period interval from 500 days to 2500 days are very complicated and strange. There are discernible signals in quasi-biennial and quasi-quadrennial frequencies. By adjusting the values of periods for a best fit of data, the solutions are as following.

$$\begin{aligned}\delta\psi &= +0.07_{mas} sin(2\pi(t-51544.5)/758.1) \quad -0.14_{mas} cos(2\pi(t-51544.5)/758.1)\\ &\quad +0.04_{mas} sin(2\pi(t-51544.5)/1500.2) \quad -0.11_{mas} cos(2\pi(t-51544.5)/1500.2)\\ \delta\epsilon &= -0.07_{mas} sin(2\pi(t-51544.5)/762.9) \quad -0.02_{mas} cos(2\pi(t-51544.5)/762.9)\\ &\quad +0.01_{mas} sin(2\pi(t-51544.5)/1858.1) \quad -0.07_{mas} cos(2\pi(t-51544.5)/1858.1)\end{aligned}$$

Geophysical or physical studies will play the key role to reconcile whether they are counterparts of the PFCN and ICW related to the motion of the solid inner core or not.

REFERENCES

Charlot. P., Sovers, O. J., Williams, J. G., & Newhall, X. X.: 1995, AJ 109, 418
Dong Danan & Zheng Dawei:1985, Annals of Shanghai Observatory 7, 13
Gwinn, C. R., Herring, T. A., & Shapiro, I. I.: 1986, JGR 91, 4755
Herring. T.A., Buffett, B.A., Mathews, P.M., & Shapiro, I. I.: 1991, JGR 96, 8259
Herring. T. A., Gwinn, C. R., & Shapiro, I. I.: 1986, JGR 91, 4745
Kinoshita, H., & Souchay, J.: 1990, Celest. Mech., 48, 187
Mathews, P.M., Buffett, B.A., Herring, T.A., & Shapiro, I.I.: 1991, JGR 96, 8219
Mathews, P.M., Buffett, B.A., Herring, T.A., & Shapiro, I.I.: 1991, JGR 96, 8243
McCarthy, D. D. (ed.): 1996, IERS Tech. Note 21: IERS Conventions. Observatoire de Paris, Paris, p.25
McCarthy, D. D., & Luzum, B. J.: 1991, AJ, 102, 1889
Rochester, M. G., Jensen, O. G., & Smylie, D. E.: 1974, Geophys. J. R. astr. Soc. 38, 349
Sasao, T., & Wahr, J. M.: 1981, Geophys. J. R. Astron. Soc. 64, 729
Seidelmann, P. K.: 1982, Celest. Mech. 27, 79
Souchay, J., Feissel, M., Bizouard, C., Capitaine, N., and Bougeard, M.: 1995, A&A 299, 277
Souchay, J., Feissel, M., and Ma, C.: 1996, A&As 116, 473
Souchay, J. & Kinoshita, H.: 1996, A&A 312, 1017
Vondrak, J.: 1977, Bull. Astron. Inst. Czech 28, 84
Walter, H. G. & Sovers, O. J.: 1996, A&A 308, 1001
Williams, J. G., Newhall, X. X., & Dickey, J. O.: 1991, A&A 241, L9
Zheng Dawei & Dong Danan: 1986, Acta Astronomica Sinica 27, 368
Zhu, S. Y. & Groten, E.: 1989, AJ 98, 1104

EXCITATION OF NUTATION AS DEDUCED FROM RESULTS OF THE RECENT ATMOSPHERIC REANALYSIS PROJECT

Aleksander Brzeziński, Christian Bizouard†, and Sergei Petrov‡

Space Research Centre, Polish Academy of Sciences
Bartycka 18A, 00-716 Warszawa, Poland
Fax.: +48(39)121273, e-mail: alek@cbk.waw.pl

Nearly diurnal variations in the equatorial components of the atmospheric angular momentum are expected to excite minor but well measurable nutational motions of the earth's pole. These motions include the *free core nutation* (FCN) of variable amplitude between 100 and 500 microarcseconds (μas), as well as a contribution to the amplitudes of some important constituents of the lunisolar nutation. We investigate this problem using a 29-years long homogeneous series of the 4-times daily EAM (*effective angular momentum*) estimates based on results of the common U.S. NCEP/NCAR reanalysis project. The most important atmospheric contributions are found for the following nutation constituents: prograde annual (77 μas), retrograde annual (53 μas), prograde semiannual (45 μas), and for the constant offset of the pole ($d\psi \sin\varepsilon_o = -86$ μas, $d\varepsilon = 77$ μas); among them only the prograde semiannual component is driven mostly by the wind term of the EAM function while in all other cases the pressure term is dominating. Comparison with the VLBI corrections to the IAU 1980 nutation model and taking into account the ocean tide contribution, yields a good agreement for the prograde annual and semiannual nutations, that is our estimation receives an important observational confirmation. We also investigated time variability of the atmospheric contribution to the nutation amplitudes by performing the sliding window least squares analysis of both the atmospheric excitation and VLBI nutation data. Almost all detected variations of atmospheric origin can be attributed to the pressure term, the largest being the in-phase annual prograde component (about 30 μas) and retrograde one (as much as 100 to 200 μas). It seems that there are physical reasons limiting the precision of classical modeling of nutation to the level of 0.1 mas. Comparison with the VLBI data shows significant correlation for the retrograde annual nutation after 1989, while for the prograde annual term there is a high correlation in shape but the size of the atmosphere-driven variations is about 3 times less than deduced from the VLBI data. Our comparison yields a considerably better agreement with the VLBI nutation data when using the EAM function without the IB correction for ocean response, which indicates that this correction is not adequate for nearly diurnal variations. From the spectral estimation near the FCN frequency we conclude that 1) the wind contribution is not important for the FCN excitation balance, and 2) the pressure term of the atmospheric angular momentum has enough power to explain the whole observed FCN oscillation.

---

† on leave from: Paris Observatory, Paris, France

‡ on leave from: Institute of Applied Astronomy, St.Petersburg, Russia

Extended version of this paper is submitted to Journal of Geodesy.

REGIONAL ATMOSPHERIC ANGULAR MOMENTUM CONTRIBUTIONS TO POLAR MOTION EXCITATION

Jolanta Nastula [1] and David Salstein [2]

1- Space Research Center of the PAS, Bartycka 18a, 00-716 Warsaw, Poland

2- Atmospheric and Environmental Research Inc., 840 Memorial Drive, Cambridge, Ma 02139-3794, USA

Abstract We focus on a regional analysis of the equatorial components of the Effective Atmospheric Angular Momentum (EAAM) functions that are responsible for excitation of polar motion. These functions are computed from the NCEP/NCAR 40-Year Reanalysis Project data both globally and in 108 geographic sectors for the period from 1968 to 1997. We investigate regional contributions of these atmospheric angular momentum functions to the short period oscillations of geodetically-determined polar motion directly and to the global EAAM excitation functions themselves. We examine two excitation terms in parallel, both excluding and including the inverted barometer (IB) formulation which adjusts the atmosphere to account for an isostatic equilibrium response from the ocean to overlying pressure; the IB formulation tends to decrease effective atmospheric variability. In the case of pressure terms without IB the largest contributions to the equatorial components of EAAM functions originate in the South Pacific, North Atlantic and North Pacific regions. However, application of the IB correction may result in the dominance of Eurasia and North America instead, with nearly all southern hemisphere contributions disappearing. Oscillations of the polar motion excitation function are mainly coherent with variations of the pressure term of the EAAM excitation functions over northern mid-latitude land areas. Distinct oscillations appear to occur in two frequency bands: 25 - 75 days and 75 - 125 days, in both prograde and retrograde directions which correspond to counterclockwise and clockwise polar motion, respectively. Coherence and cross-spectral analyses are performed to determined degree of agreement and common amplitude, respectively. We examine the atmospheric functions with respect to one important region in Eurasia and note a propagating horizontal influence within the atmosphere.

Analysis and Results

We estimate the atmospheric influence on polar motion by comparing the function that must excite the Earth (here, the geodetic polar motion excitation function) with that produced by the atmosphere (here, the atmospheric excitation function). Perturbations of polar motion are expressed by variations of equatorial components χ_1 and χ_2 or can be considered as the complex-valued $\chi = \chi_1 + i\chi_2$. Here the χ_1 and χ_2 components of geodetic excitation function are determined by the Wilson and Haubrich time domain deconvolution formulas with IERS (EOP 90 C04) pole coordinate data. Atmospheric excitation functions for polar motion are global values from a 29-year series of the NCEP/NCAR Reanalysis calculated and stored by AER, including pressure (with and without IB correction) and wind (w) terms (Salstein and Rosen, 1997). Long period oscillations of atmospheric and geodetic excitation functions are removed using the Butterworth high-pass filter with a 150-day cutoff period.

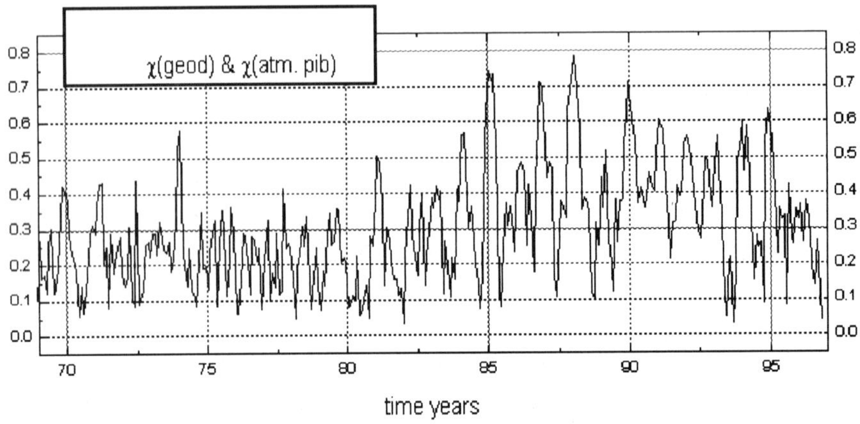

Fig. 1 Correlation coefficients between the complex χ components of geodetic and atmospheric excitation functions, computed over quarter-year intervals, starting each eighth of a year, since 1968, with about 90 independent points in each interval. A correlation coefficient of 0.33 is significant at the 99% confidence level.

The correlation coefficient between χ component of geodetic and atmospheric excitation functions, varies with time and is significant after 1983. The lack of agreement is likely due to inaccuracies in the data sets being used, and particularly the polar motion data, observed by less accurate methods before about that date. The quality of atmospheric data is more likely to be constant since 1968. Given the significance of the atmospheric-geodetic relation in the later period, we conclude that the atmospheric data, may be considered by themselves in the whole period since 1968. Correlation coefficients in Fig.1 reach the highest value 0.6-0.8. Maxima of the all correlation coefficients occur during the northern winters.

Next, regional χ_1 and χ_2 components of atmospheric excitation function of pressure, both modified and not modified by the IB correction, were computed in 108 equal-area sectors, by placing meridional boundaries every 30° of longitude and of zonal boundaries at 6.4, 19.5, 33.7, 51.1 and 90 °N and °S (Fig. 2); (Salstein and Rosen, 1989).

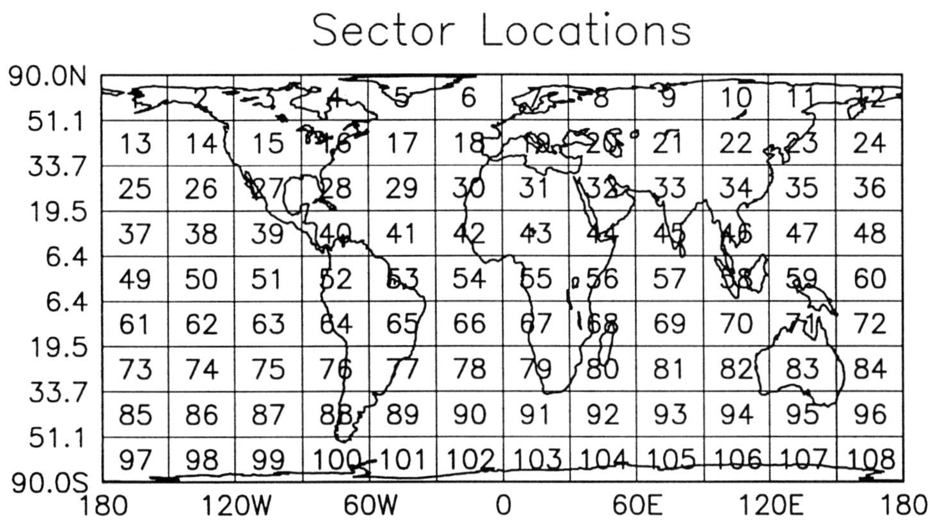

Fig. 2 Equal-area sectors for regional values of atmospheric excitation function (Salstein and Rosen, 1989).

Because the regional wind term variations are very noisy (Nastula *et al.*, 1997) we focus on pressure contributions to both the polar motion and global atmospheric excitation functions.

Figure 3 presents cross-spectra and coherences between regional variations of atmospheric excitation function in numbered sectors (see map) and either (i) polar motion excitation function during 1983-1997 or (ii) the global atmospheric excitation function, during 1968-1997. The Fourier Transform Band Pass (FT BPF) filtering method for complex-valued time series was used (Kosek, 1995). The maximum regional coherences reach 0.5-0.65 and are lower than those previously obtained between global atmospheric and geodetic excitation functions (Nastula, 1995). Such coherence behavior is probably caused by relatively higher noise levels in the regional atmospheric excitation function than in the global one. Figures 3a -3f all show evidence of distinct frequency bands at: 25 - 75 and 75 -125 days in common variations of regional and global phenomena, especially when the regional functions are compared with those of the whole study period, 1968-1997. Generally prograde parts of cross-spectra and coherences are stronger than those of retrograde, except in the case where the IB correction is included in the atmospheric functions. Equal amplitudes of prograde and retrograde parts of coherences and cross-spectra correspond to quasi-linear polarization of the EAAM function with large amplitudes in χ_2 and small ones in χ_1. Eurasian mid-latitude regions (sectors from 18 to 22) appeared to be especially prominent for polar motion excitation for both important spectral bands (Figs. 3 a-d). There is some evidence that atmosphere over the East Indian Ocean is coherent with polar motion (Fig. 3b) but seems not to contribute significantly to its power (Fig. 3a). Inclusion of the IB correction into the pressure term of regional atmospheric excitation function increases of coherence level, however, decreases the amplitudes of common oscillations in northern hemisphere and nearly eliminates them in the southern ones (Figs. 3c and 3d).

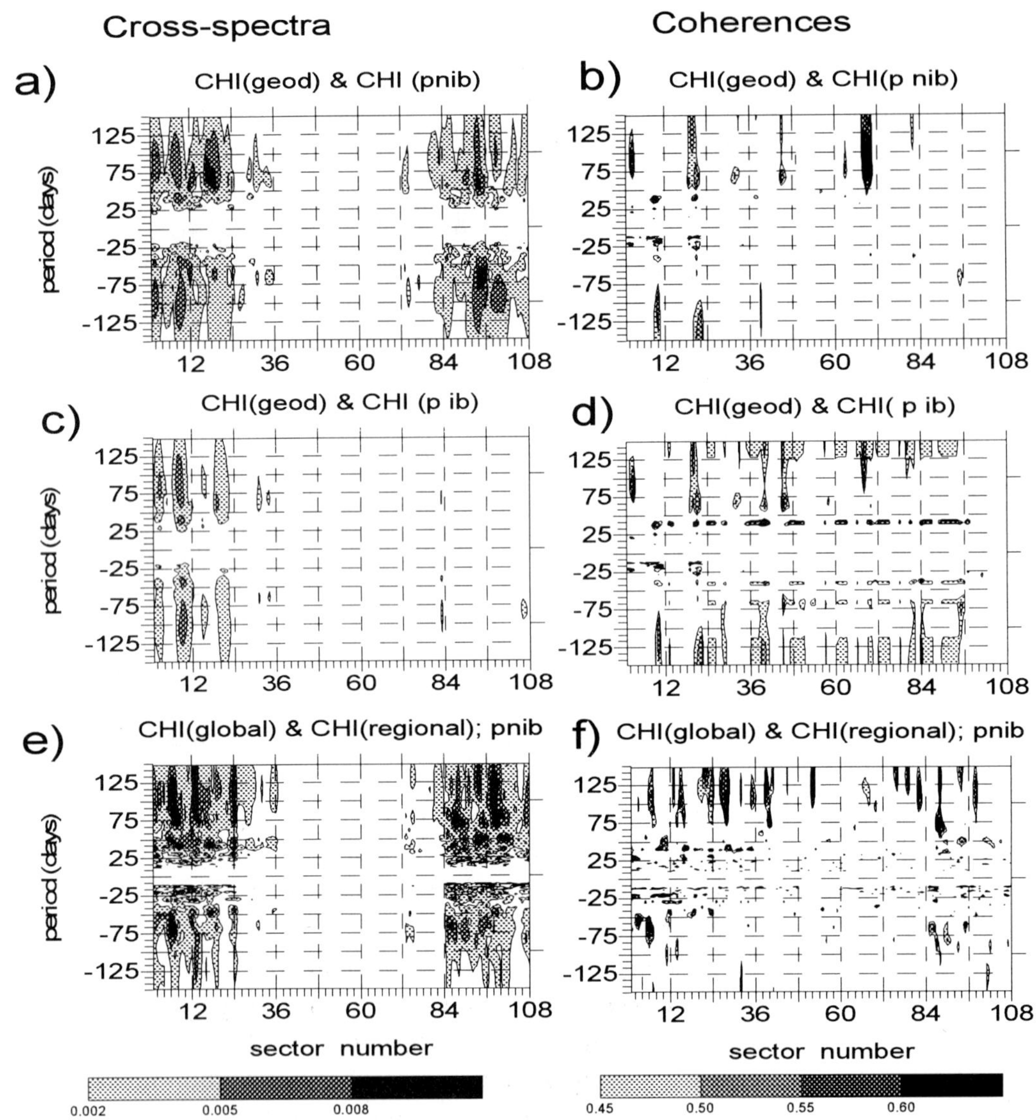

Fig.3 (a, c, e) cross-spectra and (b, d, f) coherences between regional variations of atmospheric excitation function and either (a, b, c, d) polar motion or (e, f) global atmospheric excitation functions for the case without the IB correction (p nib) and with it (pib). Sectors are numbered according the map in Fig. 2, cross-spectra are in rad*10^{-7}.

The South Pacific (sectors 84 - 88), North Pacific (sectors 13 - 14, 24), and North Atlantic (sectors 17 - 18) appear to be important as they relate to the global atmospheric excitation but do not appear to contribute significantly to polar motion excitation (Figs. 3e and 3f), likely due to their over ocean location decreased by the IB relation. Considering both cross-spectra and coherences we can conclude that sectors from 18 to 24 (51.1° N - 33.7° N Lat.; 30°W - 120° E Lon.), covering mid-latitude Europe and Asia are most important for excitation of short period variations of polar motion.

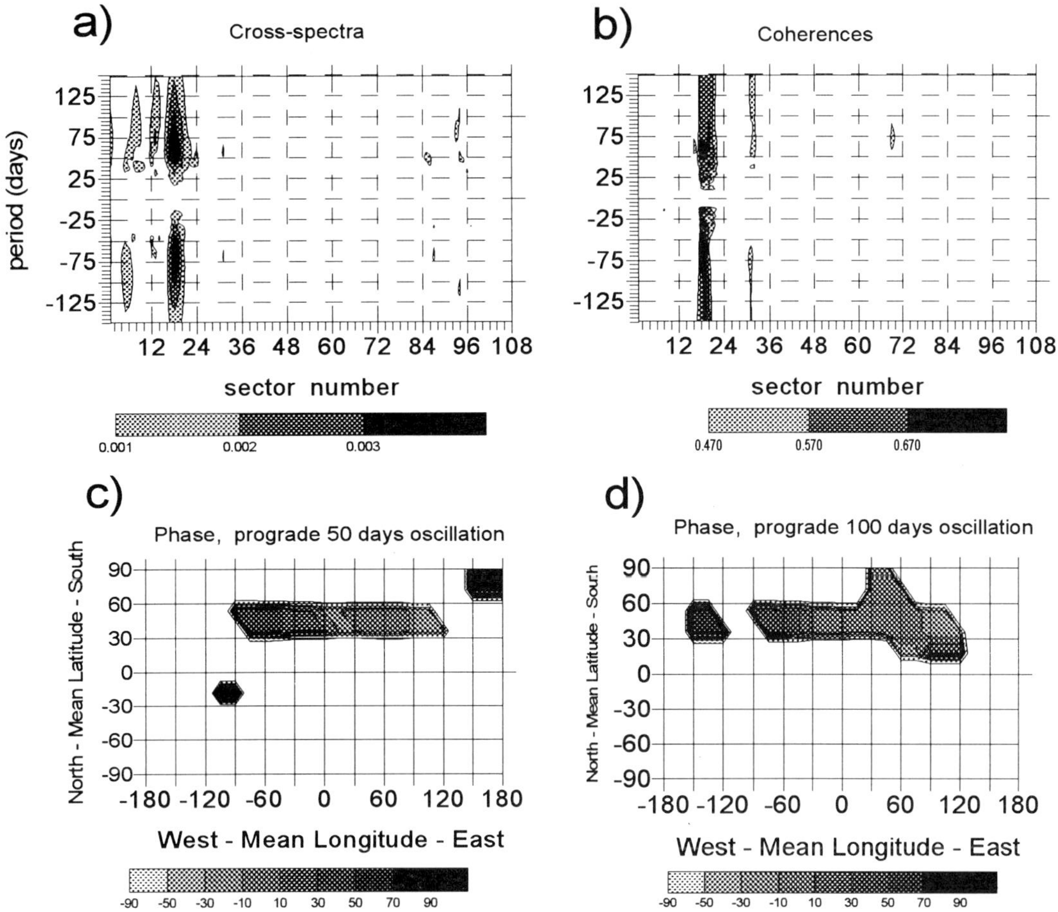

Fig. 4 (a) cross-spectra, (b) coherences and (c, d) phases between variations of atmospheric excitation function in each sector and the special region in Eurasia. Phases are plotted only if the corresponding coherence exceeds 0.3.

Atmospheric variability in the regions to the west as well as to the south of this "special" region in Eurasia are coherent with fluctuations there (Figs. 4a and 4b). The fluctuations are positive in phase with the special region indicating that atmosphere waves propagate from the west, namely from North America and the North Atlantic towards the special region, for both the 50-day and 100-day frequencies, through with slightly different characteristics (Figs. 4c and 4d). The most power of time-variable spectra of the area is concentrated in two spectral ranges: from 25 to 75 days and from 75 to 125 days (Fig. 5). Oscillations in the first, higher frequency, range peak in 1979 (very strong), 1982, 1985, 1990 and 1992. Those in the second, lower frequency, range peak in 1982, 1989 and 1997. Prograde oscillations are generally stronger then retrograde ones.

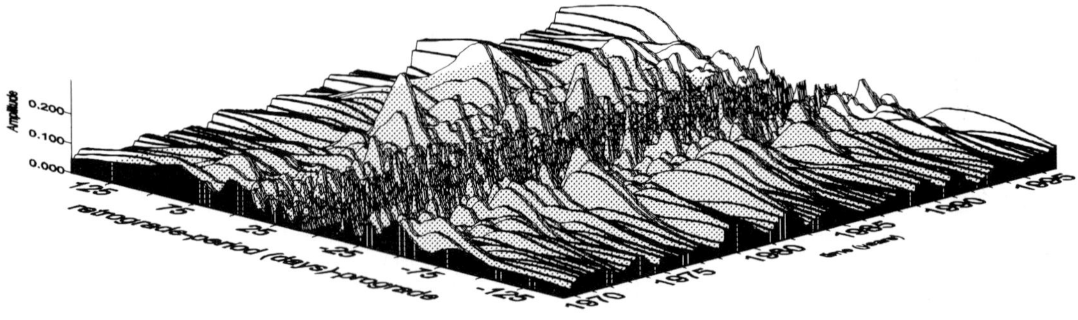

Fig. 5 Time - variable spectra of the special region atmospheric excitation functions variation showing fluctuations of amplitude for the period 1968 -1997 in the prograde (+) and retrograde (-) parts of spectrum. Units are rad*10^{-7}.

Conclusion

We can use regional atmospheric analysis to determine the origin of polar motion excitation based on coherence and cross-spectral analysis between regional atmospheric and either geodetic or global atmospheric excitation function. We find that the Eurasia region is particularly important for exciting high frequency polar motion.

Acknowledgments

The research reported here was supported under the Polish State Committee for Scientific Research through project 2 P03C 00309 and under the NASA Mission to Planet Earth Program through grant NAGW-2615.

References

Kosek, W., 1995, Time Variable Band Pass Filter Spectrum of real and complex-valued polar motion series, *Artificial Satellites, Planetary Geodesy*, **Vol. 30**., No 1, 27 - 43.

Nastula, J., 1995, Short periodic variations of polar motion and hemispheric atmospheric angular momentum excitation functions in the period 1984 - 1992, *Ann Geophysicae*, **13**, 217 -225.

Nastula, J., W. Kosek, and B. Kolaczek, 1997, Analyses of zonal atmospheric excitation functions and their correlation with polar motion excitation functions, accepted for publication in *Ann. Geophysicae.*

Salstein, D. A., and R. D. Rosen, 1989, Regional contributions to the atmospheric excitation of rapid polar motions, *J. Geophys. Res.*, **94**, 9971 - 9978.

Salstein, D. A., and R. D. Rosen, 1997, Global momentum and energy signals from reanalysis systems, *8th Conference on Climate Variations*, 2-7 February 1997, Long Beach, California, American Meteorological Society, Boston, Massachusetts, 344 -348.

TIME-FREQUENCY CHARACTERISTICS OF THE PRESSURE TERM OF THE EAAM EXCITATION FUNCTIONS IN SOME GEOGRAPHIC REGIONS

J. Nastula, O. Kotreleva*, B. Kołaczek, W. Kosek

Space Research Center, PAS, Warsaw, Poland
* on leave from Main Astronomical Observatory, RAS, St. Petersburg, Russia

Abstract

Regional values of the Effective Atmospheric Angular Momentum (EAAM) functions were computed from the Japan Meteorological Agency Global Objective Analysis data for the period 1988–1997. Time-variable amplitude spectra of the pressure terms (including or not the inverted barometric (IB) correction) of the complex-valued equatorial components of EAAM functions χ_1, χ_2 were computed for the selected geographic regions covering mid-latitude Europe, Asia with part of Siberia, North America and parts of northern Atlantic, and Pacific. These spectra of short period variations of pressure terms of χ_1, χ_2 show distinct time variations of amplitudes of the prograde and retrograde oscillations with periods of about 40–60 and 90–120 days. Amplitudes of these oscillations depend also on latitude and longitude. The computed coherence between regional and global pressure terms of the χ_1, χ_2 and between these regional terms and polar motion exciatation functions show that the regional pressure terms over Eurasia in the latitude belt $45° - 60°$ N give the most important contribution to polar motion excitation.

Analyses and Results

Research in the area of interactions between the atmosphere and polar motion is facilitated by using the equatorial χ_1 and χ_2 components of the Effective Atmospheric Angular

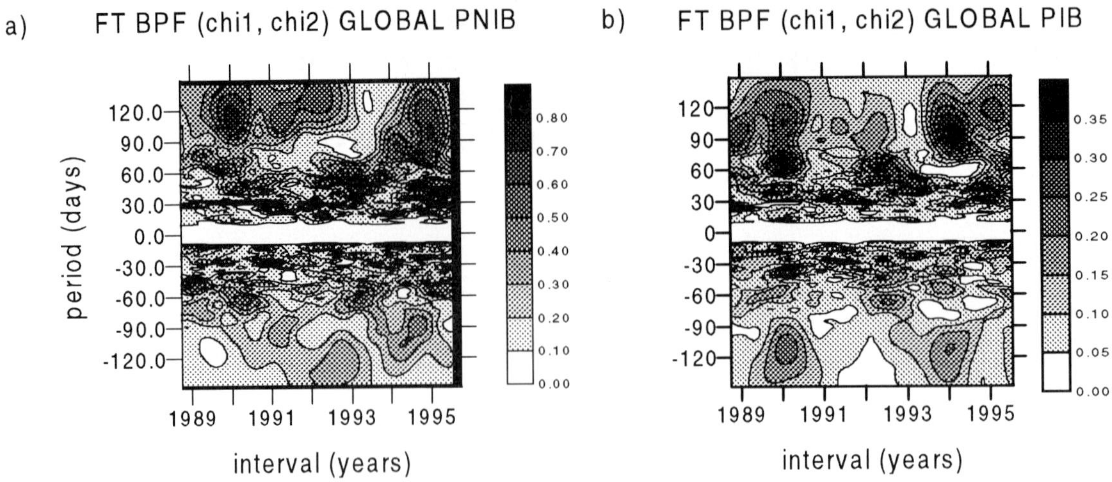

Fig. 1. Time variable spectra of global EAAM pressure terms with and without IB.

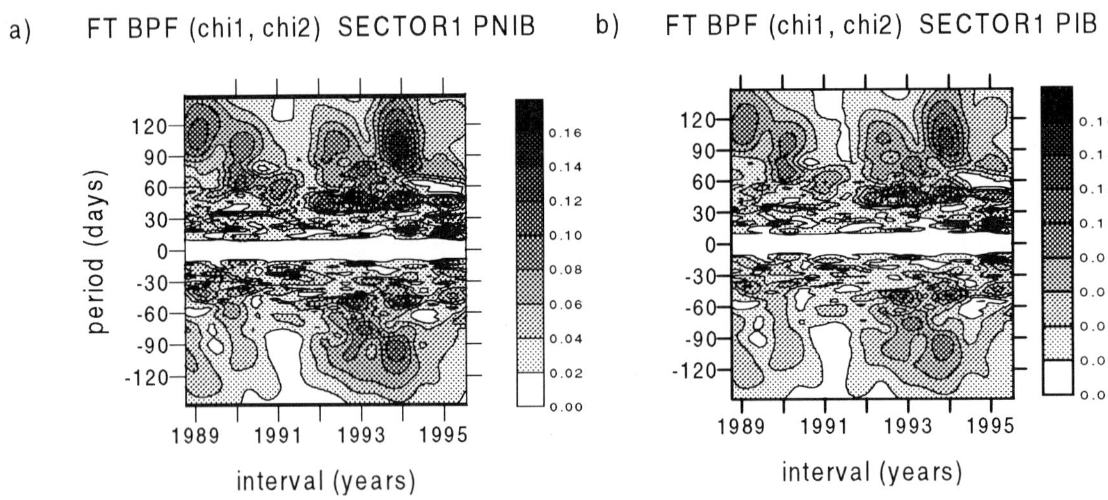

Fig. 2. Same as in Fig. 1 but for sector 1.

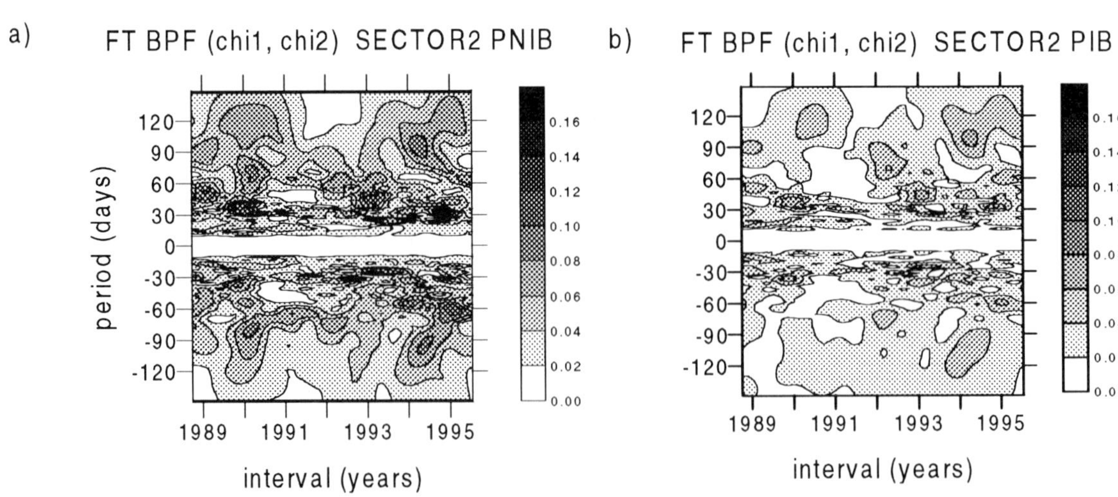

Fig. 3. Same as in Fig. 1 but for sector 2.

Momentum (EAAM) functions. Regional values of the EAAM functions were computed from the Japan Meteorological Agency Global Objective data for the period 1988–1997 (Nastula *et al.*, 1997; Nastula, 1997). Computations and analyses were made for several geographical regions responsible for the short period polar motion excitations (Nastula, 1997; Salstein and Rosen, 1989) (Table 1).

Table 1. Boundaries of regions.

No	latitude	longitude
1	45 – 60° N	0 – 90° E
2	45 – 60° N	75 – 180° E
3	45 – 60° N	60 – 165° W

Long period oscillations were removed from the data using the Butterwoth high-pass filter with a cutoff period of 150 days.

Time variable spectra of short period variations of pressure terms of global and regional χ_1, χ_2 were computed using the Fourier Transform Band Pass Filter (FTBPF) (Kosek, 1995; Kosek *et al.*, 1997) (Figs. 1–3, 5). Power of short period variations of pressure terms of global χ_1 and χ_2 is concentrated at the prograde oscillations with the major diffuse maxima for the period of 90 to 120 days in years 1990/1991 and 1994/1995, corresponding to two epochs of the El Niño events (Fig. 1). Similar maxima in the retrograde parts have smaller amplitudes than in prograde parts. The spectra show also prograde oscillations with periods of 40 to 60 days with maxima in 1990, 1992, and very strong ones in 1993, 1995. We can see also several maxima for prograde oscillation with a 30-day period. The spectra patterns are more clear for the pressure terms with the IB correction, although amplitude maxima are smaller.

Regional short period prograde oscillations of pressure terms of χ_1 and χ_2 are the highest over Eurasia in the latitude belt ranging from 45° to 60° N (Figs. 2, 3). Similarly as in the case of global terms addition of the IB corrections to the pressure terms of χ_1, χ_2 decreases the spectra amplitudes and clarify the general pattern of maxima distribution of prograde oscillations. It is necessary to stress the large similarity of time variable spectra of pressure terms of χ_1, χ_2 over the Eurasia to the global ones. It means that the pressure terms of χ_1, χ_2 variations over this region give the most important contribution to the global variations of these terms. The coherences between the global and regional pressure terms of χ_1, χ_2 over Eurasia confirm the similarity of variations of global and regional pressure terms of χ_1, χ_2 in the considered region especially in years when amplitudes of these terms are high (Fig. 4). Coherences show diffuse maxima of prograde and retrograde oscillations with periods of 90–120 days in 1990, 1993–1994 and some peaks for oscillations with periods of 50 days and 60–70 days.

The time variable amplitude spectra of regional pressure terms of χ_1, χ_2 over North America are not similar to the spectra of the global data (Fig. 5). However there are maxima in the prograde part of the spectra for periods of 90–120 days in 1991 and 1995.

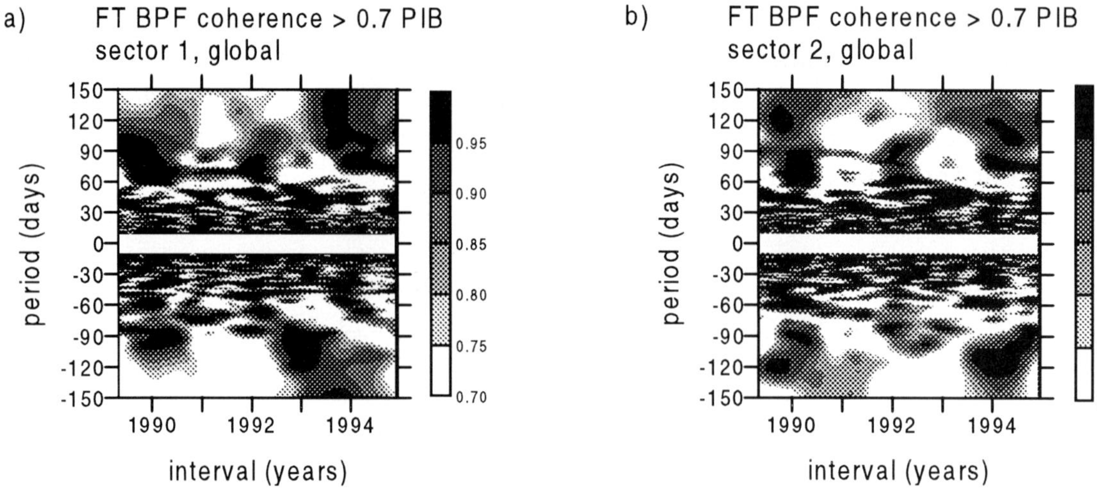

Fig. 4. Time variable coherence between the regional and global EAAM pressure terms with the IB correction.

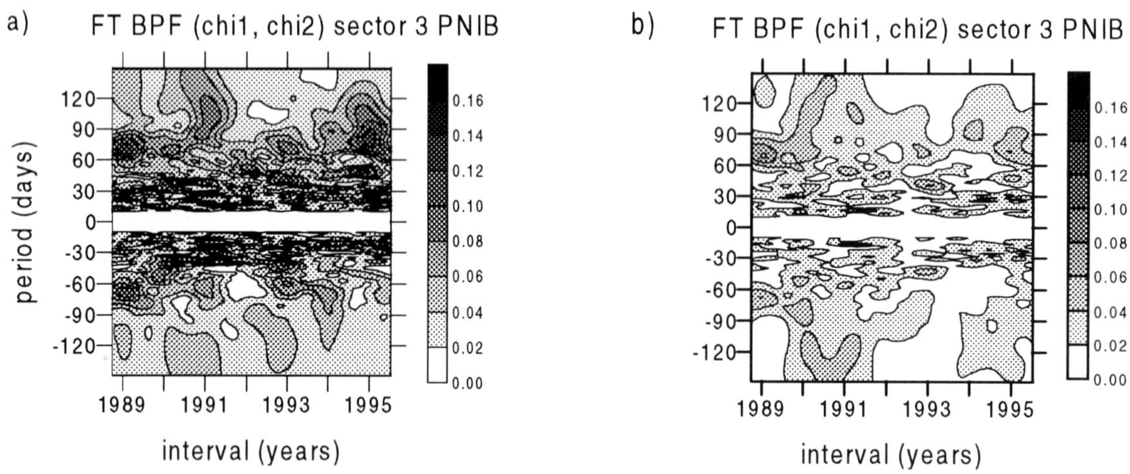

Fig. 5. Same as in Fig. 1 but for sector 3.

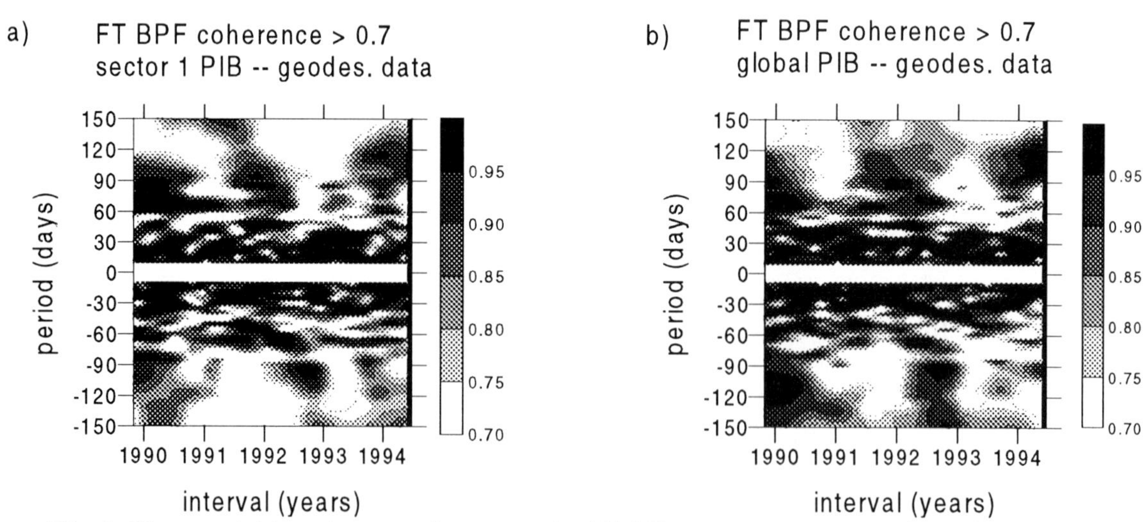

Fig. 6. Time variable coherence between the EAAM pressure terms and geodetic excitation functions.

The important contribution of the short period variations of regional pressure terms of χ_1 and χ_2 over Eurasia to such global term variations means also that the short period variations of polar motion are excited by them (Nastula, 1997). To show this effect we computed coherences between equatorial components of global and regional pressure terms of χ_1, χ_2 and equatorial components of geodetic excitation functions ψ_1, ψ_2 (Fig. 6). The geodetic excitation functions were computed from the IERS(EOP 90C04) pole coordinate data using the modified Wilson and Haubrich formula (Wilson *et al.*, 1976) and were filtered in the same way as the atmospheric excitation functions.

The coherences between global or regional pressure terms of χ_1, χ_2 over Eurasia and geodetic excitation functions ψ_1, ψ_2 are similar. There are maxima of prograde coherence in years 1990–1991, 1992 and 1994 for oscillations with periods from 60 to 120 days. The time variations of coherences in spectral range from 10 to 60 days are very changeable. The maxima of these coherences correspond well to the maxima of the time variable spectra of the equatorial components of the regional pressure terms of χ_1, χ_2 over Eurasia. All these facts imply that the regional pressure terms over Eurasia in the latitude belt ranging from 45° to 60° N give the most important contribution to the polar motion excitation, especially in the years of the coherence maxima.

Conclusions

Time variable spectra of the global and regional complex valued pressure terms (including or not the IB) of equatorial components of EAAM functions over selected geographical regions in mid-latitude Europe, Asia and North American continent for the period 1988–1996 were computed.

These regional spectra were compared with the global ones in the period range 10–150 days. These spectra of regional pressure terms over Eurasia region are very similar to the global spectra, but the maximum amplitudes are the strongest over the latitude belt ranging from 45° to 60° N. This implies that short period oscillations of the global pressure terms of χ_1, χ_2 in the spectral range of 10–150 days are generated, probably, mainly over Eurasia region in a latitude belt 45° – 60° N. The time variable coherences computed between the regional and global pressure terms of χ_1, χ_2 confirm these results.

Maxima of oscillations with period about 90–120 days are the highest in the epochs of EL Niño events in 1990/1991 and 1994/1995. Coherences between the global geodetic excitation function and global and regional pressure terms of χ_1, χ_2 show that polar motion excitations come mostly from the regional variations of pressure terms of χ_1, χ_2 over Eurasia in a latitude belt ranging from 45° to 60° N.

Acknowledgments

The support of the Polish State Committee for Scientific Research (project 2 P03C 00309) and the US-Polish Maria Skłodowska Curie Fund (project PAN/SMI 92-97) are all gratefully acknowledged. The authors are grateful to Prof. K. Yokoyama and Prof. S. Manabe from Japan National Astronomical Observatory for their discussions and help in computations of regional atmospheric data.

References

Kosek, W., 1995, Time variable band filter spectrum of real and complex-valued polar motion series, *Artificial Satellites, Planetary Geodesy*, **30**, 27-43.

Nastula, J., 1997, The regional atmospheric contributions to the polar motion and EAAM excitation functions, *Proc. International Symposium on Gravity, Geoid and Marine Geodesy*, 1996, Tokyo, Japan, (accepted).

Nastula, J., Kosek, W., and B. Kołaczek, 1997, Analyses of zonal atmospheric excitation functions and their correlation with polar motion excitation functions, *Ann. Geophys.*, (accepted).

Salstein, D. A., and R. D. Rosen, 1989, Regional contributions to the atmospheric excitation of rapid polar motions, *J. Geophys. Res.*, **94**, 9971-9978.

Kosek W., D. D. McCarthy, B. J. Luzum, 1997, Possible improvement of Earth orientation forecast using autocovariance prediction procedures, *J. Geodesy*, (accepted).

Wilson, C. R. and R. A. Haubrich, 1976, Meteorological excitation for the Earth's wobble, *Geophys. J. Royal Astr. Soc.*, **46**, 707-743.

THE SPACETIME GRAVITATIONAL FIELD OF A DEFORMABLE BODY

E. W. Grafarend
Department of Geodetic Science
University Stuttgart
Geschwister-Scholl-Str. 24/D
D-70174 Stuttgart
Germany

The high resolution analysis of orbit perturbations of terrestrial artificial satellites has documented that the *eigengravitation* of a massive body like the Earth *changes in time*, namely with periodic and aperiodic constituents. For the spacetime variation of the gravitational field the action of internal and external volume as well as surface forces on a *deformable massive body* are responsible. Free of any assumption on the symmetry of the constitution of the deformable body we review the incremental spatial ("Eulerian") and material ("Lagrangean") *gravitational field equations*, in particular the source terms (two constituents: the divergency of the displacement field as well as the projection of the displacement field onto the gradient of the reference mass density function) and the *'jump conditions'* at the boundary surface of the body as well as at internal interfaces both in linear approximation. A *spherical harmonic expansion in terms of multipoles* of the incremental Eulerian gravitational potential is presented. Three types of *spherical multipoles* are identified, namely the dilatation multipoles, the transport displacement multipoles and those multipoles which are generated by mass condensation onto the boundary reference surface or internal interfaces. The *degree one term* has been identified as non-zero, thus as a *"dipole moment"* being responsible for the varying position of the deformable body's *mass centre*. Finally for those deformable bodies which enjoy a *spherically symmetric constitution* emphasis is on the *functional relation* between Green functions, namely between Fourier- / Laplace-transformed volume - *versus* surface Love-Shida functions ($h(r)$, $l(r)$ *versus* $h'(r)$, $l'(r)$) *and* Love functions $k(r)$ *versus* $k'(r)$ expressed in *a box*. The *functional relation* is numerically tested for an active tidal force / potential and an active loading force / potential by *tables* proving an excellent agreement with experimental results.

In print, Journal of Geodesy

Temporal Variation of the Geopotential: Processes and Interactions Among the Earth's Subsystems

J. O. Dickey, D. Dong and R. S. Gross
Jet Propulsion Laboratory, 4800 Oak Grove Drive, Pasadena, CA 91109-8099
tel 818–354–3235; fax: 818-393-6890; e-mail: jod@logos.jpl.nasa.gov

Abstract

Seasonal variations in the Earth's gravitational field are investigated through the analysis of LAGEOS I satellite laser ranging measurements and are compared with those produced by atmospheric mass redistribution as inferred from global surface pressure data from the National Centers for Environmental Prediction (NCEP) reanalyses. The effect of oceanic tides and groundwater are considered as well. Focusing on the even zonal harmonics, atmospheric pressure fluctuations and ground water are shown to be the dominant cause of the observed zonal gravitational field variation at the annual period. At the semi-annual period, the modeled effect of the self-consistent equilibrium ocean tide dominates. The geographical distribution of the seasonal atmospheric variations are addressed. The potential use of LAGEOS for studies of polar ice sheet variation and the need for good atmospheric pressure data in this region are stressed.

Introduction

The Earth is a complex system with dynamical subsystems (such as the overlying fluid hydrosphere and atmosphere, underlying metallic core, and mantle) with complicated interactions among them (such as the melting of glaciers, sea level rise, and post-glacial rebound). Changes in the inertia tensor of the solid Earth are brought about by interfacial stresses, the gravitational attractions associated with astronomical objects and mass redistributions in the Earth's fluid and solid region. As the Earth's gravitational field changes only in response to net mass redistribution, observations and analysis of the Earth's time varying global gravitational field permits the isolation and study of the changing mass distributions and provides insight into the processes that cause them (for a review, see *Nerem et al.*, 1995 and *NRC*, 1997).

Data and Analyses Procedures

LAGEOS I laser ranging measurements during 1980–1994 are analyzed utilizing the GEODYN software package to perform numerical integration of satellite orbits and to construct normal matrices for every monthly orbit segment. We follow the procedure described in detail in *Dong et al.*, 1996 with the exception that the SPACE94 Earth Orientation Series [*Gross*, 1996] is used. Figure 1a shows the recovered gravitational field coefficient ΔC_{even} (the linear combination of even zonal Stokes coefficients with each term as a function of time).

For atmospheric pressure loading we calculate the spherical harmonic coefficients of the NCEP (National Centers for Environmental Prediction) reanalysis (*Kalnay et al.*, 1996) gridded (2.5°x2.5°) global surface pressure data, spanning 1980–1994 at 6-hour intervals, under both the non-inverted barometer (NIB) and inverted barometer (IB) assumptions for the response of the oceans to atmospheric pressure variations. The benefits of utilizing reanalyses results are several; the use of a single atmospheric model and the more complete edited data (as opposed to those that conform to the normal operational constraints) permit a more robust analysis. Monthly mean values of each pressure harmonic coefficient series were formed over the same time intervals as the monthly LAGEOS solutions. Individual Stokes coefficients predicted from atmospheric pressure were computed using Equation 2 from *Dong et al.* (1996). For comparison with the observed C_{even}, the same linear combination of spherical harmonics was formed using the time-dependent linear combination coefficients. Figure 1b displays the atmospheric pressure predicted C_{even} under the IB assumption. In addition, gridded values were calculated to enable a regional analysis of both pressure amplitude and phase (the convention is given in Table 2).

A self-consistent equilibrium ocean tide model [*Ray and Cartwright*, 1994, appendix B] has been used to compute the effect on the gravitational field of the annual and semiannual ocean tides. For continental surface water we have utilized the results of *Chao and O'Connor* [1988] who have computed the effect of annual and semiannual variations in continental surface water on the zonal (through degree 4) Stokes coefficients. They considered the effects of changes in snow cover, soil moisture stored in the root zone from rainfall and snowmelt, and surface water run-off that has not yet returned to the ocean. Variations in groundwater stored below the root zone were not included due to lack of accurate global estimates of this quantity. Figure 1c shows the modeled ocean tidal and surface water effects on ΔC_{even}.

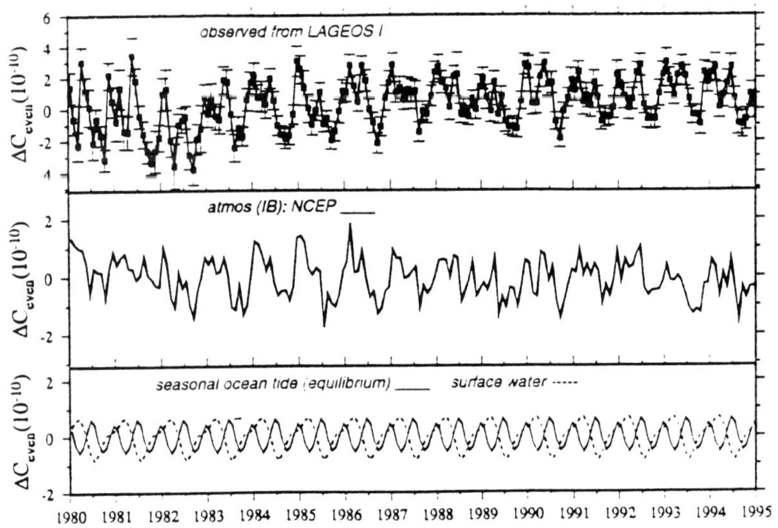

Figure 1. (a) Monthly C_{even} series spanning 1980-1994 recovered from LAGEOS I SLR data (the mean value has been removed). The error bar represents a 1σ formal uncertainty. (b) Monthly C_{even} calculated from the NCEP reanalysis (1980-1994) atmospheric surface pressure data using the same linear combination of the spherical harmonic coefficients to which the observations are sensitive. Solid line: NIB model; dashed line: IB model. (c) Monthly C_{even} reanalyses calculated from an equilibrium seasonal ocean tide model (solid line) and the surface water (dashed line) results of *Chao and O'Connor* [1988].

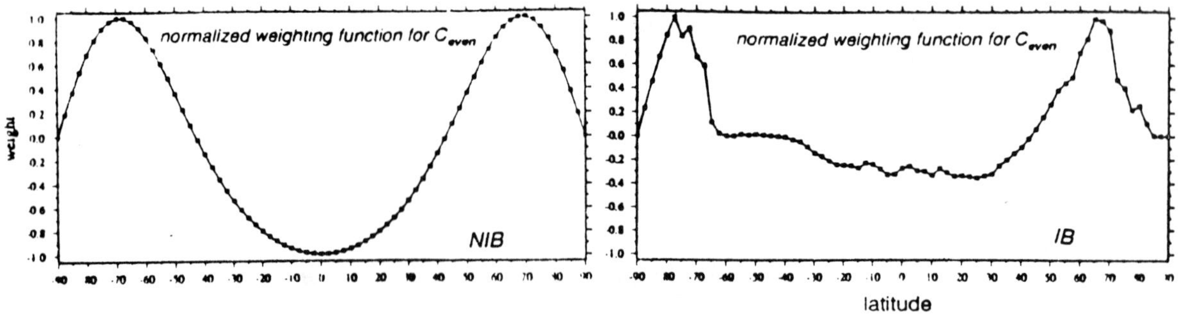

Figure 2. The normalized weighting function for C_{even} for the non-inverted barometer (NIB—left figure) and the inverted barometer (IB—right figure) case as a function of latitude.

Seasonal Global Results

The recovered monthly C_{even} solutions from the LAGEOS I data (1980–1994) are dominated by annual and semiannual variations with a significant secular trend and some indication of interannual variations (Fig. 1a). To focus on seasonal variations and the more robust results from both the NCEP and LAGEOS analyses, the observed and modeled monthly C_{even} time series from 1985 to 1994 illustrated in Figure 1a have been highpass-filtered with a cutoff period of 2 years. Table 1 lists the correlation and variance explained between the LAGEOS-observed and predicted C_{even} series. To quantitatively compare observations with predictions at the annual and semiannual frequencies, a weighted least-squares fit for a mean, trend and sinusoidal terms at these frequencies is made to the series. The comparisons are listed in Table 2.

Comparisons from Table 1 indicate that on timescales of 1 month to 2 years, and when the atmospheric pressure effect is computed under the IB assumption, the three mechanisms together (A+O+W) can account for 78.4% of the observed variance, and have a correlation of 0.89 with the observations (note definitions in Table 1). The strong preference for the IB assumption has been shown by *Dong et al.* (1996) and is consistent with numerous studies that have demonstrated the validity of the IB assumption on these timescales.

Table 1. Correlation and Variance Explained Between Observed and Modeled C_{even} (1985–1994)

	Inverted Barometer Assumed	
Series Considered	Corr.	Var*.
A†	0.83	59.0%
A + O	0.91	75.1%
A + O + W‡	0.89	78.4%

*The amount of the variance of the observed series explained by the modeled series is computed as: $(\sigma_o^2 - \sigma_{o\text{-}m}^2)/\sigma_o^2$ where σ_o^2 and $\sigma_{o\text{-}m}^2$ are the variance of the observed and residual (observed minus modeled) series, respectively.
† Notations: A atmospheric pressure; O equilibrium ocean tides; W surface water.
‡ Surface water results are from *Chao and O'Connor* [1988].

At the annual frequency, atmospheric pressure and surface ground water effects are the dominant contributors. The best agreement with the observations is obtained when all effects (A+O+W) are considered with closure seen at the 1σ level in amplitude and at the 2σ level in phase. At the semiannual frequency, the ocean tide is the main contributor; the best agreement is obtained with the observations when only the atmospheric pressure and ocean tidal effects are considered (results are within 1σ for the phase and with 20% of amplitude unaccounted for). Adding the predicted effects of surface water worsen the agreement with the observations, indicating that (a) the semiannual component of the surface water variations may not be as well determined as is the annual component, (b) other semiannual series may be erroneous, or (c) some important excitation sources are missing.

Table 2. Annual and Semiannual Variations of C_{even} from LAGEOS Observations and as Predicted from Atmospheric Pressure, Ocean Tidal, and Surface Water Fluctuations (1985–1994)

	Annual		Semiannual	
Series Considered	Amp.* (10^{-10})	Phase* (degrees)	Amp. (10^{-10})	Phase (degrees)
Observed from LAGEOS I	1.05 (0.06)†	26.5 (3.3)	0.93 (0.06)	116.1 (3.7)
O#	0.09	267.7	0.55	110.7
W‡	0.63	59.3	0.26	266.3
A (IB)	0.57 (0.06)	22.5 (6.3)	0.19 (0.06)	108.9 (9.6)
A (IB)+O	0.54 (0.06)	13.9 (6.7)	0.73 (0.06)	110.2 (4.9)
A (IB)+O+W	1.08 (0.06)	38.4 (3.4)	0.57 (0.06)	122.2 (7.1)

* Amplitude A and phase ϕ are defined by $A\sin[\omega(t-t_o) + \phi]$ where ω is the frequency and t_o is January 1, 1985.

\# Notations are the same as Table 1.

† The quoted uncertainties given in parentheses for the LAGEOS results are the 1σ formal errors in the fit. For the atmospheric series, the quoted uncertainties are the rms scatter about the fit, which are also used for the summed series since no uncertainties are available for the ocean tidal or surface water effects.

‡ From *Chao and O'Connor* [1988].

Seasonal Regional Analysis

The global mean surface pressure (the atmosphere's moment of inertia variation) has an annual cycle of ~ 0.45 mbar (e.g. *Trenberth*, 1981) that is predominantly caused by seasonal variations in water vapor content, reaching its maximum in July (during Northern Hemisphere summer). Figure 3a displays the mean annual pressure amplitude under the non-inverted barometer assumption when the data from 1980-96 are considered, with Figure 3b showing the corresponding phase. The IB case (not shown) would simply have the values over the ocean replaced by a single averaged value. Several mass redistributions are clearly visible. The effects of topography, for example, over the Andes, the Tibetan Plateau, and Greenland are evident both in phase and amplitude. The effect of distribution of land and ocean is also visible; note the high magnitude over the north Pacific and Asia, being roughly 180° out of phase. In addition, a characteristic difference exists between the northern and southern hemispheres. This seesaw is especially seen in phase and clearly visible when one examines the difference between July and January (*NRC*, 1997) results from the difference in the heating of the northern and southern hemispheres and the land and oceans (*van den Dool* and *Saha,* 1993). Figure 2a illustrates the sensitivity of LAGEOS I measurements of C_{even} to mass changes occurring uniformly over both land and oceans (2a) and to mass changes occurring over just land (2b). For LAGEOS I, in general, there are two areas of high sensitivity, the near polar and equatorial regions (Figure 2a). However, for mass changes over land, LAGEOS I is particularly sensitive to polar regions. Note the large sensitivity to variations south of 65° south and 50° north for Figure 2b. The semiannual terms (not shown) are roughly a factor of 3 smaller with maxima over the polar regions and the north Pacific with a complex phase structure. Applying the weighting fit (Figure 2) results in an effective C_{even} of the atmosphere that is very much dominated by the near polar regions.

The atmosphere is the Earth's best measured fluid; hence, it is critical that the best atmospheric data be available in order to unravel the effects of the Earth's other subsystems. A key issue in sea level and global change research is the rate at which Greenland and Antarctica are gaining or losing ice. Even the sign is uncertain. With accurate atmospheric pressure data and post-glacial rebound models, ice sheet mass variations can be deduced from time variable gravity measurements.

Acknowledgments

We are grateful to the Space Geodesy Branch at Goddard Space Flight Center for providing the GEODYN software and the LAGEOS normal point data, R. D. Ray for providing the self-consistent equilibrium ocean tidal admittance grid data, M. K. Cheng, R. S. Nerem, D. Rowland, M. Torrence and M. Watkins for helpful discussions. This work was performed at the Jet Propulsion Laboratory, California Institute of Technology, under contract with the National Aeronautics and Space Administration.

References

Chao, B. F., and W. P. O'Connor, Global surface-water-induced seasonal variations in the Earth's rotation and gravitational field, *Geophys. J.*, **94**, 263–270, 1988.

Dong, D., R. S. Gross, and J. O. Dickey, Seasonal Variations of the Earth's Gravitational Field: An Analysis of Atmospheric Pressure, Ocean Tidal, and Surface Water Excitation, *Geophys. Res. Lett.*, **23**, 725-728, 1996.

Gross, R. S., et al., Combinations of Earth Orientation Series: SPACE94, COMB94, POLE94, *J. Geophys. Res.*, **101**, 8729-8740, 1996.

Kalnay et al., The NCEP/NCAR 40-year reanalysis project, *Bull. Amer. Met. Soc.*, **77**, 437-471, 1996.

Kaula, W. M., Theory of Satellite Geodesy, 124 pp., Blaisdell, Waltham, Mass., 1966.

NRC, Satellite Gravity and The Geosphere: Contributions to the Study of the Solid Earth and its Fluid Envelope, National Academy Press, Washington, D.C., 1997.

Nerem, R. S., C. Jekeli, and W. M. Kaula, Gravity field determination and characteristics: Retrospective and prospective, *J. Geophys. Res.*, **100**, 15,053-15,074, 1995.

Ray, R. D., and D. E. Cartwright, Satellite altimeter observations of the Mf and Mm ocean tides, with simultaneous orbit corrections, in *Gravimetry and Space Techniques Applied to Geodynamics and Ocean Dynamics*, edited by B. E. Schutz, A. Anderson, C. Froidevaux, and M. Parke, pp. 69–78, American Geophysical Union Geophysical Monograph 82, IUGG Volume 17, Washington, D.C., 1994.

Trenbert, K. E., Seasonal variations in global sea level pressure and the total mass of the atmosphere, *J. Geophys. Res.*, **86**, 5238-5246, 1981.

Van den Dool, H. M., and S. Saha, Seasonal redistribution and conservation of atmospheric mass in a general circulation model, *J. Climate,* **6**, 22-30, 1993.

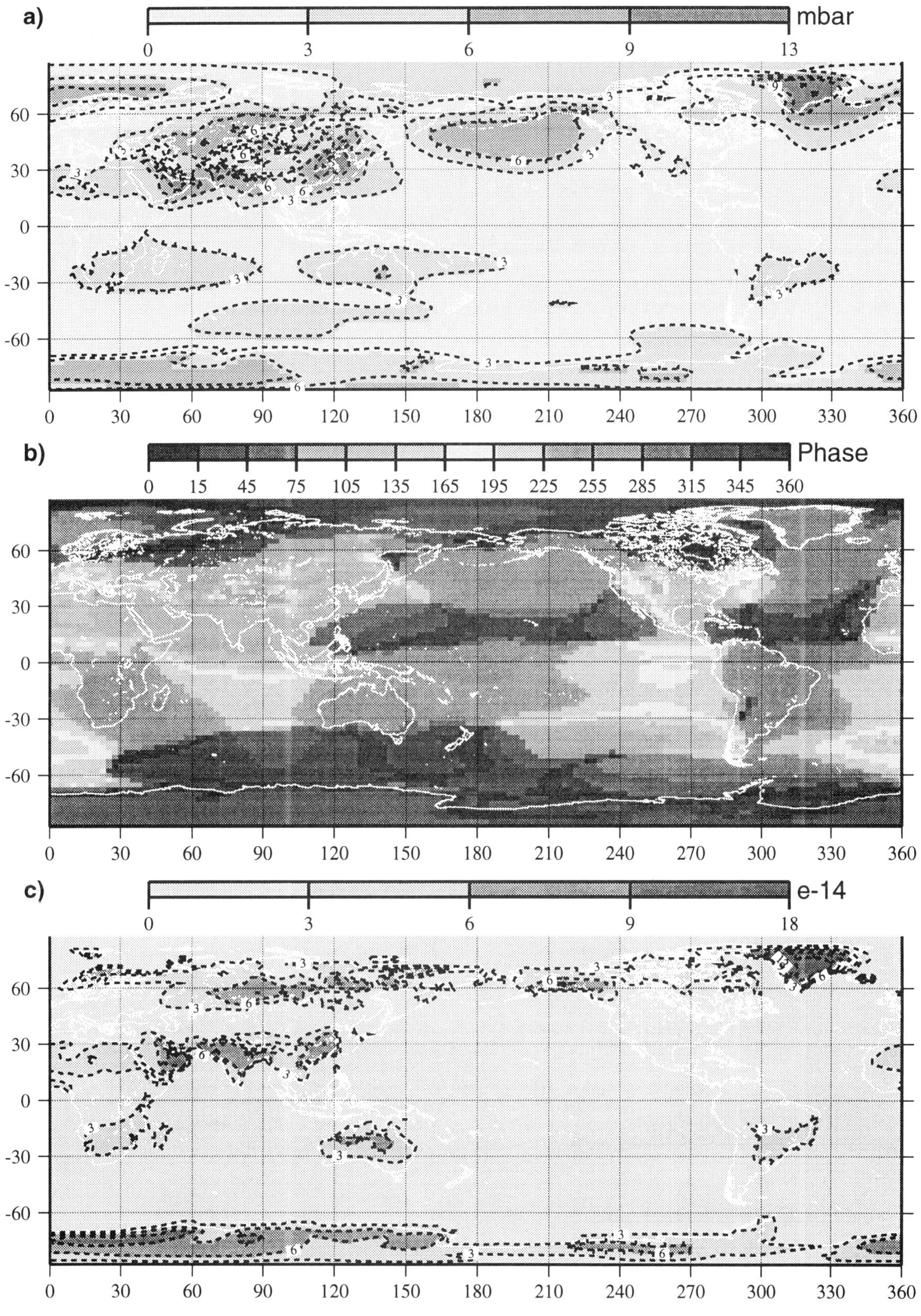

Figure 3. Annual regional pressure variation for the NIB (a) and NIB (c) case and the corresponding phase (b and d) for the recent NCEP reanalysis (*Kalnay et al.*, 1996). The effective C_{even} annual amplitude (e).

MONITORING GEOIDAL POTENTIAL ON THE BASIS OF TOPEX/POSEIDON ALTIMETER DATA AND EGM96

Milan Burša[1], Jan Kouba[2], Karel Raděj[3], Scott A.True[4], Viliam Vatrt[3], Marie Vojtíšková[3]

[1] Astronomical Institute, Czech Acad.Sci., Prague, CR; [2] Geodetic Survey Div., Natural Resources Canada, Ottawa, Canada; [3] Topographical Service of the Army of the Czech Republic, Prague, CR; [4] National Image and Mapping Agency, St.Louis, MO, U.S.A.

Abstract

The geopotential scale factor $R_o = GM/W_o$ (GM - the adopted gravitational constant) and/or geoidal potential W_o have been determined from the EGM96 geopotential model, the POCM 4B sea surface topography model and three years (1994-1996) of TOPEX POSEIDON (T/P) altimetry data: $R_o = (6\,363\,672.589 \pm 0.050)$ m; $W_o = (62\,636\,855.72 \pm 0.50)$ m^{2}s^2. The 5 cm uncertainty in altimetry calibration limits the actual accuracy of the solution. However, the formal rms of R_o is less than ±1mm which makes the monitoring W_o and R_o promising. Continuous W_o / R_o monitoring has been initiated in early 1997 and monthly values of W_o and R_o are presented and discussed.

Introduction

With the availability of high precision TOPEX/POSEIDON and ERS-1 altimeter data (Nerem et al., 1993, Ménard et al., 1994, Le Traon et al., 1995), the determination of geoidal geopotential W_o and/or the geopotential scale factor $R_o = GM/W_o$ has entered a new era. Since they are not limiting the actual accuracy of the solution, the geocentric gravitational constant (Ries et al., 1992) $GM = (398600441.8 \pm 0.8) \times 10^6$ m^{3}s^{-2} and the mean angular velocity of the Earth's rotation (IAG SC3 Rep., 1995) $\omega = 7292115 \times 10^{-11}$ rad s^{-1} can be adopted. The T/P and ERS-1 altimeter data made available to us by Groupe de Recherches on Géodésie Spatiale, Toulouse, AVISO Altimetry Project (AVISO, 1995) and the sea surface topography (SST) model POCM 4B (360 × 360) based on a global circulation model (Rapp et al., 1996) made it possible to significantly refine the previous GEOSAT altimeter data solutions (Burša et al., 1997). The following values were obtained from the first year (Oct. 92 - Dec. 93) of T/P and ERS-1 altimetry:

$$R_o = (6\,363\,672.58 \pm 0.05) \text{ m}; \quad W_o = (62\,636\,855.8 \pm 0.5) \text{ m}^2\text{s}^{-2}.$$

Although the formal *rms* of R_o is ±0.001 m, the actual accuracy of the solution is limited by the 5 cm uncertainty in the T/P calibration (Ries, personal comm., 1996). The solution is only slightly dependent on the geopotential model (Table 1) and/or on the degree of the harmonics retained (Table 2).

Table 1. Geopotential (W_0) and geopotential scale factor (R_0) determined from OSU91A, EGM96 and the geopotential models distributed by Sideris (1995; 1996).

Geopotential model	W_0 [m^2s^{-2}]		R_0 [m]	
OSU91A	62 636 855.87	± 0.014	6 363 672.574	± 0.001
EGM-X01	62 636 855.83	± 0.016	6 363 672.578	± 0.002
EGM-X02	62 636 855.89	± 0.012	6 363 672.572	± 0.001
EGM-X03	62 636 855.87	± 0.015	6 363 672.574	± 0.001
EGM-X04	62 636 855.86	± 0.015	6 363 672.575	± 0.002
EGM-X05	62 636 855.84	± 0.010	6 363 672.577	± 0.001
EGM96	62 636 855.84	± 0.010	6 363 672.577	± 0.001

Table 2. Dependence of W_0 and R_0 on degree $\bar{n}$ of harmonics retained in the EGM96 geopotential model.

$\bar{n}$	W_0 [m^2s^{-2}]		R_0 [m]	
6	62 636 863.27	± 0.46	6 363 671.822	± 0.047
90	62 636 855.90	± 0.04	6 363 672.571	± 0.004
180	62 636 855.90	± 0.02	6 363 672.571	± 0.002
240	62 636 855.86	± 0.01	6 363 672.575	± 0.001
300	62 636 855.84	± 0.01	6 363 672.576	± 0.001
360	62 636 855.84	± 0.01	6 363 672.577	± 0.001

Testing of the EGM96 Geopotential Model

The accuracy of EGM96 was tested using 20768 T/P altimetry testing sites over the oceans, 1830 testing sites in the U.S.A., 963 testing sites in Canada (Mainville, 1997), 220 testing sites in the Central Europe and at 25 sites in the Baltic region (Chen and Kakkuri, 1996). The results are summarized in Table 3.

The Testing Geopotential Model (TGM) methodology (Burke et al., 1995) is based on the difference δW between the geopotential value at the testing site computed using the function

$$W = W(GM,\ \omega,\ J_2,\ W_o,\ x_j,\ H_q)$$

and the geopotential value computed from the tested geopotential model:

$$\delta W = W - W(model);$$

where x_j stands for geocentric coordinates, H_q for the Molodensky normal height of the testing site and $J_2 = -(1082626.7 \pm 0.1) \times 10^{-9}$ is the second zonal Stokes parameter in the tide-free system (IAG SC3, 1995).

Table 3. EGM96 radial distortions (δR) of the equipotential surface passing through the testing sites.

Testing Area	mean [m]	rms [m]
Ocean Altimetry (20768 sites)	-0.001 ± 0.001	±0.141
U.S.A. (1830 sites)	0.518 ± 0.010	±0.438
Canada (963 sites)	0.532 ± 0.011	±0.340
Central Europe (220 sites)	0.088 ± 0.013	±0.189
Baltic region (25 sites)	-0.150 ± 0.064	±0.318

The TGM methodology can be applied in the mean, tide-free or zero-tide systems with identical results. The only condition is that all the data must be consistent in regards the tide system. If the mean-tide system is applied, then the following second Stokes parameter should be used: $J_2 = -(1082666.7 \pm 0.1) \times 10^{-9}$. However, in that case the geopotential values computed from EGM96, which is in the tide-free system, should be corrected for the direct and indirect zero-frequency tidal terms. Note that fundamental constants GM, ω and W_o are invariant in regards to the tide system. Also invariant is the quasigeoidal height $\zeta_q = (MN)$ (Fig.1) as well as the space position of point N defined by condition $U(N) = W(M)$.

Figure 1. Testing site M situated on the Earth's surface over continents and islands

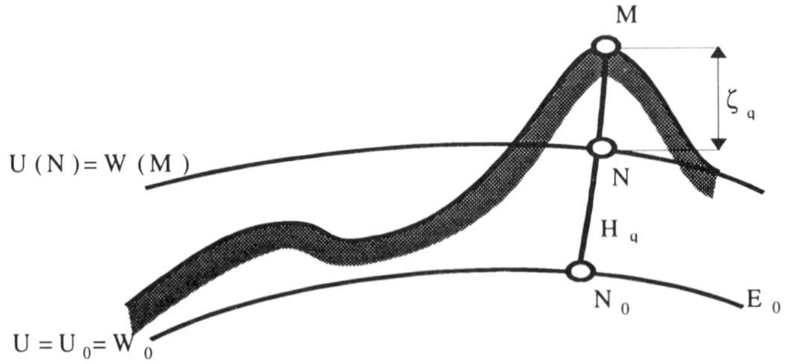

However, the normal potential

$$U_{mean} = U_{tide\text{-}free} + (1+k)\,\Delta W$$

and the heights (both normal and orthometric) differ:

$$H_{mean} = H_{tide\text{-}free} - (1+k)\,\Delta W / g\,,$$

where ΔW stands for the direct zero-frequency tidal term due the Moon and the Sun, k is the Love number and g is the gravity at M. Analogously, the same relation holds for the ellipsoidal height $h = H_q + \zeta_q$. However,

$$h - H_q = \{U(N) - U(M)\}/\gamma(N) = \{W(M) - U(M)\}/\gamma(N) = \zeta_q$$

does not depend on the tide system.

Instead of δW the radial distortion δR of the equipotential surface passing through the testing site can be used in TGM. Distortions δW and/or δR are due to the geopotential model tested if the errors of the four fundamental constants, the geocentric coordinates x_j and the Molodensky normal heights H_q of the testing site are small enough. The uncertainty in the adopted fundamental constants contribute to δR as follows: GM (±13 mm), ω (±1 mm), J_2 (±0.6 mm), W_o (±50 mm). The errors in x_j and H_q can reach up to a few centimeters and decimeters, respectively.

At the testing sites M's situated on the sea surface (i.e. the sea surface topography) the testing values required are as follows (Fig. 2): the altimetric height h of M and its ellipsoidal coordinates, as well as the sea surface topography height h_{SST} of M related to the equipotential surface $W = W_o$. Note that h_{SST} is identical to the Molodensky normal height of M, i.e. $h_{SST} = (M_o M) = (N_o N)$.

Figure 2. Testing site *M* situated on the sea surface

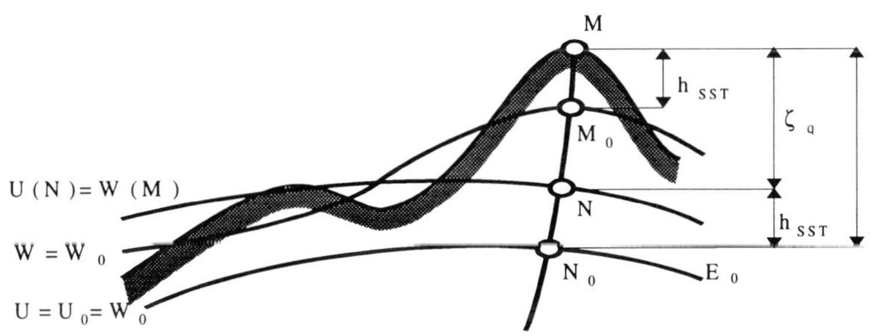

The Molodensky normal heights H_q required by the TGM methodology may not, however, be generally available. That is why, the heights available should be transformed as accurately as possible into the Molodensky normal heights. If the Helmert orthometric heights, H_{or} are based on the simplified Prey reduction, the exact relation is:

$$H_q - H_{or} = (g - \gamma_m + 0.0424\, H_{or})H_{or}/\gamma_m , \qquad (1)$$

here g is the actual gravity at the testing site, γ_m stands for the integral mean value of the normal gravity along the arc of normal plumb-line $(N_o N)$ (Fig. 1).

This height conversion was applied at the 1830 U.S. and at 963 Canadian testing sites. Figures 3 graphically depicts corrections (1) for the U.S.A. The height corrections (1) applied at Canadian test sites show similar behaviour as in Figure 3.

The non-zero biases in δW and in δR, i.e. the mean values computed over TGM regions connected to the same vertical datum (a tide gauge station) should be expected and they can be interpreted as vertical datum shifts with respect to a global reference.

Figure 3. Differences between the Molodensky normal heights and the Helmert orthometric heights for the U.S.A. (1830 sites); also shown is an average curve.

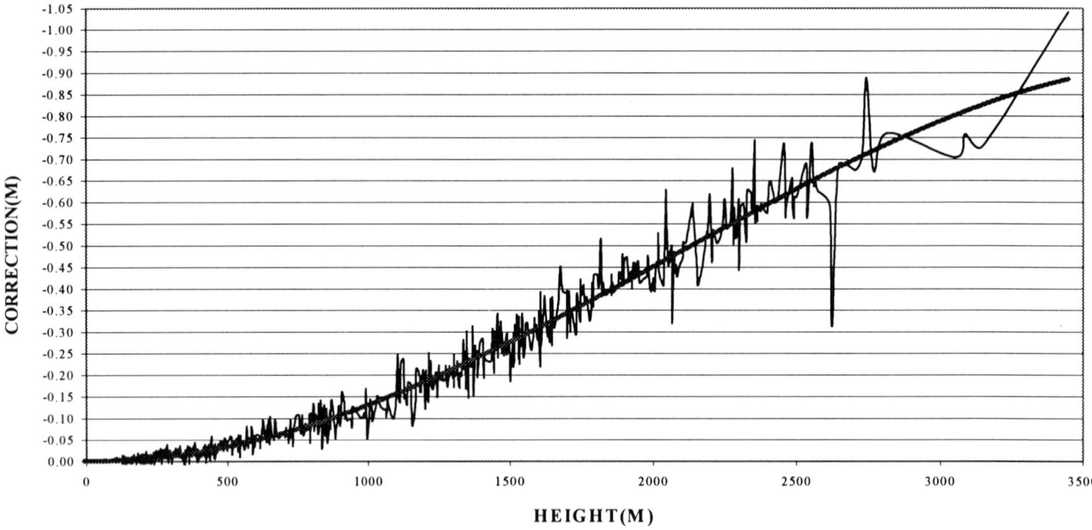

Monitoring of W_o and R_0

There are three quantities which, if varying, give rise to the variations in W_o and R_o : GM, ω and the volume (τ) defined by surface $W=W_o$. The error in GM contributes to the geopotential error by about ± 0.13 m^2s^{-2} and ± 13 mm; the seasonal variations in ω ± 0.01 m^2s^{-2} and 1mm, respectively. The error due to GM is not negligible, however, if it is a constant in time, it does not affect the monitoring of W_o / R_o variations. The 2-5 cm calibration altimeter error limits the actual accuracy of W_o and R_o. The increase in the Earth's mass due to celestial body impacts is too small to influence W_o significantly, i.e. $d(GM)/dt = 6.7 \times 10^{-4}$ m^3s^{-2}/year.

Long-term variations in τ may be caused by physical characteristics of oceans, e.g. the salinity of water. The monitoring of W_o and R_o during a few years should help with interpretation of $d\tau/dt$. However, there may also be long-term variations in the altimeter calibration and satellite orbits which may mask the real signal in $d\tau/dt$. The careful analysis of the long-term variation in W_o / R_o should help to clarify its origins.

Recently, we have commenced permanent monitoring of W_o and R_o. The initial results for the period of Aug 14, 1993 - May 15, 1997 are shown in Figure 4. Here, an annual signal with maxima in February and minima in August and with an amplitude of about 0.1 m^2s^{-2} (10 mm) is clearly visible. This highly periodic behaviour and other smaller systematic variations seen in Figure 4 are likely related to seasonal and systematic variations in SST.

Figure 4. Monthly variations in W_0 (62363800 m^2s^{-2} subtracted) and R_0 (6363600 m subtracted)

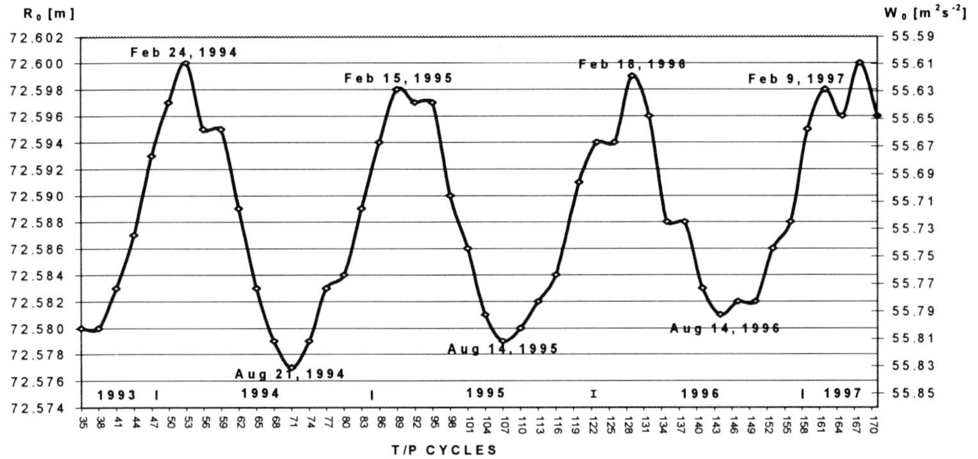

Conclusions

A continuous monitoring, based on T/P altimetry data, of the geoidal geopotential W_o and geopotential scale factor R_o have been initiated. The formal *rms* of W_o and R_o are about ± 0.016 m^2s^{-2} (± 1.6 mm) for monthly and about ± 0.005 m^2s^{-2} (± 0.5 mm) for yearly solutions, respectively. The yearly solutions are the same as the mean values computed for the three year period of Jan. 1, 1994 to Dec. 31, 1996: $W_o = (62636855.72 \pm 0.003)$ m^2s^{-2} and $R_o = (6363672.589 \pm 0.0003)$ m. However, the TOPEX altimeter calibration error limits the actual accuracy of W_o and R_o to about ± 0.5 m^2s^{-2} and 5cm, respectively (Ries, personal comm., 1996). The W_o / R_o changes due to the differences in recent geopotential models as well as the degree of harmonics retained are not significant. The corresponding values of the semi-major axis of the level ellipsoid are as follows:

tide-free system: a = (6378136.581 ±0.050) m;
zero-tide system: a = (6378136.609 ±0.050) m;
mean-tide system: a = (6378136.708 ±0.050) m.

There are annual variations with an amplitude of about 0.1 m^2s^{-2} in W_o and 10 mm for R_o. The origin of the annual and other systematic variations should be searched in the seasonal variations of sea surface topography heights and, eventually also in T/P altimetry data and/or orbits. The permanent W_o / R_o monitoring should enable investigation of the long-term variations $d\tau/dt$ in the volume defined by surface $W = W_o$ as well as unmodeled long-term variations in T/P satellite orbits.

Acknowledgments

The authors wish to express their sincere thanks to Dr. John C. Ries for his interest and the personal communications concerning the TOPEX altimeter calibration error, and to

Dr. Patrick Vincent, who made the TOPEX altimeter data available and who kindly instructed us during its treatment.

References

AVISO,: 1995, 'AVISO User Handbook: Reduced GDRs', AVI-NT-011-311-CN, *Edition 1*, 3G, pp. 17.

Burke, K.F., S.A.True, M. Burša, K. Raděj: 1996, 'Accuracy Estimates of Geopotential Models and Global Geoids' Proceedings of Symposium No. 116 held in Boulder, CO, USA, July 12, 1995; Edited by R.H. Rapp, A.A. Cazenave and R.S. Nerem Springer Verlag 1996, pp. 50-60.

Burša, M., K. Raděj, Y. Šíma, S.A. True, V. Vart: 1997, 'Determination of the Geopotential Scale Factor from TOPEX/POSEIDON Satellite Altimetry', Studia geoph. et geod. 41 (1997), pp. 203-216.

Chen, R., J. Kakkuri,: 1996, 'Results of the Baltic Sea Level 1993 GPS Campaign', 95:2 Final Results of the Baltic Sea Level 1993 GPS Campaign, Research Works of the SSG 5.147 of the International Association of Geodesy, 1995; edited by Juhani Kakkuri, pp. 21-30.

IAG SC3 Final Report, Travaux de L'Association Internationale de Géodésie, Tome 30, Paris 1995, pp. 370-384.

Le Traon, P.Y., P. Gaspar, F. Ogor, J. Dorandeu: 1995, 'Global adjustment of ERS-1data using TOPEX/POSEIDON as a reference', Techn. Report contract CNES/CLS GO/94/522, pp. 12.

Mainville, A.: 1997, 'The February 1997 GPS on BM data file from Canada', A distribution data set for 1482 stations in Canada, Geodetic Survey Division, Natural Resources Canada, Ottawa, Canada, May 9.

Ménard, Y., E. Jeansou, P. Vincent: 1994, 'Calibration of the TOPEX/POSEIDON altimeters at Lampedusa: Additional results at Harvest', Journ. Geophys. Res., 99, No. C12, 24, pp. 487-504.

Nerem, R.S., B. H. Putney, J.A. Marshall, F.J. Lerch, F.J., E.C. Pavlis, S.M. Klosko, S.B. Luthcke, G.B. Patel, R.G. Wiliamson, N.P. Zelensky: 1993, 'Expected orbit determination performance for the TOPEX/POSEIDON Mission', IEEE Trans. Geosci. Remote Sensing, 31, pp. 333-354.

Rapp, R.H., C. Zhang, Z. Yi : 1996, 'Analysis of Dynamic Ocean Topography Using TOPEX Data and Orthonormal Functions', Journ. Geophys. Res.- Oceans, 101, No. C10, pp. 22 485-22 494.

Ries, J.C., R.J. Eanes, C.K. Shum, M.M. Watkins: 1992, 'Progress in the Determination of the Gravitational Coefficient of the Earth's', Geophys. Res. Letters}, 19, No. 6, pp. 529-531.

Sideris, M.: 1995 and 1996, 'Earth's Models EGM X01-05', Distributed for members of Special Working Group (IAG) set up for testing the new GSFC/DMA Geopotential Models.

Absolute positioning, tectonic plate motions and polar motion determination with the DORIS system.

J-F Crétaux, L. Soudarin, F. Bouille, and A. Cazenave, GRGS, UMR5566 CNES/CNRS,
18 av Ed Belin, 31400 Toulouse, France.

Abstract

DORIS data collected between 1993 and 1996 from the 3 satellites Spot2, Spot3, and Topex/Poseidon have been analysed in order to provide precise positions and velocities of the ground beacons of the DORIS network, and daily values of the polar motion. Recent improvements in the processing have provided solutions at the centimeter level precision for the beacon position, and at the 1-milli-arc-second level for the polar motion parameters. Furthermore, angular velocities and Euler poles of absolute and relative motions between eight major tectonic plates have also been computed from the station velocities determined over the 4-year time span.

Introduction

The DORIS (Doppler Orbitography and Radiopositioning Integrated by Satellite) system has been developed in the early 90s by the CNES (Centre National d'Etudes Spatiales) in France to be placed onboard the altimeter satellite Topex/Poseidon launched in 1992. Prior to Topex/Poseidon, DORIS has been operational from the launch of the Spot-2 satellite in January 1990. In September 1993 DORIS had also been placed onboard the remote sensing satellite Spot-3. In order to reach the performance initially expected for its first mission which was precise orbit determination of the Topex/Poseidon satellite, a worldwide network of permanent transmitting beacons has been deployed with the collaboration of the IGN (Institut Géographique National). The DORIS system allows the orbit of Topex/Poseidon to be computed with a precision of 1-2 cm (Nouel et al., 1994). DORIS is also used for absolute positioning of the ground beacons. Since the launch of Spot-2 in 1990, the DORIS performances for absolute positioning have been regularly improved, from a precision of 10 to 20 cm obtained during the first year of the system life time to 1 cm nowadays. In the mean time the number of beacons has increased from 29 to 50 now, providing a dense and homogeneous network covering eight major tectonic plates. This global coverage is also suitable to observe the variations with time of the Earth rotation axis position. The improvement of the geodetic results obtained with DORIS have been made possible by improvements in the data analysis. The characteristics of the system and the geodetic performances comparable to GPS, SLR, and VLBI have led in 1994 the IERS (International Earth Rotation Service) to include DORIS as a new technic for the ITRF (International Terrestrial Reference Frame) determination and the computation of Earth rotation parameter time series.

Data analysis

DORIS is an uplink Doppler system working on two frequencies (400 Mhz and 2 Ghz) transmitted by ground beacons. The DORIS data used for orbit determination and ground beacon position computation are derived from the 2 Ghz frequency signal. The 400 Mhz signal is used only for the calculation of the ionospheric propagation delay. The processing of DORIS data is performed by the GINS/DYNAMO software based on a semi-dynamical method developed at GRGS (Groupe de Recherche de Géodésie Spatiale). It consists of computing the satellite's orbit, beacons positions and velocities, and Earth orientation parameters, in a single inversion, together with selected parameters required to improve the acceleration models (used to describe the satellite orbit) and the measurement corrections. The acceleration model includes geopotential, lunisolar and planetary accelerations, atmospheric drag, solid Earth and ocean tides, direct and Earth-reflected radiation pressure, and relativistic acceleration. For the wet tropospheric and the ionospheric corrections, the CNET1 model (Berrada-Baby, 1987) and the dual frequency measurements were used respectively. Corrections due to Earth tide and ocean loading were applied to the initial coordinates of the stations.

We recently investigated various ways of improving the DORIS performances. This includes improved model for the satellite orbits (use of the JGM-3 geopotential model (Tapley et al 1994), and the recent FES-95 ocean tide model (Le-Provost et al, 1997), and use of a new set of beacons coordinates). This has led to substantial improvements in the accuracy of both absolute positioning (by 20%) and polar motion (by 30 to 40%). Moreover for Topex/Poseidon we now used 3-day (instead of 1-day) arc length for orbit determination, which is a subcycle of the repeat period. This reduces noise in polar determination by nearly 50%.

A two-step analysis was performed. In the first step, orbits are computed over 1 or 3-day arcs. At this step, the geodetic parameters (beacons positions and Earth polar coordinates) are kept fixed at their initial values while empirical parameters of the force model and orbital parameters are adjusted. These parameters are then used as initial values for the second step where a new orbit is computed over the same arc length, as well as the normal equation from which station positions and velocities, and Earth orientation parameters are derived. Then the orbital parameters are backsubstituted and the daily normal equations accumulated over a monthly or a multi-year basis using weights, computed by an Helmert method consisting of estimating the components of the variance matrices of the residuals. Then inversion of the normal equation is performed and the solution (positions and velocities of the DORIS beacons) derived.

The Doppler data used in this study consist of 4 years (1993-1996) of DORIS measurements for Spot-2 and Topex/Poseidon and 3 years (November 1993-November 1996) for Spot-3. The estimated average internal precision of the Doppler measurements (expressed as radial velocity between satellites and ground beacons) is 0.51, 0.48, 0.46 mm/sec for Topex/Poseidon, Spot-2, and Spot-3 respectively.

Positioning results

Twenty four bimestrial solutions and fourty eight monthly solutions for site positions, referred to 1993.0, have been produced from January 1993 to December 1996 as well as a mean 4-year solution combining all normal equations of the three satellites. The coordinate system was stabilized by fixing the 3 coordinates X, Y, and Z, of the Toulouse station. The polar motion coordinates were also kept fixed at their IERS values for this study. The formal standard deviations are at the level of a few millimeters. For each bimestrial solution, translation, rotation and scale factor were computed in order to put them in the same reference frame. Comparisons between bimestrial solutions and the 4-year solution are then performed allowing an estimate of the repeatability (or internal coherence) of the absolute position solutions. As all bimestrial solutions are referred to the same date by using the Nuvel-1 tectonic plate model (DeMets et al, 1990), any slope observed in the month to month positions represents a discrepancy between DORIS and Nuvel-1. Figure 1 presents histogram of the differences between monthly and bimestrial solutions (expressed in geographical coordinates) and the 4-year solution for the whole network. The values remain in the range of 1-1.5 cm. We have estimated the position repeatability for each station individually. For 90% of the stations, it remains at the centimeter level or less, while for the remaining 10% stations, variations are between 1 and 2 centimeters.

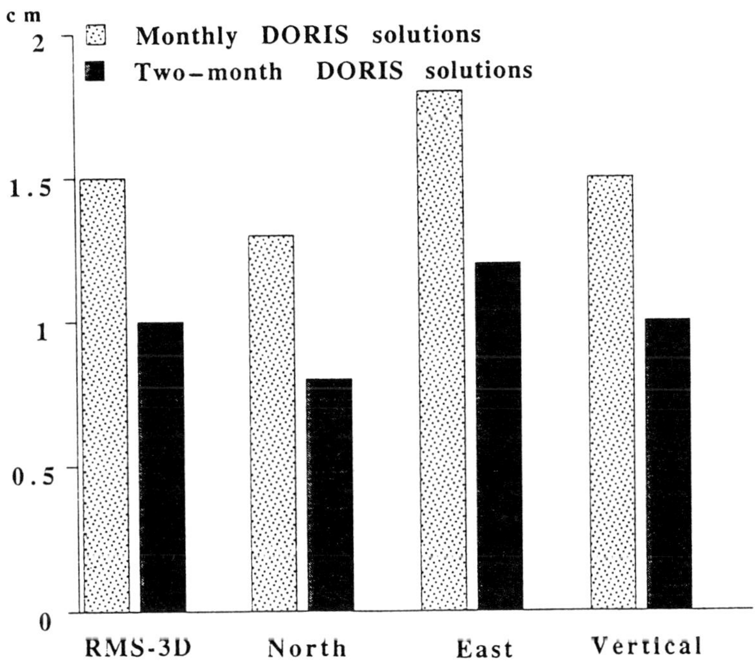

Figure 1: Precision of the DORIS beacon coordinates.

Tectonic plate motions

Due to the homegeneous coverage of its network and the continuous data collection since 1990, DORIS is becoming increasingly useful for the study of tectonic plate motions. We present below a synthesis of results obtained recently on large scale tectonic motions (detailed results can be found in Crétaux et al, 1997). Table 1 shows absolute plate motions in terms of Euler vectors and angular velocities, for eight major tectonic plates as well as predictions from the Nuvel-1 and Nuvel-1A models. We chose at least 2 beacons per plate sufficiently far away from plate boundaries. Figure 2 shows the horizontal velocities of these beacons obtained with DORIS and their corresponding Nuvel-1 predictions. This figure indicates clearly that in terms of absolute velocities, the DORIS determinations are close to kinematic model predictions.

Table 1 Absolute DORIS plate rotation vector: comparison with Nuvel-1 and Nuvel-1A (The number of sites used for each plate are indicated in parenthesis).

Plate		Φp (deg)		λp (deg)		Ω (deg/m.a)	
*****************		*****************		*****************		*****************	
Eurasia	(4)						
DORIS		47.5	+/-6.5	-112.2	+/-5.3	0.26	+/-0.02
Nuvel-1 / Nuvel-1A		50.6		-112.4		0.24	0.23
Africa	(4)						
DORIS		50.1	+/-2.9	-76.3	+/-8.1	0.33	+/-0.02
Nuvel-1 / Nuvel-1A		50.6		-74.0		0.30	0.29
North-America	(4)						
DORIS		4.7	+/-10.3	-77.0	+/-3.5	0.20	+/-0.02
Nuvel-1 / Nuvel-1A		-2.5		-86.0		0.22	0.21
South-America	(3)						
DORIS		-30.6	+/9.5	-127.5	+/-13.7	0.15	+/-0.03
Nuvel-1 / Nuvel-1A		-25.4		-124.6		0.12	0.11
Nazca	(2)						
DORIS		43.5	+/-6.4	-101.6	+/-4.3	0.71	+/-0.07
Nuvel-1 / Nuvel-1A		47.8		-100.2		0.78	0.74
Antartica	(5)						
DORIS		65.3	+/-11.1	-114.2	+/-7.2	0.24	+/-0.05
Nuvel-1 / Nuvel-1A		63.0		-115.9		0.25	0.24
Pacific	(2)						
DORIS		-60.0	+/-3.7	91.2	+/-20.9	0.68	+/-0.06
Nuvel-1 / Nuvel-1A		-63.0		100.7		0.67	0.64
Australia	(4)						
DORIS		32.3	+/-3.2	37.4	+/-3.0	0.63	+/-0.04
Nuvel-1 / Nuvel-1A		33.8		33.2		0.68	0.65

For six of the eight plates studied, we observe for the location of the Euler pole a very good agreement with the Nuvel-1 and Nuvel-1A models (differences <5 degrees in the north and east directions). For the north-American plate, the difference with the model is due to the Yellowknife station which presents a DORIS velocity in a direction sligthly different from that of Nuvel-1. If this beacon is removed from the computation, observed differences with the kinematic model are significantly reduced (<5 degrees). The error observed for the Pacific plate is due to the lack of stations on this plate. Only two beacons have regularly observed during the 4 years and even if the velocities of these beacons are close to model, small differences could conduct to high differences in the rotation pole determination.
Table 1 also shows that the DORIS velocities are in a better agreement with Nuvel-1A than Nuvel-1, for the Australia, Antartica, Nazca and North-America plates. Hence DORIS plate velocities are closer to the Nuvel-1 model for half of them while they are closer to the revisited Nuvel-1A model for the other half. It is probably premature however to conclude anything from this result.

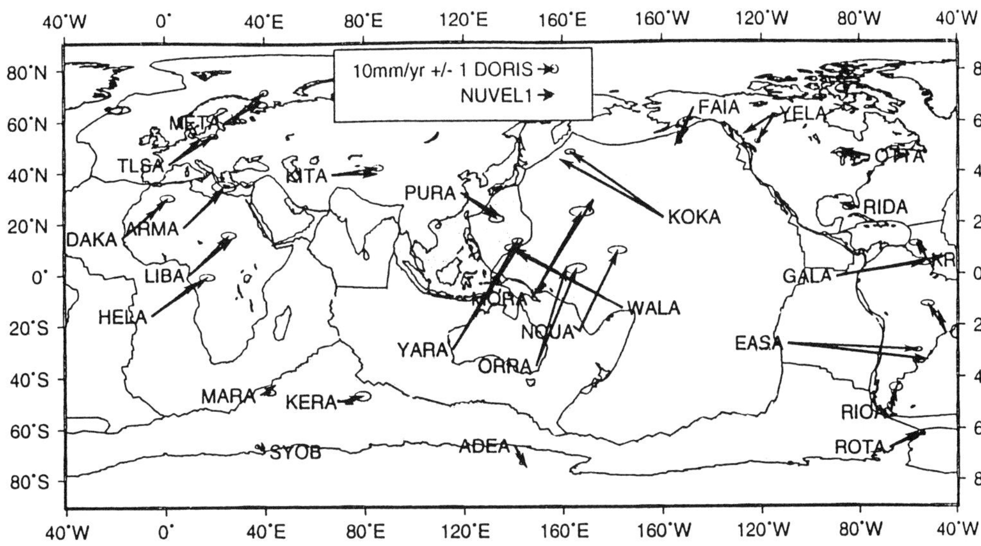

Figure 2: DORIS and Nuvel-1 absolute horizontal velocities.

Earth polar motion determination

In the dynamical orbit determination, daily solutions for the Xp and the Yp coordinates of the pole axis are determined. Each daily solution is based on multi-satellite data. The accuracy currently obtained for the polar motion is one milli arc second in the two directions. Results are still less precise than for GPS or VLBI, but recent improvements obtained with DORIS, principally due to the availibility of three satellites in orbit allow us to except more accurate results in the future when more satellites carrying the DORIS system will be launched.

Future evolution of the system

It still remains residual errors that could be decreased in the near future. One can distinguish two types of errors: orbital errors, and errors due to geometry of the satellite passes. One way of decreasing orbital errors is to improve the geopotential model by incorporating DORIS data in a tailored version. Gravitational errors are indeed very important for the Spot satellites. DORIS data (as well as laser and GPS data) are presently used in the computation of a new version of the GRIM model (calculated by GRGS and GFZ (GeoForschungsZentrum) in Germany).

The fact that at present onboard receivers cannot collect several transmissions simultaneously reduces the number of measurements available for each ground beacon. In order to increase the positioning precision, a new generation of receivers are planned. They will be able to observe two signals of two nearby beacons simultaneously.

The last major limitation is the few number of satellites currently equipped with a DORIS receiver. As we also saw in the past, the availability of new satellites equipped with DORIS receivers, especially satellites in different orbital inclination and altitude will allow to improve, once again, geodetic results. This was demonstrated when adding the Topex/Poseidon data in the processing. Topex/Poseidon is on a different orbit than the other two satellites and thus provides different information which could not be obtained with the Spot satellites only, even if processing several years of data. Fortunately the future looks very promising for DORIS. In the beginning of 1998, DORIS will placed onboard the remote sensing satellite Spot-4, then onboard the European satellite Envisat (1999) dedicated to study the Earth environment. DORIS will also support altimetry onboard the American-French satellite Jason to be launched in 1999. All these projects will assess the perenity of the system and certainly allow more accurate geodetic performances.

References

Berrada-Baby, H., P. Golé, and J. Lavergnat, Effets de la troposphère sur les mesures de distances Terre-satellite. Application au projet DORIS, *Note Technique*, CRPE, 1987

Boucher, C., Z. Altamini, M. Feissel, P. Sillard, Results and analysis of the ITRF94, *IERS Technical Note* 20, 1996.

Cazenave, A., J-J. Valette, & C. Boucher, Positioning results with DORIS on SPOT-2 after first year of mission, *J. Geophys. Res.*, 97, 7109-7119, 1992.

Crétaux, J-F., L Soudarin, F. Bouille, and A. Cazenave, Current tectonic plate motions and crustal deformations using DORIS data, submitted to *J. Geophys. Res.*, 1997.

Le Provost, C., F. Lyard, J. M. Molines, M. L. Genco and F. Rabilloud, A hydrodynamic ocean tide model improved by assimilating a satellite altimeter derived data set, in press to *J. Geophys. Res.*, 1997.

DeMets, C., R. G. Gordon, D. F. Argus & S. Stein, Current plate motions, *Geophys. J. Int* 101, 425-478, 1990.

DeMets, C., R. G. Gordon, D. F. Argus & S. Stein, Effect of recent revisions to the geomagnetic reversal time scale on estimates of current plate motions, *Geophys. Res. Lett.*, 21, 2191-2194, 1994.

Nouel, F.,et al., Precise CNES orbits for Topex/Poseidon: Is reaching 2 cm still a challenge? *J. Geophys. Res.*, 99, 24405-24419, 1994.

Sahin, M., P. A. Cross, and P. C. Sellers, Variance component estimation applied to satellite laser ranging, *Bulletin Géodésique*, 66, 284-295, 1992.

Soudarin, L., and A. Cazenave, Global Geodesy using doris data on SPOT-2, *Geophys. Res. Lett.*, 20, 289-292, 1993.

Soudarin, L., and A. Cazenave, Large-Scale tectonic plate motions measured with the DORIS space geodesy system, *Geophys. Res. Lett.*, 22, 469-472, 1995.

Tapley, B. D., et al., The JGM-3 gravity model, *Ann. Geophys.*, 12, Suppl. 1, C192, 1994

GEODETIC BASELINE ADJUSTMENT BY VLBI TECHNIQUE

B. GHEZALI, S. KAHLOUCHE

National Center of Spatial Techniques

BP 13 Arzew - 31200 - ALGERIA

Abstract :

The elaboration of a VLBI geodetic data processing named **"GLORIA/Solve"**, and the mobile VLBI observing session analysis of the IRIS program, allowed the validation of one part of the VLBI geodetic and astrometric chain analysis GLORIA established by the IERS. The preliminary data processing were developed for the tropospheric effects and clock offset, with the selection of the estimated parameters and the definition of their domain of validity. The residues, after model adjustment on the measurements , have an RMS of 170 picoseconds, and the precision on the baseline components is in the order of 5 cm. In comparison with the ITRF92 reference solution, the centimetric results obtained presents a decimetric deviation on the baseline components. These deviations were essentially caused by the IRIS network configuration, the model used and the restricted number of observations.

Key words : Geodetic baseline - VLBI - Troposphere offset - Clock offset - Adjustment.

Introduction :

Under the aegis of the central office of the IERS, a research team with the participation of the CNTS, contributed to the elaboration of a project on the development of a VLBI geodetic and astrometric chain analysis called GLORIA.

The work done in the CNTS needs to setting up and validate the chain analysis GLORIA, carried mainly on the pre-processing and the adjustment by least squares, of VLBI parameters (clock and tropospheric parameters , station coordinates,...). The VLBI data processing **"GLORIA/Solve"** consists of two main steps :

- *Pre-processing :* localisation of the clock offset and determination of the optimal time interval on whom the tropospheric delay will be estimated.
- *Processing :* adjustment of the parameters (clock difference and shift, wet tropospheric delay, station coordinates and the baseline components).

I. Model of VLBI delay :

The model of VLBI delay of the GLORIA version was conceived in order to assure a precision of 10 ps.

In general, the delay τ issued from the correlation is constituted by the following terms:

$\tau = \tau g + \tau hor. + \tau inst. + \tau ion. + \tau trop. + \tau grav. + \tau struct.$

with :

- τg : geometric delay (main term).
- $\tau hor.$: delay caused by the clock difference and shift, delay due to clock offset.
- $\tau inst.$: propagation delay between the receiving antenna and the recorder.
- $\tau ion.$: ionospheric delay.
- $\tau trop.$: tropospheric delay.
- $\tau grav.$: delay caused by the wave deviation in the presence of a mass.
- $\tau struct.$: delay due to the effects of the complex morphology of the extragallactic radiosources.

Previously to the scalar product between the baseline vector $\vec{B}$ and the radiosource vector $\vec{S}$ $(\vec{S}.\vec{B})$, it is necessary to link the terrestrial conventional frame [RTC] on whom are referred the stations of observation, to the celestial conventional frame [RCC] on whom are positioned the sources.

In order to express this transformation we have adopted the classic astronomical modelisation :

$$[\,\mathbf{RCC}\,] = P(t)\ N(t)\ S(t)\ W(t)\ [\mathbf{RTC}]$$

P(t) : Precession; N(t) : Nutation; S(t) : sideral time; W(t): pole coordinates .

The baseline vector $\vec{B}$ varies in the time because it is liable to some physical effects as:

- Movement of the plates (1 to 15 cm/ year),
- Terrestrial tides (max. 30 cm),
- Oceanic overcharge (3 cm, to 100 km of the coasts),
- Atmospheric overcharge (a few mm),
- Polar tide (< 1 cm).

II- Presentation of VLBI chain analysis : "GLORIA"

The processing of the VLBI observations for the geodesy and the astrometry do generally in three steps:

1. The first step consists in the correlation into deferred time of the signals recorded on a tape in order to obtain the delay (τ) and the speed of phase delay ($\dot{\tau}$). This operation is done by some specialized centers as the National Geodetic Survey of the United States (NGS), who collected the VLBI measurements processed in the CNTS.

2. The second step consists to calculate τ and/ or $\dot{\tau}$ from the model as well as the partial derivatives with regard to the estimate parameters. This step correspond to the first part of the GLORIA software called "GLORIA/O-C", and established by the IERS.

3. A third step where we minimize by the least squares method deviation between the observed value ($\tau obs.$ and/or $\dot{\tau}$ obs.) and the computed value ($\tau cal.$ and/or $\dot{\tau}$ cal.), adjusting some number of

the parameters which were separated in two categories : the main or global parameters (earth rotation parameters, stations and sources coordinates) and the disturbing or local parameters (tropospheric delay, clock offsets).

This step corresponds to the second part of denominated GLORIA software "GLORIA/Solve", and was developed in the CNTS. It consists of :

- a pre-processing used to locate the clock offsets and to determine the interval of optimal time for the adjustment of the tropospheric parameters,
- and a processing used to adjust the main parameters which are the the radiotelescope positions (case of the geodetic applications) .

The used observations, available at the CNTS from an IERS convention of collaboration, are constituted by an IRIS-A program session with a length of 24 hours between May 11 and May 12, 1992. During this period, a set of 518 observations was made on 28 sources from 04 stations distributed on two tectonic plates (EURA and NOAM).

The results obtained during this analysis focused only on the processing of the observing relations of band synthesis delay and are of the level of precision permitted by this mobile VLBI observing campaign.

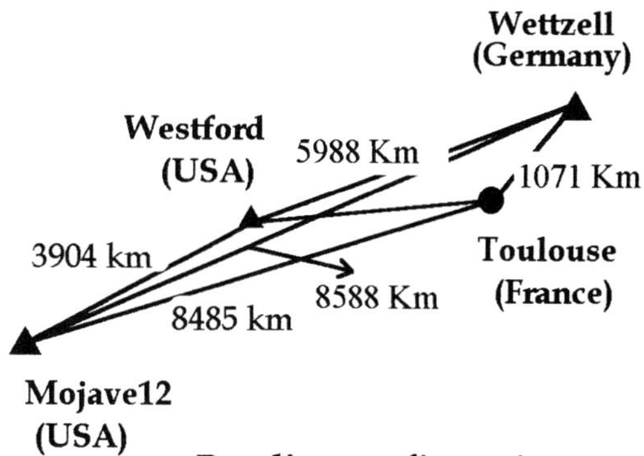

Baseline configuration
▲ : Primary (Colocation) station,
● : Secondary station.

III. Computation of model : " GLORIA/O-C " .

The software "GLORIA/O-C," developed by the IERS, corresponds to the first part of GLORIA which provided the computed delay for each observation (τcal.) and the partial derivatives with regard to the estimate parameters. In the setting of this work, based on the modelisation of the IERS, the models adopted are the following:

- **Motion of the tectonic plates :** *model NUVEL NNR-1 (Standard IERS 1992).*
- **Terrestrial Tides:** *theory of Wahr (1981) (Standard IERS 1992).*
- **Moon and Sun ephemeris :** *analytic theories ELP2000-85 and VSOP82.*
- **Oceanic Tides:** *results of Scherneck (1991).*
- **Gravitational Delay (Sun, Earth):** *model constructs by (Eddington 1923; Shapiro 1964; Richter et Matzner 1983).*
- **Passage of terrestrial frame to the celestial frame :** *classic astronomical modelisation.*

- **Troposphere:** *models of wet and dry zenith delay (Saastamoinen, 1972 and 1973); models of mapping functions : wet components (Chao, 1974) and dry (CFA-2.2, Davis and [al]. 1985).*
- **Ionosphere:** *use of the two frequencies.*
- **Precession, Nutation,** *conventional System of astronomical constant (IAU, 1976).*
- **PPN parameter (γ):** *equal to 1 in general relativity.*

These different models were adopted in order to reach a precision of 10 picoseconds on the delay τ.

IV. Processing the VLBI data : " GLORIA/Solve ".

IV.1. Pre-processing :

- ***Detection of the clock offsets:***

In the processing, the behaviour of each clock is represented like a linear model. On the 518 raw observations, 134 (26%) were excluded because they were considered either as erratic (points out of tolerance) or as rough quality (provided by the correlator).

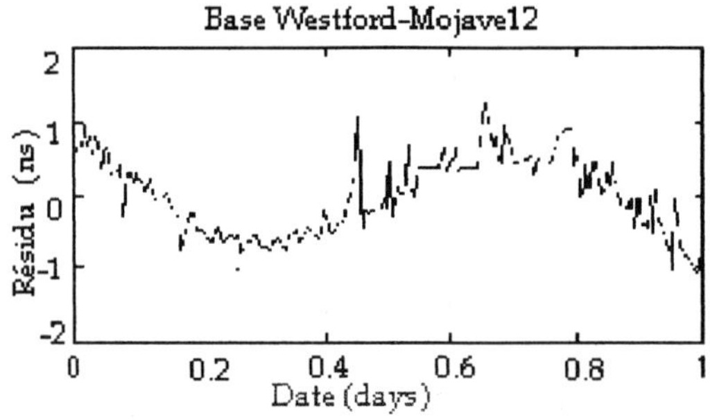

Distribution of the residues in the case of 02 clock offsets on the base Westford-Mojave12.

The residues, after model adjustment on the measurements , have an RMS of 534 ps. After residues analysis, 06 clock offsets were detected for all the session.

- ***Determination of the optimal time interval of troposphere adjustment:***

The analysis method used consists to determine, for each *i* station, the interval of average time ΔTi of adjustment of the tropospheric parameters from the tracing of the dry component considered as a reference profile and, to compare the RMS of this solution with the one of the solution which uses an adjustment on some regular time interval (2h, 4h, ...).

In order to limit the risk of overparametrisation, we selected the interval of 06 hours which results are analogous to those obtained for 04 and 05 hours.

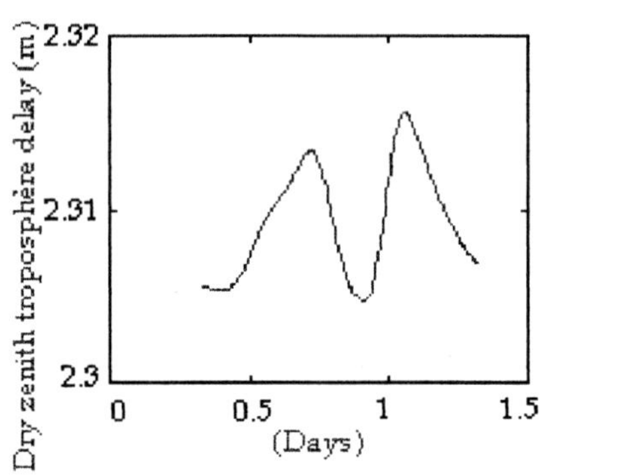

IV.2. Adjustment :

The preliminary results of the processing data and the station coordinates and baselines adjustment are provided by the following table :

Total number of observations	518
Number of stations	04
Time interval of troposphere adjustment	06 hours
Reference station	WESTFORD
Number of clock offsets	06

- ***Adjustment of the station coordinates :***

The percentage of the excluded observations is in the order of 27%. The residues, after model adjustment on the measurements , have an RMS of 170 ps for the delay (≈ 5 cm). The mean precision is respectively for the components X, Y and Z of :

$\bar{\sigma}_x$=3.6 cm; $\bar{\sigma}_y$= 3.0 cm; $\bar{\sigma}_z$= 4.6 cm.

The X and Y components are well determined than the Z component. This phenomenon was caused by the very flattened IRIS network geometry. The precision obtained on the positions of the radiotelescopes is for every station:

σ(Wettzell) = 5.4 cm ;

σ(Mojave12) = 5.6 cm ;

σ(Toulouse) = 9.3 cm.

- ***Adjustment of the baseline components :***

The accuracy obtained on the transatlantic distances, containing the Toulouse mobile station, is very important (13 cm for the biggest baseline `Toulouse-Mojave12'). In the same way, the deviation which expressed the difference between the calculated distances and ITRF92 is very important (30 cm) for the transatlantic baselines containing Toulouse. On the other hand, the baseline `Mojave12-Westford', which does not contain the mobile station and is composed of two stations, located on the American plate, is determined with a good accuracy (6 cm) and approximates the ITRF92 solution. The baseline `Toulouse - Wettzell' which is a short distance, in spite of the fact that it contains a mobile station, is estimated with a good accuracy (8 cm) and its deviation with the ITRF92 solution is only of 16 cm.

IV.3. Results analysis :

A re-processing of the observing campaign (TOULOUS2) excluding the Toulouse mobile station (WETWESMO) was carried out, but as expected , it appears useless since it does not have improved the results in a significant way.

Nevertheless, the exclusion of Toulouse, confirmed that the observations of this mobile station are more noisy than those of the fixed stations taking into account the size of the antenna. The percentage of the excluded observations is more reduced (≈ 27% for the session TOULOUS2 and ≈19% for the session WETWESMO).

The reasons which could be in the origin of the systematism on the station of Toulouse were probably caused by the restricted number of observations and the reduced quality of measurements collected by this station. An other source of error results from the geometry of the IRIS network which is composed by stations situated near the same parallel; it is generally used for the determination of the earth rotation parameters. The transatlantic baselines of IRIS sessions including the mobile stations, could create some problems on the observations.

The residues, after model adjustment on the measurements , have an RMS of 170 ps. The precision is in the order of 160 ps (≈ 5 cm) on the baseline components and of 130 ps (≈ 4 cm) on the coordinates of the stations.

Conclusion :

The results obtained during the elaboration of a VLBI geodetic data processing called "GLORIA/Solve", are of the level of precision provided by this mobile VLBI observing session. They allowed to test the delay model developed by the IERS and assure the validation of one part of the VLBI chain analysis GLORIA. The deviation between the solution obtained and the ITRF92 solution is essentially caused by the restricted number of observations and the IRIS network configuration which is intended for earth rotation parameters determination.

The adjustment of the geodetic baselines by the VLBI technique can be improved by the introduction of a sufficient number of observations , an appropriate method of detection of the Clock offset as the wavelet transform, and by the integration of the delay rate in the modelisation.

Acknowledgements :

The authors acknowledge the Central Bureau of the I.E.R.S., particularly Pr. M. FEISSEL and Dr. A.M. GONTIER, for their help on the VLBI data availability of the IRIS session and the modelisation part of GLORIA/O-C.

Bibliography :

- **EVEN P., 1992 :** *Analyse de données VLBI pour la détermination des paramètres d'orientation de la terre,* Enginner report - Paris Observatory.
- **GHEZALI B., 1996 :** *Traitement des données VLBI pour des applications géodésiques,* Magister thesis - CNTS - Arzew- Algeria.
- **GONTIER A. M., 1992 :** *Orientation de la terre par mesure VLBI ; Contribution à la chaîne d'analyse de VLBI astrométrique GLORIA,* Doctoral thesis - Paris Observatory.
- **REBAI N., 1992 :** *Traitement de la campagne Européenne de VLBI mobile 1989 et contribution à un nouveau réseau géodésique de surveillance géodynamique Europe-Afrique au niveau de la Méditerranée occidentale,* Doctoral thesis- Paris Observatory.

Acronyms :

- **GLORIA:** *GLObal Radio Interferometric Analysis.*
- **ITRF :** *IERS Terrestrial Reference Frame .*
- **GLORIA/O-C :** *part of GLORIA for processing the delay model.*
- **GLORIA/Solve :** *part of GLORIA for the geodetic parameter adjustment.*
- **[RCC] :** *the celestial conventional frame.*
- **[RTC] :** *the terrestrial conventional frame.*

ADJUSTMENT OF A EUROPEAN COMBINED GPS-SLR-VLBI FUNDAMENTAL NETWORK

Klaus Kaniuth, Horst Müller and Herbert Tremel
Deutsches Geodätisches Forschungsinstitut, Abt. I (DGFI/I)
Marstallplatz 8, D-80539 München, Germany

Axel Nothnagel
Geodätisches Institut, Universität Bonn (GIUB)
Nußallee 17, D-53115 Bonn, Germany

ABSTRACT

This paper presents the results of the first steps to set up a rigorous adjustment of a European network including GPS, SLR and VLBI data by combining these on the normal equation level. The network consists of about 20 permanently operating GPS stations, of which ten provide colocations with SLR systems, and seven stations are colocated with VLBI.

The epochs of the processed GPS data are selected to coincide with the European VLBI sessions covering a period of more than two years allowing to solve for site velocities. The SLR data set comprised two years of continuous Lageos observations. The pre-processing of the individual data sets is done with the technique specific software systems available at the involved institutions; the combination of all subsets of normal equations and the final solution is performed with the program ACCSOL at DGFI/I. A comparison with the ITRF94 velocity field is included.

INTRODUCTION

Each of the three space techniques, the Global Positioning System (GPS), Satellite Laser Ranging (SLR) and Very Long Baseline Interferometry (VLBI) has since many years demonstrated its capability of determining station positions and velocities in regional and global networks approaching accuracies of some millimeters. Indepently achieved solutions are regularly combined for establishing and maintaining an International Terrestrial Reference Frame (ITRF, e.g. Boucher et al. 1996). However, these combinations so far do not sufficiently account for the fact that there are parameters common to all techniques, such as site velocities, and tropospheric delays shared by GPS and VLBI.

An extremely successful common adjustment of GPS and VLBI observations from a dedicated single epoch experiment has been achieved by Andersen et al. (1995). Recently Andersen (1996) reported on a combined analysis of GPS, SLR and VLBI yielding a unique epoch solution.

In this paper, we have tried to perform an adjustment of GPS, SLR and VLBI by combining normal equations of all three techniques and then solving the accumulated system including site velocities. The applied strategy and the obtained results are discussed in the sequel.

DATA SETS AND PRE-ANALYSIS

We have selected for this analysis data of a continental European network including about 20 permanent GPS stations as well as the SLR and VLBI systems presently in operation. Ten SLR stations and seven out of nine VLBI systems are colocated with GPS. The composition of the network can be seen in figure 2 in which the GPS stations are labeled with their four characters IGS identifications. The sites occupied by SLR or VLBI are as follows:

SLR BOR1, GRAS, GRAZ, HERS, KOSG, MATE, METS, POTS, WETT, ZIMM

VLBI MADR, MATE, MEDI, NOTO, NYAL, ONSA, WETT

In recent years the European geodetic VLBI network performed 24 hours sessions regularly six times per year. As not all of the available data sets include all VLBI stations we have selected those six sessions between February 1995 and June 1997 which provide a reasonable spacing in time and a good occupation in terms of the number of participating telescopes. All the stations performed MarkIII dual frequency X/S band observations, and each session comprised about 200 group delay measurements per baseline.

The GPS periods processed for this combined adjustment were selected to coincide with the VLBI sessions. Each period included between two and six days of 24 hours observations amounting to a total of $2.6 \cdot 10^6$ double difference phase measurements. Again, like with VLBI not all the processed periods did contain the complete set of stations.

In the case of SLR, unlike VLBI and GPS, we have not analyzed dedicated short periods of tracking data. Instead, a global network comprising 28 stations and covering the whole two years period between January 1995 and December 1996 has been processed. Even then the data distribution is rather heterogeneous compared to GPS. The SLR data set included about 10900 Lageos 1 and 9420 Lageos 2 passes. The two years period was covered by estimating 24 monthly orbit arcs.

The pre-analysis of the three data sets has been done with the technique specific software packages available at the involved institutions using the same models and identical reference frame parameters. For example, the Earth orientation series was taken from IGS and applied to all three techniques. The VLBI observations were processed at GIUB with the Calc 8.2/Solve software (Ryan et al. 1993). In addition to the positions at epoch 1996.0 and the linear site velocities common to all three techniques, nutation offsets relative to the 1980 IAU model, polynomials modelling the station clocks, and one hour piecewise linear functions accounting for the wet tropospheric delay were adjusted.

In the case of GPS, the Bernese software version 3.4 (Rothacher et al. 1993) including some modifications implemented at DGFI was used. The classical Saastamoinen zenith delay model and the mapping function developed by Niell (1996) were applied for the a priori calibration of the tropospheric delay, and residual zenith path delays were estimated in the adjustment every two hours. The combined IGS orbits were fixed in the GPS processing. With regard to this analysis the most important modification is the direct estimation of velocities.

The SLR data processing was performed using the DGFI orbit computation and geodetic parameter estimation software system DOGS-OC/CS. In addition to the six orbital elements solved per monthly arc, solar radiation pressure and along track acceleration parameters were estimated for each 15 days interval.

After reducing technique specific session or arc dependent parameters the GPS, SLR and VLBI normal equations were saved for the common adjustments. In case of the SLR network, all position and velocity parameters refering to stations outside Europe were pre-eliminated as well. Unfortunately, it was not possible to save also the troposphere parameters in the VLBI normal equations at this stage of the analysis. Thus, the present solution includes only station positions and site velocities.

COMBINATION OF NORMAL EQUATIONS

DGFI's program ACCSOL was used for accumulating the GPS, SLR and VLBI normal equations. The strategy applied in this program and the basic idea to postpone the application of all fiducial informations and any constraints to this final stage of the analysis are displayed in figure 1, and the resulting velocity field is presented in figure 2. The approach can be characterized as follows:

- The local ties between the GPS, SLR and VLBI reference points (table 1) mainly available from the ITRF 94 documentation (Boucher et al. 1996) were included as individually weighted observation equations.
- The datum of the combined network is defined by refering the fiducial point Wettzell 7224 to the ITRF 94 position at epoch 1996.0. The velocity of Wettzell according to ITRF 94 was applied as high weight pseudo-observation.
- The individual velocities of colocated GPS, SLR and VLBI stations were constrained to identity by applying automatically defined condition equations.
- In those cases where the daily GPS networks had been divided into subsets the troposphere estimates at the sub-network junction points were constrained to equality.

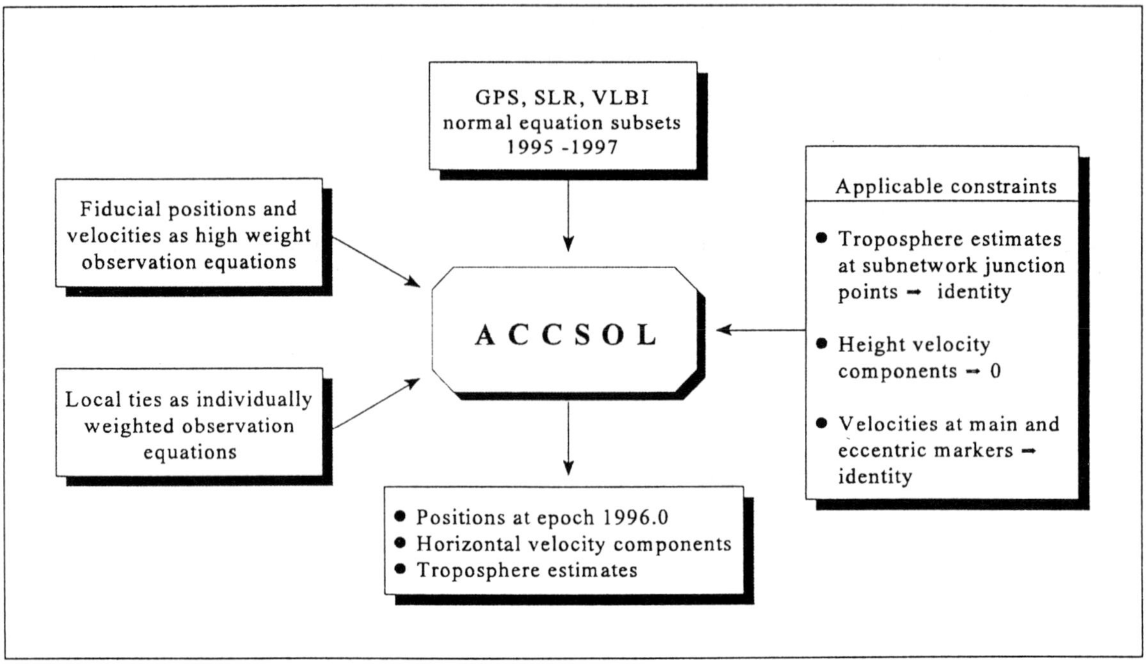

Fig. 1 Strategy for accumulating GPS, SLR and VLBI normal equations and solving the combined system applying fiducial informations and constraints

Table 1 Available local ties between reference points of colocated systems

Site	Reference points Technique / ID		Technique / ID		Δx [m]	ΔY [m]	ΔZ [m]
Wettzell	GPS	WETT	GPS	WTZR	2.105	.981	- 1.994
	VLBI	7224	GPS	WTZR	40.802	118.398	- 61.317
	SLR	8834	GPS	WTZR	- 3.828	- 68.200	15.517
Onsala	VLBI	7213	GPS	ONSA	52.631	- 40.464	- 43.865
Herstmonceux	SLR	7840	GPS	HERS	6.505	10.278	- 3.945
Matera	VLBI	7243	GPS	MATE	10.946	42.246	- 38.203
	SLR	7939	GPS	MATE	- 15.173	- 24.827	24.964
Noto	VLBI	7547	GPS	NOTO	- 34.462	60.954	- 4.994
Borowiec	SLR	7811	GPS	BOR1	25.767	- 72.908	- .324
Zimmerwald	SLR	7810	GPS	ZIMM	13.680	6.012	- 6.242
Madrid	VLBI	1565	GPS	MADR	-134.246	159.664	164.275
Medicina	VLBI	7230	GPS	MEDI	30.917	- 3.409	- 54.521
Grasse	SLR	7835	GPS	GRAS	6.007	- 33.777	- 8.036
Metsahovi	SLR	7805	GPS	METS	- 24.252	35.529	23.204
Potsdam	SLR	7836	GPS	POTS	50.091	95.219	- 40.438
Graz	SLR	7839	GPS	GRAZ	- 2.558	8.516	- 1.321

We do not show velocity estimates for HFLK, KIRU and NYAL in figure 2, because these stations are included in a few GPS data sets only, and ITRF 94 does not provide velocities as well. The average precision estimate of the displayed horizontal velocity components is 0.3 mm/year.

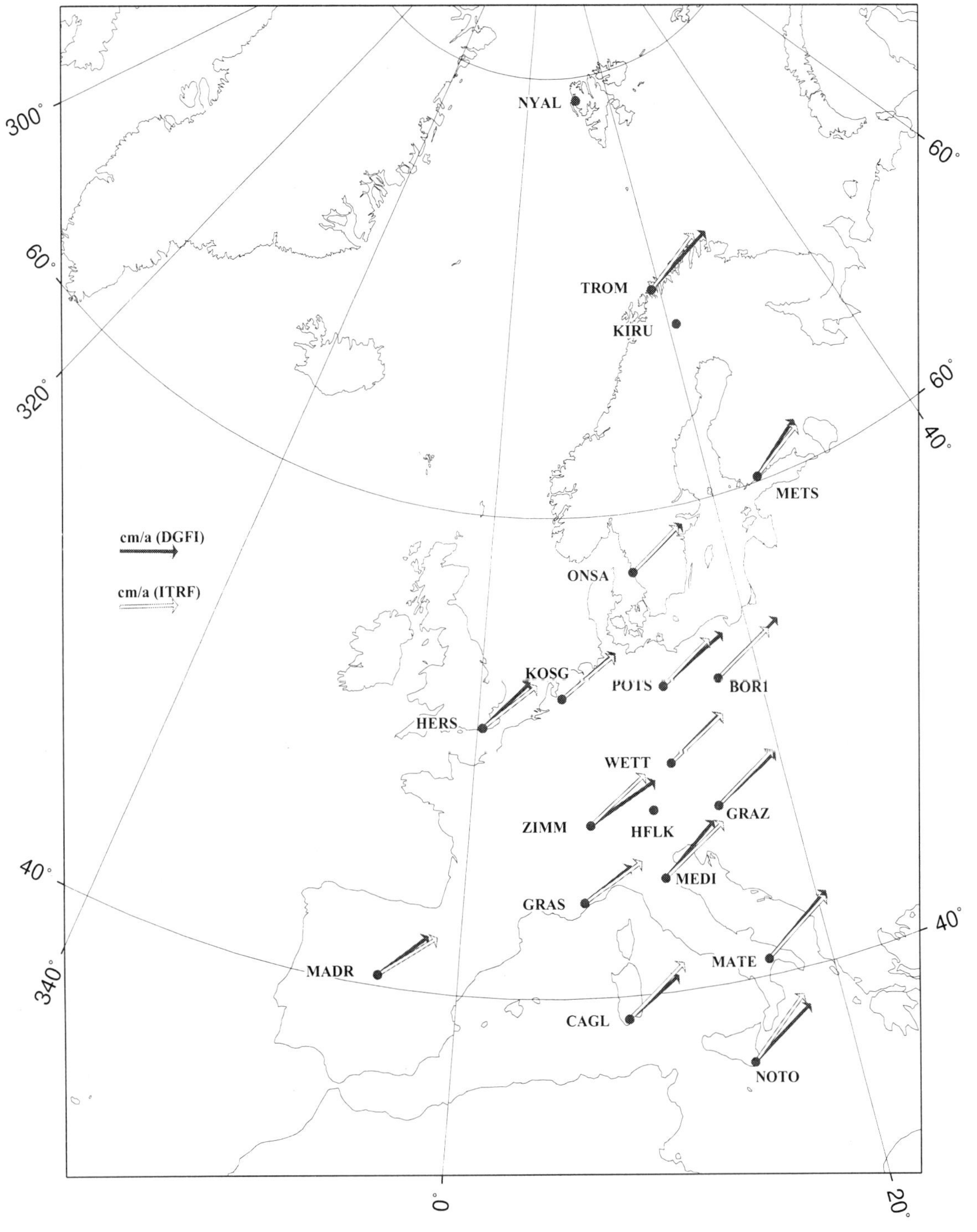

Fig. 2 Horizontal velocity field resulting from the combined adjustment in comparison to ITRF 94

CONCLUSION

We have made a first attempt to perform a common adjustment of GPS, SLR and VLBI by combining normal equations of these three techniques and solving for positions at epoch 1996.0 and annual velocities. Unfortunately, at this stage of the analysis the troposphere parameters could not be saved in the reduced VLBI normal equations; therefore, the analysis does not include a comparison of GPS and VLBI troposphere estimates.

SLR contributed least to the solution due to the heterogeneous data distribution. The relative weight of GPS and VLBI was set such as to allow both techniques to contribute equally to the adjustment result. Considering that the formal errors of the resulting velocities are not at all measures of accuracy, the comparison with the ITRF 94 velocity field shows rather good agreement.

Further progress in performing combined adjustments of GPS, SLR and VLBI data sets can be expected from

- identifying and accounting for remaining inconsistencies between the technique related models and reference systems,
- applying a rigorous variance component estimation procedure to the different data sets, and
- including all common parameters such as tropospheric delays in the individual reduced normal equation systems.

REFERENCES

Andersen, P.H., O. Kristiansen and N. Zarraoa (1995): Analysis of Data from the VLBI-GPS Collocation Experiment CONT 94. In Beutler, G. et al. (Eds): GPS Trends in Precise Terrestrial Airborne and Spaceborne Applications. Proceedings IAG Symposium 115, Springer, 315-319.

Andersen, P.H. (1996): Status Report on the Combination of VLBI and other Space Geodetic Data. In Elgered, G. (Ed.): Proceedings of the Eleventh Working Meeting on European VLBI for Geodesy and Astrometry, Onsala, Sept. 1996, 172-179

Boucher, C., Z. Altamimi, M. Feissel and P. Sillard (1996): Results and Analysis of the ITRF 94. IERS Technical Note 20.

Niell, A.E. (1996): Global Mapping Functions for the Atmospheric Delay at Radio Wavelength. J. Geophys. Res. 1101, B2, 3227-3246.

Rothacher, M., G. Beutler, W. Gurtner, E. Brockmann and L. Mervart (1993): The Bernese GPS Software Version 3.4. Astronomical Institute, University of Berne.

Ryan, J.W., C. Ma and D.S. Caprette (1993): NASA Space Geodesy Program - GSFC Data Analysis - 1992: Final Report of the Crustal Dynamics Project VLBI Geodetic Results 1979-1991. NASA Technical Memorandum 104572, Greenbelt.

COMBINATION OF VLBI, SLR AND GPS DETERMINED STATION VELOCITIES FOR ACTUAL PLATE KINEMATIC AND CRUSTAL DEFORMATION MODELS

Hermann Drewes
Deutsches Geodätisches Forschungsinstitut, Abt. I (DGFI/I)
Marstallplatz 8, D-80539 München, Germany

ABSTRACT

Recent space geodetic station velocity solutions of different techniques are combined in a common global adjustment to estimate lithospheric plate rotation vectors and to model regional crustal deformations. Kinematic datum parameters (3D rotation vectors) and relative weights are estimated for each technique in the combination procedure. The overall kinematic datum is realized by a "no net rotation" constraint. The result provides an Actual Plate KInematic Model (APKIM) including twelve major plates (only Cocos plate is missing) and the regional deformation pattern in four plate boundary zones (Mediterranean, Japan, California, South American Andes).

INTRODUCTION

Global station coordinate and velocity solutions are nowadays provided routinely (in general annually) by various analysis centers for the different space geodetic techniques. They are combined by the International Earth Rotation Service (IERS) to realize the IERS Terrestrial Reference Frame (ITRF) in terms of epoch station coordinates and individual station velocities in a common reference system (e.g. Boucher et al. 1996). In the present paper we shall restrict to the analysis of station velocities.

We are using data sets from three space geodetic techniques: GPS, SLR and VLBI. In a first step we compare one identical technique (GPS). The corresponding data sets are those of the International GPS Service for Geodynamics (IGS) Analysis Centers CODE (Rothacher 1996), GFZ (Gendt 1996), and JPL (Heflin 1996). In a second step we are comparing different space techniques using the mentioned GPS solution of JPL, the SLR solution CSR96L01 (Eanes and Watkins 1996) and VLBI solution GLB1024 (Ma 1996). In the third step we combine the different techniques' results for crustal dynamics studies.

COMPARISON OF SINGLE TECHNIQUE (GPS) VELOCITY SOLUTIONS

The coordinate and velocity solutions of the IGS Global Network Associate Analysis Centers (GNAAC) are officially using the same terrestrial reference system realized by a frame of 13 fixed or tightly constrained station coordinates and velocities (Kouba 1996). The results should therefore be directly comparable. In table 1 we present the r.m.s. deviations of such a comparison computed from 52 identical stations included in the three solutions. It is obvious that the deviations are significantly greater than the internal precision (formal errors) of the individual solutions which are all less than 1 mm/a. There is no significant difference of deviations in the velocity components North, East, Height.

To check the identity of the kinematic datums we compute a least squares fit of the velocities V in identical stations of the individual data sets i, k with respect to each other by reducing a global rotation $\underline{\Omega}$ applied to the positions $\underline{X}$.

$$\underline{V}_i = \underline{V}_k - \underline{\Omega} \times \underline{X} \tag{1}$$

The resulting r.m.s. deviations after the datum reduction are listed in table 1, too. We find a significant improvement of the fit and we also see that the height component fits significantly worse than the horizontal (N, E) components, as is well known for GPS results. The maximum relative datum effect on station velocities is 5 mm/a in the comparisons with CODE solution and 8 mm/a between GFZ and JPL solutions.

Table 1: Weighted r.m.s. deviations of different GPS station velocity solutions [m/a]

Compared	Original Data			Datum Reduced		
Solutions	North	East	Height	North	East	Height
CODE vs. GFZ	± 0.0069	± 0.0041	± 0.0058	± 0.0050	± 0.0039	± 0.0058
GFZ vs. JPL	± 0.0071	± 0.0047	± 0.0051	± 0.0036	± 0.0029	± 0.0051
CODE vs. JPL	± 0.0030	± 0.0039	± 0.0043	± 0.0023	± 0.0033	± 0.0043

COMPARISON OF VELOCITY SOLUTIONS FROM DIFFERENT TECHNIQUES

Traditionally there are two space geodetic techniques other than GPS providing station velocities, namely Satellite Laser Ranging (SLR) and Very Long Baseline Interferometry (VLBI). We are using here those solutions with the longest covered time interval which are the University of Texas SLR solution CSR96L01 (Eanes and Watkins 1996) with 72 station velocities from 20 years of observations between 1976 and 1996 and the Goddard Space Flight Center VLBI solution GLB1014 (Ma 1996) including 78 station velocities derived from observations from 1979 to 1995. These data sets are analysed together with the above mentioned JPL 1996 GPS solution providing 173 station velocities from only three and a half years of observations (1993 to 1996).

The kinematic datum of these solutions is always defined by the geophysical plate motion model NNR NUVEL-1A. It is realized, however, in a quite different way by fixing the velocities of some individually selected stations. Due to this nonuniform fixing we have different datums in all the solutions. The r.m.s. deviations after the datum removal by a relative global rotation between each two solutions after (1) is also displayed in table 2. The datum parameters are given in table 3.

Table 2: Weighted r.m.s. deviations of station velocity solutions from different techniques [m/a]

Compared	Original Data			Datum Reduced		
Solutions	North	East	Height	North	East	Height
VLBI vs. SLR	± 0.0050	± 0.0027	± 0.0080	± 0.0018	± 0.0012	± 0.0080
SLR vs. GPS	± 0.0057	± 0.0041	± 0.0068	± 0.0022	± 0.0023	± 0.0068
VLBI vs. GPS	± 0.0022	± 0.0029	± 0.0075	± 0.0016	± 0.0017	± 0.0075

COMBINING THE VELOCITY SETS FOR CRUSTAL MOVEMENT MODELS

Velocities of geodetic ground stations reflect all kinds of crustal movements: rigid lithospheric plate motions, regional (inter or intra plate) deformations, and local distortions. The comparison and interpretation of the observed velocities has thus to include all these effects. Furthermore we learn from the previous chapters that we have to count for an observation datum effect, too. The complete observation equation for estimating crustal movements from geodetically observed station velocities $\underline{V}$ at stations with position vectors $\underline{X}$ may then be written as

$$\underline{V}_k = \underline{\Omega}_n \times \underline{X}_k + \underline{S}^{-1} \times \underline{f} + \underline{\Omega}_{obs} \times \underline{X}_k. \qquad (2)$$

$\underline{\Omega}_n$ is the geocentric rotation vector of rigid plate n, $\underline{S}$ is the stiffness matrix of the deformable crust (resulting from the rheology, i.e., rock density, elasticity, viscosity, etc.), $\underline{f}$ are the active tectonic forces, and $\underline{\Omega}_{obs}$ is the rotation vector (datum) of the technique dependent observation data set.

Estimation of rigid plate rotation vectors from space geodetic observations has been described in various publications (e.g., Drewes 1982, Drewes 1990). The general question is whether to use 3D geocentric velocities (V_X, V_Y, V_Z) or only 2D velocities at the Earth's surface in latitudinal and longitudinal direction (V_φ, V_λ). The rigid plate model does not allow vertical motions; on the other hand the vertical component of positions determined by space geodesy is the weakest one (see above). Using 3D velocities would thus strongly be affected by the greater height errors. We therefore decided to use the approach with ellipsoidal coordinates as presented in the mentioned publications.

We need at least two observed station velocities on the same plate to estimate the three components of the rotation vector. Eleven of the twelve major plates included in the current plate models (e.g., DeMets et al. 1990) are nowadays covered by this minimum number of stations, only the Cocos plate is missing. On the other hand, all the Eurasian continent from the Atlantic to the Pacific is considered as one rigid plate in the geophysical models. Our analysis of geodetic observations indicates, however, that there is a significantly different plate rotation in the western (Europe) and the eastern (East Asia) part. We therefore split the Eurasian plate into western Eurasia and East Asia and come up with twelve plate rotation vectors. They are given in table 4.

The modelling of regional plate boundary deformations is done using two different approaches. The Finite Element Method (FEM) uses geophysical models of rheology and forces to describe the continuous deformation. The application of the FEM for geodetic purposes has been described in various publications (e.g., Drewes 1993). We are here using homogeneous, isotropic elastic material.

The second approach is the least squares vector prediction using empirical auto-covariance functions of the latitudinal and longitudinal velocities, and cross-covariance functions between latitudinal and longitudinal velocities. It can be shown that the auto-covariances and cross-covariances play the same role as the elastic linear and shear moduli (Young's modulus and Poisson ratio). The results of the least squares prediction and the FEM modelling are very close to each other.

We model four deformation zones along plate boundaries of those shown by Gordon (1995) and covered by geodetic stations: the mediterranean area, Japan, California and the South American Andes. The deformation field in the Mediterranean from least squares prediction, for example, is shown in Drewes (1997).

The global datum of the combination approach is defined by the condition of no net rotation:

$$\int \underline{V} \; d\sigma = \underline{0} \quad (\sigma = \text{Earth surface}) \tag{3}$$

It is realized by summing the velocities in a global equal area 1° x 1° grid (which is identical to a 1° x 1° geographical grid weighted by the cosine of latitude) and constraining this sum to zero by a common global rotation.

The datum parameters (rotation vectors) of the individual techniques data sets (VLBI, SLR, GPS) are estimated as unknowns in the adjustment procedure relative to the global (no net rotation) datum. The result is shown in table 3 in a relative way and opposed to the direct comparison of data sets (see above). There is in general a good agreement between both methods of datum estimation. The weights of the individual techniques are determined by an a posteriori variance estimation in the common adjustment of the crustal movement model. They are GLB1014 : CSR96L01 : JPL96GPS = 0.2 : 2.7 : 0.03.

Table 3: Datum parameters of individual techniques data sets from direct comparison and from the combination model [deg/Ma]

Compared Techniques	Direct Comparison			Combination Model		
	Ω_X	Ω_Y	Ω_Z	Ω_X	Ω_Y	Ω_Z
VLBI vs. SLR	- 0.0430	- 0.0165	0.0254	- 0.0452	- 0.0248	0.0259
SLR vs. GPS	- 0.0624	- 0.0159	- 0.0148	- 0.0678	- 0.0122	- 0.0076
VLBI vs. GPS	- 0.0232	- 0.0027	- 0.0354	- 0.0226	- 0.0106	- 0.0335

INTERPRETATION OF PLATE KINEMATIC MODEL

The estimated rotation vectors of the actual plate kinematic model (APKIM) from the combination of observations using different space geodetic techniques is opposed in table 4 to the geophysical model NNR NUVEL-1A (Argus and Gordon 1991). There is generally a good agreement (within the 95 % confidence interval) but some vectors are significantly different (e.g., Antarctica, Eurasia, Asia, Pacific). The deviation has only an effect of a few millimeters per year in the station motions, it is, however, significant because of the high accuracy of the model.

Table 4: Plate rotation vectors of the actual plate kinematic model (APKIM) from space geodetic observations and from geophysical model NNR NUVEL-1A

Plate Name	APKIM 8.80			NNR NUVEL-1A		
	Φ[°]	Λ[°]	ω[°/Ma]	Φ[°]	Λ[°]	ω[°/Ma]
Africa	53.1 ± 1.8	269.6 ± 3.6	0.283 ± 0.006	50.6	286.0	0.291
Antarctica	50.1 ± 4.2	220.5 ± 4.3	0.250 ± 0.023	63.0	244.2	0.238
Arabia	55.5 ± 19.3	359.5 ± 65.3	0.505 ± 0.551	45.2	355.5	0.546
Australia	33.8 ± 0.6	36.8 ± 1.2	0.633 ± 0.005	33.8	33.2	0.646
Caribbea	30.0 ± 5.9	274.7 ± 13.4	0.426 ± 0.235	25.0	267.0	0.214
Eurasia	57.9 ± 1.2	258.4 ± 1.9	0.270 ± 0.004	50.6	247.7	0.234
Asia	11.9 ± 6.3	285.2 ± 2.7	0.522 ± 0.072	50.6	247.7	0.234
India	43.5 ± 18.0	43.2 ± 33.2	0.702 ± 0.405	45.5	0.3	0.545
Nazca	28.7 ± 28.4	255.3 ± 2.7	0.736 ± 0.257	47.8	259.9	0.743
N. America	- 2.6 ± 1.0	273.7 ± 0.4	0.187 ± 0.003	- 2.4	274.1	0.207
Pacific	-62.7 ± 0.3	93.9 ± 1.4	0.699 ± 0.006	-63.0	107.3	0.641
S. America	-19.4 ± 5.1	210.1 ± 12.3	0.127 ± 0.010	-25.3	235.6	0.116

CONCLUSION

Although all space geodetic coordinate and velocity solutions are nominally using the same terrestrial reference system (ITRF and NNR NUVEL-1A) there are considerable datum discrepancies in the station velocity fields derived from different techniques and also between different solutions using the same technique. One has to reduce these datum effects when combining different data sets for crustal deformation studies.

Nearly all major tectonic plates are now covered by at least two space geodetic stations providing station velocities (except Cocos plate). The estimated plate rotation vectors from space geodetic data are in general very similar to, but (because of their exactness) sometimes significantly different from the geophysical models. Deformation zones along plate boundaries are more extended than formerly expected. They can be modelled as continuous deformation. The geodetic "no net rotation" datum differs from that of NNR NUVEL-1A because the geophysical model does not include deformation zones.

REFERENCES

Argus, D.F., R.G. Gordon: No-net-rotation model of current plate velocities incorporating plate motion model NUVEL-1. Geophys. Res. Lett. (18) 2039-2042, 1991.

Boucher, C., Z. Altamimi, M. Feissel, P. Sillard: Results and analysis of the ITRF94. IERS Paris, Technical Note No. 20, 1996.

DeMets, C., R.G. Gordon, D.F. Argus, S. Stein: Current plate motions. Geophys. J. Int. (101) 425-478, 1990.

Drewes, H.: A geodetic approach for the recovery of global kinematic plate parameters. Bull. Géod. (56) 70-79, 1982.

Drewes, H.: Global plate motion parameters derived from actual space geodetic observations. Springer, IAG Symposia, No. 101, 30-37, 1990.

Drewes, H.: A deformation model of the Mediterranean from space geodetic observations and geophysical predictions. Springer, IAG Symposia, No. 112, 373-378, 1993.

Drewes, H.: Global plate kinematics and plate boundary deformations. Terra Nostra (4/97), 4 pp, 1997.

Eanes, R.J., M.M. Watkins: Earth orientation and site coordinates from the Center for Space Research solution CSR96L01. Internet ftp csr.utexas.edu, 1996.

Gendt, G.: GFZ 1996 GPS station coordinate and velocity solution. personal communication, 1996.

Gordon, R.G.: Present plate motions and plate boundaries. In: T.J. Ahrens (ed.), Global Earth Physics, A Handbook of Physical Constants. AGU Reference Shelf 1 66-87, 1995.

Heflin, M.B.: IGS Mail No. 1393.

Kouba, J., Y. Mireault: IGS analysis coordinator report. IGS 1995 Annual Report 45-76, 1996.

Ma, C.: GSFC 1995 Annual Report. Internet ftp gsfc.nasa.gov, 1996.

Rothacher, M.: CODE coordinate and velocity results CM33V031.SNX. Internet ftp ubeclu.unibe.ch, 1996.

GEODESY DEMONSTRATION EXPERIMENT (GEDEX) FOR SPACE VLBI: STATE OF THE ART AND SOFTWARE DEVELOPMENT

M. N. Kulkarni[1], J. Adam[2], I. Fejes[3], S. Frey[3], B. Kampes[4], Y. Zheng[5], X. Hu[6]

ABSTRACT

The aim of the GEDEX, which is being designed by an international team of scientists, is to explore the feasibility of Geodetic & Geodynamic applications of Space VLBI. In this paper, the outline of this experiment has been presented. A brief description of the Space VLBI technique has been provided, followed by a summary of the significant theoretical developments. Salient features of the VSOP mission, and an overview of the various aspects of GEDEX have been presented. The state of the art of software for Geodetic SVLBI has been outlined, and the different steps in software development for the GEDEX have been described in detail. Conclusions drawn from the work done so far, and recommendations for future work have been presented.

1. INTRODUCTION:

One of the challenges facing the geodesists in the 21st century is the precise definition, practical realization, and interconnection of different references frames, for which new observational techniques, models and procedures are constantly being explored and implemented. With the launching of the first Space Very Long Baseline Interferometry (SVLBI) satellite: the VLBI Space Observatory Programme (VSOP) of Japan, in February 1997, to be followed by the proposed launch of the second satellite: RADIOASTRON of Russia in 1999, the new astro-geodetic technique of Space VLBI has become a reality. This unique technique, which provides a direct tie between the Conventional Inertial reference System (CIS) and the Conventional Terrestrial reference System (CTS), has potential applications in Geodesy and Geodynamics. The aim of the GEodesy Demonstration EXperiment (GEDEX) for VSOP is to explore the feasibility of using this promising technique for some important Geodetic and Geodynamic applications, including the interconnection of the reference frames. An international team of scientists, working under the auspices of the FÖMI Satellite Geodetic Observatory (SGO), Penc, Hungary, is in the process of designing the GEDEX.

---

1. Survey of India, Dehradun, India, 2. TU of Budapest, Hungary, 3. FÖMI SGO, Hungary, 4. TU of Delft, The Netherlands, 5. Z.Z.Inst.for Svy.& Mpg.,China, 6. Centre for Astro-Geod.Res.,China.

2. THE SPACE VLBI: SYSTEM DESCRIPTION

Space VLBI is an extension of the ground based VLBI in to the space, with either one or both the VLBI antennas mounted on Earth orbiting satellite(s). These antennas observe the signals from the extragallactic sources simultaneously with the ground VLBI network, yielding new observables like the space baselines delay and delay rates, along with ground VLBI delay and delay rates. Due to the extension of the baseline beyond the physical dimensions of the Earth, made possible by SVLBI, and the direct tie between CIS and CTS provided by these new observables, this unique technique has potential for some important applications in Geodesy and Geodynamics. The first SVLBI satellite: VSOP of Japan has already been launched in Feb. 1997, with the second: RADIOASTRON of Russia, presently under preparation (see Kulkarni (1992)).

3. THE VSOP MISSION:

The VSOP Mission is divided into four phases during which different kinds of mission operations will be conducted, viz. the Launch Phase, the Spacecraft Checkout Phase, Space-Ground Checkout Phase, and the Observational Phase. The Mission is designated as Astrophysics Mission, however, the observations obtained will also be utilized for Geodetic and Geodynamic applications (see VSOP Internet homepage: http://www.vsop.isis.ac.jp/).

4. VSOP GEDEX PROJECT:

In response to the first VSOP Announcement of Opportunity, the proposal on GEDEX was submitted by the SGO and accepted by the VSOP Proposal Review Committee. The proposal was ranked among the top 10% by the Scientific Review Committee, hence, it is "highly likely to be scheduled". The GEDEX is a concept demonstration experiment using the VSOP satellite and the co-observing ground VLBI stations in the following areas of Geodesy and Geodynamical research:

(i) Connection of the terrestrial and extragalactic celestial reference frames, as the observation equations contain the radio source and ground station coordinates and the EOP's, which are the basic components of transformation between these two systems.

(ii) An improvement in the orbit determination accuracy of SVLBI satellites, using the SVLBI delay and delay rate observables, which would also provide a tie between the dynamically defined inertial reference frame of the SVLBI satellite orbit and the terrestrial and celestial reference frames.

(iii) A direct determination of the geocentric position of the ground VLBI stations participating in the SVLBI experiment, and checking their translation with respect to other reference frames, including estimation of the geocentre offsets.

The input data expected for the GEDEX Project is the ground and space VLBI delay and delay rates, and also the time tagged range and range rate data from the telemetry stations, including the two way reference frequency Doppler shift data. Unfortunately, only single frequency VLBI data will be available. In addition to this data, the precise geocentric ITRF coordinates of the ground VLBI and telemetry stations will be required, which can be obtained from the global VLBI solutions and from the IERS. The standard reconstructed orbit data will be obtained from the Orbit Determination Groups. The strategy used in the analysis will depend on which geodetic parameters are to be estimated, and which can be fixed as a priori information. The focus will be on the three goals outlined above, viz. the estimation of the EOP's, determination of the orbit with the highest possible accuracy, and estimating the geocentre offsets. The ground network and the SVLBI data will be handled separately in the first phase of the estimation process, and in the second phase, a combined solution will be implemented to yield the best possible estimates of the geodetic parameters being estimated. The output of the analysis will answer the question: how effectively can the SVLBI data be used for geodetic and geodynamical applications in the future.

In GEDEX, using specially designed observation sequences of highly compact extra-galactic radio sources, the consistency of the terrestrial frame and its geocentric origin defined by the IERS Terrestrial Reference Frame (ITRF) locations of the participating ground VLBI stations and the satellite orbit will be analyzed. New analysis tools and software will be developed and tested. It is expected that by using real SVLBI data, new insights will be obtained concerning the SVLBI applications, even though significant improvements in the accuracy of estimation of the parameters may not be expected, due to several limitations and observation constraints. The experience gained through this experiment will certainly be useful in planning and design of the next generation SVLBI missions, including the future Geodetic SVLBI satellite missions. This experiment may also provide some explicit benefits for astronomy experiments, by testing the limit of the orbit determination accuracy of the SVLBI satellite. The method and procedures developed could later be used where high precision orbit is necessary.

5. GEDEX SOFTWARE DEVELOPMENT:

As no standard procedures presently exist for the kind of data analysis involved in the GEDEX, considerable amount of software development is required to be carried out. For this purpose, an international team of scientists working under the auspices of the SGO has been formulated. The GEDEX Working Group has initiated the geodetic management and development of the GEDEX software during their collaborative work at the SGO during the summer of 1997, and the work is presently under progress.

The first and most important task was to identify the input/output data and system requirements of the GEDEX Project from the user point of view. In consultation with various SVLBI experts, these have been finalized at the SGO. The input data, as identified in the User Requirements, is required to be obtained, along with its detailed format specifications, stored in a data base specially established and managed for this purpose, converted to the format acceptable to the processing software, and input to it. For this purpose, interfaces have to be developed. This task is especially difficult for this project due to two reasons: firstly, the non-standard data requirements of the task, which make the data format specifications and the data itself difficult to obtain, and secondly, the numerous data formats involved in the available real data and the input data to the processing software, which necessitate the development of many interfaces for data transfer. This task is presently underway at the SGO.

The different modules of the GEDEX software are shown in a flow chart in Figure 1. The GEDEX software system will consist of the following main modules: data transfer interfaces, data input modules, orbit estimation module, geodetic parameter estimation module, error analysis module, orbit and data simulation modules, software testing module, and data output modules. The geodetic parameter estimation software being developed and tested will be based on the OCCAM/SPVK software, with modifications/augmentations incorporated from other software like OSUSVLBI, to include estimation of EOP's and geocentre offsets, and a detailed error analysis. The software will be fully tested for simulated data, before using real data.

The output data products from the GEDEX software system will be stored in a database accessible to the scientific community for further research, in user-friendly data formats. These will also be distributed to the SVLBI Research Groups for their scrutiny, evaluation and suggestions, for future developments. The main products will be the improved SVLBI satellite orbits, the estimated EOP's, and the geocentre offsets.

6. CONCLUSIONS:

The new astro-geodetic technique of Space VLBI has the potential for some important applications in Geodesy and Geodynamics. With the feasibility of these applications having been established through the theoretical investigations in this area, as outlined above, it is important to study them for the real data, which will soon be available from the VSOP Mission. The GEDEX Project is the first step in this direction. With this experiment, it is hoped, that the doors to this new unique technique will open for the geodetic community. Even though the configuration of the VSOP is not suited to geodetic applications, and hence, high accuracy in geodetic parameter estimation may not be achievable, this experiment will form the basis of future investigations of SVLBI technique. Thus, the method, procedures and software developed for the GEDEX, and the experience gained, will prepare the geodetic community for the possible launch of Geodetic SVLBI missions, hopefully in the near future.

7. ACKNOWLEDGEMENTS:

The GEDEX Project was conceived, and the software development group was organized at the FÖMI SGO, under the guidance of Dr. Istvan Fejes, Head, SGO. The financial support received by the FÖMI SGO from the Hungarian Space Office under contract no. TP 17, and from the European Union under contract no. ERBCIPDCT925072 is gratefully acknowledged. The authors would also like to express their gratitude to the SVLBI experts from different organizations, who have provided valuable suggestions and information to the GEDEX Project.

10. REFERENCES:

- **Ádám, J.(1995)**: Report of IAG Special Study Group 2.109 - Application of Space VLBI in the Field of Astrometry and Geodynamics. Presented at the XXI IAG/IUGG General Assembly, Boulder, Colorado, USA, July.
- **Kulkarni, M.N. (1992)**: A Feasibility Study of Space VLBI for Geodesy and Geodynamics. *Report No. 420*, Dept. of Geodetic Science and Surveying, The Ohio State Univ., Columbus, Ohio, June.
- **Kulkarni, M.N. and J. Ádám (1993)**: A Sensitivity Analysis for Orbit Determination of Space VLBI Satellites for Interconnection of Reference Frames. *Acta Geod. Geoph. Mont. Hung.*, Vol. 28(3-4), pp. 441-456(1993).
- **Zheng, Y. (1992)**: The Application of Space VLBI for Astrometry, Geodynamics and Space Physics (in Chinese). *Ph.D. Thesis*, Shanghai Observatory, Chinese Academy of Sciences, Shanghai, August.

FIGURE 1: FLOWCHART OF GEDEX SOFTWARE DEVELOPMENT MODULES

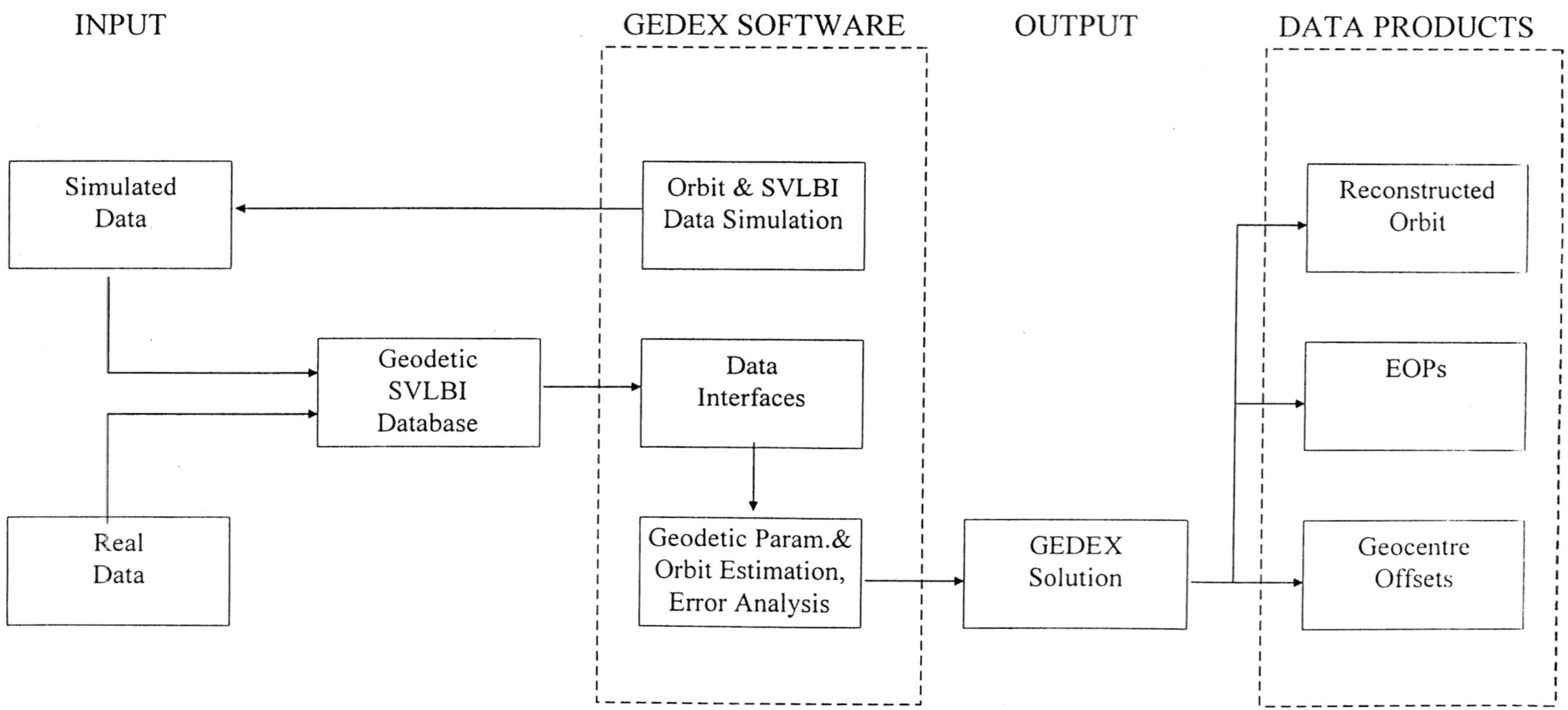

AN INTEGRATED CRUSTAL MOVEMENT MONITORING NETWORK IN CHINA

J.Y.Chen
National Bureau of Surveying & Mapping(NBSM)
Baiwanzhuang, 100830, Beijing, China

Abstract

A integrated crustal movement monitoring network in China is being established and will be fully completed in 2000. The goals for the network are to reduction and mitigate the natural hazards and try to make a acceptable environment for the sustainable development in China. Besides, The network is useful for the improvement of the earthquake prediction, geodynamic study and kinematic monitoring of earth surface spheres (atmosphere, ionosphere ocean, crust) in China. It consists of 25 GPS permanent stations 56 GPS fundamental stations, 1000 regional stations, 2 fix VLBI, 4 fix SLR, 1 mobile VLBI and 2 mobile SLR. At the same period, the absolute and relative gravimetry and leveling will also be carried out in the network.

I.Scientific Significance

1. Important basis for earthquake prediction

China continent lies on the west Pacific and Eurasia earthquake zones. Since 1988, crustal movement in mainland of China has entered into the fifth active period of strong earthquake of the century. It is predicted that a series of strong earthquake at or higher than magnitude 7.0 will be occurred in the next 10 years. As well known that the pregnant and occurrence of big earthquake are a process of accumulation, concentration, strengthening of crustal stress in a large area and finally the sudden release of the strain energy. The process mentioned above inevitably accompanies with corresponding crustal movement and deformation therefore, it is very important for the earthquake prediction to monitor the crustal movement and its change in a large area, which are caused by the change of crustal stress field. Base on the monitoring results the zone of crustal stress concentration, and the further strengthening part of crustal stress, which might be epicentral area of a big earthquake in the future, can be found and determined. As crustal movement and earthquake is usually close with each other, thus the crustal movement monitoring has been an important basis and method for the earthquake prediction . It is obvious that GPS combined with SLR and VLBI can be taken as main technique for the monitoring of the crustal movement in China. The establishment of a crustal movement monitoring network in China (CMMNC) will play an important role in the improvement of the prediction for earthquake and other geological disasters, e.g. landslides, land falling, land subsidence, volcanic eruption etc. .

2. Fundamental basis for geodynamics

China continent lies in the southeast of Eurasia plate, which is an important component of Eurasia land-crust plate. The east part of China continent is affected by dive and diminish of west Pacific ocean plate that caused a series of marginal sea expansion and subsidence basin related with back arc expansion. The west and northwest part of China continent are affected by the tectonics collision between India plate and Qinghai-Xizang block, The recent crustal movement in China is characterized by the rapid uplift of Qinghai-Xizang plateau and strike-slip or adverse strike-slip along large active zone. Qinghai-Xizang plateau is the most active area of horizontal and vertical crustal movement in the China continent, and it is also the most active area of earthquake in the world continents. Its horizontal and vertical movement rates reach 1-4 cm/a; and 0.5-1 cm/a respectively. It is to say that the modern crustal movements in China continent and its surrounding area are constrained by two typical kinds of boundaries (i.e. Type A : land-land crust collision and Type B : ocean- land crust dive) in the modern tectonics, i.e. it is not only of significance for globe tectonics, but also with the characteristics of special evolution. Hence the crustal movement monitoring in China is a case study in the global geodynamics .

3. Kinematic monitoring for earth surface spheres

CMMNC will facilitate traditional geodesy towards the kinematic geodesy. According to the continuous observed GPS signals, the kinematic change of vapor components in troposphere can be obtained, which can be served for weather broadcast and meteorological study. Meanwhile based on these signals, the kinematic data for the electron density in ionosphere can be also tracked, so that it can be used for the relevant study on atmosphere physics in upper atmosphere and micro wave communication. Combining GPS data with satellite altimeter information, time-space kinematic change of sea level can be detected.

The data obtained form GPS, SLR, VLBI observations are also valuable for the determination of earth rotation.

General speaking, the establishment of CMMNC will push forward the development of kinematic monitoring technology and relevant scientific theory on the three spheres of earth surface in China.

II. Structure of CMMNC

CMMNC is characterized of a network in a large area with high accuracy and high time-space resolution. CMMNC consists of GPS (main part), VLBI and SLR, as well as precise gravimetry and leveling. CMMNC is composed of four components, e.g. fiducial, fundamental, regional network, as well as data transmission, analysis and processing system.

1. Fiducial network is comprised by 25 GPS permanent stations in combination with VLBI or SLR.

2. Fundamental network is comprised by 56 GPS regular remeasurement stations.

3. Regional network is comprised by 1000 GPS remeasurement stations.

4. Data transmission and analysis and processing system, including one data center and three data subcenters.

III. Establishment and Accuracy Index of Fiducial Station

1. The fiducial station take the monitoring of block tectonic movement as their main objective. It must be ensured, that each of the six blocks in China continent, except Hainan block, should have at least three fiducial stations. Considering the importance of Beijing region, the densification of fiducial stations appropriately in the region is necessary. Besides, the optimum distribution of fiducial stations for the GPS orbit determination should be also taken into account as far as possible. Now 25 fiducial stations in total planed(see Fig.1).

2. In addition to the continuous GPS observation, the ability of real time data transmission at the fiducial station is also absolutely necessary. so fiducial station must be equipped with modern or satellite data transmission devices, computer, weather observation equipment and so on.

3. At the 25 GPS fiducial stations, three VLBI stations (among them a mobile VLBI station is taken as a fixed one), four SLR stations, two mobile SLR stations are coincided.

4. Absolute gravimetry should be carried out at each fiducial station.

5. Accuracy Index of fiducial stations is as follows. The measurement accuracy of annual change rate for GPS side between adjacent stations is ±1-3 mm. The accuracy for the determination of annual change rate of the baselines between VLBI stations is ±2 mm. The accuracy for coordinate determination of fixed and mobile SLR is ±1-3 cm, and ±3-5 cm respectively. The accuracy for absolute gravimetry is ±3-5 microgal.

VI. Establishment and Accuracy Index of Fundamental Network

1. The data of the crustal relative movement and deformation along their boundary zone among 6 main tectonic blocks in China continent will be obtained by the regular remeasurement of fundamental stations and the continuous observation of fiducial stations in CMMNC.

2. The number of GPS fundamental and fiducial stations on each main tectonic blocks of China continent should not be less than 10 (except Hainan block), and it should be densified appropriately in some important and active regions and also their boundary zone. 56 GPS fundamental stations are planed in CMMNC(see Fig.2) for regular remeasurement (once every 2-3 years).

3. The accuracy for the determination (each time) of GPS baseline between adjacent fundamental stations is ±3-5 mm for horizontal component and ±10-15 mm for vertical one. During the measurement, each fundamental station will be occupied for 7 days and 24 hours per day.

V. Establishment and Accuracy Index of Regional Network

1. GPS regional network is established mainly on the active tectonic zones, along earthquake belts and in some important areas, It serves directly the earthquake prediction.

2. The separation among regional stations in the area of stable tectonic activity is about 250 to 350 km; In strong tectonic activity areas, and the important areas, the distribution of stations can be densified, and distance can be shorten to 30-50 km, or even shorter. Regional network is composed of 1000 GPS observation stations(see Fig .3).

3. During the measurement in each regional station should be occupied for 4 days and 24 hours per day. The accuracy of baseline between adjacent regional stations is ±3-5 mm for horizontal component and ±15 mm for vertical component.

VI. Data Transmission, Analysis, and Processing System

Data Transmission, analysis, and processing system is constituted by a data center and three data subcenters. The data center is composed of three systems: data transmission system, data base system , data processing and analysis system.

1. Data transmission system

Data transmission system is a communication network for acquiring the continuous observation data at GPS fiducial stations as well as the data from fundamental and regional stations. It is mainly used for inter -connection with GPS data communication in CMMNC via computer network. Meanwhile the system is also used for the information sharing and exchange of GPS data and crustal movement information, domesticly and internationally.

Data transmission system is composed of control center, communication frame lines among fiducial stations, data collection node for fundamental and regional stations, mobile data communication, power supply etc.

2. Data Base System

Data base system consists of basic information (geography, geodesy, geophysics, seismology, et.) and GPS data sub-database, SLR data sub-database, VLBI data sub-database, gravity data sub-database and leveling data sub-database.

3. Data processing and analysis system

Data processing and analysis system consists of following three sectors. GPS precise ephemeris calculation sector; crustal movement and comprehensive analysis sector; and earthquake prediction and evaluation sector.

VII. Real-time kinematic Monitoring System of Crustal movement(wide area differential GPS system)

The real time kinematic monitoring system for crustal movement is for short term earthquake prediction in high hazard area. The system is semi-mobile, so is very useful for surveying and investigation in an area before and after an earthquake. The system contains two parts: wide area differential GPS data processing center, and wide area differential GPS data broadcast center.

VLBI, SLR and GPS observations in the Key Stone Project

Yasuhiro Koyama, Ryuichi Ichikawa, Tadahiro Gotoh, Mamoru Sekido,
Tetsuro Kondo, Noriyuki Kurihara, Fujinobu Takahashi,

Kashima Space Research Center, Communications Research Laboratory,
893-1 Hirai, Kashima, Ibaraki, 314 JAPAN

Jun Amagai, Toshimichi Otsubo, Hideyuki Nojiri,
Kouichi Sebata, Hiroo Kunimori, Hitoshi Kiuchi, Akihiro Kaneko,
Yukio Takahashi, Shin'ichi Hama, Yuko Hanado, Michito Imae,
Chihiro Miki, Mizuhiko Hosokawa, and Taizoh Yoshino

Communications Research Laboratory
4-2-1 Nukui-Kita, Koganei, Tokyo, 184 JAPAN

Abstract

A space geodetic observation network has been established around Tokyo, Japan under a project name of Key Stone Project by Communications Research Laboratory. Three space geodetic methods, i.e. Very Long Baseline Interferometry, Satellite Laser Ranging, and Global Positioning System, are involved in the project. As of September, 1997, VLBI and GPS observation facilities at all four stations are operational, whereas developments of SLR observation facilities are in course of final alignment procedures. Daily VLBI observations began in January 1995 with a single baseline between Koganei and Kashima, and the full network observations with four stations began in September 1996. Observations and data analysis of VLBI measurements are fully automated and the analysis results are produced shortly after all observations of an experiment session finished. GPS observations at four sites began in July 1997 and the automatic data collection and analysis system are under developments.

1. Introduction

Communications Research Laboratory (CRL) has been establishing a compact space geodetic observation network around Tokyo, Japan under a project which was named as Key Stone Project. The Key Stone Project network is consist of four observation sites at Koganei (Tokyo), Kashima (Ibaraki), Miura (Kanagawa), and Tateyama (Chiba). The geographic locations of these four sites are shown in Figure 1.

At each site, observation facilities of three space geodetic techniques, i.e. Very Long Baseline Interferometry (VLBI), Satellite Laser Ranging (SLR) and Global Positioning System (SLR), are collocated. An 11m diameter fully steerable cassegrain

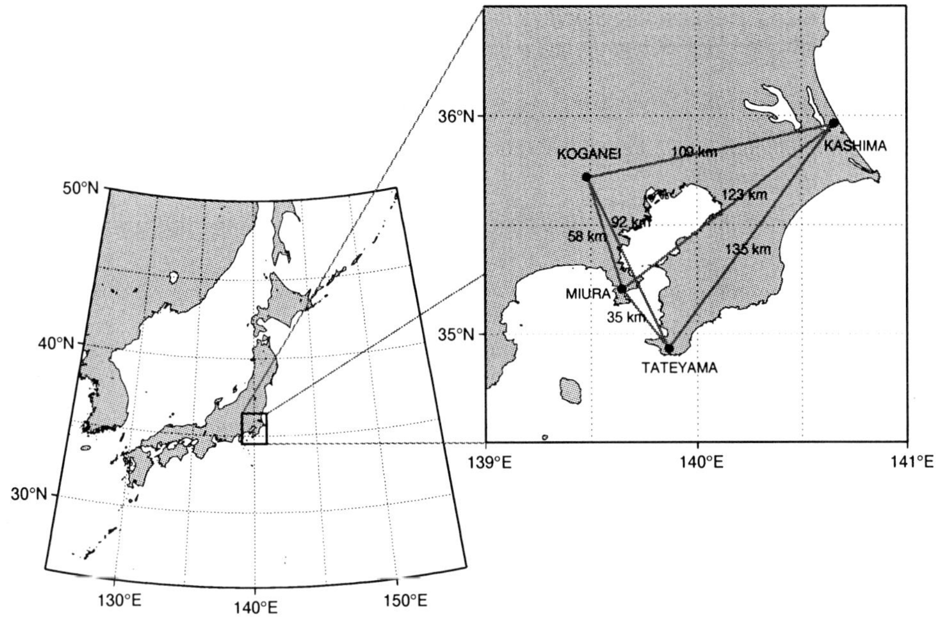

Figure 1. Geographic locations of four observation sites of the Key Stone Project space geodetic network.

antenna for VLBI observations, an optical telescope with an aperture of 75cm in diameter for SLR observations, and a GPS antenna mounted on top of a stable pillar are placed within about 100m from each other. As of September, 1997, VLBI and GPS observation facilities at all four stations are operational. Daily VLBI observations began in January 1995 with a single baseline between Koganei and Kashima, and the full network observations with four stations began in September 1996. Observations and data analysis of VLBI measurements are fully automated and the analysis results are produced shortly after all observations of an experiment session finished. GPS observations at four sites began in July 1997 and the automatic data collection and analysis system are under developments. The first signal of the laser echo from Lageos satellite was detected in December 1996 with the SLR observation system at Kashima. A lot of engineering data were then collected at Kashima and at Koganei in February, 1997 and final alignment procedures of the laser and optical system are on the way at present. The regular SLR observations will begin after automated operation software is completed.

2. VLBI, SLR, and GPS observation and data analysis system

One of the objectives of the Key Stone Project space geodetic network is to measure precise site positions with three space geodetic techniques and to investigate dynamical behaviors which might be seen in the series of estimated site position data

Table 1. Estimated VLBI site positions R on January 1, 1997, and velocities V defined in the ITRF94 coordinate system.

Site		X (mm)	Y (mm)	Z (mm)
Koganei	R	-3941937398.4 ± 5.8	3368150858.3 ± 4.7	3702235261.4 ± 5.0
	V	7.5 ± 0.7	-5.2 ± 0.6	-15.8 ± 0.6
Kashima	R	-3997505622.1 ± 6.1	3276878350.2 ± 4.8	3724240665.9 ± 5.2
	V	5.1 ± 2.3	-3.6 ± 2.0	-19.8 ± 2.5
Miura	R	-3976129918.1 ± 6.9	3377927833.6 ± 5.7	3656753813.7 ± 5.9
	V	24.2 ± 1.1	-6.7 ± 0.9	-15.5 ± 1.0
Tateyama	R	-4000983352.7 ± 5.7	3375275900.1 ± 4.7	3632213145.2 ± 5.0
	V	33.5 ± 1.9	-13.3 ± 1.6	-24.0 ± 1.7

before and during seismic activities. Because of this objective, the observations of VLBI, SLR, and GPS are required to be performed as frequently as possible. To achieve this requirement, observations and data analysis processing have been automated as much as possible.

In the VLBI system, almost all procedures from observations to data analysis are automated (Koyama 1997) and it will be fully automated when the developments of the real-time VLBI data correlation software (RKATS) is completed in the near future (Kiuchi *et al.* 1996). The VLBI observations are performed with two frequency bands at S-band and X-band. Before June 1997, the total data rate of observations were 64 Mbps and the observed data were recorded on ID-1 standard magnetic tapes. These tapes are then shipped to Koganei station everyday and usually delivered next day. The recorded data are processed with a VLBI correlator system developed for the Key Stone Project. Simple operations were required to change observation tapes in the digital mass storage system at each observation station and to ship recorded tapes. Since June 1997, on the other hand, the observations are performed at the total data rate of 256 Mbps and the observed data are transfered to the VLBI correlator at Koganei station with high speed data communication line which can transfer digital data at 2 Gbps using Asynchronous Transfer Mode (ATM). After data correlation processing and following band-width synthesis processing (Kondo *et al.*, 1996), Mark-3 format databases are created and site positions are estimated along with other parameters such as clock offsets and tropospheric delay. In the data analysis, model calculations are performed by a software called CALC which has been developed in Goddard Space Flight Center of National Aeronautics and Space Administration and the least square estimates are performed by a software called VLBEST which has been developed in CRL. The estimated results are made available to public over the Internet via anonymous FTP and World Wide Web accesses.

The estimated site positions and their rates of change are shown in Table 1. Site positions are given on the epoch of January 1, 1997 in the ITRF94 coordinate system based on results of Key Stone Project VLBI experiments and a joint VLBI experiment with 34m antenna station at Kashima on May 1, 1997.

Observations of GPS are performed with the Ashtech Z-X113 data acquisition unit and the Rev. B (Koganei) or Rev. D (Kashima, Miura, and Tateyama) antenna placed on a stable pillar. Data are collected at 30 seconds of intervals for 24 hours a day, and the observed data stored in a PC system are transferred to a GPS data analysis workstation through the computer network configured for the VLBI observation system. Observed data are analyzed by using Bernese Version 4.0 software which has been developed in Bern University. To make the entire process from data transfer to data analysis automated as much as possible, the software developments are under way.

SLR observation system consist of an optical telescope with a 75cm primary mirror and a 15cm secondary mirror in a Coude configuration, and a laser optic system. The telescope is placed in a fully covered dome equipped with a high quality glass window through which the ranging observations are performed. Thus the telescope can be operated in a temperature controlled stable environment and is protected against precipitation or harsh weather without an operator attending to the telescope system. By using eye-safe laser system of 1.5 micron of wavelength in addition to the main high power laser transmitting system, any possible hazards are automatically protected. The remote control and data transfer are done over 128kbps computer network independently configured for the SLR observations. The automated operation software and data analysis which are now being developed will allow regular extensive SLR observations in the near future.

3. Ground Survey Measurements

Ground survey measurements were performed with conventional measurement facilities using laser transmissions to measure relative position of three reference points of VLBI, SLR and GPS at each site. The results are tabulated in Table 2. The results obtained from VLBI and GPS observations are also tabulated for comparison.

Table 2. Positions of SLR and GPS reference points seen from VLBI reference point obtained by (1) ground survey measurements and (2) VLBI and GPS observations.

	Site		Eastward (mm)	Northward (mm)	Upward (mm)
(1)	Koganei	SLR	17420.8	−31359.6	−1975.0
		GPS	3493.0	−23843.2	−1769.9
	Kashima	SLR	12387.9	76693.2	7725.0
		GPS	18285.7	−24052.4	−4471.4
	Miura	SLR	16603.7	−74138.4	−2240.0
		GPS	−17032.1	6364.4	...
	Tateyema	SLR	−37205.4	−15229.3	−2056.0
		GPS	−34012.5	−19654.7	...
(2)	Koganei	GPS	3481.1 ± 1.7	−23854.5 ± 1.7	−1741.6 ± 8.6
	Kashima	GPS	18282.3 ± 1.6	−24069.7 ± 1.8	−4429.7 ± 9.1
	Miura	GPS	−17049.2 ± 2.0	6352.5 ± 2.2	−4441.4 ± 10.3
	Tateyama	GPS	−34011.9 ± 1.8	−19674.0 ± 1.6	−259.9 ± 8.6

Reference points of VLBI and SLR are defined as intersection of azimuth and elevation axes of the antenna and the telescope, respectively. Reference point of GPS is defined as a phase center of the antenna. The vertical components of the GPS reference point seen from the VLBI reference points at Miura and Tateyama stations are not yet available since the vertical offsets of GPS antenna phase centers from the top of pillars are not measured yet. VLBI results are obtained from daily regular VLBI observations and a joint VLBI experiment with four stations of the Key Stone Network and 34m antenna station at Kashima. GPS results were obtained from a data analysis of observations on April 21 and 22, 1997 at Kashima, Koganei, and at Tsukuba observed by Geographical Survey Institute, and another five day observations from July 28, 1997 at four stations of Key Stone Project network. The GPS results are obtained based on the ITRF94 global coordinate through a IGS site at Tsukuba (TSKB).

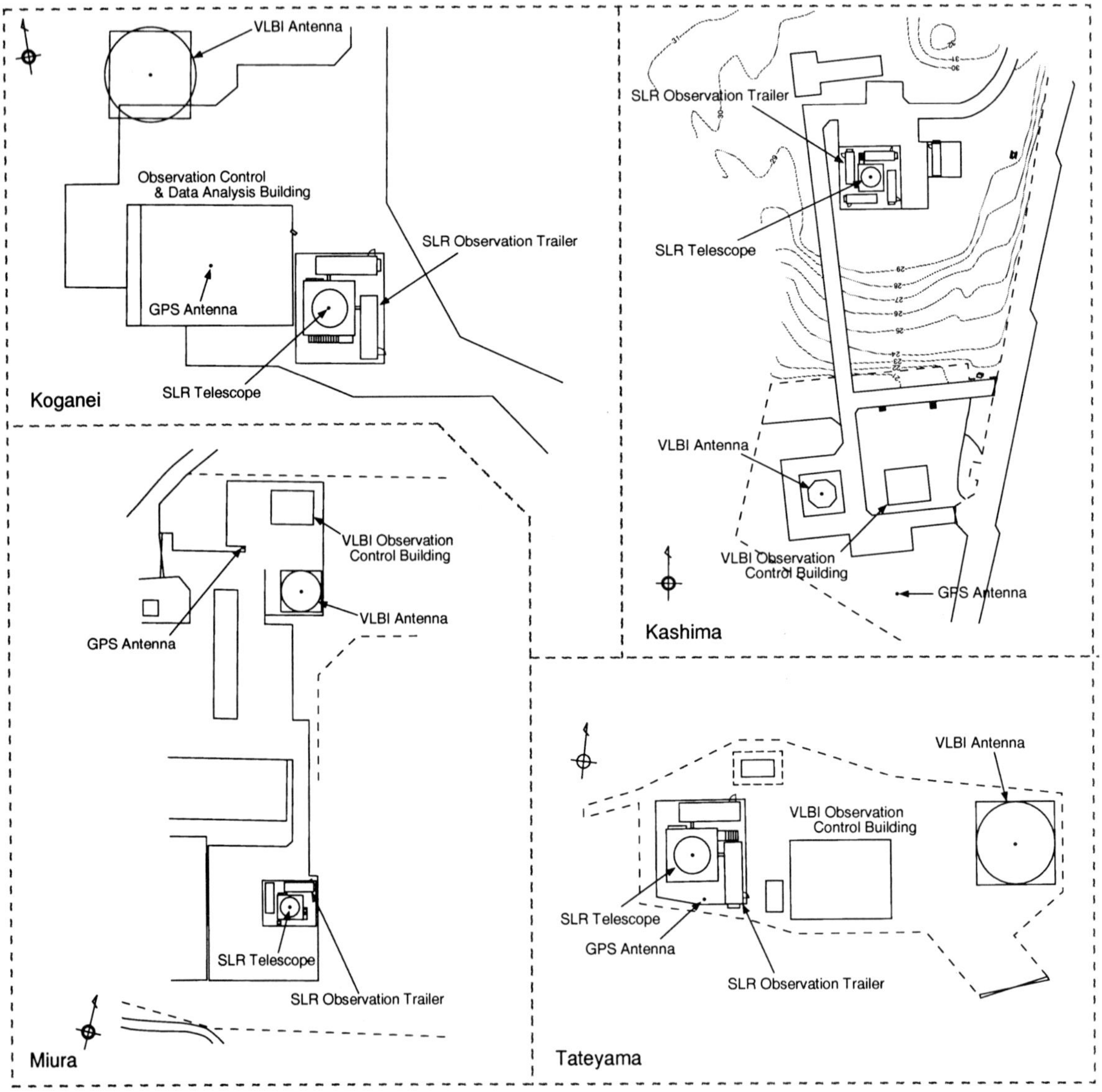

Figure 2. **Site map of each observation site.**

The comparison between ground survey results and two space geodetic measurements of VLBI and GPS showed an agreement within $2cm$ in horizontal components, and $5cm$ in a vertical component. The other comparison will be made when SLR results are obtained and the reasons of the discrepancies will be investigated to obtain better agreements. The precise ground survey measurements will be very useful for the collocation studies to tie different space geodetic techniques. Locations of reference points of VLBI, SLR, and GPS are shown in site maps in Figure 2.

4. Concluding Remarks and Future Plans

The system of the Key Stone Project space geodetic network has been described. The network is relatively compact with largest distance of 135km considering that space geodetic techniques are often applied to much longer distances. Because of its compactness and remote operation capabilities, however, the Key Stone Project can be considered as a unique and ideal test-bed of the technical developments and system improvements. Regular and extensive observations of VLBI, SLR, and GPS will be compared with each other to improve consistencies and accuracies. The network will be regularly tied with global space geodetic networks by joint VLBI experiments and interchanges of observation data of SLR and GPS.

Acknowledgments

The authors would like to express appreciations to colleagues in Geographical Survey Institute for GPS observations at Tsukuba and data correlation of a joint VLBI experiment of Key Stone Project and 34m antenna at Kashima. We would also like to appreciate colleagues in Goddard Space Flight Center of National Aeronautics and Space Administration for use of many involved softwares, and members of Communications Research Laboratory at present and in the past for the various system developments.

References

Kiuchi, H., M. Imae, T. Kondo, M. Sekido, S. Hama, T, Yamamoto, H. Uose, and T. Hoshino, "Real-time VLBI of the KSP", proceedings of the Technical Workshop for APT and APSG 1996, p. 125, 1996

Kondo, T., H. Kiuchi, and M. Sekido, "KSP Correlator and Data Reduction System", proceedings of the Technical Workshop for APT and APSG 1996, p. 188, 1996

Koyama, Y., "Automated Remote Operation System and Data Analysis System for the Key Stone Project", proceedings of the Technical Workshop for APT and APSG 1996, p. 139, 1996

Kunimori, H., C. Miki, J. Amagai, H. Nojiri, and T. Otsubo, "Automated SLR Observation System for Keystone Project", proceedings of the Technical Workshop for APT and APSG 1996, p. 144, 1996

COOPERATION OF SIXTEEN CEI (CENTRAL EUROPEAN INITIATIVE) COUNTRIES IN GEODESY AND GEODYNAMICS

Prof. Dr.-Ing. habil. Janusz Śledziński
Chairman and International Coordinator of the CEI Section C "Geodesy"
Chairman of the IAG Subcommission "Geodetic and Geodynamic Programmes of the CEI"
Co-Chairman of the Project CERGOP

Warsaw University of Technology
Institute of Geodesy and Geodetic Astronomy, Warsaw, Poland
00-661 Warszawa, Pl. Politechniki 1
Tel.: (4822) 6228515; Fax: (4822) 6210052; E-mail: sledzinski@gik.pw.edu.pl

1. COOPERATION LINKS BETWEEN IAG AND CEI. IAG SUBCOMMISSION "GEODETIC AND GEODYNAMIC PROGRAMMES OF THE CENTRAL EUROPEAN INITIATIVE (CEI)".

In February 1996 a new IAG (International Association of Geodesy) Subcommission was founded to initiate the cooperation links between CEI Section C and IAG. The Subcommission "GEODETIC AND GEODYNAMIC PROGRAMMES OF CEI" was created within the Commission VII "Recent crustal movements" (chaired by Prof. Dr. T. Tanaka, Japan), Section V "Geodynamics" (chaired by Dr. M. Feissel, France). The Chairman of the Subcommission was elected Prof. Dr. J. Śledziński (Poland).

The following charter duties of the Subcommission are the following: (1) coordination and/or integration of the international geodetic and geodynamic programmes supported by IAG and CEI, (2) creation of close links between running projects of IAG and those of CEI (e.g. CEI CERGOP - Central Europe Regional Geodynamics Project and IGS and EUREF, use of CEI permanent GPS stations within IGS and other programmes for maintenance of the ETRF and ITRF, etc.), (3) initialisation of common geodetic and geodynamic projects for the region of Central and Eastern Europe, (4) fostering the cooperation among universities and research centres from Central Europe and Western countries in the field of geodesy and geodynamics, promoting actions contributing to the development of innovative technologies and participation of CEI scientists in international IAG research programmes.

Members of the Subcommission and their responsibilities are: Janusz Śledziński (Poland) - Chairman of the Subcommission; Jozsef Adam (Hungary) - links between national geodetic/geodynamic networks and EUREF; Kazimierz Czarnecki (Poland) - cooperation between European universities (links between CEI and FIG or IAG similar bodies, the Subcommission scientific secretary; Istvan Fejes (Hungary) - links between CERGOP and other geodynamic IAG projects; Jan Hefty (Slovakia) - problems of coordinate systems, EUREF, ITRF and ETRF; Iginio Marson (Italy) - interregional gravimetric connections, gravimetric projects of IAG/CEI; Peter Pesec (Austria) - geodynamic use of CEI permanent GPS stations within IGS and other IAG programmes; Ewald Reinhart (Germany) - links/coordination and fostering a collaboration among projects of EU (e.g. ESA) and those of IAG and CEI; Jerzy Rogowski (Poland) - geodynamic use of CEI permanent GPS stations within IGS and other IAG programmes; Jaroslav Šimek (Czech Republic) - problems of homogeneity and time changes of precise levelling networks and vertical datum in Central Europe.

2. BRIEF HISTORY OF THE CEI.

In November 1989 the Foreign Ministers of Austria, Hungary, Italy and Yugoslavia at the conference in Budapest founded an organisation named QUADRAGONALE. A few months later, in April 1990 (Vienna) former Czechoslovakia joined this organisation forming the PENTAGONALE and in July 1991 at the conference of Prime Ministers in Dubrovnik Poland was admitted creating the HEXAGONALE. In July 1992 the HEXAGONALE was renamed as Central European Initiative. Violent political development in Europe, break-up and civil war in Yugoslavia, disintegration of Czechoslovakia, formation of new countries in the region of Eastern and Southern Europe, all these events caused considerable changes in organisation and international cooperation within the Central European Initiative. The current (1.09.1997) status of the CEI membership is the following: Albania, Austria, Belarus, Bosnia & Herzegovina, Bulgaria, Croatia, Czech Republic, Hungary, Italy, Macedonia, Moldova, Poland, Romania, Slovakia, Slovenia and Ukraine.

The main objectives of the CEI cooperation are to strengthen the stabilisation within the region of Central Europe, to promote all-European integration processes and to help the Central and Eastern European countries in entering the integrated world by adjusting their multi-lateral relations to Western European standards.

The Central European Initiative is a loose grouping of states with no legal status. It was agreed that a Summit (Heads of Governments and Foreign Ministers) would be held once a year in October and a meeting of Foreign Ministers of the member countries would be organised every year in April. Working Groups constitute the basic structural component of the CEI. They plan and approve initiatives, agreements and projects which they promote and complete in cases requiring financing, or which they pass to the National Coordinators or Foreign Ministries for approval and financing. There are sixteen CEI Working Groups at the moment: (1) Environment, (2) Transport, (3) Small and medium size enterprises, (4) Media, (5)Telecommunications, (6) Culture, education, youth exchange, (7) Science and technology, (8) Migration, (9) Energy, (10) Disaster relief, (11) Tourism, (12) Statistics, (13) Agriculture, (14) Civil defense, (15) Minorities, (16) Experts on vocational training, ad-hoc. The Earth sciences and, in particular geodesy, are represented in the Working Group "Science and Technology." The programme of the Working Group on Science and Technology includes presently nine endorsed scientific and technological projects (committees): Centres of Excellence Physics of Matter, Technology Transfer Centre, Earth Science, Experimental Mechanics, Meteorology, Astronomy and Astrophysics, Space, Industrial Technologies and Automation, Parallel Computing.
The work of the Committee for Earth Science is divided into three sections: Section A: "Geology", Section B: "Geophysics", Section C: "Geodesy".

3. GEODETIC RESEARCH PROGRAMME OF THE CEI SECTION C "GEODESY"

The work of the Section C is organised by specific working groups and projects. Now the following structure is accepted: (1) Working Group on Interconnection of Geodetic Networks; (2) Working Group on GIS/LIS; (3) Working Group on University Education Standards; (4) Working Group on Satellite Navigation Systems; (5) Project CERGOP; (5) Project UNIGRACE.

4. GPS EUREF CAMPAIGNS IN CENTRAL AND EASTERN EUROPE.

The interconnection of horizontal control networks of CEI countries is recognised in the programme of activities of the Section C "Geodesy" as one of the main tasks. The best way to fulfil this task is the cooperation with the IAG Subcommission EUREF. It is stressed in several Section C resolutions that the national geodetic services of CEI countries should intensify and finish the interconnection of geodetic control networks as soon as possible within the actions of EUREF organised by the Institut für Angewandte Geodäsie, Frankfurt/Main, Germany in cooperation with CERCO Commission VIII (Comité Europées de la Cartographie Officielle) and national geodetic services and institutions/agencies. The unified EU geodetic system have adopted so far the following 11 CEI countries: Austria, Bulgaria, Croatia, the Czech Republic, Hungary, Italy, Macedonia, Poland, Romania, Slovakia and Slovenia. It is to be stressed that in almost all these countries there are established national geodetic satellite control networks and the works are already completed.. It is also recommended that transformation parameters from the ETRF to the national coordinate systems used in particular countries and vice versa should be available as soon as possible for each CEI country. Five countries have not yet entered to the unified EU geodetic system. The EUREF campaigns are planned in Albania and Bosnia & Herzegovina. No plans are reported from Belarus, Moldova and Ukraine.

5. NETWORK OF PERMANENT GPS STATIONS IN CEI COUNTRIES.

The great importance of permanent stations for practical geodetic works in every country and for research has been fully recognized in the research programmes of the Central European Initiative (CEI). The programme of Section C recommends that each member country should establish and maintain at least one permanent station. A special CEI Study Group (CSG.5) was up to recognise the situation as regards the existing and planned permanent GPS stations in CEI countries. The current situation indicates that there are 29 permanent stations in operation in Central Europe (19 stations involved in the maintenance of the EUREF), and that the next 24 stations are to be established within the next two years. The situation should be assessed as very favourable and satisfactory.

6. CONCISE OUTLINE OF THE PROJECT CERGOP.

The area of Central Europe has a particular position from the point of view of geotectonics. This results from the fact that three geological units of various age contact on the territory of Central Europe: East European Precambrian Platform, Palaeozoic Platform of the Central and Western Europe and South European Alpine Orogeny. The Teisseyre-Tornquist's zone separating two above mentioned Platforms is of prime importance for the regional geodynamics of Central Europe. Also investigations of the geodynamic interactions in the Carpathian Orogenic Belt, Pannonia Basin and the Bohemian Subalpine Region may contribute to better knowledge of geodynamics of Central Europe.
Project CERGOP (Central Europe Regional Geodynamics Project) was initiated in 1993 by the scientists from FÖMI Satellite Geodetic Observatory Penc (Hungary), the Institute of Geodesy and Geodetic Astronomy of the Warsaw University of Technology and the Space Research Centre of the Polish Academy of Sciences. It was approved for realisation by the CEI member countries in May 1993 in Książ Castle, Poland. Eleven countries participate in the CERGOP Project.: Austria, Croatia, Czech Republic, Germany, Hungary, Italy, Poland, Romania, Slovakia, Slovenia, Ukraine.
The main objectives of the Project are the following: (1) to integrate the geodynamic research in the region of Central Europe based on high accuracy space geodetic surveys and an integrated

geodynamic network; (2) to foster the international cooperation among research groups of participating countries; (3) to provide a precise geodetic frame - so called Central European GPS Reference Network (CEGRN) - for studies on geodynamics of Central European areas of Pannonia Basin, Bohemian Massif, Teisseyre-Tornquist Zone, Carpathian Orogenic Belt and Subalpine Region; (4) connection of local geodynamic networks established on the territory of participating countries; (5) to collect satellite observations for studies and interpretation of geodynamic interactions in Central Europe.

The main coordinating body of the Project is the International Project Working Group (IPWG) headed by the Project Chairman Dr. István Fejes (Hungary) and the Project Co-Chairman Prof. Dr. Janusz Śledziński (Poland). Each participating country delegates to the IPWG one representative, the National Investigator. Project CERGOP is partially financed by the EU programme COPERNICUS. It was agreed that the conferences of the International Project Working Group (IPWG) would be organised twice a year to discuss the progress of scientific research and plan of next actions within the Project.

Ten study groups were formed to carry out research in particular fields. They are:
CSG1. Investigation of tropospheric delays, CSG2. CERGOP site quality monitoring, CSG3. CERGOP reference frame, CSG4. Standardisation of data and processing centres, CSG5.CERGOP GPS Permanent and Epoch stations, CSG6. CEGRN and height determination, CSG7. CERGOP gravity network, CSG8. Geotectonic analysis of the region of Central Europe, CSG9. cancelled, CSG10. Monitoring of recent crustal movements in Eastern Alps with GPS, CSG11.Threedimensional plate kinematics in Romania.

Establishment of a Central European Regional GPS Reference Network (CEGRN) was proposed as the first action within this Project CERGOP. The design required that the sites and monumentations of the CEGRN stations would satisfy optimal conditions for GPS observations as well as identify and fulfil the geodynamical needs. For technical reasons the number of sites of the CEGRN was limited to 31. It was also decided that Graz Lustbühel Computing Centre (Austria) would serve as the CERGOP Data Centre. There are eight CERGOP Processing Centres which process the observations of the CEGRN campaigns. These centres are: Graz (Austria), IGS Evaluation Centre IG&GA Warsaw (Poland), FÖMI SGO Penc (Hungary), VUGTK Pecny (Czech Rep.), IfAG (Germany) and STU (Slovakia)), ISA Matera (Italy) and University of Zagreb (Croatia). The first GPS zero-epoch observation campaign (named also "pilot project campaign") of the CEGRN'94 was organised from May 2nd to May 6th, 1994 and the CEGRN'95 campaign was performed from 29 May to 3 June 1995. The third CEGRN campaign (a second COPERNICUS CEGRN'96 campaign) was organised from 10th to 15th June 1996 and the fourth CEGRN'97 campaign from 4 to 10 June 1997. Now the new phase of the Project CERGOP is being discussed within the CEI Section C.

7. THE OVERLAP BETWEEN OTHER INTERNATIONAL PROJECTS.

There are some geodynamic projects running in Europe that indicate or may indicate some interferences with the CERGOP. These projects are: WEGENER, EXTENDED SAGET, BALTIC SEA LEVEL PROJECT, EUROPROBE, EUREF, IGS. As an example of the interference of two overlapping projects we can mention the CERGOP and EXTENDED SAGET which was initiated by the Institute of Geodesy and Geodetic Astronomy of the Warsaw University of Technology in 1991 as an extension of the Polish Project SAGET launched at this Institute in 1986. There is a significant coincidence of scientific aims of both projects; however, there are also very essential differences. The EXTENDED SAGET network covers much more extended area and an unlimited number of points can be incorporated to this network. This gives an excellent opportunity to all participating institutions to connect new established points to the ITRF coordinate system. The

CERGOP campaigns are envisaged to be organised till 1997; EXTENDED SAGET campaigns are thought as long-term action and will be performed every year at least in the whole of current decade. So far, the EXTENDED SAGET campaigns were organised in the following periods: (1) 1992, September 7-11; (2) 1993, August 2-6; (3) 1994, May 2-6; (4) 1995, May 29 - June 3; (5) 1996, June 10-15; (6) 1997, June 4-10. In order to have more points related to the same epoch of observation the last four campaigns of EXTENDED SAGET fully overlapped the period of CEGRN campaigns. The following conclusions may be pointed out when comparing both projects: (1) The same standards of GPS observations are observed in both CEGRN and EXTENDED SAGET campaigns, (2) EXTENDED SAGET network includes stations of Scandinavia and Mediterranean Region. CEGRN is limited only to CEI countries; only some regions of Germany, interesting from tectonic point of view, are included, (3) EXTENDED SAGET campaigns give the possibility to connect to ITRF system new points that may be currently needed, (4) Both networks (projects) can coexist. Campaigns of both projects may supplement each other.

8. PROJECT UNIGRACE.

A new project on Unification of the Gravity Systems in Central and Eastern Europe was lately launched by the CEI Section C. The Project will provide benefits for (1) geodesy for the definition of a unique height system and for the determination of the geoid in Europe as well as for sea level variation studies, (2) metrology to transfer to Central and Eastern Europe the gravity standard for precise determination of weight and related instrumentation and to improve the testing methods and the level of agreement between laboratories, (3) geophysics for the definition of a unique European reference for geophysical investigations.

Ten absolute gravity intraplate and seven tide gauge stations will be measured in 12 countries: Austria (Graz), Bulgaria (Sofia, Burgas), Croatia (Zagreb, Dubrovnik), Czech Republic (Pecny), Germany (Wettzell, Rostock), Finland (Metsähovi), Hungary (Penc), Italy (Trieste), Poland (Józefosław, Władysławowo), Romania (Gilau, Constanţa), Slovakia (Modra Pesky), Slovenia (Ljubljana). As intraplate stations mainly permanent GPS stations were chosen, as tide gauge points the stations with long series of sea level observations. Twelve countries will participate in the Project, two non-CEI countries (Finland and Germany). The gravity observations will be made by five absolute meters from Austria, Finland, Germany, Italy and Poland. The meters will be compared at the stations in Sèvres, Wettzell and in Józefosław. Two observation campaigns in 1998 and 1999 will be organised. Four review UNIGRACE conferences are planned. The Project was accepted by the European Union INCO COPERNICUS Programme. Works will start very soon.

9. WORKING GROUP ON UNIVERSITY EDUCATION STANDARDS.

The Section C Working Group on University Education Standards was founded at the Fourth Section Review Conference in Warsaw in February 1995. The general charter duties of this Working Group are the following: (1) unification of university programmes and curricula, (2) compatibility of programmes of universities from Central and Eastern Europe and those from Western countries, (3) exchange of scientists and lecturers, organisation of postgraduate studies, summer/winter schools, etc. (4) postgraduate and doctoral studies, (5) exchange of student groups with the aim of field training in surveying and geodesy, diploma works, scientific visits etc., (6) cooperation with similar Working Groups of IAG and FIG, (7) initiation of international projects (TEMPUS, PHARE, COPERNICUS etc.). Several very important and valuable actions have been undertaken by this Working Group last time. First action was organisation of the international symposium "Education in GPS Application to Geodesy and Geographic Information Systems (GIS)". The symposium was held in September, 18-20, 1996, in Ljubljana, Slovenia. Another

International Symposium "DGPS in Engineering and Cadastral Measurements - Education and Practice" was organised in Ljubljana, Slovenia, 25-27 August 1997. A Summer School "Education in GPS Application to Geodesy and LIS/GIS" to be held in Grybów, Poland is planned for August 1998.

10. COOPERATION LINKS BETWEEN CEI AND EGS.

Cooperation between CEI Section C and the EGS (European Geophysical Society) was discussed during the XXI General Assembly of the EGS in The Hague, The Netherlands in May 1996. It was decided that since there are many projects of common interest of both CEI and EGS, during the next XXII General Assembly of the European Geophysical Society, Vienna, Austria, 21-25 April 1997 a special symposium devoted to geodetic and geodynamic programmes of the CEI would be held. Prof. Dr. J. Śledziński (Poland) was appointed the Convener of the Symposium. Papers on EUREF campaigns in CEI countries, CERGOP, UNIGRACE, activities of the CSGs, contribution of permanent GPS in CEI countries to international programmes were presented at this symposium. CEI Section C will take an active part in all next EGS General Assemblies. In 1998 the XXIII General Assembly of the European Geophysical Society will be held in April in Nice (France) and a symposium "Geodetic and Geodynamic Achievements of the CEI Section C" will be he organised.

11. REFERENCES.

I.Fejes, M.Barlik, I.Busics, W.Pachelski, J.Rogowski, J.Śledziński, J.B.Zieliński. The Central Europe Regional Geodynamics Project. Proceedings of the International Seminar "GPS in Central Europe", Penc, Hungary, 27-29 April 1993.

J. Śledziński. CERGOP Project as a multilateral cooperation of CEI countries to set up a GPS monitoring network in Central Europe (invited paper). Paper presented at the Third Round Table Session of the Earth Science Committee, Working Group on Science and Technology of the Central European Initiative (CEI), Trieste, Italy, 12-13 December 1994.

J.Śledziński. Programme of research of CEI (Central European Initiative) countries in geodesy and geodynamics. Paper presented at the 8th International Technical Meeting ION GPS-95, Palm Springs, California, 12-15 September 1995.

Janusz Śledziński. Central European Initiative (CEI) as the organisation of international cooperation in geodesy and geodynamics. Paper presented at the XXI General Assembly of the European Geophysical Society, Session: G1: "Open Session on Geodesy", The Hague, The Netherlands, 6-10 May 1996.

Janusz Śledziński. Geodetic and geodynamic research projects realised in international cooperation of fifteen CEI (Central European Initiative) countries. Paper presented to the Geodetic Seminars at: Geophysics Directoriate, Phillips Laboratory (PL/GRIM), Hanscom AFB, MA; Massachusetts Institute of Technology, Cambridge, MA; Naval Research Laboratory, Washington DC; NASA Goddard Space Flight Center, Greenbelt, MD; Defense Mapping Agency (DMA) Headquarters, Fairfax, VA; DMA Geodesy and Geophysics Department (SMWD), St. Louis, MO; Institute of Navigation, Kansas City, MO. September 1996.

Janusz Śledziński. GPS techniques for geodesy and geodynamics. International cooperation of fifteen countries of the Central European Initiative. Paper presented at the 9th International Technical Meeting ION GPS-96, Kansas City, Missouri, USA, September 17-20, 1996. Proceedings ION-96, Kansas City, Missouri, USA.

Janusz Śledziński. Central European Initiative (CEI) - Organisation and Programme of the International Cooperation in Geodesy and Geodynamics. Paper presented at the XXII General Assembly of the European Geophysical Society, Session G14 "Geodetic and Geodynamic Programmes of the Central European Initiative (CEI)", Vienna, 21-25 April 1997.

NO RELATIVE MOVEMENTS BETWEEN SOUTH AMERICAN AND SCOTIA PLATES IN TIERRA DEL FUEGO

D. Del Cogliano, R. Perdomo, J. L. Hormaechea,
C. Santarossa and E. Zunino
Facultad de Cs. Astronómicas
y Geofísicas - Univ. de La Plata
CONICET
Argentina

Abstract:

In December, 1993, a twenty points precise GPS network for cartographic and geodetic purposes was first established in the argentine portion of the Tierra del Fuego Island. The northern part of the network (15 points) is placed on the Southamerican plate (SAM), three points are situated on the Scotia plate and two points are located near the main fault. Redundant baselines were measured in such a way that each point is directly related to at least three neighbor points. During December, 1994 and December, 1995, the network was partially remeasured (12 common points) to detect tectonic movements between SAM and Scotia plates. TRIMBLE 4000 SSE receivers were used during the 1993 campaign, and Ashtech Z12 receivers were used in 1994 and 1995 campaigns. Each time, the individual baselines were processed using the GPSurvey software, and the Trimnet program was used to compensate the redundant baselines. Precise IGS orbits were used in all cases. The estimated accuracy of each individual network was 1 cm in latitude and longitude and 2 cm in ellipsoidal height.

The repetitivity of the coordinates between the results of the different campaigns, after a Helmert transformation is better than 1 cm in latitude and longitude and about 2 cm in ellipsoidal height. Due to the relatively small distances between the extremes of the network (about 150 km), the spatial coordinates were transformed to plane Gauss Krüger coordinates and a two dimension transformation was performed (a scale factor, one rotation angle and X, Y origin offsets). The repetitivity of the plane coordinates is better than 1 cm. In conclusion, no movements were detected in this region during the last two years, to this level of accuracy.

1. Introduction:

The Tierra del Fuego island is cut by a fault zone running East-West that represents segments of the South American (SAM) - Scotia (SCO) transform plate boundary (Dalziel, 1984; Klepeis, 1994).

Several groups are interested in the region and very valuable informations were obtained from the project SCARP proposal (Lawver et al., 1994), and from recent geological investigations of local groups (Olivero et al, 1995).

In December 1993, a GPS network was first established in the Argentine portion of the island. The La Plata University group carried out the measurements and calculations. The main purpose of this network was to serve as reference frame for cartographic and cadastral services. However, the good quality of the results and the fact that some points are located on the SAM plate while other points are on the SCO plate, suggested the authors to re measure the network to determine relative movements between both plates in the region.

2. The network:

Twenty points distributed over the whole province were measured in 1993. Fifteen points are located in the northern region (SAM plate), three are clearly situated on the SCO plate, while other two are close to the fault zone (Perdomo et al, 1994). The typical distance between adjacent points is 40 km.

The monuments are concrete pillar completely buried excepting for a 25 cm sides parallelepiped on the top, with a bench mark in the center.

The origin of the network is the EARG (Estación Astronómica Río Grande) GPS pillar. A Doris beacon is operating in EARG since 1989 and a GPS permanent receiver is also working since 1995, in cooperation with Potsdam GeoForschungZentrum (GFZ).

3. The campaigns:

The first campaign was carried out in December 1993 using three Trimble 4000SSE receivers (borrowed from GFZ). Each session lasted about 2.5 hs. producing two independent baselines. Twenty points were occupied and 43 independent baselines were measured.

The second campaign (December 1994) and the third one (December 1995) were measured using only 2 Ashtech Z12 receivers. Thirteen points were occupied and 24 baselines were measured. Only the northern portion of the 1993 network was not re measured.

4. Processing strategies:

The software used for processing the individual baselines of the three campaigns was GPSurvey 2.00. IGS precise orbits were used in all the cases.

The adjustments were calculated using Trimnet with the only condition of one fixed station (EARG). The error ellipses are very similar for different campaigns. From them, it is possible to estimate the errors: about 1 cm in both Latitude and Longitude and 2 cm in ellipsoidal height.

5. Discussion of the results:

The comparison between the results obtained in 1994 with those of the 1993 campaign shows an excellent agreement and may be taken as a validation of the first ones. So the following discussion is centered in the comparison between 1995 and 1993 results.

The simple differences between coordinates of the same points have the following mean values and standard deviations:

Table 1

	Latitude differences	Longitude differences	height differences
	[m]	[m]	[m]
mean value	+0.013	-0.003	+0.022
standard dev.	0.008	0.010	0.022

The individual deviations about the mean value show that a transformation was necessary to take into account the slight differences introduced by the system (reference frame, orbits).

As a first approximation, a seven parameters transformation was used, but even when better residuals were obtained, the highly correlated values obtained for the parameters clearly show that this solution had no sense. The main reason surely is the smallness of the region covered by the network: the length of the more distant points is about 150 km.

Besides, the fact that the height individual determinations and the comparisons between different campaigns are less precise than those for Latitude and Longitude, suggested that further analysis might be made using plane coordinates.

The geodetic coordinates were converted into plane coordinates (Gauss - Krüger projection) and a simple but more realistic transformation was performed: one scale factor, one rotation and a two components vector for the origin.

The residuals obtained after these transformation are the following:

Table 2

	X dif. [m]	Y dif. [m]
EARG	+0.004	-0.001
Despedida	+0.001	+0.000
Viamonte	+0.003	-0.008
Guazu Cue	+0.000	-0.001
PteJusticia	-0.007	-0.003
Yehuin	-0.001	+0.009
EOLO	-0.005	-0.001
Ran.Hambre	+0.005	+0.008
CADIC	-0.003	-0.004
Almanza	+0.008	+0.001
Moat	-0.005	-0.002

The stations are ordered from North to South, so that the last three are located on the SCO plate. In the adopted argentine convention, the X Gauss - Krüger coordinate is measured along the meridian.

6. Conclusions

The Tables 1 and 2 show that:

- The agreement between the coordinates obtained in 1993 campaign with those obtained later is better than the estimated accuracy of the individual networks !!

- It seems clear that in this region, the horizontal movements between South American and Scotia plates is less than 1cm/year.

References:

- Dalziel, I., 1984, The Scotia Arc: an international geological laboratory: Episodes, vol. 7, n 3.
- Klepeis, K., 1994, The Magallanes and Deseado fault zones: major segments of the South American - Scotia transform plate boundary in southernmost South America, Tierra del Fuego: Journal of Geoph. Res.
- Lawver, L., Dalziel, Y., Schutz, B., Bevis, M. & Smalley, R., 1994, Scotia Arc GPS Project (SCARP): a proposal submitted to the National Sc. Foundation.
- Olivero, E., Malagnino, E. & Gagliardini, D., 1995, Interpretación preliminar del sistema de fracturas del este de Tierra del Fuego basada en imágenes ERS-1: SELPER, Rev. Tec. de Int. Iberoamericana y Mundial, vol.11, 1-2.
- Perdomo, R., Hormaechea, J.L. & Del Cogliano, D., 1994, Informe final del trabajo "La red geodésica de la Tierra del Fuego": Dirección de Geodesia, Prov. de Tierra del Fuego.

A NUMERICAL MODEL FOR THE STUDY OF THE LOAD EFFECT OF THE WATER IN THE AREA OF THE RIVER PLATE.

M. Cristina Caamaño[1,2], Claudia L. Ravazzoli[2], Rodolfo Rodríguez[3]

[1] Becaria de Perfeccionamiento, C.O.N.I.C.E.T., Argentina.

[2] Facultad de Ciencias Astronómicas y Geofísicas, Universidad Nacional de La Plata. Paseo del Bosque s/n, (1900) La Plata, Argentina.

[3] Departamento de Ingeniería Matemática, Universidad de Concepción. Casilla 4009, Concepción, Chile.

Abstract.

We present a simplified geophysical model to evaluate the deformation field of the local lithosphere during a marked swell of the River Plate. The influence of this phenomenon on the displacement field is analyzed. The differential equations of elastic equilibrium were solved by means of a three dimensional finite element method with appropiate boundary conditions. The results obtained indicate that very high precision measurements would be neccesary to quantify these effects. This, in turn, would allow us to improve the estimation of the elastic parameters of the area.

Introduction.

The study of the deformation of the Earth by surface tractions is a classic problem of geophysical interest as a tool for the study of its mechanical behaviour. As it is well known, large masses of water produce important loading tilts in the crust which extent their effect far from the loads. In some cases this effect may also cause temporal fluctuations of the absolute gravity which can be registered in continental areas with high precision gravimeters.

The River Plate, one of the widest in the world, frequently shows an important increase of its volume due to strong south easterly winds and other meteorological effects. These phenomena present particular characteristics such as very high levels of water (on the order of the oceanic tide) and temporal aperiodicity, which are observed by a continuous network of tide gauges around the river.

The aim of this work is to evaluate the load effect of water during the forementioned swells. Using the available geological information, we developed a simplified numerical model to compute the deformation of the lithosphere in the area of the River Plate under the hypothesis of linear elastic behaviour. The differential equations of static equilibrium for the displacement field were solved by using a finite element method. Unlike the classic convolutional method, valid for uniform halfspaces (Farrell 1972), this approach allowed us to consider geological strata of variable thickness. It should be mentioned that finite element modelling was previously used in this context for instance, by Beaumont and Lambert (1972) and Sato and Harrison (1990).

Definition of the three-dimensional model.

In order to build up our simplified three-dimensional model of the local lithosphere, the first step was to define a plane two-dimensional domain extending from 29° 45' S to 39° 45' S and from 51° 29' W to 61° 29' W, according to Figure 1. The dimensions of the area were chosen so as to contain the whole surface of the river and to include points sufficiently far from it as well. This was necessary to ensure (as much as possible) the validity of the boundary conditions applied for the numerical solution of the problem.

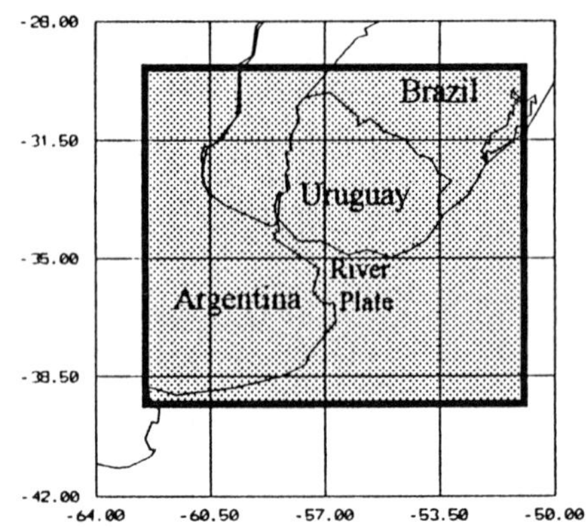

Figure 1. Location of the model (geographical coordinates)

To include the main geological features of the region in our model, we looked for information in the geological and geophysical literature on the subject (Urien 1981; Sprechmann et al. 1981; Bracaccini 1972). From that analysis we can briefly say that the crust in the area consists of a wide variety of Cretaceous-Cenozoic sedimentary fill with seismic p-wave velocities lower than 5 km/sec, lying on a basement composed of Precambrian and Paleozoic methamorphic and Upper Jurassic volcanic rocks with seismic velocities higher than 5 km/sec. Then, for the local crust we adopted the simple scheme of a *sedimentary layer* of variable thickness (from approximately 0.025 to 6 km) over a *basement* extended up to 33 km depth (Barrio 1993). To delineate the top of the basement (see Figure 2) we also used some borehole information taken from Barrio (1993). For simplicity, the effects of the curvature and topography were not taken into account. Also, we considered a plane bottom for the basement (i.e. a plane Mohorovicic discontinuity). The *lower lithosphere* was included as a third layer, extending from 33 to 115 km depth, according to PREM model (Dziewonski and Anderson 1981). In Figure 3a we show a scheme of the three-layer model.

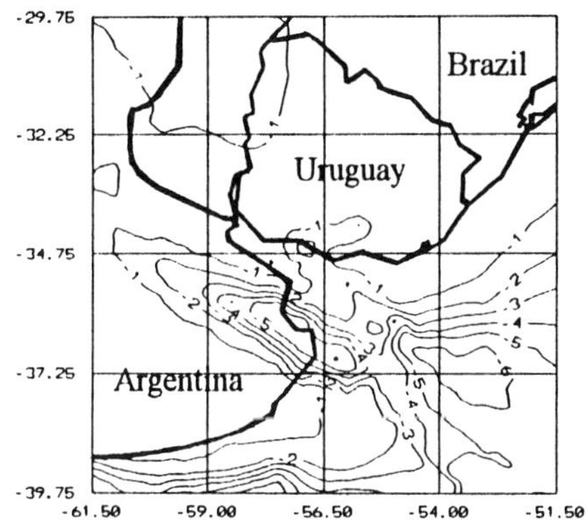

Figure 2. Contour map showing basement depth in km measured from top surface of the model (geographical coordinates)

To assign the elastic parameters to sediments and basement we used the forementioned p-wave velocities measured by Bracaccini (1972) in seismic refraction surveys. From that data and using Nafe-Drake relations (Ludwig et al. 1970) to estimate the corresponding s-wave velocities, we established the following range for the elastic parameters λ and μ (Lamé constants) for the sedimentary layer: λ^S_{min} = 6.15 GPa, μ^S_{min} = 0.824 GPa and λ^S_{max} = 18.33 GPa, μ^S_{max} = 21.44 GPa. Using the same procedure, we obtained $\lambda^B \doteq \mu^B$ = 34 GPa for the basement, but for the numerical tests we also used $\lambda^B_{low} = \mu^B_{low}$ = 10 GPa, according to one of the values given by Melchior (1983) for the crust. The mean elastic properties adopted for the lower lithosphere are λ^L = 85.12 GPa and μ^L = 67.26 GPa and were calculated from PREM model. Finally, we must mention that for the shallow water of the Argentine sea in the area of study, we considered the same elastic properties of the sediments.

Resolution of the numerical problem and analysis of results.

To evaluate the load effect of water during an important swell of the river we computed the static deformation of an elastic non-gravitating three-dimensional bounded domain[1] subjected to the action of surface forces. The equilibrium equations for the displacement field were solved using a finite element method. To do this, we first constructed a plane

[1] It should be pointed out that due to the large horizontal extension of the domain compared to its vertical size, a plate model would probably be more efficient to solve this problem.

two dimensional mesh of triangular elements with the least possible distortion (Figure 3b). In some cases the location of the nodes was chosen to coincide with the position of the tide gauges situated around the river. Next, we generated three-dimensional mesh with three layers (subdomains), as described in the previous section. The total number of elements was equal to 5076 with 1192 nodes.

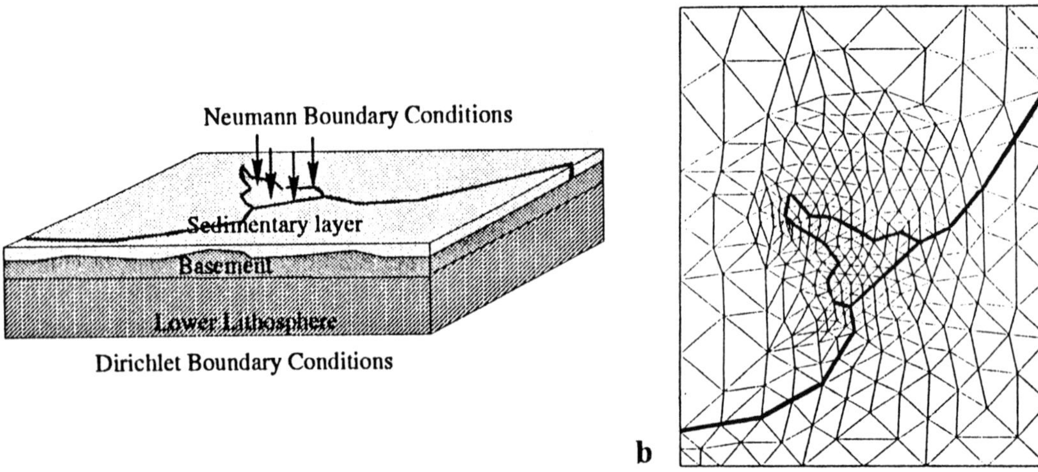

a b

Figure 3a,b. a Schematic representation of the three dimensional model. **b** Surface two-dimensional mesh

The load effect of water was simulated by means of a Neumann type boundary condition on the top, representing a distribution of surface forces with the real shape of the river (Figure 3a). The pressure at each point was computed using water height data measured during an ordinary swell due to meteorological causes. On the bottom boundary (i.e. the lithosphere-astenosphere contact), we used a Dirichlet boundary condition (zero displacement) and on the lateral faces we tested the algorithm with both types of condition (zero displacement and zero tractions).

In Figures 4a, b, c we plot the vertical component of the displacement field at the surface for different combinations of the above mentioned elastic parameters. The contours in Figure 4d display the same magnitude as in Figure 4a calculated for a lower water height. From these results we observe that the area of significant deformation is located immediately under the loads and extends up to approximately 20 km from the coast line. Within this area the shape of the contours is mainly controlled by the volume distribution of the loads and the elasticity of the basement (lower values of λ^B, μ^B result in greater displacements). In the remainder, the deformation pattern depends only on the elastic modulus of the basament. The greatest value of displacement (approx. 1.56 cm) is controlled by the elastic properties of the sediments. As can be seen in Figures 2 and 4 this occurs at points within the Salado basin, where the sedimentary thickness and the water heights reach their maximum values simultaneously.

Similarly, at the top of basement and at Mohorovicic discontinuity we observe that the pattern of the displacement field is related to the volume distribution of the loads and the elasticity of the basement. In Figure 5 we show the amplitude decrease of the vertical displacement with depth along a representative cross-section.

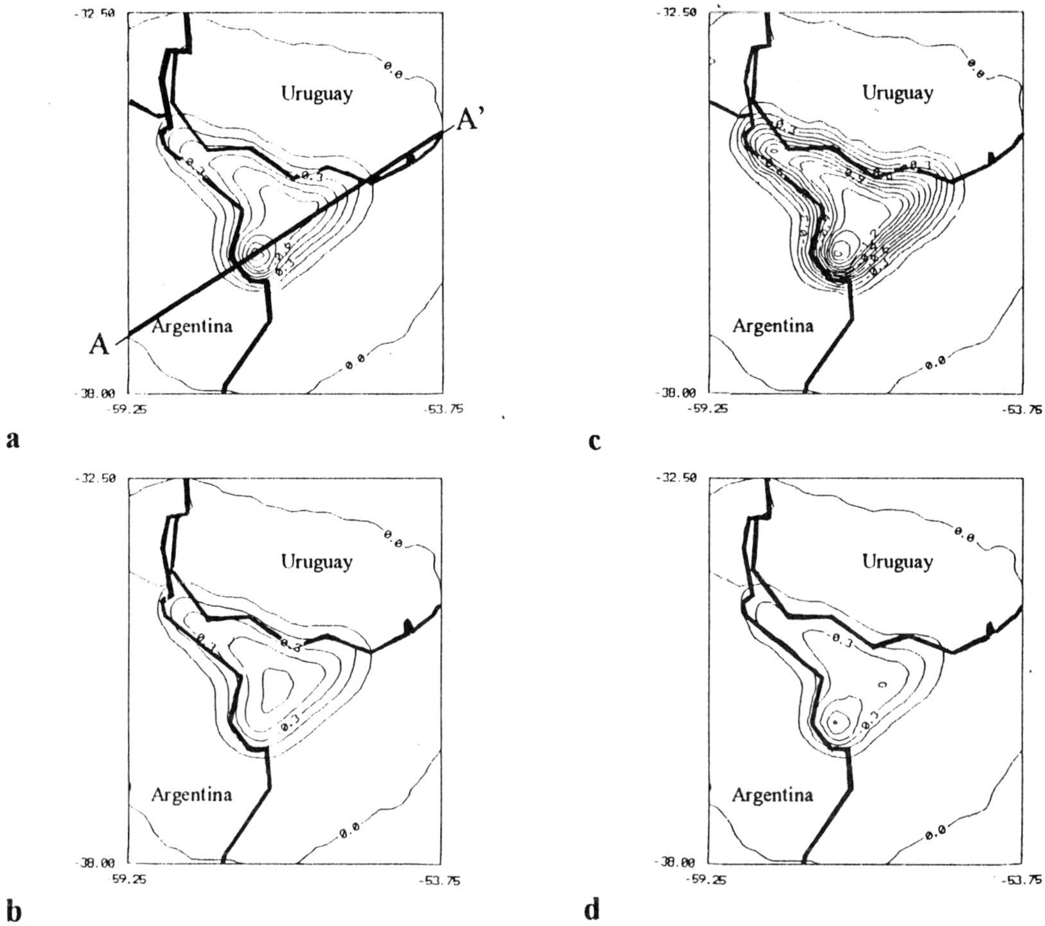

Figure 4a-d. Numerical model estimates of the vertical component of the displacement field at the surface measured in cm, **a** for λ^{S}_{min}, μ^{S}_{min} and λ^{B}, μ^{B}, **b** for λ^{S}_{max}, μ^{S}_{max} and λ^{B}, μ^{B}, **c** for λ^{S}_{min}, μ^{S}_{min} and λ^{B}_{low}, μ^{B}_{low}, **d** for λ^{S}_{min}, μ^{S}_{min} and λ^{B}, μ^{B} and for a lower water height. The graph was restricted to the area where there exist significant values of displacement

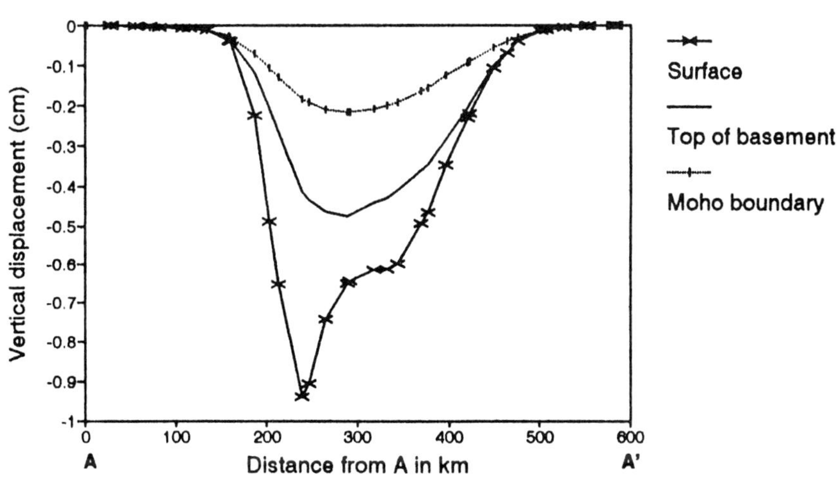

Figure 5. Vertical displacements at different depths along the A-A'profile (indicated in Figure 4a)

As expected, the results obtained using Dirichlet and Neumann boundary conditions on the lateral sides of the domain do not show meaningful differences in the neighborhood of the loads.

As a conclusion we may state that, despite of its simplicity, the numerical model presented here gives a reasonable approximation of the static deformation of the local lithosphere during an important swell of the River Plate. Thus we consider that the combination of this model with high precision GPS and tidal gravity measurements could help not only to achieve a better understanding of the deformation mechanisms but to improve the estimation of the elastic properties in the surrounding of the river as well.

Acknowledgements.

We would like to thank Ing. Juan Carlos Usandivaras for many stimulating discussions concerning this work. We also wish to extend our thanks to Prof. Paul Melchior, Dr. Dalla Salda and Ing. Donofrio for providing us useful bibliography and data.

References.

Barrio L (1993) Modelado bi y tridimensional de campos potenciales: metodología y aplicación. Tesis Doctoral. Facultad de Ciencias Astronómicas y Geofísicas, Universidad Nacional de La Plata, Argentina

Beaumont C, Lambert A (1972) Crustal structure from surface load tilts, using a finite element model. Geophys J R astr Soc 29: 203-226

Bracaccini O (1972) Cuenca del Salado. Geología Regional Argentina. Academia Nacional de Ciencias de Córdoba, pp 407-417

Dziewonski M A, Anderson D L (1981) Preliminary reference earth model. Phys Earth Planet Interiors 25: 297-356

Farrell W E (1972) Deformation of the Earth by surface loads. Rev Geophys Space Phys 10: 761-797

Ludwig W J, Nafe J E, Drake C L (1970) Seismic refraction. In: Maxwell A E (ed) The sea, Vol. 4, Part 1, Wiley-Interscience, New York, pp 53-84

Melchior P (1983) The tides of the planet Earth. Pergamon Press, Oxford

Sato T, Harrison J C (1990) Local effects of tidal strain measurements at Esashi, Japan. Geophys J Int 102: 513-526

Sprechmann P, Bossi J, Da Silva J (1981) Cuencas del Jurásico y Cretácico del Uruguay. In: Volkheimer W, Musacchio E A (ed) Cuencas sedimentarias del Jurásico y Cretácico de América del Sur. Comité Sudamericano del Jurásico y Cretácico, Vol. 1, Buenos Aires, pp 239-270

Urien C M (1981) The basins of Southeastern South America (Southern Brasil, Uruguay & Eastern Argentina) including the Malvinas Plateau and Southern South Atlantic paleographic evolution. In: Volkheimer W, Musacchio E A (ed) Cuencas sedimentarias del Jurásico y Cretácico de América del Sur. Comité Sudamericano del Jurásico y Cretácico, Vol. 1, Buenos Aires, pp 45-125

CRUSTAL DEFORMATION ALONG THE CARIBBEAN - SOUTH AMERICAN PLATE BOUNDARY DERIVED FROM THE CASA GPS PROJECT

Klaus Kaniuth, Hermann Drewes, Klaus Stuber, Herbert Tremel
Deutsches Geodätisches Forschungsinstitut, Abt. I (DGFI/I)
Marstallplatz 8, D-80539 München, Germany

Napoleón Hernández
Servicio Autónomo de Geografía y Cartografía Nacional
Edif. Camejo, Esquina Camejo CSB, Piso 2, Oficio 216, Caracas 1010, Venezuela

Melvin Hoyer and Eugen Wildermann
La Universidad del Zulia, Apartado Postal 10311, Maracaibo, Venezuela

Hans-Gert Kahle, Alain Geiger and Christian Straub
ETH Hönggerberg, HPV G 52, Ch-8093 Zürich, Switzerland

ABSTRACT

The Central and South America (CASA) GPS project aims at monitoring crustal deformations in the complex tectonic zone of the Caribbean, Cocos, Nazca and South American plate boundaries. Since 1988 several GPS campaigns have been carried out, unfortunately not always covering the whole area simultaneously.

Main emphasis is laid in this paper on the Caribbean-South American plate boundary. We present our first estimates of deformations derived from GPS observations acquired between 1988 and early 1997. Our analysis comprises all the data available in the Venezuelan part of CASA but includes also a number of stations in adjacent countries, mainly in Colombia. The main characteristics of the applied processing strategy is the simultaneous estimation of epoch positions and site velocities by defining the velocity components already on the observation equation level. The precision of the estimated horizontal velocity components varies between 0.15 and 0.55 mm depending on the number of occupations. The results do not indicate abrupt changes of motions at the Boconó - El Pilar fault, but suggest a smooth continuous crustal deformation.

INTRODUCTION

During an international workshop held in Maracaibo/Venezuela on March 17, 1987, the decision for establishing a Central and South America (CASA) GPS project was made by various institutions from Venezuela, Colombia, USA and Germany. The first GPS campaign was performed by about 30 participating institutions in January/February 1988 including some 25 stations in Costa Rica, Panama, Colombia, Venezuela and Ecuador (Kellogg and Dixon 1990).

Partial repetitive GPS observations were carried out several times in different regions of the project area. The Venezuelan part (Drewes et al. 1989) was for the first time reobserved and considerably extended with regard to the network area and to the number of stations in 1993 (Drewes et al. 1995). Another complete observation of that part was carried out in 1996. A number of CASA sites has been occupied within other GPS campaigns in 1994, 1995 and 1997.

The present analysis comprises all the GPS observations acquired in Venezuela at the mentioned epochs. Including data collected at some stations in Colombia and Costa Rica allows a common interpretation of the deformations in the entire CASA region in a unified reference system. However, in this paper emphasis is laid on estimating the crustal deformations along the Boconó - El Pilar fault system in Venezuela.

MAIN GEODYNAMIC FEATURES

The geographical area of the CASA project is characterized by the geodynamic interaction of four major tectonic plates: Cocos, Nazca, Caribbea and South America (Wadge and Burke 1983, Mann and Burke 1984). The dominant feature in the eastern part is the Boconó - El Pilar fault system forming the southern boundary of the Caribbean plate (see figure 1).

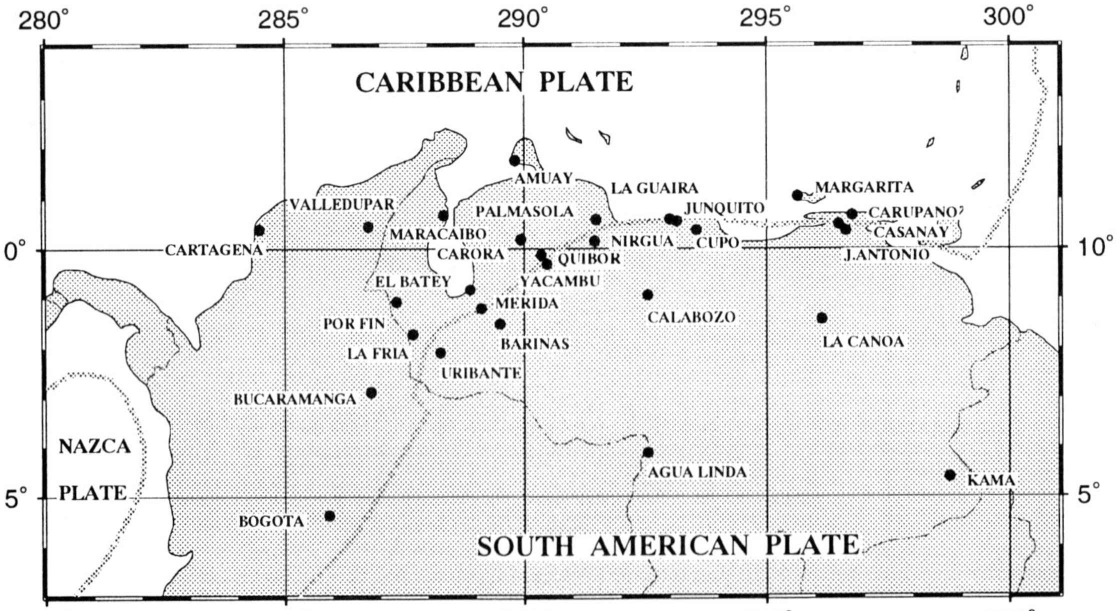

Fig. 1 CASA area and sites included in this analysis

The recent tectonics along this fault system has been studied from geological (e.g. Schubert 1984) and seismological (e.g. Suarez and Nábělek 1990) investigations. A relative motion of some 15 mm/a between the Caribbean and South American plates in this area can be derived from global plate models showing a right-lateral slip with a convergent component.

Geodetic monitoring of recent crustal movements along the Boconó fault was already done before the beginning of the CASA project by several local or regional networks in the Venezuelan Andes (Henneberg and Schubert 1986). A present-day strike-slip movement is indicated from these terrestrial measurements. Results of crustal movements in the western part from GPS data of the CASA project are discussed by Kellogg and Vega (1995).

DATA ANALYSIS

In addition to the sites displayed in figure 1 some IGS stations outside the CASA area are included to provide fiducial information for fixing the terrestrial reference frame. Table 1 summarizes the main characteristics of the individual campaigns and demonstrates that the data set is very heterogeneous with regard to receiver types involved, availability of fiducial points, number of stations occupied, and session length.

Table 1 Features of the GPS campaigns included the CASA solution

Epoch	Stations		GPS Receiver Systems Involved	Observations	
	CASA	Fid.		Days	Hrs/Day
1988.1	5		Texas Instruments 4100	6	2 + 5
1993.2	21	1	Leica 200, Trimble SST	4	7
1994.1	12	4	Trimble SSE/SST, Rogue/TurboRogue	5	8 - 24
1995.4	7	5	Leica 200, Trimble SSE, Rogue/TurboRogue, Ashtech Z-XII	10	24
1996.2	25	4	Leica 300, Trimble SST, Rogue/TurboRogue	5	8
1997.4	7	4	Leica 300, Trimble SST, Rogue/TurboRogue	3	8

For estimating site velocities we do not follow the concept of deriving these from sets of epoch coordinates; instead we define the velocity parameters, as linear functions of time, already in the GPS observation equations (Kaniuth et al. 1994). This strategy has been realized in context with the Bernese Software (Rothacher et al. 1993) which we use for data processing and has been successfully applied (e.g. Kaniuth et al. 1995). Depending on the number of observations and solve for parameters the campaigns were processed in subsets, saving the normal equations for external accumulation using DGFI's program ACCSOL. These normal equation subsets include station positions, velocities and troposphere parameters. Only at this final stage fiducial information, local ties and any constraints are applied. This strategy is displayed in figure 2.

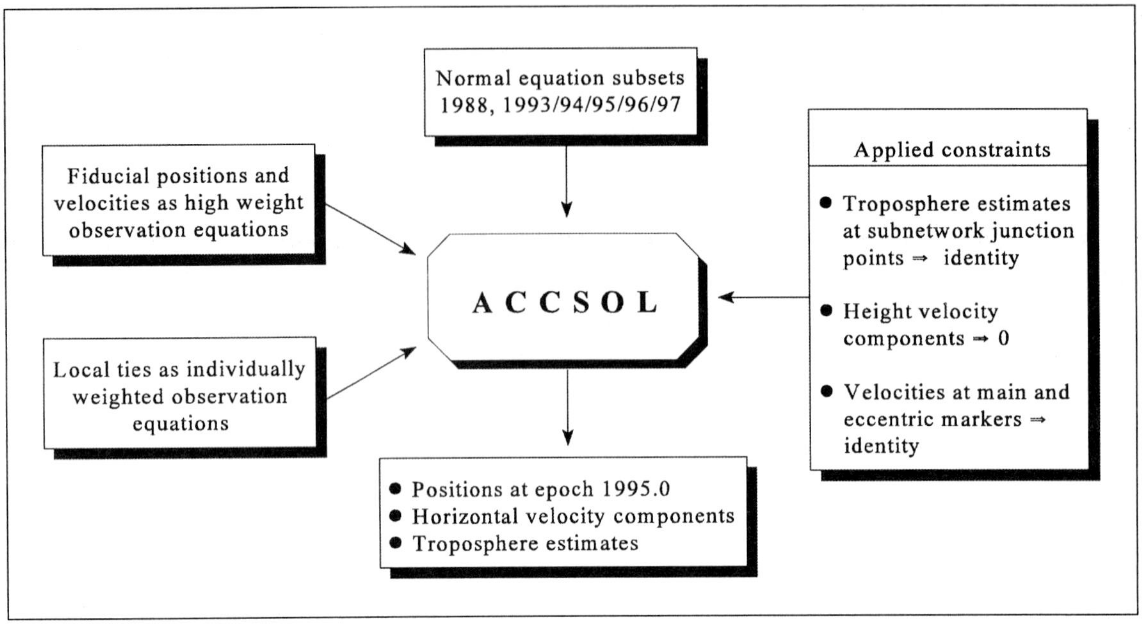

Fig. 2 Accumulation of CASA normal equations using program ACCSOL

The main features of the data processing and modelling are summarized in the sequel. The ionosphere free linear carrier phase combination L3 is processed in the double difference mode, code measurements are only used for receiver clock synchronization. The observation sampling rate was 30 seconds in all campaigns, and the elevation angle cutoff was set to 15°. Ambiguities are partly resolved according to statistical tests. In order to account for receiver antenna phase center variations, offsets of the mean phase center as well as elevation dependent corrections were applied according to Rothacher and Mader (1996). The tropospheric delay was predicted using Saastamoinen's (1973) zenith delay model and the mapping function proposed by Neill (1996), both applying standard atmosphere data; relative to this a priori calibration residual delays were estimated every two to three hours. We have not attempted to improve the satellite orbits from the regional data, because since 1994 the combined IGS orbits were available, and in 1993 orbits generated by the Center for Orbit Determination in Europe (CODE) could be used. The five stations occupied in 1988 are located close together, and therefore we do not expect any degradation due to the poor quality of the broadcast ephemerides at that time.

For establishing a reference frame for our regional network we have not used ITRF 94; instead we have selected the latest CODE (1997) global GPS solution including data up to spring 1997 and providing position and velocity estimates for the five fiducial points BOGT, CRO1, FORT, KOUR and RCM5.

RESULTS AND CONCLUSION

Considering that fixing positions at a reference epoch and velocities of five fiducial points would generate an overconstrained solution, we have applied weights allowing the fiducials to vary some millimeters in all components. The precision of the estimated annual velocities of the regional network points varies between 0.15 and 0.55 mm depending on the number of occupations. Figure 3 displays the horizontal velocity components of all sites occupied at least twice at epochs covering by at least 2.5 years time interval.

We clearly see the general north-northwest motion of stations on the South American plate. This motion is turning eastward on the Caribbean plate. Obviously there is no abrupt change of motions at the Boconó - El Pilar fault, but there is a smooth variation indicating a continuous crustal deformation. Station Bogotá in the southwest of the area displays the motion of the North Andes Block in between the adjacent plates.

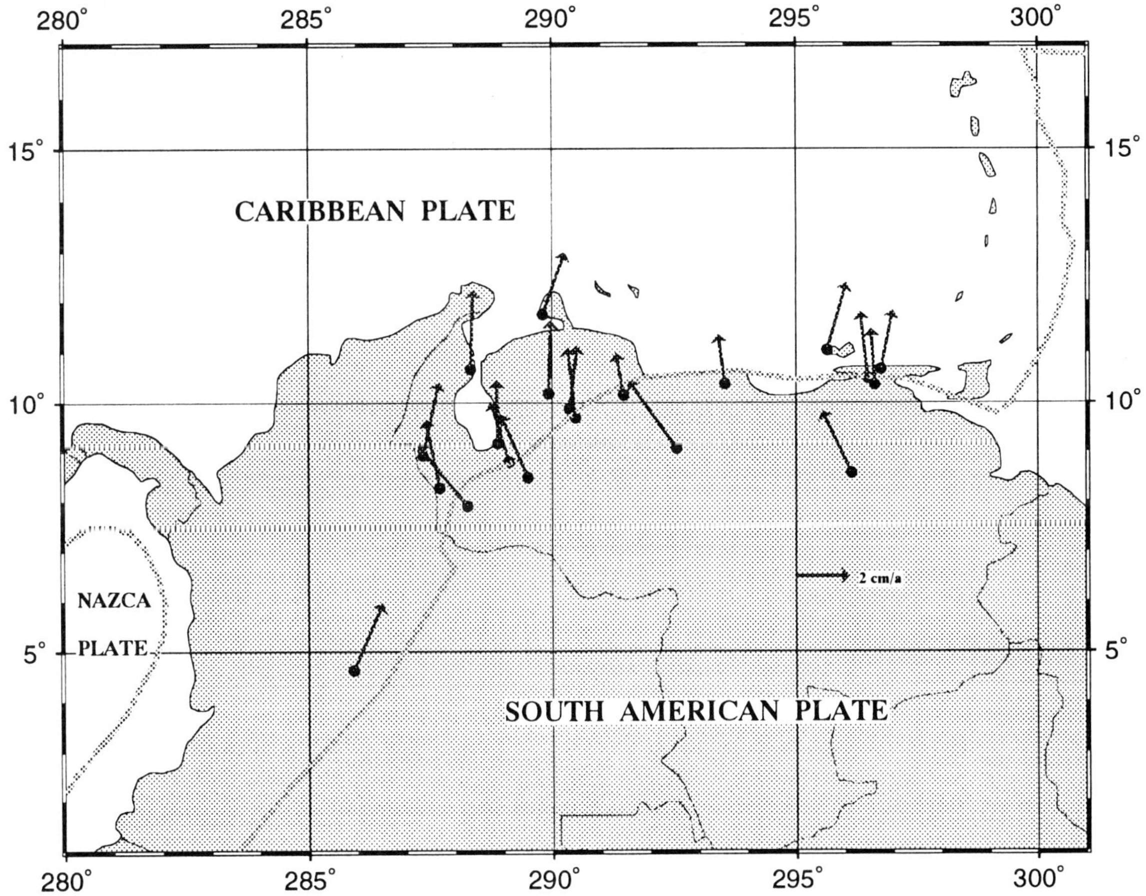

Fig.3 Velocity field along the Caribbean - South American plate boundary from the Venezuelan part of the CASA project

REFERENCES

CODE (1997): Center for Orbit Determination in Europe, University of Berne, Solution SC/SSV 97 P01.

Drewes, H., K. Kaniuth, K. Stuber, H. Tremel, H.G. Kahle, Ch. Straub, N. Hernandez, M. Hoyer and E. Wildermann (1995): The CASA`93 GPS Campaign for Crustal Deformation Research Along the South Caribbean Plate Boundary. J. Geodynamics 20 (2),129-144.

Drewes, H., Ch. Reigber, K. Stuber, M. Suarez, H. Tremel, H. Henneberg, M. Hoyer, O. Chourio and S. Rekkedal (1989): The Venezuelan Part of the CASA/UNO GPS Project. Manuscripta Geodaetica 14, 339-344.

Henneberg, H.G. and C. Schubert (1986): Geodetic Networks Along the Caribbean-South American Plate Boundary. Tectonophysics 130, 77-94.

Kaniuth, K., H. Drewes, H. Tremel, K. Stuber, S. Zerbini, L. Pezzoli and G. Corrado (1995): Crustal Deformations in the Calabrian Arc Area from Five Years of GPS Measurements. In Proc. of Istanbul-94, 1st Turkish Int. Symp. on Deformations, 781-790.

Kaniuth, K., K. Stuber and H. Tremel (1994): Multi-Epoch Adjustment of a European GPS Control Network. In Gubler, E. and H. Hornik (Eds): Report on the Symposium of the IAG Subkommission for the European Reference Frame (EUREF) held in Warsaw 8-11 June 1994. Veröff. Bayer. Komm. Int. Erdmessung 54, 220-227.

Kellogg, J.N. and T.H. Dixon (1990): Central and South America GPS Geodesy - CASA UNO. Geophys. Res. Lett. 17, 195-198.

Kellogg, J.N. and V. Vega (1995): Tectonic Development of Panama, Costa Rica and Colombian Andes: Constraints from Global Positioning System Geodetic Studies and Gravity. Geol. Soc. Am. Special Paper 295, 75-90.

Mann, P. and K. Burke (1984): Neotectonics of the Caribbean. Rev. Geophys. 22, 309-362.

Niell, A.E. (1996): Global Mapping Functions for the Atmospheric Delay at Radio Wavelengths. J. Geophys. Res. 101, B2, 3227-3246.

Rothacher, M., G. Beutler, W. Gurtner, E. Brockmann and L. Mervart (1993): The Bernese GPS Software version 3.4. Astronomical Institute, University of Berne.

Rothacher, M. and G. Mader (1996): Combination of Antenna Phase Center Offsets and Variations; Antenna Calibration Set IGS_01. Int. GPS Service for Geodynamics (IGS).

Saastamoinen, J. (1973): Contributions to the Theory of Atmospheric Refraction, Part II: Refraction Corrections in Satellite Geodesy. Bull. Geod. 107, 13-34.

Schubert, C. (1984): Basin Formation Along the Boconó-Morón-El Pilar Fault System, Venezuela. J. Geophys. Res. 89, 5711-5718.

Suarez, G. and J. Nábèlek (1990): The 1967 Caracas Earthquake: Fault Geometry, Direction Rupture Propagation and Seismotectonic Implications. J. Geophys. Res. 95, 17459-17474.

Wadge, G. and K. Burke (1983): Neogene Caribbean Plate Rotation and Associated Central American Tectonic Evolution. Tectonics 2, 633-643.

THE AQABA EARTHQUAKE OF NOVEMBER 22, 1995 AND CO-SEISMIC DEFORMATION IN SINAI PENINSULA, DEDUCED FROM REPEATED GPS MEASUREMENTS

F. KIMATA*; A. TEALEB; H. MURAKAMI***;N. FURUKAWA****; S. MAHMOUD**;H. KHALIL**; K.O. SAKR** AND A.M. HAMDY****

Abstract:

The newly developed techniques of space geodesy, especially the Global Positioning System (GPS), allow the monitoring of ongoing crustal deformation and provide a powerful tool for analysis of the regional seismotectonic regime and characterization of current strain accumulation and release.

Recently, the northern region of the Red Sea, the southern part of the Gulf of Suez and the Gulf of Aqaba-Dead Sea Fault are a tectonically seismoactive zone. Mainshocks, aftershock sequences as well as earthquake swarms are associated. Moreover, this tectonic zone is capable of producing large earthquakes that could cause extensive damage in the region. Large earthquake was occurred in March 31, 1969 at Shedwan island, to the south of the Gulf of Suez and north Red Sea, with a magnitude 6.9 and the last large earthquake was occurred in November 22, 1995 in the Gulf of Aqaba with a magnitude 7.0.

In order to monitor the crustal deformation in Sinai Peninsula and around the regions of the northern Red Sea-Gulf of Suez and Gulf of Aqaba a GPS geodetic network consisting of 12 geodetic points was established in 1994. The initial geodetic measurements were performed in April 1994 using GPS receivers type Trimble 4000 SSEs. The geodetic measurements were repeated in April 1995 and April 1996 using the same type of GPS receivers. Twenty-eight baselines ranging from 46 to 166 km were determined from the six-to eight-hours observation using precise ephemerides.

A southwest displacement of 166 mm was recorded at Dahab geodetic point, 26 km southwest of the epicenter of the November earthquake, in the period April 1995-April 1996, and a northeast displacement of 32 mm was observed at Ras-Mohamed, 50 km southwest of the epicenter. These displacements show a left lateral fault and it agree well with the focal mechanism of the November earthquake (strike-slip mode with the tension axis of 65 degree from north), and aftershocks distribution.

N.B. Under publication in "ACTA Geodetica, 1997" Journal of the Institute of Geodesy and Geophysics Sopron, Hungary.

Fault Model of the 1995 North Sakhalin Earthquake Based on SAR Interferogram

S. Ozawa, M. Tobita, S. Fujiwara, H. Nakagawa, K. Nitta, M. Murakami
Geographical Survey Institute
Kitasato-1, Tsukuba-shi, Ibaraki-ken, 305 Japan
M. Shimada
National Space Development Agency of Japan
1-9-9. Roppongi, Minato-ku, Tokyo, 106 Japan
P. A. Rosen
Jet Propulsion Laboratory
Pasadena, CA91109, USA

Abstract

A large (Mw=7.0) earthquake occurred on May 27, 1995 at the northern part of Sakhalin Island. Field workers reported right-lateral, strike-slip seismic fault as long as 35km, running NNE on the surface, which maximum lateral displacement is about 8.1m.

We report crustal deformations associated with the earthquake using SAR interferometry (InSAR) and a fault model we constructed from this result. JERS-1 L-band (24cm) SAR data of one month (April 28) before and two weeks after (June 11) the mainshock were used to detect the crustal deformation. The fringes due to topographic component were removed from the interferogram by using 3-pass method. A fault model was constructed by non-linear least square method using this crustal deformation data.

The resulting interferogram shows characteristic features. First, around the eastern side of the southern end of the fault, the uplift with a pair of adjacent peaks clearly appears, one of which is as large as 70 cm. Second, the eastern side of the north end of the fault subsided about 1m. These structures could not be observed in detail by any other means than InSAR, such as GPS observation data.

A model of the seismic fault was analyzed from the SAR interferogram. The model indicates that the southern end of the seismic fault extends more than the one that appears on the ground surface.

INTEGRATION OF GPS ALGERIAN SITES IN WEST MEDITERRANEAN GEODYNAMICAL STUDIES : CASE OF TYRGEONET PROJECT

Salem KAHLOUCHE [(1)], **Said TOUAM** [(1)], **Marco ANZIDEI** [(2)]

[(1)] National Center of Spatial Techniques
Geodetic Laboratory- BP 13 ARZEW - 312OO - ALGERIA
[(2)] Istituto Nazionale di Geofisica
Via di Vigna Murata, 605 - 00143 ROMA - ITALIA

Abstract

The western Mediterranean, particularly the North African region, is characterized by an important geodynamical activity due essentially to the collision between the Eurasian and African plates which gave rise to some earthquakes of high intensity (El Asnam, 1980). In the setting of geodynamical surveillance project of the Tyrrhenian sea (TYRGEONET) by the ING of Italy, two Algerian stations (Arzew and Algiers) were integrated in the 1995 and 1996 GPS campaign observations. The processing, with the Bernese Software, of data collected during 7 days, on 7 sites of the network (Arzew, Algiers, Madrid, Cagliari, Lampedusa, Matera and Noto) allowed to show a few displacements confirming the results provided by global geodynamical models based on the geological observations. The final solution was obtained by the adjustment of the daily GPS baselines with regard to the fixed station Matera in the ITRF92 System.

The analysis of the results obtained between the two observing campaigns showed that, between the Algerian stations, the relative movement is due probably to the presence of the seismic zones (Arzew fault, El Asnam active area) between the two stations. In the North of Africa, for bringing to light the slow geodynamical movements (5 mm / year), observations have to be performed over a long period (every 5 years).

Finally, in the elaboration of the error assessment, the tests performed on Arzew site showed an important variation (cm) of the phase center of antennas. This is due to the influence of the Ashtech Z-12 antennas orientation.

I - introduction

In response to the international interest caused by the project ' Crustal Dynamics' of the NASA since 1985, some observations were performed in Central and Eastern Mediterranean from interferometry, laser and GPS measurements.

The relative movement of African and Eurasian tectonic plates, the presence of particularly seismic regions (Algeria, Greece, Italy) on one hand, the emergence of the spatial geodesy techniques on the other hand , contributed to the make use of projects for a Geodynamical purpose.

At the local level, the Algerian North constitutes a vast seismic zone and the surveillance networks are until now connected to the national geodetic network expressed in the North Sahara system.

The project called WEGENER-MEDLAS (Working-group of European Geoscientists for Establishment of Networks for Earthquake Research - MEDiterranean LASer project), which started with the establishment of seismic surveillance network, was expanded to the survey of the temporal variation of mean sea level and to the scientific aspects of geodesy (precise positioning , pole motion, etc.).

From 1989, some approaches were undertaken for the installation of mobile VLBI station in North Africa in order to well known the tectonic motions in the Mediterranean basin. A project entitled WEMED (WEstern MEDiterranean) carried on a GPS network in Western Mediterranean had then risen and its various aspects were discussed during the symposium held in Nice (France) in October 1991. Nevertheless, there was not any approval of the South Mediterranean countries to this project.

II- Geodynamical Context of the Western Mediterranean

The continental lithosphere is an area of important distortions produced under the effect of continental collision.

It is well known that the African and Eurasian plates are animated by a relative motion which is translated by a closeness rising from the West to the East in order of a half centimeter by year at the level of Algiers to reach on average the centimeter at the Sicily meridian. This relative shifting is accompanied with a rotation of the two plates whose axis is being located in the vicinity of Rabat: we remark then an almost non existent motion .

Some major distortions characterize the global active tectonics on either sides of the Mediterranean. They are materialized by some folds, some faults and by great accidents in the main orogenic areas : Rif, Atlas, Betic Cordillera, Alps, Apennines, Pyrenees. With these zones of important distortions, alternate some continental blocks, with few deformations: the High plateaus (Algeria), the Corsica-Sardinia block, the Sicily and the Iberian Meseta.

III- TYRGEONET Network: extension to Africa

The TYRGEONET project (TYRhenian GEOdetic NETwork), initially mounted by the ING (Istituto Nazionale di Geofisica) and the University of Bologna for the oceanography and the geodynamical surveillance of the Italian peninsula knew an extension toward several riversides country of the Mediterranean among them Algeria.

The TYRGEONET network is constituted about fifty stations of which some are determined by the VLBI and SLR techniques: Cagliari, Madrid, Matera, etc.

In the setting of the extension of the project to North Africa, two Algerian institutions contribute to the project, each one by the observation of one site: the CNTS (National Center of Spatial Techniques) and the INC (National Institute of Cartography). The project allowed to determine 2 points (Arzew and Algiers) in the WGS-84 system (linked to the VLBI and SLR stations) and offers the possibility to densification by the determination of points in the inner regions of the country, and therefore, improve the accuracy of the geodetic networks.

IV- Observation campaigns 1995-1996

The GPS observations of each campaign have taken one week and are recorded in 24 h daily files; the interval of recording is 30s for the whole of the receivers used.

During the counting, we proceeded to an examination of the GPS observations done at each station of the network (period, quality of the observations,etc.). Depending on these parameters and of the baseline length, one summit of network by day of observations was fixed.

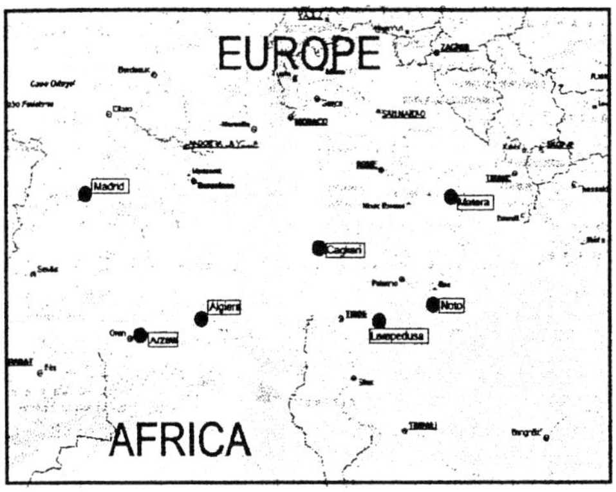

Fig. 2: Stations processed in the course of this study

Station	J.D 144	J.D 145	J.D 146	J.D 147	J.D 148	J.D 149
Cagliari						
Noto						
Lampedusa						
Madrid						
Matera						
Algiers						
Arzew						

a/

J.D 157	J.D 158	J.D 159	J.D 160	J.D 161	J.D 162	J.D 163	J.D 164

b/

Tab. 1*:Sites used and observation period: **a/ 1995** Campaign ; **b/ 1996** Campaign*

V- Processing the TYRGEONET Data.

All the available GPS data are processed with the Bernese software Version 3.5 developed at the Institute of Astronomy of the University of Bern (Swiss).

The pseudorange measurements are used in the linear combination L3 (free of the ionosphere) in order to estimate the initial coordinates of the receiver in the WGS-84 system and the synchronization of the receiver clock with the GPS time.

The precise ephemeris of the satellites and the information files which relates to the satellite clock are included in the data processing of the TYRGEONET 1995 and 1996 campaigns.

We selected one fixed summit and one constituted network of GPS baselines for the construction of a network in the shape of pivot for each daytime of observations.

A 20° minimal angle of elevation is required by Bernese in order to reduce all effects caused by the multipath error and the atmospheric noise.

This standard model uses the values for the pressure, the temperature and the relative humidity obtained by extrapolation of the standard values of sea level to the altitude of the stations.

The use of the initial coordinates allowed to resolve the ambiguities (linear combination L5) for each session and campaign. At this step of pre-processing, the ionospheric models was introduced in order to correct the observations in L1 and L2.

The ambiguities being initially known, the solution L3 elaborated for each day with the resolution of the tropospheric parameters.

The final coordinates for each campaign are obtained by the adjustment of the solutions on the daily GPS baselines obtained after having fixed the station of Matera with its coordinates in the ITRF92 system.

VI- Influence of the antenna phase center

In order to determine the role of orientation of the antennas, some baseline tests were measured in the CNTS with the use of three Ashtech Z-12 receivers. Below are presented the results of one of among the whole, long of 11,211 m.

The choice of the very short bases were imposed by the in order to don't submerge the variation of the position of the phase center with the noise of measurement which would be relatively important for a long base.
Four sessions of observations of 30 mn each were conducted in the same conditions of work (previous optimization of the constellation by the choice of the times of observations, session duration), with proceeding at the end of each session by a rotation of the antenna of 90°.

Sess.	GDOP & *Nb.Sat*	Az. ant.	Wide Lane	L1	L2
A	3.9 *5*	0°	11.209	11.211	11.212
B	2.4 *7*	90°	11.192	11.204	11.212
C	2.5 *7*	180°	11.207	11.207	11.212
D	3.5 *6*	270°	11.215	11.208	11.210

Tab. 2: Baselines obtained in' Wide Lane,' L1 and L2 modes in function of the antenna orientation (Azimuth).

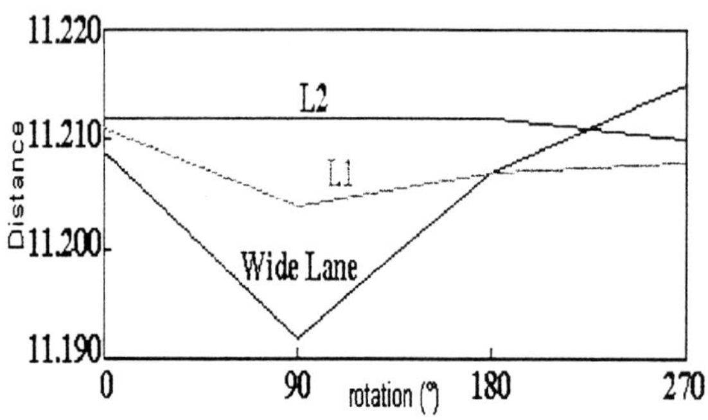

***Fig.** 3: Influence of the antenna orientation and the processing mode on the base measurement.*

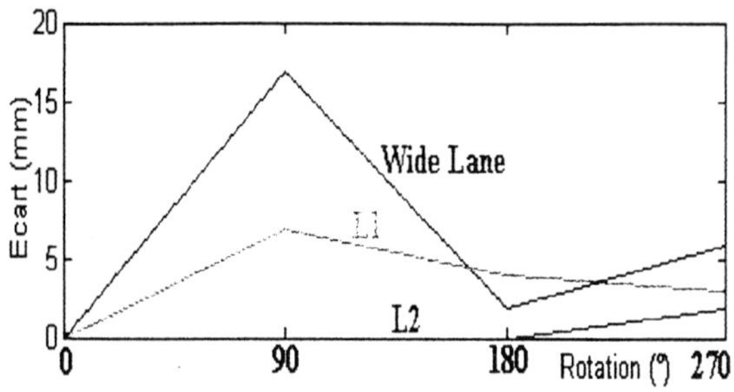

***Fig.** 4: Deviation in function of antenna orientation and processing mode of the base measurement.*

The relative accuracy of the GPS technique for a short base is millimetric when taking all the useful precautions such as the orientation of the antennas toward a same direction and the choice of adequate mode of processing.
The illustrations shown above allow to affirm that the center of phase is not confounded with the center of the antenna; the results obtained depend on the orientation of the antenna
The results obtained with monofrequency mode are more coherent when it comes to processing very short baselines .

VII- Comparison of the two campaign: results analysis

The statistical analysis conducted to separate the sites in two groups, one group formed of stations not having presented some significant differences of coordinates and therefore, called Datum, the other regrouping the points of which the coordinates changed statistically.
The used program is the Demo-Package, developed at the University of Rome (Crespi Mr., 1996); it permits the determination of the distortions and the rigorous statistical analysis of results taking into account the covariance matrix of the parameters.

Site	Δφ mm	Δλ mm	Δh mm	range 2D mm	Az. (°)
Cagliari	-4	1	-12	4	170
Noto	1	3	2	3	74
Lampedusa	-2	-2	19	19	-138
Arzew	0	-10	10	10	- 89
Algiers	1	-2	-1	2	- 63
Madrid	2	-4	-20	4	- 65
Matera	-3	6	-71	6	115

Tab.3: Variation of the station coordinates, error ellipses parameters and shifting vectors.

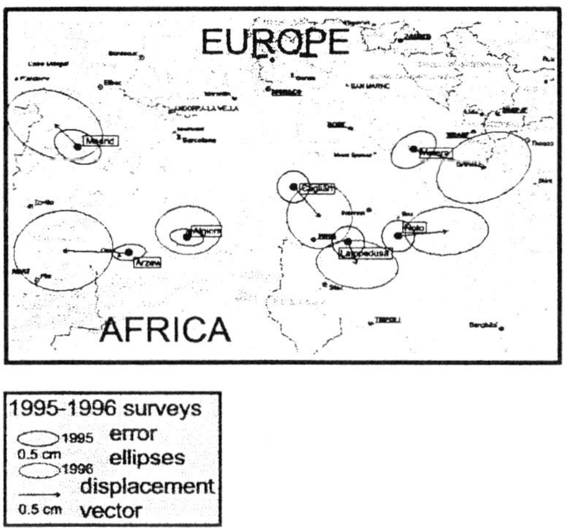

Fig.5: GPS Network, error ellipses derived from 1995-1996 campaigns 1995-1996 and correspondent vectors shifting.

VIII- Interpretation of the distortions

Though we have not detected some significant distortions of the terrestrial crust between the 2 campaigns of observation (1995 and 1996), it is remarkable through the variations of coordinating that the 2 stations located in Algeria (Arzew and Algiers), could provide some preliminary data as for their geodynamical behavior showing in this way, a relative opposite motion translating an extension. This could be in relation with the presence of the seismic active area located in the region of Arzew (Arzew fault) and the reputed El Asnam area for its catastrophic earthquakes (10 October 1980, M= 7.3) and located between the 2 stations.

The stations of Matera and Cagliari present the evidence of significant shifting, while the other stations of the network do not present a significant relative movement.
The station of Grasse (South of the France) would have contributed to the optimization of the network configuration and to the improvement of the analysis if data were available.

IX- Conclusion

The processing of the GPS data of TYRGEONET project were done using the powerful software (Bernese and Demo-Package) and the files of precise ephemeris.
The exhaustive modeling of different intervening parameters in the processing of the TYRGEONET network allowed to obtain interesting results. Nevertheless, the introduction of meteorological parameters measure on stations would have allowed, by substitution with the global model, best results. Some serious investigations could be taken and contribute to the geodynamical studies of the Mediterranean Western. Besides, the important effects of the antenna orientations showed through the tests done on the site of Arzew/ CNTS, must be taken in account, notably with the use of various means (different antennas) during the TYRGEONET campaign.
The centimetric bias between the phase center and the center of the antenna must be introduced in the final adjustment of the network in order to benefit of all the GPS power and guarantee the quality of the results for the geodynamical applications.
Considering the accuracy obtained for the trans-mediterranean baselines, it is obvious that a GPS observing campaigns spaced from 4 to 5 years are more adequate for the determination of significant shifting of tectonic plates. this would allow to make

the centimetric amplitude of noise of the measurement less than the shifting signal.

The analysis of the results obtained from a participation to the 2 observing campaigns (1995 and 1996) let us believe to an extension of the network and provide results which oppose the geological global models which forecast a compression of the African and Eurasian plates. The extension of the period of observation (4 to 5 years) or the use of more accurate techniques as the VLBI, would allow to confirm the global motion of geodynamical blocks of the western Mediterranean basin. Finally, the processing of the TYRGEONET data allowed to set in Algeria 2 points (Arzew and Algiers) in the WGS-84 system with a centimetric absolute accuracy. The most precise available points until then are the 11 points established at the time of the ADOS African campaign where the results in the NSWC-9Z2 system were published in 1987 (general Assembly of the IAG), with a final precision of 0.8 m on average.

Acknowledgments

The authors are thankful to Institut National de Cartographie (Algeria) for its cooperation during the 1995 and 1996 surveys.

References

Ambrosius B. A. C., Rebaï N., Wakker K. F., 1991. *First results of WEGENER-MEDLAS data analysis*. Global and regional geodynamics. I.A.G Symposia 101. Springer-Verlag.

Anzidei M. et al., 1995. *Crustal deformation analysis applied to GPS observations in the frame of the TYRGEONET Project*. XXI IUGG General Assembly, Boulder.2-14 July 1995.

Anzidei M, Baldi P., Casula G.,.... 1996. *Repeated GPS Surveys across the Ionian Sea: Evidence of crustal deformations*. Geophysics J. International N° 127.

Barlier F. et al; *Projet de Campagne GPS en Méditerranée Occidentale. Preliminary version* . presented at the IV Symposium on the Geodesy in Africa - Tunis. May 1990

Benouar D., Laradi N., 1996. *A reappraisal of the sismicity of the Maghreb Countries-Algeria, Morocco, Tunisia*. Natural Hazards 13: 275-296.

Bonnin J., Olivet L., 1986. *Seismic Hazard in Mediterranean Regions*. Kluwer Academic Publishers for Commission of the European Communities.

Chen R., 1992. *Processing of the 1991 GPS Campaigns in Central Finland for crustal deformation studies*. Reports of the Finnish Geodetic Institute.

Crespi M., 1996. *Demo-Package. Version 1.0. I.T.S.* Area Topografia. Faculta di Ingegneria. Universita di Roma.

Kahlouche S., Touam S., Anzidei M., 1997. *Apport du GPS à l'étude géodynamique de la Méditerranée Occidentale*. International seminar 'the use of the spatial tool for the prevention of the major risks' CNTS- Algeria .5-6 May 1997.

Rebaï N. et al., 1996. *Crustal deformations in the Mediterranean area computed from SLR and GPS observations*. J. Geodynamics. Vol. 21 N° 1.

Meghraoui M., Doumaz F., 1996. *Earthquake-induced flooding and paleoseismicity of the El Asnam, Algeria, fault-related fold*. Journal of Geophysical Research. Vol. 101.

Touam S., 1997. *Contribution du GPS à l'étude géodynamique de la Méditerranée Occidentale*. Magister Thesis - C.N.T.S / Algeria.

MULTI-PARAMETRIC EXPERIMENT FOR OBSERVING CRUSTAL DEFORMATIONS IN SOUTHERN BRAZIL

S.R.C. de Freitas[1], M.C. Santos[1], J. Cordini[1] & E. Marone[2]
[1]Universidade Federal do Paraná, Curso de Pós-Graduação em Ciências Geodésicas, P.O. Box 19011, 81.531-990 Curitiba, PR, Brazil
sfreitas@cce.ufpr.br, mcsantos@geoc.ufpr.br, cordini@geoc.ufpr.br
[2]Universidade Federal do Paraná, Centro de Estudos do Mar, Av. Beira Mar, SN, 83.255-000 Pontal do Sul, PR, Brazil
maroneed@cce.ufpr.br

Abstract

The periodic actions provoked by the tidal potencial on portions of continents, oceans and atmosphere cause direct and indirect effects on the earth's crust (Melchior, 1983). The induced deformations cannot be neglected when dealing with high-accuracy geodetic positioning, specially in view of the diurnal and semi-diurnal tidal components. The dynamic effects of earth tides, and the loading coming from variations in atmospheric pressure and the indirect effect of the oceans on these frequencies (Farrel, 1972), have significant amplitudes in both low and middle latitudes. A discussion on the validity of a simple application of response models for correcting positions is pertinent.

An experiment was devised in order to verify how adequate predicted corrections of tides are. This experiment was carried out over a 315 km long continental profile, along the S-N direction, in Southern Brazil. Three sites were stablished on this profile carrying out simultaneous multi-parametric observations, being one site located on seashore (in Imbituba), the second 50 km away from the sea (Blumenau) and the third one 80 km inland (in Curitiba). In each one of them it was installed a dual-frequency GPS receiver, a gravimeter for observation of gravimetric tides, and atmospheric pressure and temperature sensors. In addition, a tide gauge was installed at the coastal site aiming to control local ocean effects. As a first step of the study, we established an initial multiparametric analysis for the coastal station, trying to find the local admitance among some envolved parameters. For this purpose, we used hourly observed values from gravity tide, ocean tide and air pressure. Results of this analysis, including interactions and instrumental effects, are discussed in this paper. In a next step, we intend to extend the analysis to the other two stations of the network and to include GPS hourly positions to check the crustal deformation and compare them with both predicted and observed gravity tides, corrected from air pressure and ocean load.

Description of the experiment and its objectives

The recent level of precision achieved both in position by space techniques and gravity tides by modern gravimeters, has opened new horizons on the study of geodynamical effects of solid and ocean tides on inland horizontal and vertical positioning. De Freitas et al. (1995) found a strong difference between the predicted and observed tidal gravity response in a tidal gravity E-W profile envolving 15 stations around the latitude 22°S. The gravimetric factors corrected from the ocean loading using Schwidersky (1980) cotidal maps, show a decreasing of about 2% from the Atlantic coast to the Pacific coast. Recents ocean tidal models from TOPEX-POSEIDON (JPL, 1996) confirm most of these behaviors (Melchior & Francis, 1996). They indicate that the predicted earth's response to tides used to the differential positioning on medium baselines should be investigated accouting for observed tidal to model geodynamical effects. On the other hand, vertical positioning, since shore references must be also invetigated due to coastal effects in the ocean topography (Zikoski et. al. 1992), and also consequence of local ocean tide and solid tide.

The permanent GPS station Curitiba/UFPR -"PARA", which took part in the SIRGAS Project, has the longest time series of continuos GPS observations in South America and nowadays it contributes to the Brazilian Network for Continuos Monitoring of GPS (RBMC - see Fortes, L.P.S, in this proceedings). The Curitiba tidal station is located close by. This station is the fundamental tidal gravity station for South America, in operation since 1983 linked to the International Center for Earth Tides (ICET of IAG) network. These facilities in Curitiba, with well determined position and gravimetric factors, allow the stablishement of an experimental 315 km N-S profile, with three multi-parametric stations envolved: the Imbituba coastal station, where the vertical brazilian datum is placed; an intermediate station in Blumenau with the purpose to check the relative behavior; and, the 80 km inland Curitiba station (fig. 1 and tables 1-2).

Two main objectives were stablished, considering the experimental profile: a) to understand the coastal effect on the vertical datum and coastal positioning; b) to check the tidal implications over baselines with usual lenghts in Brazilian high precision relative GPS positioning, mainly for envolved coastal stations.

In the experiment simultaneous observations of position (dual frequency GPS with a 30s rate) and gravity tide (with a 1 min. sampling rate) were collected at the three stations between April/09/1997 and May/21/1997, along with aditional air pressure and temperature measurements (with a 1min. sampling rate) in order to correct the tidal gravity observations from the main enviromental effects. In the same period a tide gauge at the Imbituba station was kept in operation (with a 10 min. sampling rate). Considering the possibilities of the experiment, the following steps in the research have been defined:

1) determination of tidal gravity response in Blumenau and Imbituba stations;
2) determination of local enviromental effects on tidal observations;
3) precise determination of GPS position in Imbituba and Blumenau;
4) determination of an hourly time series of GPS position in the three stations;
5) determination of the local admitance between ocean tide and gravity tide in Imbituba;
6) determination of geocentric position of the tide gauge in Imbituba;
7) modelling of ocean loading using different ocean models for Imbituba and Blumenau;
8) determination of admitances among position and tides;
9) determination of ocean topography in the vertical brazilian datum;
10) determination of tidal effects on relative positioning along the profile.

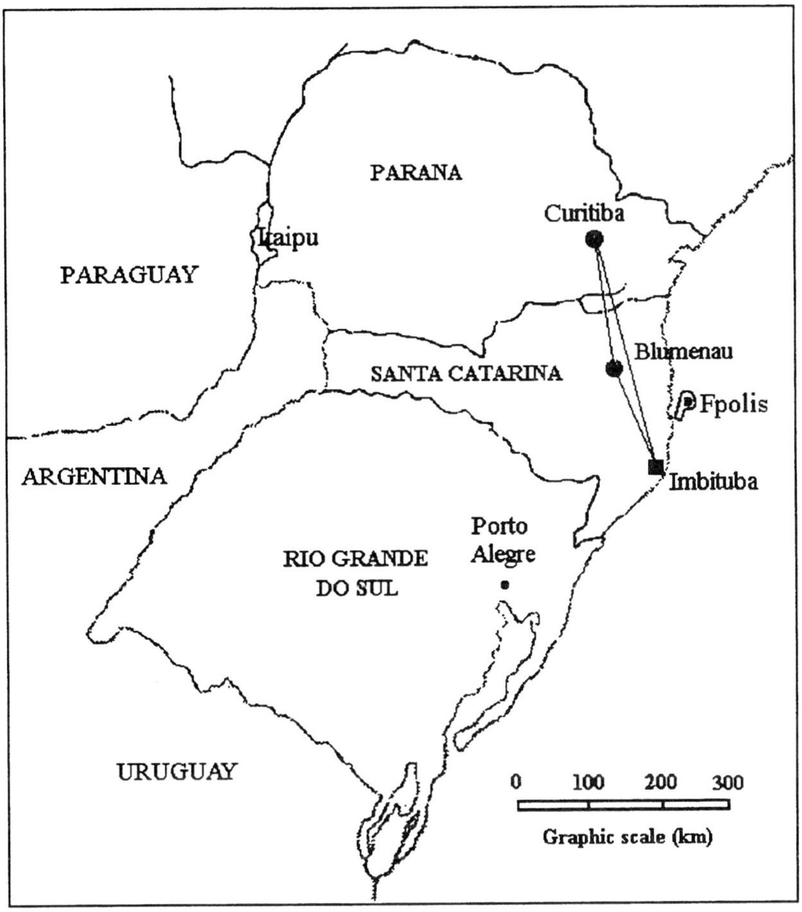

Fig. 1: Stations used in the experiment developed between April/09/1997 and May/21/1997 (● - Dual frequency GPS + Gravity tide + Air pressure + Temperature measurements; ■ - Dual frequency GPS + Gravity tide + Ocean tide + Air pressure + Temperature measurements).

Table 1: Station's position (GPS)

Station	φ	λ	**H (m)**	**d → Sea (km)**
Curitiba	25° 26′ 54.1291 S	49° 13′ 51.4368 W	925.7591	≈ 80
Blumenau	26° 53′ 30.0420 S	49° 05′ 00.8291 W	26.7192	≈ 50
Imbituba	28° 14′ 11.8106 S	48° 39′ 21.8820 W	11.7957	≈ 0.200

Table 2: Linear distances between stations (km)

Curitiba → Blumenau	160.592
Curitiba → Imbituba	314.170
Blumenau → Imbituba	154.895

Initial analysis and results in the Imbituba station

Even though the tidal gravity observations which started in Imbituba in April covering the whole period (fig. 2) of simultaneous multi-parametric observations, the time series are not enough long for a trustworthy determination of the local tidal parameters to allow an analytical treatement. For this purpose, its is necessary, at least, a six-month long continuous series for comparison with predicted tide and to correct it from the modeled ocean tide. Only after that, plus performing a final calibration of the gravimeter in the fundamental station, it will be possible to determine the true coastal effect and to estimate the ocean topography. However, it was already possible to determine provisory local tidal parameters (table 3 presents the M2 components, the main wave in Imbituba) allowing the generation of a local synthetic gravity tide (fig. 3) as an initial aproach in the analysis. The analysis of gravity tides are conduced considering the vectors for each tidal componente:

$$\mathbf{B} = \mathbf{A} - \mathbf{R} \qquad \text{and} \qquad \mathbf{X} = \mathbf{B} - \mathbf{L}$$

where **A** is the observed gravity tide, **R** is the theoretical gravity tide, **B** the tidal residual, **L** is the ocean indirect effect and **X** the final residual.

Considering the observed air pressure (fig. 4), it was possible to generate the gravity tide residuals and to correct the ocean tide corrected from air pressure (fig. 5). The regression between the tidal residuals and observed ocean tide, allowed the determination of provisory final gravity tide residuals (fig. 6). These residuals permit an evaluation of the effect from ocean loading. The the mean tidal gravity residuals has the value of -0.21 ±69.35 nm/s^2 and the mean final gravity tidal residuals has the value of 0.1 ±9.99 nm/s^2. An improved treatement must be donne considering the air pressure effects and temperature variations on the 1 min. gravity data.

The whole GPS observations are now being reduced to hourly coordinates to compare them with the other time series in the same period. The end of tidal gravity observations is expected for October/1997. This acquisition must be followed by a three-month long period for final calibration of gravimeters in Curitiba, allowing the computation of improved tidal parameters for Blumenau and Imbituba. The connection between the GPS station and the tide gauge is now being provided. Only with these final procedures, it will be possible to compute the coastal effects and the geodynamical effects on the baselines.

Table 3 - Tidal parameters for M2 wave at the Imbituba station (Ampl. in nm/s^2)

A	RMS	Amp.fc	RMS	Ph. °	RMS	B	B ph. °	L	L ph. °
689.2	3.0	1.1831	0.0052	1.07	0.25	18.5	44.1	21.7	71.4

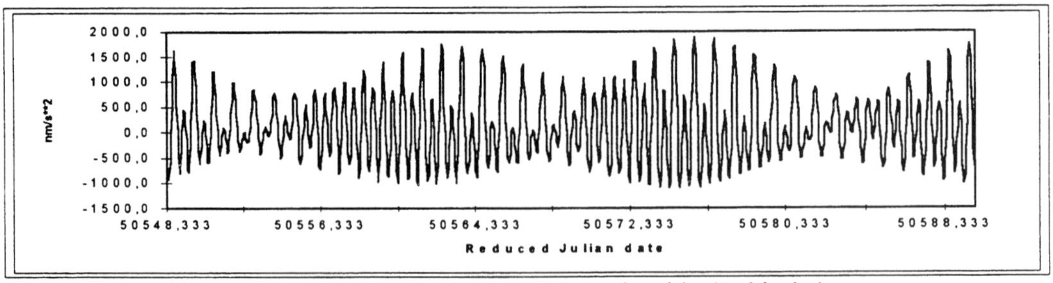

Fig 2 - Time series of observed gravity tide (Imbituba).

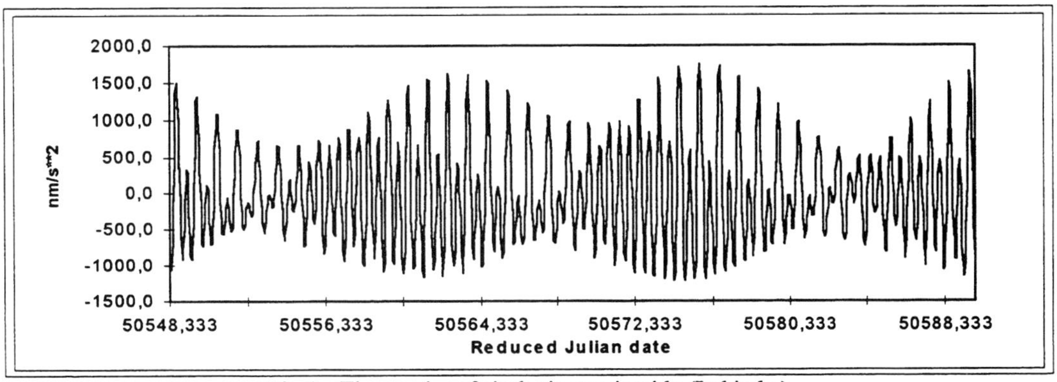

Fig 3 - Time series of sinthetic gravity tide (Imbituba).

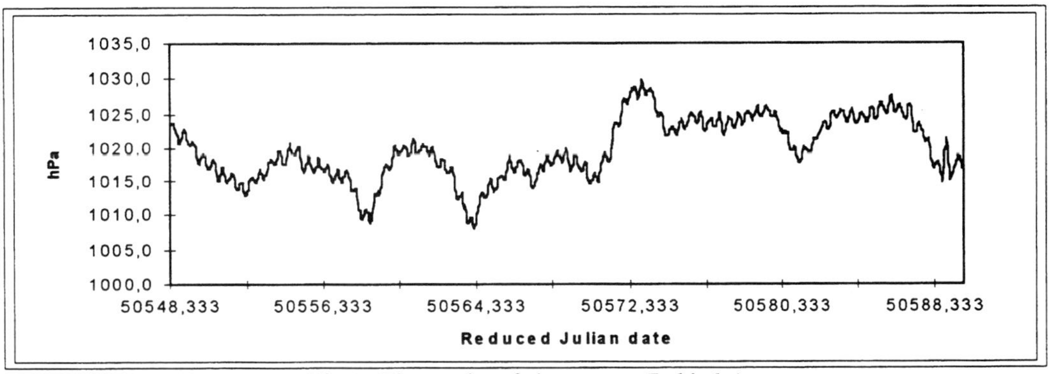

Fig 4 - Time series of air pressure (Imbituba).

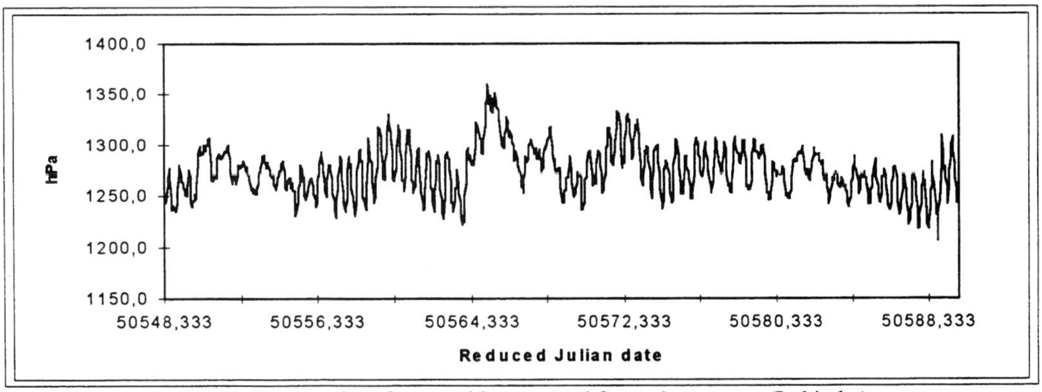

Fig 5 - Time series of ocean tide corrected from air pressure (Imbituba).

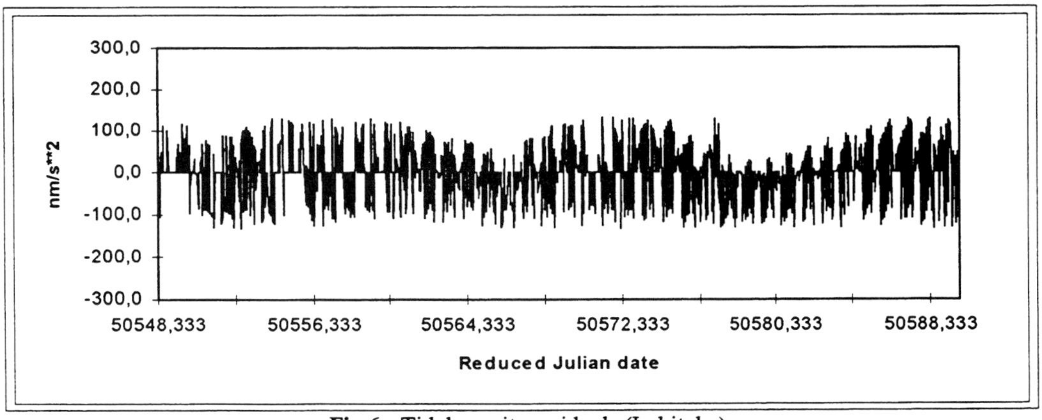

Fig 6 - Tidal gravity residuals (Imbituba).

Conclusions

The present project is not concluded yet. The present results are only preliminary as a consequence of the long series needed to determine the tidal parameters and the large quantity of multi-parametric data still to be crossed. However, its observational part is nearly over and the conditions of execution have been satisfactory in light of the usual accepted standards.

The partial computation represents an provisory aproach for checking the instrumental resolution considering the necessary accuracy of the collected data. The tidal residuals at the Imbituba station are well explained by ocean loading.

The analysis of the preliminary results indicates the consistency of the experiment to determine geodynamical effects on positioning. When the mean tidal gravity residuals is comparated with the mean value of corrected final tidal gravity residuals, it is possible to evaluate the improvement in the observations when introduced air pressure and ocean loading corrections.

Aknowledgements

The autors would like to thank the scientific and field support from CNPq (Proc. 524030/96-0), CAPES, Observatoire Royal de Belgique, ICET, Compania Docas de Imbituba and Fundação Universidade da Região de Blumenau, and to the collegues Prof. S.H.S. Schwab, Eng. E. Saatcamp, Eng. M. Saldanha and Geog. M.S. Kungenfluss by the kindly help for the developement of the project.

References

DE FREITAS, S.R.C; MANTOVANI, M.S.M.; MELCHIOR, P. & SHUKOWSKY, W., 1995. Some implications of the gravity tides in the South Americam Plate. In: Earth Tides, ed. H.T. Hsu, Science Press, New York, p. 387-393.

FARREL, W.E., 1972. Deformation of the Earth by surface loads. Rev. Geophys. Sp. Phys., 10:761-797.

JPL, 1996. A collection of Global Ocean Tide Models from TOPEX-POSEIDON. Pasadena. CD-ROM.

MELCHIOR, P., 1983. The tides of the planet Earth. 2nd ed., Pergamon, Oxford, 640pp..

MELCHIOR, P. & FRANCIS, O., 1996. Comparison of recent Ocean Tide Models using ground-based tidal gravity measurements. Marine Geodesy, 19:291-330.

SCHWIDERSKY, E.W., 1980. On charting global ocean tides. Rev. Geophys. Sp. Phys., 18:243-268.

ZIKOSKI, D.B.; RICHARDS, J.H. & YOUNG,G.M., 1992. Results of the general adjustment of the North American vertical Datum of 1988. Surveing and Land Inf. Sys., 52(3):133-149.

The determination of motion rates of blocks in the Qinghai-Tibet Plateau by GPS measurements

Jingnan Liu Caijun Xu Chenghua Song Dingbo Chao
Wuhan Technical University of Surveying and Mapping,
39 Luoyu Road, Wuhan, 430070, P.R. China.

Abstract:
Space geodesy provides an efficient technique to determine contemporary crustal deformations. At present, GPS is one of the best methods to study the crustal movements in the Qinghai-Tibet plateau. Two GPS campaigns were carried out in this area in 1993 and 1995 by Wuhan Technical University of Surveying and Mapping. The GPS measurements show that the Qinghai-Tibet plateau moves at about 33.4mm/a to Siberia northeastward at present, which is in agreement with the kinematic model derived from geophysical evidence. It can also be deduced that the motion rate and azimuth of Bayan Har terrane, Qangtang terrane, Lhasa-Gangdise terrane and Himalayan terrane are about 17.3mm/a, N35°E; 25.7mm/a, N33°E; 38.9mm/a, N29°E; 46.0mm/a,N27°E relative to Siberian block (Siberia) respectively.
Key words : Kinematic model, The crustal motion in the Qinghai-Tibet plateau, GPS measurements.

1. Introduction

The Qinghai-Tibet Plateau is located at collision zone between the Eurasian and Indian plates. The collision and press of two plates result in uplifting of the Qinghai-Tibet plateau, they also cause horizontal strong tectonic movement . A great amount of studies about the tectonic movement in the Qinghai-Tibet plateau are presented (Molnar, Armijo ,et al.). The crustal movement rates and directions are mainly based on the geological examination, palaeomagnetic survey, satellite photography, focal-mechanism solution etc., these results are chronological average, and they haven't verified by modern geodetic data before now. Space geodesy provides an efficient technique to determine contemporary crustal deformations. At present, GPS is one of the best methods to study the crustal movements in the Qinghai-Tibet plateau . Two GPS campaigns were carried out in this area in 1993 and 1995 by WTUSM. The motion rates of blocks in the area are discussed with GPS measurements in the paper.

2.GPS data acquisition and processing

The sketch of GPS monitoring sites in the Qinghai-Tibet plateau shows in Fig.1. The GPS monitoring sites are mainly deployed at Tibet block ,Himalayas block and Qaidam block. The GPS -chain network across the Honxil-Bayan Har ,Jinsha Jiang, Nujiang-Lancang Jiang and Yarlung Zangbo Jiang rift systems. The shortest baseline and longest baseline in the network are 62km and 1185km respectively, the mean length of baselines is 364 km. The first GPS campaign took place from July 18 to August 24, 1993, and 3 Rogue SNR 8000 GPS receivers were used in 12 sites(refer to fig.1), the second GPS campaign took place from June 5 to July 5 , 1995, and 4 Rogue SNR 8000 GPS receivers were used in 15 sites. There were two observation sessions at every sites in each campaign, one was in daytime ,the other was at night, every observation session kept 9 hours.The GPS data are analyzed with the GAMIT software, the GPS baseline solutions are obtained in the ITRF91 frame with help of the NGS precise ephemeris. Here, a fiducial strategy was implemented, fixing the coordinates of Lhasa station to its a priori value(ITRF91).The GPSADJ software improved for scientific studying are used to process the baselines of GPS network(ITRF94).. The coordinates of Golumd site are fixed as a fiducial strategy because the Golmud site is motionless compare to other monitoring sites repeated. The mean relative accuries of baselines adjusted in two campaigns are 2.8×10^{-8} and 1.6×10^{-8} respectively.

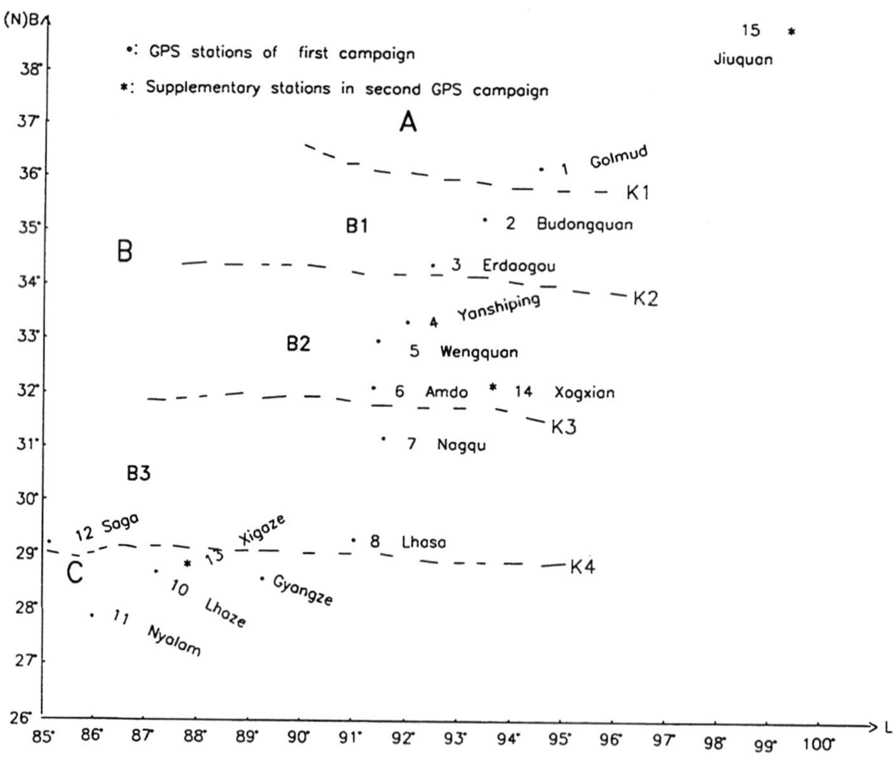

Fig.1.The sketch of GPS monitoring stations in the Qinghai-Tibet plateau

K1:Honxil-Bayan Har rift system K2:Jinsha Jiang fault B:Tibet block C:Himalayan block B1:Bayan Har terrane
K3:Nujiang-Lancang Jiang fault K4:Yarlung Zangbo Jiang fault B2:Qangtang terrane B3:Lhasa-Gongdise terrane

3.The horizontal crustal movement

It is hard to give verification of uplifting of Qinghai-Tibet plateau by GPS vertical components because data noise,tropospheric errors and remaining satellite orbit errors. This paper is mainly to analyze the horizontal displacements provided by GPS measurements.

Let B_i^{95} and B_i^{93} represent the latitude of monitoring site I in 1995 and in 1993,L_i^{95} *and* L_i^{93} are the longitude of monitoring site I in 1995 and in 1993. Then ,we have:

$$\begin{cases} \delta B_i = B_i^{95} - B_i^{93} \\ \delta L_i = L_i^{95} - L_i^{93} \end{cases} \quad (1)$$

The displacements of the ith site in N-S direction and in E-W direction could be expressed as follows:

$$\begin{cases} \delta N_i = M_i^{\circ} \delta B_i \\ \delta E_i = N_i^{\circ} \delta L_i \cos B_i \end{cases} \quad (2)$$

here M_i° *and* N_i° are radius of curvature of meridian and radius of curvature in the prime vertical .

Thus ,the displacement and direction of horizontal crustal movement are obtained:

$$\begin{cases} \Delta_i = (\delta N_i^2 + \delta E_i^2)^{1/2} \\ A_i = arctg \, \delta E_i / \delta N_i \end{cases} \quad (3)$$

The displacements and their accuries, azimuths of movement of GPS monitoring sites relative to Golmud site are figured by equation (1)~(3). They are listed in Tab.1.

Table 1.Horizontal displacements, movement rates and their azimuths of GPS monitoring sites from 1993 to 1995

NO.	δN (cm)	δE (cm)	Δ, M_Δ (cm,cm)	A (°)	motion rate (cm/a)	U- Test	
						$u = \Delta / M_\Delta$	remark
2	0.499	0.683	0.846,0.349	54	0.43±0.18	2.48	significant
3	1.066	1.330	1.704,0.699	51	0.87±0.35	2.49	significant
4	1.759	1.765	2.492,0.699	45	1.27±0.35	3.64	significant
5	2.132	1.836	2.813,1.201	41	1.44±0.60	2.40	significant
6	2.708	1.863	3.287,1.186	35	1.68±0.59	2.83	significant
7	3.203	1.603	3.582,1.158	27	1.83±0.58	3.16	significant
8	4.446	1.689	4.756,1.164	21	2.43±0.58	4.17	significant
9	5.271	2.229	5.723,1.163	23	2.92±0.58	5.02	significant
10	5.776	3.379	6.692,1.272	30	3.42±0.64	5.37	significant
11	6.951	3.959	8.000,1.302	30	4.09±0.65	6.28	significant
12	6.389	4.948	8.081,1.336	38	4.13±0.67	6.18	significant

The displacements are driven from two independent GPS campaigns and they are approximately coincide with normal distribution. The displacements are significant and reliable by U-test. Since Golumd station is located in Qaidam block, and Qaidam block moves about 11.3~15.4 mm/a with north by east 25° relative to Siberian block(Ding and Lu,1986),the movement rate of Golmud site relative to Siberian block(Siberia) can be figured as follows:

$$\begin{cases} V_N = 11.3 \times \cos 25^\circ = 10.2(mm/a) \\ V_E = 11.3 \times \sin 25^\circ = 4.8(mm/a) \end{cases}$$

Considering this fact, the displacements and movement rates of GPS monitoring sites relative to Siberian block could be obtained and they are showed in Fig.2.The GPS measurements indicate that the Qinghai-Tibet present-day rate of movement is about 33.4mm/a north by east 30° in the Siberian reference frame The motion rates of GPS sites relative to Golmud site or Siberian block(Siberia) show increase tendence from north to south. It can be deduced that the motion rate and azimuth of Bayan Har terrane, Qangtang terrane, Lhasa -Gangdise terrane and Himalayan terrane are about 17.3mm/a, north by east 33°,25.7mm/a,north by east 33°,38.9mm/a ,north by east 29° and 46.0m/a ,north by east 27° relative to Siberian block(Siberia) respectively.

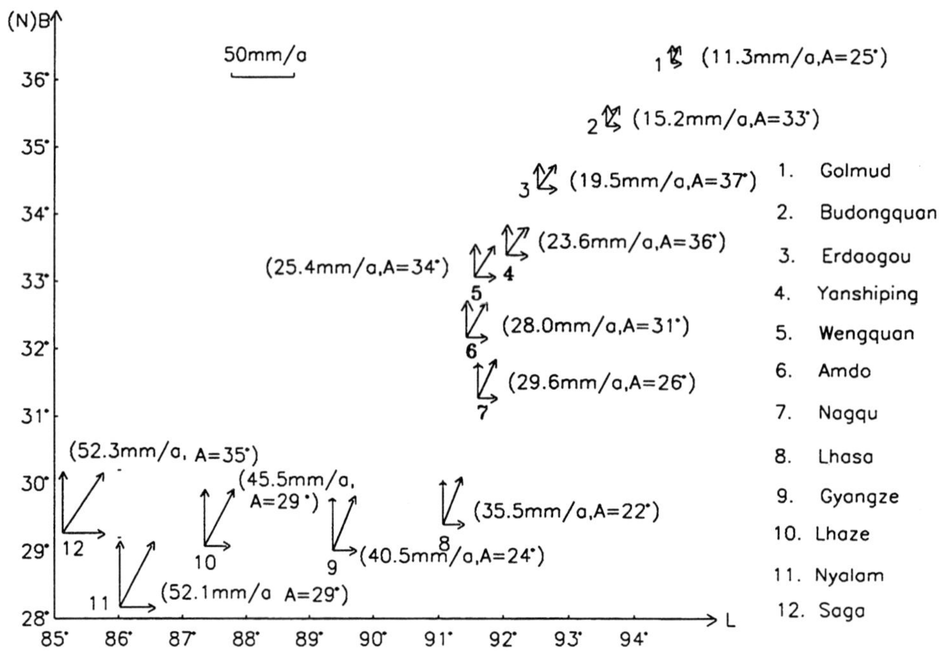

Fig.2. The horizontal components of motion vector solution for GPS station relative to Eurasia (Siberia)

4.Kinematic model of GPS measurements in comparison with geological model.
A kinematic model of blocks in western China, relative to Siberia, derived from slip-rates of Neogenic active faults by Ding and Lu (1986) The movement rates of Tibet block and Himalayan block relative to Siberian block , which are in comparison with the rates from GPS measurements, are listed in the Tab.2.
The velocity of field of present-day deformation in central Asia is modeling using a set of four rotating blocks (Siberia,tarim,tibet,india) on a spherical earth by J.P.Avouace and Tapponnier(1993).The euler vectors of tibet block is
ϕ_N=44.2°±5°,λ_E=64.4°±12°,ω=0.89°/Ma±0.35°/Ma in Siberian reference frame. However, the motions as given by this model are averaged over at least a number of million years, they are chronological average . They are also in comparison with the velocities from GPS measurements and shown in Tab.2 and Fig.3.

Table 2. Horizontal motion in the Qinghai-Tibet Plateau from GPS measurements in comparison with geological kinematic model

	Vn (mm/a)	Vn (mm/a)	Vn (mm/a)	Ve (mm/a)	Ve (mm/a)	Ve (mm/a)	rate, azimuth	rate, azimuth	rate, azimuth
	GPS	Ding, Lu	J.P. Avouac	GPS	Ding, Lu	J.P. Avouac	GPS	Ding, Lu	J.P. Avouac
TB	24.4	28.1	32.1	14.8	10.0~ 20.0	24.3	28.5, 31	31.9, 28.0	40.3, 37.0
HM	40.8	42.3	28.0	21.1	/	29.1	46.0, 27	42.3, 10.0	40.4, 46.0
QT	28.9	35.2	30.1	16.5	/	26.7	33.4, 30	/	40.4, 42.0

*.TB: Tibet Block HM: Himalayas Block QT: Qinghai-Tibet Plateau

The results in the tab.2 and Fig.3 show that the present-day deformation rate provided by GPS measurements is consistent with the average rate of chronology by geological model

6. Conclusion remarking.
The GPS measurements indicate that Tibet block, Himalayan block and Qinghai-Tibet Plateau present-day rates of movement are respectively about 28.5mm/a,N31° E,46.0mm/a N27° E and 33.4mm/a,N30° E in the Siberian reference frame. they are in agreement with the geological kinematic model. These result indicate that the collision between Eurasian plate and Indian plate cause east-west crustal extension in the center of Tibet .However, the present-day kinematic model from GPS measurements is regional model because there are only 15 sites distributed mainly along the Qinghai-Tibet highway in the GPS-chain network .It is necessary to get a more reliable kinematic model , further GPS campaign in this area should be carried out.

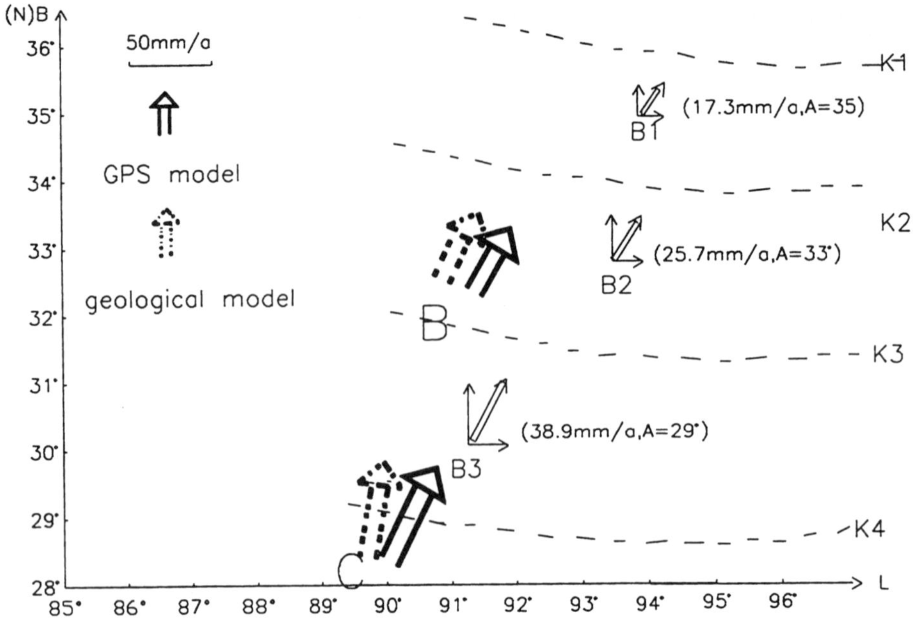

Fig3.Motion rates of blocks in the Qinghai–Tibet plateau from GPS and geological model

K1:Honxil–Bayan Har rift system K2:Jinsha Jiang fault B:Tibet block B1:Bayan Har terrane C:Himalayan block

K3:Nujiang–Lancang Jiang fault K4:Yarlung Zangbo Jiang fault B2:Qangtang terrane B3:Lhasa–Gangdise terrane

Acknowledgments: The authors would like to express their hearty gratitude to Mr. Shi Chuang, Mr. Jiang Weiping and Mr. Liu Zhizhao for their efforts on GPS baselines solutions and network solutions of GPS sites , and to the National Science Foundation of China .

References

1.R. Armijo et al., Late Cenozonic Right- Lateral Strike-slip Faulting in Southern Tibet,JGR.vol.94,No.B3, 2787-2838,1989

2.J.P.Avouace and P.Tapponnier, Kinematic Model of Active Deformation in Central Asia, Geophysical Research Letters,vol.20, No.10,895-898, 1993.

3.Editorial Board for Lithosphere Dynamics Atlas of China, Lithosheric Dynamos Atlas of China, China Cartographic Publish House, Beijing ,China,142-153,1990.

4.Ding G.Y. ,Lu Y.Z. A preliminary analyze of present-day motion in the intraplate of China ,Chinese Science Bulletin, 18, 1412-1415,1986

5.P. Molnar and Deng Q ., faulting associated with large earthquakes and the average rate of deformation in central and eastern Asia, J.G.R. Vol.89,6203-6227,1984

6.P. Molnar et al., Fault plane solutions of earthquakes and active tectonics of the Tibetan Plateau and its margins. Geophys. J.Int .99,123-153,1989.

Detection of Crustal Deformations by SAR Interferometry in Combination with Dense GPS Array

S. Fujiwara, M. Tobita, S. Ozawa, H. Yarai, H. Nakagawa, K. Nitta, M. Murakami
Geographical Survey Institute
Kitasato-1, Tsukuba-shi, Ibaraki-ken, 305 Japan
M. Shimada
National Space Development Agency of Japan
1-9-9. Roppongi, Minato-ku, Tokyo, 106 Japan
P. A. Rosen
Jet Propulsion Laboratory
Pasadena, CA91109, USA

Abstract

We aim to improve the accuracy of space-borne synthetic aperture radar (SAR) interferometry making advantages of Japanese nationwide GPS array. In order to detect crustal deformations in high accuracy without delay, the Geographical Survey Institute is constructing dense GPS array that consists of 900 permanent GPS stations all over the country. The average interval of the stations is about 25km. Though this is one of the densest GPS arrays in the world, the surface change observed by GPS is the data "at a point". So it can miss detecting the surface changes associated with geological activities at the most important spot, such as deformations near an epicenter. To supplement this disadvantage, we employ SAR interferometry (InSAR). Though the InSAR measurements are less accurate than the GPS measurements, it provides better spatial resolution of about 100m, measuring dislocations at millions of points over areas that can not be covered by GPS. Since approximately nine (9) GPS stations are included in a SAR image of about 76km by 76km square obtained by JERS-1, we can use them to constrain the amount of surface change from InSAR result. We analyzed by this technique a fault model of an earthquake (M=6.3) that occurred in Kagoshima-ken, Japan on March 26, 1997. Though the maximum amount of detected deformations is as small as 5cm, we have obtained a unique solution by using no other data than GPS and InSAR.

MODELLING VERTICAL CRUSTAL DEFORMATION IN TIME AND POSITION DOMAINS, INFERRED FROM GPS LEVELLING

Quanwei Liu and Teuvo Parm
Institute of Geodesy and Cartography, Department of Surveying, Helsinki University of Technology, P. O. Box 1200, FIN-02015 HUT, FINLAND

ABSTRACT

The principle and application of the vertical crustal deformation model from the integration of Global Positioning System (GPS) and levelling in the time and position domains are discussed in this paper. Many attempts are made to address the height datum inhomogeneity problem between the GPS ellipsoid height and the levelling orthometric height. Corresponding observation and computation strategies are designed. A special statistical technique is imposed to test the validity of the model developed. The numerical example shows the effectiveness and feasibility of the approach.

1. INTRODUCTION

Geodetic spirit levelling measurements, along with water level and tide gauge records have traditionally been used to monitor vertical crustal deformation. Geodetic levelling is of low efficiency in terms of cost and observation speed, especially when the stable areas are far from the deformation area. Now there is GPS technology, which provides fast, convenient, low cost and comparable accuracy for height information over large areas. However, height measurements determined by GPS are presently not quite as accurate as those obtained by levelling over short distances. To take full advantage of the two techniques of monitoring vertical crustal deformation, a model based on the integration of the GPS technique with levelling will be discussed. Its application will be demonstrated by a numerical example.

2. VERTICAL DEFORMATION MODELS IN TIME AND POSITION DOMAINS

Crustal deformation is a function of not only time, but also position. The models which describe crustal deformation should be relevant to both the time and the position factors. With such models, the results of crustal deformation can be interpolated and extrapolated. The models can provide a panoramic view of the whole deformation area when evaluating deformation. It also becomes convenient to predict the deformation of any point within the deformation area during and even beyond the surveying periods, as long as the horizontal coordinates of the point are known.

The general model for vertical crustal deformation in the time and position domains can be expressed as

$$D(x,y;t-t^0) = B(x,y,t-t^0)\alpha, \tag{1}$$

where D is the vertical deformation of a point within the deformation area of interest, (x, y) are the horizontal coordinates of the point in a local coordinate system, t is the present time, t^o is the initial time selected; B is a matrix of base functional values to be selected and α is the vector of unknown coefficients to be estimated. Different suitable base functions may be used to approximate the deformation, so as to produce different models. For details about different models the reader is referred to e.g. Holdahl (1978); Vanicek and Sjöberg (1987), for the recently developed combining model, referred to Liu and Parm (1997).

In order to estimate the coefficients in Eq. (1) and, moreover, the deformations by applying the least squares approximation technique, the model has to be connected with observation equations. Here we consider a pair of points, *k* and *l*, as an example to show the observation equations for the orthometric height differences:

at the initial epoch: $$\Delta H_{kl}(t_0) + V_{kl}(t_0) = H_l(t_0) - H_k(t_0), \qquad (2)$$

at the *m* th epoch:

$$\Delta H_{kl}(t_m) + V_{kl}(t_m) = H_l(t_0) - H_k(t_0) + [B(x_l, y_l; t_m - t_0) - B(x_k, y_k; t_m - t_0)]\alpha, \qquad (3)$$

where $\Delta H_{kl}(t_0)$ and $\Delta H_{kl}(t_m)$ are levelled height difference observations between points *k* and *l* at the initial epoch t_0 and the following *m* th epoch t_m, respectively; $V_{kl}(t_0)$ and $V_{kl}(t_m)$ denote their corresponding residuals; $H_l(t_0)$ and $H_k(t_0)$ represent orthometric heights of points *l* and *k* at the initial epoch, respectively.

3. THE INTEGRATION MODEL OF GPS AND LEVELLING

In the above model, observations are orthometric height differences determined by levelling measurements. From a theoretical point of view, the model is also valid if only the height differences determined by GPS measurements are used. However, from a practical point of view, these two different kinds of measurements have their own advantage and disadvantage and are appropriate to different particular needs. GPS is convenient and economical to use to connect deformation areas with stable points. This is very important in assessing deformation. Levelling is appropriate for areas where detailed measurements are needed. Inspired by the work done by Chrzanowski *et al.* (1989) who successively used the GPS technique to replace levelling for monitoring ground subsidence in oil fields along the east coast of Lake Maracaibo in Venezuela, in the following, we will combine the two techniques to take full advantage of the two kinds of measurements to detect the vertical crustal deformation in a practical example.

Levelled heights (orthometric heights *H*) refer to the geoid, which is defined as the gravity equipotential surface which best approximates mean sea level over the entire earth. GPS heights (ellipsoidal heights *h*) refer to a geocentric ellipsoid which is a fictitious surface introduced as a mathematical approximation of the geoid. The non-parallelism of the geoid surface leads to the spatial separation and the relative slope between the geoid and the ellipsoid. The former known as the geoid undulation (*N*) can be in the order of tens of metres, and the latter known as the deflection of the vertical can range up to one minute of arc in extreme cases (Collier and Croft, 1997).

From Heiskanen and Moritz (1964), the two types of heights can be connected together, to an adequate level of approximation (< 1 mm), by the geoid undulation *N*:

$$H = h - N. \qquad (4)$$

Since the observations in vertical deformation detection involve only height differences between two stations, Eq. (4) can then be reduced to

$$H_i - H_j = (h_i - h_j) - (N_i - N_j),$$

that is $$\Delta H_{ij} = \Delta h_{ij} - \Delta N_{ij}, \qquad (5)$$

where ΔN_{ij} is the relative geoid undulation between the two points.

If we neglect the temporal variations of the geoid, ΔN is needed only in one epoch to build the observation equation (5). We can design an overlapping epoch with a large number of sites where both the GPS ellipsoid heights and levelling orthometric heights are observed. Therefore, among these sites, both the levelled orthometric height differences ΔH_{ij} and the GPS ellipsoid height differences Δh_{ij} are available. With these two kinds of height differences we can evaluate the geoid undulation differences ΔN_{ij} among these sites by using Eq. (5). Then, we can directly construct the observation equations for the GPS ellipsoid height differences. Here, we take a pair of points, *k* and *l*, as an example to show the observation equation for the ellipsoid height differences:

at the initial epoch it reads $$\Delta h_{kl}(t_0) - \Delta N_{kl} + V_{kl}(t_0) = H_l(t_0) - H_k(t_0), \qquad (6)$$

at the following *m* th epoch, if we neglect the temporal variation of the geoid, and if the levelling is surveyed at the same pair of GPS sites, it is

$$\Delta h_{kl}(t_m) - \Delta N_{kl} + V_{kl}(t_m) = H_l(t_0) - H_k(t_0) + [b(x_l, y_l; t_m - t_0) - b(x_k, y_k; t_m - t_0)]\alpha, \quad (7)$$

where $\Delta h_{kl}(t_0)$ and $\Delta h_{kl}(t_m)$ are observed GPS ellipsoid height differences between points *k* and *l* at the initial epoch t_0 and the following epoch t_m, respectively; $V_{kl}(t_0)$ and $V_{kl}(t_m)$ denote their corresponding residuals; $H_l(t_0)$ and $H_k(t_0)$ represent orthometric heights of points *l* and *k* at the initial epoch, respectively; other notations have the same definition as in the previous equations.

For those sites where actual levelling or GPS data are not available, we have to predict the geoid undulation differences. The geoid undulation differences are continuous in both the time and position domains. We may approximate the geoid undulation differences (ΔN), at a point *(x, y)*, by selecting a best-fitting surface in the position domain, for instance, by the following polynomial, if the temporal variations of the geoid are again neglected as in Eq. (7) (Vanicek and Krawisky, 1986):

$$\Delta N(\Delta x, \Delta y) = \sum_{i=0}^{n_y} \sum_{j=0}^{n_x} x^j y^i \kappa_{ji}, \quad (8)$$

where $\Delta N = N - N_0$, with N_0 representing the geoid undulation at the origin of the local coordinate system; the orders n_x, n_y of the polynomial depend on the smoothness of the geoid and the accuracy required in the area in question.

The coefficients in Eq. (8) can be determined through application of the least squares approximation techniques. The input data for the adjustment are the known geoid undulation differences and their corresponding horizontal coordinates obtained at the overlapping epoch. After producing the geoid undulation differences at every GPS independent baseline, we can build the observation equations for all GPS height differences as in Eqs. (6) and (7).

With all observation equations ready, we can integrate models (2), (3), (6) and (7) to evaluate the coefficient matrices of the models with their variance-covariance matrices by applying the least squares approximation technique.

To test the significance of the coefficients α contained in Eqs. (3) and (7), a rigorous statistical technique has to be applied to obtain a reliable result. To do this, we can form a test statistic:

$$t = \frac{\alpha_i}{\sigma_{\alpha_i}} \sim t(f), \quad (9)$$

where σ_{α_i} is the standard deviation of α_i obtained from the resultant coefficient matrices and variance-covariance matrices and *f* is the number of redundant observations. α_i is accepted or rejected depending on whether the value of *t* is less or greater than *t(f)* at a chosen significance level.

4. NUMERICAL EXPERIMENT

The data used in this practical example has been observed since 1988 by spirit levelling. In 1994 a GPS network of 14 sites was established and was re-observed in 1995 with the same geometry as in 1994. Data of the levelling from 1988 to 1995 and of GPS from 1994 to 1995 were used in this study. The 1994 GPS campaign served as the overlapping campaign in order to derive directly the geoid undulations. Therefore, the GPS sites were all visited by geodetic levelling in this year. Part of the levelling was replaced by GPS in 1995. GPS data in 1994 and 1995 were separately processed using the Bernese Version 3.4 post-processing program (Rothacher *et al.*, 1993). By comparing the height differences obtained by GPS and levelling in 1994, we obtained the geoid undulations. The levelled height differences from 1988 to 1994 as well as that of 1995 (those which were not replaced by GPS) were called data 1, while the levelled height differences from 1988 to 1994 and the integration of levelled height differences with the GPS height differences in 1995 were called data 2. Then the observation equations were constructed according to Eqs. (2) and (3) for data 1, and according to (6) and (7) for data 2. The next step was to separately process the two types of constructed observation equations corresponding to data 1 and data 2 by using the model described in section 2 to evaluate the vertical deformations through the applications of the least squares principle to obtain result 1 and result 2, corresponding to data 1 and data 2, respectively. Fig. 1 represents the vertical deformation rate (1994-1995) achieved by levelling data only; while Fig. 2 displays the vertical deformation rate (1994-1995) obtained by the integration of GPS and levelling. Fig. 3 shows the differences of vertical deformation between levelling and the integration of GPS and levelling. The statistical information of these differences is presented in Table 1.

One point we had to bear in mind when we processed data 2 is that the observations in Eqs. (6) and (7) in 1995 are correlated with the observations in Eqs. (2) and (3) for the levelled height differences in 1994, because the geoid undulation differences contained in Eqs. (6) and (7) were derived by using the 1994 levelled height differences. However, the effect of this correlation on the determination of deformation is considerably small (<0.5mm) according to our numerical test, therefore, it was neglected in this study to reduce the computation burden.

Fig. 1 and Fig. 2 supply us an opportunity to compare the deformations evaluated by the two methods in detail. Fig. 3 shows the differences of the deformations by the two methods. Table 1 indicates that the maximum differences between the deformation results obtained by levelling and by the integration of GPS and levelling can reach up to -5 mm to 4 mm in extreme cases. As an average value, though, they generally remain below -2 mm to 1 mm. The differences between the deformation results by levelling and by the integration of GPS and levelling approaches are about 2.0 mm and 1.6 mm in terms of the root-mean-squares (rms) and standard deviation (σ), respectively. The comparison results in Table 1 and the inspection of Fig. 1, Fig. 2, and especially Fig. 3, obviously shows that the differences between the deformation results evaluated by levelling and that by the integration of GPS and levelling is considerably small.

method	max	min	mean	σ	rms
levelling (data1)	10 mm	-70 mm	-44 mm	2.9 mm	4.2 mm
GPS integration (data2)	7 mm	-68 mm	-39 mm	4.2 mm	5.1 mm
levelling-GPS integration	4 mm	-5 mm	-2 mm	1.6 mm	2.0 mm

Table 1: The statistical results of deformation rate (1994-1995) by levelling and GPS levelling integration

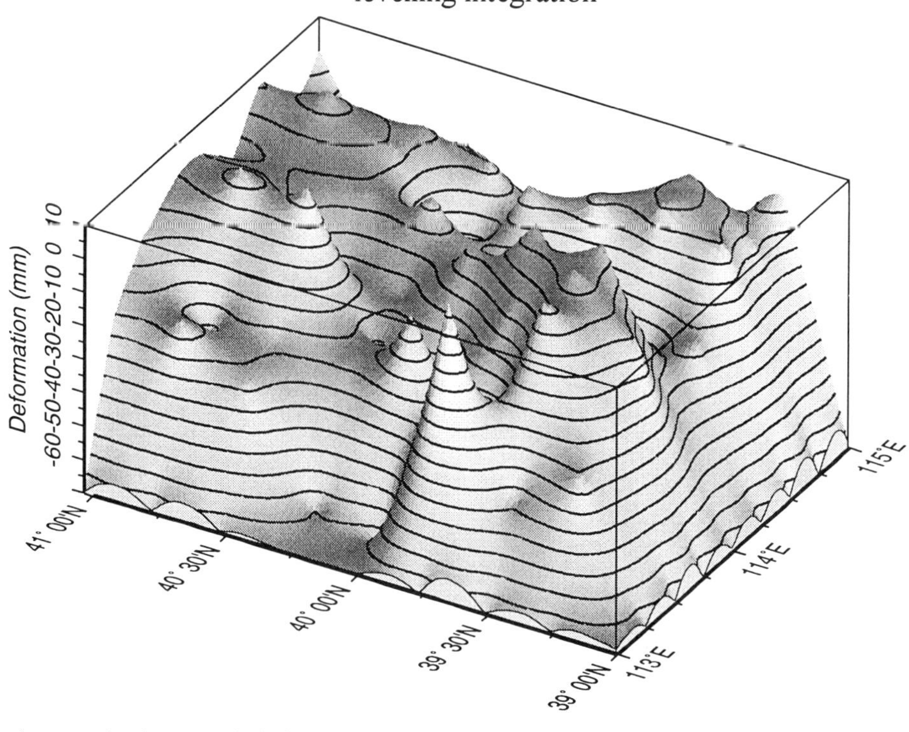

Fig. 1. The vertical crustal deformation rate (1994-1995), represented by both the contour (contour interval: 5 mm) on the surface and the surface itself, determined by the levelling data (1988-1995).

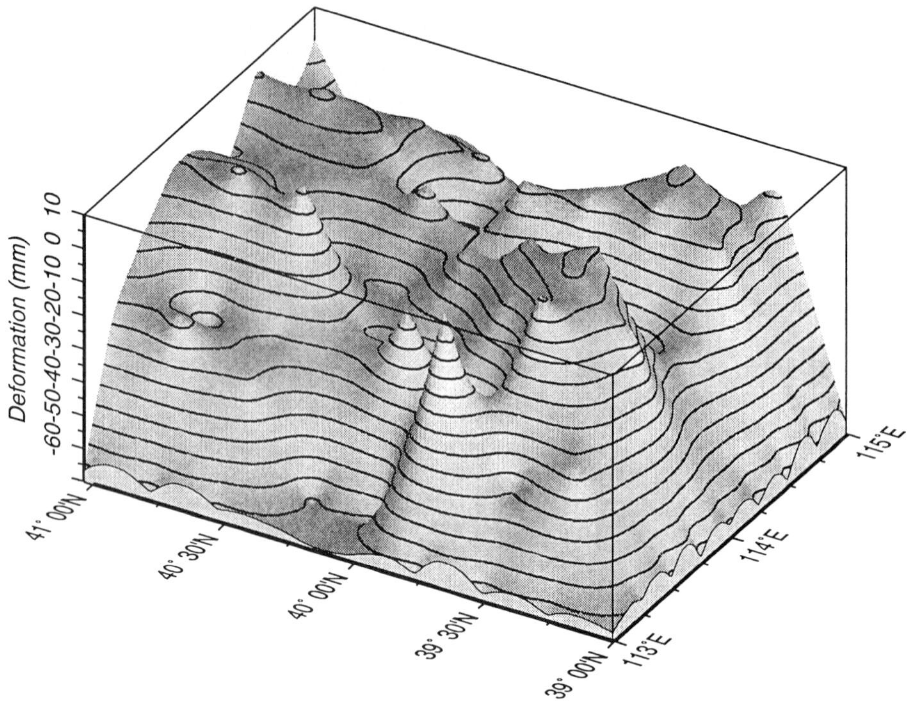

Fig. 2. The vertical crustal deformation rate (1994-1995), represented by both the contour (contour interval: 5 mm) on the surface and the surface itself, determined by the integration of GPS (1994-1995) and levelling data (1988-1995).

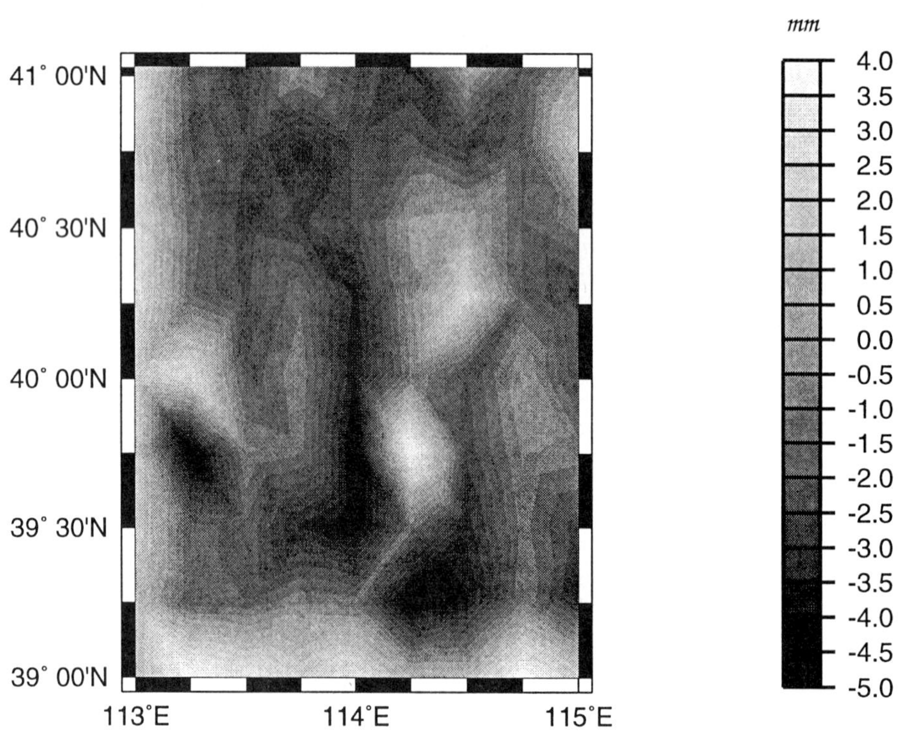

Fig. 3. Differences of the vertical deformation, represented by the degree of greyness in the region, between the methods determined by levelling and the integration of GPS and levelling.

On the other hand, according to our analysis in economical aspect, the yearly average cost for the levelling from 1988 to 1994 was approximately six thousands US Dollars, as a comparison, the total cost for the levelling and GPS measurements in 1995 was only approximately four thousands US Dollars. As a result, the integration method had a 33% cost saving in one campaign due to the fact that less levelling is needed to be done.

5. CONCLUSION

We have researched the principle and applications of the integration model of GPS and levelling to detect vertical crustal deformation in this paper. To obtain a homogeneous solution, procedures have been put in place to address the height datum problem between the GPS ellipsoid height and the levelled orthometric height. Corresponding observation and computation strategies have been discussed. From the theoretical research and the practical application of the technique on a real world project, we can conclude that the use of the integrated model improves significantly the observation efficiency, reduces the high levelling cost, with little sacrifice of accuracy.

REFERENCES

Chrzanowski, A., Chen, Y. Q. and Leal, J. (1989), Modelling of Ground Subsidence from a Combination of GPS and Levelling Surveys. Global and Regional Geodynamics. International Association of Geodesy Symposium No. 101, pp. 182-191, Edinburgh, Scotland.

Collier P. A. and M. J. Croft (1997), Heights from GPS in an Engineering Environment (Part I), Survey Review, Vol. 34, No. 263.

Denker, H. and Wenzel, H.-G. (1987), Local geoid determination and comparison with GPS results. Bulletin Geodesique, 61.

Gubler (1984), The Determination of Recent Crustal Movements from Precise Levelling Data, A Review, Workshop on Precise Levelling, Eds. H. Pelzer and W. Niemeier, Ferd. Dummlers Verlag, Bonn.

Hardy, R. L.(1978), The Application of Multiquadratic Equations and Point Mass Anomaly Models to Crustal Movements Studies. NOAA Technical Report Nos. 76, NGS 11, U.S.A.

Heiskanen, W. A. and Moritz, H. (1967), Physical Geodesy. Freeman, San Francisco.

Holdahl, S. H. (1978), Models for Extracting Vertical Crustal Movements from Levelling Data. Proc. of the 9th GEOP Conference, An International Symposium on the Applications of Geodesy to Geodynamics, Dept. of Geodetic Science Report No. 280, The Ohio State University, Ohio, U.S.A.

Holdahl, S. H. and R. L. Hardy (1979), Solvability and Multiquadratic Analysis as applied to Investigations of Vertical Crustal Movements. Tectonophysics 52.

Liu Q. and T. Parm (1996a), Optimization of the Combining Models for Vertical Crustal Deformation. Proc. of the 8th FIG International Symposium on Deformation Measurements. Hong Kong, June 25-28.

Liu Q. and T. Parm (1996b), On the Vertical Deformation models, Proc. of the 8th FIG International Symposium on Deformation Measurements. Hong Kong, June 25-28.

Liu Q. and T. Parm (1997) A New Approach on the Time Dependent Models of Vertical Crustal Deformation. Bollettino di Geodesia e Scienze Affini, Vol. 56, No. 3.

Moritz H. (1989), Advanced Physical Geodesy, Second edn. Wichmann.

Rapp, R. H. and Pavlis, N. K. (1990), The development and analysis of geopotential coefficient models to spherical harmonic degree 360. Journal of Geophysical Research, 95 (B13).

Rothacher, M., G. Beutler, W. Gurtner, E. Brockmann and L. Mervart. (1993), Bernese GPS Software Version 3.4.

Schwarz, K. P., and R. Forsberg (1987), Orthometric Heights without Levelling. Journal of Surveying and Mapping, Vol. 113, No. 1.

Vanicek, P. and E. Krakiwsky (1986), Geodesy: The concepts, Second edn. Amsterdam, Netherlands: North-Holland.

Vanicek, P. and L. E. Sjöberg (1987), A Note on Vertical Crustal Movement Determination Techniques. Department of Geodesy, The Royal Institute of Technology, Technical Report No. 9, Stockholm, Sweden.

COMPARISON OF CRUSTAL DEFORMATIONS OBSERVED WITH GPS AND STRAINMETERS/TILTMETERS

Torao TANAKA*, Takehide NAKANO*, Yoshinobu HOSO*, Kazuro HIRAHARA**, Ashraf El-Kutb MOUSA*,Taiichi HAYASHI*, Akira SUEMINE* and Sei YABE*

* Disaster Prevention Research Institute, Kyoto University, Gokasho, Uji, Kyoto-fu 611, Japan

** Graduate School of Science, Nagoya University, Furou-cho, Chikusa-ku, Nagoya, 464-01, Japan

Abstract

Daily solutions from GPS observations can give uniform tectonic deformation and relative motions of stations more than 100km apart to each other. More precise wet delay correction is necessary in order to monitor local or the order of 10km deformations. On the other hand observation with strainmeters and tiltmeters show irregular changes which are probably associated with local stress accumulation or irregular concentration, but it is not easy to distinguish the irregular change caused by tectonic stress from that caused by meteorological or ground water disturbances. Combined observations with GPS and high sensitive strainmeters/tiltmeters are important to make clear the nature of observed irregular changes, and further to get information relating to the occurrence of earthquakes of M7 class and accordingly to approach the long-term earthquake prediction.

1. Introduction

Since introduction of three Minimac 2816TM receivers in Disaster Prevention Research Institute (DPRI), we extended our monitoring network of crustal meovements in the south western Japan, and now we have 9 GPS stations for continuous recording. In 1994 the Geographical Survey Institute of Japan (GSI) deployed a dense GPS monitoring network with about 600 stations and started nationwid monitoring of the movements of the Japan Islands. Now GSI issues deformation maps of Japan Islands on approximately yearly base. It is actually marvelous progress to see the deformation of Japan Islands at the present moment which we couldn't see until ten years ago when the conventional geodetic techniques were the main tools in this field.

On the other hand, strainmeters and tiltmeters with high sensitivity have been used for monitoring earth tides and ground movements and recorded irregular changes near the earth' s surface. Much of them are caused by meteorological disturbances such as rainfalls, surface temperature, atmospheric pressure and so on, but local movements due to the tectonic force are also considered to be included naturally. Previously, such irregular or annual movements within the term of a few years or shorter couldn't be observed with the conventional terrestrial techniques, so that there were no ways to intercompare the movements oberved with strainmters/tiltmeters with regional or nation-wide deformations which are to be observed with geodetic methods.

In this paper we compare the two years records observed with continuous GPS monitoring and strainmteres and tiltmeters, and discuss the possibility of observation of irregular crustal movments and application of these data to long-term earthquake prediction.

2. GPS Observations

We are carrying out continuous GPS observations at six stations in the Kinki district as shown in Fig.1. We use Ashtech Z-12 receivers and analyze every 30 second data by the Bernese software Ver.3.4. In the present analysis we used the IGS Ephemeris to calculate every day solution referred to the GSI point TSKB about 500km distant from Kinki district. Movements of the GPS stations obtained by Nakano (1996) are also shown by the arrows in Fig.1. Daily solutions of Ikeda (IKED) are given in Figs.2 for the two years of 1994 and 1995. As seen in the figure, approximately constant movements and annual change are

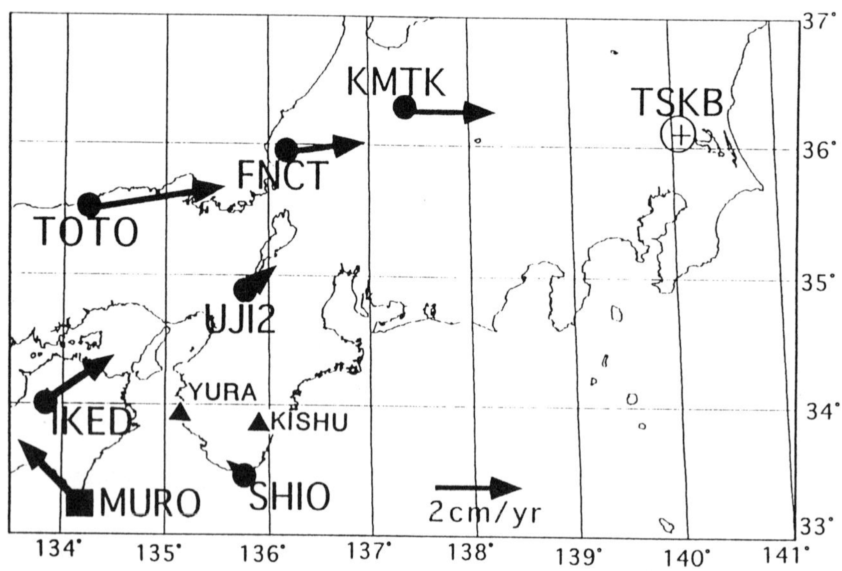

Fig.1. GPS stations (solid circles), and underground stations with strainmeters and tiltmeters (solid triangles) run by DPRI in Kinki district, southwestern Japan. Station movements observed by GPS referring to the GSI station TSKB are shown by arrows. (Nakano,1996)

remarkable. The distance from TSKB to IKED is 600km, so that the linear movement of the station corresponds to the strain of $2x10^{-8}$ if the strain is assumed to be uniform between the two stations.

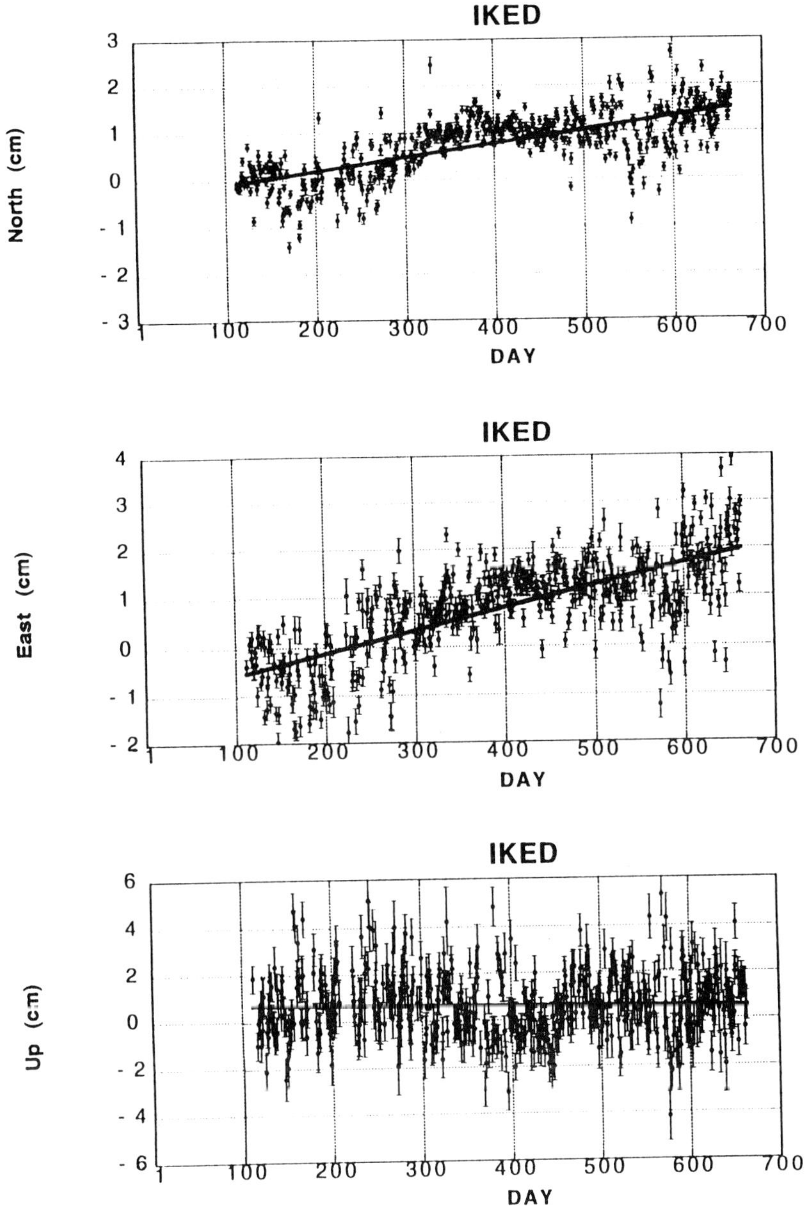

Fig.2. Daily solutions of the vector of IKED referred to TSKB. The solid lines are obtained by the least squares fitting. Error bars of + 3 are shown. (Nakano,1996)

3. Observations with Strainmeters and tiltmeters

DPRI is continuing crustal movement obervations at 11 stations in the Kinki district by using strainmeters and tiltmeters. Here we show the results from two stations, Yura and

Kishu. Their locations are shown in Fig.1. The results are given from Fig.3. Annual changes and approximately linear trend with irregular fluctuation are prominent, although observational faults and disturbances cause some indefinite situations. The amplitudes of annual changes reach 10^{-6} . Secular strain and tilt rates are also of the order of 10^{-6} a year and two orders of magnitude large compared with the strain calculated from the GPS monitoring.

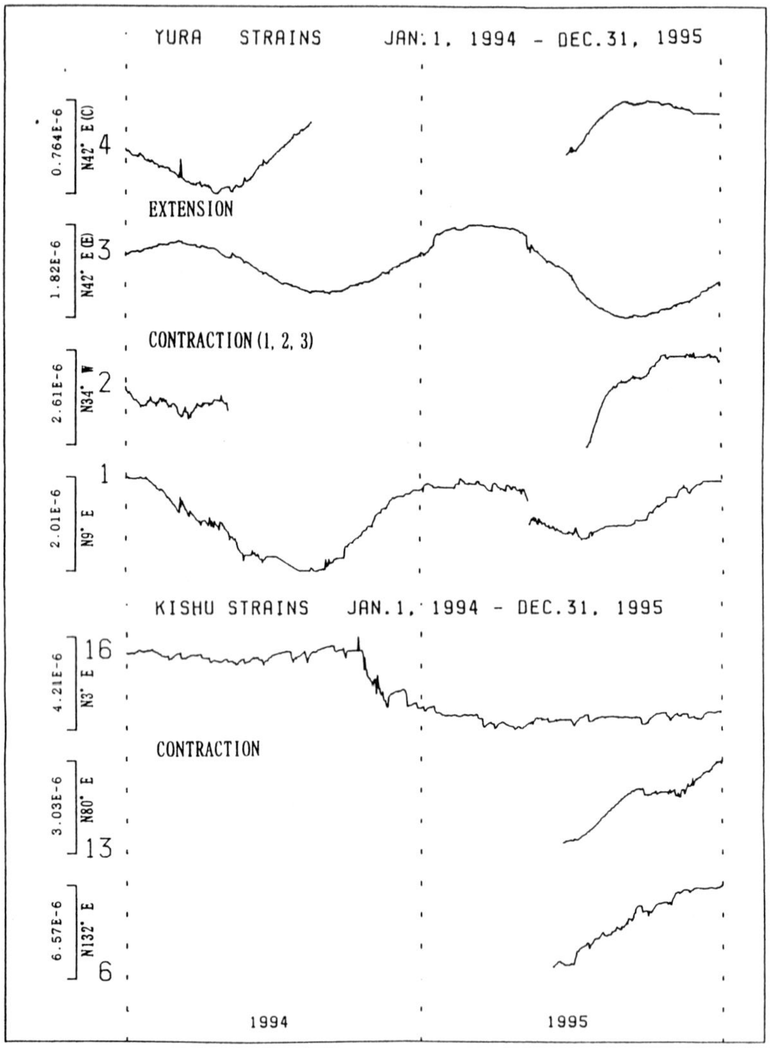

Fig.3. Ground strains observed at Yura and Kishu in 1994 and 1995.

4. Annual Changes

Baseline solutions referred to TSKB in Fig. 2 show remarkable annual changes which give maximum and minimum values in January - February and July - August as well as large scattering in the warm seasons. Fig.4 shows the excess path delay calculated from routine radiosonde observations by Japan Meteorological Agency (JMA). The excess path delay by water vaper generally shows maximum delay in August and minimum in February in southwest Japan. While the sharp increase of the delay is seen in August on the radiosonde data, GPS baseline solutions have no such sharp changes in the summer, though the scattering is very large. In the case of GPS solutions the zenith delay were estimated by using a default air model in the present analysis (Nakano ,1996) and both the dry and wet delays have been reduced to some extent. Thus the annual change seen in Fig.2 is interpreted as the residual or error of the approximate correction.

On the other hand, the annual changes seen on the strain and tilt results have maximum and minimum values around March and October which correspond well to those of the temperature in the underground tunnel. The surface temperature also gives the maximum and minimam values around August and January. Accordingly the annual changes of the

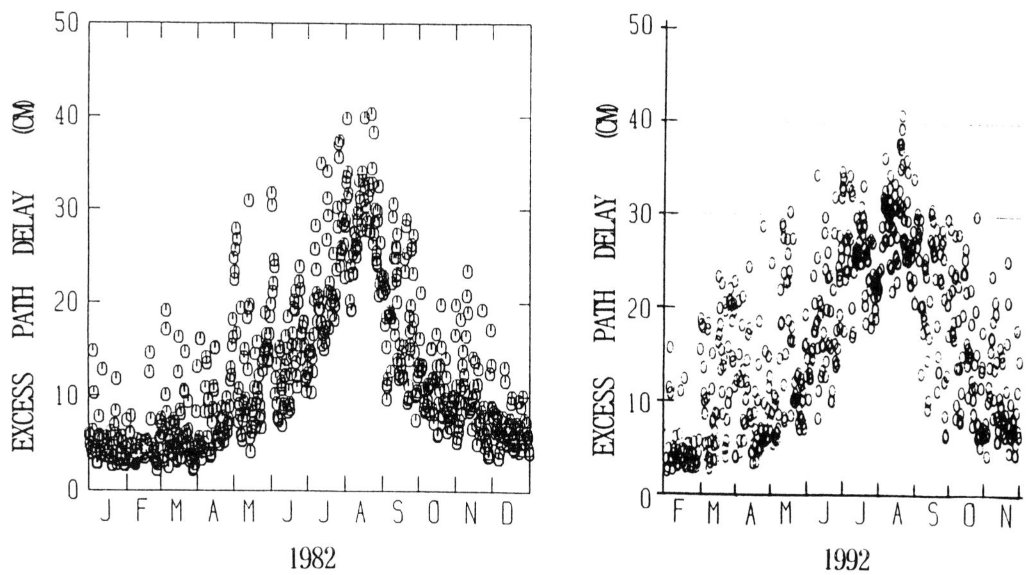

Fig.4. Wet excess path delay calculated from data by routine radiosonde observations by JMA at SHIO.

ground strains and tilts are not caused by the thermal deformation at the earth surface, but by the temperature in the tunnel or thermal deformation at some depth under the ground where the annual change of the temperature is delayed due to the propagation delay of heat.

The annual movement of GPS pillars will reach a few mm at least due to the thermal effect on the pillar itself and the ground surface, and occasionally due to change of ground water level. Though it is difficult to distinguish the real movement of the pillar on the present observations since the phase of the wet delay and the air temperature are same, it is necessary to correct the propagation delay to detect a few mm movement of the pillar.

5. Detection of Irregular Movements

The mean rate of strain change in Japan Islands is approximately 2×10^{-7} a year. Since the strain rates at Yura and Kishu are far large compared with the tectonic strain rate, they are considered as very local strains caused by meteorological and/or ground water disturbances. This kind of disturbances decrease in deeper part or at more suitable circumstances. As examples we refer here to the result from Matsushiro in Fig. 5. As seen in the figure long-

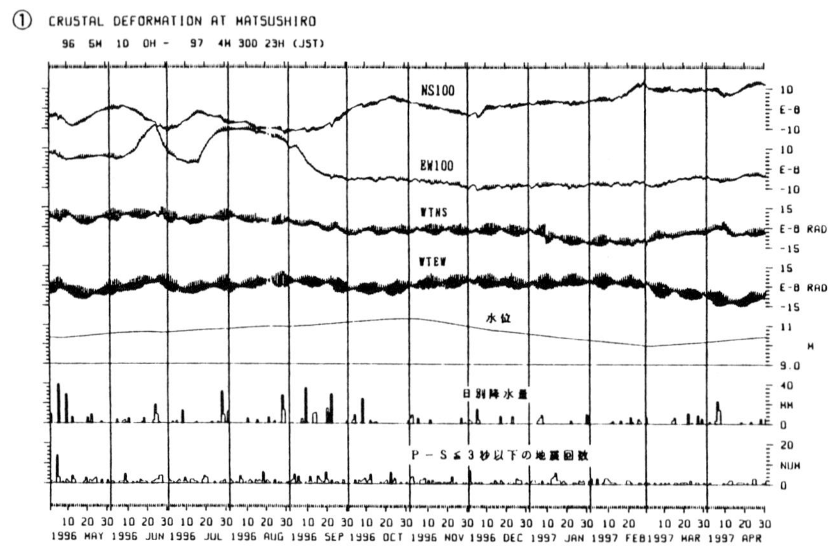

Fig.5. Strain and tilt at Matsushiro of JMA. (after Matsushiro Seismological Observatory, JMA, 1997)

term changes are order of 10^{-7} which correspond well to the magnitude of the average strain rate.

It is to be noted that there can be seen irregular strain changes of the order of 10^{-7} in Fig. 5, and they are considered to relate to the temporal fluctuation of stress accumulation, detection of which will be very imortant for long-term prediction of earthquakes.

Fig.6 is the result from GPS observations of aftereffect caused by the Hyogoken Nanbu Earthquake by Nakano and Hirahara (1997). Since the baseline length is about 40 km, the scatter is small compared with Fig.2, but it reaches about 1cm which corresponds to 2~ $3x10^{-7}$ of strain. Therefore the irregular changes which are expected to be detected are just the same order of the data scattering in Fig.6. The fault length of M7 class earthquake is approximately 40km. In order to detect strain changes associated with the expected irregular stress accumulation by GPS which will generate an earthquake of M7, accuracy of detecting a few mm changes of baseline length of the oder of 40km is required.

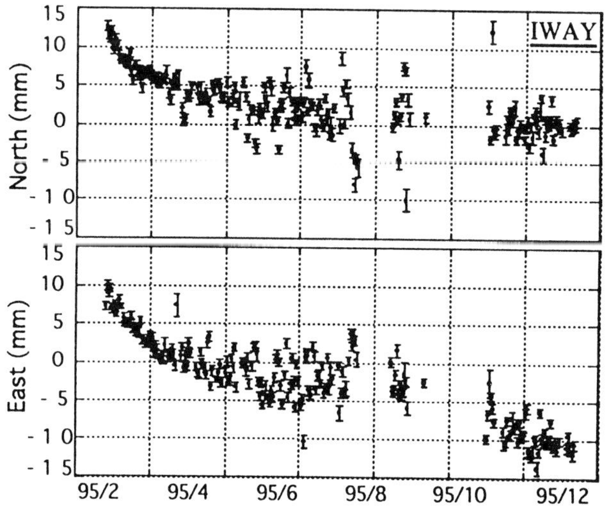

Fig.6. Postseismic displacement at IWAY relative to MITU site about 40km to northwest . (after Nakano and Hirahara, 1997)

6. Conclusions

GPS can detect uniform tectonic movements of regional scale by establishing networks larger than 100km, but it is still difficult to monitor local or irregular strains which will be associated with local stress accumulation. Monitoring such local irregular strains together with strainmetrs and tiltmeters will be very effective to understand the local tectonics and long-term earthquake prediction. In order to attain a few mm accuracy in GPS observations improvement of wet delay correction is necessary under the actual weather condition.

References

Matsushiro Seismological Observatory,JMA (1997): Seismic activity in northern part of NaganoPrefecture and observation of crustal movement at Matsushiro (November 1, 1996 - April 30, 1997), Report of the Coordinating Committee for Earthquake Prediction, Vol.58,435-439. (in Japanese)

Nakano,T.(1996): Postseismic deformation following the Hyogo-ken nambu earthquake observed with GPS, Master Thesis of graduate School of Sciences, Kyoto University.

Nakano,T. and K.Hirahara(1997): GPS observations of postseismic deformation for the 1995 Hyogo-ken Nanbu earthquake, Japan, Geophysical Res. Lett., 24, 5, 503-506.

PRESENT-DAY KINEMATICS OF CENTRAL ASIA DERIVED BY JOINT INVERSION OF GEOLOGICAL AND GEODETIC DATA

Ruizhi Chen[1] and Junyong Chen[2]
[1]The Finnish Geodetic Institute, Geodeetinrinne 2, 02430, Masala, Finland
[2]National Bureau of Surveying and Mapping, Baiwanzhuang, Beijing, 100830, China

Abstract

Joint inversion of geological and geodetic data is performed with a combination procedure of a genetic algorithm and the quasi-Newton method for Central Asia. The geological data includes geological fault rates and stress orientations, while the geodetic data include two sets of GPS results: one set from the repeated measurements of the Chinese A-order GPS network and the other from two GPS networks in southwest China. Although a simple 2-D elastic model is used in the forward computation, the best-fit model fits the geological and geodetic data reasonably. The prediction errors are ±33.1 degrees for stress orientations, ±9.4 mm for change rates of the GPS baselines obtained from the Chinese A-order GPS network and ±8.3 mm for relative displacement vectors obtained from the GPS networks in southwest China. The estimated slip rates are generally less than the geological estimations. The most significant one occurs in the central part of the Altyn Tagh fault. Our estimation is less than 15 mm/yr, while the geological estimation is 30 mm/yr. The deduced deformation model shows that the regional component dominates the main part of the deformations. The northward push of the India plate is absorbed by about 50% by the convergence along the Himalayan arc. For eastern Tibet, a compression of about 20 mm/yr occurs in the east-west direction , while an extension of about 10 mm/yr occurs in the north-south direction. The characters and magnitudes of the movements agree very well with the north-trending right-lateral shear model proposed by England and Molnar (1990).

The paper has been submitted to *Geophysical Journal International* for review.

APPLICATION OF THE ANOMALY OF ASTROMETRIC OBSERVATIONS TO EARTHQUAKE PREDICTION STUDY

Hu Hui[1) and Han Yanben[2)]

1. Yunnan Observatory, Academia Sinica, Kunming, 650011, China
2. Beijing Astronomical Observatory, Academia Sinica, Beijing 100080, China

Abstract. The article reports on a method of providing valuable information for earthquake prediction by the anomalous observations in the residuals of astronomical time and/or latitudes measured at Yunnan Observatory and gives some primary results.

Key words: astrometry, time-latitude residual, anomaly, earthquake prediction

1. Introduction

Since the occurrence of a major earthquake in Tangshan region of China in 1976, it has been verified from many examples that before a strong earthquake, there appears an abnormal fluctuation of the residuals of astronomical time and/or latitudes near the shock region (Li Zhisen et al., 1978). It shows that the accurate optical astrometric observations are used not only for the astronomic purposes, but also for the research of the local geophysical phenomena, like the prediction of earthquakes. After an earthquake of magnitude 6.1 which occurred in Luquan County, Yunnan Province, China, on April 18, 1985 (Han Yanben et al., 1987), the authors made a model for the prediction of a strong earthquake based on the detailed analyses of the relation between the strong earthquakes and the residuals of time and latitude obtained with the photoelectric astrolabe at the Yunnan Observatory, and have regularly provided the reports of the time-latitude residuals for the departments concerned, like the Seismology Bureau of Yunnan Province since November in 1985(Hu Hui et al.,1989). So far, the authors successfully provided the valuable information for seven strong earthquakes by using the model.

2. Analytic Method

The data of astronomic time and latitude residuals include rich geophysical information. Different information can be abstracted by means of different analytical methods, so the important thing is to use the optimum method. In order to cleanup the time and latitude residuals, one should first optimize the reduction of time and latitude observations so as to increase their accuracies. The authors reduce all the observations according to project MERIT standards (i.e. standards of a programme of international collaboration to monitor earth rotation and intercompare the techniques of observation and analysis) (Melbourne et al., 1983), and two preliminary calculations are carried out with the data obtained from 1981 to 1986. The first is made to normalize the data within a group of stars, and the correction for the individual position of each star is found so that the calculated values of time and latitude of a group having not enough stars are basically equivalent to those of a complete group. The second calculation is made to find the group correction after making the correction for the individual star positions (Debarbat and Guinot, 1970). The application of the group corrections not only eliminates the zonal differences of the catalog but also decreases the seasonal effects because a fixed group of stars is observed in the same season of every year. Thus the reduction of observations is based on the normalization of the data within each star group and adding the group correction.

The calculation of time and latitude residuals is mainly carried out according to the following steps:

1. The initial calculation of the residual of time determination RT_i, and the residual of latitude determination RF_i

$$RT_i = (UT1 - UTC)_i - (UT0 - UTC)_i - (X_i sin\lambda - Y_i cos\lambda) tan\varphi / 15 \quad (1)$$

$$RF_i = \Delta\varphi_i - X_i cos\lambda - Y_i sin\lambda, \quad (2)$$

where $\Delta\varphi_i$ stands for the daily mean value of the determined values of the latitude variation in the observing day and, λ and φ for the adopted values of geographical coordinates of the photoelectric astrolabe at Yunnan Observatory.

2. The calculation five-day mean value and the zeroed mean value.

It is found from the comparison and analysis that the average time interval of 5 days is more advantageous to the preservation of the information of the warning signs of an earthquake than the averaging time intervals of 15 days or a month do. Therefore, the five-day weighted averages of $RT_i andRF_i$ are respectively calculated and then the two sequences of $RT_i andRF_i$ are respectively reduced to zero, i.e.

$$R_o = \frac{1}{N} \sum R_i \quad (3)$$

$$R_{oi} = R_i - R_o. \quad (4)$$

3. Removal of the annual and semiannual terms

According to the analyses made by astronomers for many years, (Kimuna. H., 1902) there generally exist local terms with annual or semiannual harmonic variations in the observations of time and latitude obtained with the optical astrometric instrument at all the stations over the world. In order to remove the local terms the conventional harmonic analyses were universally adopted.

The practical method includes two steps:

a) The RT and RF are respectively fitted into the sinusoidal curve by means of the following formula

$$Asin(2\pi t + \alpha) + Bsin(4\pi t + \beta), \quad (5)$$

where t is calculated from the initial date of the calculated year.

b) The local term of the harmonic variation is deduced respectively from the sequences of RT and RF, thereby obtaining the purified time and latitude residual sequences RT and RF.

4. The five-point sliding average

In order to remove the short-term accidental disturbance the five-point weighted sliding average is taken with the formula for each sequence as follows:

$$S_i = \frac{1}{9}(F_{i-2} + 2F_{i-1} + 3F_i + 2F_{i+1} + F_{i+2}), \tag{6}$$

where F represents RT or RF.

5. The raise of the criterion of anomalies

According to the standard deviation formula

$$\sigma = \sqrt{\frac{\sum s^2 - \frac{(\sum s)^2}{N}}{N-1}}, \tag{7}$$

the standard deviations σ of the RT and RF sequences are respectively calculated and those greater than or equal to 2σ, or greater than or equal to 3σ, are respectively used for the warning indications of earthquakes.

3. Preliminary Practice of The Forecasts

According to the analyses of the earthquakes in Yunnan region and the fact that the Xikang-Yunnan plate moves mainly in the direction of south-southeast, the forecasting model should principally depend on the residuals in latitude. The real-time treatment of the observed data, obtained with the instrument, has been carried out by means of the computer program of the prediction model and prediction values of BIH/IERS since March 1987. When astronomic observations are available, they are reduced twice a month, and the curve of the residuals is drawn by a IBM computer. Then a report on the behavior of the residuals is sent to the departments concerned, such as the Seismology Bureau of Yunnan Province and the Kunming Seismology Office, and numbered uniformity since November 1985. Until December 1989, four reports of abnormal residuals have been issued. It so happened that four moderate or strong earthquakes occurred in the regions surrounding the Yunnan Observatory during the time span when the astronomical observations were made (Hu Hui et al., 1990).

In 1990-1992 years the work on the automation of the instrument has been carried out, with a photon counting detector being also added to it, and therefore the sections of continuous observed data are seriously short. But after the recover of continuous observation of the instrument in October 1992, four strong earthquakes were precisely monitored. The specific case of four anomalies is shown in Table 1.

4. Discussion

1. The local vertical line is taken as a datum line of the optical astrometric instrument. The angle between Earth's rotation axis and local vertical line is determined by optical astrometric instrument, and so the residuals, which is the part after drawing Earth Rotation Parameters from the observations, include the variational component of the local vertical line. This is called the astronomical time and latitude residuals. The variation of the vertical before earthquakes occurrences is due to the

Table 1. Prediction for anomaly of residuals and the Parameters used to test earthquakes in 1990s.

No.	date	place	φ		λ		M	D	A	St	Dt	Sφ	Dφ	Y/N	T(d)
			^{o}	$'$	^{o}	$'$			^{o}						
1	1993.1.27	Puer	22	56	101	05	6.3	291	216	$>3\sigma$	$-$	$>3\sigma$	$+$	Y	90
2	1994.1.11	Myitkyina	25	05	97	19	6.7	551	269	$>2\sigma$	$-$	$>3\sigma$	$-$	Y	140
3	1995.7.12	Menglian*	22	00	99	18	7.3	476	225	$>2\sigma$	$+$	$>3\sigma$	$-$	Y	140
4	1996.2.3	Lijiang	27	18	100	06	7.0	360	310	$>2\sigma$	$+$	$>3\sigma$	$-$	Y	140

*The earthquake occurred practically within the boundaries of Burma outside Menglian County, Yunnan Province.

λ and φ are the epicenter position of the earthquake.

M is the magnitude of the earthquake.

A and D denote respectively the azimuth (reckoning from north to east in the range of 0° to 360°) and distance (km) of the epicenter for Yunnan Observatory.

St and Sφ note respectively the anomalous quantities of RT and RF.

Dt and Dφ note respectively the anomalous directions of RT and RF.

Y/N denotes the anomalous synchronism of RT and RF.

T(d) is the advance days of the anomaly of RT and RF with respect to the earthquake.

motion of the groundmass (like the movement of a fault block of the intraplate) in a seismic region during the pregnancy of earthquake may be the main reason of the anomalous fluctuation. It reflects practically the change in the gravity field around the observatory, and therefore it is also a method for monitoring the change of gravity

2. The practice of the forecasts above has proved again that the anomalies of the residuals of astronomical time and/or latitudes before strong earthquakes are true. It may be a significant precursory of medium/short-term earthquake. The comprehensive analyses should be made by combining observations obtained with astrometric instruments with those from other geodetic techniques.

3. The difficulties in the prediction of a strong earthquake by means of time and latitude residuals are that the observations made with optical instruments are seriously affected by the weather, because the astronomic observation can be only carried out at clear night. The rainy and cloudy in Kunming from June to October.

In addition, one cannot predict by means of the residuals of time and latitudes where an earthquake will happen, how far it will be from the instrument and which actual magnitude it will have. How large is the distance range of earthquakes corresponding to anomalies in the time and latitude residuals? What is the minimum earthquake magnitude that can be detected? Further research should be made to answer these questions, and others.

References

Debarbat S. and Guinot B. 1970, La Methode des hauteur egales en Astronomie, Paris

Han Yanben, Hu Hui, Du Hongrong, 1987, *Kexue Tongbao*, Vol.32, No.17, 1205-1207.

Hu Hui, Kan Rongju, Wang Rui, Cai Xin, Chen Cuixian, 1989, *Astron. Astrophys.*, Vol. 224, 321-322

Hu Hui, Kan Rongju, Wang Rui, Cai Xin, Chen Hongming, 1990, *Chinese Science Bulletin*, Vol.35, No.14, 1229.

Kan Rongju, et al., The exploration of the stress fields and active characteristics of modern tectonics in Southwest China, *Acta Geophysic Sinica*, No.2, Vol.20, 96-109(1977)

Kan Rongju et al., The topographical section from Zhefang to Malong in Yunnan Province, Seismology Press, 1992, in Chines

Kimura, H., 1902, *Astron. J.*, **12**, 49.

Li Zhisen, Zhang Guodong, Zhang Huanzhi, Liu Xuejun, 1978, *Acta Geophysical Sinica*, **4**, 278-291.

Melbourne, W., Anderle, R., Feissel, M., King, R., McCarthy, D., Smith, D., Tapley, B., Vicente, R., 1983, Project MERIT standards, USNO Circular nr167.

INTERDISCIPLINARY STUDIES OF ASTRONOMICAL FACTORS AND EARTHQUAKES IN CHINA

Han Yanben[1)], Li Zhian[2)] and Hu Hui[3)]

1) Beijing Astronomical Observatory, Academia Sinica, Beijing 100080, China
2) Department of Astronomy, Beijing Normal University, Beijing 100875, China
3) Yunnan Observatory, Academia Sinica, Kunming 650011, China

Abstract. The history and present condition of study of the seismicity in astronomical observations and methods in China and initial results are introduced in this paper. Significance and problems in the study are discussed. China is a country in which there are many earthquakes and a new period of active seism is beginning, great attention should be paid to the study of astronomy on seismicity.

Key words: Astronomy, Earthquake, Seismicity

1 Introduction

Humanity is facing varied natural disaster, and seism is one of the most serious disasters. Accurate earthquake prediction is an effective means to reduce the effects of this type of disaster, but the form of an earthquake is a very complicated physical process and modern science has not recognized the features of seismicity completely and profoundly, so the study of seismicity, especially earthquake prediction, is a difficult scientific problem. Of course, this study is a subject of seismology. Some studies have show that many non seismic elements also affect seismicity. So, it needs the participation and cooperation of some other fields of scientific study to solve this difficult scientific problem. In recent years, people have recognized that the study of seism requires the cooperation of astronomers since astronomical observations and studies include not only the motion and variation of the Earth but also the astronomical environment of the Earth. So the effects of astronomical elements should be not ignored. Actually, many results have show that there are some relative features between astronomical factors and seismicity and some of them are closely related. This subject of the interdisciplinary study should be called astro-seismology.

2 The History of Research of Astro-seismology in China

There are some records show that there is a correlation between earthquakes that occurred in ancient China and certain astronomical phenomena. Only since the end of the last century and the beginning of this century have scientists noticed the relationship. Some considered that seismic events are random, so the correlation had not been considered or were negated when others tried to study the regularity (Divison 1938; Jeffreys 1938). In this condition, of course, people did not pay more attention to the study of possible correlations between the formation and the occurrence of earthquakes and the motion of the Earth and astronomical events happening in the solar system. Interdisciplinary studies among some scientific subjects has been considered along with the development of modern science. Possible correlations between seismic events and astronomical factors, such as variations of the Moon's phase, the celestial tidal force and solar activity, had been noticed since the 1950's (Gutenberg et al 1954; Benioff 1951). In the 1930's, a Chinese geologist, Professor Li Siguang, remarked upon the relationship between the movement of the Earth's crust and the variation of the Earth's rotation, and then he found the geological mechanisms (Li 1979). In the middle of 1950's, Wang Jiayin advanced a point of view that there are some correlations between earthquakes and sunspot and the variation in the strength of the Earth's magnetic field, and in the beginning of the 1960's he suggested that the celestial gravitation and tidal effect in earthquake prediction should be considered synthetically (Wang 1963).

In March 22, 1966, a major earthquake(M=7.2) occurred in the Xingtai region located in Hebei province of China. An active period of seism had begun in the north China. Because of this, the Chinese government began to pay great attention to the study and prediction of seismic events. So the study of astro-seismology was developed in China. At the end of the 1960's, some astronomers of the Beijing Observatory and Department of Astronomy of Beijing Normal University conjoined to analyze and study the relationship between astronomical factors and earthquakes. Similar studies were also developed at the Shanghai, Yunnan and Purple Mountain observatories, the National Marine Bureau and the Geological Institute of the Chinese Academy of Sciences. The studies dealt with the variations in the Earth's rotation, polar motion, variations in the relative number of sunspot, variations in the position of the Sun and Moon and the phase of the Moon and so on, and some valuable initial results were obtained (Xiao et al. 1973; Li et al. 1973; Earthquake Prediction Group of National Marina Bureau 1974).

3 Essential Study Substances of Astro-seismology

An earthquake is the result of the strenuous motion of the internal medium of the Earth bursting. The birth of a seism is a long and complicated physical process, and the process starts with the accumulation of stress in rock due to the effects of various internal and external elements. Then the rock bursts suddenly and an earthquake happens when it can not bear the tectonic stress. Because the Earth exists among heavenly bodies, the motion and variation of its lithosphere must be affected by its astronomical surroundings and by the movement and changes in the Earth itself. The astronomical factors may affect the accumulation process of stress in internal rock formations in the Earth. These effects may modulate the alternations of active and tranquil periods of seismicity, and even trigger the occurrence of an earthquake.

Therefore, astro-seismology will mainly study the relationship between the variations of astronomical factors and conditions and seismicity, and study the physical excitation mechanism which affects seismicity. On this basis, the study will make use of astronomical observations and extend and forecast data to offer valued information for the study of seismicity and earthquake prediction in order to improve the accuracy of the prediction.

4 The Main Results of Astro-seismology Study in China

In former years, the main content of astro-seismologic research was the relationship between astronomical factors and seismicity and to find their interrelation regularity. The studies have involved many astronomical factors and conditions and some valuable initial results, which are followed with interest by seismologists, were obtained. Some results have been taken as reference in studies of seismicity and earthquake prediction. It can be said that astro-seismological studies has became an important part of seismic studies in China. The main results may be the following:

4.1 The relationship between seism and variation of the Earth's rotation

If the Earth's rotation is constant, the Earth should be in a quasi-elliptic state of of revolution and this rotation would not cause any sudden change of crustal movement. However, the rate of the rotation is variable. This will cause additional stress along latitudinal and longitudinal directions and affect the normal crustal movement process. It was a contending problem that the additional stress should start up a earthquake or affect it. At present, some studies have shown that the energy caused brought by variations in the spin rate is not enough to bring up a major earthquake, but the variation of the rotation may be one of the main reasons for the modulating and conditions which excite an earthquake. In studying some relationships between the long, middle and short-term variations of the rotation of the Earth and earthquakes, some scientists discovered that there are high correlations between the variation of energy released by intermediate and deep earthquakes which have occurred since the beginning of the century and the fluctuation of stress caused by variations of the rotation. Statical analysis of the yearly frequency of global earthquakes of $M \geq 8$ and $M \geq 7$ have been made, and it has been discovered that there is a close correlation between ten year rotational variations (Zhang 1979; Zheng et al. 1995). The frequency of large earthquake is higher when the rotation slows down, and it is lower when the rotation grows faster (Gao et al. 1981; Li 1984). The additional force of longitudinal direction and the Earth's deformation will be the largest at $\pm 45°$ of latitude and the smallest at the poles when the rotation changes, and we can discover that there are many earthquakes in the area between $35° - 45°$ latitude in the northern hemisphere and the peak of seismic frequency is there (Li 1984). Analysis of seismic data in China showed that higher monthly frequencies existed in March and August, and lower ones in June and November. Earthquakes may be affected by the seasonal variations of the Earth's rotation (Li 1984). During studies of seismicity of different areas and tectonic belts in China, it was discovered that there is close relationship between seismicity and this variable rotation (Li et al. 1973). This can help us to recognize the modulation and excitation of earthquakes from variations in the rotation from the point of view of mechanics.

The relationship between polar motion and seism also is an interesting problem and some phenomena has shown this relationship but some studies have shown that the excitation of the polar motion is very small.

4.2 The relationship between seism and solar activity

Solar activity may possibly affect some geophysical processes, such as earthquakes, through a magnetic or electric phenomena. The relationship between the relative number of sunspot and earthquakes was noticed. Then it was discovered that the seismicity in the northern hemisphere closely related to the periods of about 11 and 22 years. The frequency spectrum of seismic data in China and neighbouring countries showed the seismicity exists in periods of about 11 and 21.3 years (Liu et al. 1982; Xu et al. 1980), and the seismic strength was discovered to be stronger in many areas when the sunspot number peaks or bottoms (Liu et al. 1984; Han 1989). It was also observed that the frequency of earthquakes of $M \geq 7$ varied depending on whether it was day or night. This may be related to the relative position between the Sun and Earth. The magnetic storm is one of the phenomena associated with strong solar activity. Some scientists suggest that magnetic storms may modulate and excite seism. According to an analysis of observations of the Beijing Magnetic Observatory, among 168 records of magnetic storms, 127 earthquakes occurred in the same day or before or after the day of each storm (Du et al. 1989). An analysis of earthquakes($M \geq 7.5$) in the region around the Pacific showed that seismicity is stronger along the eastern Pacific when the Kp index is large, but it is weaker along the western Pacific. The result showed that the affect may be related to the characteristic of the geological structure (Zhang et al. 1989). Five major earthquakes ($M \geq 7$) in the north of China and 11 of 16 ($M \geq 7$) in the western area of China (including two earthquakes of $M \geq 8$) happened when the Kp index was high (Zhang et al. 1989). Magnetic storms generally are related to solar flares with the strongest energy, so scientists consider that the excitation of the magnetic storm should be larger than the excitation of variations of sunspot for the occurrence of an earthquake (Jiang 1985).

4.3 The relationship between seism and tidal forces

The relationship between seism and tidal forces had been researched early and extensively. The Xingtai earthquake (March 22,1966, M=7.2) which occurred the first day of the lunar month and Tangshan earthquake (July 28,1976, M=7.8) which occurred the second day of the lunar month are two very well known examples that the occurrence of earthquakes is affected by the tidal force. Study and analysis of major earthquakes ($M \geq 6$) occurring in Hebei province of China showed that the earthquakes in the three days around the first day and the fifteenth day of the lunar month were about 2.5 times the normal frequency (Ren 1982). The frequency of major earthquakes ($M \geq 5$) in the north China occurring in the three days around the first day and fifteenth day of the lunar month is about twice the normal frequency (Du et al. 1989). There is a close correlation between the earthquakes in Yunnan province and the Moon's position. The frequency of earthquakes is high when the Moon is in the last quarter and the first day of the lunar month. Earthquake statistics ($M \geq 6.2$) in the 19th and 20th centuries show that the frequencies of earthquakes occurring near these two lunar aspects are about 6.2 and 4.5 times the normal frequency respectively (Ren 1982).

The relationship between earthquakes and variations in the components of tidal forces for different seismic belts have also been studied and some interesting results have been obtained. The analysis of major earthquakes in the southwest of China in the part thirty years showed that there is a close relationship between the date of the earthquakes and certain components of the tidal forces. This suggests that tidal forces of the Moon and Sun probably can excite seism (Li et al. 1994; Han et al. 1994).

4.4 Studying seismicity with observations of time and latitude

At the end of the 1970's and the beginning of the 1980's, a phenomenon that anomalies appearing in residuals of universal time and latitude (RT and RF) before major earthquakes was found and confirmed preliminarily by astronomers of the Beijing Observatory (Li et al. 1978; Han et al. 1986). Then the study was done in depth. Similar study was also developed at the Yunnan and Shanghai observatories (Hu et al. 1989).

The main regularity of the phenomenon may be summarized as follows:

(1) The anomalies usually become apparent in the RT and/or RF of optical astrometric instruments before the occurrence of a major earthquake within the distant of 300 or 400 km (sometimes it is about 500 km);

(2) The beginning time of the anomalies are several weeks or months before the occurrence of the earthquake;

(3) The fluctuations of residuals of some instruments installed in the regions without major earthquakes were relatively small.

(4) The long-term variation of the RF was also related to the seismicity around the instruments (Tian 1989).

Some studies showed that the large variations of the local vertical before major earthquakes due to the motion of groundmass may be the main reason for the anomalies in the residuals and the rich geophysical information is included in the data. So the residuals is from high precision instruments can be used not only in the studies of variations of the vertical and gravitational field, but also in the study of seismicity and provide information for earthquake prediction (Zhang 1981; Han et al. 1989).

There were major earthquakes in the region around the Yunnan observatory. One Mark-2 PA is working in the observatory. The prediction information for several earthquakes occurring in the region has been provided according to the anomalies of the residuals of the instrument (Hu et al. 1988). Of course, the observations from one instrument can not show the position of an earthquake, but a network including several instruments may be helpful for determining the position of an earthquake. We are trying to set up an observational network including three instruments in the Beijing Observatory.

5 Conclusion and Prospects

Because the study of seismicity in astronomy is a new interdisciplinary subject, there is a lot of problems: (1) The formation and occurrence of an earthquake is a very complicated process and it is not completely understood. (2) At present, research is not organized and focused and there is a short age of systematic study. The study especially lacks the synthetical analysis to the overall background of earthquakes, various factors and local features. (3) It is lacking in depth of study of the physical excitation mechanisms.

The study of seismicity and earthquake prediction is a difficult scientific problem that needs some interdisciplinary studies. Some astronomical conditions and phenomena may be important factors that induce and excite seism. The study of astro-seismology already is one of the important contents of astrogeodynamics. The eastern region of China, faces the Pacific seismic belt, and the western region of China connects with the seismic belt of Eurasia. In the both of these seismic belts earthquakes occur at a high frequency. The works of Chinese seismologists considered that a new period of active seismicity is already coming to the world and China. Astronomer should pay more attention to this study.

Attention should be paid to the following tasks: (1) To organize some astronomers and geophysicists and choose several hopeful subjects to make in depth studies, especially, with regard to the above-mentioned subjects (the Earth's rotation, solar activity, celestial tidal forces and residuals of astronomical time and latitude) and strive to get systematic results. (2) To improve methods of the study and consider geological structure in choosing seismic samples, and combine a part with the whole, and systematically study the relationship between the variations of astronomical factors and seismicity at different time-scales. First obtain large samples of information and then recognize if there is a close correlation between astronomical factors and seismicity. Especially, research the possibility of exciting seism, and study in depth the regularity and physical mechanisms of these phenomena from dynamic point of view.

To study astro-seismology is not only to understand seismicity profoundly, but also to deeply understand the movement and variable processes of the Earth (e.g., the variation of Earth's rotation, polar motion and the solar-terrestrial relationship), and to push relevant studies and observational work of astronomy forward. Development of the work related to the relationship between residuals of optical observations of time-latitude and earthquakes broke through traditional recognition about the astrometric local terms. Astronomers discovered that these observations contain rich geophysical information, especially the variation of the local vertical. With regard to the relationship between solar activity, variation of the Earth's rotation and seism, one assumes that solar activity may directly affects seismicity, and perhaps the solar activity firstly affects the variation of Earth's rotation then affects seismicity. To study this subject in depth will increase the knowledge of the relationship between solar activity and the Earth's rotation and the mechanisms of the variation of the Earth's rotation. Of course, earthquake prediction is a difficult problem that can not be solved only by the study of astro-seismology. The joint efforts of astronomers and seismologists can play an active and proper role in the solution of this difficult problem.

Reference

Benioff, H.: 1951, *Bull. Geol. Soc. Amer.*, **62**

Divison, C., *Studies in the Periodicity of Earthquakes*, T. Murby, London, 1938

Du Pinren et al., An Introduction to Astroseismology, Seismic Press, 1989 (in Chinese)

Earthquake Prediction Group of National Marina Bureau: 1974, *Scientia Geologica Sinica*, No.2 (in Chinese).

Gao Xu et al.: 1981, *Earthquake*, No. 3 (in Chinese)

Gutenberg, B. et al., *Seismicity of the Earth and Associated Phenomena*, Princeton Univ. Press, 1954.

Han Yanben, *Progress in Comprehensive Study of the Mutualrelations among Cosmos, Earth and Life*, Chinese Science and Technology Press, 1989 (in Chinese)

Han Yanben et al.: 1994, *Astrophysics (Publ. of BAO)*, No. 25

Han Yanben et al.: 1986, *Chinese Science Bulletin*, **31**, 1244 (in Chinese)

Han Yanben et al., *Comprehensive Study of the Mutualrelations Among Cosmos, Earth and Life*, Chinese Science and Technology Press, 1989 (in Chinese)

Hu Hui et al.: 1989, *Astron. and Astrophys.*, **224**,321

Hu Hui et al.: 1988, *Publ. of the Yunnan Astronomical Observatory*, No.1, (in Chinese)

Jeffreys, H.: 1938, *Beitr. Geophys.*, **53**, 111

Jiang Boqin: 1985, *Acta Seismological Sinica*, **7**, 452 (in Chinese)

Li Qibin et al.: 1973, *Acta Geophysical Sinica*, **16**, No. 3 (in Chinese)

Li Siguang, *An Introduction to Geology*, Science Press, 1979 (in Chinese)

Li Yuanjun: 1984, *North China Earthquake Sciences*, **2**, No. 3 (in Chinese)

Li Zhian et al.: 1994, *Journal of the Beijing Normal Univ.*, **30**, No. 3 (in Chinese)

Li Zhisen, Zhang Guodong et al.: 1978, *Acta Geophysical Sinica*, **21**, 278 (in Chinese)

Liu Defu et al.: 1982, *Journal of Seismological Research*, No. 1 (in Chinese)

Liu Defu et al.: 1984, *Earthquake*, No. 5 (in Chinese)

Ren Zhenqiu: 1982, *Chinese Science Bull.*, No. 10 (in Chinese)

Tian Jing: 1989, *Journal of Graduate School, USTC*, **6**, 92 (in Chinese)

Wang Jiayin, *Historical Data of Geology in China*, Science Press, 1963 (in Chinese)

Xiao Xinghua et al.: 1972, *Publ. of the Beijing Astronomical Observatory*, No. 2 (in Chinese)

Xu Daoyi et al., *Motion of Heavenly Bodies and Earthquake Prediction*, Seismic Press, 1980 (in Chinese)

Zhang Huanzhi, *Collected Works of Astro-geodynamics*, Shanghai Astronomical Observatory Press, 1979 (in Chinese)

Zheng Dawei et al.: 1995, *Acta Seismologica Sinica*, **17**, No. 1 (in Chinese)

Zhang Baoshu et al., *Comprehensive Study of the Mutualrelations Among Cosmos, Earth and Life*, Chinese Science and Technology Press, 1989 (in Chinese)

Zhang Guodong: 1981, *Acta Seismological Sinica*, **3**, 152 (in Chinese)

The Ice-Age Geodynamics of Antarctica

W. R. Peltier
Department of Physics, University of Toronto
Toronto, Ontario, Canada M5S 1A7

Abstract

During the last deglaciation event of the current ice-age, significant deglaciation of the Antarctic continent is expected to have occured, primarily of West Antarctica. Detailed analyses of the magnitude of the geodetic signatures that would be expected to characterize this event at present, using a complete gravitationally self-consistent theoretical theory of the global process of glacial isostatic adjustment, suggest that these signatures should be easily observable. They include the current rates of displacement in both the "vertical" and "horizontal" as well as the time rate of change of the gravitational acceleration measured on the solid surface of the planet.
The interpretation of measurements of this kind is complicated by the fact that significant present day melting of Antarctic ice could also be occuring although this is by no means established. It is clear, however, that some ongoing melting of land ice, in excess of that from small ice-sheets and glaciers, seems to be necessary to account for the approximately 1.8 mm yr^{-1} globally averaged rate of sea level rise that seems to be occuring at present, a signal that has often been attributed to global warming.
I will present a new set of analyses of this range of issues using recently refinements of the mantle viscosity structure and deglaciation in an attempt to demonstrate the utility of the geodetic constraints that may be invoked to help resolve the outstanding issues.

References

Peltier, W. R. (1998): *Postglacial variations in the level of the sea: Implications for climate dynamics and solid earth geophysics*. Review of Geophysics (in press).
Peltier, W. R. (1998): *The inverse problem for mantle viscosity*. Inverse Problems (in press).

Elevation changes in Antarctica 1992-1996: Implications for Ice Sheet Mass Balances

D. J. Wingham, A. L. Ridout and R. J. Arthern,
Mullard Space Science Laboratory, Department of Space and Climate Physics, University College London, Holmbury St. Mary, Dorking RH5 6NT, U. K.

Abstract

The uncertainties in the mass balance of the ice sheet of Antarctica of 5×10^{14} kg yr^{-1} are the largest uncertainties in the causes of the observed rise in sea level. The uncertainty in the mass balance of Antarctica has increased in time as the estimated role of ice shelf bottom melting has grown. The past few years have seen dramatic improvements in the force modelling of altimeter satellites and improved correction of the surface and volume scattering contributions of ice sheet altimeter echoes. In consequence, we have been able to constrain the elevation change 1992-1996 of 60% of the Antarctic ice sheet to 0,5±0,7 cm yr^{-1}. The possible sources of error from the satellite orbit, the time-variant scattering at the ice sheet surface, and the echo travel-time corrections are all examined, and it is concluded these are too small to be significant to the five-year elevation trend. The relationship between the fluctuations of elevation and mass is examined through numerical modelling of the densification of firn at the surface of the ice sheet and it is concluded that, at a single location, time-variant densification, due to fluctuations in surface mass balance and surface density, may cause the elevation time-series to misrepresent the mass change by as much as ±2 cm yr^{-1}. However, evidence from deep Antarctic cores, and from atmospheric models, point to a spatial correlation length of these surface fluctuations of less than 1.000 km. On this basis, we estimate the elevation time series to track the mass fluctuations to within ±0,6 cm yr^{-1}. We thus conclude that 60% of the sheet has been in balance to within 7% in the period 1992-1996. Extended over Antarctica this is a mass balance uncertainty of 140 Gt yr^{-1}, a substantial reduction on the previous estimate.

Reference

Wingham, D. J.; Scharoo, R.; Arthern, R.; Ridout, A. and Shum, C. K. (1998): *Antarctic Elevation Change 1992-1996.* (in preparation).

The SCAR GPS Campaigns: Accurate Geodetic Reference in Antarctica

R. Dietrich[1], R. Dach[1], J. Perlt[1], H.-W. Schenke[2], T. Schöne[2], M. Pohl[2], G. Soltau[3], G. Engelhardt[3], H.-W. Mikolaiski[3], G. Seeber[4], F. Menge[4], W. Niemeier[5], H. Salbach[5], K. Lindner[6], H.-J. Kutterer[6], M. Mayer[6]

[1] Technische Universität Dresden, Institut für Planetare Geodäsie
D–01062 Dresden
[2] Alfred–Wegener–Institut für Polar- und Meeresforschung Bremerhaven
Columbusstr., D–27568 Bremerhaven
[3] Bundesamt für Kartographie und Geodäsie, Außenstelle Leipzig
Karl–Rothe–Str. 10, D–04105 Leipzig
[4] Universität Hannover, Institut für Erdmessung
Schneiderberg 50, D–30167 Hannover
[5] Technische Universität Braunschweig
Institut für Geodäsie und Photogrammetrie
Gau"s–Str. 22, D–38106 Braunschweig
[6] Universität Karlsruhe, Geodätisches Institut
Englerstr. 7, D-76128 Karlsruhe

1 Introduction

During the XXIII Meeting of the Scientific Committee on Antarctic Research (SCAR) held in Rome in 1994 the Working Group on Geodesy and Geographic Information (SCAR WG-GGI) established a working program for the next two years. It was decided that SCAR Epoch GPS Campaigns should be organized as a continuation of former Epoch Campaigns.

The objectives of GPS Epoch Campaigns, as a part of the Working Group project GIANT (Geodetic Infrastructure for Antarctica), can be described as follows:

- Linking of Antarctica with the Global Terrestrial Reference Frame ITRF (IERS – International Earth Rotation Service – Terrestrial Reference Frame) with highest accuracy
- Measurement of the relative rates and directions of motion of the Antarctic Plate with respect to the adjoining plates and microplates
- Determination of the relative motion of crustal blocks within the Antarctic Plate

- Unification of the vertical datum, determination of the height of the mean sea level at tide gauge stations
- Determination of the vertical motion of the Antarctic lithosphere due to changes of the ice and ocean loading.

2 Observations

The SCAR 95 Epoch GPS Campaign took place 1995 from January 20, 00:00 UT until February 10, 24:00 UT.

The campaign data were completed by the data of permanent tracking sites on the southern hemisphere (stations of the International GPS Service for Geodynamics – IGS – and regional stations). Figures 1(a) and 1(b) provide an overview of geographical distribution of the stations.

The main observation standards were:

- providing daily data sets of 24 hours starting at 0:00 UT
- using only dual frequency geodetic GPS receivers without squaring technique (if possible)
- data sampling interval: 15 seconds (if possible)
- tracking all satellites above 10 degrees elevation.

To ensure a long-term use of the obtained results it was strongly recommended to pay special attention to a stable monumentation of the GPS markers.

In 1996 the SCAR 96 Epoch GPS Campaign was carried out. The schedule was the same as in 1995.

All data and documentation sheets were collected and filed at the *Institut für Planetare Geodäsie* (Technische Universität Dresden, Germany).

3 Data Analysis

In order to get an optimum accuracy and reliability of the results as well as realistic estimates of the accuracy of the results, six research groups processed the data. Different GPS analysis software packages were used:

- TU Dresden: Bernese GPS Software Version 3.5 (DRES)
- AWI Bremerhaven: GAMIT/GLOBK (AWI)
- BKG Leipzig: Bernese GPS Software Version 3.5 (IfAG 1) and GIPSY (IfAG 2)
- Universität Hannover: GEONAP (HANN)
- TU Braunschweig: GEONAP (BRAUN)
- Universität Karlsruhe: Bernese GPS Software Version 3.5 (KARL).

All groups used the same input data:

- GPS observations from the stations of the SCAR95 Epoch GPS Campaign
- GPS observations from several IGS stations

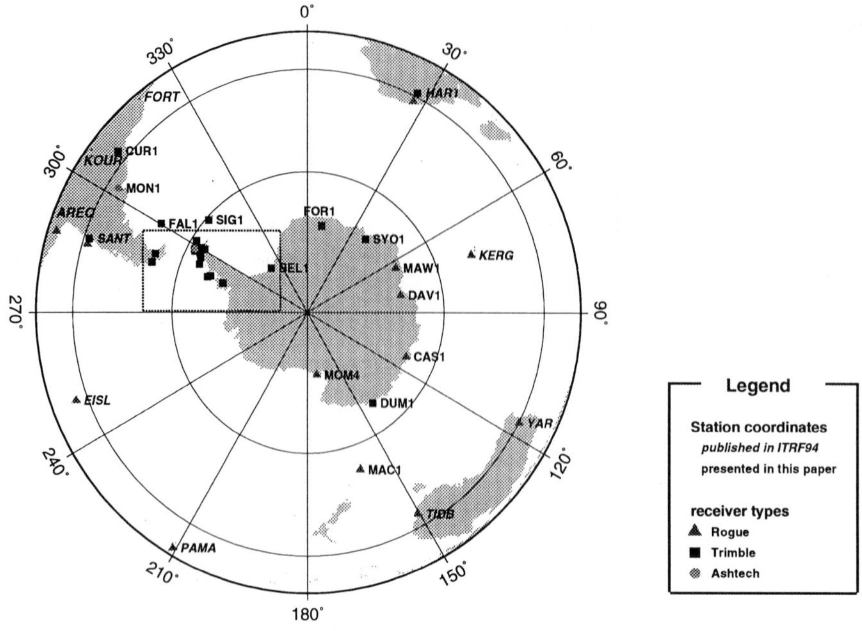

(a) Map of the network

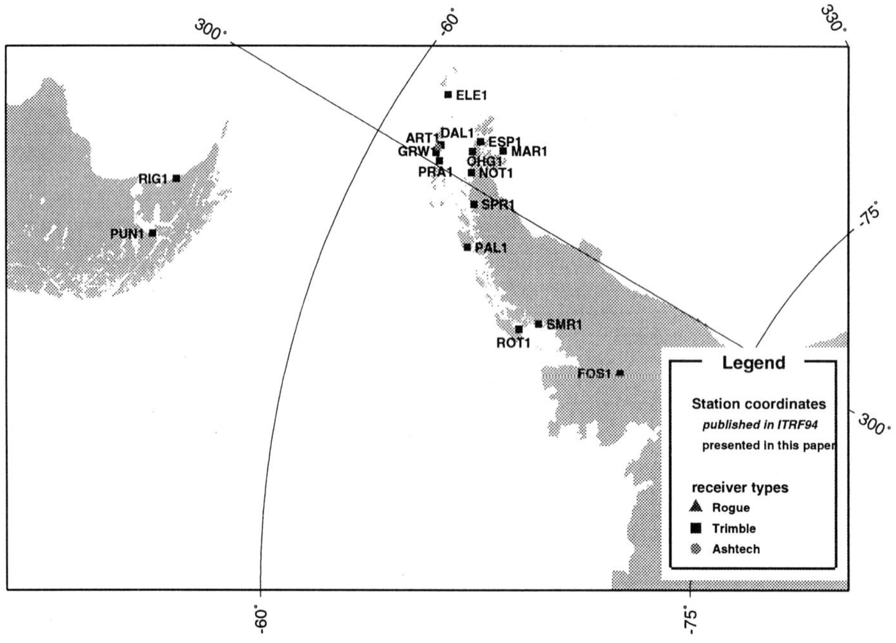

(b) Map of the Antarctic Peninsula

Figure 1: Stations of the SCAR 95 Epoch GPS Campaign including permanent IGS tracking sites

Table 1: Mean of squared residuals from Seven-Parameter-Transformations between the individual GPS solutions of SCAR 95 Epoch GPS Campaign in mm

		AWI	IfAG1	IfAG2	HANN	BRAUN	KARL
DRES	dN	7.7	9.5	9.6	9.6	10.5	5.4
	dE	5.7	6.8	9.7	10.0	12.0	6.6
	dU	23.1	30.5	12.5	20.8	20.2	36.0
AWI	dN		6.6	3.7	9.2	7.9	8.1
	dE		5.8	4.9	8.9	9.6	9.1
	dU		31.0	21.0	31.6	20.6	32.3
IfAG1	dN			7.8	9.5	9.0	7.1
	dE			8.1	9.4	11.2	7.2
	dU			26.4	39.7	23.9	17.6
IfAG2	dN				10.4	9.4	9.1
	dE				11.9	11.4	12.2
	dU				26.7	21.1	30.4
HANN	dN					6.2	7.8
	dE					9.0	10.8
	dU					24.8	44.5
BRAUN	dN						9.4
	dE						12.2
	dU						26.1

- final precise orbits from IGS
- polar motion parameters from Bulletin B of IERS.

The calculations were performed in the reference frame ITRF94 (cf. Boucher et al., 1996), epoch 1995.1 for observations of 1995 and epoch 1996.1 for observations of 1996. Because of larger discrepancies between the GPS velocities and the NUVEL-Ia model[1] it was decided to use only stations with GPS-based published velocities. The coordinates of these stations were fixed in the analysis.

On the other hand, the selection of the different software options to be applied for the GPS analysis was left to the free decision of the operators:

- selection of fiducial points (at least four stations)
- setting of the elevation mask
- modeling of the influence of troposphere and ionosphere.

The aim was to get the best solution using optimum strategies of each software, and to get realistic accuracy measures when comparing the solutions.

4 Results

The results obtained by the different groups are provided – as described – in the datum ITRF94 (epoch 1995.1). The full covariance matrices of the solutions were not used

[1]See the residual graphs of ITRF94 in the IGN data base where ITRF94 was published.

in the processing of the combined final solution because these matrices resulted from different software packages. Therefore, they are not directly comparable. All solutions were introduced with the same weight in producing the final combined solution.

Different approaches were discussed and numerically tested for combining the seven solutions. These comprised:

- the arithmetic mean of the station coordinates,
- medians of the station coordinates,
- Helmert transformations between all solutions and a final optimum fit of the resulting coordinate set to the ITRF stations, and
- coordinate adjustment with an additional transformation parameter estimation.

It turned out that the resulting combined station coordinates for the SCAR stations correspond at mm-level for arithmetic mean as well as for the other approaches. In

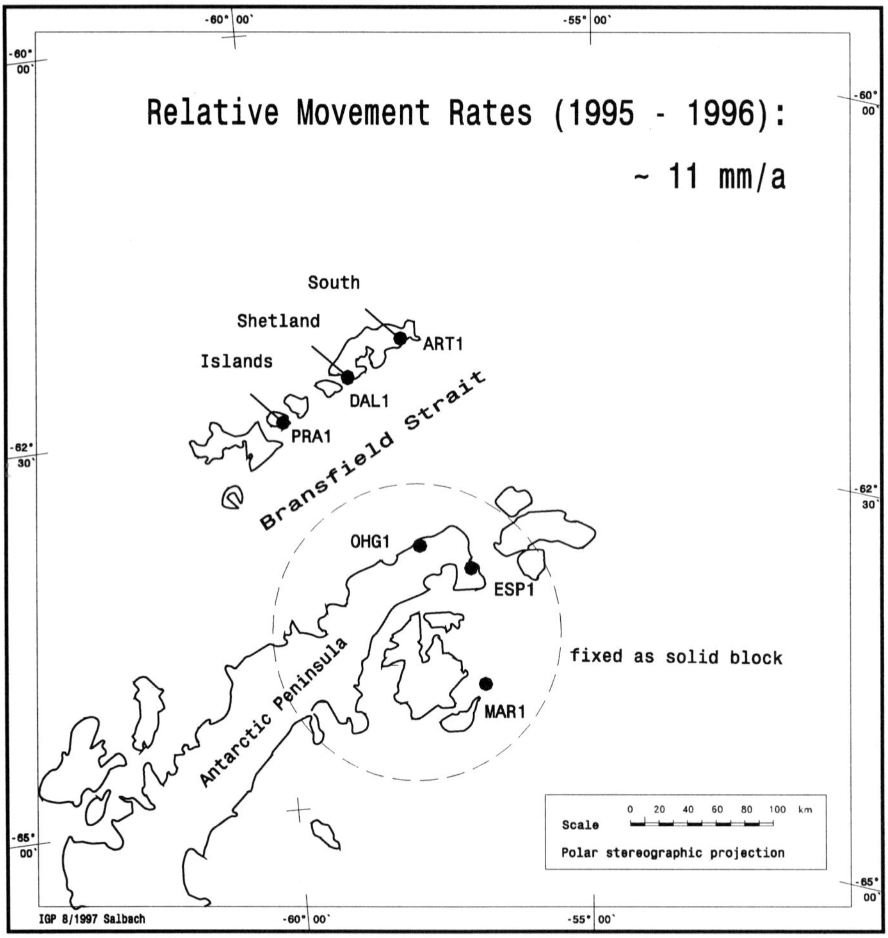

Figure 2: Relative Movement of the South Shetland Islands in relation to the Antarctic Peninsula from the analysis of the combined GPS solutions of the SCAR 95 Epoch GPS Campaign and the SCAR 96 Epoch GPS Campaign

favour of simplicity we decided to use the averaging of station coordinates of the individual solutions to get the combined solution. The station coordinates of the different solutions are compared in Table 1. The final solution as well as other details are published in (Dietrich, 1996). It is important to note that within the subnetwork of the Antarctic Peninsula the accuracies are smaller by a factor of about 3.

The analysis of the deviations from the mean leads to the following conclusions:

- There are no significant differences in the accuracy measures of the individual solutions.
- The r.m.s. values for one individual solution (computed from the deviations from the mean) can be summarized as follows:
  North Component: ± 9 mm, East Component: ± 8 mm, Up Component: ± 20 mm
- The accuracy of the combined solution is probably better by a factor of about 2.

The results of the 1996 campaign were of similar accuracy. The application of the obtained results to study tectonic deformations is shown in Fig.2, as an example for the Antarctic Peninsula region.

5 Conclusions

The SCAR GPS Campaigns contribute to link Antarctica to the ITRF and provide a good basis for geodynamic research.

Acknowledgment

This research was funded by BMBF grant 03PL013.

The SCAR GPS Campaigns could only be successful, because many institutions supported this work scientifically and logistically, and because a lot of people contributed with a high commitment to the realization of field observations.

Special thanks are directed to the SCAR Working Group on Geodesy and Geographic Information, which acted as a scientific focus within the international research community in Antarctica.

Finally we like to thank the IGS Central Bureau for supplying GPS data through their international services.

References

Boucher, C., Z. Altamimi, M. Feissel and P. Sillard (1996): *Results and Analysis of the ITRF94.* IERS Technical Note 20, Observatoire de Paris.

Dietrich, R. (ed.) (1996): *The German Antarctic Project GAP95, German Contributions to the SCAR95 Epoch Campaign.* DGK, series B, number 304.

Uruguayan Contribution to the Antarctic Geodesy

Subiza, W., Rovera, H., Melconian, J.
Servicio Geográfico Militar (SGM)
Av. 8 de Octubre 3255, Montevideo 11600, Uruguay
E-mail:geografi@seciu.uy

1. Abstract

In the austral summer of 1986, the SGM started systematic geodetic and topographic surveys in order to support various scientific projects around Artigas Base. Maps of different parts of King George and Nelson Islands were made at several scales.

In 1995 the Uruguayan Antarctic Institute / Military Geographical Survey joined the SCAR Epoch GPS 95 Campaign, observing two stations with Ashtech Z-12 receivers, „Montevideo" in Uruguay and „Artigas" at Artigas Base, King George Island. After that, the 1996 and 1997 campaigns were carried out. During the last campaign another station, „Dallman" at Jubany Base (Argentina), was added. Some local results are presented here.

Since December 1987 gravimetric projects have been carried out by the Military Geographical Survey in order to support geological researches. Three airtrips were started from Punta Arenas gravity station (Chile) with two LaCoste & Romberg gravimeters, and a local network was set up.

In March 1997 an earth tide station was established at Artigas Base as a contribution to the SCAR Epoch GPS Campaign and for the regional study of geoid models. This station will be working continuously over a period of two years. It shows the international cooperation in scientific research, following the Antarctic Treaty's spirit.

2. The Beginning

In 1985, the first Uruguayan Antarctic Station was established by the Uruguayan Antarctic Institute at King George Island (South Shetlands).

In order to support geological, geochemical and biological research the Military Geographical Survey started to set-up a geodetic network and to determine its orientation. During February 1986 astronomic observations of the Sun as well as local ocean tide measurements were performed and recorded at a geodetic datum point. Artigas Station and its surrounding areas were surveyed.

Later on, a classical triangulation network was set-up. More than 25 references were constructed and determined, which allowed to re-fix the link to the two Doppler points 13103 and 13104 at Marsh Station (Chile). Horizontal angels were observed using Wild T2 theodolites. Infrared instruments enabled a practical and quick distance determination.

Afterwards, the vertical datum was defined by means of an ocean tide gauge AANDERAA WLR 7. These measurements were performed by the Oceanographic, Hydrographic and Meteorological Survey for the Navy in summer time. A precise leveling connected the tide gauge reference point with the datum point. The difference was only a few centimeters compared with previous determinations. Due to well known classical survey difficulties at these latitudes, it was necessary to use the trigonometric leveling in some cases. Vertical angles were observed simultaneously from two extremes. The triangle closure errors are less than 5 centimeters.

Supported by this network several topographic products have been launched: Maps of Nebless Point (scale 1:5.000, with a Wild RDS theodolite); Armony Point at Nelson Island; South-West of King George Island (scale 1:5.000, in February 1986). The northern area of Marion Cove was derived from the original network. A coordinate transfer was made using a Wild T2 and an infrared distance meter. During March and April 1988 also detailed field surveyings were carried out by means of a Wild RDS theodolite (scale 1:4.000).

At Collins Glacier, the small dome was represented in order to assist a glaciological project within the scope of an agreement with the National Committee for Antarctic Research of the People´s Republic of China. During November and December 1992 a total station was used to produce a 1:10.000 scale chart as well as to measure ice profiles.

At Fildes Peninsula (King George Island) a photogrametric flight was performed . The original aerial photo scales are 1:15.000 and 1:5.000, and the final product will be at a scale of 1:10.000 with contour intervals of 10 m.

3. GPS Observations

The first GPS receivers were bought in 1991. Two Magellan 5000 PRO C/A code receivers with submeter kits have been tested in our country and afterwards in the Antarctic classical network. The results allowed to readjust the original network.

Two years later, SGM got Ashtech Z-XII receivers with dual frequency and code capabilities. Experiences showed that satellite receivers are suitable to solve positioning problems in Antarctica. Applying static techniques for more than 1 hour and pseudo-kinematic observations for less than 4 minutes of observation time per point allowed accurate, rapid and cheap surveys for almost any purpose.

Table 1. Artigas Station Control Network (units: meter)

From	To	Dist.	95-96	95-97
ART1	RC10	97.019	+0.002	+0.002
ART1	RC12	247.871	+0.005	-
ART1	CO13	272.060	+0.007	-0.004
RC10	RC12	164.663	0.000	-
RC10	CO13	189.218	+0.006	-0.006
RC12	CO13	24.62	+0.003	-

The SGM joined the SCAR 95 Epoch GPS Campaign. A suitable station was selected near the Artigas station scientific facilities (ART1). A small-range control network consisting of five stations was also determined in order to monitor local displacements at cm level. Table 1 shows the control measurements carried out during the 95-97 SCAR Epoch Campaigns.
The annual control measurements proved a good repeatability and the stability of the Artigas station. It is measured with an approximately 3-4 times repetition of each single vector. The accuracy of them is below 1 cm at 2 sigma confidence. Table 2 shows the output of the analysis using Geolab adjustment software version 2.4d for the three stations observed in 1997.

Table 2. Geolab 2.4d output for 1997 control network confidence regions

Artigas ART1's Control Network 1997

GeoLab V2.4d WGS 84 UNITS: m, DMS Page 0011

2-D and 1-D Relative Station Confidence Regions (95.000 percent):

FROM	TO	MAJ-SEMI	AZ	MIN-SEMI	VERT.	DIST.	PPM
ART1	CO13	0.004	3	0.003	0.007	272.064	15.06
ART1	RC10	0.003	3	0.002	0.006	97.017	35.97
CO13	RC10	0.003	179	0.003	0.006	189.224	18.46

At the same time, another reference point was occupied at the South American continent, Montevideo (MONT or MON1), which was also determined during the SIRGAS Campaign 1995. This fact will give an extra connection between both continental projects. During three years, from January 20 to February 10, the SGM has been working on the project, and there is the will to continue in the future. Problems of power supply, transportation and a small buffer capacity affected the observing time. Nevertheless, the observation time was always longer than required by the campaign specifications.
Last February, another point has been occupied for seven days: Dallman/Jubany (DALL1) at Jubany Station (Argentina). Figure 1 shows the observation time at each campaign.
The processing of the observations was carried out at the SGM with the aim to control the quality of the recorded data, and to get precise WGS84 coordinates for cartographic and positioning purposes. The Campaign 95 was processed using the precise IGS ephemeris and the standard Ashtech software Prism, version 2.0. The Campaign 96 was processed using broadcasted ephemeris and the same software.
We know that the position accuracy depends on how the software handles different errors contributing to the measurement. The results are affected by the accuracy of satellite ephemeris and clocks, by vehicle dynamics, multipath effect, noise and of course atmospheric delays especially on long baselines like in our continental project, and the ionospheric and tropospheric conditions in Antarctica.

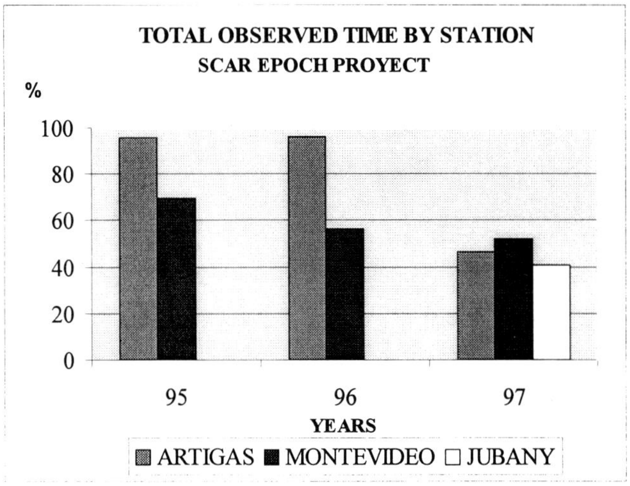

Figure 1: Observation Time at GPS Stations
(Data sampling interval: 1995/96 - 30 min, 1997 - 15 min)

Software limitations enabled us to work only with a part of the data. Meteorological information could not be completely used. Despite of all efforts, our PRISM software unfortunately did not fix any single cycle slip. Observing the differences in Table 3 & 4, and considering that we are talking about a 3000 kilometer vector, we will agree on the quality of the observations, system and equipment.

Table 3. Geocentric coordinate differences

Results	Δ X (m)	Diff.	Δ Y (m)	Diff.	Δ Z(m)	Diff.
Epoch 95 *	1367888.140	-	1800246.167	-	1990367.710	-
Local 95 Prec.	1367888.023	-0.117	1800246.085	-0.082	1990367.886	+0.176
Local 96 Broad.	1367888.020	-0.120	1800245.885	-0.282	1990367.893	+0.183

* ITRF 94 (Epoch 1995.1)

Table 4. Processed Cartesian vectors, lengths differences to SCAR Epoch 95 solution

Results	Vector (m)	Diff	RMS	Observations
Epoch 95	3012236.353	-		ITRF 94 Epoch 1995.1
Local 95	3012236.253	-0.010	0.091	31 sessions, Prec. Eph., All solutions
Local 95	3012236.410	-0.057	0.068	10 sessions, Prec. Eph., 80-96 Ratio Confid.
Local 96	3012236.169	-0.184	0.113	21 sessions, Broad. Eph., All solutions
Local 96	3012236.221	-0.132	0.066	15 sessions, Broad.Eph., 55-82 Ratio Confid.

4. Gravity Observations

Precise gravity determinations have been carried out with the help of three air links from Punta Arenas 9814-72 (President Ibáñez Airport) and IAGS continental station to Marsh Airstrip (TN1), King George Island, using two LaCoste & Romberg model G (# 013 and 703) gravimeters.

The operation was accomplished on December 07, 11, and 13, 1987. A F-227 Fairchild airplane, belonging to the Uruguayan Air Force, was used to transport both instruments. During the flights they were specially conditioned by the surveyors in order to minimize the drift induced by plane vibration.

The gravity differences Punta Arenas – King George were:

gravimeter G-013: + 8942.63 μms^{-2}

gravimeter G-703: + 8942.67 μms^{-2}

These allowed to estimate new calibration coefficients (scale factors): 0.99955 for G-013 and 1.00038 for G-703.

In the summer of 1988 a local gravity network was established. 23 closed polygons with a total of 65 stations (50 at Fildes Peninsula, 14 at Marion Cove and 1 at Nelson Island) were set-up. Bouguer and Free-air anomalies charts were produced.

During 1991 new links were observed between Fildes and Nelson. Two years later Armony Point (Nelson) was densified with 35 new stations. New anomalies charts were published. Last year another station was established at King Sejong (Korean Station) in cooperation with the Korean Antarctic Research Program.

5. Earth Tides Monitoring

This year, within the scope of the SCAR WGGI and under the auspices of the IAG Earth Tide Commission, we started to operate an earth tide station at King George Island.

The Project objectives are:

1- To contribute to a *precise determination of geoidal and ellipsoidal models* in the Antarctic region.
2- To *improve and verify theoretical earth tides.*
3- *Gravity data analysis* of the period, using new values registered at the station.

Last January, with the authorization and collaboration of N.I.M.A. (National Imagery Mapping Agency, former D.M.A.), a LCR gravimeter #062 was modified at the Royal Observatory of Belgium, Bruxelles. A Maximum Voltage Retroaction (MVR) feedback system was added. The main advantage of the system is the fact, that it applies a force to balance the permanently changing gravity, thereby keeping the gravimeter beam at the reading line. The gravity tide signal is then proportional to the applied feedback force. In that way, errors introduced by the measuring screw are eliminated.

The station (WGS84 ϕ = - 62° 10'52".48170, λ= - 58 ° 54' 34".59224, H= 37.09 m MSL) is set-up on a bedrock 650 m north of Artigas station. The gravimeter is located on a concrete pillar (40 by 40 cm) inside a cargo container, which was specially reconstructed to an observation facility. Main operation difficulties are to maintain stable environmental conditions, esp. concerning pressure and temperature. In addition, we found quick and

frequent changes of atmospheric masses, seismic disturbances, etc.
We expect to continue the operation of the station at least for 2 years. We also like to reactivate local oceanic tide and GPS satellite observations as soon as possible.
Data preprocessing and analysis are being carried out between SGM, Parana Federal University of Brazil and the Royal Observatory of Belgium.

6. Conclusions

Uruguay has been making strong efforts to contribute to local and international scientific projects. Cartographic, geophysical, gravimetric, geodetic and geodynamic projects have been carried out or are in progress.
Our country has a long tradition in cooperation and *offering the complete sharing of scientific data on an exchange basis* and we are open to continue this. *International scientific cooperation is the challenge of our time* to countries which do research in Antarctica in order to economize resources. You may have heard these words many times, but you know how difficult it is to turn theory into practice. Ideas, good will and imagination are often not enough.

7. Acknowledgement

The logistic support by the Instituto Antártico Uruguayo (IAU, Av. 8 de Octubre 2958, Montevideo 11600, Uruguay, e-mail: antartic@iau.gub.uy) is thankfully acknowledged, as well as the good cooperation within the joint IAU-SGM programs.

8. References

B.Murphy, J.Manning, T.Morrison .1990 „The transition to GPS: The Australian experience with satellite position in Antartica".
Dietrich, R. et al. The SCAR 95 GPS Campaign: Objectives, Data Analysis and Final Solution. XXIV Seminar of SCAR WGGI, England 1996.
Dietrich, R. The Geodetic Antarctic Project GAP95. German Contributions to the SCAR 95 Epoch Campaign, Munich 1996.
Torge, W. 1989 „Gravimetry".
W. Subiza. 1995/996. „Aplicaciones de la Geodesia Satelital en la Antártida". Informes de las campañas IAU/SGM.
J.Acuña, A.Bacci, J.Lazo, G.Sanguinetti. 1988. „Relevamiento Topográfico en las Islas Shetlands del Sur".
Ashtech INC.1990. User Manual
E. Forbes.1990. Observaciones del nivel del mar frente a la Base Científica Antártica „Artigas" (Isla Rey Jorge).
SCAR Working Group on Geodesy and Geographic Information , Circular letter No.6. October 1996.

GPS for ice sheet movement monitoring and grounding line detection

Capra A.*, Frezzotti M. **, Mancini F.*, Radicioni F. *, Vittuari L.***

***DISTART - University of Bologna**
Viale Risorgimento 2, 40136 - Bologna - Italy
**** ENEA - C.R. Casaccia - AMB/CLIM/MOP**
Via Anguillarese 301 - 00060 – Roma - Italy
***** Istituto di Ingegneria Ambientale - University of Perugia**
Via Duranti 1, 06125 – Perugia - Italy

Introduction

The velocity field of outlet glaciers at the grounding line is a critical parameter, together with ice thickness, in determining the ice discharge rate for the Antarctic Ice Sheet. Moreover the position of the grounding line is very sensitive to changes in the mass balance of the ice sheet. Data on ice velocities have been obtained with remote sensing and GPS measurement techniques on the David Glacier-Drygalski Ice Tongue, Priestley and Reeves Glaciers. A measurement technique particularly suitable for glacier monitoring is the satellite GPS technique, due to the environmental conditions and the accuracy that can be achieved in positioning. GPS static, fast-static and kinematic applications have been used since 1989 and repeated five times for ice stream velocity field determination, floating glacier undulation determination under sea tidal effects and grounding line detection (Capra et al., 1996,Frezzotti et al., 1997). The GPS measurements made on the David Glacier –Drygalski Ice Tongue are presented with particular attention to the method used and data analysis. On two floating glaciers, Drygalski and Hells Gate (a small ice shelf that seems to be particularly sensitive to changes), different in size and shape, GPS techniques were used for the floating glacier undulation determination under sea tidal effects (Bondesan et al., 1994) and kinematic profiles were obtained for grounding line detection (Capra et al., 1996). The GPS instruments used were the Geodetic L1/L2 Trimble model 4000 SSE and all the data were processed with the Geotracer software v. 2.25 and GPSurvey software v.2.0.

David Glacier-Drygalski Ice Tongue monitoring

The most important ice tongue, in term of climatic influence, near the Italian Base in Antarctica (Terra Nova Bay - Victoria Land) is the David Glacier and its ice tongue Drygalski. The David Glacier is the largest glacier in Victoria Land, which drains from the

Talos and Circe Domes, covering an area of 224,000 km^2 and moves across bedrock well below sea level. Velocity measurements were made at different time intervals with the comparison of sequential satellite images (Frezzotti-et al., 1997) and GPS measurements were performed between 1989 and 1994. Satellite velocity measurements are more abundant and are even possible for areas with widespread crevasses, while GPS can give more precise values of ice movements. So it is possible to monitor extended areas by locating some GPS station to verify and to support the solutions. Comparison between GPS and satellite tracking data in areas with tie-points have shown that std. dev. in the average velocity measured may be estimated at +/- 15-20 m per year, considering the 28.5 m pixel dimension (Spot XS). An accuracy of 15-20 cm can be considered for GPS measurements and an average velocity was found of 1.5 to 2.5 m per day for the David Glacier and the Ice Tongue points (Frezzotti et al., 1997). GPS surveys may provide positions in static or kinematic modes, one receiver remains on a fixed location (reference station), on rocky outcrops, and the other receivers are located on the ice, to control the co-ordinate variations over time. For logistic reasons, due to morphology of the area (long distances, long ice tongues without rock outcrops on the sides), only one fixed point was used. This was located on Huges Bluff (HB) (Fig. 1).

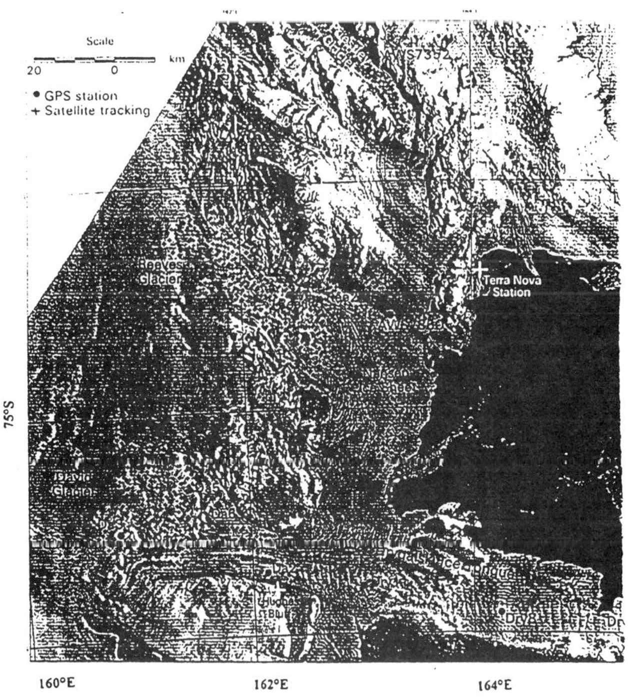

Fig. 1 - GPS stations and satellite tracking points on David-Glacier Drygalski Ice Tongue monitoring

More mobile receivers were located with contemporary acquisition, consisting of aluminium stakes of about 13 cm of diameter, 3 m long and driven into the ice by about 1.5 m. The measurement stations on the ice were located at 30 to 100 km from the fixed point which made it possible to carry out different tests on the distance effect on the results of GPS data processing. In order to verify the effect of movements of points during the measurement session, kinematic profiles of 12 hours were processed and compared with the planimetric solutions obtained with a static survey at the same period. The results showed that the static positions, in the case of linear movements, were in the middle of the processed kinematic period, taking into consideration the precision of the method (5-10 cm) for different distances (Frezzotti et al., 1997). To verify the same results on height, where variation is due to tidal undulations, the static solutions in the half time were compared with the kinematic ones for different intervals on points Dry2 and Dry1 at different tidal levels, one being closer to the grounding line than the other (Tab.1 and Fig.2).

Tab.1 – Height from static solutions and from kinematic profile relative to the half-time.

	Interval 1		Interval 2	
	Single base	kinematic	Single base	kinematic
Dry 2	36.52 m	36.65 m	36.41 m	36.52 m
Dry 1	6.84 m	6.93 m	6.80 m	6.91 m

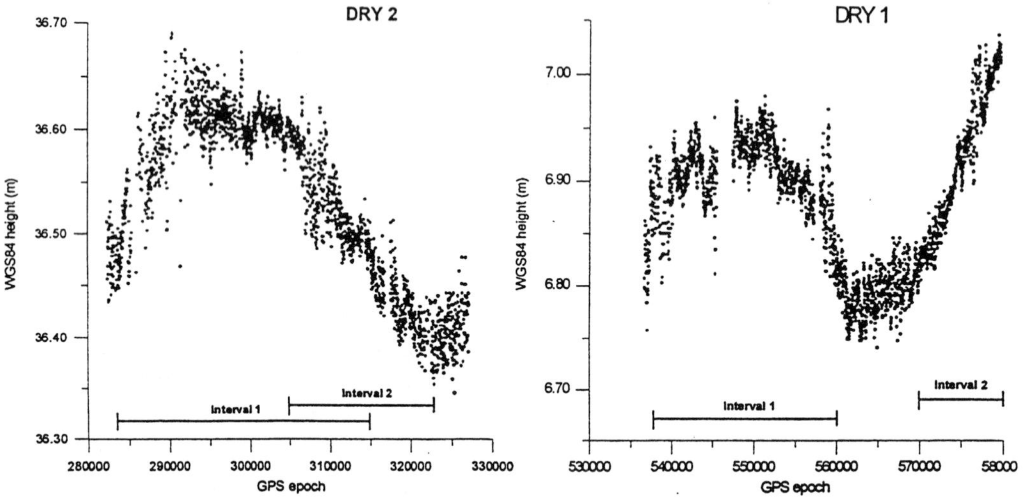

Fig.2 – Kinematic altimetric profile for Dry1 and Dry2 points in 1993 measurements.

The static solution is different from the kinematic one for the middle of the period. If the interval is symmetrical with respect to the curve, due to tidal effect (Int.1 for Dry1 and Dry2), the shape of the height static solution reflects the average value of heights of the kinematic profile of the interval. While the interval is decreasing (Int. 2 of Dry2) or increasing (Int. 2 for Dry1), the height of static solution differs from the average height value of the kinematic profile. The next test was carried out on different points to verify whether the planimetric position obtained from data relating to increasing acquisition period remain centred on the position determined at the half time of the period. The points are located at different distances: Dry2 at about 15 km, Dry1 at about 30 km and Dry8 at about 85 km. The solutions for the more distant point, Dry8, were not obtained for the 12 hour interval, probably due to the difficulty of processing long distance baselines and the movement of the point during the measurement period. The values differ from one another by an amount equivalent to the method precision for the points at different distances, except for Dry 8, probably for the reasons mentioned above; therefore the hypothesis can be accepted. The solutions appear divergent for the 12 hour period, even if the st. dev. is even smaller. This is probably due to the very high number of observation in GPS set (2880) which stabilises the rms solution, even with significant point position variations. The static baseline for a minimum of 3 to 6 hours period acquisition would appear to provide a solution of sufficient repeatability, considering the difference in values and the distances. Another test was performed by processing the triangle formed by the independent bases HB-Dry1-Dry8, in static mode, which provided superabundant observation throughout successive sessions. Even if this is not really a static network since points Dry1 and Dry8 are moving.

Tab.2 – Difference between the co-ordinates obtained with single baseline solutions, centred at the same time for three different points. Data referred to the 1993 measurements.

	Dry1 (std. dev. in mm for the two solutions)					
	1.5 - 3	1.5 - 6	1.5 - 12	3 - 6	3 - 12	6 - 12
Δ Lat. (m)	-0.0219 (1.5 – 0.9)	-0.0360 (1.5 – 0.8)	-0.0702 (1.5 – 0.9)	-0.0141 (0.9 – 0.8)	-0.0483 (0.9 – 0.9)	-0.0342 (0.8 – 0.9)
Δ Long. (m)	0.0025 (1.9 – 1.1)	-0.0258 (1.9 – 0.9)	-0.1122 (1.9 – 1.0)	-0.0283 (1.1 – 0.9)	-0.0014 (1.1 – 1.0)	-0.0864 (0.9 – 1.0)
Δ h (m)	-0.0440 (2.4 – 1.7)	-0.0260 (2.4 – 1.3)	0.0220 (2.4 – 1.5)	0.0180 (1.7 – 1.3)	0.0660 (1.7 – 1.5)	0.0480 (1.3 – 1.5)

	Dry2 (std. dev. in mm for the two solutions)					
	1.5 - 3	1.5 - 6	1.5 - 12	3 - 6	3 - 12	6 - 12
Δ Lat. (m)	0.0354 (2.9 – 1.5)	0.0735 (2.9 – 1.1)	0.0858 (2.9 – 0.7)	0.0381 (1.5 – 1.1)	0.0504 (1.5 – 0.7)	0.0123 (1.1 – 0.7)
Δ Long. (m)	0.0405 (1.7 – 1.0)	0.0365 (1.7 – 1.0)	0.2596 (1.7 – 0.7)	-0.0040 (1.0 – 1.0)	0.2191 (1.0 – 0.7)	0.2230 (1.0 – 0.7)
Δ h (m)	0.0250 (2.2 – 1.7)	0.0140 (2.2 – 1.5)	-0.0860 (2.2 – 1.1)	-0.0110 (1.7 – 1.5)	-0.1110 (1.7 – 1.1)	0.1000 (1.5 – 1.1)

	Dry8 (std. dev. in mm for the two solutions)					
	1.5 - 3	1.5 - 6	1.5 - 12	3 - 6	3 - 12	6 - 12
Δ Lat. (m)	0.0603 (3.6 – 2.3)	0.1068 (3.6 – 1.7)	–	0.0465 (2.3 – 1.7)	–	–
Δ Long. (m)	0.0262 (1.9 – 1.5)	0.0404 (1.9 – 1.4)	–	0.0142 (1.5 – 2.4)	–	–
Δ h (m)	0.0040 (3.2 – 2.5)	0.0290 (3.2 – 2.2)	–	0.0250 (2.5 – 2.2)	–	–

The tests on superabundant measurements showed that the precision was of the same size as those obtained with single base static solution. The apparent advantage offered by superabundant measurements in the search of gross errors, can not be guaranteed under these dynamic conditions, because the base is also affected by movements over time (Frezzotti et al.,1997).

Hells Gate Ice Shelf grounding line detection

GPS fast static and kinematic measurements were applied to obtain the altimetric variation for several stations over the ice surface due to tidal undulation. The curves obtained showed a similar pattern, they were in phase and presented the same amplitude as the tidal curves provided by the tide-gauge recordings, taking into account the accuracy of the method (Bondesan et al., 1994). The discrete distribution of measured points did not facilitate grounding line detection for the large area that had to be investigated, so further experiments on GPS kinematic surveying were carried out with an antenna located on sledges (Capra et al., 1996). After an accurate determination of the time of high and low tide, kinematic profiles were made for the maximum and minimum times. The comparison between the altimetric profiles obtained at the different tidal times makes it possible to distinguish buoyancy from grounding areas, if a displacement between the profiles occurs at different times. Two profiles of different areas, at a distance of around 20 km, were surveyed at the same time. The first one, quasi-linear, was processed and the results showed that a portion of the profile referred to floating ice, while another portion of the profile seemed to refer to grounding ice (Capra et al., 1996). The first profile, 27 km long, was subsequently completely processed using the GPSurvey OTF module. Figure 3 shows the profile height differences determined. The oscillation of the solutions are due to the precision of OTF algorithm and to the inaccuracy generated by the fact that the height comparison is not made exactly at the same point in the two outward and return journey, even if the topographic variation is small (about 15-20 cm for the distance travelled at velocity of 20 km/h). The point of separation between the floating and grounding ice appears clearly at a distance of about 20 km from the starting point. Previous processing of the second profile with a zig-zag pattern showed the altimetric differences of homologous points along the track (fig.4 a).

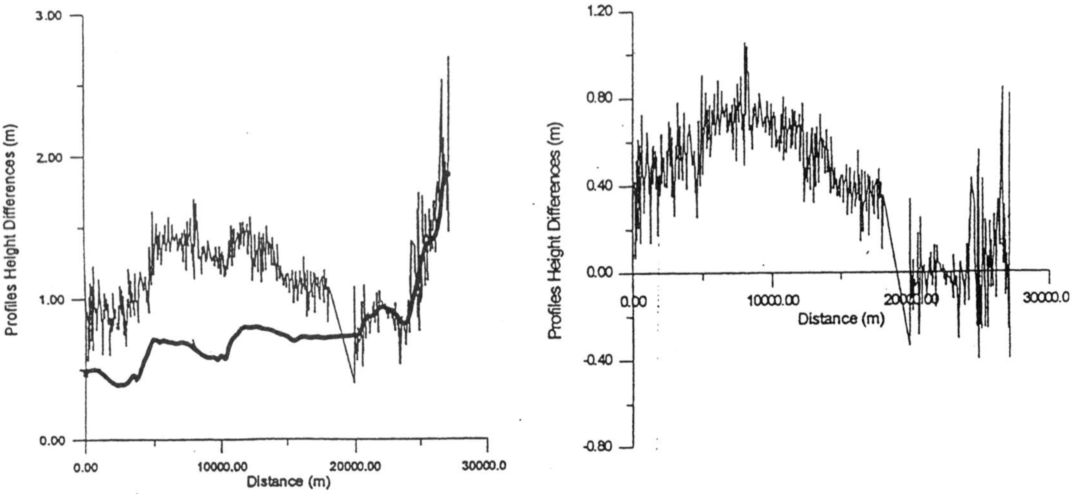

(a) (b)

Fig.3 – Height differences between the ice altimetric profile (a) and the GPS kinematic profiles.

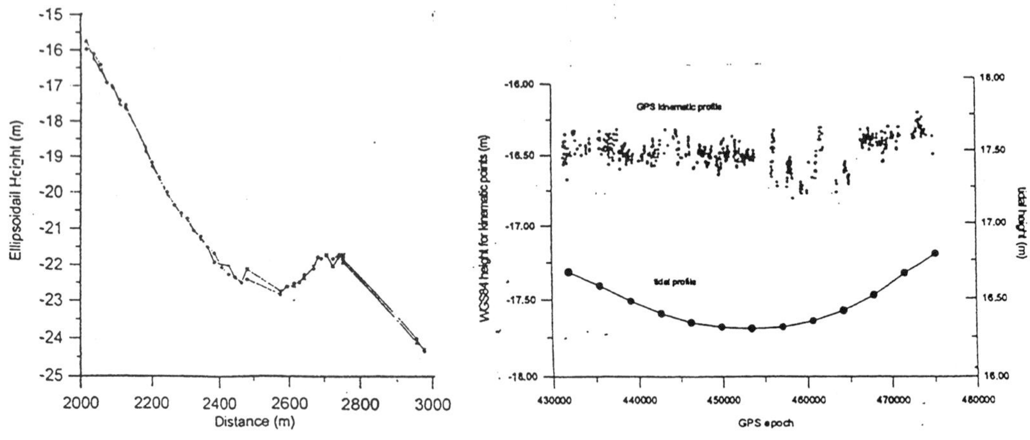

(a) (b)

Fig.4 – (a) comparison between the different GPS altimetric profiles made in 1995-96.
(b) GPS altimetric profile for 12 hours acquisition and corresponding tidal curve obtained from tide gauge.

The differences were scattered, but the mean value was centred around zero, so, considering the expected differences due to tidal effects, it can be hypothesised that the profile was probably made on grounding zone. To verify these preliminary results, in the following Antarctica campaign 1996-97, a kinematic continuous survey was carried out on a fixed station for 24 hours, approximately at the start of the profile (fig.4 a), where the ice appeared to be grounded. The GPS kinematic data were processed with the GPSurvey OTF software. The 12 hour altimetric GPS profile and the respective tidal curve are shown in fig.4 b. The GPS height shows a scattered solution, due to measurement errors, but of the same order of method precision (about 10 cm), while the tidal curve presents an amplitude of about 60 cm. It appears to confirm the previous results and the usefulness of GPS kinematic surveying to detect the separation between floating ice and grounded areas.

Conclusion

Processing of GPS data acquired on David Glacier-Drygalski Ice Tongue clearly shows that the vertical oscillation due to tidal effects and planimetric movements do not influence the result of static processing for planimetric co-ordinates determination or its precision when the ice movements are more than 50 cm per day. More attention needs to be given to the data set used, the length of acquisition and the time distribution, especially for height determination using static solutions The tests make it possible to determine that co-ordinates values sufficiently reliable to be used as comparisons may be obtained from the single base static solution with observation periods suitable for the base length, considering the position obtained as the one relative to the period half time. Previous experiments carried out on different Antarctica Ice tongues and Ice sheets have shown the validity of GPS fast static and kinematic techniques for the determination of ice undulation under tidal effects. The last experiments on the Hells Gate Ice Shelf seem to confirm the usefulness of kinematic GPS surveys to detect the separation between floating ice and grounding areas.

Acknowledgements

To Dr. O.Al-Bayari for collaboration in kinematic GPS data processing.
Research was carried out in the framework of a project on Glaciology and Paleoclimatology of the *Programma Nazionale di Ricerche in Antartide*, and financially supported by ENEA through a co-operation agreement with *Università degli Studi di Milano*, and by C.E.C. under grants ENV4-CT95-0124.

References

Bondesan A., Capra A., Gubellini A., Tison J.L. (1994): "On the use of Static GPS Measurements to record the tidal response of a small antarctic ice shelf (Hells Gate ice shelf- Victoria Land)". Geografia Fisica e Dinamica Quaternaria, 17, pp.123-129.

Capra A., Gubellini A., Radicioni F., Vittuari L. (1996): "Italian Geodetic activities in Antarctica". On "Italian Geophysical Observatories in Antarctica", edited by A.Meloni e A.Morelli, Editrice Compositori, Bologna

Capra A., Radicioni F., Vittuari L. (1996): Kinematic GPS profiles and navigation in Antarctica. Int. Archives of Photogrammetry and Remote sensing. 31, B1. ISPRS 1996 Comm. I, WG2, Vienna, 31-35.

Frezzotti M., Capra A., Vittuari L. (1997): Comparison between glacier ice velocities inferred from GPS and sequential satellite images. Presented at Antarctica and Global Change: interactions and impacts, Hobart, Tasmania, Australia, 13-18 July 1997. In press on Proceedings of the Congress..

Comparison of Altimetric and Ship borne Marine Gravity over Ice-free and Ice-covered polar Seas

O. B. Andersen, R. Forsberg, P. Knudsen,
Kort- og Matrikelstyrelsen DK-2400, Copenhagen NV, Denmark.

S. Laxon,
University College London, MSSL, Holmbury St. Mary, Great Britain.

D. McAdoo,
NOAA, Geosciences Lab, Silver Spring, MD 20910, USA.

Various recent marine altimetric gravity fields based on geodetic mission altimetry are compared with ship borne gravity in four regions around Greenland. This is done over different states of the ocean surface ranging from ice free open ocean to permanently ice covered ocean. Distinguishing between results obtained over these different states is important because, usually, the accuracy of the altimetric products decreases with increasing coverage of ice and proximity to coast.
Initial results have standard deviations with marine surveys of 4-3 mGal or better in the open and partly ice-covered ocean ranging up to 8 mGal in permanently ice-covered regions. At intermediate wavelengths (50-500 km), where valuable information about regional geologic structures is found, the various gravity fields agree remarkably well.

Introduction

Marine gravity field information may be obtained from satellite altimetry using different methods. Several new gravity fields based on the geodetic mission altimetry from the ERS-1 and GEOSAT geodetic mission have been released. These new gravity fields have excellent spatial resolution and provide an unprecedented view of tectonic fabric in the sea floor.
In ice-covered seas the standard altimeter data products are corrupted due to complex radar echoes confusing the onboard data processor. However, data from ice-covered seas may be re-processed using the full waveform records enabling altimetric marine gravity fields to be derived over polar seas (Laxon & McAdoo, 1994).
These polar marine gravity fields significantly extend the coverage of gravity fields

derived from satellite altimetry (e.g. Andersen & Knudsen, 1995, Sandwell, 1994) which are only assumed to be accurate over global, ice-free and partly ice-covered oceans.
In this investigation the focus will be on the most recent global gravity field derived from geodetic mission satellite altimetry having very high spatial resolution. Comparisons with marine gravity field observations in different ice-free oceans have recently been presented by, e.g. (Rapp and Yi, 1997, Behrend et al., 1996, and Andersen & Knudsen, 1997).

Gravity fields.

Four marine gravity fields were included in this investigation. These are the global marine gravity fields by Andersen and Knudsen, by Hwang et al. and by Sandwell and Smith. To investigate the importance of using altimetry over ice-covered ocean regions the polar marine gravity field derived from retracked altimetry by Laxon and McAdoo has also been included. Each marine gravity field, the methodology, its coverage and resolution is described below

Andersen & Knudsen Global Marine Gravity field.
Gravity anomalies were derived from crossover adjusted sea surface heights using Fast Fourier Transformation (FFT). Recently, the method used by Andersen and Knudsen (1995, 1996) has been fine tuned to obtain an optimal gravity field from a combination of GEOSAT and ERS-1 geodetic mission altimeter data. The Andersen and Knudsen gravity field is the only marine gravity field in this study that uses the direct sea surface height observations by the satellites. All other gravity fields use the slopes between neighbouring along track observations to derive the gravity anomalies. Therefore, this field is sensitive to sea surface variability. However, much effort was done in minimising this effect by interpolating the data onto regular grid using local collocation in which residual ocean variability was considered. The conversion of the heights into gravity anomalies was subsequently carried out using FFT. This gravity field has a spatial resolution of 1/16° in both latitude and longitude and covers marine regions between the 82° parallels. This comparison used the version (October 1996) derived using a remove restore technique with respect to the joint NASA GSFC and DMA EGM96 geoid model complete to degree and order 360 (Lemoine et al., 1996).

Laxon and McAdoo polar marine gravity field.
A new altimetric marine gravity field of the polar oceans covering the entire Southern Oceans and the Arctic Ocean up to 81.5N has been derived from the ERS-1 geodetic mission and ERS-1/2 tandem mission. The fields include poorly charted areas covered by both seasonal and persistent sea ice. To retrieve gravity measurements over ice-covered regions requires reprocessing of the full waveform telemetry data set to correct for errors in surface elevation measurements occurring over ice. The gravity anomalies are derived from along track sea surface slopes using FFT (McAdoo and Marks, 1992). The resolution of this gravity field is 1/30° in latitude and longitude and the version of April 1997 has been used.

Sandwell and Smith Global Marine Gravity field.

This field is version 7.2 of the global marine gravity field by Sandwell and Smith (1997). The marine gravity anomalies have been gridded on a Mercator projection giving it a variable latitude spacing and a constant longitude spacing of 1/30°. Therefore, the provided software program "interp_ship" was used to interpolate towards the positions of marine gravity observations. Version 7.2 used in this study was derived from the following data sources: All ERS-1 GM data (two 168-day cycles Ocean Product)all GEOSAT/GM data, stack of 62 repeat cycles of GEOSAT/ERM, and a stacks of 16 repeat cycles of ERS-1 35-day repeats. Applied methods are described in Sandwell (1984, 1992). This marine gravity field only covers regions up to the 72° parallels, and will consequently only be used in some comparisons.

Hwang et al. Global marine gravity field.

Global marine gravity anomalies have been determined from Seasat, GEOSAT, ERS-1 and TOPEX/POSEIDON altimeter data. The inverse Vening Meinesz formula with a 1D FFT method was used to compute gravity anomalies from gridded north-south and west-east geoid gradients, in a remove-restore procedure with the EGM96 gravity model as the reference field. Coverage of the marine gravity field is within the 82° parallels similar to the Andersen and Knudsen gravity field. The resolution of the global marine gravity field is 1/30° in both latitude and longitude. Applied method is described in Hwang et al. (1997). For this investigation the version gmgr9706 has been provided.

Ship Gravity Data:

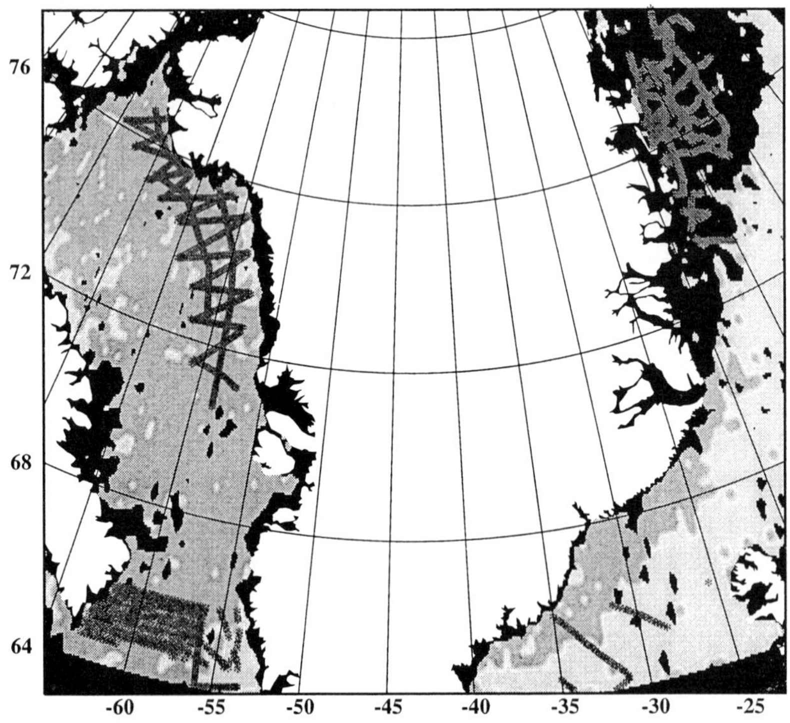

Figure 1. Location of the four surveys and their location in different states of the ocean. Dark grey is mostly icefree ocean. Light grey is partly icecovered ocean, grey is permanently icecovered ocean.

Marine gravity data from four different surveys around Greenland were selected. These marine gravity data were collected from various oil exploration surveys and are consequently confidential data.

The location of these four marine surveys is shown in figure 1. In this figure the altimetry from the Quick look ocean product (QLOPR data) of the ERS-1 Phase C mission was used to define the different states of the ocean. Areas with no or sparse observations are defined as permanently ice-covered regions. Areas with observations only during the arctic summer define the partly ice-covered regions. Similarly, areas with all or most observations define mostly ice-free regions.

The first survey having 777 observations is located to the Northeast of Greenland in permanently ice-covered areas between 73°N - 80°N and 10°W - 20°W. The second survey has 803 observations and is located to the west of Greenland. This is in the Baffin Bay between 70°N - 77°N and 55°W - 75°W which is a partly ice-covered region. The third survey is located to the Southwest of Greenland in the Davis Strait in a similar partly ice-covered region. This survey is located between 64°N - 66°N and 53°W - 63°W and has 649 observations. Compared with the second survey this survey has a much smaller gravity signal. Finally, the fourth survey is located south of the Denmark Strait in the Atlantic ocean. This is to the Southeast of Greenland between 64°N - 67°N and 25°W - 35°W. This survey has only 72 observations.

In the following these four surveys will be called the Northeastern, the Northwestern, the Southwestern and the Southeastern surveys. Schematically the states of the ocean in each surveys can be displayed like:

Northwest: Partly ice covered	Northeast: Permanently ice covered
Southwest: Partly ice covered	Southeast: Mostly ice free

The four areas have different magnitudes of the variations of the gravity field. The mean and standard deviations of the marine observations are displayed along with the minimum and maximum values in the parentheses. All values are in mGal.

Northwest: 6.9 ± 58.0 mGal (-95.9, 234.3)	Northeast: -9.4 ± 42.3 mGal (-85.0, 85.6)
Southwest: 19.3 ± 39.0 mGal (-77.5, 97.3)	Southeast: 37.2 ± 24.5 mGal (-3.5, 71.0)

Comparisons with marine data

Statistics of gravity differences for the four regions and four models are displayed below. These are given as mean offset calculated as observations minus interpolated gravity field. Similarly, standard deviations are given along with minimum and maximum differences in brackets.

Andersen and Knudsen global marine gravity field:

Northwest: -0.4 ± 5.7 mGal (-18.3, 30.3)	Northeast: -3.2 ± 17.4 mGal (-54.2, 48.3)
Southwest: 0.4 ± 4.5 mGal (-18.7, 17.5)	Southeast: 1.0 ± 2.9 mGal (-6.3, 6.3)

Sandwell and Smith global marine gravity field:

Northwest: Not covered	Northeast: Not covered
Southwest: -3.4 ± 5.22 mGal (-14.3, 16.1)	Southeast: 1.4 ± 3.8 mGal (-4.9, 8.5)

Hwang et al. global marine gravity field:

Northwest: -2.7 ± 18.0 mGal (-35.2, 94.4)	Northeast: -2.7 ± 17.8 mGal (-54.8, 59.9)
Southwest: 1.1 ± 4.5 mGal (-14.3, 16.1)	Southeast: 1.4 ± 2.7 mGal (-4.9, 8.5)

Laxon and McAdoo polar marine gravity field:

Northwest: -5.1 ± 8.2 mGal (-36.0, 22.1)	Northeast: -12.7 ± 7.9 mGal (-16.8 7.9)
Southwest: -3.4 ± 8.1 mGal (-30.0 27.6)	Southeast: -1.5 ± 6.4 mGal (-17.6, 9.1)

Discussion

Overall the comparisons show that the accuracy of the altimetric products in polar regions decreases with increasing coverage of ice and proximity to coast.

The standard altimetry does relatively poorly in permanently ice-covered regions due to lack of data having standard deviation around 18 mGal. In such regions the retracked products are extremely valuable (standard deviation around 8 mGal).

In all regions it appears that the retracked gravity field suffers from a bias that makes it less accurate than the global marine gravity field in ice-free and partly ice-covered regions. However, the polar marine gravity field is still very accurate, and has a relatively constant standard deviation around 8 mGal or better in any state of the ocean. All global marine gravity fields are very accurate in partly ice covered and ice-free regions. However, the Hwang et al global gravity field suffers from very large standard deviation in the Bay of Baffin compared with the gravity field by Andersen and Knudsen. This is probably caused by differences in the editing of data.

Compared with ship gravimetry the altimetric derived gravity field ranges from an accuracy of 4-5 mGal or better in mostly ice-free and partly ice-covered regions to 8 mGal or better in permanently ice covered regions.

Acknowledgement. The authors would like to thank scientists and organisations for providing gravity field observations and models for this comparison.

References

Andersen, O. B. and P. Knudsen, Global Marine Gravity Field from the ERS-1 and GEOSAT Geodetic Mission Altimetry, J. Geophys Res. In press, 1997.

Andersen, O. B., P. Knudsen & C.C. Tscherning, Investigation of methods for global gravity field recovery from dense ERS-1 geodetic mission altimetry. In: Global Gravity Field and its temporal variations. Ed.s: Rapp, Cazenave & Nerem, IAG Symposia no. 116, p. 218 - 226, Springer Verlag, Berlin Heidelberg, 1996.

Andersen, O. B., and P. Knudsen, Global gravity field from the ERS-1 geodetic mission, Earth Observation Quarterly, 47, ESRIN, 1-5, 1995.

Andersen, O. B., and P. Knudsen, Altimetric gravity field from the full ERS-1 geodetic mission, Proceedings EGS general meeting, Springer Verlag, Berlin Heidelberg, 1996.

Behrend, D. H. Denker and K. Schmidt, Digital gravity data sets for the Mediterranean Sea derived from available maps, Bull d.Inform. 78, BGI, Touluse Cedex, Fr., 32 - 41, 1996.

Hwang, C. , E.-C. Kao, and B. Parsons, 1997. Global marine gravity anomalies from Seasat, GEOSAT, ERS-1 and TOPEX/POSEIDON altimeter data, paper submitted GJI.

Laxon, S.W., and D. McAdoo, Satellite Altimetry. Science, 256, pp. 621-624, 1994.

Laxon S. W., and D. McAdoo, Polar Marine Gravity Fields from ERS1, Proceedings of the 3rd ERS symposium, Florence, 17-21 March, 1997.

Lemoine, F. G., et al., The development of the NASA GSFC and DMA joint geopotential model, in (J. Segawa, H. Fujimoto and S. Okubo (eds) Gravity, Geoid and Marine Geodesy, IAG Symposia 117, Springer-Verlag, Berlin, 1997.

McAdoo and Laxon, 1997, Antarctic Tectonics : Constraints from a new ERS-1 Satellite Marine Gravity Field, Science, 276 , 5312, pp. 556-561, 1997.

McAdoo, D.C. & Laxon, S.W. Marine Gravity from GEOSAT and ERS-1 Altimetry in the Weddell Sea. in Weddell Sea Tectonics and Gondwana Breakup, Eds. B.C. Storey, E.C. King and R.A. Livermore, The Geological Society, London, 155-164, 1996.

Rapp, R. H., Yi Y. Role of ocean variability and sea surface topography in the recovery of the mean sea surface and gravity anomalies from satellite altimeter data. J. of Geod., 71(10), p. 617-629, 1997.

Sandwell, D. T., Antarctic marine gravity field from high-density satellite altimetry, Geophys. J. Int., v. 109, p. 437-448, 1992.

Sandwell, D. T., A detailed view of the South Pacific from satellite altimetry, J. Geophys. Res., v. 89, p. 1089-1104, 1984.

Sandwell, D. T and W. H. F. Smith. Marine gravity anomaly from Geosat and ERS-1 Satellite Altimetry, J. Gephys Res. 102, p. 10039-10054, 1997.

Italian geodetic network as reference frame for geodynamic purposes (Terra Nova Bay - Victoria Land - ANTARCTICA)

Capra A. *, Radicioni F., Vittuari L.***

***DISTART - University of Bologna**
Viale Risorgimento n.2, 40136 - BOLOGNA - ITALY
**** Istituto di Ingegneria Ambientale - University of Perugia**
Via Duranti 1, 06125 - PERUGIA - ITALY

Introduction

An Italian geodetic network has been monumented and surveyed using a GPS satellite technique several times since 1988. The network can be divided into a general network, or reference frame, and a detailed network, smaller than the general one, to control the crustal deformations of the Mt. Melbourne volcano area. Some experiments were carried out to study the effects of some possible causes of variation in the GPS reference system, interesting for geodynamic purposes. In particular, tests were made on the use of precise ephemeris instead of broadcasting and different GPS data processing software, Bernese v.4.0 and Geotracer v.2.25. In 1995-96 absolute positioning was made for the network reference station, with the satellite Doris system, and the ITRF co-ordinates were obtained to compare them with the same ITRF co-ordinates obtained using GPS international campaigns.

Geodetic network

The Italian geodetic network (fig.1) has been monumented and surveyed using GPS satellite technique since the 1988-89 expedition. The network was established in order to supply a reference system for several studies: crustal deformations in the Mt. Melbourne area, global geodynamics, glacier movement monitoring, mapping, etc.

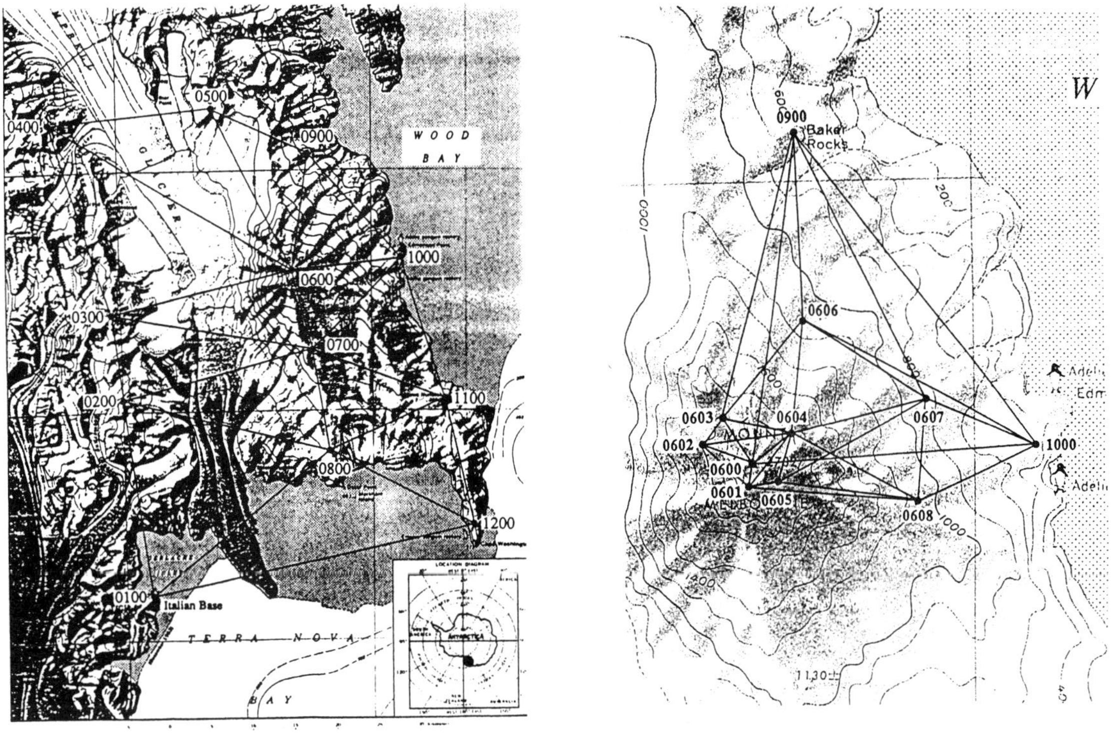

Fig.1 – Geodetic network and detailed network for crustal deformation control.

The network consists of 20 stations covering an area of 2000 km^2 and can be divided into a general network of 12 stations, with an average baseline length of about 30 km, and a detailed network of 8 stations with an average baseline length of 7 km, located inside the general net and distributed around the volcanic cone of Mt. Melbourne. The frame network was completely surveyed twice during the 1990-91 and 1993-94 campaigns and the deformation control network was surveyed three time during the 1990-91, 1993-94 and 1995-96 campaigns. The GPS receivers used were the geodetic L1/L2 Trimble model 4000 STD and SSE. The network reference station (0100) co-ordinates were determined with high precision within the ITRF system with data processing by NOAA, as part of a co-operative aerophotogrammetric GPS assisted test. The results of adjustments did not reveal any appreciable variations in the coordinates in relation to the GPS method accuracy (Al Bayari et al., 1996 b, Capra et al., 1996,). Studies and research were also made on GPS signals and data processing to determine the effects of environmental conditions and different software (Capra et al., 1995, Al Bayari et al., 1996 a).

Precise and broadcast ephemeris effects

Precise ephemeris generally gives better results than broadcast ephemeris. Unfortunately, there are few, unevenly distributed permanent GPS tracking stations in the Southern hemisphere and in Antarctica. So tests were carried out to assess the improvement in accuracy provided by precise ephemeris instead of broadcasting. The deformation control network was surveyed in 1995-96 campaign and the same data were processed using broadcast and precise ephemeris with Geotracer software v.2.25.

Tab. 1 – Network adjustment solutions with broadcast and precise ephemeris. Lat. and Long. are in DMS and the height in m. Seven parameter Helmert Transformation residuals for the geocentrical components of the two solutions are shown (RMS: 12 mm)

Station		Broadcast ephemeris (std. dev. in mm)		Precise ephemeris (std. dev. in mm)		Helm.Trans. residuals (mm)
	Lat	74 41 55.6965		74 41 55.69441		
0100	Lon	164 06 10.57020		164 06 10.58752		
	h	73.054		73.085		
	Lat	74 23 41.60117	(3.7)	74 23 41.60120	(4.3)	-1.5
0300	Lon	163 58 14.51610	(4.6)	163 58 14.51573	(5.2)	-2.3
	h	1972.529	(5.0)	1972.575	(5.7)	-0.6
	Lat	74 21 00.00033	(3.1)	74 21 00.00029	(3.6)	-2.9
0600	Lon	164 41 26.31569	(3.8)	164 41 26.31607	(4.4)	-7.7
	h	2671.942	(4.4)	2672.005	(5.0)	-24.2
	Lat	74 13 55.42188	(4.8)	74 13 55.42185	(5.3)	-1.3
0900	Lon	164 44 00.87181	(6.4)	164 44 00.87044	(7.3)	6.0
	h	602.325	(6.1)	602.347	(6.1)	24.0
	Lat	74 20 57.66700	(3.9)	74 20 57.66718	(4.3)	5.7
1000	Lon	165 05 09.71748	(5.0)	165 05 09.71621	(5.5)	6.2
	h	211.317	(5.4)	211.317	(5.8)	31.7
	Lat	74 37 42.00658	(5.0)	74 37 42.00680	(5.8)	5.6
1200	Lon	165 25 30.39116	(6.4)	165 25 30.39085	(7.5)	0.1
	h	363.298	(8.2)	363.315	(9.6)	-10.2
	Lat	74 21 26.42150	(3.2)	74 21 26.42153	(3.7)	-0.6
0601	Lon	164 42 01.74517	(4.1)	164 42 01.74463	(4.7)	1.3
	h	2589.251	(4.5)	2589.312	(5.0)	-23.2
	Lat	74 20 40.71341	(3.3)	74 20 40.71341	(4.0)	-1.5
0602	Lon	164 36 30.09969	(3.8)	164 36 30.09958	(4.4)	-4.3
	h	1975.742	(4.5)	1975.797	(5.3)	-15.1
	Lat	74 20 03.05853	(3.6)	74 20 03.05859	(4.0)	1.2
0603	Lon	164 38 21.25562	(4.0)	164 38 21.25520	(4.8)	-1.1
	h	1924.936	(4.7)	1924.984	(5.4)	-7.8
	Lat	74 20 16.17871	(3.4)	74 20 16.17869	(3.9)	-2.0
0604	Lon	164 44 28.58446	(4.0)	164 44 28.58388	(4.6)	0.3
	h	1935.340	(4.6)	1935.389	(5.3)	-10.6
	Lat	74 21 13.24709	(3.8)	74 21 13.24720	(4.3)	1.3
0605	Lon	164 44 50.74075	(4.8)	164 44 50.74032	(5.6)	-0.2
	h	2197.136	(5.3)	2197.193	(6.1)	-20.2
	Lat	74 17 26.75338	(4.1)	74 17 26.75339	(4.3)	-0.7
0606	Lon	164 42 22.80297	(4.4)	164 42 22.80204	(4.9)	3.5
	h	984.389	(5.0)	984.413	(5.4)	18.4
	Lat	74 19 12.01626	(3.9)	74 19 12.01631	(4.4)	1.2
0607	Lon	164 58 02.84641	(4.5)	164 58 02.84573	(5.2)	0.6
	h	664.638	(5.1)	664.656	(5.9)	17.9
	Lat	74 21 39.95080	(3.6)	74 21 39.95078	(4.1)	-1.3
0608	Lon	164 50 18.35508	(4.4)	164 50 18.35433	(5.0)	2.2
	h	1191.735	(4.9)	1191.772	(5.6)	-1.3

To make the results easier to interpret, all the co-ordinates differences presented in this paper were transformed into metric values, adopting the values of 30.9 m for 1” latitude and 8.4 m for 1” longitude computed in this area. The comparison revealed that the coordinates std. dev. are of approximately the same size, even if the values for precise ephemeris are greater. The station co-ordinate values are significantly different of 2-3 cm for latitude and longitude and from 3 to 7 cm for the height, but a relevant systematic effect is present, probably due to reference system variations, as showed by the Helmert transformation residuals of the two solutions (tab.1). After the transformation, there is still a small difference, close to the accuracy of the solution, in the planimetric components, while there is an appreciable difference in vertical component. The greater difference in height (very close to the Z component at this latitude) is probably due to the fact that in Antarctica the satellites are relatively low above the horizon, so the variation in the orbit affects the overall height value of the station co-ordinates. In previous experiments significant differences for the co-ordinate values of network adjustment were obtained with different software (Capra et al., 1995). Geotracer software guaranteed a sufficiently stable solution, but it was decided to carry out a new test. During the 1996-97 campaign, measurements were taken on some stations of a sub-network (stations 0100, 0300, 0600 and 1200) with sessions lasting about 2 hours. The data processing and the network adjustment were carried out with the Geotracer and Bernese software and broadcast and precise ephemeris. The solutions adopted were Lc for Geotracer and L3 for Bernese, together with a standard tropospheric model. The results confirmed a small but significant difference between the broadcast solutions. This difference is probably also due to the fact that the Bernese software includes a clock module in case of broadcast ephemeris. A difference of about the same value was found with the precise ephemeris solutions with both software, whereas with Bernese there was a similar value for broadcast and precise ephemeris. The effect of using precise instead of broadcast ephemeris seems to have exactly the same effect with the two software solutions. The st. dev. of solutions are also presented, the values demonstrate the good quality of the results, although the parameters are generally under-estimated.

Tab.2 Comparison between the network solutions obtained with the Geotracer and Bernese software in 1996-97 measurements with broadcast and precise ephemeris (mm). (In brackets: std dev.).

		96-97 Geotracer (Broadcast eph)	Geotracer-Bernese (Broadcast eph)	Bernese Broadcast-Precise	Geotracer-Bernese (Precise eph)
0300	Lat	74 23 41.60166	8.8 (4.8,1.1)	2.3 (1.1,1.1)	11.2 (3.5, 1.1)
	Lon	163 58 14.51384	39.3 (5.3,1.4)	1.2 (1.4,1.5)	34.3 (4.6, 1.5)
	h	1972.502	18 (6.7,2.4)	2.9 (2.4,2.3)	19.0 (4.6, 2.3)
0600	Lat	74 21 00.00091	1.1 (3.3,0.7)	3.6 (0.7,0.7)	6.9 (3.7, 0.7)
	Lon	164 41 26.31504	-10.6 (3.3,0.9)	2.4 (0.9, 0.9)	-12.5 (4.7,0.9)
	h	2671.934	-25.4 (4.0,1.5)	5.4 (1.5,1.3)	-12.0 (5.0,1.3)
1200	Lat	74 37 42.00652	0.2 (1.6,1.5)	4.1 (1.5,0.6)	8.1 (3.3, 0.6)
	Lon	165 25 30.38866	-9.7 (2.1,1.5)	-5.5 (1.5,0.8)	-11.6 (4.0,0.8)
	h	363.301	-35 (2.8,2.5)	-4.8 (2.5,1.1)	-31.2 (4.5,1.1)

Doris measurements

In 1995-96, the absolute positioning of the network reference station was established with the Doris system. The co-ordinates of the 0100 station, obtained with a semi-dynamic method in the ITRS (International Terrestrial Reference System) of IERS (ITRF92), were determined. The Doris measurements were made between 15 November 1995 and 29 December 1995 with the Spot2 and Spot3 satellites. Data for five days were processed and the accuracy found was 3-4 cm in absolute positioning. In table 3 the absolute co-ordinates with the Doris measurements are shown and are compared with the ITRF co-ordinates obtained with GPS.

Tab.3 – Doris and WGS84 absolute co-ordinates of station 0100 in ITRF system.

	0100 NOOA co-ordinates	0100 Doris ITRF 92 co-ordinates	Doris-NOOA (mm)
Lat	74 41 55.6965	74 41 55.69441	-64.5
0100 Lon	164 06 10.57020	164 06 10.58752	145.5
h	73.054	73.085	31

The differences are - 6.45 cm in latitude, 14.55 cm in longitude and 3.1 cm in height. The difference between the ITRF co-ordinates is very small and it appears particularly interesting with respect to the ITRF co-ordinate absolute stability, obtained with two completely different satellite systems. Deeper analysis will be made, also using the next 0100 station co-ordinates, from the SCAR GPS Epoch 1995, 96 campaign solutions. The adjusted co-ordinates for the sub-networks, mentioned above, were compared with the same networks surveyed in the 1995-96 and 1996-97 campaigns, using different reference co-ordinates. As expected, there was little difference between the two adjusted co-ordinate values, always less than the method accuracy, taking into consideration a constant shift, corresponding to the difference in the 0100 station co-ordinate values. This shows the absence of network distortion even with an average baseline value of 40 km.

Tab 4 – Differences in adjusted co-ordinate values by NOAA and by Doris in the 0100 reference station network co-ordinates. Lat. and Long. are in DMS, height is in m.

	95-96 co-ordinates from 0100 NOOA	Doris - NOOA 95-96 (mm) (std. dev.in mm)	96-97 co-ordinates from 0100 NOOA	Doris - NOOA 96-97 (mm) (std. dev.in mm)
Lat	74 23 41.60120	-60.6 (7.5,7.5)	74 23 41.60166	-63 (4.8, 4.1)
0300 Lon	163 58 14.51716	136 (10.2,9.8)	163 58 14.51384	142.2 (5.3,4.7)
h	1972.581	24.0 (10.0,9.9)	1972.502	33.0 (6.7,5.6)
Lat	74 21 00.00047	-69.5 (7.8,7.9)	74 21 00.00091	-65.5 (3.3,2.6)
0600 Lon	164 41 26.31507	149.8 (10.1,9.9)	164 41 26.31504	140.8 (3.3,2.6)
h	2672.000	45.0 (11.4,11.2)	2671.934	32.0 (4.0, 3.1)
Lat	74 37 42.00687	-73.2 (30.1,26.2)	74 37 42.00652	-67.7 (1.6, 1.3)
1200 Lon	165 25 30.39033	148.9 (34.1,32.1)	165 25 30.38866	143.2 (2.1,1.6)
h	363.350	42.0 (45.5,38.2)	363.301	33.0 (2.8, 2.2)

Conclusions

The results revealed differences between the values obtained with precise and broadcast ephemeris, using the Geotracer and Bernese software, as found in previous experiments with Antarctica GPS data. Even if the use of precise ephemeris in the last 1996-97 campaign seems to give smaller variations between the different software solutions, the results require a deeper analysis of GPS data processing to assess the effects of precise ephemeris on adjusted co-ordinates values. At the moment the data are also processed with broadcast ephemeris for two reasons: the first is that the network-adjusted station comparisons started in 1990-91 and at that time, and for the 1993-94 campaigns, acceptable precise ephemeris values were not available; the second one is to verify whether the actual precise ephemeris could, at Antarctica latitudes, provide a really improvement in network adjustment results. Today it is common opinion that the difference in precision between broadcast and precise ephemeris is relatively small. As a result of a project for a permanent GPS tracking station at 0100, started in the 1996-97 expedition, the station will be active in February 1998 (Capra et al., 1997). This station will be included in the SCAR GPS permanent station network in Antarctica for geodynamics studies and will be included in the IGS network for precise ephemeris determination. This is particularly important as it is located in a continent where there are still only 4 permanent GPS tracking stations. The computation of precise ephemeris through a larger number of evenly distributed permanent GPS stations in Antarctica, the values from the SCAR GPS Epoch 1995-96 campaign and the use of the 0100 GPS tracking station could provide more stable reference system co-ordinates and thus make it possible to determine which algorithms and software are able to give the most accurate and stable solutions. This is fundamental for crustal deformation control in the Mt. Melbourne area and for regional and global geodynamics studies.

Acknowledgements

Research was carried out in the framework of a project on Geophysics and Geodesy of the *Programma Nazionale di Ricerche in Antartide*, and financially supported by ENEA, Italy. Thanks to Dr.Francesco Mancini for scientific collaboration.

References

Al Bayari O., Capra A., Unguendoli M.,Vittuari L.(1996 a). "Actual and simulated results in GPS networks". Reports on Surveying and Geodesy, editor M.Unguendoli, Pubblicazione del DISTART, Università di Bologna, pp.51-63, Edizioni Nautilus, Bologna.

Al Bayari O., Capra A.,Radicioni F., Vittuari L.(1996 b). "GPS network for the deformation control in the Mount Melbourne area (Antarctica) : preliminary results of third measurements campaign". Reports on Surveying and Geodesy, editor M.Unguendoli, Pubblicazione del DISTART, Università di Bologna, pp.64-72, Edizioni Nautilus, Bologna.

Capra A., Gubellini A., Vittuari L.(1995)."Reti GPS per il controllo di deformazioni : analisi dei risultati ottenuti con differenti programmi di elaborazione dati".Bollettino della SIFET, n.4.

Capra A., Gubellini A., Radicioni F., Vittuari L.(1996). "Italian Geodetic activities in Antarctica". On "Italian Geophysical Observatories in Antarctica", edited by A.Meloni e A.Morelli, Editrice Compositori, Bologna.

Capra A., Gandolfi S. (1997). Italian GPS permanent station in Antarctica (Terra Nova Bay-Victoria Land). Proceedings of Workshop "Standard e Servizi della Rete GPS", 10-11 April 1997, Matera, Italy.

Ice-Ocean-Solid Earth Interactions in Dronning Maud Land / Antarctica: A Geodetic Approach to Solve Open Questions

R. Dietrich, R. Dach, W. Korth, R. Metzig and J. Perlt
Technische Universität Dresden
Institut für Planetare Geodäsie
Germany

R. Hartmann, W. Winzer
Jena-Optronik GmbH
Germany

1 Introduction

The coastal regions in Antarctica are sensitive parts for mass balance studies of the Antarctic ice sheet. On the one hand they have a maximum snow accumulation compared to the relatively dry inland regions (see e.g. Giovinetto et al., 1990), on the other hand the grounding line, where the continental ice starts to swim up, is the boundary for mass balance studies. The ice flow across the grounding line is the main output quantity of the Antarctic ice sheet. Therefore, it is easy to conclude, that the determination of the location of the grounding line is an important condition for mass balance studies.

The fact, that the floating ice shelves participate in the vertical tidal motions of the ocean, whereas the grounded ice parts do not, can be used to separate these two ice bodies from each other with the help of geodetic observations. We have carried out our research in the coastal region of Dronning Maud Land where we have ice-free parts (e. g. Schirmacher Oasis) as well as the Nivlisen Ice Shelf north of the grounded inland ice (Fig. 1).

2 Data and Results

During the last years several observations have been carried out to determine ice surface velocities along geodetic traverses (see Korth and Dietrich, 1996), which cover both inland ice and the ice shelves. Ocean tides were determined from a one year pressure gauge record in an epishelf lake north of the oasis (Dietrich et al., 1995, 1997).

In 1995 GPS observations were carried out at different locations of the ice shelf. The base station was established on bedrock in Schirmacher Oasis. Having already collected one day of data it was possible to detect the shelf ice motion (Fig. 2) performing a kinematic GPS data analysis. This concerns both the horizontal motion (mainly in northern direction) and the tidal induced vertical motion. Predicted height variations based on the tide gauge observations fit very well to the actually observed height changes. Therefore, kinematic GPS is an useful geodetic tool to decide whether the observer is located on floating or on grounded ice.

The application of interferometric SAR is a new and very efficient method for grounding line location (see e.g. Goldstein et al., 1993 and Rignot, 1996). During the ERS-1 / ERS-2 tandem mission in 1996 the first ERS radar scenes of the working area could be recorded at the Japanese Antarctic Station Syowa (see Doi et al., 1997). As one example we show the analysis of tandem scenes from May 20th resp. 21st, 1996 (see Fig. 1). The amplitude image (Fig. 3, top) shows very clearly the Schirmacher Oasis (left side) and flow lines of the ice. A large blue ice area (dark part of the image) can be separated from the accumulation areas (snow surface in light).

During the Antarctic winter time catabatic winds use to blow and, thus, change the snow surface. In spite of that fact, there is a sufficient coherence also in these areas (Fig. 3, centre). This confirms the fact, that the radar signal is not only reflected from the surface, but also from upper snow and ice layers. In the interferogram (Fig. 3, bottom) good fringes appear over the whole image. Because the orbit positions of the satellites are very close to each other (normal component of baseline about 9 m), the dominating source of the fringes are surface displacements. The vertical displacements of the ice shelf due to ocean tides are especially interesting: The grounding line appears in the interferogram as a fringe belt. The expected tidal displacement, using the known local ocean tide constituents, agrees very well to the observed one (Table 1). If one converts the height difference of about 16 cm into fringe numbers, one will expect about 5 fringes at the grounding line, which fits very well to Fig. 3. One should, of course, take into account the additional horizontal displacement, which may – depending on flow direction – slightly increase or decrease the number of fringes.

In this way SAR interferometry provides a very good possibility to locate the grounding line and even to study the displacement processes at the grounding line in more detail, e. g. the extension of the deformation area.

3 Conclusions

The interferometric SAR data can of course also be used to determine horizontal surface motion. Knowing additionally the ice thickness at the grounding line one can compute the ice output of the Antarctic area under investigation as an important component of the ice mass balance.

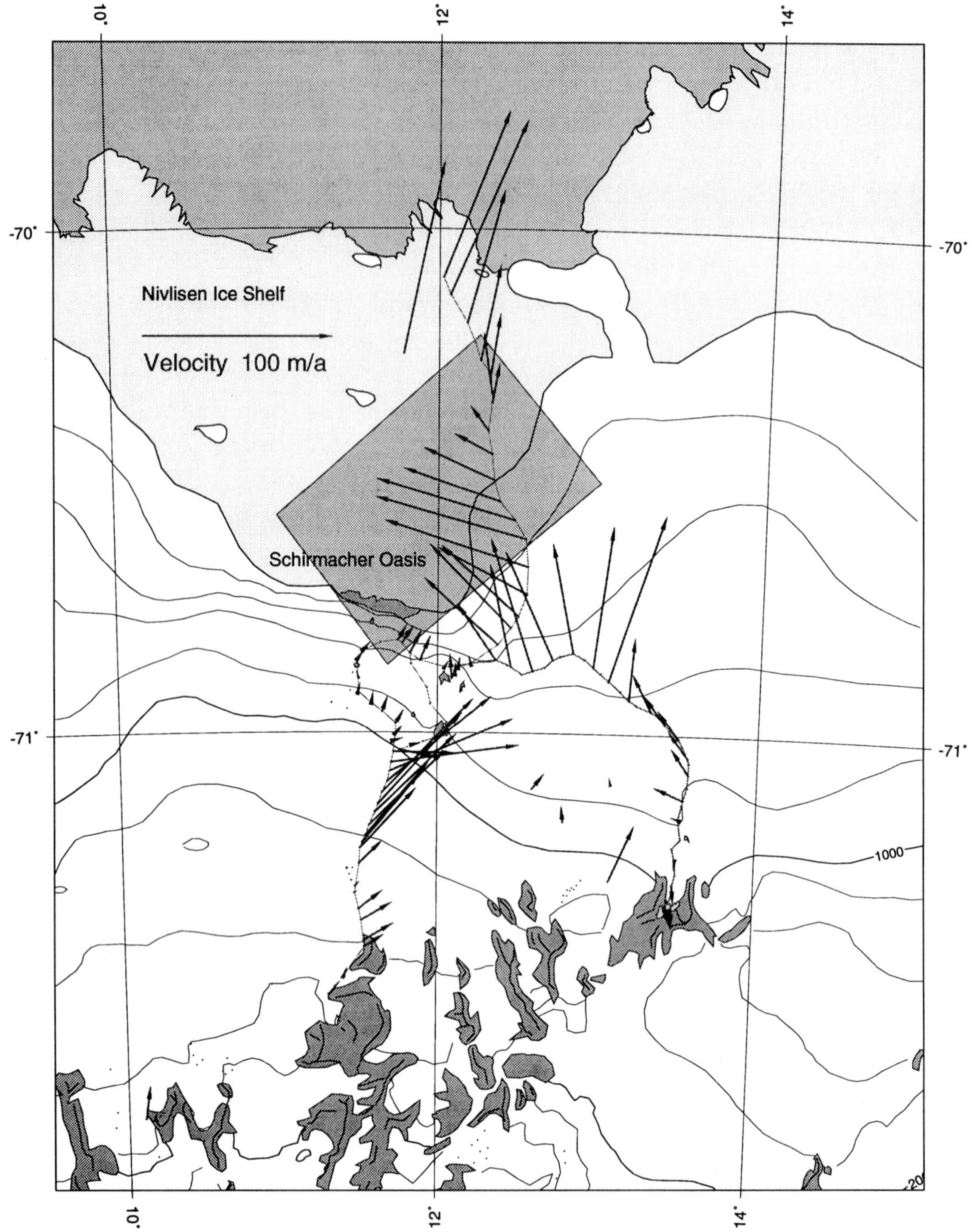

Figure 1: Region around Schirmacher Oasis with ice surface velocities. The position of the SAR images below is shaded.

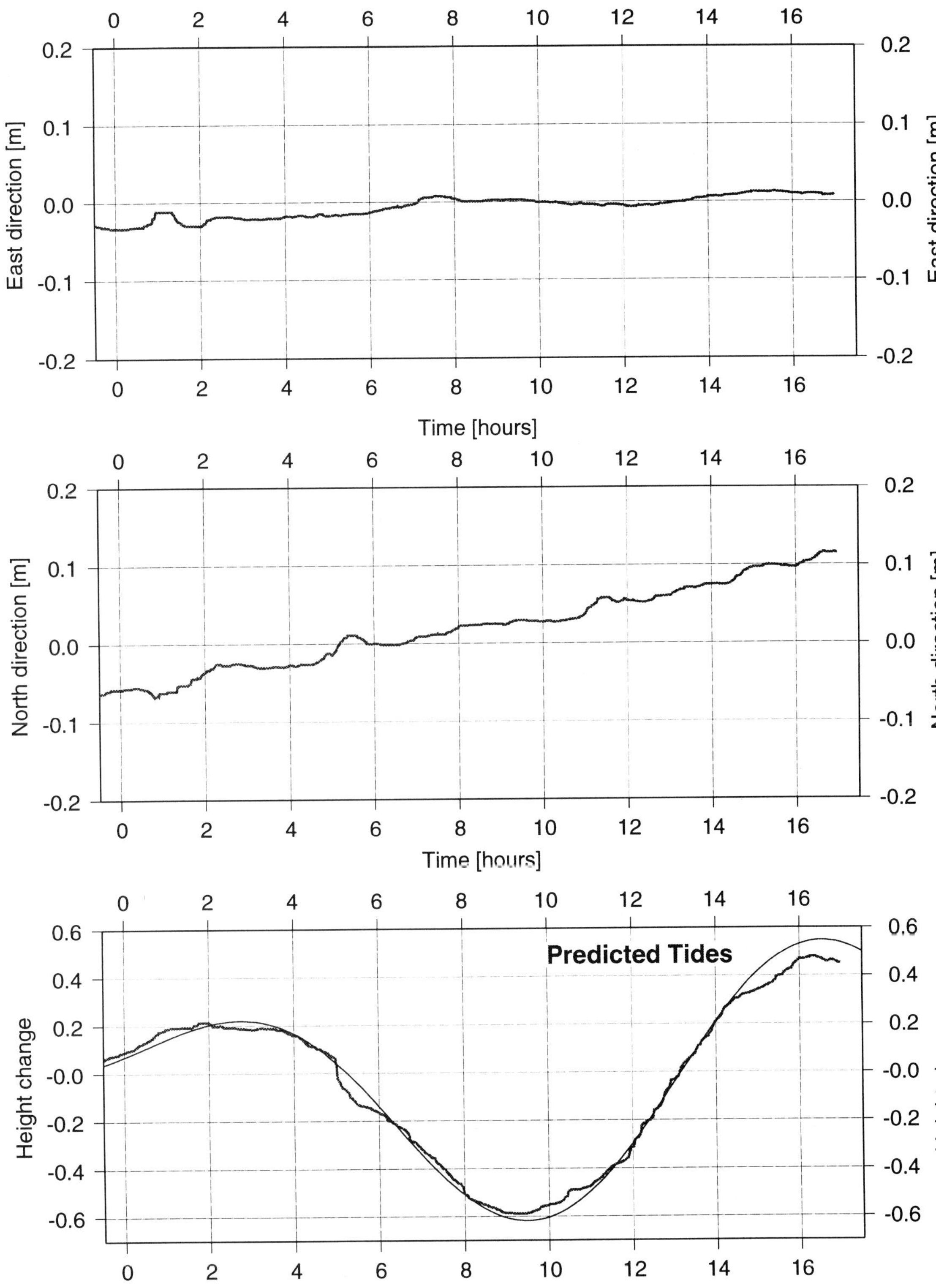

Figure 2: Kinematic GPS-Positioning at the ice shelf with tidal signal in the height component

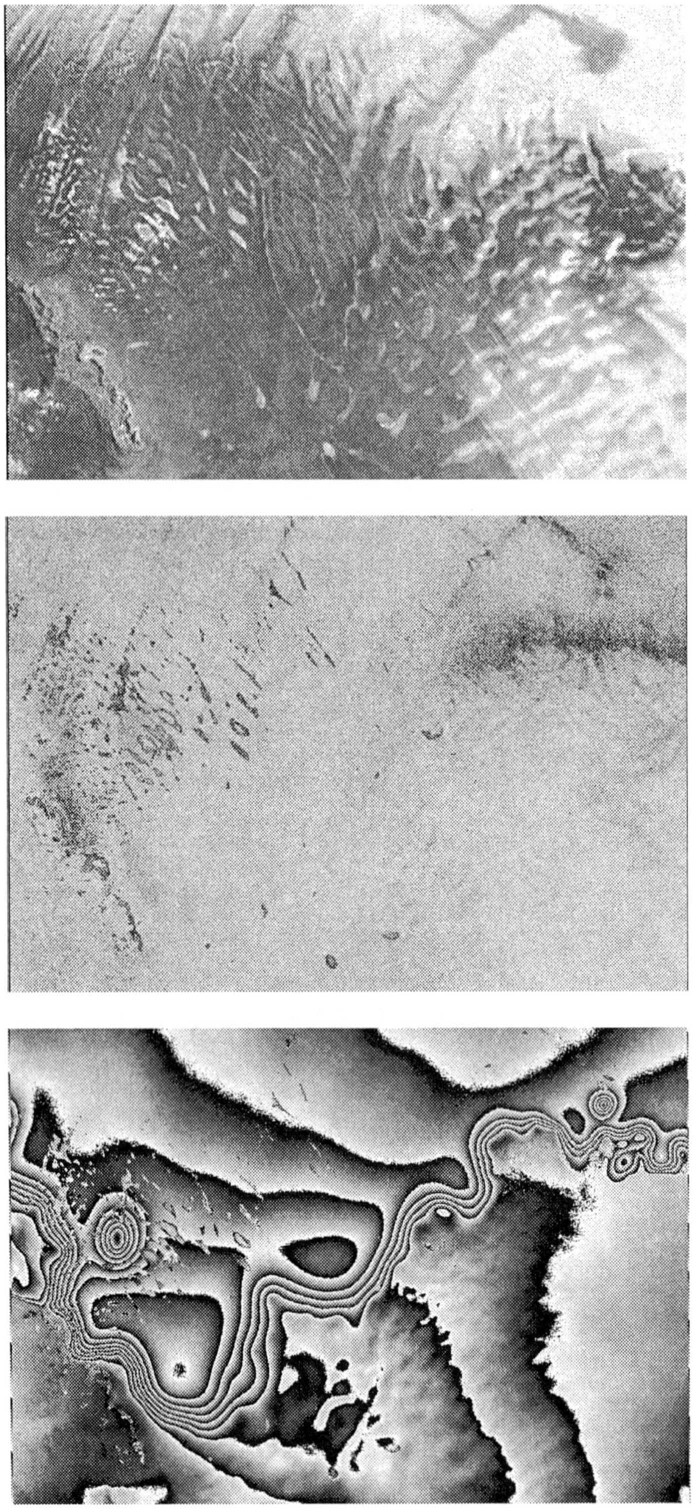

Figure 3: ERS-1 radar amplitude image (top); phase coherence image (centre); flattened interferogram combining orbits 25354 (ERS-1) and 5681 (ERS-2) (bottom)

Table 1: Tidal signal in the region of the grounding line (ERS-1/ERS-2 tandem mission)

Satellite	ERS-1	ERS-2
Epoch Predicted tidal signal at epoch of observation	20.05.1996 23:58 $h_1 = -85.3$cm	21.05.1996 23:58 $h_2 = -69.2$cm
Height difference Conversion into fringe numbers (pure tidal displacement)	$\Delta h = h_2 - h_1 = 16.1$cm $N = cos\,\eta \frac{\Delta h}{\lambda/2} \approx 5.4$ λ: wavelength ($\lambda = 5.6$cm) η: incidence angle ($\eta \approx 20°$)	

Acknowledgment

The authors would like to thank K. Shibuya and K. Doi (NIPR Tokyo, Japan) for the support in recording ERS-1/ERS-2 SAR tandem data at Syowa station.
This research was carried out with the help of BMBF grants 03PL016B and 03PL013.
SLCIs copyright by European Space Agency 1997.

References

Dietrich, R., G. Liebsch, H.-J. Dittfeld, and G. Noack (1995): Ocean tide and attempt of Earth tide recordings at Schirmacher Oasis / Dronning Maud Land (Antarctica). In *Proceedings of 12th Int. Symposium on Earth Tides,* Science Press, Beijing, New York.

Dietrich, R., R. Dach, W. Korth, J. Polzin, and M. Scheinert (1997): Gravimetric Earth Tide Observations in Dronning Maud Land / Antarctica to Verify Ocean Tidal Loading. In *Proceedings of 13th Int. Symposium on Earth Tides,* July 22–25, Brussels 1997. (in press)

Doi, K., K. Shibuya, T. Ozawa, and H. Nakagawa (1997): *SAR Data Acquisitions at Syowa Station in Antarctica.* This volume.

Giovinetto, M.B., N.M. Waters, and C.R. Bentley (1990): *Depending of Antarctic Surface Mass Balance on Temperature, Elevation, and Distance to Open Ocean.* Journal of Geophys. Research, 95(D4), 3517–3531.

Goldstein, R.M., H. Engelhard, B. Kamb and R.M. Frolich (1993): *Satellite Radar Interferometry for Monitoring Ice Sheet Motion: Application to an Antarctic Ice Stream.* Science, 262(5139), 1525–1530.

Korth, W. and R. Dietrich (1996): *Ergebnisse geodätischer Arbeiten im Gebiet der Schirmacheroase / Antarktika 1988–1993.* DGK Reihe B, Heft 301, München.

Rignot, E. (1996): *Tidal motion, ice velocity and melt rate of Petermann Gletscher, Greenland, measured from radar interferometry.* Journal of Glaciology, 42(142), 476–485.

The Contribution of the Geodetic Observatory O'Higgins to the Realization of a Geodetic Reference Frame in Antarctica

J. Ihde[1], A. Reinhold[1], H. Seeger[2], G. Soltau[1] , V. Thorandt[1], R. Wojdziak[1]
[1]Bundesamt für Kartographie und Geodäsie* , Außenstelle Leipzig;
Karl-Rothe-Straße 10-14, D-04105 Leipzig, Germany
[2]Bundesamt für Kartographie und Geodäsie*, Richard-Strauss-Allee 11,
D-60598 Frankfurt am Main, Germany

On the development of O'Higgins

Since 1948 O'Higgins is a permanently manned Antarctic military base managed by the Chilean Army, located near the northern end of the Antarctic Peninsula on a small offshore island. At the end of the eighties the Federal Republic of Germany developed a combined SAR receiver and geodetic VLBI antenna to be placed in the neighbourhood of the Weddell Sea to collect radar data of the European Remote Sensing Satellites (ERS). The project was managed by the Bundesamt für Kartographie und Geodäsie (BKG), the German Aerospace Research (DLR) and the German Institute for Polar and Marine Research (Alfred-Wegener-Institute) under BKG's leadership.

The location of the receiver/radiotelescope was selected to fit both ERS and VLBI requirements (stable foundation on bedrock, temporarily free of ice). An already existing station with some logistics was desirable. By a kind agreement of the Chilean Government the installation of the scientific observatory at O'Higgins and the hosting the German crew during periodical visits, could be realized.

The first SAR image from ERS-1 was received successfully on October 9, 1991, the first VLBI experiments were performed during January/February 1992. The SAR receiver system is operated by DLR, which in addition is responsible for the management and logistics of the German part of O'Higgins. The responsibility for all geodetic observations and the care and integration of further geodetic equipment is in the hands of BKG (Reinhold, et al. 1996). Both DLR and BKG organize combined sessions (mostly) twice a year lasting for up to eight weeks each. From the beginning of the geodetic activities at O'Higgins BKG contributed also to GPS-campaigns, organized by the Scientific Committee for Antarctic Research (SCAR).

---

* formerly Institut für Angewandte Geodäsie (IfAG)

Since January 1992 the VLBI equipment at O'Higgins participated in internationally scheduled VLBI experiments. Some results will be presented within this paper. Since February 1995 O'Higgins is involved in the International GPS Service for Geodynamics (IGS) with a Turbo Rogue permanently operating GPS receiver. At the same time an underwater pressure tide gauge has been installed. In order to be able to relate the different observing systems to each other a precise geodetic control network has been established.

The VLBI equipment

The VLBI equipment is the most important geodetic tool of the station as VLBI is the only method for precisely connecting terrestrial stations with the celestial reference frame and for estimating all parameters of Earth with respect to this frame. The construction of the VLBI-antenna is able to resist wind speeds up to 300 km/h, the surface of the reflector and the subreflector can be heated to keep the system operationable even during heavy snowing.

Icing of the antenna is excluded by a special layer. The vertical axis of the telescope is conventionally set parallel to the local vertical. The second axis is tilted against the vertical axis by 45°, and the collimation axis is again tilted against the second axis by an angle of 45°. Keeping the vertical axis fixed the collimation axis is describing a cone when the second axis is rotated (Figure 1).

The radiation from the extragalactic radio sources (quasars) enters the feedhorn and is observed in the operational frequency bands from 8.0 to 8.6 GHz (X-Band) and from 2.0 to 2.3 GHz (S-Band) and gets separated in different conductors. The necessary electronic amplifiers are cooled down to a minimum temperature of 20 K by liquid Helium. The received signal stream is amplified by about 30 dB and transformed to a more convenient frequency region by mixing with intermediate frequencies in the range of 0 to 500 MHz.

The local oscillator of the receiver, the different converters, sampler and formatter are all phase-locked to the stable 5 MHz frequency of an H-maser. The electronics for VLBI (data acquisition rack, recorder, time and frequency rack, H-maser, and the controlling computer) are placed in a container. The time and frequency system consists of a Cesium standard, H-Maser, and a GPS time receiver. The VLBI data are stored on magnetic tapes.

The following comprises some parameters of the antenna:

– diameter of cassegrain radio telescope antenna (main reflector)	9 m
– diameter of subreflector	1,38 m
– focal length	3,6 m

As in other geodetic VLBI equipment the O'Higgins VLBI system is a Mark III type which operates simultaneously in S- and X-band subdivided into 28 separated bands of 2 MHz width each. The frequency reduction to 2 or 4 MHz and the mixing of the originally observed frequency band with the local oscillator frequency are performed independently in each of the 28 channels. The sampling frequency is 4 MHz, this means two samples within a reduced period (Kovalevsky 1995).

VLBI activities at O'Higgins

The first successful VLBI recordings at O'Higgins have been performed in January/February 1993. Besides contributing within IERS to the development and maintaining of ICRF and ITRF and the estimation of EOPs the geodetic VLBI experiments run within the framework of the project DOSE (Dynamics of the Solid Earth), which was initiated by NASA and is executed in worldwide cooperation. Since 1992 O'Higgins was involved in 41 international experiments (status: August 1997).

An important condition for VLBI is the close cooperation between the participating observatories, based on a sophisticated scheduling which is substantial for high precision results. The schedule (made by NASA) combines the operating telescopes at different regions of the earth to radiointerferometers pointing at identical extragalactic radio sources simultaneously.

VLBI results

The VLBI data processing yields station coordinates and station velocities in the ITRF, the coordinates of the observed radio sources in the ICRF, polar motion, the offsets of the earth rotational velocity and offsets from the conventional theory of nutation. After ceasing the optical astrometric methods for earth orientation VLBI is the only tool to relate ITRF to ICRF. BKG has set up a team for processing VLBI data in cooperating closely with the VLBI group of the University of Bonn (Geodetic Institute).

	Coordinates and velocities in mm	σ [mm]	correlation matrix X	Y	Z	X vel	Y vel	Z vel
X	1525833.069	3.2	1.000					
Y	-2432463.886	3.6	-0.136	1.000				
Z	-5676174.545	3.8	-0.227	0.145	1.000			
X vel	0.013	1.8	-0.571	-0.026	-0.187	1.000		
Y vel	0.018	1.9	-0.025	-0.512	0.262	0.020	1.000	
Z vel	-0.004	0.9	-0.271	0.431	-0.324	0.496	-0.858	1.000

Table 1: Free network station coordinates for O'Higgins from the global solution VLBI-ifag95.01, velocities are given in m/year

The applied software is CALC/SOLVE/GLOBL (Caprette et al., 1990), developed at the Goddard Space Flight Center (GSFC-NASA). In the framework of the GPS campaign SCAR95 (cf. next chapter) special emphasis has been laid on a comparison of the VLBI results (station coordinates) with those resulting from the GPS Antarctic reference net-

Figure 1: View on the geodetic observatory O'Higgins. From left to right: 2 radoms of INMARSAT antenna, PRARE test installation on pillar K4, radom of TURBO ROGUE antenna, VLBI antenna

work. For this comparison all available global VLBI data (collected since 1984) have been mixed with the data of 14 VLBI experiments at O'Higgins (1993: 7 experiments, 1994: 3 experiments, 1995: 4 experiments).

The combined data (in total 562 experiments of 28 stations) resulted in a parameter set of a free network solution (cf. Table 1). The solution (VLBI.ifag95.oh01) was finally related to the global coordinates of ITRF94 (epoch 1993.0) by a least squares transformation (7-parameter transformation), minimizing the residuals between the BKG-solution and the ITRF94 coordinates at the epoch in question.

results (epoch 1995.1)	x [m]	y [m]	z [m]
VLBI.ifag95.oh01	1525832.989 ± 0.003	-2432463.652 ± 0.003	-5676174.527 ± 0.003
GPS.scar95.all Trimble GPS receiver centered to VLBI	1525833.000 ± 0.007	-2432463.664 ± 0.007	-5676174.534 ± 0.011
ITRF94 VLBI solution	1525832.983 ± 0.010	-2432463.638 ± 0.012	-5676174.503 ± 0.016

Table 2: Comparison of results from VLBI and GPS (transformed with NUVEL 1 A to the epoch 1995.1 of the GPS SCAR95 campaign)

The coordinates of the geodetic reference point of the VLBI antenna can be compared with the results from the SCAR95 GPS campaign (cf. next chapter). The last line (Table 2) contains the coordinates of the VLBI antenna published by the IERS Central Bureau. The results are in good agreement with each other.

GPS-activities at O'Higgins

In order to improve the geodetic connection of O'Higgins with its environment as well as to establish a reference network in Antarctica German geodesists created such a network, observed and processed the collected data (Dietrich, 1996). These German activities have been integrated in the program of SCAR WG-GGI.

Within this network (25 stable marked stations in Antarctica and 12 additonal stations) there is a subset of about 12 stations on the Antarctic Peninsula, which are of specific interest for O'Higgins: Repeated observations in future will enable to detect differential motions in this district. The observations within this network have been performed in 1995. A final coordinate set was ready in 1996. The precision of the final solution is in the order of 1 cm for the horizontal components and 2 cm for the vertical component of the coordinates related to the frame of ITRF94 and the epoch 1995.1. A comparison of the GPS with VLBI results for O'Higgins with VLBI was given in the last chapter and is confirming the precision estimates. The first repetition of the observations is scheduled for 1998. Permanent GPS observations at O'Higgins started in 1995. By these O'Higgins is participating in the IGS. Data transmission from O'Higgins is performed by a satellite link using the INMARSAT system.

Sea level monitoring

In 1995 an underwater tide gauge system was installed at O'Higgins. The tidal data will allow investigations of local features of the ocean tides. A second item will be analyses of long term sea level changes induced by postglacial continental uplift as well as by the melting of polar ice through possible changes of the global climatic behaviour. The latter item is also of common interest.

The direct collocation of space geodetic equipment and sea level monitoring will be an important feature of the station. The space geodetic observations yield "absolute" coordinates, this means coordinates related to the centre of the Earth´s mass. The sea level observations can directly be related to the geodetic systems due to the short distance. So the sea level recordings at O'Higgins are more or less directly connected to the Earth´s centre.

The changing ice coverage at the shore and the rocky coast make the cable connection a highly vulnerable part of the tide gauge installation. The sensors have been solidly mounted and fixed to underwater bedrock. Two pressure sensors were in operation until March 1997. Unfortunately the sea level recording is interrupted since this time, probably due to cable problems.

The PRARE Ground Unit

The PRARE system (Precise Range And Range Rate Equipment) is an allweather microwave ranging system (accuracy at the cm level) for the measurement of satellite to ground ranges and range rates (DGFI, 1991). PRARE has been designed and developed mainly at the Institute for Navigation, University of Stuttgart, and was originally dedicated for precise satellite positioning for a number of ESA's satellite missions.

PRARE provides quick information for the determination of satellite orbits, which is important for an efficient application of remote sensing with satellites (e. g. satellite altimetry, SAR). On the other hand PRARE allows the precise determination of the coordinates of ground stations (Dornier, 1991). The installation of a PRARE ground unit was done in 1996 to assist the ERS-2 satellite mission. Since that time the system operates automatically and continuously (with only short interruptions).

O'Higgins geodetic control network

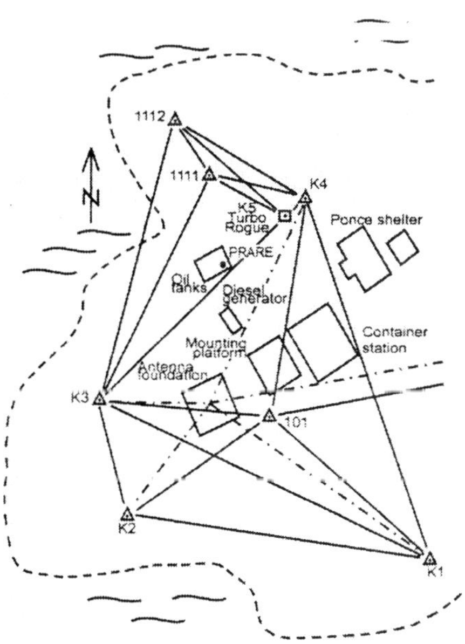

Figure 2: The local control network at O'Higgins

	dX	dY	dZ
K1	19.229	61.825	-10.501
K2	-34.111	9.814	-2.758
K3	-24.123	-15.065	8.215
K4	44.163	-17.916	29.285
K5	39.450	-17.639	28.436
101	8.086	13.987	4.750
1111	20.956	-34.586	36.002
1112	19.931	-48.620	44.347
M	0.000	0.000	0.000

M is the VLBI reference point (no 7245)

Table 3: Local coordinate differences in the control network of O'Higgins in [m]

In order to relate the different geodetic measuring systems of O'Higgins to each other a control network has been installed at the observatory. The observation points are stable monumented to allow the precise surveying of distances and angles and levelling with conventional surveying techniques. Whereas in case of the GPS antennas the corresponding reference point (this is the point to which the coordinates are related after processing the GPS data) is realized and marked by the manufacturer, the reference point of the VLBI antenna must be found from observations: it is the (virtual) crossing point between the vertical axis and the tilted second axis of the telescope. The control network is shown in Figure 2.

The difference between the coordinates of the VLBI reference point and other geodetic points are given in Table 3. Pillar K1 has been used during the SCAR95 GPS campaign in

Antarctica, pillar K4 within earlier SCAR GPS activities. On K5 the Turbo Rogue antenna is positioned. The PRARE antenna has not yet been precisely surveyed within this local network.

Prospects

In January and February 1995 some gravimetric tests have been made on O'Higgins, to get information on the microseismic behaviour. BKG is planning longer recordings for at least some months to determine local tidal parameters for this site (earth tides) as well as absolute gravity observations.

Future activities at O'Higgins focus on continuing the VLBI, GPS, and sea level observations as well as on stabilizing and increasing the precision. Direct observations of a baseline between O'Higgins and Syowa by VLBI are in discussion for 1998. To ensure a valuable contribution to ITRF and to an Antarctic reference network it is necessary to continue the observations over a longer period.

Acknowledgement

The geodetic operations, especially the VLBI campaigns and the realization of the logistic part, have only been possible in and due to the good cooperation with the colleagues from DLR. The support of the Goddard Space Flight Center of NASA in integrating the station O'Higgins into international projects and analyzing the VLBI observations is gratefully acknowledged.

References:

Caprette et al. (1990): NASA Technical Memorandum 100765. - NASA Goddard Space Flight Center, Greenbelt (MD), USA.

DGFI (1991): The ERS-1 PRARE System; Information on occasion of the launch of ERS-1. - Darmstadt.

Dietrich (1996): The Geodetic Antarctic Project GAP95 - a German Contribution to the SCAR95 GPS Campaign. - DGK (B) No. 304, München.

Dornier (1991): PRARE Ground Stations. - Information on occasion of the launch of ERS-1; Darmstadt.

Hase et al. (1993): The ERS/VLBI Station in Antarctica. - In: Proceedings of the 9th Working Meeting on European VLBI for Geodesy and Astronomy at Bad Neuenahr; Bonn.

IERS (1995): IERS-Missions and Goals for 2000. - Paris.

Kovalevsky (1995): Modern Astrometry. - Springer, Berlin-London-New York.

Reinhold et al. (1996): Status of the ERS/VLBI Station O'Higgins (Antarctica). - Working Meeting on European VLBI for Geodesy and Astronomy at Onsala. - Leipzig.

TOWARDS A NEW FREE AIR ANOMALY MAP OF THE ANTARCTIC PENINSULA

Phil Jones (British Antarctic Survey, High Cross, Madingley Road, Cambridge, CB3 0ET, United Kingdom).

ABSTRACT

Since 1959 the British Antarctic Survey (BAS) has undertaken reconnaissance gravity surveys of the Antarctic Peninsula. Approximately 1500 on-rock and about 600 on-snow station values were measured. Continuing climatic, topographical and logistic constraints resulted in an uneven distribution of stations, with a concentration of measurements on the coast and very few on the ice covered spine of the peninsula.

An even sampling of the gravity field over two thirds of the Antarctic Peninsula has now been obtained from a BAS airborne gravity survey over areas with sparse land station coverage. A total of 10,700 line km of data were flown during the 1996/97 season using a modified LaCoste and Romberg S meter deployed in a Twin Otter aircraft. Thc frcc air anomaly field was recovered with an accuracy of 5 mGals for along line wavelengths greater than 9 km and gridded wavelengths greater than 20 km. Ice thickness and topographical data were recorded from simultaneous radio echo soundings.

INTRODUCTION

Recently it has been recognised that airborne gravimetry can make a significant contribution to the determination of a precise geoid (Schwarz and Li, 1995). It is also recognised that airborne gravimetry represents the only realistic way of determining the gravitational field for wavelengths of 10 - 50 km in unsurveyed parts of the world, such as the Antarctic (Schwarz and Li, 1995). Whilst the long-term goal of determing the gravity field at such wavelengths over the whole of the Antarctic continent requires a large-scale survey such as the one conducted in Greenland (Brozena *et al.*, 1992), the denser line spacings required over mountainous regions (Schwarz and Li, 1995) can be achieved more efficiently using light aircraft. The Antarctic Peninsula (Figure 1) is a prime example of such a mountainous area.

Despite the fact that BAS has been obtaining recconnaissance land gravity data on the Antarctic Peninsula for nearly 30 years, the data distribution remains uneven and is non existent over large parts of the peninsula. Therefore during November and December 1996, an airborne gravity survey was flown over regions with sparse land coverage, with the aim of obtaining uniform coverage over the majority of the Antarctic Peninsula.

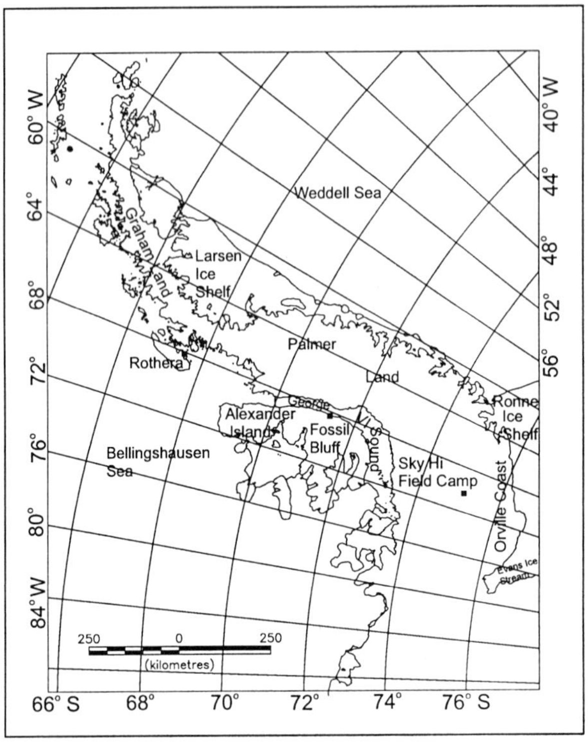

Figure 1. Antarctic Peninsula location map. Squares mark the three airborne gravity base stations from which the survey was flown.

LAND GRAVITY MEASUREMENTS ON THE ANTARCTIC PENINSULA

Since 1959 BAS have obtained approximately 1500 on-rock and about 600 on-snow station gravity values (Figure 2). Most of the measurements were made using either Worden geodetic or LaCoste and Romberg meters. A more detailed description of all the reconnaissance gravity surveys of the Antarctic Peninsula is given in Renner *et al.*, (1985).

Gravity surveys in the Antarctic Peninsula have been liable to a degree of error which places the accuracy of the measurements at reconnaissance level. Elevation control represents the major element of error in the measurements, with some recorded errors of up to 20 m. The vast majority of errors in station height are less than 10 m. The total error due to incorrect determination of instrumental drift, calibration factor and latitude correction, range from 0.2 mGals to 2 mGals (Renner *et al.*, 1985).

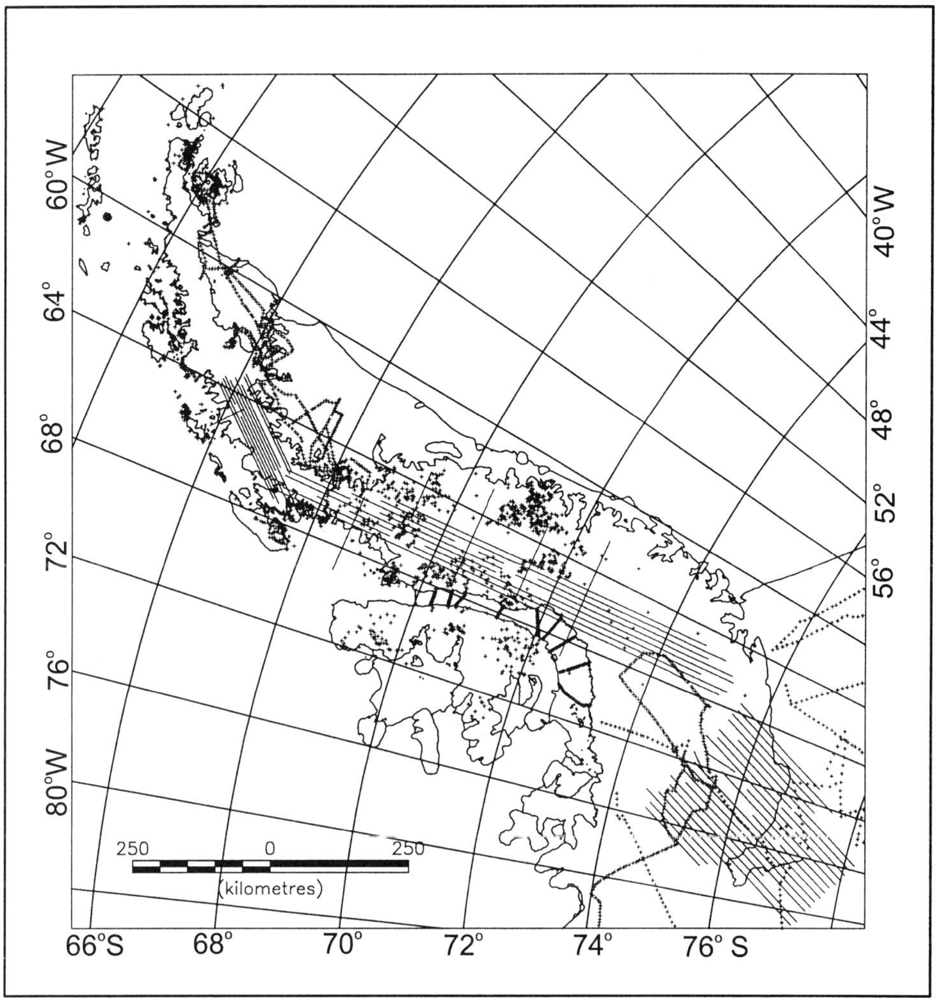

Figure 2. Location map of gravity measurements on the Antarctic Peninsula. The dots show positions of land gravity stations. The thin solid lines show airborne gravity flight lines. The thick lines show the coast line and the grounding lines of ice shelves.

THE AIRBORNE GRAVITY SURVEY

Continuing climatic, topographical and logistic constraints have ensured an uneven distribution of land gravity stations, with a concentration of measurements on the coast and very few on the ice-covered spine of the peninsula (Figure 2). The airborne gravity data will be used primarily to study regional scale crustal features. Therefore, when logistical and safety considerations constrained the period of acquisition, the line spacing was kept to a maximum of 10 km (a denser line spacing than that meeting current geoidal requirements in the region; Schwarz and Li, 1995) but the survey was restricted to regions where the land data coverage was relatively sparse. The survey consisted of 10,700 line km of data, with a line spacing of 5 km in Graham Land, and 10 km in Palmer Land, see the solid lines in figure 2. Figure 2 also shows, as a series of solid lines south of Sky Hi field camp (Figure 1), a previously obtained airborne gravity survey (Jones and Johnson, 1995).

The gaps in the solid lines in Figure 3 represent regions where the airborne gravity field was so corrupted by turbulence, that the data were not used in this preliminary analysis of the free air gravity field. Although survey height was kept constant along each line, the need to both avoid turbulence and optimise radio echo sounding returns, required varying the flight height from line to line. For the purposes of this preliminary examination of the airborne gravity survey, all the data were upward continued to 3850 m above the reference ellipsoid, the maximum flight height used during the survey.

A LaCoste and Romberg air/sea gravimeter was used to measure the gravity signal. The meter has been kindly loaned to BAS by the Hydrographic Office of the Royal Navy and was then modified by ZLS for use in an aircraft. Trimble geodetic receivers and antennas were used to make differential, dual frequency, carrier phase, GPS measurements of the aircraft's motion. Ice thickness data were obtained using an in-house built, radio echo sounding system. In-field processing indicated that ice-bottom returns over most of the survey area were obtained. The equipment was deployed in a BAS De-Havilland Twin Otter aircraft and flown without the aid of an autopilot.

Standard processing steps were used to obtain free air anomalies and are described in Jones and Johnson (1995). Cross-over analysis of the along line free air anomaly data was undertaken on the 28 cross overs present in the data set. Before the analysis took place, a low pass filter with a 9 km cut off was applied and the data were upward continued to a uniform height. The absolute average value and standard deviation of the absolute miss-ties at the intersections were 5.4 mGals and 4.5 mGals respectively. These miss-ties are relatively large compared to those of other airborne gravity surveys (Klingele *et al.*, 1997). However, this is only a preliminary analysis of the survey and I hope to improve the errors with a more careful analysis of the data.

Figure 3 shows free air anomalies from the airborne gravity survey of the Antarctic Peninsula. Despite the fact that the map shows the field a few km above the ellipsoid, none of the data used in its production are from more than 1500 m above the terrain surface. Anomaly amplitudes range from -60 to 180 mGals and there is a good alignment of features between profiles. A number of features of the map may be related to major geological features, although the geophysical significance is difficult to gauge without a digital elevation map. Figure 3 clearly shows a relatively broad free air anomaly high over the elevated spine of the peninsula in Graham Land and northern Palmer Land. The other major feature is the free air anomaly low between latitudes 71° 30' S and 73° 30' S, which may be related to extensional features seen in the vicinity of George VI Sound.

FUTURE WORK AND CONCLUSIONS

In order to obtain a free air anomaly field of the whole of the Antarctic Peninsula region at ground level, work on two fronts is needed. It is clear from Figure 2, that there are large gaps in the gravity data coverage and plans have been made to obtain airborne gravity data over the whole of the Antarctic Peninsula with a 10 km line spacing in the next 5 years.

A free air anomaly map over the majority of the Antarctic Peninsula can be obtained by

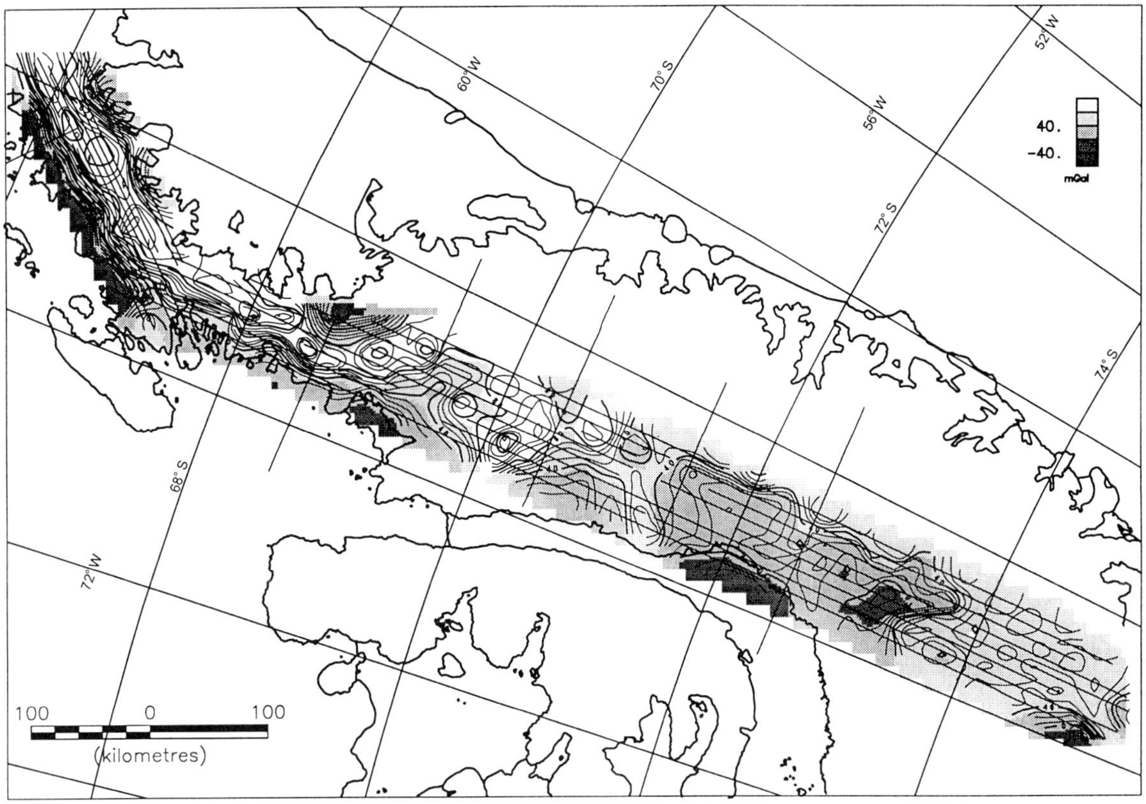

Figure 3. A map of the free air anomaly field derived from the airborne gravity survey of the Antarctic Peninsula. All the data have been filtered with a 20 km low pass filter, upward continued to a height of 3850 m above the reference ellipsoid and subsequently gridded using a minimum curvature routine onto a 10 km grid. The contour interval is 10 mGal. The thick lines show the coastline and grounding lines of the ice shelves.

integrating the current data sets. This would require downward continuation of the airborne gravity data, using a suitable methodology (e.g. Forsberg and Kenyon, 1995). In order to facilitate this, a digital elevation model of the region needs to be constructed from radio echo soundings obtained during this and previous surveys (Crabtree *et al.*, 1985; Renner *et al.*, 1985). Finally the free air anomaly field on the land needs to integrated with the satallite derived field over the Weddell and Bellingshausen seas (McAdoo and Laxon, 1997).

Airborne gravity surveys from light aircraft represent the most efficient way of obtaining a precise geoid over the mountainous region of the Antarctic Peninsula. Preliminary analysis of an airborne gravity survey indicates that the gravitational field can be determined with an accuracy of a few mGals for wavelengths greater than 20 km at a height of less than 2.5 km from the source. Although airborne gravity coverage does not extend to the whole peninsula, a combination of the currently available land, airborne and satellite derived data will yield a free air anomaly field over the majority of the region.

ACKNOWLEDGEMENTS

Comments by Bryan Storey, Mike Thomson and Martin Vermeer helped improve the final manuscript. In addition, the author would like to thank the staff at Rothera scientific station for assistance during data acquisition. The contribution of BAS pilots Greg Harris and Pete Buckley is gratefully acknowledged.

REFERENCES

BROZENA, J.M., CHALONA, M., FORSBERG, R. and MADER, G. 1992. The Greenland Aerogeophysics Project. *EOS Trans. American Geophys. Union*, **73**.

CRABTREE, R.D., STOREY, B.C. and DOAKE, C.S.M. 1985. The structural evolution of George VI Sound, Antarctic Peninsula. *Tectonophysics*, **114**, 431-442.

FORSBERG, R. and KENYON S. 1995. Downward continuation of airborne gravity data. (*In* SCHWARZ, K.P., BROZENA, J. and HEIN, G. *eds. Proceedings of IAG Symposium on Airborne Field Determination, at the IUGG XXI General Assembly Boulder Colorado, July 1995*, Special Report No. 60010 of the Department of Geomatics Engineering at the University of Calgary, 2500 University Drive North-West, Calgary, Alberta, Canada, T2N 1N4)

JONES, P.C. and JOHNSON, A.C. 1995. Airborne gravity survey in Southern Palmer Land, Antarctica. (*In* SCHWARZ, K.P., BROZENA, J. and HEIN, G. *eds. Proceedings of IAG Symposium on Airborne Field Determination, at the IUGG XXI General Assembly Boulder Colorado, July 1995*, Special Report No. 60010 of the Department of Geomatics Engineering at the University of Calgary, 2500 University Drive North-West, Calgary, Alberta, Canada, T2N 1N4)

KLINGELE, E.E., COCARD, M., KAHLE, H.-G. and HALLIDAY, M. 1997. Kinematic GPS as a source for airborne gravity reduction in the airborne gravity survey of Switzerland. *Journal of Geophysical Research*, **102 B4**, 7705-7715.

McADOO, D. and LAXON, S. 1997. Antarctic tectonics: constraints from ERS-1 satellite marine gravity field. *Science*, **276**, 556-560.

RENNER, R.G.B., STURGEON, L.J.S. and GARRETT, S.W. 1985. Reconnaissance gravity and aeromagnetic surveys of the Antarctic Peninsula. *Brit. Antarct. Surv. Sci. Rep.*, **110**, 50pp.

SCHWARZ, K.P. and LI, Y.C., 1995. What can airborne gravimetry contribute to geoid determination?. (*In* SCHWARZ, K.P., BROZENA, J. and HEIN, G. *eds. Proceedings of IAG Symposium on Airborne Field Determination, at the IUGG XXI General Assembly Boulder Colorado, July 1995*, Special Report No. 60010 of the Department of Geomatics Engineering at the University of Calgary, 2500 University Drive North-West, Calgary, Alberta, Canada, T2N 1N4)

Regional Geoid Improvement Based on Surface Gravity Data

W. Korth and R. Dietrich
Technische Universität Dresden, Institut für Planetare Geodäsie, Germany

G. Reitmayr and V. Damm
Bundesanstalt für Geowissenschaften und Rohstoffe, Hannover, Germany

1 Introduction

A major problem for accurate geoid determination in Antarctica is the lack of gravity data in most parts of the continent. There are very dense data available for other regions of the world but almost none for the Antarctic continent. For that reason global geopotential models are imprecise for Antarctica (Sjöberg and Fan, 1993). The differences between different global geoid models may reach more than five meters in the continental part of Antarctica (fig. 1). As a consequence, orthometric heights calculated from ellipsoidal GPS heights may have errors of that magnitude.

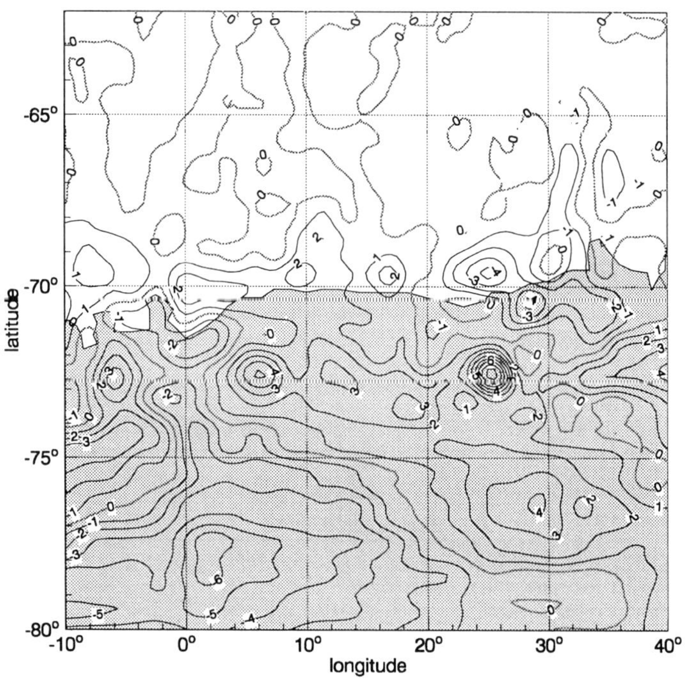

Figure 1: Differences of geoid undulations for the Atlantic sector of Antarctica computed from high resolution geopotential models EGM96 (Lemoine et al., 1996) and GFZ93a (Gruber and Anzenhofer, 1993). Contour line interval 1 meter. The border between white and gray indicates the ice shelf edge.

A regional geoid improvement is possible if regional gravity data are available. Such data can be used to compute the residual disturbing potential referring to a given global model. The global geoid model as well as the effect of topography, including the ice masses, can be handled by the remove-restore technique.

2 Surface gravity data

In the region of the Schirmacher Oasis in central Dronning Maud Land geophysical and geodetic fieldwork has been carried out for more than two decades. First gravity data along profiles have been collected by russian scientists (Frolov and Koryakin, 1967), (Kogan and Stroev, 1972). Those data are included in the global data set that was used for the computation of the actual global models of the gravity field.

Since 1991 and especially during the GeoMaud Expedition in 1996 new terrestrial gravimetric measurements have been carried out by german geophysicists. The survey covers an area of about $270 \cdot 270\,\mathrm{km}^2$. Altogether 238 gravity points were measured on a nearly regular grid with a separation of approx. 30 km in latitude and 14 km in longitude (Reitmayr, 1996) and

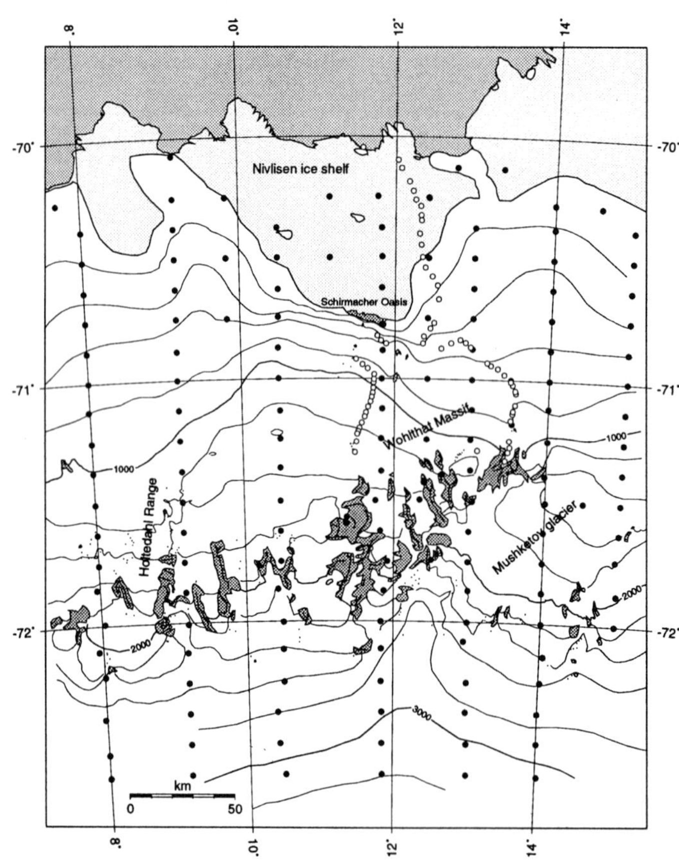

Figure 2: Distribution of the gravimetric points in the working area (open circles: measurements till 1995 along traverses, filled circles: helicopter supported measurements during GeoMaud Expedition).

along three traverses (Fritzsche, 1997). Figure 2 shows the distribution of the gravity data. In addition to gravity itself the ellipsoidal height (by differential GPS) and the thickness of the glaciers have been measured at each of these gravimetric points.

3 Ice thickness data

Very dense ice thickness and bedrock topography data have been measured during the GeoMaud Expedition for the cental part of the working area by an airborne radio echo sounding survey (RES) (Damm and Eisenburger, 1996). These data were used for computing a detailed digital terrain model including the surface and the sub-ice bedrock topography as well. The model was completed by use of height and ice thickness data of the gravimetric points outside the central part of the area. Figure 3 shows the distribution of the RES survey points.

The variation of the ice thickness in the ice shelf area is rather low with values between 150 m near the ice edge and up to 700 m at some parts of the grounding line. High variations

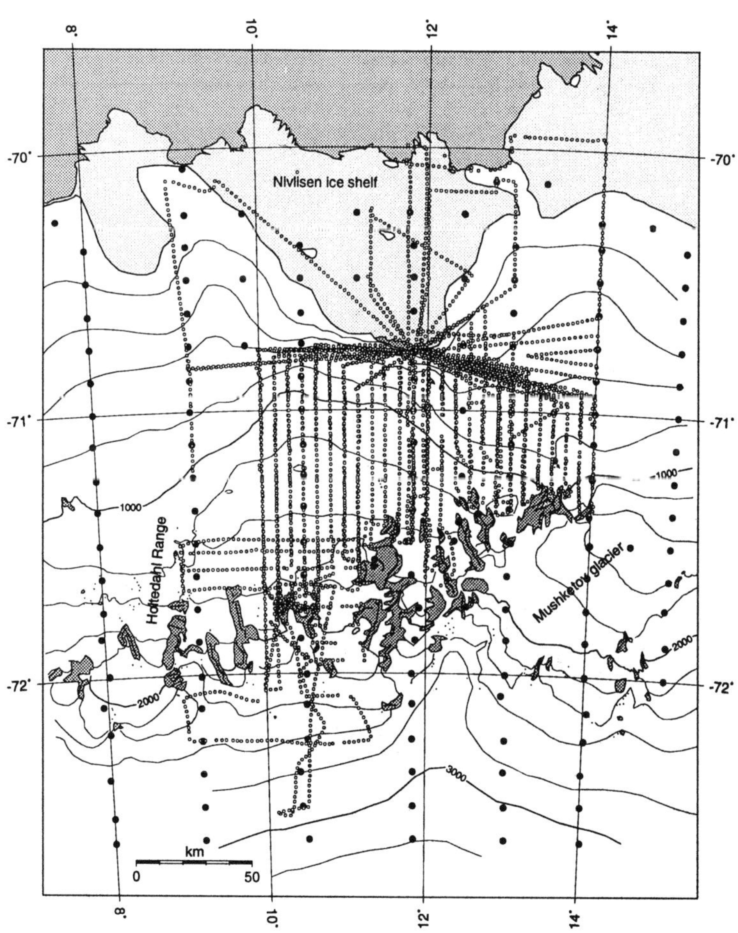

Figure 3: Distribution of the RES survey points in the working area (open circles: airborne measurements, filled circles: gravimetric points with ice thickness measurements).

of the ice thickness in the region between the Schirmacher Oasis and the southern mountain ranges between 700 m and 2500 m indicate a very rough morphology of the sub-ice bedrock. At some places there are deep throughs with sub-ice topography some 100 m below sea level. The Wegener Inland Ice further to the south is more than 1000 m thick in general.

4 Model for the disturbing potential

The geoid computation was carried out by a point mass representation of the disturbing potential T in the region:

$$T = \sum_{i=1}^{n} G\mu_i \frac{1}{l_i} \tag{1}$$

(G... gravitational constant; μ_i... mass; l_i... distance between point mass and point of observation)

The use of gravity data and ellipsoidal heights of the points of measurement leads to gravity disturbances. The relation between gravity disturbances δg and the disturbing potential can be expressed as (Heiskanen and Moritz, 1967, p. 85):

$$\delta g = -\frac{\partial T}{\partial r} = -G \sum_{i=1}^{n} \mu_i \frac{\Delta z}{l_i^3} \tag{2}$$

(Δz... height difference between point mass and point of observation in respect to the topocentric coordinate system of the point of observation)

In the framework of the point mass adjustment the masses and their positions were determined by iterative nonlinear optimization. The algorithm is based on a method published by Barthelmes and Dietrich (1991). For an optimal regularization of the algorithm a method proposed by Scales (1985) was used.
The global potential model EGM96 (Lemoine et al., 1996) and topographic masses (solid rock as well as ice bodies) were handled by the remove-restore technique.

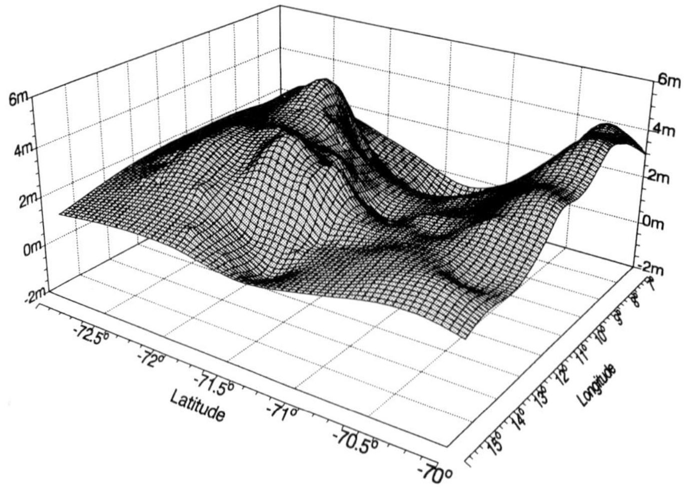

Figure 4: Regional geoid (EGM96 and topographic effects removed).

5 Results

A geoid model was computed for the region around the Schirmacher Oasis applying a point mass model for the disturbing potential T. 94 masses were used for the approximation of T. The depths of the point masses were limited to a minimum of 1 km and a maximum of 34 km in order to restrict the wavelength of the potential caused by the point masses (fig. 4).

The absolute level of the geoid was verified through a comparison with the ellipsoidal height of the mean sea level determined by tide gauge measurements and GPS. The difference between the ellipsoidal height measured by GPS and corrected for sea surface topography and for ocean tides and the geoid undulation computed from the model is less than 2 centimeters.

The deviations of the resulting geoid undulations from the model EGM96 vary between -1 and 4 meters. The estimated point to point accuracy is $\pm 5 \ldots 15$ cm and the accuracy of the absolute level is ± 0.5 m.

No additional gravity measurements were available in the vicinity of the area. The resulting influence on the geoid undulation errors was estimated to be in the order of $\pm 5 \ldots 10$ cm for the central part and $\pm 20 \ldots 30$ cm at the edges of the model.

The computed geoid model for the region of the Schirmacher Oasis is shown in figure 5. The undulations in the area studied are rather strong. They vary between 13 m and 24 m in North-South direction mainly. An East-West striking conspicuous minimum is located around 71°S. The values increase rapidly towards the Polar Plateau in the South, where an almost circular maximum between 12° and 14°E shows up.

References

Barthelmes, F. and Dietrich, R. (1991). Use of point masses on optimized positions for the approximation of the gravity field. In *Determination of the Geoid, Present and Future.* Springer-Verlag, New York, Inc.

Damm, V. and Eisenburger, D. (1996). *Bericht über hubschraubergestützte Radareisdickenmessungen im zentralen Queen-Maud-Land während der Antarktisexpedition GeoMaud 1995/96.* Bundesanstalt für Geowissenschaften und Rohstoffe, interner Bericht, Archiv-Nr.: 114890, August 1996 (unpublished).

Fritzsche, D. (1997). Gravimetrische und Eisdickenradarmessungen in der Region der Schirmacheroase. pers. comm.

Frolov, A. I. and Koryakin, E. D. (1967). Investigations by the gravimetric method of subglacial relief in the Lazarev Station region. *Bull. Sov. Antarct. Exped.*, pages 103–107.

Gruber, T. and Anzenhofer, M. (1993). The GFZ 360 Gravity Field Model. In *Proc. Of Sess. G3 of XVIII European Geophysical Society General Assembly. Wiesbaden 1993.*

Heiskanen, W. A. and Moritz, H. (1967). *Physical Geodesy.* W. H. Freeman and Company, San Francisco.

Kogan, A. L. and Stroev, P. A. (1972). Gravimetricheskie issledowanie v rayone Sovietskich Antarkticheskich stanciey Lazarev i Novolazarevskaya (Gravimetric investigations in the region of Antarctic stations Lazarev and Novolazarevskaya). *Trudy gos. astron. Inst. im. P. K. Sternberga XLIII, Moskva*, pages 3–7.

Lemoine, F. G., Smith, D. E., Kunz, L., Smith, R., Pavlis, E. C., Pavlis, N. K., Klosko, S. M., Chinn, D. S., Torrence, M. H., Williamson, R. G., Cox, C. M., Rachlin, K. E., Wang, Y. M., Kenyon, S. C., Salman, R., Trimmer, R., Rapp, R. H., and Nerem, R. S. (1996). The development of the NASA GSFC and NIMA joint geopotential model. In *Proc. of the Int. Symp. on Gravity, Geoid, and Marine Geodesy, (GraGeoMar 1996), Tokyo, Sep. 30 – Oct. 5, 1996.*

Reitmayr, G. (1996). *Bericht über gravimetrische Messungen und sie begleitende GPS-Ortsbestimmungen im Königin Maud Land, Antarktis, während der Expedition GeoMaud.* Interner Bericht der Bundesanstalt für Geowissenschaften und Rohstoffe, Archiv-Nr. 114900, Hannover (unpublished).

Scales, L. E. (1985). *Introduction to non-linear optimization.* MacMillan Publishers Ltd., London and Basingstoke.

Sjöberg, L. E. and Fan, H. (1993). The Antarctic Gravity and Geoid Situation. Presented paper, General Meeting of the IAG, Beijing.

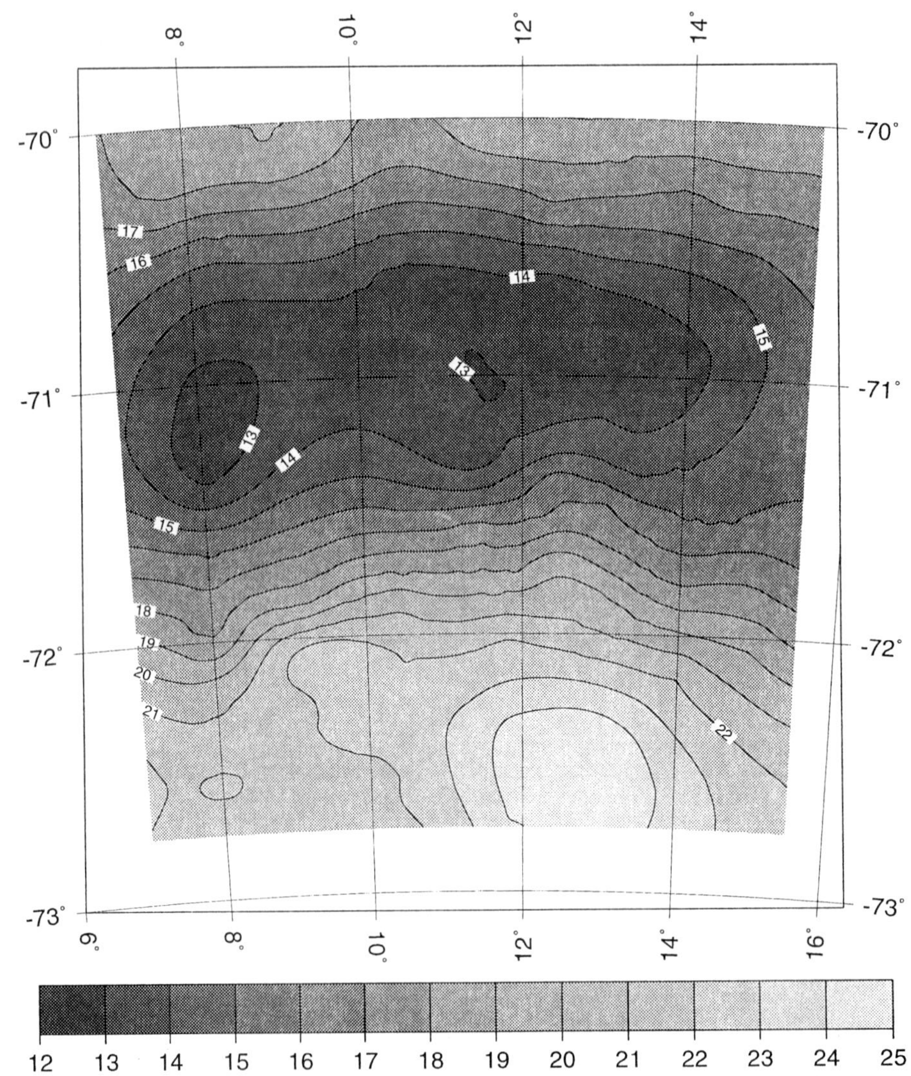

Figure 5: Computed geoid model for the region of the Shirmacher Oasis. (Contour line interval 1 meter.)

SAR Interferometry for Improved Terrain Corrections in Surface and Airborne Gravimetry

Cecilia Svenné Nielsen, René Forsberg and Simon Ekholm
National Survey and Cadastre, Dept. of Production and Research
Rentemestervej 8, DK-2400 Copenhagen NV, Denmark
Phone: +45 35 87 52 43, fax: 45 35 87 50 52, e-mail: csn@kms.min.dk

Johan Jacob Mohr
Danish Center for Remote Sensing, Dept. of Electromagnetic Systems
Technical University of Denmark, B-348, DK-2800 Lyngby, Denmark

Abstract

Derivation of Digital Elevation Models (DEM's) by ERS SAR interferometry is a promising method for obtaining the necessary elevations for computing accurate terrain corrections in remote, poorly surveyed areas. InSAR DEM's are especially useful due to the inherent high spatial resolution. In the poster the quality of a SAR derived DEM of an East Greenland area is evaluated based on comparisons to geodetic control points and airborne laser altimetry.

The impact on terrain corrections for both surface and airborne gravity surveys is evaluated by comparison to computations based on previous DEM's. Results show that effects can often be 10 mGal ($1 \cdot 10^{-4}$ ms^{-2}) or more and especially affecting airborne gravimetry. Derivation of SAR DEM's are, however, not without problems since atmospheric effects, phase unwrapping errors, shadowing and foreshortening / lay-over may produce systematic errors and data voids especially in areas with very steep slopes.

Introduction

Derivation of high quality Digital Elevation Models (DEM's) by different remote sensing techniques is a promising method for obtaining the necessary elevations for computing accurate terrain corrections in remote, poorly surveyed areas.

The National survey and Cadastre - Denmark (KMS) is involved in projects dealing with the evaluation of different observation methods and their capability of detecting subtle changes (both in time and space) in elevation. The techniques studied are primarily:

Radar) data, and ERS-1/2(launched by ESA) SAR data since 1991 using an S/X-band parabola antenna whose diameter is 11 meters. SAR data acquired at Syowa Station have been often used to investigate relations between intensity of backscattering and surface features characteristics of ice sheet and sea ice(e.g. Takahashi et al.1995).

Recently, interferometric SAR technique (InSAR) have been applied to generate digital elevation models(DEM) (Zebker 1986) and to detect surface crustal displacements(Massonnet et al. 1994). There are also some investigations to estimate topography and displacements of ice sheet on Antarctica and Greenland by InSAR (Goldstein et al. 1993, Kwok et al. 1996).

The topographic maps around Syowa Station obtained from aerial photographs are confined around coastal area from logistic difficulties, even though precise and detailed topographic maps are highly required for specific field observations and logistic supports during the Antarctic operation. Digital elevation model of ice sheets in Antarctica are also expected to provide basic data for estimating changes of ice mass in the Antarctic region.

Our purpose to generate DEM around Syowa Station by means of InSAR is to provide basic topographic map data for the missions of JARE and to monitor ice sheet movements in relations to environmental changes. We also aim at detecting crustal uplifts associated with post glacial rebound.

In this paper, we introduce SAR data acquisitions at Syowa Station and some topics obtained from JERS-1 SAR analyses.

2. Data Acquisition System

The satellite data receiving system at Syowa Station consists of a 11meter-diameter dish antenna, S-band and X-band receiving systems, and a herical-scan digital recorder to record on D-1 cassette tapes. The antenna is covered by a radome.

We usually plan the JERS-1 and/or ERS-2 observation schedule six weeks before the acquisitions. We send the JERS-1 plan to NASDA/EOC(Earth Observation Center) and the ERS-2 plan to ESA. When the acquisition request is accepted, NASDA/EOC and/or ESA send NIPR the schedules two weeks before the acquisitions and NIPR relays precise acquisition times and orbit information of satellites to Syowa Station one or two days before the acquisitions.

Data tapes recorded at Syowa Station have to be transported to Japan every April by icebreaker Shirase. Then they are sent to NASDA/EOC and quality of the data is checked. After the quality check, some of data with good quality are processed and intensity images and CEOS formatted SAR raw data for ERS-1 and JERS-1 are produced by NASDA.

The number of acquired passes for each year from 1991 to1996 for ERS-1,ERS-2, and JERS-1 are shown in Table 1.

We can also acquire JERS-1 SAR data with onboard memories mainly for assessing the icebreaker Shirase's route instead of downloading at Syowa Station. The onboard recorded data are usually downloaded at NASDA/EOC(Hatoyama, Japan) or Alaska SAR Facility(Fairbanks, Alaska). The data can be processed within two or three months after the acquisitions.

Table 1 The number of acquired passes of JERS-1& ERS－1／2 data at Syowa Station

	JERS-1		ERS－1	ERS－2
sensor	SAR	OPS	AMI	AMI
1991			63 passes	
1992	34 passes	5 passes	45	
1993	8	6	43	
1994	21	4	55	
1995	11*	9 *	54	
1996	15*	5*	76	72 passes
total	89	29	336	72

* There is no data from Feb. 1995 to Jul. 1996, because of a trouble in the recording system of Syowa. Station.

3. Acquisitions of ERS-1/2 Tandem Mission Data

We have acquired the data pairs of the ERS-1 and ERS-2 tandem mission from February to June, 1996. The obtained data will be used to generate DEM of Antarctica and Greenland and to detect motions of ice sheets by interferometric SAR processing. Some results on detection of ice motions using ERS-1/2 tandem data have already been reported (Thiel et al. 1997).

The number of the acquired passes at Syowa Station in the tandem period are shown in Table 2, and the areal coverages are also shown in Figure 1.

The observed tandem data already arrived at NIPR in April, 1997, and they were sent to NASDA/EOC for data validation. The data will be catalogued and make available to the scientific community.

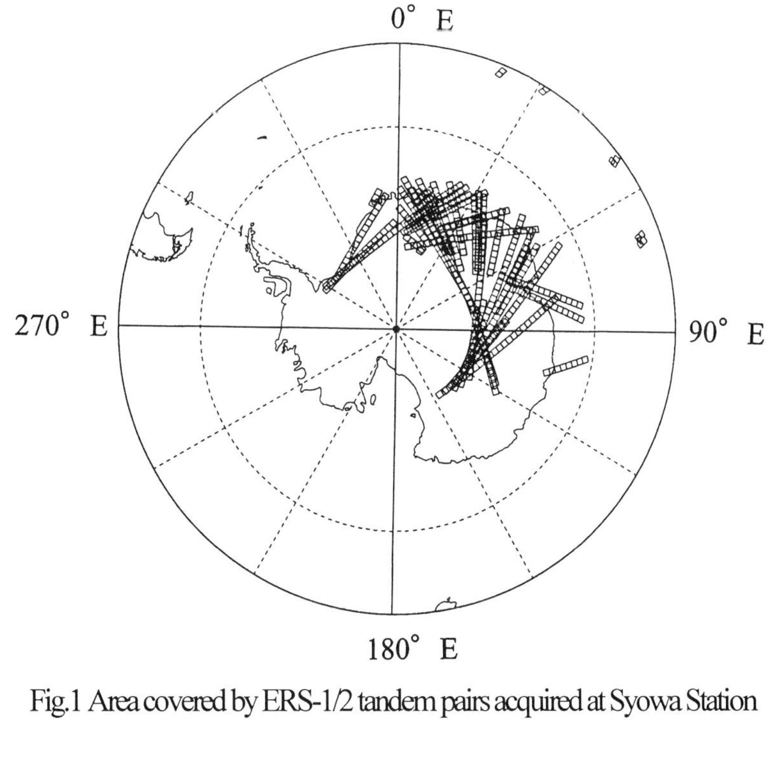

Fig.1 Area covered by ERS-1/2 tandem pairs acquired at Syowa Station

Table 2 The number of acquired passes during the tandem mission

ERS-1	ERS-2	Tandem Pairs
76 passes	62 passes	49 passes

4. Examples of Observed JERS-1 SAR Data

We show here examples of SAR intensity images and interferogram observed with JERS-1.

The inteferometric processing was carried out using two JERS-1 SAR image data which were observed on November 26, 1996 and January 9, 1997. The CEOS formatted raw data were provided by NASDA/EOC. The observed area corresponds to a rectangle in Figure 2. The area index number of the scenes defined by NASDA is 171-414. The area covers the Amundsen Bay, located in western Enderby Land.

Figures 3 and 4 are intensity images of the SAR data observed on November 26, 1996 and January 9, 1997, respectively. Figure 5 is an interferogram, which is superimposed on intensity image map. Many portions of fringes detected in Figure 5 are considered to include topographic effects, where we can refer to Figure 6 of the topographic map around the Amundsen Bay produced by the Division of National Mapping, Department of Minerals and Energy, Australia. Preliminary analysis shows that topographic fringes in Figure 5 indicate good agreement with the topographic map.

5. Summary

JERS-1 and ERS-1/2 SAR data are acquired by the 11 m antenna at Syowa Station, Antarctica. The recorded data tapes are transported to NIPR every April by icebreaker Shirase. The SAR data are validated at NASDA/EOC. Data with good quality are catalogued and made available by NASDA/EOC to users according to request.

Our purpose is to generate digital elevation models for logistic support of JARE operations, estimation of ice sheet movements, and detection of crustal movements associated with post glacial uplift.

An interferogram was obtained for the area around the Amundsen Bay where geological survey and GPS measurements will be carried out at the end of 1997. We have also plans to perform verification experiments using GPS and radar reflector sites to check produced DEM and estimated ice sheet movements.

Acknowledgment

NASDA has copyright of the data shown here for JERS-1. Processing to generate intensity images and CEOS formatted raw data for JERS-1 and ERS-1 were done by NASDA/EOC according to the interface documents between NASDA and NIPR. We thank greatly NASDA/EOC, RESTEC and ESA for their cooperation concerning SAR data acquisitions at Syowa Station and data productions.

Fig. 2 Observed Area

Fig.3 SAR intensity image observed on November 26, 1996. ©MITI/NASDA

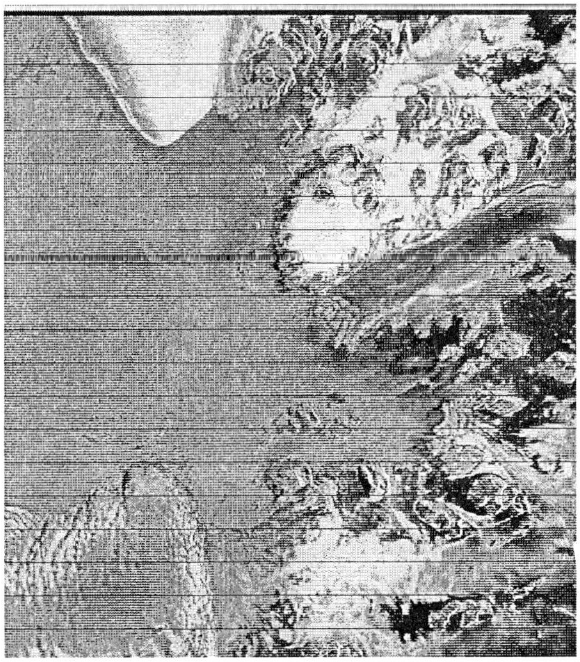

Fig.4 SAR intensity image observed on January 9, 1997. ©MITI/NASDA

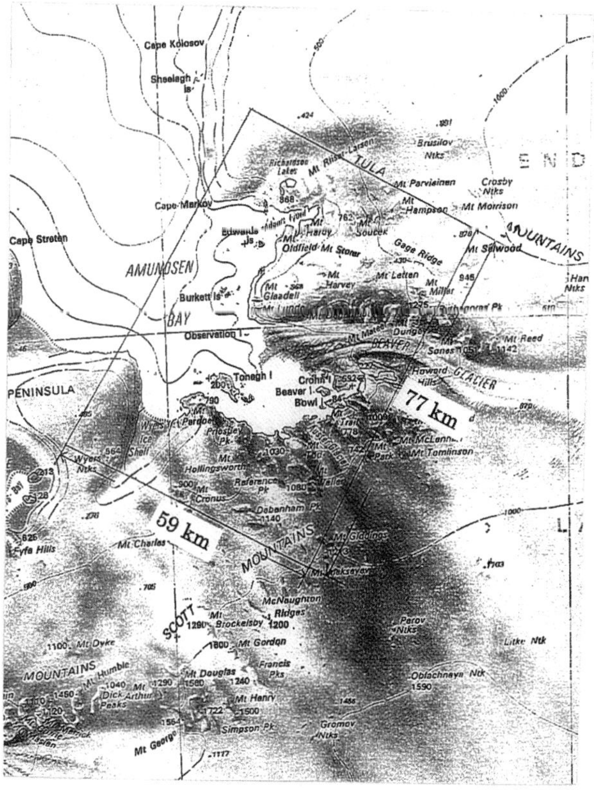

Fig.6 Topographic Map around Amundsen Bay.

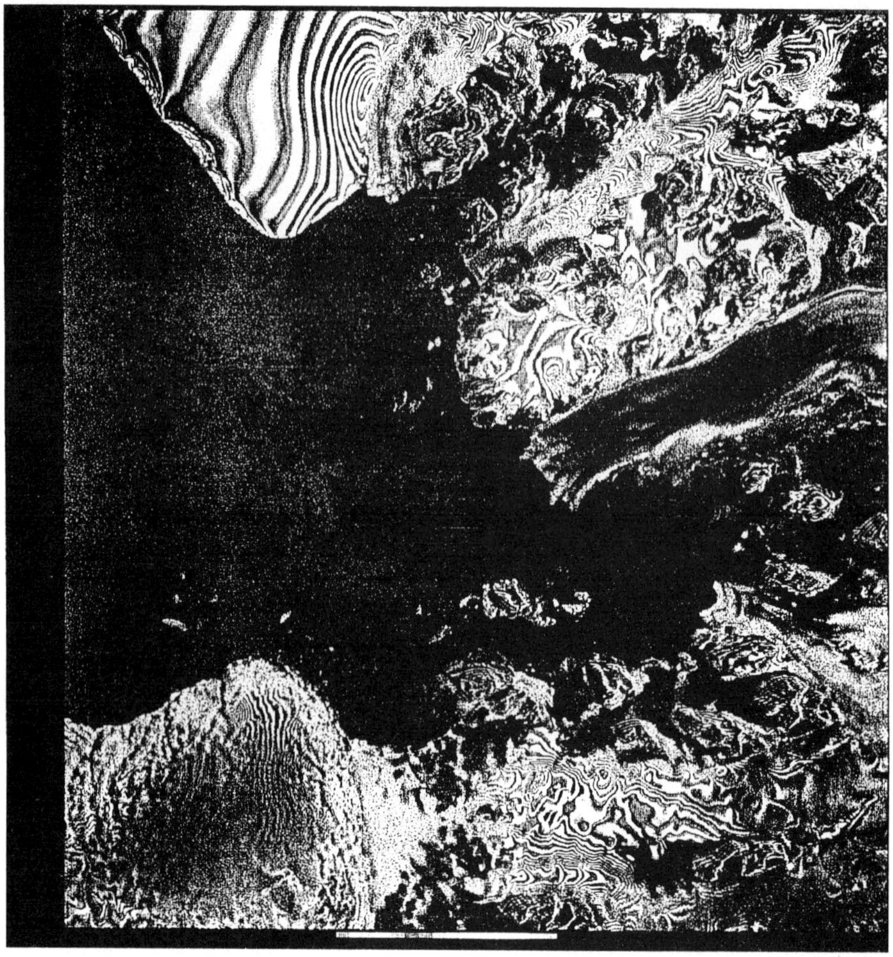

Fig . 5 Interferogram derived from JERS-1 SAR data observed on Nov. 26,1996 and Jan. 9, 1997.

References.

Goldstein, R. M., H. Engelhardt, B. Kamb, and R. M. Flolich, Satellite Radar Inteferometry for Monitoring Ice Sheet Motion: Application to an Antarctic Ice Stream, Science, 262, 1525-1530,1993.

Kwok, R., M. Fahnestock, Ice Sheet Motion and Topography from Radar Interferometry, IEEE Trans. Geosci. Remote Sensing, 34, 189-200, 1996.

Massonnet, D., K. Feigl, M. Rossi, and F. Adragna, Radar Interferometric Mapping of Deformation in the Year after the Landers Earthquake, Nature, 369, 227-230, 1994.

Takahashi, A., Y. Fujii, K. Cho, F. Nishio, T. Furukawa, O. Watanabe, Application of Synthetic Aperture Radar(SAR) Imagery on Antarctic Glaciology, Antarctic Record, 39, 205-232, 1995.(in Japanese)

Thiel, K., X. Wu, P. Hartl, ERS-Tandem-Interferometric Observation of Volcano Activities in Iceland, Proceedings of the Third ERS symposium, Florence, 1997.

Zebker, H. A., R. M. Goldstein, Topographic Mapping from Interferometric Synthetic Aperture Radar Observations, J. Geophys. Res., 91, 4993-4999, 1986.

CRUSTAL UPLIFT AROUND SYOWA STATION, ANTARCTICA

Katsutada Kaminuma
National Institute of Polar Research (NIPR)
9-10, Kaga-1, Itabashi, Tokyo 173, Japan
e-mail : kaminuma@nipr.ac.jp
fax : +81-3-3962-5741

Abstract
There is a great deal of evidence concerning crustal uplift, after deglaciation, in the vicinity of Syowa Station (69.0°S, 39.6°E) from tide gauge data, seismic evidence, elevated beaches, marine terraces, etc. The geomorphological and tide gauge data show that crustal uplift is going on around Syowa Station. Local earthquake activity corroborates the crustal uplift, and it is an intermittent phenomenon. Sea level falling of 4.5 mm/y was found using data in 1975-1992. This falling rate is consistent with geomorphological data. A route for repeat leveling survey was established in East Ongul Island. No appreciable change was observed for the last 14 years. The observations from repeat leveling measurements suggest no significant changes, which further supports the idea that the crustal uplift is not a tilt trend movement but a block movement.

1. Introduction
The Antarctic coastal area and surrounding islands are places where crustal uplift has occurred after deglaciation. Characterization of crustal movement and estimation of sea level change have been accomplished using a variety of classical high precision methods at Syowa Station (69°S, 39°E) which is located on East Ongul Island in Lützow-Holm Bay, East Antarctica. Kaminuma (1996) summarized a process of crustal uplift.

To detect crustal movement around Syowa Station, a leveling survey was established on East Ongul Island in 1982, and repeat measurements over the route were done in 1992, 1996 and 1997. However there was no significant height change of leveling at Syowa Station during 14 years (Kaminuma and Kimura, 1997).

In this paper, a dynamics of crustal uplift around Syowa Station is discussed using the geomophological, oceanic tide, seismic and leveling data.

2. Sea level Change
Two sea level falling rates at Syowa Station were obtained using tide gauge data. Odamaki et al. (1991) obtained a trend of sea level falling at a rate of 9.5 mm/y using the

International Association of Geodesy Symposia, Vol. 119
Forsberg et al. (eds.), Geodesy on the Move

tide gauge data from 1981 to 1987. This sea level change is measured relative to a land based bench-mark on East Ongul Island. Considering the rise of global mean sea level at the rate of 1-2 mm/y (Warrick, 1993), the falling rate of mean sea level at Syowa Station should be over 10 mm/y. This value is too big as the value of crustal uplift on East Ongul Island to compare with the uplift rate estimated from heights of elevated beaches as mentioned in the section 2.

Kaminuma and Akamatsu (1992) noted that the larger inferred rate of uplift is an intermittent movement. The crust around Syowa Station must be uplifted during 1981-1987. This idea is supported by local earthquake activities as mentioned in the section 4 in this paper.

Michida et al. (1995) estimated the mean sea level change at Syowa Station using 18 years of data in 1975-1992. However the data in 1977-1980 were excluded in the calculation of the sea level change, because the reliability of the data in that period was very low or on account of instrumental troubles. They obtained a trend of sea level falling with a rate of 4.5 mm/y. This value is consistent with the uplift rate estimated from the heights of elevated beaches.

3. Elevated beaches

Elevated beaches and the emergent marine deposits represent important data/information for estimating vertical crustal movement, sea level change, ice advance and retreat, and hence environmental change in the polar regions. East Ongul Island is 4 km off of the Antarctic Continent across the Straits of Ongul. Evidence of past glaciation is observed throughout the Ongul Islands: Erratic boulders, glacial scour, and various glacial deposits etc. can be found in the snow free area. Shell fossils have been found on elevated beaches on East Ongul Island and marine deposits have been identified on beaches from 0.5 to 16 m asl (above sea level) (Hayashi and Yoshida, 1994). Crustal uplift following deglaciation is continuing in the Ongul Islands at present.

Holocene elevated beaches, marine terraces and their emergent marine deposits have been reported in the east coast of the Lützow-Holm Bay (e.g. Hayashi and Yoshida, 1994; Igarashi et al., 1995). These elevated beaches and marine terraces have been formed by the relative lowering of sea level caused by the crustal uplift. Generally the highest elevation of elevated beaches in the coastal area of the Antarctic Continent is around 20 m asl and the rate of crustal uplift caused by deglaciation in Antarctica is estimated to be 2.5 mm/y on average, the possible maximum being 5-6 mm/y during last several thousand years (e.g. Yoshida and Moriwaki, 1979).

The elevated beaches, hence the crustal uplift, are a reflection of the regional isostatic rebound. The Antarctic ice sheet still extends to the coastal areas at present; however, crustal uplift coincident with deglaciation appears to be continuous since Holocene.

4. Local Earthquakes

No large earthquakes of magnitude greater than 5 have been located in the Antarctic Continent; smaller magnitude earthquake activities are detected by the worldwide seismic network and local earthquakes are located by some local seismic networks in the Antarctic. A tripartite seismic array had been operated at Syowa Station in 1987-1989 for studying the local seismicity. Ten local earthquakes were located by the tripartite array during the 29 months from June 1987 to October 1989 . The annual number of local earthquakes counted on the routine observation seismograms is shown in Fig. 1. A process for the local earthquake activity in the vicinity of Syowa Station was suggested by Kaminuma and Akamatsu (1992) as follows:

1) Local earthquakes around Syowa Station are inferred to be caused by the tectonic stress which is accumulated by the crustal uplift after deglaciation, as the earthquakes are located in the coastal and offshore areas .

2) The rate of stress accumulation is very small, so that only small, micro/ultra-micro earthquakes occur in the coastal and offshore areas.

3) The crustal uplift occurs only for a few years during one decade/more. The occurrence of earthquakes corresponds with the intermittent crustal uplift (see Fig. 1).

This process can explain the discrepancy between the 10 mm/y falling of sea level and 3-6 mm/y uplift of elevated beaches. The uplift rate of elevated beaches with a rate of 3-6 mm/y is the mean value during the last 5000-6000 years, while the rate of sea level falling from tide gauge data is averaged over several years. If crustal uplift occurs intermittently, the rate of the uplift might be larger than 3-6 mm/y as presented by Odamaki et al. (1991) and Michida et al. (1995). The sea level fall of 10 mm/y appears to correspond to the crustal uplift in the 1980s.

The period of local earthquake activity around Syowa Station corresponds to the same period when the 10 mm/y sea level fall was obtained. The local earthquake activity must indicate that the crustal uplift occurs intermittently. The crustal uplift occurred during the last few years of the 1980's.

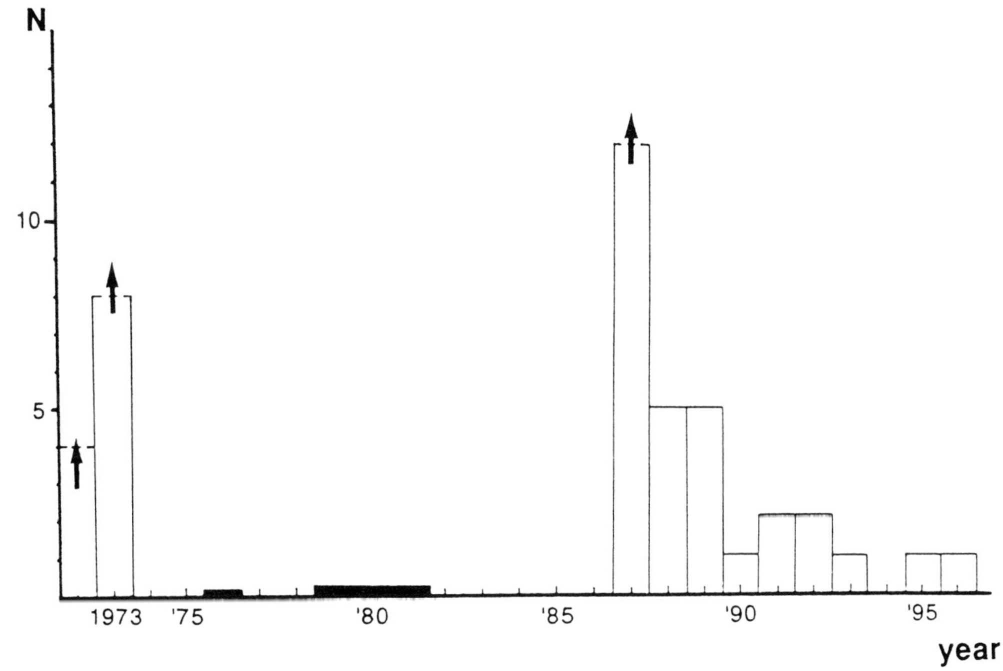

Fig. 1 Annual numbers of local earthquakes around Syowa Station. Arrows indicate that the exact number of earthquakes is more than that shown in the figure.

5. Leveling Survey

A route for repeat leveling survey was established on East Ongul Island in 1982. The leveling measurements were repeated along the route in 1992, 1996 and 1997 as shown in FIg. 2 (Kaminuma and Kimura, 1997; Kaminuma and Ootaki, under preparation). The leveling measurements were carried out according to the specification of the Geophysical Survey Institute (GSI) of Japan. The discrepancy of forward and backward leveling, and closing error of the measurement circuit, were obtained within the limitation of the first order leveling of GSI.

Figure 3 shows height changes between measurement in 1996 and 1997, and 1982. The cause of large height change at Bench Mark (BM) 1032, BM23-16 etc. are estimated to artificial noise because these bench marks are located along the road in the working area and one of the station huts. No significant relative height change is discernible in East Ongul Island during the last 15 years.

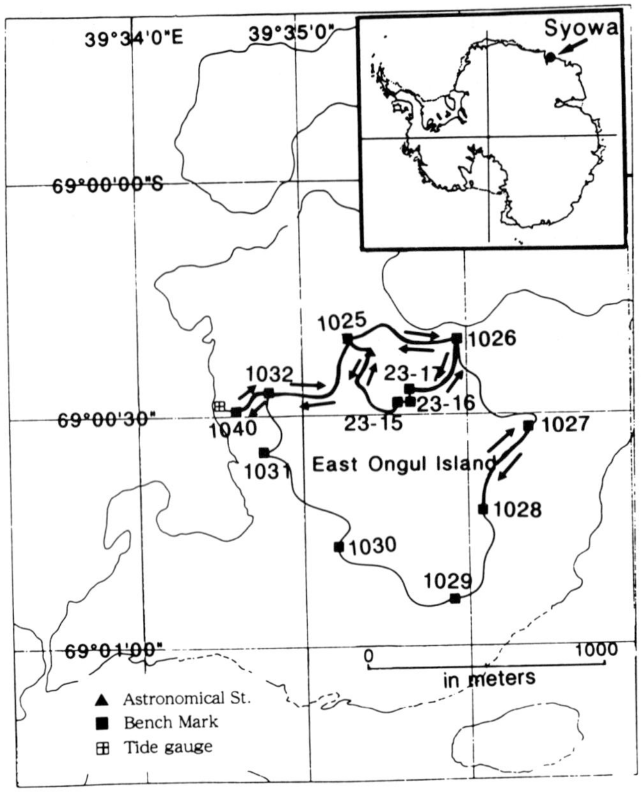

Fig. 2 Leveling route on East Ongul Island. Arrows show the Survey route in 1996.

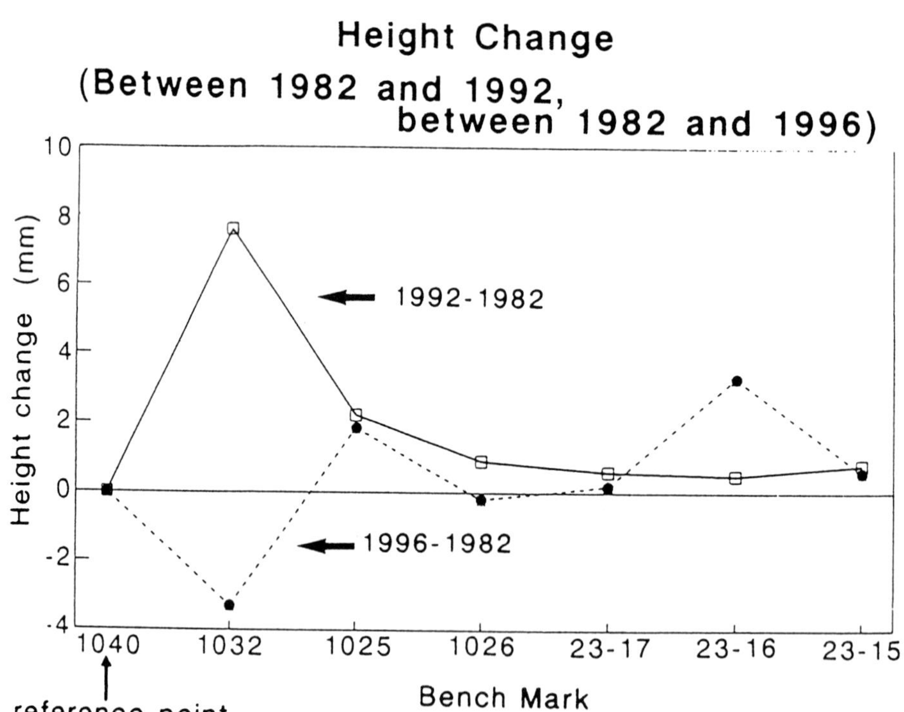

Fig. 3 Height changes between the measurements in 1992 and 1982, and between the measurements in 1996 and 1982.

6. Discussion and Conclusion

Crustal movements around Syowa Station are summarized as follows based on elevated beaches, oceanic tide, local earthquake activities and leveling survey :

1) Elevated beaches around Syowa Station show that crustal uplift after deglaciation is still going on at present.

2) A trend of sea level falling at a rate of 4.5 mm/y was obtained from oceanic tide data during 18 years in 1975-1992, and a falling rate of 9.5 mm/y was also obtained from oceanic tide data during 7 years in 1981-1987. These two falling rates are indicated that the falling is an intermittent phenomena.

3) As the location of epicenters are in the coastal and offshore areas, local earthquakes are inferred to be caused by tectonic stress accumulated by crustal uplift after deglaciation. The occurrence of local earthquakes is intermittent.

4) The crustal uplift occurs only for a few years during one decade/more, because the occurrence of earthquakes corresponds with the intermittent crustal uplift (see Fig. 1).

5) As there was no significant height change of leveling during 15 years in 1982-1997, crustal uplift is a block movement (Fig. 3).

6) Estimating from all data mentioned above, crustal uplift in the Ongul Islands is not a tilt trend movement but represents a block movement as shown in Fig. 4.

If the crustal uplift with a rate of 5 mm/y continues for ten years, the total uplift obtained will be 5 cm. The total amount of the vertical crustal uplift would produce a change of about 15 μGal in gravity. Repeated measurements of absolute gravity over many years might detect the absolute value of vertical crustal movement.

Observations of oceanic tide, seismic monitoring and gravity using a superconducting gravimeter (SCG) and LaCoste Romberg gravity meter type D and G are continued at Syowa Station. Crustal uplift after deglaciation continues at present and next intermittent uplift is expected to occur in early of the 2000's. The SCG observation is believed to be provide the highest sensitivity and highest resolution data for study of vertical crustal movement. Using the absolute gravimeter together with the SCG will provide significant insight into the absolute elevation change.

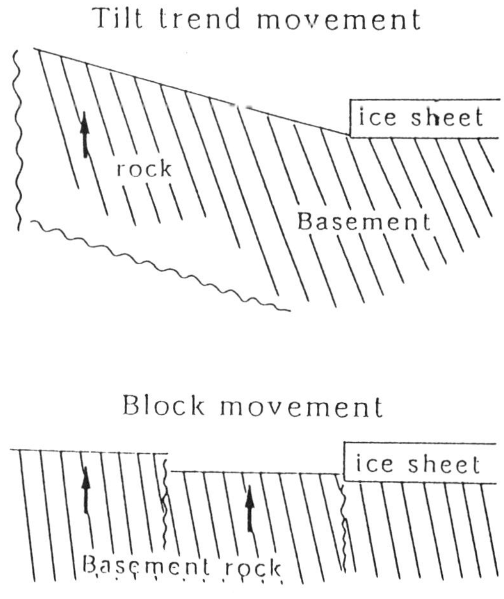

Fig. 4 A scheme of crustal uplift in the coastal area. Tilt trend movement (upper) and flock movement (bottom).

Acknowledgment
The author expresses his thanks to two connivers of the symposium for giving him the opportunity to present the paper. The manuscript was prepared by Ms. M. Minegishi of National Institute of Polar Research.

References

Akamatsu, J., Yoshikawa, S. and Kaminuma, K., Preliminary report of local seismic activity around Syowa Station, East Antarctica. Proc. NIPR Symp. Antarct. Geosci., 2, 1-6, 1988.
Hayashi, M. and Yoshida, Y., Holocene raised beaches in the Lützow-Holm Bay region, East Antarctica, Mem. Natl Inst. Polar Res., Sepc. Issue, 50, 49-84, 1994.
Igarashi, A., Harada, N. and Moriwaki, K., Marine fossils of 30-40 km in raised beach deposits, and late Pleistocene glacial history around Lützow-Holm Bay, East Antarctica. Proc. NIPR Symp. Antarct. Geosci., 8, 219-229, 1995.
Kaminuma, K., On the possiblity of detecting absolute crustal uplift at Syowa Station, Antarctica, Proc. NIPR Symp. Antarct. Geosci., 9, 16-23, 1996.
Kaminuma, K. and Akamatsu, J., Intermittent micro-seismic activity in the vicinity of Syowa Station, East Antarctica. Recent Progress in Antarctic Earth Science, ed. by Y. YOSHIDA et al. Tokyo, Terra Sci. Publ., 493-497, 1992.
Kaminua, K. and Kimura, I., Leveling survey on East Ongul Island, Antarctica and its implications, Proc. NIPR Symp. Antarct. Geosci., 10, 19-25, 1997.
Michida, Y., Namiki, M. and Iwanaga, Y., Mean sea level change at Syowa Station in these 20 years. Dai-15-kai Nankyoku Chigaku Shinpojiumu Puroguramu no Kôen Yôshi (The 15th Symposium on Antarctic Geosciences Program and Abstract). Tokyo, Natl Inst. Polar Res., 10-12 (in Japanese), 1995.
Odamaki, M., Mochida, Y., Noguchi, I., Iwanaga, Y., Ikeda, S. and Iwamoto, K., Mean sea-level observed at Syowa Station, East Antarctica., Proc. NIPR Symp. Antarct. Geosci., 5, 20-28, 1991.
Warrick, R. A., Climate and sea level change : A synthesis. Climate and Sea Level Change : Observations, Projections and Implications, ed. by R. A. Warrick et al. Cambridge, Cambridge Univ. Press, 3-12, 1993.
Yoshida, Y. and Moriwaki, K., Some consideration on elevated coastal features and their dates around Syowa Station, Antarctica. Mem. Natl Inst. Polar Res., Spec. Issue, 13, 220-226, 1979.

GEODETICAL AND GEOPHYSICAL OBSERVATIONS AT SYOWA STATION, ANTARCTICA

KatsutadaKaminuma and KazuoShibuya
NationalInstituteof PolarResearch(NIPR)
9-10, Kaga-1,Itabashi,Tokyo 173,Japan
e-mail : kaminuma@nipr.ac.jp
fax : +81-3-3962-5741

Abstract

A Japanese Antarctic Station Syowa (69°S, 39°E) was established in East Ongul Island, Lützow-holm Bay, East Antarctica for the International Geophysical Year (IGY) in 1957. Since that time, the following geodetical and geophysical observations have been carried out in the vicinity of Syowa Station : 1. Gravity : a) Continuous recording of gravity for detecting earth tides using LaCoste-Romberg gravity meter type D and G. b) Continuous observation by a superconducting gravimeter was started in March 1993; 2. Oceanic tide : Continuous recording since 1966; 3. Geomagnetism : a) Base line measurment. b) Continuous recording of long-period and short-period geomagnetic variations, and total intensity; 4. Seismology : Continuous recording by the three-component of long-period, short-period and broadband digital seismographs; 5. GPS / DORIS becon : Tracking continuously.

Syowa Station was selectied one of the 36 subset (A) stations of the International Absolute Gravity Basestation Network (IAGBN) by IAG. The measurments were made three times using four different absolute gravimeters in 1992, 1993 and 1995.

VLBI measurements were made among Syowa Station, Kashima (35.8°S, 140.7°E), Japan and Tidbinbilla (35.2°S, 149°E), Australia in 1990. Measurements will be done again in 1998. A route for repeat leveling survey was established in East Ongul Island in 1982. The leveling measurements were repeated in 1996 and 1997.

1. Introduction

Station Syowa (69°S, 39°E) was established in East Ongul Island, Lützow-holm Bay, East Antarctica for the International Geophysical Year (IGY) in 1957. Since that time except four years in 1962-65, Syowa Station has been occupied by Japanese Antarctic Research Expedition (JARE) and carried out various scientific observations. Geophysical observations such as geomagnetic, seismic etc. have been continued more than three decades. Oceanic tide observation has continued since 1966. Observations and measurements such as VLBI, GPS, DORIS etc. using space geodetic techniques started in the early 1990's.

East Ongul Island is a small island only one by one kilometer in size. Despite of its small size, many different geodetic and geophysical observations have been carried out around Syowa Station as shown in Fig. 1, and many significance results have been obtained.

International Association of Geodesy Symposia, Vol. 119
Forsberg et al. (eds.), Geodesy on the Move

An overview of the present status of the geodetic and geophysical observations at Syowa Station are summarized and some scientific results are introduced briefly in this paper.

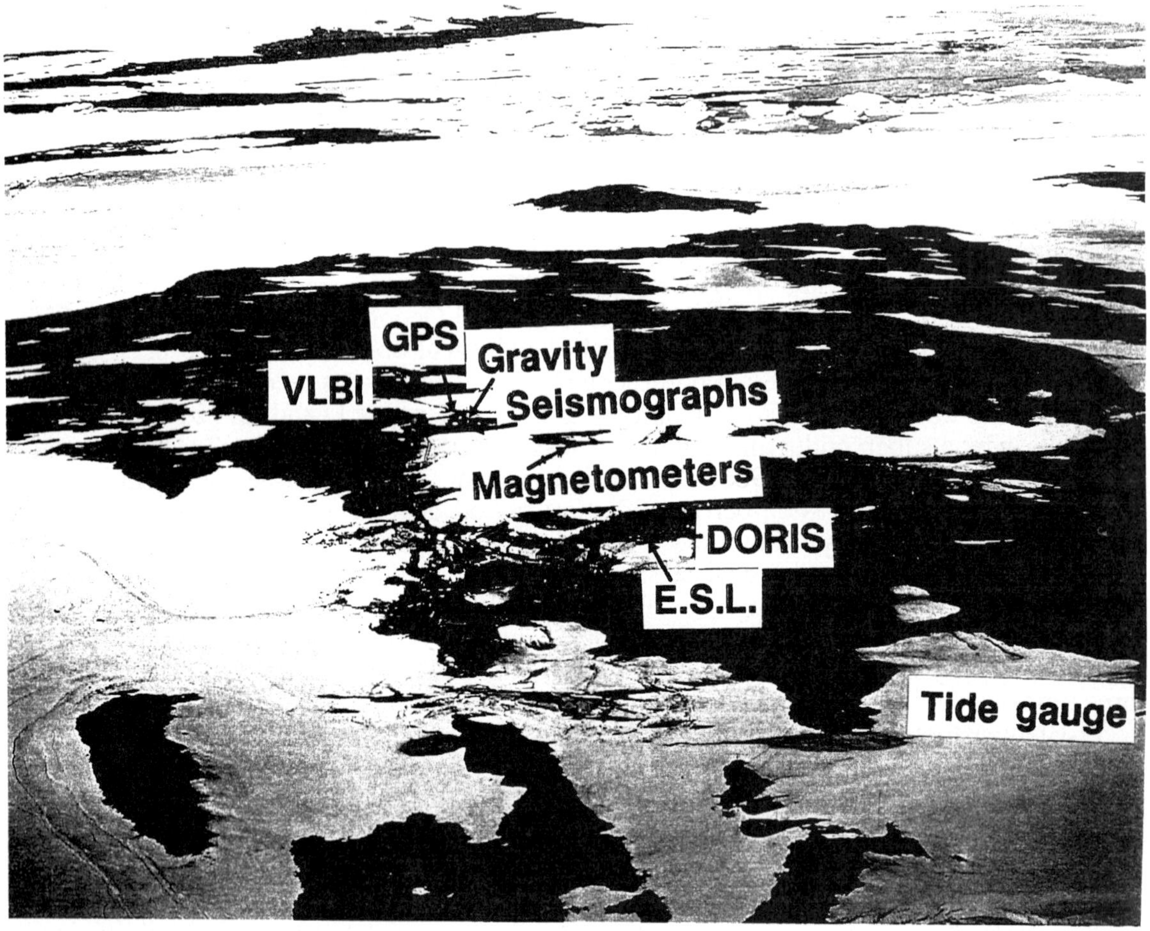

Fig. 1 A view of Syowa Station, East Ongul Island. The photograph was taken from north-west side of the island.

2. Gravity

The first pendulum gravity station was established using the Geographical Survey Institute (GSI) gravimeter for establishing a gravity control point at Syowa Station in 1962 (Harada et al., 1963). The gravity survey measurements in the field around Syowa Station have been carried out since that time.

In 1987, Syowa Station was selected as one of the 36 subset (A) stations of the International Absolute Gravity Base Stations Network (IAGBN). A gravity hut was constructed to improve the gravity measurements at Syowa Station in 1991. The absolute gravity measurements were made three times in the gravity hut using four different absolute gravimeters in 1992, 1993 and 1995 (Nakagawa et al., 1994; Kaminuma et al., 1996).

At the workshop organized by National Insitute of Polar Research (NIPR) in January 1996, it was concluded that the following absolute gravity value by the FG5 gravimeter

(#104) be adopted as the value at Syowa Station instead of the value by the GA60 gravimeter which was reported before. The main reason to adopt the value by the FG5 gravimeter (#104) is the high reliability of the FG5 gravimeters and is already a consensus choice of the international gravimetric community.

$$g = 982524.327 \pm 0.004 \text{ mgal}$$

SD of a single measurement : 0.015 mgal

$$\varphi = 69°00'27.035'' \text{ S}$$

$$\lambda = 39°35'06.372'' \text{ E}$$

$$h = 21.492 \text{ m}$$

Gravity is recorded continuously for detecting earth tides using LaCoste-Romberg gravity meters type D and G in the hut.

Continuous observation by a superconducting gravimeter (SG) was started in March 1993 (Sato et al., 1995). This is the first installation of SG not only in the Antarctic but in the southern hemisphere. Preliminary analyses of one year tidal data confirmed reasonableness of the previously obtained diurnal and semidiurnal tidal gravimetric factors and a 10 % larger M 2 δ factor than the theoretical value. The recorded data include high quality free oscillation signals from the Hokkaido Nansei-Oki earthquake (Ms = 8.2) on July 12, 1993, the earthquake of South of Mariana Island (Ms = 8.2) on August 08, 1993 and the deep earthquake of Bolivia (Ms = 6.8, depth = 637 km) on June 09, 1994. Polar motion is under analysis. Hourly sampling data of three years (from March 22, 1993 to January 28, 1996) is shown in Fig. 2.

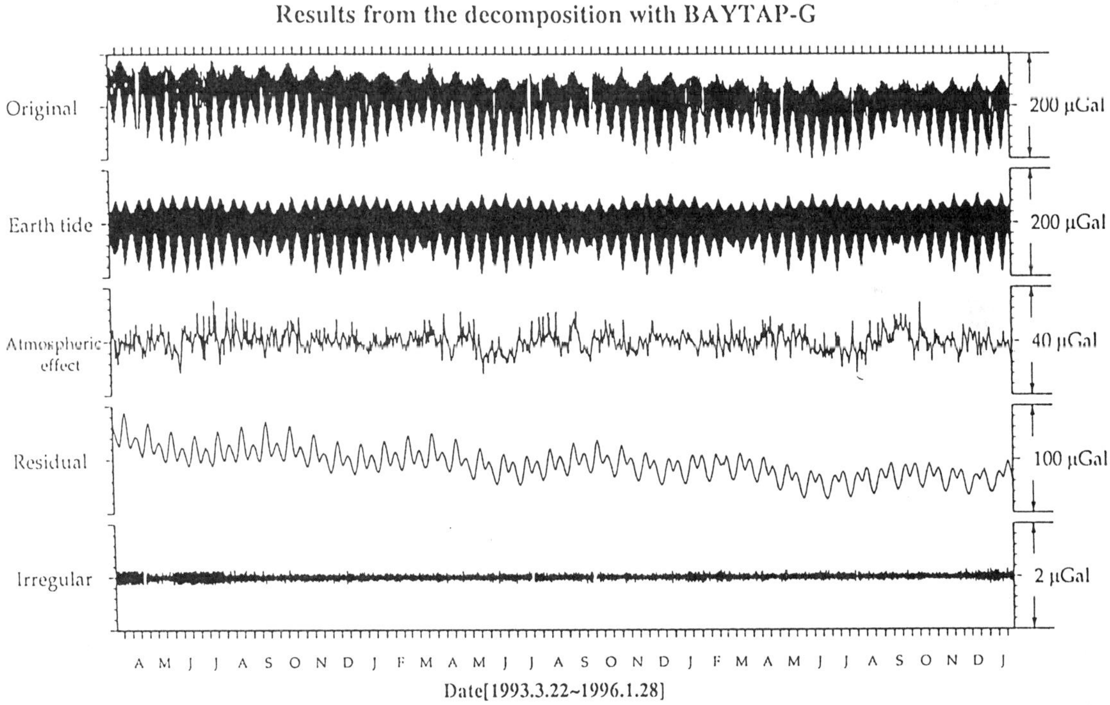

Fig. 2 Hourly sampling interval records of SG at Syowa Station from Mar. 22, 1993 to Jan. 28, 1996.

3. *Oceanic tide*

Continuous tidal observations have been made at Syowa Station since 1966 using a mechanical pressure sensor tide gauge. Syowa Station is the only station in the coastal area of the Antarctic Continent where tidal observations have been continued for more than a quarter of a century; normally observations are disturbed by sea ice and tide gauges are not installed at most coastal stations in the Antarctic. The observations at Syowa Station had also problems in monitoring mean sea level up to the end of the 1970's. A new tide gauge using a strain gauge sensor was installed at Syowa Station in 1981, and thereafter the mean sea level has been monitored continuously.

Monthly mean sea level was therefore available using the new tide gauge data from 1981 to 1987. Sea level falling at a rate of 9.5 mm/y was obtained by Odamaki et al. (1991). Recently Michida et al. (1995) estimated the mean sea level change at Syowa Station using 18 years of data in 1975-1992. They obtained a trend of sea level falling with a rate of 4.5 mm/y. This value is consistent with the crustal uplift rate estimated from the geomorphological data.

4. Geomagnetism

The first measurement of the absolute geomagnetic intensity at Syowa Station was made in 1959. Continuous recordings of long-period and short-period geomagnetic variations, and total intensity were also started in 1959.

A geomagnetic observation hut was constructed in 1966. The absolute geomagnetic measurements have been made in the hut mostly once a month since that time. Secular variations of total intensity, inclination and declination of geomagnetic field are obtained over more than 30 years since 1966. The observations of geomagnetic variations and total intensity using three component proton magnetometers are also continued since 1966.

5. Seismology

The seismic observation at Syowa Station was started in 1959. The three-component short-period and long-period seismographs were installed at Syowa Station in 1967 (Kaminuma, 1969). The ability of earthquake observation at Syowa Station made at one of the stations of the World Wide Seismic Station Network (WWSSW) which was established in the western countries by The United States Geological Survey (USGS) for detecting underground nuclear explosions in the 1960's. A new seismic vault was constructed in 1970. The detection capability of earthquakes at Syowa Station was increased.

The seismic observations at Syowa Station have been operated using nine seismographs, three-component short-period and long-period seismographs and broadband digital seismographs, since 1989.

An intraplate earthquake occurred near Syowa Station as shown in Fig. 3. This earthquake is the first event which locates within 400 km from Syowa Station, and of magnitude larger than 4. Three component digital seismograms of the earthquake are given in Fig. 4.

6. Space geodetic techniques

A multipurpose parabola antenna with 11 m in diameter was constructed at Syowa Station in 1989. The first VLBI (Very Long Baseline Interferometer) experiment in Antarctica was made using the antenna. The experiment connected Antarctica with other continents (Kurihara et al., 1991). Kashima (35.8°N, 140.7°E) in Japan and Tidbinbilla (35.2°S, 149.0°E) in Australia participated in the experiment in conjunction with Syowa Station. The baseline lengths among three stations obtained by the VLBI experiments are as follows :

Syowa Station	- Kashima	11,391,620.78 m
	- Tidbinbilla	6,735,680.19 m
Kashima	- Tidbinbilla	7,436,721.48 m

The second VLBI experiment at Syowa Station will be done during 1998.

The GPS (Global Positioning System) receivers have been installed at Syowa Station in 1990. Since that time, continuous monitoring of GPS signals is going on.

The DORIS (Doppler Orbitography and Radiopositioning Integrated by Satellite) beacon antenna was also installed at Syowa Station in 1994. Signals are received continuously since that time. The accuracy of the antenna position increases year by year.

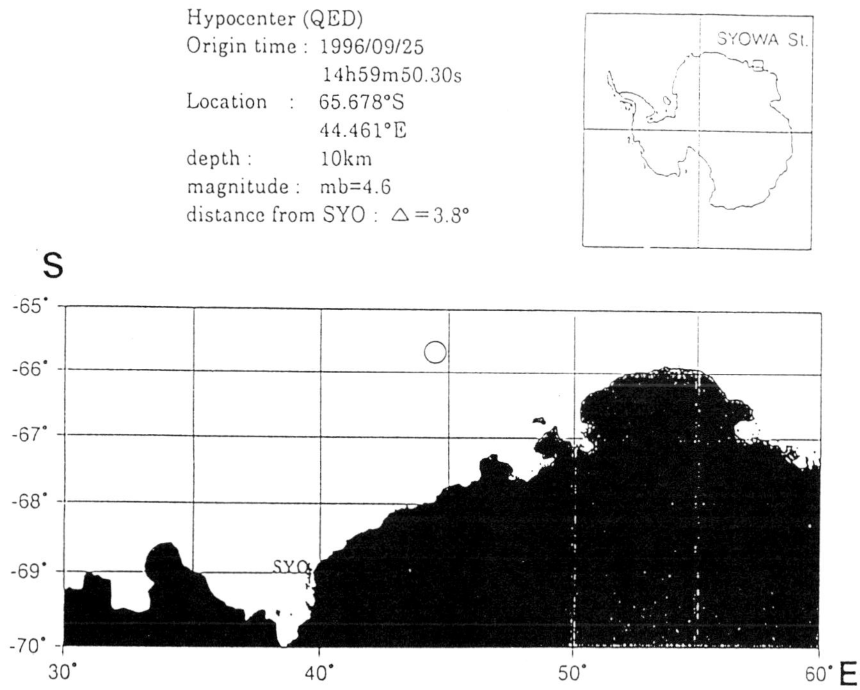

Fig. 3 Location of an intraplate earthquake occurred near Syowa Station.

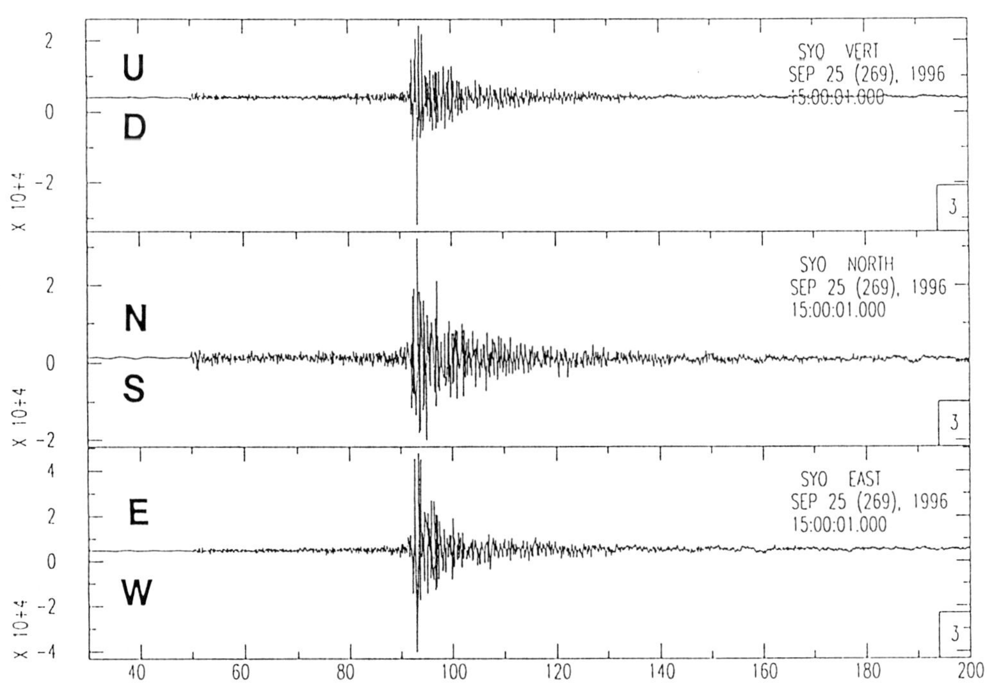

Fig. 4 Three component broadbabd seismograms of the earthquake shown in Fig. 3.

Acknowledgments

In this review, a number of results by members of JARE, especially Drs. Y. Nogi, K. Kanao and K. Doi of NIPR are referred to. The authors' thanks are due to these colleagues. The manuscript was prepared by Ms. M. Minegishi of NIPR.

References

Harada, Y., Kaminuma, S. and Murata, I., Pendulum determination of the gravity differences between Tokyo, Mowbray and Syowa Base, Antarctic Rec., 17, 35-50, 1963.

Kaminuma, K., The seismological observation and the earthquake detection capability of Syowa Station, Antarctica, Bull. Earthq. Res. Inst., 47, 453-466, 1969.

Kaminuma, K., Tsukahara, K. and Takemoto, S., Absolute gravity value measured at Syowa Station, Antarctica, Bull. d' Inform., Bureau Gravimetrique International, 80, 26-29, 1997.

Kurihara, N., Kondo, T., Takahashi, Y. and Ejiri, M., The results of test VLBI experiments with the Syowa Station in Antarctica, J. Communi. Res. Lab., 38, 605-611, 1991.

Nakagawa, I., Shibuya, K., Kaminuma, K., Fujiwara, S., Watanabe, K., Murakami, M., Ishihara, M., Tsubokawa, T., Hanada, H. and Yokoyama, K., Absolute gravity measurements at Syowa Station during the Japanese Antarctic Research Expedition. Bull. d' Inform., Bureau Gravimetrique International, 75, 41-56, 1994.

Sato, T., Shibuya, K., Tamura, Y., Kanao, M., Ooe, M., Okano, K., Fukuda, Y., Seama, N., Nawa, K., Kaminuma, K., Ida, Y., Kumazawa, M.and Yukutake, T., One Year observations with a superconducting gravimeter at Syowa Station, Antarctica, J. Geodet. Soc. Japan, 41, 75-89, 1995.

AUTHOR INDEX